Essentials of
Foye's Principles
of Medicinal
Chemistry

Essentials of Foye's Principles of Medicinal Chemistry

Thomas L. Lemke, PhD
Professor Emeritus
College of Pharmacy
University of Houston
Houston, Texas

S. William Zito, PhD
Professor of Pharmaceutical Sciences
College of Pharmacy and Allied Health Professions
St. John's University
Jamaica, New York

Victoria F. Roche, PhD
Professor of Pharmacy Sciences
School of Pharmacy and Health Professions
Creighton University
Omaha, Nebraska

David A. Williams, PhD
Professor Emeritus of Chemistry
School of Pharmacy
MCPHS University
Boston, Massachusetts

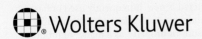 Wolters Kluwer

Philadelphia • Baltimore • New York • London
Buenos Aires • Hong Kong • Sydney • Tokyo

Acquisitions Editor: Matt Hauber
Senior Product Development Editor: Amy Millholen
Editorial Assistant: Brooks Phelps
Marketing Manager: Michael McMahon
Production Project Manager: Marian Bellus
Design Coordinator: Joan Wendt
Manufacturing Coordinator: Margie Orzech
Prepress Vendor: Aptara, Inc.

Copyright © 2017 Wolters Kluwer

All rights reserved. This book is protected by copyright. No part of this book may be reproduced or transmitted in any form or by any means, including as photocopies or scanned-in or other electronic copies, or utilized by any information storage and retrieval system without written permission from the copyright owner, except for brief quotations embodied in critical articles and reviews. Materials appearing in this book prepared by individuals as part of their official duties as U.S. government employees are not covered by the above-mentioned copyright. To request permission, please contact Wolters Kluwer at Two Commerce Square, 2001 Market Street, Philadelphia, PA 19103, via email at permissions@lww.com, or via our website at lww.com (products and services).

9 8 7 6 5 4 3 2 1

Printed in China

Library of Congress Cataloging-in-Publication Data

Names: Lemke, Thomas L., author. | Zito, S. William, author. | Roche,
   Victoria F., author. | Williams, David A., author.
Title: Essentials of Foye's principles of medicinal chemistry / Thomas L.
   Lemke, S. William Zito, Victoria F. Roche, David A. Williams.
Description: Philadelphia : Wolters Kluwer, [2017] | Includes index. |
   Abridgement of: Foye's principles of medicinal chemistry / edited by
   Thomas L. Lemke, David A. Williams ; associate editors, Victoria F. Roche,
   S. William Zito. 7th ed. c2003.
Identifiers: LCCN 2016000560 | ISBN 9781451192063
Subjects: | MESH: Chemistry, Pharmaceutical
Classification: LCC RS403 | NLM QV 744 | DDC 616.07/56–dc23
LC record available at http://lccn.loc.gov/2016000560

This work is provided "as is," and the publisher disclaims any and all warranties, express or implied, including any warranties as to accuracy, comprehensiveness, or currency of the content of this work.

This work is no substitute for individual patient assessment based upon healthcare professionals' examination of each patient and consideration of, among other things, age, weight, gender, current or prior medical conditions, medication history, laboratory data and other factors unique to the patient. The publisher does not provide medical advice or guidance and this work is merely a reference tool. Healthcare professionals, and not the publisher, are solely responsible for the use of this work including all medical judgments and for any resulting diagnosis and treatments.

Given continuous, rapid advances in medical science and health information, independent professional verification of medical diagnoses, indications, appropriate pharmaceutical selections and dosages, and treatment options should be made and healthcare professionals should consult a variety of sources. When prescribing medication, healthcare professionals are advised to consult the product information sheet (the manufacturer's package insert) accompanying each drug to verify, among other things, conditions of use, warnings and side effects and identify any changes in dosage schedule or contraindications, particularly if the medication to be administered is new, infrequently used or has a narrow therapeutic range. To the maximum extent permitted under applicable law, no responsibility is assumed by the publisher for any injury and/or damage to persons or property, as a matter of products liability, negligence law or otherwise, or from any reference to or use by any person of this work.

LWW.com

CCS0416

Dedication to William O. Foye

William O. Foye, Sawyer Professor of Medicinal Chemistry at the MCPHS University (formerly Massachusetts College of Pharmacy), Boston, MA, was born in 1923 in western Massachusetts. He received his BA (1944) in chemistry from Dartmouth College and PhD in Organic Chemistry (M. Carmack) from Indiana University in 1948. He served in the U.S. Navy during World War II as a chemical warfare instructor. He joined DuPont (Delaware) as a research scientist, then in 1950 joined the School of Pharmacy at the University of Wisconsin as assistant professor of pharmaceutical chemistry. In 1955, he moved to MCPHS University, Boston as Professor of Chemistry, where he brought a new vision of pharmaceutical chemistry (medicinal chemistry) to the pharmacy curriculum. As department chair, he advocated for organic medicinal chemistry in the pharmacy curriculum.

The impetus for a new text in medicinal chemistry, grounded on Alfred Burger's two-volume Medicinal Chemistry, came from Dr. Norman Doorenbos, College of Pharmacy University of Maryland (Baltimore), who had made arrangements with Lea & Febiger (forerunner of Lippincott Williams and Wilkins) for publishing a companion text to Wilson & Gisvold's *Textbook of Organic Medicinal and Pharmaceutical Chemistry*. Because Dr. Doorenbos was moving to chair the new Pharmacognosy Department at the School of Pharmacy, University of Mississippi, he relinquished the job of editing this text to Dr. Foye. During this time, a number of teachers and researchers in medicinal chemistry felt that a text on drugs that included biochemical modes of action, aimed primarily for undergraduates, should be written. Although other pharmaceutical and medicinal chemistry books had been written during this time (*The Chemistry of Organic Medicinal Products*, Jenkins and Hartung, and *Textbook of Organic Medicinal and Pharmaceutical Chemistry*, Wilson and Gisvold), these authors organized their books according to the accepted scheme of chemical classification of the more important organic medicinal compounds, their methods of synthesis, properties and descriptions, and their uses and modes of administration. Therefore, this "Principles" text provided a contemporary basis for the biochemical understanding of drug action that included the principles of structure – function relationships and drug metabolism. Dr. Foye assembled authors who were experts in their respective fields and published the first edition of *Principles of Medicinal Chemistry* in 1972.

The current authors of the 7th edition *Foye's Principles of Medicinal Chemistry* uphold the original concept for an undergraduate medicinal chemistry principles textbook to assist its readers to pull together the chemical understanding of drug action and to appreciate its practical relevance to contemporary pharmacy practice. Because medicinal chemistry students have often expressed the desire for a short and "to the point" text that clearly summarizes the most important chemical elements of therapeutically relevant drug classes, the *Essentials of Foye's Principles of Medicinal Chemistry* book was constructed to meet this need. The authors hope this Essentials book not only meets your short-term need for focused medicinal chemistry education, but also stimulates a lifelong desire to know more about how the chemicals we call drugs work.

Dr. Foye had a long and creative career in medicinal chemistry with more than 150 refereed scientific publications, book chapters, and editor of Cancer Chemotherapeutic Agents ACS Monograph Series. His research focused on numerous areas especially the SAR of antiradiation organosulfur compounds, anticancer agents, and chelation as mechanism of drug action. He was elected Fellow of the AAPS, Fellow of APhA, received the APhA Foundation Research Achievement Award in Medicinal Chemistry, and was an emeritus member of AAAS, ACS, APhA, and AAPS. He was one of the founding members of the Medicinal Chemistry Group of the Northeast Section of the American Chemical Society (1968). He retired in 1995. In addition to being Chair of the Department of Chemistry, he was also the Dean of Faculties and Dean of Graduate Studies at the MCPHS University. Dr. Foye traveled widely and was an invited participant at numerous meetings.

Not only was Dr. Foye a distinguished scientist, he was an avid fly-fisherman, outdoorsman, and environmentalist. He published several books about the western Massachusetts wilderness, *Trout Waters: Reminiscence with Description of the Upper Quabbin Valley* (1992) and *North Quabbin Wilds: A populous Solitude* (2005). He was a dedicated supporter of the arts in Boston. He married Lila Siddons in 1974 and has a son Owen and stepson Kenneth. Dr. Foye died in 2014.

David A. Williams, PhD.

William O. Foye (1923–2014)

Reviewers

Faculty Reviewers

Kimberly Beck
Assistant Professor
Butler University
Indianapolis, Indiana

Marc W. Harrold
Professor of Medicinal Chemistry
Duquesne University
Pittsburgh, Pennsylvania

Philip Kerr
Senior Lecturer
Charles Sturt University
Bathurst, Australia

John Malkinson
Senior Lecturer
University College London School of Pharmacy
London, United Kingdom

Amjad Qandil
Associate Professor
King Saud bin Abdulaziz University
   for Health Sciences
Riyadh, Saudi Arabia

Alex White
Associate Professor
Cardiff University
Cardiff, United Kingdom

Student Reviewers

William Cooper
Union University School of Pharmacy

Heather Gegenhuber
Pacific University School of Pharmacy

Letetia Jones
Hampton University

Emily Li
University of Alberta

Sebastian Nguyen
Harding University College of Pharmacy

Hemina Patel
Ferris State University

Anna Recker
Butler University

Aesha Shah
Roosevelt University

Nathan Watson
Butler University

Acknowledgments

This book was conceived of and made possible by the actions of Sirkka Howes of Lippincott Williams and Wilkins who brought the idea to us and then secured approval for the project. We also would like to acknowledge the help of Dr. Marc Harrold of Duquesne University who served as a consultant reviewing our work and made helpful suggestions to bring uniformity to the work.

Contents

Essentials of *Essentials*

The authors of the *Essentials* text want readers to glean the most from this abbreviated and focused medicinal chemistry resource. In this first chapter, we share the rationale behind the development of the book and provide guidance for its optimal use to advance chemical understanding and appreciate its practical relevance to contemporary pharmacy practice.

Why an Essentials *Text?*

Students have often expressed the desire for a short and "to the point" text that clearly summarizes the most important chemical elements of therapeutically relevant drug classes. The *Essentials* book was constructed to meet this need.

Chapter Content and Format

Each chapter of *Essentials* captures the key "take-home" chemical points of its companion chapter in *Foye's Principles of Medicinal Chemistry*, 7th edition. "Essential" concepts are bulleted to get to the heart of the chemical message with a limited amount of text.

Sections related to mechanism of action (MOA), structure–activity relationships (SAR), physicochemical and pharmacokinetic properties, metabolism, and clinical applications (most commonly, therapeutic use) are found consistently in all chapters. Additional sections appropriate to specific topics are sometimes incorporated (e.g., resistance mechanisms in the Antimicrobial Agents and Cancer Chemotherapy chapters). Figures and tables are commonly used in lieu of words to illustrate concepts. Complementary information on adverse effects, drug–drug or drug–food interactions, chemical points of interest, and unique aspects of clinical use are often captured in side boxes or summarized in Chemical Note side bars.

Each chapter concludes with a section emphasizing the clinical relevance of the content covered in the chapter, and five review questions to allow self-assessment of learning and identification of areas for further study.

Getting the Most From Each Chapter

The in-depth understanding of medicinal chemistry that underpins therapeutic decision-making requires more than memorizing facts and diagramming pathways. The authors recommend that learners first read the appropriate chapter in *Foye* to get the full chemical picture of the drug classes under discussion. The *Foye* chapters are written to be descriptive and comprehensive narratives that educate about all aspects of therapeutically relevant drug chemistry.

Armed with that significant level of understanding, the *Essentials* text can then be used to prompt recollection of crucial chemical concepts and allow assessment of the ability to summarize central aspects of, among others, chemical mechanism, SAR, and metabolic pathways. Areas of deficient understanding uncovered by a review of the *Essentials* chapter can be addressed by revisiting the appropriate chapter in *Foye*.

The section on clinical relevance that concludes each chapter was crafted to stimulate reflective thinking on how understanding of drug chemistry allows practitioners to scientifically approach patient care. Drugs are chemicals, and they behave as such in the environments encountered in vitro and in vivo. Pharmacists are the only health care professionals educated in the chemical behavior of drug molecules and are in a unique position to predict therapeutic opportunities and problems based on that understanding. This distinction gives pharmacists a very important and valued place on the health care team.

Lifelong Self-Directed Learning

Understanding the body's biochemical systems and the mechanisms of disease is advancing at an almost inconceivable rate due to innovations in molecular biology, technology (including computer-facilitated ligand–receptor modeling), biotechnology, and other relevant sciences. Human knowledge has been recently estimated to double every 13 months, with clinical knowledge doubling every 18 months. Some predict that the rate of the doubling of new knowledge will approach every 12 hours in the not-so-distant future. This underscores the importance of students approaching their studies with a healthy intellectual curiosity and embracing lifelong self-directed learning skills. The ability to search the primary literature and to understand what one reads in seminal, ground-breaking papers will be even more important tomorrow than it is today.

Drug design and discovery will follow and, in many cases lead, new knowledge acquisition. According to the FDA, in 2014, the Center for Drug Evaluation and Research gave the nod to 41 new drugs as new chemical entities (NCEs) or biologics. Of these new drugs, 31 are small molecules (including 1 ^{18}F radiopharmaceutical and 1 sulfur hexafluoride lipid microsphere) and 10 are therapeutic proteins. This is a larger figure than the average number of new drugs approved annually over the past 10 years. Of these new drugs, 41% (18) are the first in their therapeutic class to be approved, and are sure to be followed by analogs and novel entities that aim to improve on therapeutic efficacy, pharmacokinetic profile, or adverse effects risk. The diseases targeted by these new drugs run the gamut from various types of cancer to cardiovascular disease to diabetes to infection. Of the drugs approved in 2014, 41% (18) are designed to treat orphan diseases such as multicentric Castleman's disease, Gaucher's disease, and leishmaniasis. To further illustrate the rate of new drug development, Table 1.1 names some of the biologics recently approved as therapeutic monoclonal antibodies.

The point of this discussion is that every text, no matter how comprehensive, needs to be supplemented with regular research on new scientific discoveries in drug development and the self-directed evaluation of the impact of novel chemical entities on disease control, prevention, and wellness.

With this brief introduction, the authors welcome the reader to the *Essentials* text. We hope it not only meets your short-term need for focused medicinal chemistry education, but also stimulates a lifelong desire to know more about how the chemicals we call drugs work.

Table 1.1   FDA-Approved Monoclonal Antibodies (MoAbs) Therapeutic Agents 2013–2015

Generic Name	Trade Name	Type	Indication
Alirocumab	Praluent	Fully human	Hypercholesterolemia
Blinatumomab	Blincyto	Mouse	Acute lymphoblastic leukemia (ALL)
Daratumumab	Darzakex	Fully human	Multiple myeloma (IgG1κ)
Dinutuximab	Unituxin	Chimeric	Neuroblastoma
Elotuzumab	Empliciti	Humanized	Multiple myeloma
Evolocumab	Repatha	Fully human	Hypercholesterolemia
Ipilimumab	Yervoy	Fully human	Late-stage melanoma
Mepolizumab	Nucala	Humanized	Severe eosinophilic asthma
Necitumumab	Portrazza	Fully human	Metastatic squamous non small cell lung cancer (NSCLC)
Nivolumab	Opdivo	Fully human	Metastatic melanoma and advanced squamous non small cell lung cancer (NSCLC)
Obinutuzumab	Gazyva	Humanized	Chronic lymphocytic leukemia in combination with chlorambucil
Pembrolizumab	Keyruda	Humanized	Advanced or unresectable melanoma
Ramucirumab	Cyramza	Fully human	Advanced or metastatic stomach or gastroesophageal junction cancer
Pertuzumab	Perjeta	Humanized	Human Epidermal growth factor Receptor 2-positive (HER2-positive) metastatic breast cancer
Secukinumab	Cosentyx	Fully human	Rheumatoid arthritis, ankylosing spondylitis, and plaque psoriasis
Siltuximab	Sylvant	Chimeric	Multicentric Castleman's disease (MCD) in patients who are HIV negative
Vedolizumab	Entyvio	Humanized	Moderate to severe ulcerative colitis or Crohn's disease.

Introduction to Medicinal Chemistry: Drug Receptors and General Drug Metabolism

2

Learning Objectives

Upon completion of this chapter the student should be able to:

1. Define what a receptor is and its role in the pharmacological activity of a drug.
2. Discuss the physicochemical factors that govern drug–receptor bonding and affinity.
3. Given a drug structure predict possible drug–receptor binding interactions.
4. Explain the different types of drug receptors and how they produce their effects when a drug binds.
5. Discuss the different classes of drug metabolism reactions.
6. Differentiate factors that affect drug metabolism.
7. Discuss drug–drug and drug–dietary interactions.
8. Distinguish between drugs that inhibit metabolism and those that induce metabolism.
9. List the different types of phase 1 and phase 2 metabolic reactions.
10. Given a drug structure draw the possible metabolic reactions.
11. Explain the importance of membrane drug transporters to the pharmacokinetic profiles of drug molecules.

Introduction

The use of chemical compounds to treat disease has its origins in the use of plants and other natural substances by primitive cultures. These cultures were attempting to mitigate the influences of evil spirits and superstitions believed to be the cause of illnesses. Over time and much research it became evident that compounds could affect pharmacological activity in either a selective or non-selective way. Nonselective drugs interact with all body cells and compartments (general anesthetics, nitrogen mustards), and changes in their structure have little effect on their pharmacological activity. On the other hand, many drugs act selectively and small changes in their chemical structure can produce profound differences in their pharmacological activity. The selective nature of these drugs stems from their affinity for specific cellular components which have been identified as receptors (see Chapter 7 in *Foye's Principles of Medicinal Chemistry, Seventh Edition,* for a complete detailed discussion).

Drug–Receptor Affinity

What is a Drug Receptor?

- Drug receptors are biomacromolecules which are normally membrane bound but are also found in the cytoplasm.
- Contain specific regions (receptor sites) that react with complementary functional groups on their endogenous substance or a drug molecule (drug–receptor affinity).
- Drug–receptor interactions are often generally referred to as ligand–receptor interactions.
- Allow for physiological messages to be communicated from the drug and/or endogenous substances.

<table>
<tr><td>

Definitions

Agonist: A drug or endogenous substance that has affinity for a receptor and producing a biological response (intrinsic activity). The response can be full, partial, or even negative (inverse agonist). Inverse agonism occurs when the receptor has basal activity in the absence of any ligand; therefore, inverse agonists can produce an effect below the basal level.

Antagonist: A drug that has affinity for a receptor but does not activate it to produce a response. Antagonists are referred to as having affinity for the receptor, but no intrinsic activity.

</td></tr>
</table>

Affinity: The Role of Chemical Bonding

- When a drug or endogenous substance interacts with a receptor the initial attraction (affinity) is based on interactive chemical bonding forces which are either covalent or noncovalent (i.e., ionic, hydrogen bonding, or hydrophobic bonding) (Fig 2.1).
- Most drugs bind via noncovalent bonding which is reversible, relatively weak, and mainly ionic or ion–dipole. H-bonding is not very significant for affinity but plays an important role in stabilizing the drug–receptor interaction.
- Compounds interacting with a drug receptor can be classified as an agonist or an antagonist.

Covalent Bond

- Strongest drug–receptor interaction, formed when two atoms share a pair of electrons.
- Usually irreversible and drug action terminated only by eventual cleavage or receptor turnover.
- Reactions that form covalent bonds include alkylation, acylation, and phosphorylation.

Ionic Bond

- Formed when two ions of opposite charge are attracted to each other through electrostatic forces.
- Acidic functional groups on the receptor (e.g., carboxylic acid on the side chain of aspartic acid or glutamic acid) bind with basic functional groups on the drug molecule (e.g., tertiary amines on the structures of morphine, atropine).
- Basic functional groups on the receptor (e.g., the side chains of lysine, arginine, and histidine) bind with acidic functional groups on the drugs (e.g., carboxylic acids of aspirin, penicillin).

Hydrogen Bond

- The H-bond is an electrostatic dipole–dipole interaction between a hydrogen atom on an electronegative atom (i.e., –OH) and an electronegative atom (i.e., oxygen, nitrogen, or sulfur).

Type	Energy (kcal/mol)	Example
Covalent	40–140	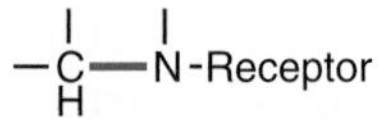
Ionic	5	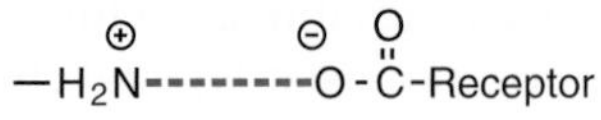
Reinforced Ionic	10	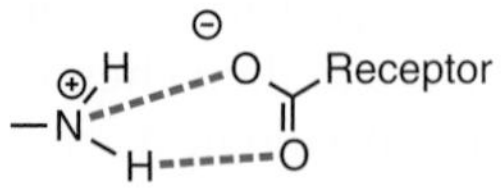
Hydrogen bond	1–7	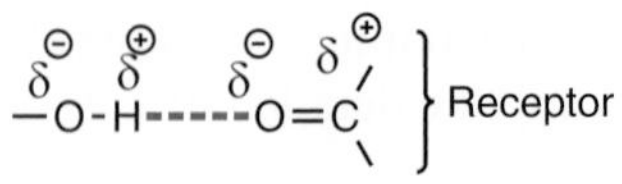
Ion–Dipole	1–7	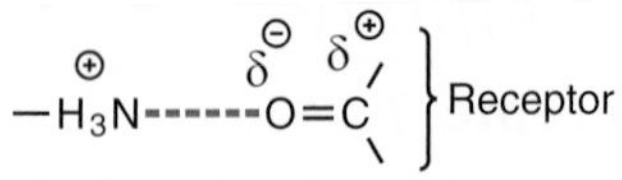
Dipole–Dipole	1–7	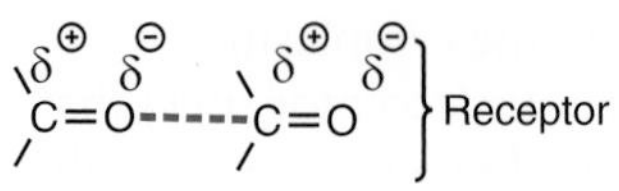
van der Waals	0.5–1	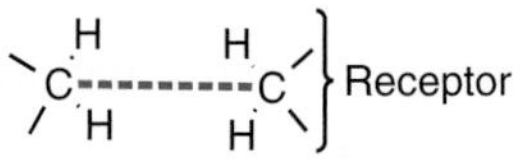

Figure 2.1 Examples of various types of drug–receptor bonds.

- The H-bond can be intramolecular (within the same molecule) or intermolecular (between two molecules). The latter is what occurs most often in drug–receptor interactions.
- Not strong enough to support drug–receptor interaction alone. However, it can be substantial when there are multiple H-bonding interactions.

Reinforced Ionic Bond
- This type of bond is formed by the combination of an ionic bond and an additional reinforcing interaction by a hydrogen bond. Can be of substantial strength (e.g., 10 kcal/mol).

Ion–Dipole Bond
- This type of bonding is between an ion and a dipole. The ionic site can be a cation or an anion functional group on the drug molecule and the dipole formed by a carbonyl group on a receptor.

Dipole–Dipole Bond
- The interaction between two dipoles, one on the drug molecule and the other on the receptor. The strength of a dipole–dipole bond can range between 1 and 7 kcal/mol.

Hydrophobic Bond Interaction
- Induced dipoles between nonpolar organic molecules as a result of the ability of water molecules to exclude (repel) hydrocarbon molecules.
- The interaction is very weak and increases inversely with the 7th power of the interatomic distance ($E = 1/\text{distance}^7$).
- Often referred to as van der Waals or London forces and are important contributions to the overall binding of drugs to the receptor.

Affinity: The Role of Conformation

Once an agonist binds to the receptor it produces a biologic response. There are a number of theories to explain how the binding causes this response including the Occupancy Theory, the Rate Theory, the Induced-Fit Theory, the Macromolecular Perturbation Theory, and the Activation–Aggregation Theory.

The Occupancy Theory
- Predicts that the biological response is directly related to the number of receptors bound (occupied) by the agonist.
- Response ceases when the drug dissociates from the receptor.
- Antagonists occupy with high affinity but do not produce a biological response.

The Rate Theory
- Predicts that the biological response is directly related to the number of times the drug binds to the receptor per unit of time.
- Thus, drugs that rapidly associate and dissociate with the receptor produce the most intense response.

The Induced-Fit Theory
- Predicts that as the drug approaches the inactive state of the receptor it induces a specific conformational change (perturbation) that leads to effective drug binding and to the biological response (Fig 2.2).
- According to this theory, antagonists induce nonspecific conformational changes that fail to produce the desired biological response.

The Macromolecular Perturbation Theory
- Combines the induced-fit and rate theories into one.
- Suggests that there are two types of specific receptor conformational perturbations: one leading to the biological response and the other to no activity.
- Therefore, the rate and ratio of their existence determines the observed biological response.
- This theory may in part explain the activity of partial agonists.

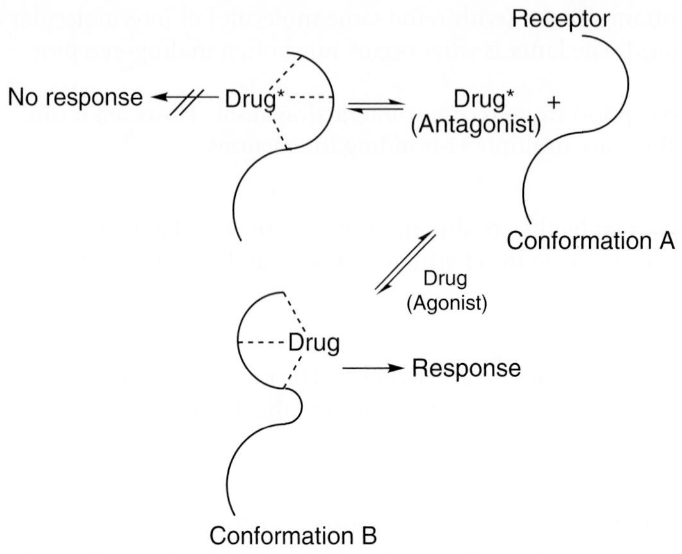

Figure 2.2 Diagrammatic representation of the drug-induced fit theory where an agonist (drug) or antagonist (drug*) interacts with two different conformations of the receptor.

The Activation–Aggregation Theory
- Postulates that the receptor is always in the state of equilibrium between active and inactive states.
- Agonists function by shifting the equilibrium to the active state while antagonists prevent the active state.
- This theory can account for inverse agonists which can produce responses opposite to that of an agonist.

Affinity: The Role of Stereochemistry
- All molecules in nature exist in three dimensions and as such their functional groups will be found in specific areas of space.
- Drug binding to receptors require that the drug molecule contain functional groups that are complementary to the functional groups on the receptor (i.e., anionic group on receptor and cationic group on the drug molecule).
- In order for the complementary functional groups to bind, they must also have the proper complementary spacial orientation.
- The orientation of functional groups in three-dimensional space is referred to as the molecules' stereochemistry.
- Molecules with the same molecular formula but differing in the arrangement of their atoms in space are called stereoisomers.
- There are three types of stereoisomers: optical, geometric, and conformational.

Optical Isomers
- Two types of optical isomers: enantiomers and diastereomers.
- Enantiomers are optical isomers that are not superimposable mirror images of one another by virtue of having asymmetric centers (chiral carbons).
- Chiral carbon atoms are attached to four different atoms or groups and each asymmetric carbon has its own absolute configuration. Usually denoted as R and S and have the ability to rotate the plane of polarized light either to the left (l) or the right (d) (i.e., optical isomers).
- Enantiomers have identical physicochemical properties (mp, bp, solubility, partition coefficient, pKa).
- A mixture containing equal amounts of the two individual enantiomers is called a racemic mixture.
- Racemic mixtures are optically inactive, having opposite rotations of plane polarized light which cancel out each other.

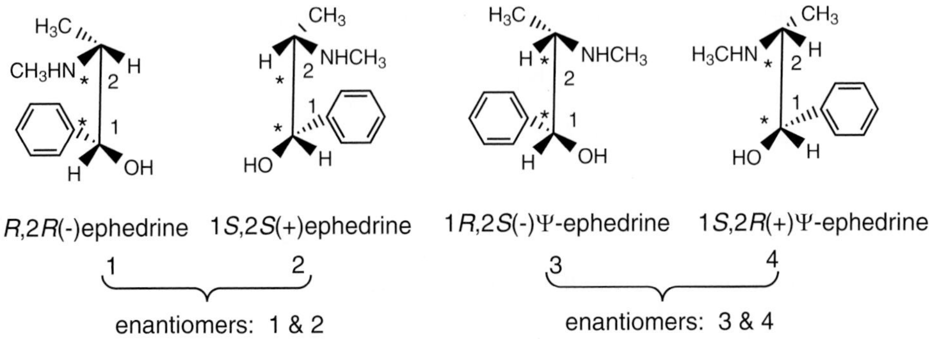

1R,2R(-)ephedrine   1S,2S(+)ephedrine   1R,2S(-)Ψ-ephedrine   1S,2R(+)Ψ-ephedrine

　　　1　　　　　　　2　　　　　　　　3　　　　　　4

enantiomers: 1 & 2　　　　　　enantiomers: 3 & 4

Figure 2.3 Example of ephedrine which has two chiral centers and, therefore, four optical isomers, two sets of enantiomers: ephedrine and pseudoephedrine.

- Many drugs are only available as racemic mixtures where only one enantiomer has functional groups in the proper orientation for maximal affinity and intrinsic activity.
- Molecules with n chiral carbons will have n^2 optical isomers; for example, two chiral carbons will have four optical isomers (two sets of enantiomers).
- Figure 2.3 depicts the optical isomers of ephedrine which contains two chiral carbons and therefore four possible enantiomers.
- Diastereomers are optical isomers that are NOT super imposable and NOT mirror images.
- They only occur when there is more than one asymmetric carbon and their physicochemical properties are different.
- In Figure 2.3, optical isomers 1 & 3, 1 & 4, 2 & 3, and 2 & 4 are diastereomers.
- Epimers are a specialized type of diastereomer in which all but one chiral carbon atom have the same configuration.

Geometric Isomers
- Refers to any system where rotation about a bond is restricted (i.e., double bonds or ring systems).
- Most commonly referred to as *cis* and *trans* isomers. Alternative nomenclature is zusammen (Z) and entgegen (E).
- Geometric isomers do not necessarily have optical activity.

Conformational Isomers
- A nonidentical spatial arrangement of atoms in a molecule resulting from rotation about one or more single bonds.
- Can occur in both acyclic and cyclic systems.
- In acyclic systems, the predominant conformer is the one with the greatest distance between the largest groups – usually the *trans, anti*. There is an exception to this when there is also a co-occurring electrostatic interaction (e.g., intramolecular hydrogen bonding).
- In cyclic systems, the predominant conformer is one where all substituents are staggered or the one that puts the largest substituent in the equatorial position (Fig. 2.4).
- Acyclic compound conformational isomers are represented by Newman projections, whereas cyclic representations depend on the number of atoms in the ring.

The Role of Stereochemistry and Drug–Receptor Interaction
- The asymmetry of the drug molecule must be complimentary to the asymmetry of the receptor.
- Figure 2.5 visually depicts how optical, geometric, and conformational isomers influence the affinity of the drug to receptor functional groups.

Dose–Response Relationships
- Dose–response curves are a representation to explain the different types of drug effects once a drug is bound to its receptor.
- In Figure 2.6, drug X is equally effective as drug Y, but drug X is more potent (less dose to give maximal response).

NEWMAN PROJECTION OF ACYCLIC SYSTEMS

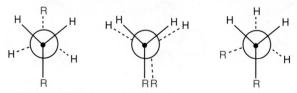

Trans, staggered, *anti*   Fully eclipsed   Skew, gauche

CONFORMATIONS OF CYCLIC SYSTEMS

Cyclopropane   Cyclobutane   Cyclopentane
planar   folded   envelop; puckered; half-chair

Cyclohexane   Cyclohexane   Cyclohexane
chair   boat   skew

Figure 2.4   Depiction of Newman projections of acyclic conformers and the different types of cyclic conformers based upon ring size.

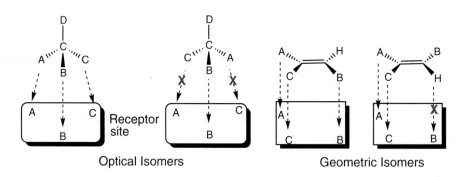

Optical Isomers   Geometric Isomers

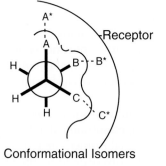

Conformational Isomers

Figure 2.5   Depiction of how optical, geometric, and conformational isomers affect the interaction of a drug molecule with its receptor.

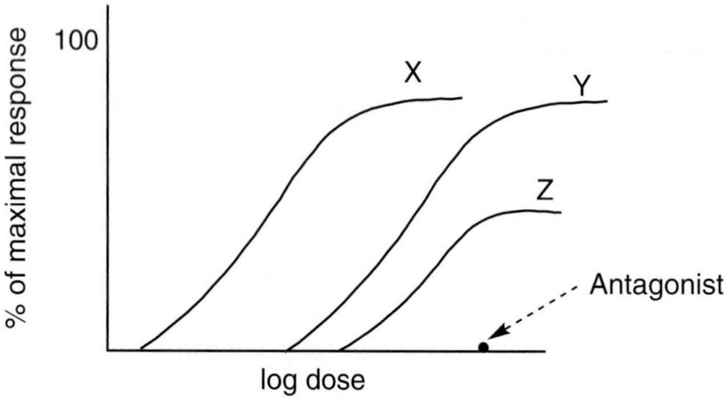

Figure 2.6   Dose–response curves representing different types of drug biological responses.

- Drug Z is less potent than either drug X or Y and less effective as well.
- Antagonists would not give any dose response curve even if drug X is subsequently administered.
- Dose–response relationships can also be expressed by the following equation:

$$D + R \underset{k_2}{\overset{k_1}{\rightleftharpoons}} [DR] \xrightarrow{k_3} E$$

Dose receptor equation

- k_1 is a measure of drug affinity for the receptor, k_2 measures lack of affinity, and k_3 measures effectiveness (intrinsic activity).
  - For agonist, $k_1 > k_2$, and k_3 is large
  - For inactive drug, $k_2 < k_1$
  - For partial agonist, $k_1 > k_2$, and k_3 is small
  - For antagonist, $k_1 > k_2$, and k_3 is 0

Drug Receptors and the Biologic Response

- Binding of a drug or endogenous substance to its receptor allows for physiological messages to be communicated from the drug and/or endogenous substances into the cells without the drug or endogenous substance having to cross through the cell membrane.
- This method of communication is referred to as signal transduction.
- Signal transduction involves a sequence of biochemical reactions inside the cell which are carried out by enzymes, proteins, and ions (especially Ca^{2+}) that are linked through second messengers, such as cyclic adenosine monophosphate (cAMP), inositol 1,4,5-triphosphate (IP_3) and diacylglyercol (DAG).
- There are four major types of receptors: transmembrane ion channels (voltage or ligand gated), transmembrane G-protein–coupled receptors, transmembrane catalytic or enzyme-coupled receptors, and intracellular cytoplasmic/nuclear receptors (Fig. 2.7).

Transmembrane Ion Channels

- There are two types of transmembrane ion channels: voltage-gated and ligand-gated.

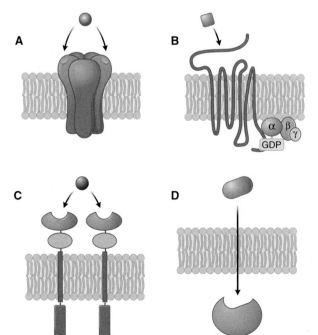

Figure 2.7 Major classes of drug receptors. **A.** Transmembrane ligand-gated ion channel receptor. **B.** Transmembrane G-protein–coupled receptor (GPCR). **C.** Transmembrane catalytic receptor or enzyme-coupled receptors. **D.** Intracellular cytoplasmic/nuclear receptor. (Reprinted with permission from Simon JB, Golan DE, Tashjian A, Armstrong E, et al., eds. Chapter 1, Drug–Receptor Interactions. In: *Principles of Pharmacology: The Pathophysiologic Basis of Drug Therapy*, 2nd ed. Baltimore: Lippincott Williams & Wilkins, 2004: 3–16)

Voltage-Gated Ion Channels

- Opened and closed by changes in the membrane potential
- Ions are Na^+, K^+, and Ca^{2+}
- Examples:
  - Neurotransmission: Sodium channels are responsible for impulse conduction in sensory nerve fibers.
  - Hormone release: Voltage-gated Ca^{2+} channels and ATP-sensitive K^+ channels work together to secrete insulin from β-cells of the pancreas.
  - Muscle contraction: Voltage-dependent calcium channels are found in pacemaker cells and skeletal, cardiac, and smooth muscles.

Ligand-Gated Ion Channels

- Action regulated by binding of a ligand to the channel.
- Ligands include acetylcholine, γ-aminobutyric acid (GABA), and glutamate.
- Action modified by many drugs (i.e., benzodiazepines for GABA receptor).
- Examples:
  - Nicotinic receptors: Binding of acetylcholine results in Na^+ influx, generation of an action potential, and activation of skeletal muscle contraction.
  - $GABA_A$ receptors: Found in the CNS. Binding of GABA results in influx of Cl^- and hyperpolarization of the responsive cells.

Transmembrane G-Protein–Coupled Receptors

- A single peptide with seven membrane-spanning regions linked to a G-protein.
- G-protein consists of three subunits: an α-subunit that binds guanosine triphosphate (GTP) and a βγ-subunits.
- Classification of particular G-proteins is based upon the characteristics of the α-subunit.
  - G_s increases both Adenylyl cyclase activity and Ca^{2+} currents.
  - G_i decreases Adenylyl cyclase activity and increases K^+ currents.
  - G_o decreases Ca^{2+} currents.
  - G_q increases phospholipase C activity.
- Binding of ligand catalyzes the exchange of GTP for G-protein–bound guanosine diphosphate (GDP).
- The GTP–G-protein then interacts with specific secondary messenger systems to elicit a biological response.
- Second messenger pathways:
  - Adenylyl cyclase (messenger = cAMP)
  - Phospholipase A_2 (messenger = arachidonic acid)
  - Phospholipase C (messenger = IP_3 and DAG)
  - Ion channels for Ca^{2+}, K^+, and Na^+
  - This process is called "Signal Transduction" because the second messenger amplifies the ligand signal causing further actions within the cell.
- Examples:
  - β-Adrenergic receptor is a G_s-type G-protein receptor. When endogenous norepinephrine binds it causes interaction with Adenylyl cyclase to form cAMP as the secondary messenger which results in intracellular actions that lead to heart muscle contraction, smooth muscle relaxation, and glycogenolysis (Fig. 2.8).
  - Histamine receptor: There are 4 subtypes – H_1 to H_4. The H_1 receptor is a G_q-type G-protein receptor. When endogenous histamine binds to the H_1 receptor it causes interaction with phospholipase C which produces the secondary messengers inositol 1,4,5-trisphosphate (IP_3) and diacylglycerol (DAG) by the hydrolysis of phosphatidylinositol 4,5-bisphosphate (PIP_2).
  - Both DAG and IP_3 phosphorylate proteins leading to a number of effects including GI smooth muscle contraction, itching, systemic vasodilation, and bronchoconstriction (Fig. 2.8).

Transmembrane Enzyme-Linked Receptors

- This type of receptor has cytosolic enzyme activity as an integral component of its structure.

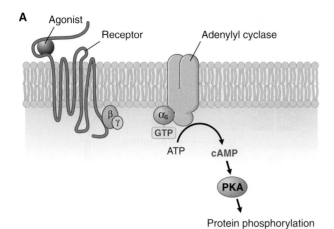

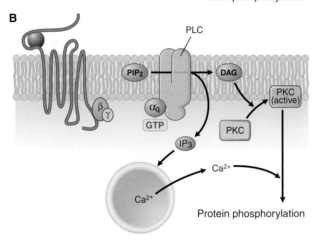

Figure 2.8 Second messenger mechanisms for cAMP and inositol triphosphate. (Reprinted with permission from Simon JB, Golan DE, Tashjian A, Armstrong E, et al., eds. Chapter 1, Drug-receptor interactions. In: *Principles of Pharmacology: The Pathophysiologic Basis of Drug Therapy*, 2nd ed. Baltimore: Lippincott Williams & Wilkins, 2004: 3–16)

- Binding of ligand can either activate (most of the time) or inhibit the cytosolic enzyme activity.
- The most common enzyme linked to a receptor is a tyrosine kinase and binding of ligand will convert the inactive kinase to its active form which then autophosphorylates the receptor, which in turn phosphorylates tyrosine residues on specific cytosolic proteins activating a cascade of intracellular effects.
- Example:
  - The insulin receptor: When insulin binds to its receptor the intrinsic tyrosine kinase activity causes autophosphorylation of the receptor itself. This then allows the receptor to phosphorylate insulin receptor substrate peptides (IRSPs) that start a cascade of intracellular activations (IP_3 and mitogen-activated protein kinase [MAPK] system) which ultimately result in a number of cellular effects including the translocation of the glucose receptor to the cell membrane.

Intracellular Cytoplasmic/Nuclear Receptors

- This type of receptor is located within the cytoplasm of the cell or is bound to the nuclear surface.
- Cytoplasmic/nuclear receptors are a single polypeptide with three functional domains.
  - The amino terminal which binds a modulator protein (transcription activation domain).
  - The carboxy terminus which contains the ligand binding site.
  - In the middle of the receptor is a binding site for DNA which contains critically placed cysteine amino acids which chelate with Zn^{2+} ions that bind to DNA.
  - When ligand binds it activates the receptor by causing the release of the heat-shock protein-90.
  - The receptor–ligand complex then dimerizes and translocates into the nucleus and binds to a DNA hormone response element (HRE), which in turn initiates translation of the target gene.

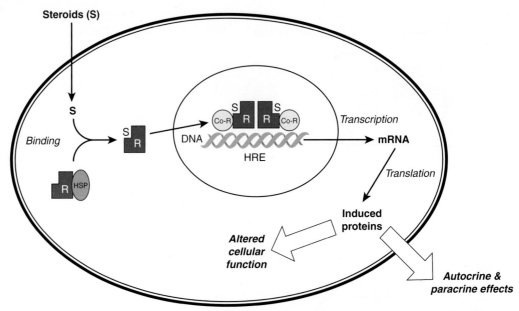

Figure 2.9 Example of intracellular cytoplasmic receptor–steroid mechanism of action (from Foye's 7th edition, Chapter 28).

- Endogenous ligands include nitric oxide, steroid hormones, and vitamin D (Fig. 2.9).

Drug Metabolism

- Humans inhale and ingest a variety of foreign substances including drugs (collectively called "xenobiotics") that can be toxic (see Chapter 4 in *Foye's Principles of Medicinal Chemistry, Seventh Edition,* for a complete detailed discussion). Protection against these xenobiotics relies on detoxification via metabolic reactions to yield less toxic compounds.
- Most drug metabolism occurs in the liver, where hepatocytes are equipped with a wide variety of nonspecific enzymes, primarily in the smooth endoplasmic reticulum.
- Extrahepatic sites of metabolism include the gut (major), kidney, lung, plasma, and skin.
- Overall, the result of metabolism is to convert the drug to a more water-soluble compound by increasing its polarity.
- The metabolism of a drug can result in four possible effects upon its action:
  - Decreased pharmacological activity (deactivation)
  - Increased pharmacological activity (activation)
  - Increased toxicity (carcinogenesis, cytotoxicity)
  - Altered pharmacological activity
- Orally administered drugs are extensively metabolized before reaching their site of action, for example:
  - First-pass metabolism: Drug absorbed into hepatic portal system and, if lipophilic enough, will gain access to hepatocytes.
  - Presystemic metabolism: Gut mucosa is rich in oxidizing enzymes and conjugating transferase enzymes which can deactivate the drug before it ever gets absorbed.

Classification of Drug Metabolism

- Drug metabolic reactions are divided into two phases: phase 1 and phase 2.
- Phase 1 metabolism serves to functionalize the drug molecule and includes:
  - Oxidations (liver microsomes): The most common metabolic reaction and is catalyzed by a family of Cytochrome P450 enzymes (CYP450).
  - Reductions (liver microsomes).
  - Hydrolyses (liver microsomes and plasma).

- Phase 2 metabolism serves to attach groups to phase 1 derivatives to mask them by the addition of a new group and generally makes them more water soluble; they include:
  - Methylation of oxygen, nitrogen, and sulfur
  - N-Acetylation of primary amines
  - Conjugation reactions with glucuronic acid, sulfate, glutathione and amino acids (glycine, glutamine)

Factors Affecting Drug Metabolism

Genetic Factors

- Individual metabolism can be affected by the genetic expression of the many metabolizing enzymes (polymorphisms).
- This can lead to differences in drug responsiveness including drug sensitivity, drug resistance, drug–drug interactions, and drug toxicity.
- The fields of pharmacogenomics and pharmacogenetics study the genomic and genetic differences between racial and ethnic groups that have different levels of gene expression and enzyme levels which can lead to very different drug responses between individuals.
  - For example, genetic polymorphisms (mutations) in one isoform of cytochrome P450 (CYP2D6) can result in poor, intermediate, or ultra-rapid metabolism of over 30 cardiovascular and CNS drugs.

Physiologic Factors

- Age is a factor since the very young and the very old have impaired metabolism.
  - In the elderly, there is a decline in drug metabolism related to decreases in hepatic blood flow, glomerular filtration rate, hepatic microsomal enzyme activity, plasma protein binding, and body mass resulting in profound changes in drug plasma concentrations and renal clearance.
  - There are a number of important metabolizing enzymes in the placenta helping to protect the fetus. However, the fetus and newborns are deficient in some microsomal enzymes, particularly glucuronide conjugation, which can lead to jaundice in newborns.
- Gender differences in metabolism between men and women, although not clearly understood, might be associated with hormonal differences which can lead to different levels of inducible oxidizing enzymes (CYP3A4). For example:
  - Side-chain oxidation of propranolol is 50% faster in males than in females.
  - Drugs that are cleared by oxidative metabolism are more rapid in males than in females (i.e., chlordiazepoxide and lidocaine).
  - Diazepam, prednisolone, caffeine, and acetaminophen are metabolized slightly faster by women than by men.

Pharmacodynamic Factors

- Dose, frequency, and route of administration, plus tissue distribution and protein binding of a drug may affect its metabolism.
  - Since there is a limited amount of glutathione available for conjugation, the dose and frequency of dosing of drugs conjugated by glutathione can saturate the system and lead to alternative paths of metabolism.

Environmental Factors

- Ingested or inhaled environmental substances can compete for metabolizing enzymes, induce or inhibit enzymes, or even poison enzymes (carbon monoxide and pesticides).

Drug–Drug Interactions

- Drug–drug and drug–food/herbal interactions are a common clinical problem.
- Some drugs act as potent enzyme inducers while others are inhibitors (Table 2.1).
- CYP450 isozymes are the enzymes most affected.
- Inhibition of metabolizing enzymes can lead to prolonged drug action and possible toxic overdose.

Table 2.1 Metabolism Inhibiting and Inducing Drugs

Metabolism Inhibitors	Example	Metabolism Inducers	Example
Macrolide antibiotics	Erythromycin	Tuberculosis drugs	Rifampin, Isoniazid
Imidazole antifungals	Ketoconazole	Anti-seizure drugs	Phenobarbital, Phenytoin
Quinolone antimicrobial	Ciprofloxacin	Environmental toxins	Polycyclic aromatic hydrocarbons
Additional drugs	Cimetidine	Additional drugs	Prednisone
	Omeprazole		Ethanol
	Ranitidine		

DRUG–DRUG INTERACTIONS

Examples of GFJ–Drug Interactions
GFJ → Statins rhabdomyolysis
GFJ → Dihydropyridines increased vasodilation
GFJ → Repaglinide hypoglycemia
GFJ → Amiodarone increased toxicity
GFJ → Clopidogrel modulated antiplatelet activity

- Induction of metabolizing enzymes decreases duration of action and causes ineffectual drug therapy.
- Most drug–drug interactions are metabolism-based, that is, competition between two drugs for the enzyme-active site.
- Some drug–drug interactions are mechanism-based in that the inhibitory effect of the drug only occurs after an activation step producing an active inhibitor.
- Grapefruit juice (GFJ) can significantly increase the oral bioavailability of drugs that are metabolized by CYP3A4 in the intestinal tract. In addition, compounds present in GFJ can inhibit intestinal P-glycoprotein (P-gp), which can also enhance drug bioavailability.

Drug–Dietary Supplement Interactions

- The present interest and widespread use of herbal remedies creates the possibility of their interaction with prescription drugs.
- Some of the more common dietary supplements that cause interactions include St. John's Wort (SJW), Echinacea, Ginko Biloba, and Kava (see Chapter 4 in *Foye's Principles of Medicinal Chemistry, Seventh Edition*).
- Some dietary supplements are hepatotoxic: Chaparral, Comfrey, Germander, Skullcap, Pennyroyal, and Valerian Root.

Phase 1 Metabolism

Hepatic Cytochrome P450 Enzyme System (CYP450)

- CYP450 enzyme system is a collection of membrane-bound enzymes located in the endoplasmic reticulum, also called mixed function oxidases or monooxygenases.
- Also found extra-hepatically: lung, intestine, skin.
- Inserts a single oxygen atom into the substrate (RH to ROH).
- Requires:
  - A heme protein: CYP450 is in a resting state when bound to Fe^{3+} (oxidized) and an active state when bound to Fe^{2+} (reduced form).
  - NADH or NADPH: cofactors of CYP450 reductase or CYP450 b_5 reductase (nonheme flavoprotein) to provide an electron to the S-CYP450 complex yielding a peroxide dianion intermediate.
  - A hydrogen atom present on the drug molecule that can be abstracted.
  - Molecular oxygen (O_2).
- Results in the activation of O_2 as an e^- acceptor; oxidation of substrate to OH and formation of one molecule of H_2O.
- Figure 2.10 gives the steps involved in CYP450 oxidation:
  - *Step a:* Drug binds to the Fe^{3+}–P450 complex.
  - *Step b:* Drug–Fe^{3+}–P450 complex is reduced to a drug–Fe^{2+}–P450 complex from an electron originating in a NADPH reductase (flavoprotein complex).
  - *Step c:* The drug–Fe^{2+}–P450 complex binds dioxygen as the sixth ligand of Fe^{2+}.
  - *Step d:* The dioxygen–drug–Fe^{2+}–P450 complex rearranges to a superoxide–drug—Fe^{3+}–P450 complex.
  - *Step e:* The superoxide–drug–Fe^{3+}–P450 complex is reduced by accepting a second electron from a flavoprotein complex and becomes a two-electron–reduced peroxy–drug–Fe^{3+}–P450 complex.

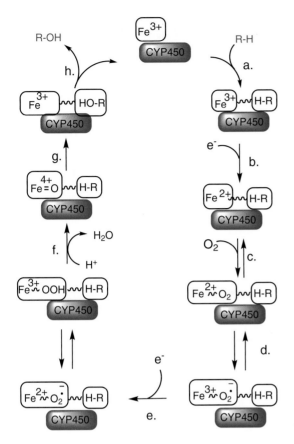

Figure 2.10 The cyclic mechanism for CYP450. The substrate is RH and the valence state of the heme iron in CYP450 is indicated.

- *Step f:* The peroxy drug–Fe^{3+}–P450 complex undergoes heterolytic cleavage of the peroxide anion to form water and a ferryl ($Fe^{4+} = O$) complex which is the catalytically active oxygenation species.
- *Step g:* Abstraction of a hydrogen atom from the drug by the ferryl complex produces a radical–hydroxide complex intermediate.
- *Step h:* Radical recombination yields the hydroxylated product and regeneration of the F^{3+}–P450 complex (*step a*).

Classification of CYP450 Family of Genes

- CYP450 is a superfamily of metabolizing genes with the root name of CYP followed by an Arabic numeral designating the family member (e.g., CYP1, CYP2, CYP3, etc.).
- The family name is followed by a capitalized letter denoting the subfamily (e.g., CYP1A, CYP2C, CYP3A, etc.), which is then followed by another Arabic numeral identifying the individual gene (e.g., CYP1A1, CYP2D6, CYP3A4, etc.).
- The primary isoforms responsible for drug metabolism are depicted in Figure 2.11. It is evident that CYP3 and CYP2 families are involved in the majority of clinically relevant drug metabolism.
  - Specifically, CYP3A4/5; and CYP2A6, 2B6, 2C8/9, 2D6, and 2E1 are the most relevant genes.
  - CYP1A1/2 are primarily involved in the metabolism of environmental substances (e.g., polyaromatic hydrocarbons [PAHs]).

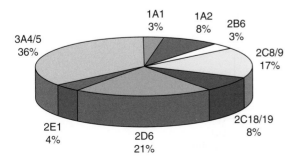

Figure 2.11 Percentage of clinically important drugs metabolized by human CYP450 isoforms.

Table 2.2 Hydroxylation Reactions Catalyzed by CYP450

Aromatic hydroxylation

$$CH_3CO-\overset{H}{N}-C_6H_5 \xrightarrow{[OH]} CH_3CO-\overset{H}{N}-C_6H_4-OH$$

Aliphatic hydroxylation

$$R-CH_3 \xrightarrow{[OH]} R-CH_2-OH$$

Deamination

$$R-CH(NH_2)-CH_3 \xrightarrow{[OH]} \left(R-C(OH)(NH_2)-CH_3\right) \longrightarrow R-CO\cdot CH_3 + NH_3$$

O-Dealkylation

$$R-O-CH_3 \xrightarrow{[OH]} \left(R\cdot O-CH_2-OH\right) \longrightarrow R-OH + CH_2O$$

N-Dealkylation

$$R-N(CH_3)_2 \xrightarrow{[OH]} \left(R-N(CH_2OH)(CH_3)\right) \longrightarrow R-NH\,CH_3 + CH_2O$$

$$R-NH-CH_3 \xrightarrow{[OH]} \left(R-NH-CH_2OH\right) \longrightarrow R-NH_2 + CH_2O$$

N-Oxidation

$$(CH_3)_3-N \xrightarrow{[OH]} \left((CH_3)_3-NOH\right) \longrightarrow (CH_3)_3-NO + H^{\oplus}$$

Sulfoxidation

$$R-S-R' \xrightarrow{[OH]} \left(\underset{OH}{R-S-R'}\right) \longrightarrow \underset{O}{R-S-R'} + H^{\oplus}$$

Sources of PAHs

PAHs are measured as benzo[a]pyrenes and their major sources are:

Cigarette smoke

Burning wood and charcoal briquettes

Wax candles

Automobile, bus, truck exhaust

- CYP450 isoenzymes oxidize compounds via a large variety of chemical reactions including hydroxylation of alkanes, aromatics, polycyclic hydrocarbons, the epoxidation of alkenes, the dealkylation of secondary and tertiary amines, conversion of amines to N-oxides, and the dehalogenation of halogenated hydrocarbons.
- Table 2.2 gives representative examples.

Azo and Nitrogen Reduction

- Azo compounds can be reduced to amines by azoreductase, an NADPH-dependent enzyme in liver microsomes (Fig. 2.12).

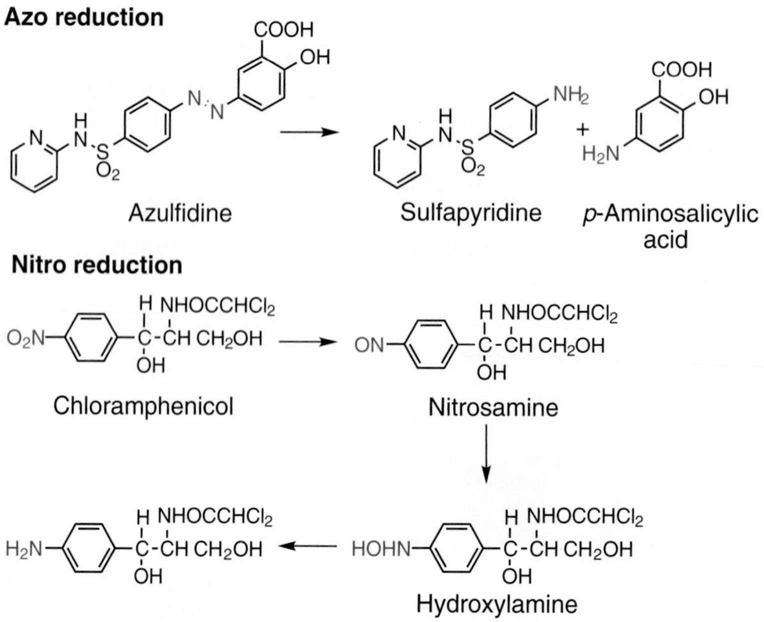

Figure 2.12 CYP450 catalyzed reduction of azo and nitro compounds.

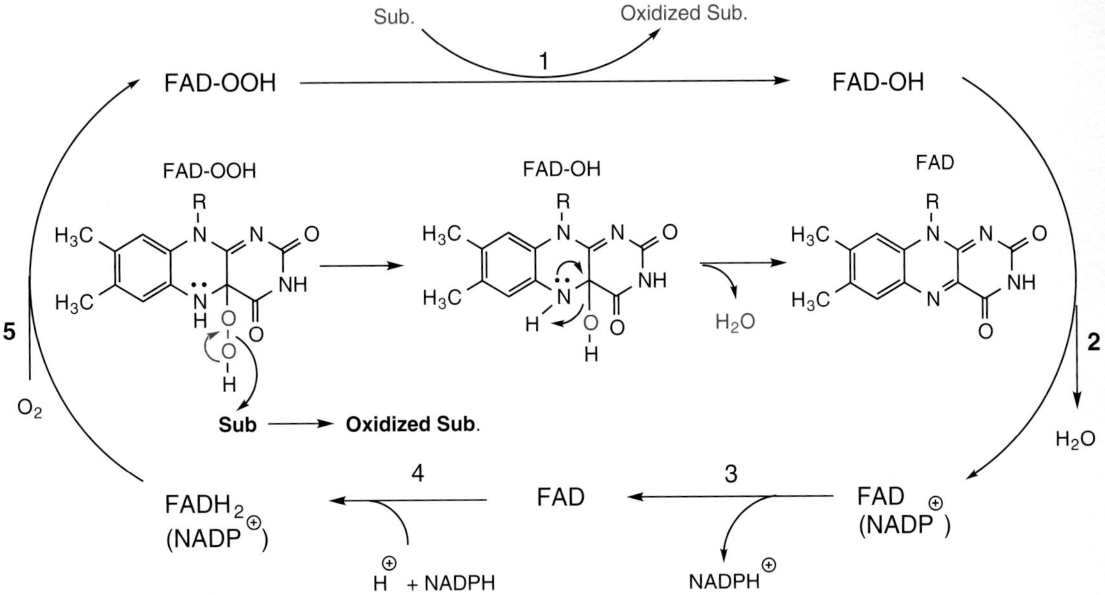

Figure 2.13 The FMO catalytic cycle. Oxygenated substrate is formed by nucleophilic attack of a substrate (Sub.) by the terminal oxygen of the enzyme-bound hydroperoxyflavin (FAD-OOH) followed by heterolytic cleavage of the peroxide (**1**). The release of H_2O (**2**) or of NADP (**3**) is rate limiting for reactions catalyzed by liver FMO. Reduction of flavin by NADPH (**4**) and addition of oxygen (**5**) complete the cycle by regeneration of the oxygenated FAD-OOH.

- Aromatic nitro compounds are reduced to aromatic primary amines by a nitroreductase.

Oxidations Catalyzed by Flavin Monooxygenases (FMOs)

- FMO primarily oxidizes drugs and other xenobiotics containing nitrogen and sulfur nucleophiles.
- FMO utilizes a two-electron–reducing equivalent of NADPH (hydride anion) to reduce one atom of molecular oxygen to water, while the other oxygen atom is used to oxidize the substrate.
- Humans express five different FMOs (FMO1 to 5) which are tissue specific with different substrate specificities.
- Figure 2.13 demonstrates the catalytic steps for FMO.

Peroxidases and Other Monooxygenases

- Peroxidases are hemeproteins closely related to CYP450 but differing in the non-porphyrin coordinating ligands (cysteine in CYP450 and histidine in peroxidases).
- Peroxidase oxidation involves the formation of a ferryl-oxo intermediate analogous to CYP450, but oxidations are limited to electron-rich substrates easily oxidized (e.g., heteroatom oxygenation, aromatization of 1,4-dihydropyridines and arylamines).

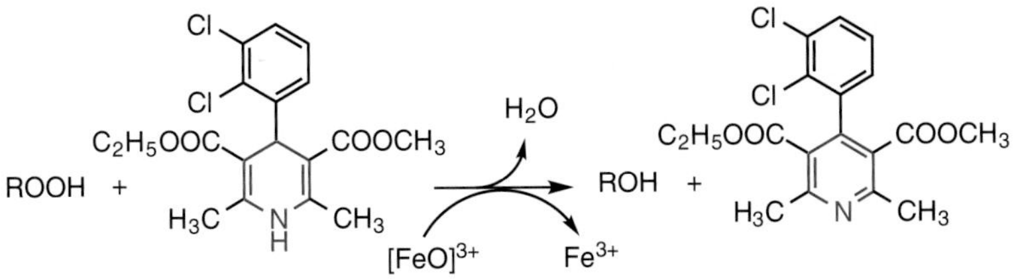

- Cyclooxygenase (COX), also known as prostaglandin synthase or prostaglandin endoperoxidase.
- Widely distributed heme peroxidase involved in the formation of prostanoids from arachidonic acid (see Chapter 31 in *Foye's Principles of Medicinal Chemistry, Seventh Edition,* for a complete detailed discussion).

- There are three COX isoenzymes: COX-1, COX-2, and COX-3, and they are inhibited by NSAIDs.
- Myeloperoxidase (MPO) is a heme peroxidase occurring in neutrophil granulocytes.
- It oxidizes chloride ion to hypochlorous acid (HOCl) which is cytotoxic to bacteria, viruses, and other pathogens.
- Other peroxidases: horseradish peroxidase (plants), lactoperoxidase (breast milk), and thyroid peroxidase (produces thyroid hormones from iodide).

Non-Microsomal Oxidations

- Oxidases and dehydrogenases found in mitochondrial and soluble fractions of tissue homogenates (Fig. 2.14).
- Oxidation of alcohols:
  - Alcohol dehydrogenases (ADHs) are a family of nicotinamide adenine dinucleotide (NAD)-specific dehydrogenases.
  - Oxidize primary and secondary alcohols to aldehydes and ketones with the reduction of NAD+ to NADH.
  - Primary pathway for the metabolism of ethanol.
- Oxidation of aldehydes:
  - Aldehyde dehydrogenases (ALDHs) are a family of polymorphic NAD-specific dehydrogenases (ALDH$_1$:cytosolic; ALDH$_2$:mitochondrial; and ALDH$_3$:stomach, cornea, and tumors).
  - Oxidize aldehydes to carboxylic acids.
  - Aldehyde oxidase (AO) is a molybdenum cofactor-specific enzyme that generates carboxylic acids and hydrogen peroxide from aldehydes in the presence of oxygen.
  - AO requires FAD in addition to molybdenum and mainly oxidizes nitrogen heterocyclic compounds.
  - Aldo-keto reductases (AKRs) are a superfamily of enzymes that are NADH/NADPH-dependent oxidoreductases.
  - AKRs metabolize sugars to sugar alcohols as well as metabolizing reactive aldehydes, prostaglandins, steroid hormones, and chemical carcinogens.
- Xanthine oxidase and xanthine dehydrogenase.
  - Both enzymes are often referred to as xanthine oxidoreductase.
  - Xanthine dehydrogenase can be converted to xanthine oxidase by reversible sulfhydryl oxidation, that is, they are interconvertible.
  - Xanthine oxidase is a molybdo-flavoenzyme that requires molybdenum, two iron–sulfur centers, and FAD.

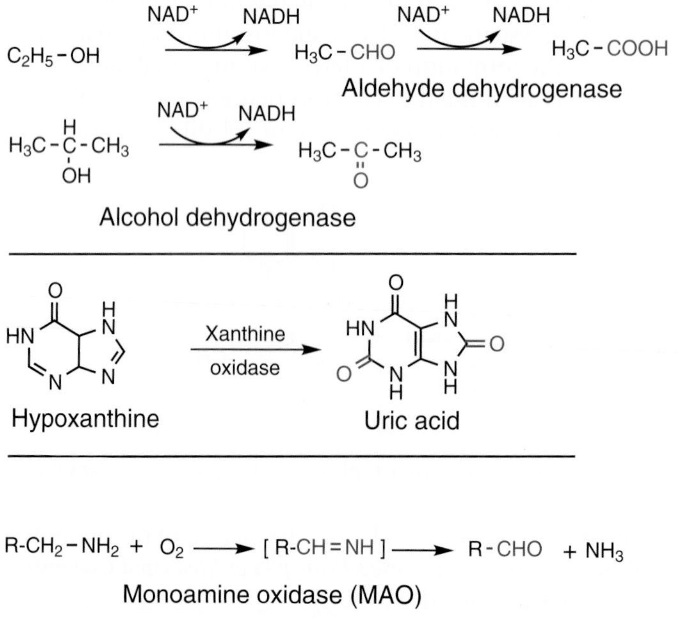

Figure 2.14 Examples of non-microsomal oxidations: ADH and ALDH, xanthine oxidase, and MAO.

- These enzymes are involved in purine metabolism converting hypoxanthine to uric acid and found primarily in the liver.
- Oxidative deamination of amines
  - Monoamine oxidase (MAO) is a mitochondrial membrane flavin-containing enzyme that catalyzes the oxidative deamination of monoamines.
  - The deamination requires oxygen to remove the amine group resulting in the formation of the corresponding aldehyde and release of ammonia.
  - The amine must be primary; however, secondary and tertiary amines with methyl substituents are also reactive.
  - The amine must be attached to an unsubstituted methylene group and substrates with α-carbon atoms are poorly or not oxidized at all.
  - MAO is important in regulating the metabolic degradation of catecholamines and serotonin in neural tissues, and hepatic MAO plays an important role in deactivating monoamines absorbed via the GI tract.
  - Diamine oxidase (DAO) is very similar to MAO; however, it catalyzes the oxidative deamination of diamines (putrescine, cadaverine, spermine) and histamine to yield aldehydes.
  - DAO is found in kidneys, intestines, liver, lung, and nervous tissue.

Hydrolysis of Esters and Amides

- Ester hydrolysis yields an acid and an alcohol. Nonspecific esterases (carboxylesterases, cholinesterase, pseudocholinesterase) are found in plasma and microsomes (Fig. 2.15).
- Amide hydrolysis yields an acid and an amine. Much slower than ester hydrolysis and found only in microsomes and lysosomes within the cell. Never in the plasma.

Phase 2 Metabolism

- Phase 2 metabolism involves drug conjugation reactions which usually renders the drug inactive and water soluble and easily excreted via urine or bile.
- The major conjugation reactions are glucuronidation, sulfonation, amino acid conjugations (glycine, glutamine), reduced glutathione, acetylation, and methylation.
- Each conjugation reaction uses specific transferase enzymes which involve adding an endogenous substrate (e.g., sulfate, amino acid, glucuronic acid) to a functional group on the drug molecule. The functional group can be initially present or created as a result of phase 1 metabolism.

Glucuronide Conjugation

- Glucuronide conjugation is the most common conjugation reaction.
- It requires glucuronic acid which is formed by the oxidation of glucose 1-phosphate.

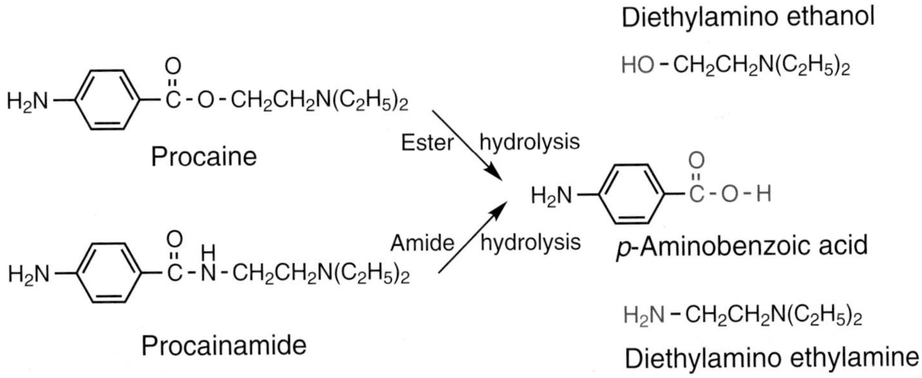

Figure 2.15 Example of ester and amide hydrolysis reactions using procaine (ester) and procainamide (amide).

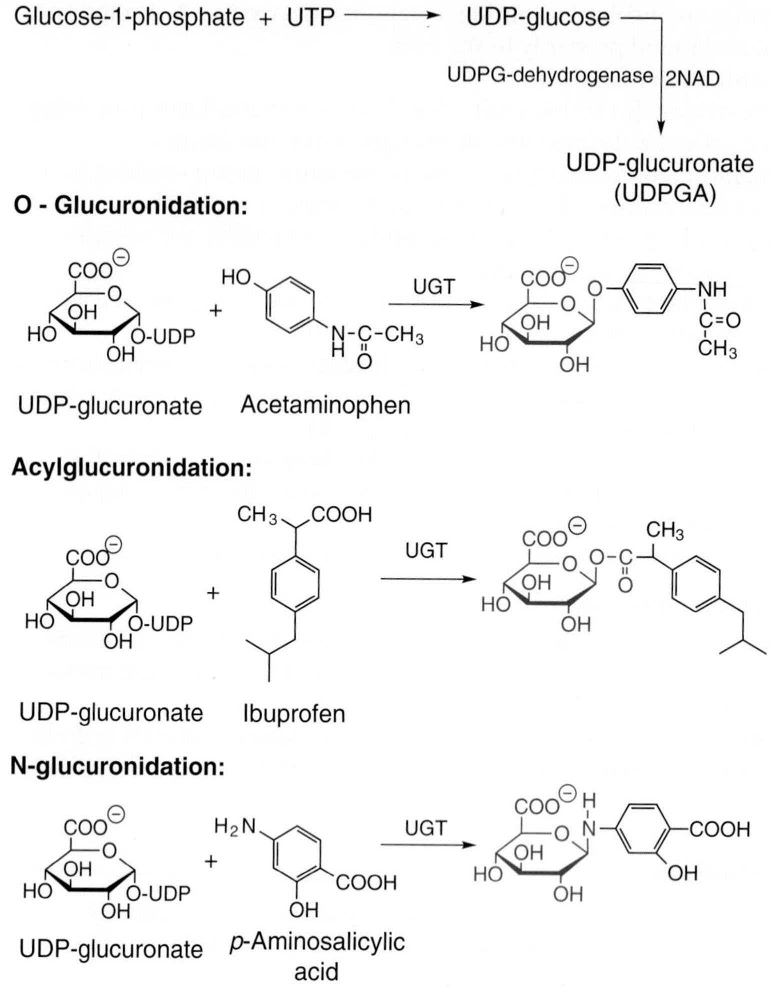

Glucose-1-phosphate + UTP ⟶ UDP-glucose

UDPG-dehydrogenase | 2NAD

UDP-glucuronate
(UDPGA)

O - Glucuronidation:

UDP-glucuronate   Acetaminophen

Acylglucuronidation:

UDP-glucuronate   Ibuprofen

N-glucuronidation:

Figure 2.16 Glucuronidation pathways catalyzed by UDP-glucuronosyltransferases (UGTs).

UDP-glucuronate   *p*-Aminosalicylic acid

- The enzyme is uridine diphosphate (UDP)–glucuronosyltransferase (UGT) which utilizes uridine diphosphate-glucuronic acid (UDPGA) as the activated cofactor (Fig. 2.16).

Sulfate Conjugation

- Sulfate conjugation is most important for the biotransformation of steroid hormones, catecholamines, thyroxine, bile acids, and phenolic drugs.
- Sulfate conjugated metabolites have a low pKa (1 to 2) and are almost totally ionized and, therefore, highly soluble and easily excreted via urine.
- Drugs are sulfonated by the transfer of a sulfonic group ($-SO_3^-$) from 3'-phosphoadenosine-5'-phosphosulfate (PAPS) by sulfotransferase (SULT).
- PAPS is formed from inorganic sulfate and adenosine triphosphate (ATP) (Fig. 2.17).

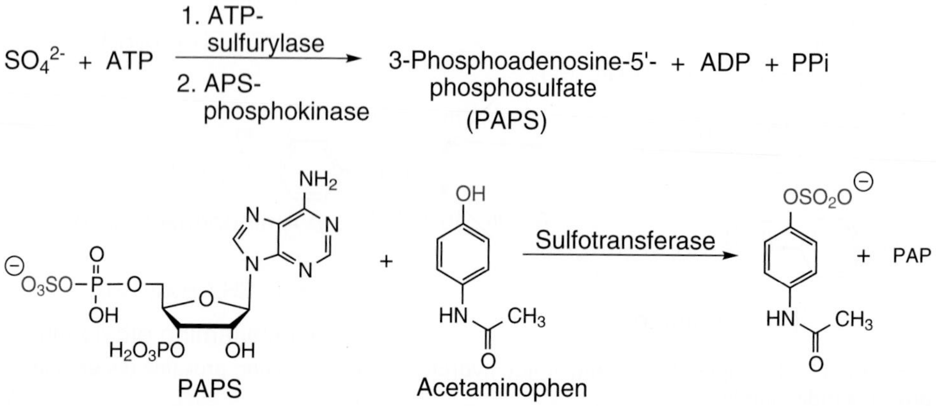

$$SO_4^{2-} + ATP \xrightarrow[\text{2. APS-phosphokinase}]{\text{1. ATP-sulfurylase}} \text{3-Phosphoadenosine-5'-phosphosulfate (PAPS)} + ADP + PPi$$

PAPS   Acetaminophen

Figure 2.17 Sulfate conjugation pathways.

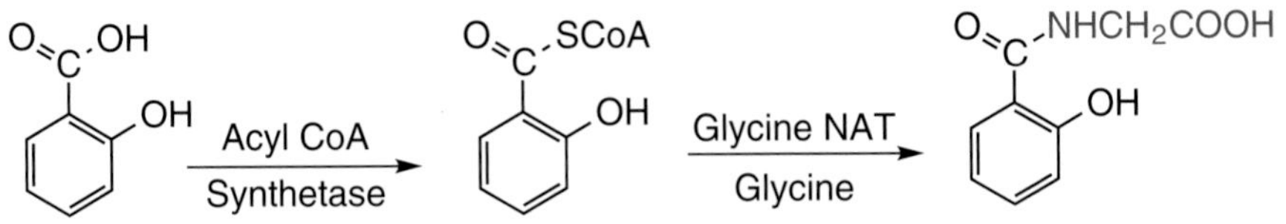

Figure 2.18 Example of amino acid conjugation.

Amino Acid Conjugations

- Amino acids will form an amide with aromatic or aralkyl carboxylic acids.
- Major amino acids are glycine and glutamine, with aspartic acid, taurine, and serine playing a minor role.
- In this reaction, the carboxylic acid group on the drug molecule is activated to an acyl coenzyme A (R-CO-S-CoA) ester by Acyl CoA synthetase and then the primary amine of the amino acid is transferred to form the amide bond by N-acyltransferase (NAT) (Fig. 2.18).

Glutathione Conjugations

- Glutathione (GSH) is a tripeptide made up of glycine, cysteine, and γ-glutamate.
- It is the only conjugating compound that does not require activation since it contains the sulfhydryl group of cysteine which can attack electrophilic centers in drug molecules, that is, GSH conjugation is a nucleophilic attack of the drug molecule by the –SH of glutathione.
- GSH conjugation is catalyzed by glutathione S-transferase (GST) and is followed by the sequential removal of glutamate (glutamyl transferase) and glycine (cysteinylglycinase) to form a mercapturic acid derivative which is easily excreted via urine (Fig. 2.19).
- Hepatic stores of GSH are readily depleted. In case of an overdose with an electrophilic poison that depletes GSH, it is necessary to treat with another SH reagent to consume the toxin. The most common SH reagent is N-acetylcysteine. It forms the identical conjugate as with GSH.

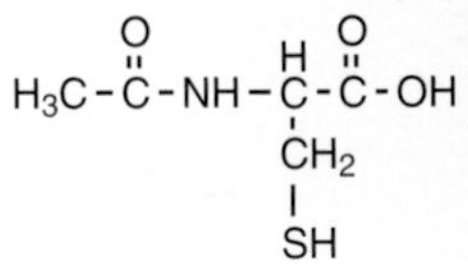

N-Acetylation of Primary Amines

- Acetylation involves the transfer of acetyl CoA to primary aliphatic and aromatic amines, amino acids, hydrazines, or sulfonamide groups.
- Secondary and tertiary amines are not acetylated.

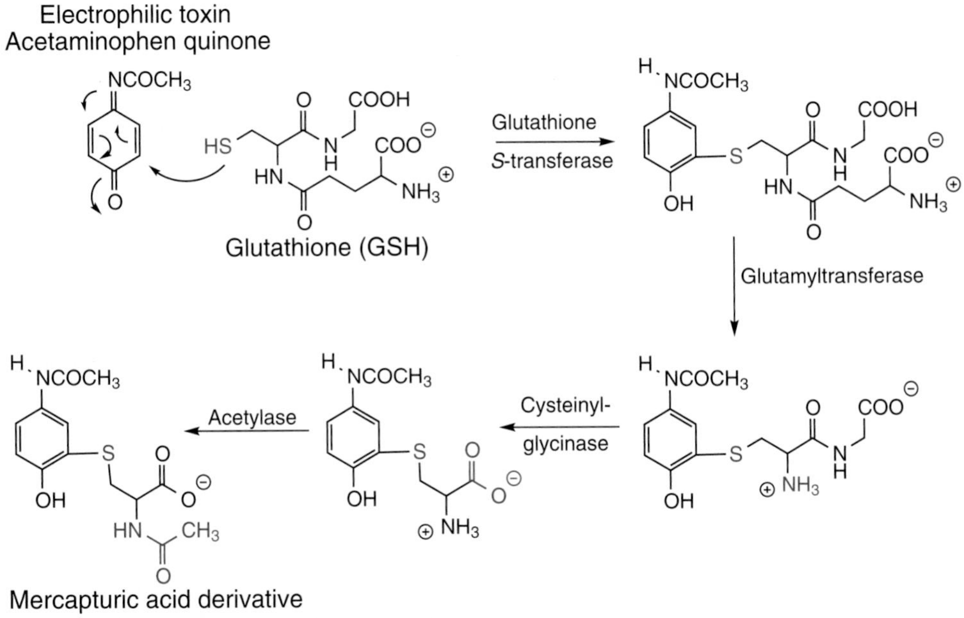

N-Acetylcysteine

Mercapturic acid derivative

Figure 2.19 Glutathione and mercapturic acid conjugation pathways.

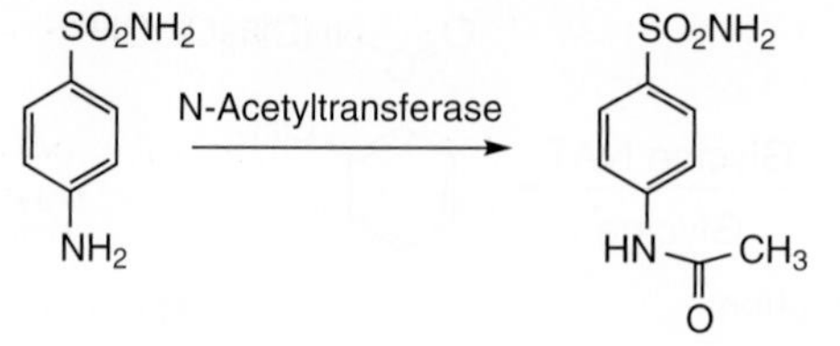

Figure 2.20 Acetylation of sulfanilamide.

Sulfanilamide   N-Acetyl sulfanilamide

- N-acetylation is catalyzed by N-acetyltransferase, which is a liver mitochondrial enzyme that uses acetyl CoA as a coenzyme (Fig. 2.20).
- Polymorphism of N-acetyltransferase results in two phenotypes: slow and fast acetylators.
- Slow acetylators are prone to drug-induced toxicities and accumulate higher concentrations of unacetylated drug.
- Fast acetylators eliminate the drug more rapidly leading to a need to adjust dose and dosing schedules.

Methylation

- Methylation can occur on O, N, or S atoms on an endogenous or drug molecule.
- It is primarily seen with endogenous compounds rather than with drugs and often leads to active metabolites (norepinephrine to epinephrine).
- Although they have specific names, methylation is catalyzed by methyltransferase enzymes that transfer a methyl group from S-adenosylmethionine (SAM).
- SAM is produced from ATP and methionine by the enzyme methionine adenosine transferase (Fig. 2.21).
- Methylation of drugs and other xenobiotics does not enhance water solubility, but usually deactivates the drug.

Membrane Drug Transporters

- Membrane transporters play an important role in the pharmacokinetic (administration, distribution, metabolism, and elimination) profiles of drugs (see Chapter 5 in *Foye's Principles of Medicinal Chemistry, Seventh Edition*, for a complete detailed discussion).

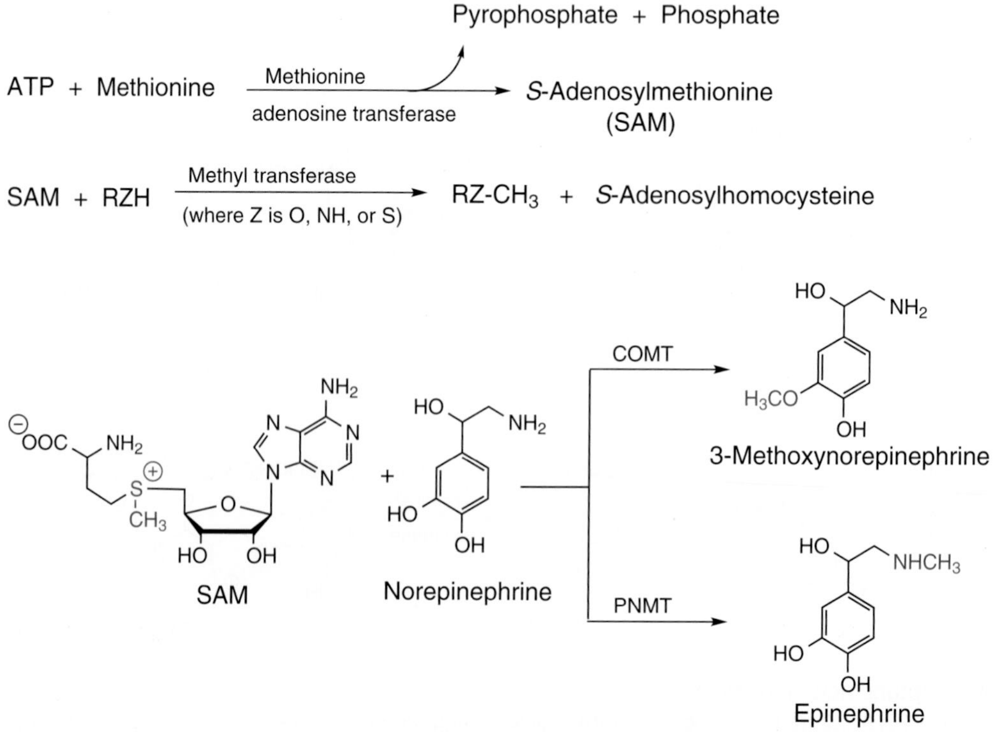

Figure 2.21 Methylation pathways. COMT, catechol methyl transferase; PNMT, phenylethanolamine methyl transferase.

- Membrane transporter interaction leads to changes in transporter function and have effects on other drug coadministered drugs.
- Membrane transporters are proteins that govern the mechanisms by which drugs get into and out of cells in order to reach system circulation as well as their sites of metabolism, storage, and action.
- Membrane transporters are found especially in epithelial tissue of the liver, intestines, and kidney, as well as the blood–brain barrier, testes, and the placenta.

Mechanisms of Membrane Transport

- There are five different mechanisms by which drugs are transported across membranes (Fig. 2.22).
  - Passive diffusion: Depends solely upon a gradient concentration across the membrane and the lipophilic character of both drug (unionized form) and the membranes.
  - Facilitated transport can be facilitative or active.
    - Facilitative transporters: Membrane proteins that move drug molecules down a concentration gradient without the use of a separate energy source.
    - Active transporters: Membrane proteins that require an energy source (usually ATP) to move drug molecules down and against a concentration gradient.
  - Paracellular transport: Drug molecules cross membranes by passing through the space between cells.
  - Transcytosis: The transport of extracellular drug molecules by their inclusion into cell surface vesicles for transport across the interior of a cell.
  - Efflux transporters: Move drug molecules out of cells usually by an active mechanism.

Classification of Membrane Transporters

- There are two superfamilies of transporters: ATP-binding cassette (ABC) and the solute carrier (SLC) superfamilies.
- The ABC type transporters are considered primary active transporters because they use ATP as an energy source.
- SLC transporters are considered secondary active transporters because they use gradients created by primary active transporters (e.g., Na^+/K^+ ATPase and Na^+/H^+ exchangers) to move drugs across membranes.

The ABC Transporters

- ABC transporters are primarily efflux transporters located on the apical side of membranes (facing the lumen) of the intestines, brain, placenta, heart, liver, and kidney.

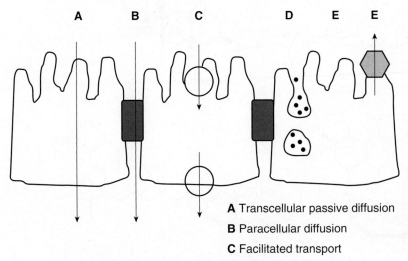

A Transcellular passive diffusion

B Paracellular diffusion

C Facilitated transport

D Transcytosis

E Active efflux

Figure 2.22 Mechanisms of membrane transport. Transport mechanisms most commonly utilized by drug therapeutic agents include passive diffusion and facilitated transport.

- ABC transporters include: P-gp, also known as multidrug resistance (MDR1) protein; breast cancer resistance protein (BCRP), and multidrug resistance-associated proteins (MRP1 to 4).

P-glycoprotein (P-gp): involved in drug transport in the intestines, kidneys, liver, brain, placenta, and in tumor cells

- Consists of 12 transmembrane-spanning domains arranged in two homologous halves and having only one ATP binding site.
- Mechanism of transport is not well understood with at least two drug binding sites and three different proposed mechanisms of transport.
- P-gp has a wide range of substrates it can transport (Fig. 2.23), making it difficult to determine SAR for P-gp–mediated transport.
- Intestinal efflux by P-gp contributes to the low bioavailability of many drugs.

BCRP: most highly expressed in the placenta where it protects the fetus by efflux of harmful xenobiotics

- Consists of six transmembrane-spanning domains with one ATP binding site.
- Also expressed in the intestines, liver, kidney, ovaries, and mammary glands.
- Attributed with causing chemotherapeutic drug resistance due to its overexpression in tumor cells.
- Associated with drug absorption, biliary clearance, and transport of drugs into breast milk.

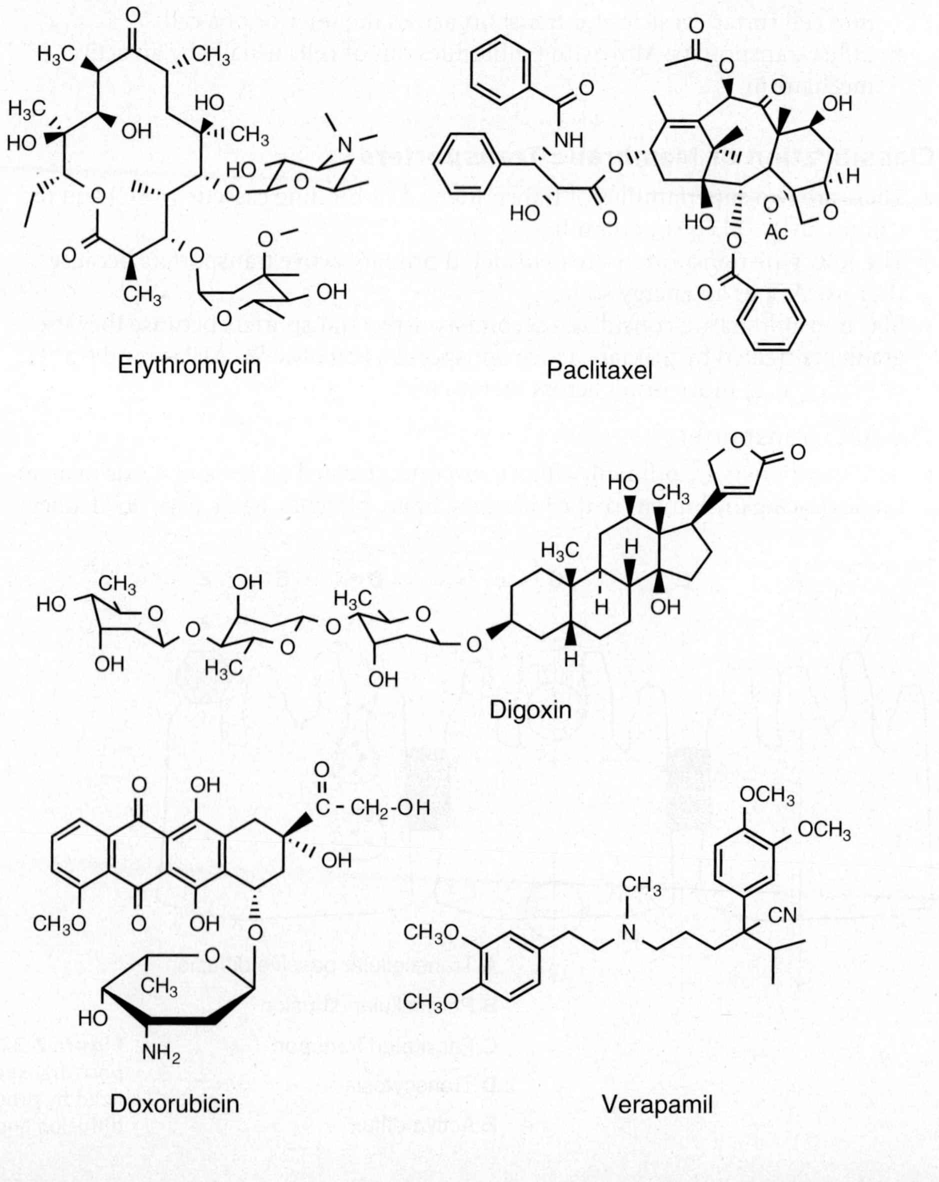

Erythromycin

Paclitaxel

Digoxin

Doxorubicin

Verapamil

Figure 2.23 Examples of P-gp substrates.

MRP1 to 4

- MRP1: It is a basolateral (facing the systemic circulation) efflux transporter that can be overexpressed in cancer cells resulting in chemotherapeutic drug resistance.
- MRP2: It is an apical efflux transporter localized in the liver canalicular membrane, the brush border of the kidney and apical membrane of the gut serving to excrete substrates into the bile, urine, and back into the lumen of the gut.
  - Major substrates are phase 2 conjugates (glucuronide, glutathione, sulfate).
  - Primary function is hepatobiliary transport and enterohepatic cycling.
- MRP3: It is located on the sinusoidal (basolateral) liver membrane effluxing primarily glucuronide conjugates out of the hepatocytes into the plasma resulting in conjugate excretion via urine.
- MRP4: Similar to MRP3 but also plays a role in brush border of the kidney in anionic drug secretion into the urine.

Table 2.3 gives the relevant drug substrates and inhibitors of the ABC transporters.

The SLC Transporters

- These transporters are secondary active transporters utilizing concentration gradients of many other substances to transport drug substances.
- SLC transporters generally facilitate drug influx, although in some physiological locations they can also facilitate efflux.
- They are physiologically expressed throughout the body and transport xenobiotics as well as many endogenous substances.
- SLC transporters include: organic anion transporting polypeptides (OATPs), organic anion transporters (OATs), organic cation/carnitine zwitterion transporters (OCTNs), multidrug and toxin extrusion transporters (MATEs), monocarboxylate transporters (MCTs), peptide transporters (PEPTs), and nucleoside transporters (CNTs).

OATPs: considered antiporters using high intracellular concentrations of glutathione as a driving force. The name implies transporting anions; however, they can also transport cationic and neutral compounds.

- OATP1B1: most clinically relevant since its substrates include the statins (3-hydroxy-3-methyl-glutaryl-coenzyme A [HMG-CoA]) inhibitors which are widely prescribed.
  - OATP1B1 is expressed on the sinusoidal membrane of the liver and is highly polymorphic which leads to decreased hepatic uptake and decreased clearance of the statins and other OATP1B1 substrates.

Table 2.3 Relevant Drug Substrates and Inhibitors of ABC Transporters

Transporter	Substrates	Inhibitors
P-gp (MDR1)	Amphotericin B, cyclosporine, daunorubicin, dexamethasone, digoxin, docetaxel, doxorubicin, erythromycin, etoposide, fexofenadine, hydrocortisone, ketoconazole, methadone, mitoxantrone, morphine, paclitaxel, phenytoin, rifampin, saquinavir, tacrolimus, verapamil, vinblastine, vincristine	Clarithromycin, cyclosporine, elacridar, quinidine, verapamil
MRP2	Ampicillin, ceftriaxone, cisplatin, daunomycin, doxorubicin, etoposide, fluorouracil, glucuronide conjugates, glutathione conjugates, irinotecan, methotrexate, mitoxantrone, olmesartan, pravastatin, ritonavir, rosuvastatin, saquinavir, SN-38-glucuronide, valsartan, vinblastine, vincristine	Cyclosporine, Probenecid, efavirenz, emtricitabine
MRP3	Fexofenadine, glucuronide conjugates, methotrexate	Efavirenz, emtricitabine
MRP4	Adefovir, furosemide, methotrexate, tenofovir, topotecan	Diclofenac
BCRP	Ciprofloxacin, imatinib, irinotecan, methotrexate, mitoxantrone, nitrofurantoin, sorafenib, topotecan	Elacridar

Table 2.4   Relevant Drug Substrates and Inhibitors of SLC Transporters

Transporters	Substrates	Inhibitors
OATP1B1	Atorvastatin, bosentan, caspofungin, cerivastatin, enalapril, fexofenadine, fluvastatin, methotrexate, olmesartan, pravastatin, repaglinide, rifampin, rosuvastatin, simvastatin, SN-38, valsartan	Atorvastatin, clarithromycin, cyclosporine, erythromycin, gemfibrozil, paclitaxel, rifampin, ritonavir, saquinavir, tacrolimus, telmisartan
OATP1B3	Digoxin, docetaxel, fexofenadine, olmesartan, paclitaxel, rifampin	Clarithromycin, cyclosporine, erythromycin, rifampin, ritonavir
OATP1A2	Digoxin, erythromycin, fexofenadine, imatinib, levofloxacin, methotrexate, pitavastatin, rocuronium, rosuvastatin, saquinavir	GFJ, rifampin, ritonavir, saquinavir, verapamil
OATP2B1	Atorvastatin, bosentan, fexofenadine, pravastatin, rosuvastatin	Cyclosporine, gemfibrozil, rifampin
OAT1	Acyclovir, adefovir, cidofovir, ciprofloxacin, lamivudine, methotrexate, penicillins, tenofovir, zidovudine	Probenecid
OAT3	Bumetanide, cefaclor, ceftizoxime, furosemide, NSAIDs, penicillins	Probenecid
OCT1	Metformin, lamivudine, oxaliplatin	Disopyramide, quinidine
OCT2	Cisplatin, lamivudine, metformin, oxaliplatin, procainamide	Cetirizine, cimetidine, procainamide
MATE1	Metformin, oxaliplatin	Cimetidine, quinidine, procainamide
MCT1	Atorvastatin, salicylic acid, pravastatin, γ-hydroxybutyric acid, valproic acid	Dietary flavonoids
PEPT1	Captopril, cefadroxil, cephalexin, enalapril, valacyclovir	Glycylproline, zinc
PEPT2	Captopril, cefadroxil, cephalexin, enalapril, valacyclovir	Fosinopril
CNT/ENT	Clofarabine, gemcitabine, ribavirin	Dipyridamole

- Other OATPs include OATP1B3, OATP1A2, OATP2B1, OATP4C1, all of which function much the same as OATP1B1 and have effects on the distribution of their drug substrates (Table 2.4).
- Figure 2.24 gives the structures of some common OATP substrates.

OATs: are highly expressed in the kidney having a significant role in the renal clearance of many anionic drugs. There are four different types of OATs.

- OAT1 is expressed on the basolateral membrane of the kidney along with OAT3. They both facilitate anion transport into renal tubule cells.
- OAT4 is expressed on the apical membrane and can transport drugs from the tubule cells into urine, but is mainly responsible for renal reabsorption.
- OAT2 is found on the sinusoidal membrane of hepatocytes facilitating the uptake of anion drugs.

OCTs: are electrogenic uniporters which use only cation concentration as a driving force. They require no cosubstrate and result in unequal distribution of ionic charges across the membrane.

- OCT1 is expressed on the basolateral membrane of the liver and OCT2 on the basolateral membrane of the kidney.
- Drug interactions are attributed to OCT-mediated transport when drugs are coadministered with inhibitors. An example is that metformin plasma concentration is increased when coadministered with the OCT inhibitor cimetidine.

OCTNs: are proton antiporters expressed in the kidney where they are localized on the apical membrane

- OCTNs share substrates with OCTs and are important in the placenta where they are responsible for transporting carnitine to the fetus.
- OCTNs also transport ipratropium and tiotropium by bronchial epithelial cells facilitating their activity via inhalation.

MATEs: Include MATE1 and MATE2-K which are extrusion transporters that use proton antiport as a driving force

- MATE1 is an apical transporter on the kidney and canalicular membrane where it acts in concert with OCTs to transport cationic substances. It has been demonstrated to be important in metformin renal excretion.

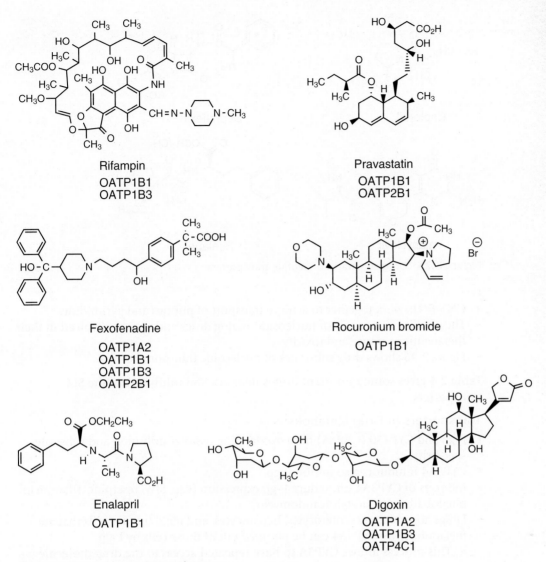

Figure 2.24 Common OATP substrates.

MCTs: are proton-coupled monocarboxylate transporters via symport.

- MCTs are driven by the proton gradient; therefore, they can influx or efflux substrates depending upon this gradient.
- Ubiquitous and play an important role in the transport of lactate and pyruvate.
- Many drugs have been identified as substrates, but clinical relevance is not clear (Table 2.4).
- Highly expressed in the intestines making this transporter a target for enhancing oral absorption.
- Overexpressed in tumor cells and is a likely target for therapeutic strategies for treating some cancers.

PEPTs: mediate the transport of di- and tripeptides via proton symport

- PEPT1 expressed on the apical membrane of the intestines is responsible for drug absorption.
- PEPT2 found on both the brush border membrane and apical membrane of the kidney where it serves to reabsorb peptides from urine in the renal cells.
- Figure 2.25 shows drug substrates for the peptide transporter.

CNTs are concentrative nucleoside transporters and the ENTs are equilibrative nucleoside transporters. CNTs are found on apical membrane of cells and are sodium dependent. ENTs are facilitative and located primarily on basolateral membranes.

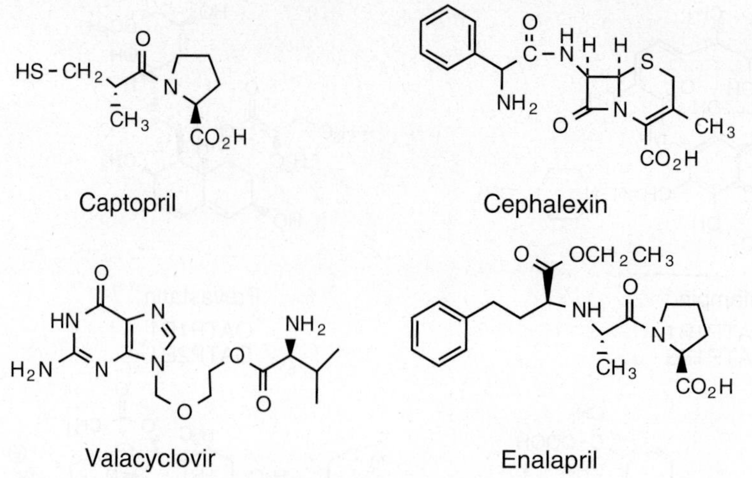

Captopril

Cephalexin

Valacyclovir

Enalapril

Figure 2.25 Drug substrates of peptide transporters.

- CNT/ENTs work together to achieve transport of purines and pyrimidines.
- They facilitate transport of nucleoside analog drugs and may be involved in their therapeutic response and toxicity.
- Figure 2.26 shows drug substrates of nucleoside transporters.

Table 2.4 gives some important drug substrates and inhibitors of the SLC transporters.

Transporters in Drug Metabolism

- P-gp and CYP450 (CYP3A) have overlapping tissue distribution and substrate specificity.
- CYP3A4 inhibitors also inhibit P-gp.
- Inducers of CYP3A4 can induce P-gp expression (e.g., SJW, reserpine, rifampicin, phenobarbital, triacetyloleandomycin).
- Drugs absorbed into enterocytes, hepatocytes, and renal tubular cells that are metabolized by CYP3A4 can be pumped out of those cells by P-gp.
  - This process allows CYP3A to have repeated access to the drug molecule resulting in the low bioavailability of CYP3A4 substrates, less accumulation of parent compound inside the cells, and more metabolite formation.

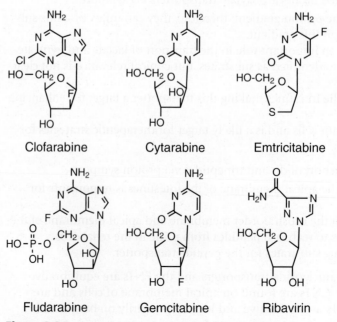

Clofarabine   Cytarabine   Emtricitabine

Fludarabine   Gemcitabine   Ribavirin

Figure 2.26 Drug substrates of nucleoside transporters.

Chemical Significance

There are many factors that govern the pharmacological effect of an administered drug. In most cases, a drug once absorbed into circulation will bind to a receptor in a sensitive tissue and affect a response. The key to receptor affinity and effect resides in the manner in which a drug's structure complements the receptor site. Understanding the physicochemical factors that influence drug–receptor interaction is a unique knowledge that is possessed by the pharmacist and can be used to advantage in the decision to use a specific drug agent to treat a specific disease.

One of the most important obstacles faced by a drug molecule is to survive its metabolism so that it can reach its site of action. Drug metabolism can affect a drug's plasma concentration and whether or not it will give the desired therapeutic outcome. In addition, knowledge of which drugs can induce or inhibit the metabolism of a coadministered drug often leads to the recommendation of effective drug therapy. The pharmacist as the chemist of the health care team is a valuable resource for the determination of appropriate drug therapy based upon their knowledge of the relationship of drug structure to possible metabolic transformation and is able to recommend dose adjustment and medication changes based upon that knowledge.

Review Questions

1 Which of the following pairs are acidic amino acids found on the receptor available for ionic bonding with basic drugs?
A. Aspartate, glutamine
B. Aspartate, glutarate
C. Arginine, asparagine
D. Lysine, ornithine
E. Aspartate, asparagine

2 Which of the following isomers are diastereomers?

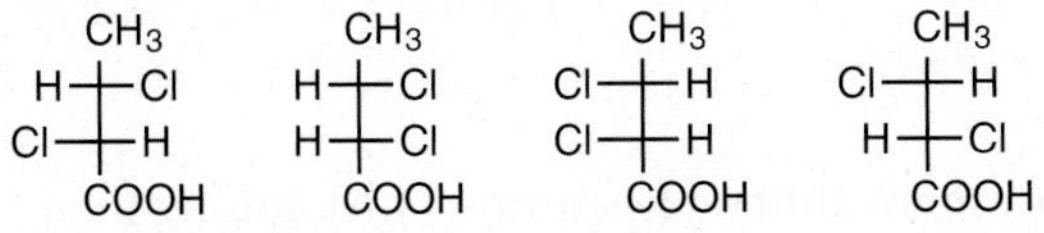

A. I and II
B. I and III
C. II and IV
D. II and III
E. Choices A and C above

3 Which of the following reactions is an example of oxidative N-dealkylation?

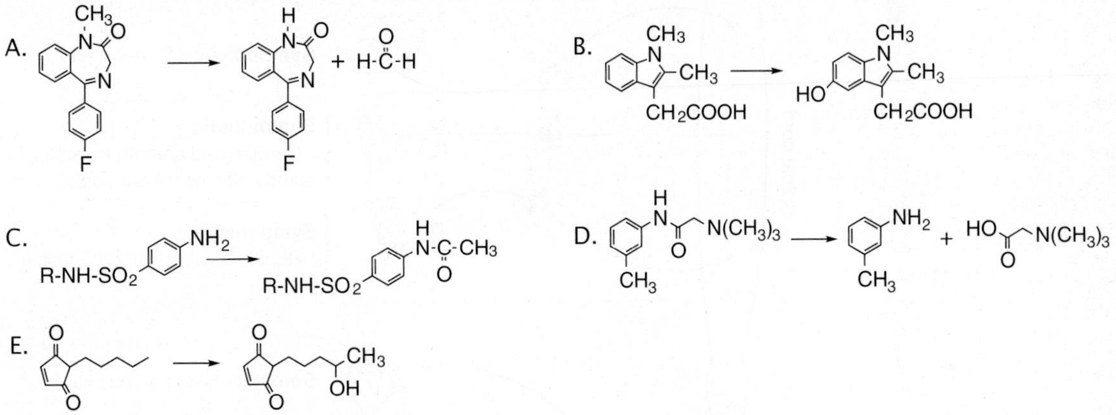

4 Which of the above reactions is an example of aromatic hydroxylation?

5 Which of the above reactions is an example of amide hydrolysis?

3

Cholinergic Agents

Learning Objectives

Upon completion of this chapter the student should be able to:

1 Discuss the role of acetylcholine (ACh) as a neurotransmitter including its biochemistry (i.e., synthesis, stereochemistry, storage, and metabolism).

2 Discuss the ACh receptors and compare and contrast the muscarinic versus the nicotinic receptors.

3 Discuss the mechanism of action of the following classes of cholinergic drugs:
   a. muscarinic agonists
   b. acetylcholinesterase inhibitors (AChEIs)
   c. muscarinic antagonists (anticholinergics)
   d. nicotinic antagonists
   e. nicotinic agonist.

4 List structure–activity relationships (SARs) of cholinergic drugs where presented: muscarinic agonists, muscarinic antagonists, neuromuscular (NM) blocking agents.

5 Identify clinically significant physicochemical and pharmacokinetic properties.

6 Discuss, with chemical structures, the clinically significant metabolism of the cholinergic drugs and how the metabolism affects their activity.

7 Discuss the clinical application of the cholinergic drugs.

Introduction

ACh is the neurotransmitter for many different neurons, chief among them being the parasympathetic nervous system in which ACh is the neurotransmitter at both the pre- and postganglionic fibers. In addition, ACh is the critical neurotransmitter in sympathetic and somatic fibers as indicated in Figure 3.1.

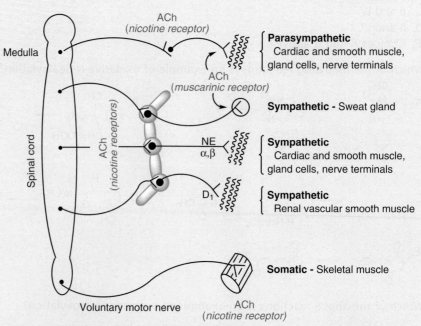

Figure 3.1 Schematic representation of autonomic and somatic motor nerves. nACh, acetylcholine nicotinic; mACh, acetylcholine muscarinic; Epi, epinephrine; NE, norepinephrine; D, dopamine.

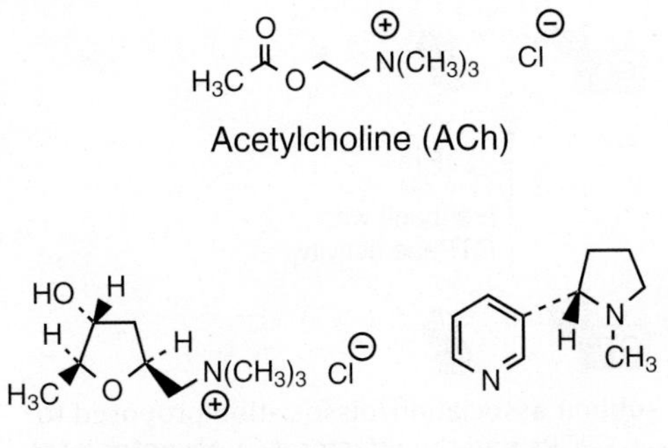

Acetylcholine (ACh)

L(+)-Muscarine chloride　　　*S*(-)-Nicotine

Figure 3.2 Structures of cholinergic agents.

Neurons that release ACh are referred to as cholinergic, as are the receptors on which these neurons synapse. The cholinergic receptors are further classified as either muscarinic (M) or nicotinic (N) receptors, based on their ability to bind the naturally occurring alkaloids muscarine or nicotine, respectively (Fig. 3.2). The natural and synthetic agonists of the cholinergic receptors are often referred to as cholinomimetic or parasympathomimetic agents and produce this effect either directly on the receptor or by inhibiting the inactivation of ACh through acetylcholinesterase (AChE)-catalyzed hydrolysis of ACh. Chemicals that exhibit affinity for the ACh receptor without intrinsic activity are referred to as cholinergic antagonists, cholinolytic, or parasympatholytic agents. The early recognition that muscarine and nicotine could mimic the actions of ACh led to the classification of receptors as either muscarinic acetylcholinergic receptors (mAChRs) or nicotinic acetylcholinergic receptors (nAChRs). The lack of selectivity and instability of ACh has led to the development of cholinergic agents with the desired selectivity (both for particular neurons and location of these neurons) and stability.

Muscarinic Receptors

- All mAChR subtypes (M_1 to M_5) are found in the CNS, while peripheral tissues may contain more than one mAChR subtype.
- Stimulation of M_1, M_3, and M_5 activates phospholipase C (PLC), while activation of M_2 and M_4 inhibits adenylyl cyclase (also named adenylate cyclase).
- Activation mAChRs involves guanosine triphosphate (GTP) and thus, mAChRs are G-protein–coupled receptors (GPCRs).
- The receptor consists of seven hydrophobic transmembrane helical domains with hydrophilic extracellular and intracellular domains (N-terminus of the GPCR protein is extracellular, and the *C*-terminus is intracellular) (Fig. 3.3).
- Following ACh binding to the AChR in the cell membrane a conformational perturbation initiates a process, which is described in Figure 3.4 involving subunit release and activation of second messengers.
- Presently, there are no mAChR agonists or antagonists with high selectivity for the various subtypes.

> **G-Protein Classification Based Upon α-Subunit**
>
> G_s increases adenylyl cyclase activity and increases Ca^{2+} currents; G_i decreases adenylyl cyclase activity and increases K^+ currents; G_o decreases Ca^{2+} currents; G_q increases PLC activity. Details of the signaling process, cellular response, function, and tissue involved can be found in Foye's 7th edition, pages 313 to 314 and Table 9.1.

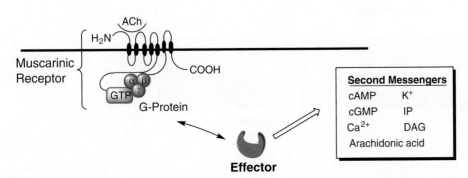

Second Messengers	
cAMP	K^+
cGMP	IP
Ca^{2+}	DAG
Arachidonic acid	

Figure 3.3 Model of signal transduction by a GPCR. This figure illustrates a proposed relationship between the binding of ACh and mAChR, G-protein, effector, and various second messengers.

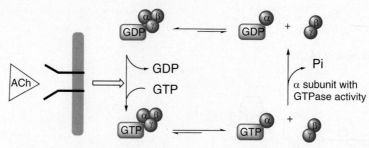

Figure 3.4 Diagram of a GTPase cycle and subunit association/dissociation proposed to control signal transduction between muscarinic GPCRs and the effector. ACh receptor interaction facilitates GTP binding to and activate the α-subunit. The α-subunit–GTP complex then dissociates from the β,γ-subunit, and each is free to activate effector proteins. The duration of separation is determined by the rate of α-subunit–mediated GTP hydrolysis.

Nicotine Receptors (nAChRs)

- nAChRs are found at the skeletal neuromuscular (NM) junction, adrenal medulla, and autonomic ganglia.
- Ganglionic neuronal (NN) and NM (somatic muscle) nAChRs are classified as ligand-gated ion channel receptors.
- nAChRs are pentameric transmembrane proteins made up of subunits: muscle type receptors are made up of α, β, ε, and δ subunits; NN types are made up of 12 different nicotine receptor subunits: α_2 to α_{10} and β_2 to β_4.
- The nAChR creates a transmembrane ion channel where ACh serves as a gate-keeper by binding with the nAChR to modulate passage of ions (i.e., K^+, Na^+) through the channel (Fig. 3.5).
- Some central and peripheral nAChR subtypes are summarized in Table 9.2 (see Chapter 9 in *Foye's Principles of Medicinal Chemistry, 7th Edition*, for a complete detailed discussion of nicotinic receptors).

Chemical Features of Acetylcholine

Biosynthesis and Processing of ACh

- ACh is biosynthesized in cholinergic neurons by the enzyme choline acetyltransferase utilizing acetyl coenzyme A (acetyl-*S*-CoA) and choline (Fig. 3.6).
- Newly biosynthesized ACh is actively transported into cytosolic storage vesicles located in presynaptic nerve endings.
- Release of ACh from storage vesicles occurs in response to an action potential, which opens voltage-dependent calcium channels affording an influx of Ca^{2+} and exocytotic release of ACh into the synapse.
- Synaptic AChE hydrolyzes ACh to terminating its action.
- Approximately half of the choline is recycled choline via the high-affinity transporter protein sodium–choline cotransporter found in cholinergic nerve endings.
- Details of the processing of ACh is shown in Figure 3.7.

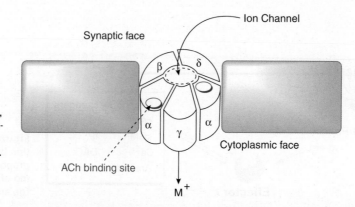

Figure 3.5 Diagrammatic representation of an ion-gate nicotinic acetylcholine receptor (nAChR) consisting of four subunits (α, β, ε, and δ) within the pentameric transmembrane protein. At the center is the transmembrane ion channel. Two ACh binding sites are shown on the extracellular domain of the α-subunits.

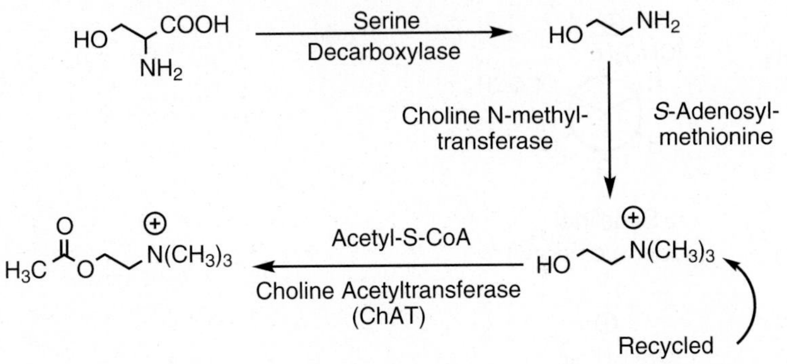

HO～COOH
|
NH$_2$ →(Serine Decarboxylase)→ HO～NH$_2$

Choline N-methyl-transferase   S-Adenosyl-methionine

H$_3$C～O～N(CH$_3$)$_3$$^{\oplus}$ ←(Acetyl-S-CoA / Choline Acetyltransferase (ChAT))← HO～N(CH$_3$)$_3$$^{\oplus}$

Recycled

Figure 3.6 Biosynthesis of ACh.

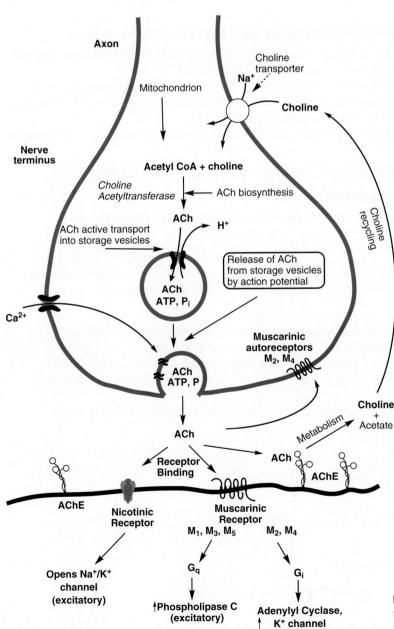

Figure 3.7 Cholinergic neuron site of biosynthesis, storage, release, and receptor activation. Following receptor binding ACh metabolism via acetylcholine esterase (AChE) and recycling of choline.

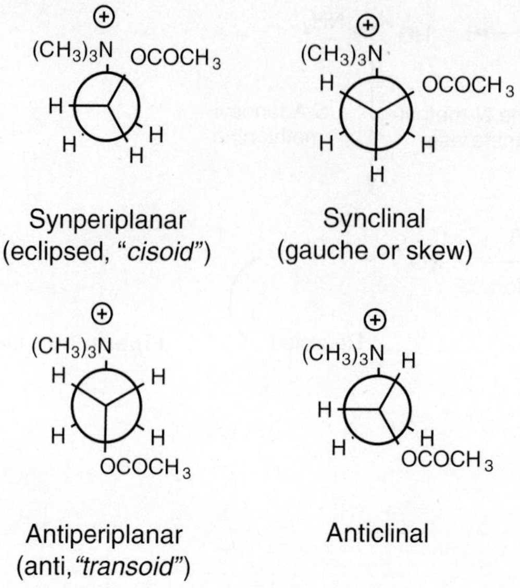

Figure 3.8 Conformational isomers of ACh.

CHEMICAL NOTE

ACh is the prototypical muscarinic and nicotinic agonist; however, it is a poor therapeutic agent due to its lack of receptor specificity and the chemical instability (i.e., ease of hydrolysis) in aqueous media, the gastrointestinal tract, and serum. ACh is also poorly absorbed across lipid membranes due to the quaternary ammonium functional group.

Stereochemical Properties of ACh and the ACh Receptors

- Although ACh is achiral, the ACh receptor exhibits chirality with respect to the binding of cholinergic agonists and antagonists.
- The stereochemistry of ACh resides in the rotation about σ bonds (i.e., conformational isomerism) and ACh can exist in an infinite number of conformations as illustrated by Newman projections (Fig. 3.8).
- ACh probably interacts with mAChRs in its less-favored anticlinal conformation.

Drugs Affecting Cholinergic Neurotransmission

Muscarinic Agonists

A limited number and role for muscarinic agonists exists today (Fig. 3.9).

MOA

- Muscarinic agonists are active by virtue of their ability to bind to the various muscarinic ACh receptor subtype M_1 to M_5 and elicit a response.

SAR

- SAR is based upon three domains of ACh (the quaternary ammonium group, the ethylene bridge, and the acyloxy group) (Fig. 3.10).
- The addition of a β-methyl group creates chiral acetyl-β-methylcholine (methacholine), which exhibits greater muscarinic potency than nicotinic potency.
- The drugs containing the chiral center are used as enantiomeric mixtures.

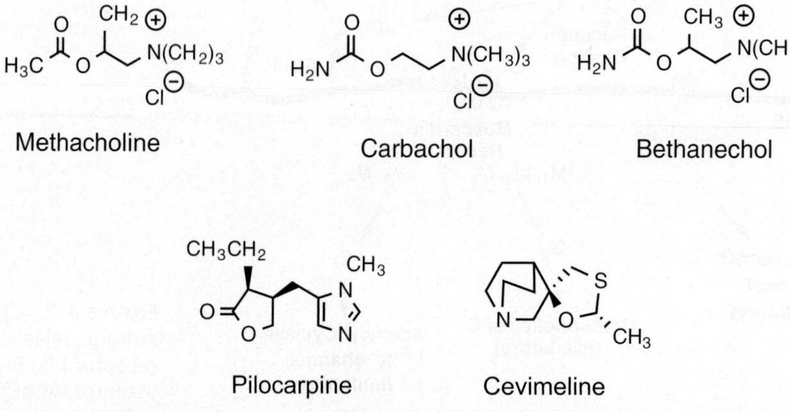

Figure 3.9 Muscarinic agonists.

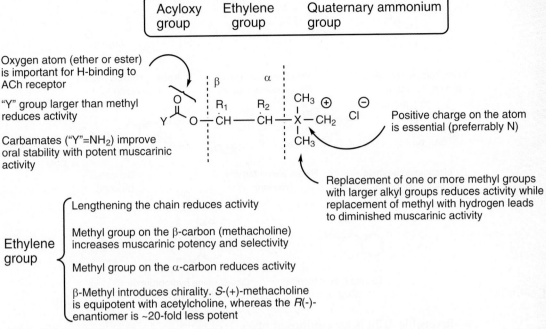

Acyloxy group | Ethylene group | Quaternary ammonium group

Oxygen atom (ether or ester) is important for H-binding to ACh receptor

"Y" group larger than methyl reduces activity

Carbamates ("Y"=NH₂) improve oral stability with potent muscarinic activity

Positive charge on the atom is essential (preferably N)

Replacement of one or more methyl groups with larger alkyl groups reduces activity while replacement of methyl with hydrogen leads to diminished muscarinic activity

Ethylene group
- Lengthening the chain reduces activity
- Methyl group on the β-carbon (methacholine) increases muscarinic potency and selectivity
- Methyl group on the α-carbon reduces activity
- β-Methyl introduces chirality. *S*-(+)-methacholine is equipotent with acetylcholine, whereas the *R*(-)-enantiomer is ~20-fold less potent

Figure 3.10 SAR limitation for muscarinic agonists.

Physicochemical and Pharmacokinetic Properties

- The quaternary and nonquaternary ammonium salts are all water soluble.
- The quaternary ammonium chlorides are poorly absorbed via the oral route because of the quaternary nature of the compound.
- The ester, lactone, and carbamates are relatively stable in aqueous media, but are still prone to metabolism via esterase.

Clinical Application

The clinical applications for muscarinic agonists are presented in Table 3.1.

Acetylcholinesterase Inhibitors (AChEIs)

The drugs shown in Figure 3.11 are all inhibitors of AChE, but some derive their therapeutic activity from additional mechanisms besides that of inhibition of AChE.

MOA

- AChE is an enzyme, which is capable of binding ACh and chemicals of a similar structure.
- Key binding sites within AChE include a tryptophan capable of cation–π bonding, two glycines with hydrogen bonding potential, and an esteratic nucleophilic serine hydroxyl group (Fig. 3.12A).
- Acetylated AChE is considered to be "inactivated" (Fig. 3.12C).
- Hydrolytic deacylation (Fig. 3.12D) rapidly regenerates the active AChE.

> **⚠ ADVERSE EFFECTS**
>
> **Acetylcholinesterase Inhibitors:** The most common adverse effects of the optically administered quaternary amine echothiophate include stinging, burning, lacrimation, blurred vision, and other eye effects. Lens opacities may also occur. The adverse effects of the other three traditional AChEIs (e.g., edrophonium, neostigmine, pyridostigmine) are associated with excessive blockage of ACh and may affect the cardiovascular, respiratory, gastrointestinal, genitourinary, or musculoskeletal systems. While similar systems can be affected in the dementia patient the adverse effects tend to be mild and consist of nausea, diarrhea, vomiting, anorexia, and ataxia.

Table 3.1 Muscarinic Agonists in Clinical Use

Drug Name	Trade Name	Clinical Application – Route of Administration
Methacholine	Provocholine	Diagnostic for asthma inhalation
Carbachol	Miostat	Treatment of glaucoma and induction of miosis – ophthalmic drops
Bethanechol	Duvoid, Uecholine	Treatment of urinary retention and abdominal distention – oral
Pilocarpine	Isopto Carpine, Pilopine HS, Salagen	Treatment of open-angle glaucoma and xerostomia, Sjögren syndrome – ophthalmic drops and oral
Cevimeline	Evoxac	Xerostomia, Sjögren syndrome – oral

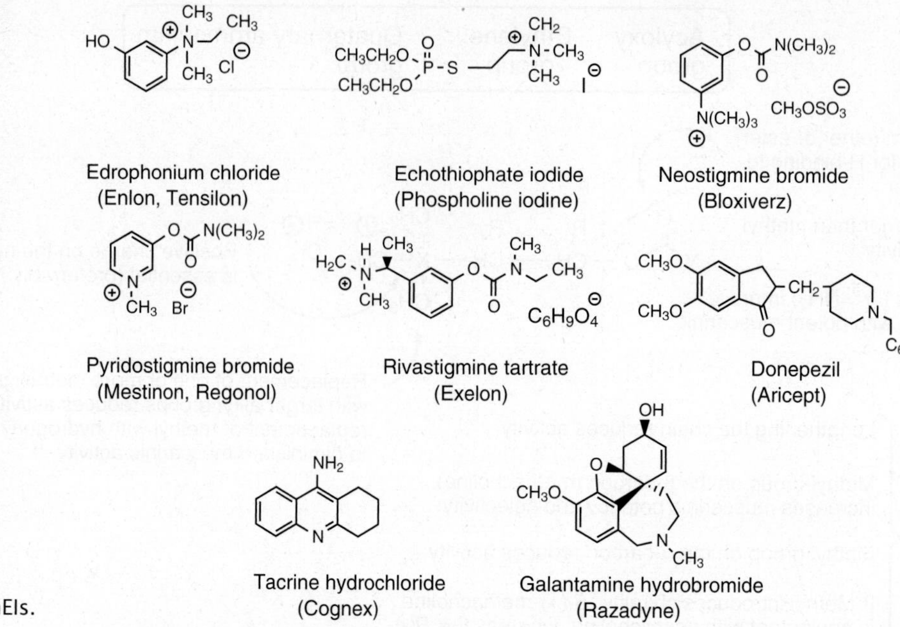

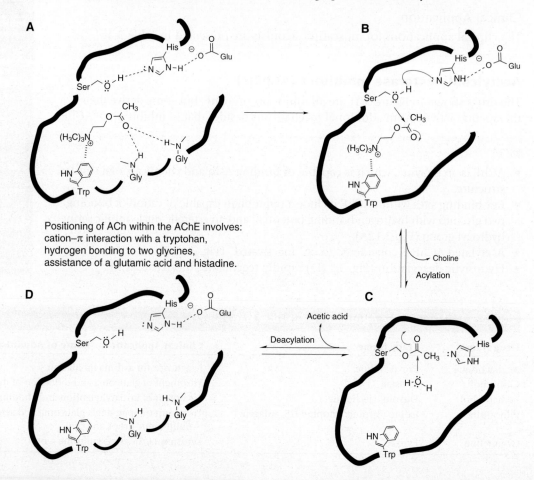

Figure 3.11 AChEIs.

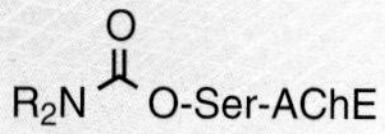

- Reversible AChEIs are composed of:
  - substrates for AChE which acylate the enzyme followed by deacylation. Most of these agents are carbamates, not esters, and form carbamylated AChE and are more resistant to hydrolysis.
  - substrates with greater binding affinity for AChE than ACh, but do not acetylate the enzyme.
- Irreversible AChEIs (echothiophate iodide) are:
  - commonly phosphate esters, which phosphorylate AChE and only slowly dephosphorylate and, as a result, have high potential toxicity.

Carbamylated AChE

Positioning of ACh within the AChE involves:
cation–π interaction with a tryptohan,
hydrogen bonding to two glycines,
assistance of a glutamic acid and histadine.

Figure 3.12 Mechanistic steps in the acetylation and hydrolysis of AChE by ACh.

Physicochemical and Pharmacokinetic Properties

- The quaternary amines (edrophonium, echothiophate, neostigmine, pyridostigmine) are either used locally, via injection, or with limited oral bioavailability (Fig. 3.11).
- The four remaining compounds are tertiary amines and are well absorbed, used systemically, and most of them enter the CNS.
- The systemic drugs are extensive metabolized (see below).

Metabolism

- Rivastigmine is a substrate for AChE and undergoes carbamate hydrolysis.
- Tacrine, donepezil, and galantamine are substrates for various CYP isoforms leading to inactive or less active products (1-hydroxytacrine retains some activity) (Fig. 3.13).

Clinical Application

- Echothiophate iodide is indicated for treatment of open-angle glaucoma for which the drug is administered as an ophthalmic solution. The drug is an irreversible inhibitor of AChE and there may be availability issues since manufacturing of the drug is limited.
- Edrophonium chloride is recommended for diagnosis of myasthenia gravis, but its short duration of action is a limiting factor for its use. On the other hand, neostigmine methylsulfate and pyridostigmine bromide are used to control myasthenia gravis, with the former being administered SC or IM while the latter is effective orally, but not well absorbed.
- The four remaining drugs rivastigmine tartrate, donepezil and tacrine hydrochloride, and galantamine hydrobromide have been approved for treatment of mild to moderate dementia including Alzheimer's type. These drugs have limited effectiveness in advanced forms of Alzheimer's disease.
- As noted above, the physicochemical properties of quaternary ammonium compounds prevent CNS penetration and limit their use to non-CNS conditions, while the tertiary amines cross the blood–brain barrier and are utilized for treatment of Alzheimer's disease.

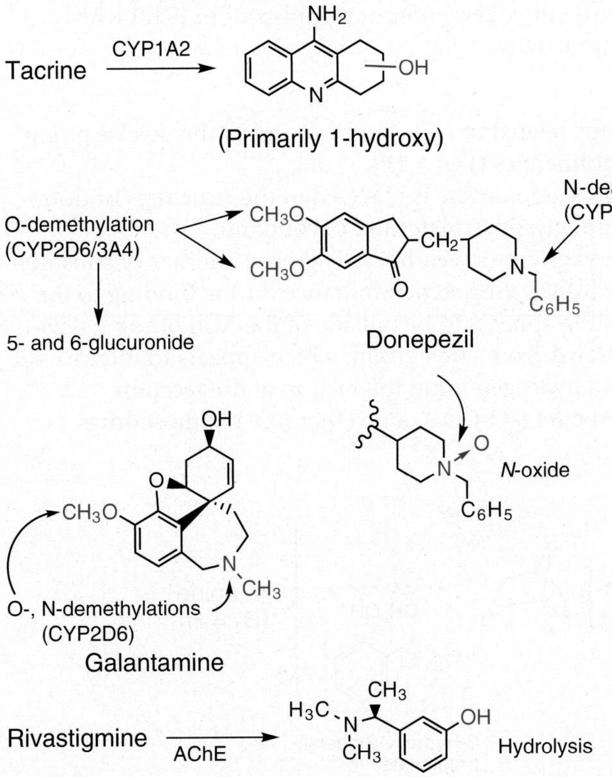

Figure 3.13 Metabolic reactions common to systemic AChE inhibitors.

CHEMICAL NOTE

A large number of organophosphate insecticides have been used for years in agriculture chemistry to protect plants from insects. These agents are effective as irreversible AChEIs with varying degrees of human toxicity. They are represented by the chemicals shown below.

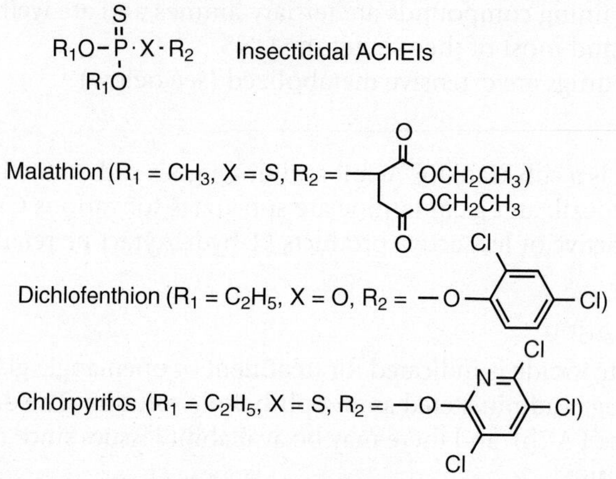

$$R_1O-\overset{\overset{S}{\|}}{\underset{R_1O}{P}}\cdot X \cdot R_2 \quad \text{Insecticidal AChEIs}$$

Malathion ($R_1 = CH_3$, X = S, $R_2 = $)

Dichlofenthion ($R_1 = C_2H_5$, X = O, $R_2 = -O-$)

Chlorpyrifos ($R_1 = C_2H_5$, X = S, $R_2 = -O-$)

ADVERSE EFFECTS

Anticholinergics:
When used systemically, the anticholinergics cause blurred vision, photophobia, dry mouth, constipation, dyspepsia, and difficulty in urination. These agents should not be used in angle-closure glaucoma and benign prostatic hyperplasia. These adverse effects are usually of less importance with inhaled or topically used drugs.

$$R_1-\overset{\overset{R_2}{|}}{\underset{|}{C}}-X-(CH_2)_n-N\big<$$

Muscarinic Antagonists – Anticholinergics

Muscarinic antagonists are commonly referred to as anticholinergics because of their ability to block the action of ACh. The earliest known anticholinergics are the alkaloids atropine and scopolamine isolated from plants in the Solanaceae family (i.e., *Atropa belladonna*, *Hyoscyamus niger*, and *Datura stramonium*) (Fig. 3.14). These drugs are still used today although a large number of structurally related anticholinergics have taken their place in many instances.

MOA

- The anticholinergics act as antagonists at muscarinic sites at postganglionic mAChR in the cholinergic nervous system often with selectivity for the M_3 receptor (Fig. 3.7).
- Some of the antimuscarinics are also active at presynaptic M_2 receptors.
- The anticholinergics are competitive reversible antagonists of mAChR with affinity, but without intrinsic activity.

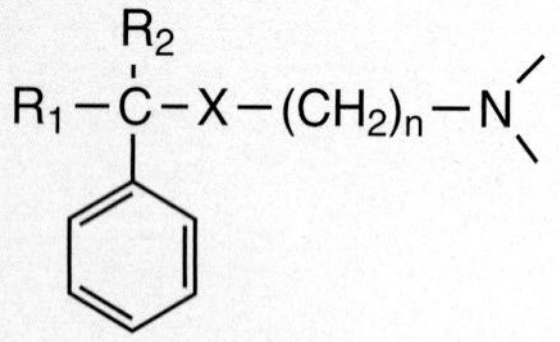

SAR

- A variety of functional groups related to ACh can be found in the diverse group of drugs classified as anticholinergics (Fig. 3.15).
- The prototype of atropine or scopolamine is reflected in the structures of tiotropium, trospium, ipratropium, glycopyrrolate, and oxybutynin.
- R_1 is commonly a second aryl group or a carbocyclic group. The size of substituents on either ring is limited suggesting steric hindrance during binding to the mAChR. Hydrophobic binding appears to be outside of the ACh binding region.
- R_2 is a tertiary hydroxyl or a hydroxymethyl group, which appears to increase the potency by participating in a hydrogen bond interaction at the receptor.
- The X substituent is either an ester (-O-CO-) or an ether (-O-) in most drugs.

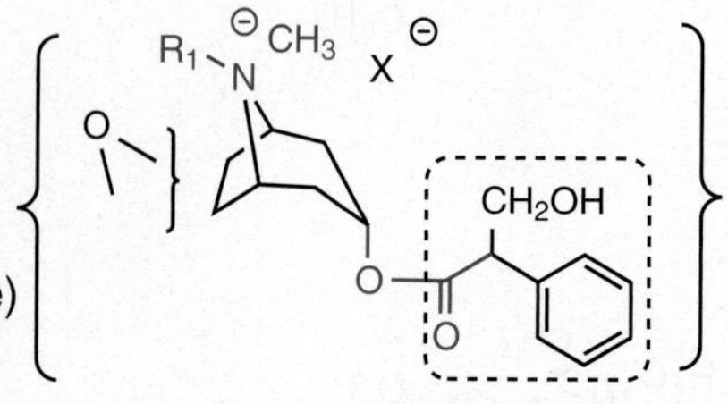

Scopolamine ($R_1 = H$, 6,7-epoxide)

Methscopolamine ($R_1 = CH_3$, 6,7-epoxide)

Atropine ($R_1 = H$)

3-Hydroxy-2-phenyl-propionate

Figure 3.14 Atropine and scopolamine are alkaloids from the plant family Solanaceae. They differ by the presence or absence of an epoxide at the 6, 7 position of the tropane ring. Highlighted are the key components of ACh.

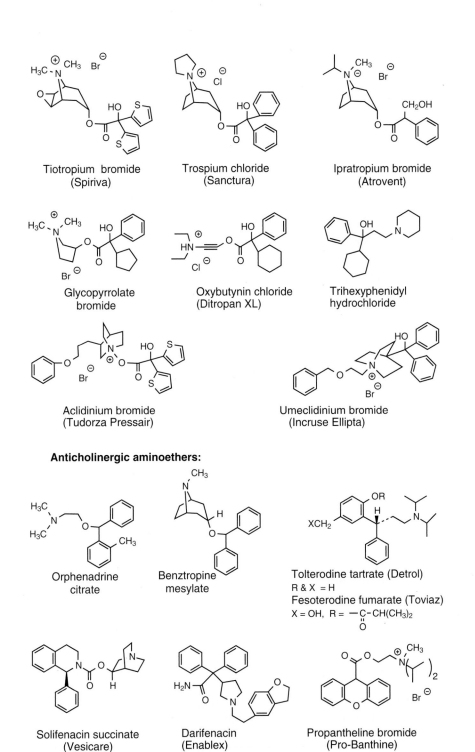

Figure 3.15 Anticholinergic drugs.

- A cationic amine (whether a protonated tertiary amine or a quaternary amine) appears to be involved in binding to the muscarinic receptor with small alkyl substituents (e.g., methyl, ethyl, or isopropyl).
- The distance between the ring-substituted carbon and the N is not critical, although two methylenes or equivalent distance is common.

Physicochemical and Pharmacokinetic Properties

- The anticholinergics are effective both orally and parenterally.
- The quaternary ammonium compounds used orally are not well absorbed, but commonly act locally affecting the respiratory or gastrointestinal tract.
- The tertiary nitrogen compounds are absorbed following oral administration and cross the blood–brain barrier.

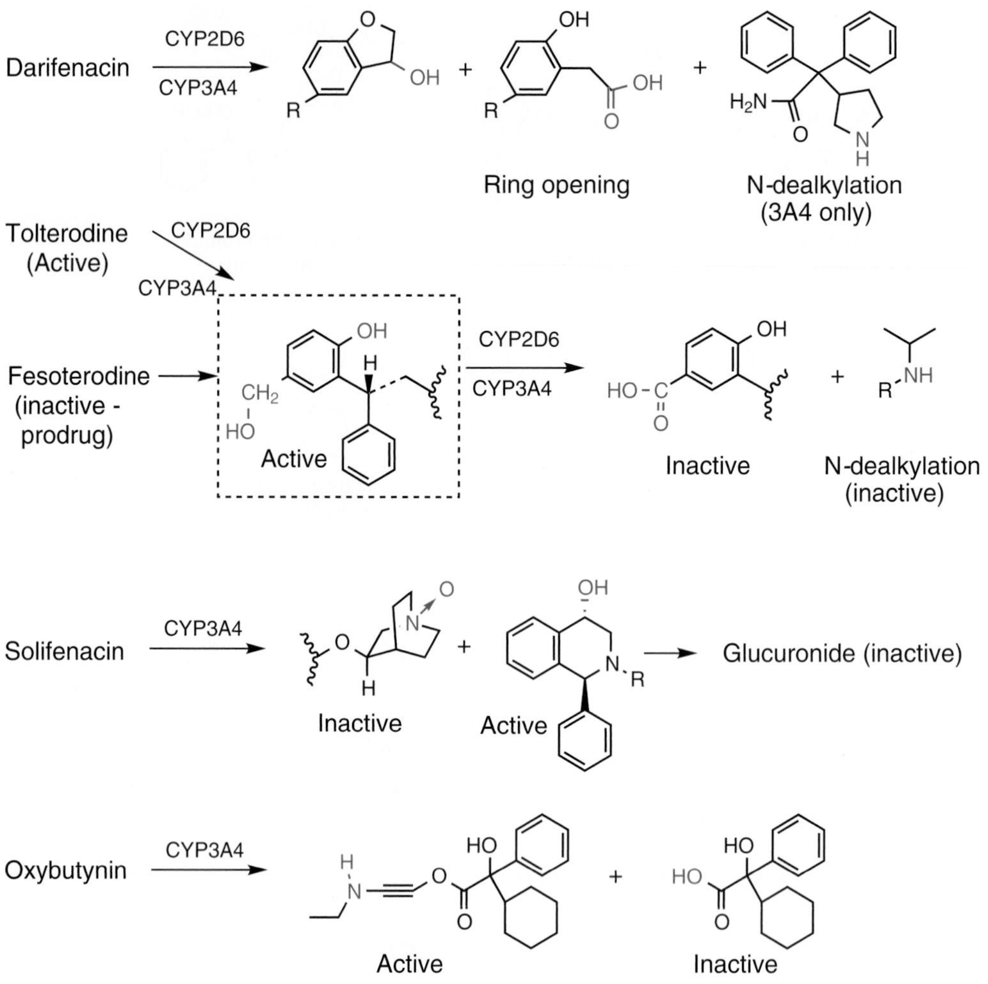

Figure 3.16 Metabolism of anticholinergics. CYP2D6 is the primary metabolizing enzyme for darifenacin, tolterodine, and fesoterodine except in poor metabolizers where CYP3A4 becomes the primary CYP metabolizer.

- Several of the tertiary nitrogen anticholinergics are absorbed transdermally (e.g., scopolamine, oxybutynin).
- Several of the quaternary ammonium drugs are available as inhalation powders and as such, their action is primarily in the lungs with poor plasma bioavailability.

Metabolism

- Most of the ester-containing anticholinergics undergo hydrolysis to inactive agents.
- N-Dealkylation reactions are common reactions resulting in active agents in some cases.
- Systemic CYP450-catalyzed reactions are shown in Figure 3.16.
- Fesoterodine is activated via hydrolysis, while active metabolites of the other anticholinergics are formed following CYP450 oxidation.
- Dose adjustments may be required when anticholinergics are combined with other drugs, which enhance or inhibit CYP-catalyzed reactions.

Clinical Applications

- A major indication for muscarinic antagonists is for treatment of overactive bladder. The commonly used agents include trospium, oxybutynin, tolterodine, solifenacin, and darifenacin.
- The anticholinergics tiotropium, ipratropium, aclidinium, and umeclidinium are used for the long-term maintenance treatment of bronchospasms (long-acting muscarinic antagonists) (LAMAs) seen in chronic obstructive pulmonary disease (COPD) as well as chronic bronchitis. The dosage form is that of an inhaled powder.
- The anticholinergics can be used as a preoperative agent through their ability to reduce salivary, tracheobronchial, and pharyngeal secretions (e.g., glycopyrrolate).
- Trihexylphenidyl and benztropine are indicated as an adjunct in treatment of parkinsonism, and possibly, control of extrapyramidal disorders.

- Scopolamine, in the form of a transdermal patch, is used for the prevention of motion sickness (e.g., nausea and vomiting).
- There are various ophthalmologic conditions for which anticholinergic agents can also be used.

Nicotinic Antagonists

The nicotine antagonists fall into two subclasses – the ganglionic blocking agents and the NM blocking agents. It is the latter class that represents the therapeutically useful drugs. These drugs resulted from the discovery that South American natives used a plant extract on their arrow tips to hunt with and that the extract contained *d*-tubocurarine, a muscle-paralyzing agent.

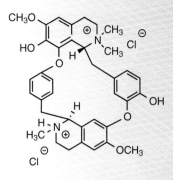

d-Tubocurarine chloride

MOA of Neuromuscular Blocking Agents

- The NM blocking agents are competitive antagonists of nicotine.
- Their action is either depolarizing (succinylcholine) or nondepolarizing (the remaining drugs shown in Fig. 3.17).
- The depolarizing NM blocking agent produces an initial muscle excitation followed by prolonged blocking action as a result of resistance to AChE.
- Nondepolarizing NM blocking agents are competitive antagonists at the α-subunits preventing ACh from binding to nAChR preventing depolarization of the NM end plate.
- Inhibition of nondepolarizing blocking agents can be reversed by AChEIs through increased concentrations of ACh at nAChR.
- It is suggested that these drugs must bind to two adjacent anionic sites simultaneously to produce their effect.

SAR

- The depolarizing NM blocking agent has a flexible structure connecting the two quaternary heads.
- The nondepolarizing agents have a more rigid structure with the quaternary and protonated heads being part of bulky groups.

> **⚠ ADVERSE EFFECTS**
>
> **Neuromuscular Blocking Agents:** Adverse reactions to most of the NM blocking agents can include hypotension, bronchospasm, and cardiac disturbances. The depolarizing agents also cause an initial muscle fasciculation before relaxation. Many of these agents cause release of histamine and subsequent cutaneous (flushing, erythema, urticaria, and pruritus), pulmonary (bronchospasm and wheezing), and cardiovascular (hypotension) effects.

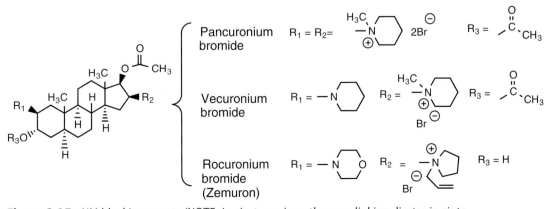

Succinylcholine chloride
(Anectine, Quelicin)

Atracurium besylate
Cisatracurium besylate
(Nimbex)

Figure 3.17 NM blocking agents (NOTE: in cisatracurium, the crosslinking diester is *cis* to the respective dimethoxybenzyl groups).

Table 3.2				**Pharmacokinetic Properties of Neuromuscular Blocking Agents**
Drug	**Onset (min)**	**Duration (min)**	**Half-life (min)**	**Mode of Elimination**
Succinylcholine	1–1.5	6–8	<1	Hydrolysis by plasma butyrylcholinesterase
Atracurium	2–4	30–40	16–20	Hofmann degradation, hydrolysis by plasma cholinesterases
Pancuronium	4–6	120–180	89–140	Renal elimination, liver metabolism, and clearance
Rocuronium	1–2	30–40	84–131	Liver metabolism and clearance
Vercuronium	2–4	30–40	65–80	Liver metabolism and clearance, renal elimination

Physicochemical and Pharmacokinetic Properties

- The NM blocking agents are not absorbed orally, but are well distributed following IM or IV administration.
- Because of their polar nature these drugs do not enter the CNS.
- The duration of action is limited usually by haptic hydrolysis of the esters present in these drugs although often, the metabolites show partial activity (Table 3.2).
- Note that these drugs fall into short-, intermediate-, and long-duration agents based upon their rate of hydrolysis and renal elimination.

Metabolism

- As indicated above the drugs are prone to enzyme-catalyzed hydrolysis.
- Atracurium is also degraded via a Hofmann elimination which occurs in the plasma and as a result, the drug's biological effects are independent of renal elimination and can be used in patients with impaired renal function (Fig. 3.18).

Clinical Application

- NM blocking agents are used primarily as an adjunct to general anesthesia.
- They produce skeletal muscle relaxation that facilitates operative procedures such as abdominal surgery and reduce the depth requirement for general anesthetics, thus decreasing the overall risk of a surgical procedure and shortens the postanesthetic recovery time.

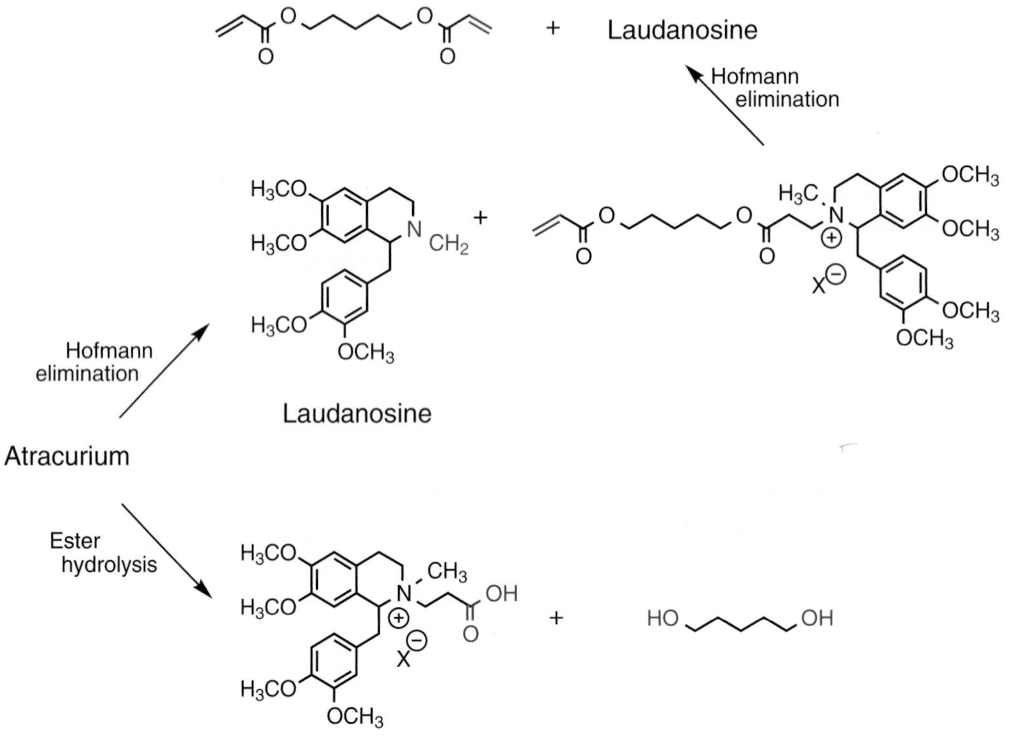

Figure 3.18 Inactivation of atracurium by Hofmann elimination and hydrolysis.

- NM blocking agents have also been used in the correction of dislocations and the realignment of fractures.
- Short-acting NM blocking agents, such as succinylcholine, are routinely used to assist in tracheal intubation.

Nicotine Agonist – Varenicline

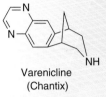

Varenicline
(Chantix)

MOA

- Varenicline is a partial agonist (efficacy relative to ACh of 13%) at nAChRs with selectivity for the $\alpha_4\beta_2$ subtype as well as weak partial agonist on $\alpha_3\beta_4$, $\alpha_3\beta_2$, and α_6-containing nAChR subunits.
- It is a full agonist on the α_7-nAChR subunit.
- Some combination of these actions may be involved in the mechanism of action of varenicline as a smoking cessation aid.

Physicochemical and Pharmacokinetic Properties

- Maximal plasma concentration is reached within 3 to 4 hours after oral administration.
- A steady-state concentration is reached within 4 days.
- Varenicline has a half-life of 24 hours.
- Oral bioavailability is not affected by food or time of administration. The drug exhibits low plasma protein binding (~20%).
- Renal excretion occurs with 81% to 92% of the drug being excreted unchanged.

Clinical Application

- Varenicline is indicated for smoking cessation, as an alternative to bupropion and nicotine patches.
- Time of treatment, with a maintenance dose of 1 mg twice daily, is 12 weeks.

> ⚠️ **ADVERSE EFFECTS**
>
> **Varenicline:** Potential adverse effects include weight gain, change in taste and appetite, mouth dryness, insomnia, headache, irritability, heartburn, vertigo, nausea, diarrhea, and tachycardia. The most common adverse events for participants receiving active-drug treatment were nausea (28%) and insomnia (22%). The FDA issued an alert on February 2008 regarding serious neuropsychiatric symptoms associated after patients taking varenicline who reported depression, agitation, and suicidal thoughts and behavior.

Chemical Significance

There is probably no better example of how important a working knowledge of chemistry can be to help understand the drugs that affect the cholinergic nervous system. The simple structure of ACh is repeated as the pharmacophore in the classes of cholinergic agents. Whether they are muscarinic agonists, AChEIs, anticholinergics, or NM blocking agents, one will commonly find a tertiary basic nitrogen or a quaternary nitrogen and an ester or ester-like functional group present. Because of these commonalities a prediction and rationalization of the physical, chemical, pharmacokinetic, and metabolic reactions all show a similarity. Obviously, there are differences between each of the classes of cholinergic agents, but a review of the chemistry allows the practitioners to immediately recognize the relationship between the drug and the cholinergic nervous system. In addition, normal processing within the cholinergic neuron and the nature of nicotinic, muscarinic, and ACh receptor sites are vital to understand the chemistry of cholinergic drugs.

Review Questions

1 Which receptor is associated with a second messenger and is classified as a GPCR?
- A. Choline transporter
- B. Nicotinic receptor
- C. Muscarinic receptor
- D. Acetycholinesterase receptor
- E. Adenylyl cyclase receptor

2 In the drug bethanechol, why is the carbamate function group used rather than the ester group found in methacholine?

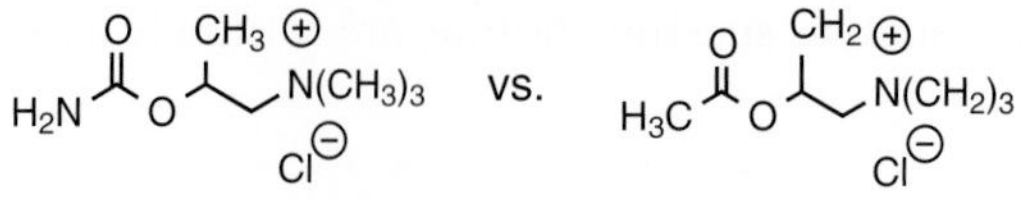

Bethanechol   Methacholine

A. Salt formation can be made on the nitrogen of bethanechol, thus improving water solubility.

B. The carbamate improves oral stability, whereas the ester is readily hydrolyzed.

C. The carbamate functional group is more lipophilic than an ester allowing for faster absorption.

D. The carbamate functional group can hydrogen bond to the receptor but the ester cannot.

E. The carbamate actually hydrolyzes faster than an ester, thus giving the advantage of a shorter acting drug.

3 Rivastigmine is active by virtue of acting as a(n):

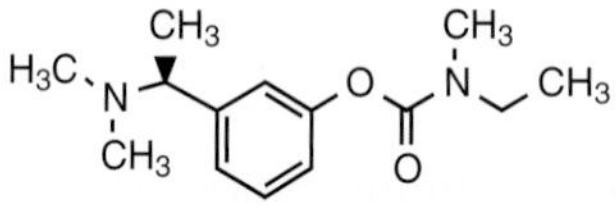

Rivastigmine

A. irreversible AChEI.

B. reversible muscarinic agonist.

C. reversible nicotinic agonist.

D. reversible AChEI which binds to the ACh esterase receptor blocking the action of ACh.

E. reversible AChEI which acetylates the acetylcholine esterase receptor.

4 Identify the active form(s) of fesoterodine shown below.

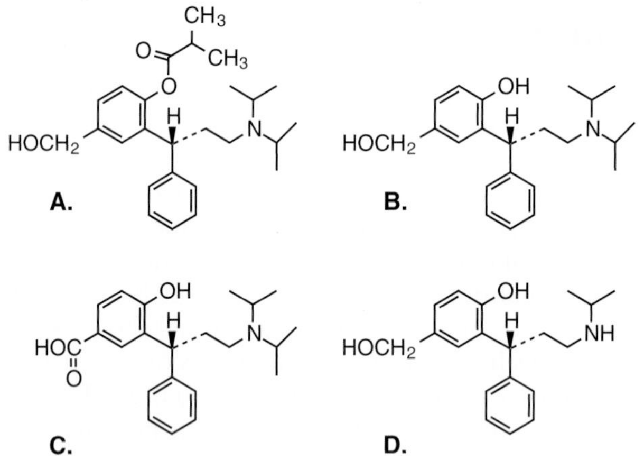

Fesoterodine and its metabolites

A. (A)

B. (B)

C. (C)

D. (D)

E. Both (A) and (B) are active

5 The clinically accepted use of atracurium besylate is:

A. skeletal muscle relaxation during abdominal surgery.

B. treatment during smoking cessation.

C. used to control myasthenia gravis.

D. treatment of overactive bladder.

E. prevention of motion sickness.

Adrenergic Agents

Learning Objectives

Upon completion of this chapter the student should be able to:

1 Describe the mechanism of action of:
   a. phenylethylamine adrenergic agonists
   b. arylimidazoline α_1-adrenergic agonists
   c. α_2-adrenergic agonists
   d. α_1-adrenergic antagonists
   e. β-adrenergic antagonists.

2 Discuss applicable structure activity relationships (SAR) (including key aspects of receptor binding) of the above classes of adrenergic agonists and antagonists.

3 Identify the clinically relevant physicochemical and pharmacokinetic properties that impact the in vitro and in vivo stability and/or therapeutic utility of adrenergic agonists and antagonists.

4 Diagram metabolic pathways for all biotransformed adrenergic agonists and antagonists, identifying enzymes and noting the clinically significant metabolism and the activity, if any, of the major metabolites.

5 Apply all of the above to predict and explain therapeutic utility of adrenergic agonists and antagonists.

Introduction

The term adrenergic comes from the word "adrenaline," a synonym for epinephrine (Epi), the N-methylated analog of the neurotransmitter norepinephrine (NE). The binding sites acted upon by this short-lived endogenous biochemical and its agonist and antagonist analogs are termed adrenergic receptors (or adrenoceptors). They are found throughout the peripheral and central sympathetic nervous systems and belong to the family of G-protein–coupled receptors (GPCRs). These receptors are broadly classified as α or β depending on their selectivity for N-substituted analogs of NE.

Adrenergic agonists and antagonists enjoy widespread therapeutic use. α-Agonists are potent vasoconstrictors that can be used for conditions ranging from annoying (e.g., cold symptoms) to life threatening (e.g., hemorrhage). Those agents that can penetrate the blood–brain barrier suppress appetite and increase alertness. β-agonists dilate bronchi and are most commonly prescribed in the treatment of asthma and related respiratory disorders. β-receptors are also found in the heart, and antagonists of both receptor classes are potent antihypertensive agents.

Endogenous Adrenergics

- NE is the adrenergic neurotransmitter produced and released from presynaptic terminals.
  - The fraction of NE not binding to pre- and postsynaptic receptors is returned to presynaptic terminals via an NE uptake-1 transporter.

- Epi, a neurohormone, is produced in and secreted by the adrenal medulla.
  - The precursor of both endogenous adrenergics is *L*-tyrosine (see Chapter 10, Figure 10.1 in *Foye's Principles of Medicinal Chemistry, Seventh Edition,* for a complete detailed discussion).

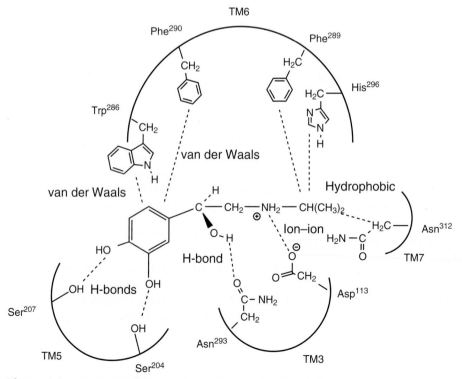

L-Tyrosine   Norepinephrine   Epinephrine

Adrenergic Receptors

- Subtypes of α- and β-adrenergic receptors are known and include α_1, α_2, β_1, β_2, and β_3. Agonists and antagonists of subtypes 1 and 2 are the most commonly used therapeutic agents.
- α_1-Receptors are predominantly postsynaptic, G_q coupled, and found in vascular smooth muscle and the CNS. Stimulation activates phospholipase C, which increases cellular levels of inositol 1,4,5-triphosphate (IP_3) and diacylglycerol (DAG).
- α_2-Receptors are predominantly presynaptic and G_i coupled. Stimulation inhibits adenylyl cyclase, which decreases intracellular cyclic adenosine monophosphate (cAMP) and inhibits the release of NE from presynaptic terminal.
- β_1- and β_2-Receptors are predominantly postsynaptic and G_s coupled. Stimulation activates adenylyl cyclase, which increases intracellular cAMP.
- β_1-Receptors are primarily myocardial. Stimulation results in increased chronotropic (heart rate) and inotropic (force of contraction) effects.
- β_2-Receptors are found in lungs, uterus, vascular smooth muscle, and skeletal muscle. Stimulation results in relaxation or dilation of these tissues. Both β-receptor subtypes are also found in the CNS.
- The human β_2-receptor has been extensively characterized and the binding residues are contained primarily within the transmembrane helices 3, 5, 6, and 7 (Fig. 4.1). Most of these binding residues are also found on the α-receptors.

Figure 4.1 Isoproterenol–human β_2-receptor binding interactions.

Adrenergic Agonist Metabolism

- Two enzymes are involved in the inactivation of both NE and Epi: monoamine oxidase (MAO) and catechol-O-methyltransferase (COMT) (Fig. 4.2).
- MAO deaminates primary amines and has a strict steric requirement for two hydrogens on the adjacent carbon (C_α). The aldehyde metabolite can be further oxidized or reduced by cytosolic dehydrogenases.
- MAO is found in neuronal mitochondria and inactivates some NE after uptake-1.
- N-substituted adrenergics (e.g., Epi) must be N-dealkylated by CYP prior to MAO deamination.
- Cytosolic COMT methylates the *m*-OH group of catecholic drugs.
- Drugs vulnerable to both MAO and COMT have short durations of action and are orally ineffective. Synthetic adrenergic agonists and antagonists resistant to one or both of these enzymes will often show a more prolonged duration and some oral activity.
- Adrenergic agonists and antagonists with one or more phenolic hydroxyls can be conjugated with glucuronic acid or sulfonated with PAPS. Conjugation in the gut and liver compromises oral activity and duration of action, respectively.

Adrenergic Agonists

MOA

- Adrenergic agonists can act by direct, indirect, or mixed mechanisms.
- Direct action occurs when an agonist binds directly to the same receptors and communicates the same chemical/physiological message as NE.
  - Direct action can be selective for one receptor subtype over another.
- Indirect action occurs when the agonist is denied access to receptors, but augments NE's action in adrenergic synapses by: (1) stimulating its release from presynaptic terminals, (2) blocking uptake-1, and/or (3) inhibiting its metabolism.
  - Indirect action is nonselective since NE stimulates all adrenergic receptors.
- Mixed action occurs when the agonist can act directly and indirectly.

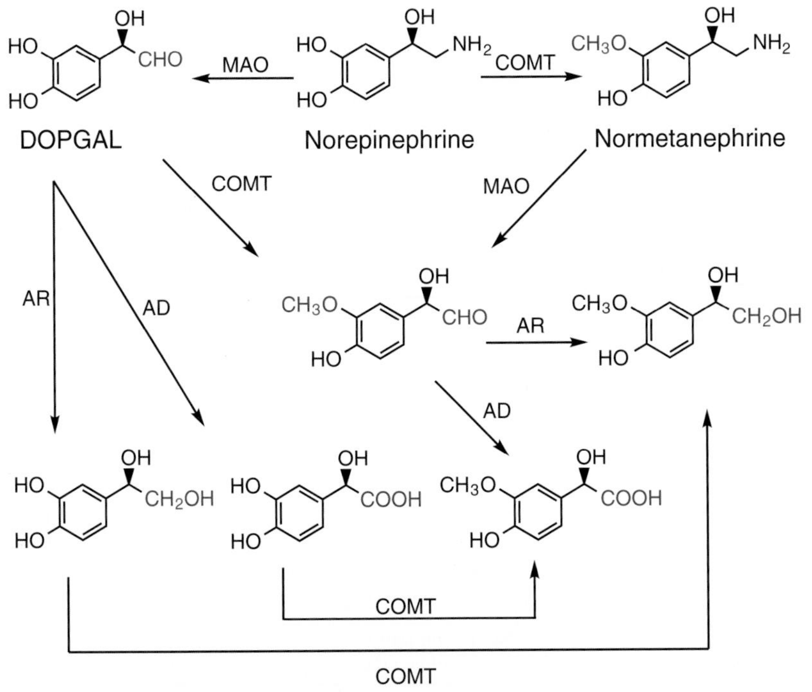

Figure 4.2 Metabolism of NE. *MAO*, monoamine oxidase; *COMT*, catechol-O-methyltransferase; *AR*, aldehyde reductase; *AD*, aldehyde dehydrogenase.

Phenylethylamine Adrenergic Agonists

SAR

- The phenylethylamine and arylimidazoline adrenergic agonist pharmacophores and general SAR overviews are provided in Figure 4.3.
- Most adrenergic agonists are primary or secondary phenylethylamines. Tertiary amines are inactive.

Amine Substituent (R_1)

- A cationic amine is essential for adrenoceptor binding.
- The amine substituent binds hydrophobically to adrenoceptors. The size of the substituent influences which receptor subtype(s) will be stimulated.
- R_1 = H: Preferential α-agonist action; used predominantly as vasoconstrictors.
  - Some β_1-agonist action (tachycardia) may be noted.
- R_1 = CH_3: Nonselective agonist action; used predominantly for their α-agonist action.
  - In the absence of other α- or β-receptor directing substituents, N-methyl agonists stimulate both receptor types.
  - The area on the α-receptor that binds the amine substituent is small. It readily accepts a CH_3, but nothing larger.
  - The corresponding area on the β-receptor is large and can easily accommodate the methyl group. Larger lipophilic groups afford higher β-affinity and activity.
- R_1 = *i*-propyl, *t*-butyl, or aralkyl: Selective β-agonist action.
  - N-*i*-propyl, *t*-butyl, and aralkyl substituted amines are devoid of α-agonist action. These substituents are too bulky to fit into the hydrophobic "methyl pocket" of α-receptors.
  - These larger, more lipophilic substituents bind with selective affinity at β-receptors.
  - The *t*-butyl and aralkyl substituents show greater β_2-selectivity, while the *i*-propyl binds at both β_1 and β_2 sites.

α-Carbon Substituent (R_2)

- R_2 = H
  - No steric hindrance to essential ion–ion anchoring at adrenoceptors.

Phenylethylamine agonists:

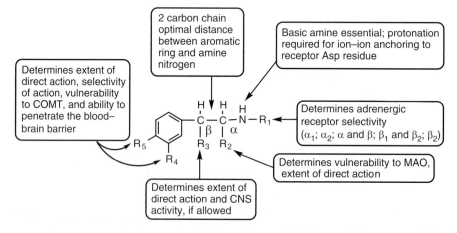

Arylimidazoline agonists:

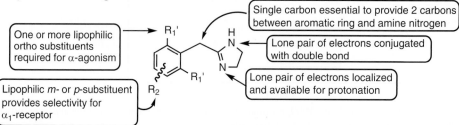

Figure 4.3 Adrenergic agonist SAR.

- MAO deamination occurs if the drug has, or can metabolically generate, a primary amine.
- $R_2 = CH_3$
  - α-Carbon substituents introduce asymmetry. R and S isomers exist.
  - Direct action at α_1-receptors is decreased due to steric hindrance to ion–ion bonding. The S isomer retains direct action at α_2- and β-receptors.
  - Affinity for presynaptic uptake-1 sites is high. Indirect action results from prolonged retention of NE in the synapse.
  - Penetration of the blood–brain barrier occurs if no phenolic hydroxyls are present. Stimulation of central α_1-receptors results in sleeplessness, agitation, restlessness, and anorexia (appetite suppression).
- $R_2 = C_2H_5$
  - β_2-Selective agonist action results, even with nitrogen substituents associated with α-preferential activity (e.g., $R_1 = H$).

β-Carbon Substituent (R_3)

- $R_3 = OH$
  - A β-OH introduces asymmetry. R and S isomers exist.
  - The R configuration of the carbinol carbon is essential for optimal direct action at adrenoceptors. The affinity-enhancing H-bond is most critical at β sites (particularly β_2) (Fig. 4.1).
  - A β-OH slows, but does not stop, CNS penetration.
- $R_3 = H$
  - Direct action decreases, especially at β-receptors. Indirect action is maintained.
  - CNS penetration in nonphenolic compounds is significantly enhanced, especially when R_1 and/or R_2 is/are CH_3.

Aromatic Substituents (R_4 and R_5)

- R_4 and $R_5 = OH$ (catechol)
  - Required for optimal direct action at all adrenoceptors, particularly β-receptors. Critical H-bonds are formed with receptor Ser residues (Fig. 4.1).
  - Catechols have low stability both in vivo (COMT metabolism) and in vitro (chemical or photo-oxidation to inactive quinones).
  - Catechols prohibit blood–brain barrier penetration because of their hydrophilicity. No CNS stimulation is observed.
- R_4 or $R_5 = OH$ (monophenol)
  - Direct action decreases as compared to catechols. Subtype specificity is dependent on substitution pattern. A m-phenol ($R_4 = OH$) retains some direct action at α_1-receptors, while a p-phenol ($R_5 = OH$) retains direct action at β-receptors.
  - Monophenols are resistant to COMT, but vulnerable to glucuronic acid conjugation or sulfonation in the gut and liver.
  - In vitro oxidation to inactive quinones can occur. Protect from light, alkaline pH, and air.
  - Monophenols are too polar to penetrate the blood–brain barrier.
- R_4 and $R_5 = H$
  - Direct action decreases, especially at β-receptors, due to loss of H-bonds with receptor Ser. Indirect action rises due to blockade of presynaptic uptake-1 sites.
  - CNS stimulation is possible, particularly in compounds lacking a β-OH and/or containing an α-CH_3.
- Other substitution patterns
  - Small H-bonding moieties at m (e.g., hydroxymethyl, formamide) retain β-agonist affinity if a p-phenol is present. β_2-selectivity results if the amine substituent is t-butyl or aralkyl.
  - A 3,5-dihydroxyphenyl (resorcinol) ring gives β_2-selective action if the N-substituent is i-propyl, t-butyl, or aralkyl. The latter two substituents result in greater β_2-selectivity.
  - The above substitution patterns confer COMT resistance.

Table 4.1   Direct Action of Ephedrine and Pseudoephedrine Isomers

Ephedrine (*Erythro* isomers)   Pseudoephedrine (*Threo* isomers)

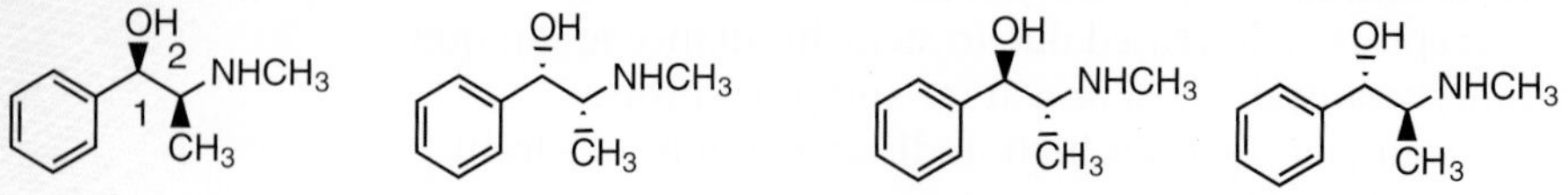

(1*R*:2*S*)   (1*S*:2*R*)   (1*R*:2*R*)   (1*S*:2*S*)

	1*R*:2*S* Ephedrine	1*S*:2*R* Ephedrine	1*R*:2*R* Pseudoephedrine	1*S*:2*S* Pseudoephedrine
Optimal *R* configuration at C1?	Yes	No	Yes	No
Optimal *S* configuration at C2?	Yes	No	No	Yes
Type of action	Mixed[a]	Mostly indirect	Mostly indirect	Mostly indirect

[a]Direct and indirect. Direct action is not optimal due to the lack of phenolic OH groups.

Stereochemistry

- Proper configuration of asymmetric centers is crucial to direct action. The carbinol carbon must be in the *R* configuration and the α-CH$_3$ should be in the *S* configuration.
- The relationship between stereochemistry and adrenoceptor affinity is exemplified by the *erythro* and *threo* isomers of ephedrine and pseudoephedrine, respectively (Table 4.1).

Physicochemical Properties

- Phenolic adrenergics are amphoteric: amine pKa 8.5 to 10 and phenol pKa 9.8 to 12.
- At pH 7.4, adrenergic agonists are monocations. The phenol(s) is/are predominantly uncharged at physiological pH.
- Log P is dependent on substitution pattern. Catecholic adrenergics with a β-OH, no α-CH$_3$, and small nitrogen substituents are the least lipophilic (e.g., NE, Epi). The most lipophilic drugs have few, if any, OH groups (e.g., amphetamine, methamphetamine).
- The negative impact of phenolic OH groups on lipophilicity is evident from the relatively high log P for ephedrine as compared to log P for Epi.

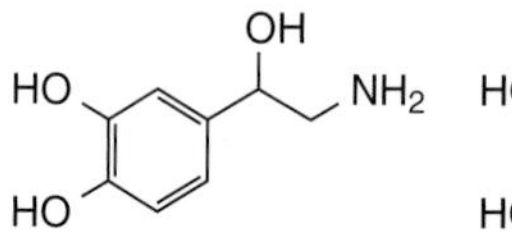

Norepinephrine
(log P = −1.26)

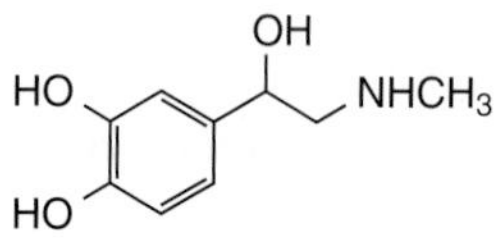

Epinephrine
(log P = −0.54)

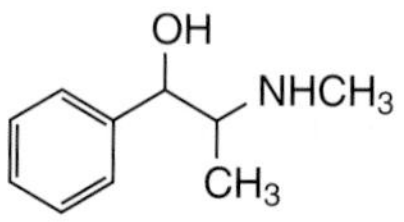

Ephedrine
(log P = 1.43)

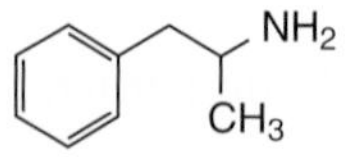

Amphetamine
(log P = 1.76)

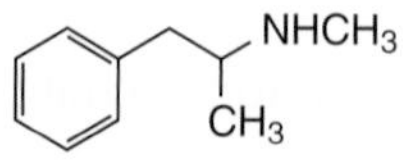

Methamphetamine
(log P = 2.20)

Arylimidazoline Adrenergic Agonists

SAR

- The phenylethylamine pharmacophore exists within the heterocyclic arylimidazoline structures. A nitrogen atom of the imidazoline ring has replaced a carbon in the "ethyl" moiety.
- One lipophilic *ortho* substituent (e.g., halogen) is required for potent α-agonist action. Two *o*-halogens augment lipophilicity and facilitate distribution to central sites of action.

- The 2-aminoimidazole segment of clonidine (a representative agonist) tautomerizes to the guanidino moiety, which has a pKa of 8.3. The electron withdrawing Cl atoms lower the pKa as compared to an unsubstituted guanidine (12.5), but the essential cationic amine still predominates at pH 7.4.
- Most agonists in this class stimulate central α_{2A}-receptors.

Phenyl"ethyl"amine   Guanidino moiety

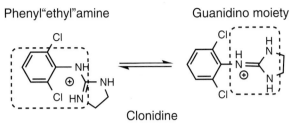

Clonidine

Clonidine tautomerization

Catecholic Agonists

Clinical Applications

- NE and Epi are used to constrict blood vessels (α_1) in life-threatening hypotensive or hemorrhagic situations (Fig 4.4). Both can stimulate the heart (β_1) in cardiac arrest and dilate bronchi (β_2) in bronchospasm.
- The *i*-propyl moiety of isoproterenol restricts its use to cardiac or bronchospasm-related emergencies.
- All three catechols are optimally direct acting, but orally inactive and of short duration due to rapid COMT and MAO metabolism.
- Dobutamine's action is stereochemically dependent, and includes α_1-agonism (*S* isomer), α_1-antagonism (*R* isomer), and β_1-agonism (*R* > *S*). The major therapeutic use is cardiac stimulation.

α-Adrenergic Agonists

Clinical Applications

Direct-Acting α_1-Agonists

- Phenylephrine's single *m*-OH provides selectivity for α_1-receptors. Its primary indication is nasal congestion (Fig. 4.5).
  - Vulnerability to MAO and conjugating enzymes in the gut limit oral bioavailability. Oral and intranasal dosage forms are marketed.
- Arylimidazoline α_1-agonists are used topically as ocular and nasal decongestants.

Mixed/Indirect-Acting Agonists

- Ephedrine and pseudoephedrine are most commonly used as nasal decongestants. Their action is predominantly indirect.
- 1*R*,2*S*-ephedrine has some direct action at adrenoceptors, and can dilate bronchi and stimulate the heart. The latter can be problematic in patients with underlying cardiovascular disease (e.g., hypertension, arrhythmia). Some CNS stimulation is possible with high-dose ephedrine.

CHEMICAL NOTE

Ma Huang
1*R*,2*S*-(-)-ephedrine is the major constituent of ephedra, a natural product that has enjoyed widespread use in traditional medicine. Ma Huang (*Ephedra sinica*) is a Chinese herb available via the internet and used for increased alertness or weight loss. Ephedrine's vasoconstriction (α_1) and cardiac stimulation (β_1) actions can cause serious adverse reactions, and fatalities from its use/misuse have been reported. The FDA has been issuing warnings on the use of Ma Huang for two decades.

CHEMICAL NOTE

Dependence Liability of Amphetamines
Stimulation of central adrenergic neurons by powerful indirect-acting agonists like amphetamine and methamphetamine produces euphoria and a sense of invincibility that, once experienced, can be very difficult to abstain from. Drug craving can be unrelenting, even after amphetamine use has been stopped. These drugs carry Black Box Warnings for dependence liability, and practitioners are urged to monitor patients for signs of illicit use or distribution of their prescription medication to others. Misuse of amphetamines results in hypertension, tachycardia, extreme anorexia, violent psychotic behavior, and the inability to regulate body temperature.

Norepinephrine bitartrate (R = H, X = $HOOC-\overset{\overset{\displaystyle HO}{|}}{\underset{\underset{\displaystyle H}{|}}{C}}-\overset{\overset{\displaystyle H}{|}}{\underset{\underset{\displaystyle OH}{|}}{C}}-COO^{\ominus}$)

Epinephrine HCl (R = CH$_3$, X = Cl$^\ominus$)

Isoproterenol HCl (R = CH(CH$_3$)$_2$, X = Cl$^\ominus$)

Dobutamine HCl (R = ... X = Cl$^\ominus$)

Figure 4.4 Catecholic adrenergic agonists.

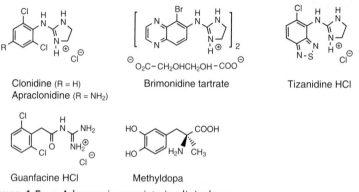

Direct Acting Phenolic α_1-Agonist
(nasal/ocular decongestant,
injectable hypertensive agent)

Phenylephrine HCl

Mixed/Indirect Acting Non Phenolic α_1-Agonists
(nasal decongestants, CNS stimulants)

Ephedrine/pseudoephedrine HCl
(R_1 = OH, R_2 = CH_3, X = $Cl^{\ominus}$, n = 1)

Dextroamphetamine sulfate
(R_1 = R_2 = H, X = $SO_4^{2\ominus}$, n = 2)

Methamphetamine HCl
(R_1 = H, R_2 = CH_3, X = $Cl^{\ominus}$, n = 1)

Arylimidazoline α_1-Agonists (topical nasal/ocular decongestants)

Tetrahydrozoline HCl   Naphazoline HCl   Xylometazoline HCl (R_1 = H)
                                                       Oxymetazoline HCl (R_1 = OH)

Direct Acting α_2-Agonists (antihypertensives, antiglaucoma, antispastic agents)

Clonidine (R = H)   Brimonidine tartrate   Tizanidine HCl
Apraclonidine (R = NH_2)

Guanfacine HCl   Methyldopa

Figure 4.5 α-Adrenergic agonists in clinical use.

- Dextroamphetamine and methamphetamine are potent CNS stimulants used in attention deficit hyperactivity disorder (ADHD) or narcolepsy. They act indirectly to facilitate stimulation of central α_1-receptors by NE.
- All agents have good oral bioavailability due to resistance to inactivating enzymes.

α_2-Agonists

- The therapeutic activity observed is dependent upon the α_2-subtype being stimulated.
- Activation of central α_{2A}-receptors by oral clonidine and guanfacine inhibits NE release and causes a profound drop in peripheral blood pressure.
  - Extended release formulations can be used in childhood ADHD.
  - An injectable clonidine formulation attenuates cancer pain. The drug can also reduce anxiety in opioid-dependent patients experiencing withdrawal.
- Apraclonidine (*p*-aminoclonidine) and brimonidine are α_{2A}-agonists used in the treatment of glaucoma.
  - Dangerous elevations in blood pressure can occur with apraclonidine use. Brimonidine is approximately 1,000 times more α_{2A}-selective and less likely to induce cardiovascular toxicity.
- Methyldopa, an α_{2A}-prodrug, is converted to the antihypertensive α-methylnorepinephrine in the CNS.
- Tizanidine, a thiadiazole clonidine analog, is selective for α_{2C}-receptors in motor neurons. It is administered orally to reduce spasticity related to injury of the brain or spinal cord.

Methyldopa and α-methylnorepinephrine

CHEMICAL NOTE

Phenylpropanolamine
In 2000, the FDA issued a public health advisory on the use of a then-readily available OTC, phenylpropanolamine (PPA), which was being used as an inexpensive weight loss aid. The predictable cardiovascular side effects of this lipophilic indirect-acting adrenergic were pronounced in some women, and a low, but significant, risk of life-threatening hemorrhagic stroke was noted. As a result, these products are no longer marketed.

CHEMICAL NOTE

Restricted Access to OTC Pseudoephedrine
Patients wishing to purchase pseudoephedrine to treat a cold must consult their pharmacist and show proof of identity before this OTC drug can be dispensed. The answer to why this must be lies in its chemistry. In a relatively simple reaction using readily available household chemicals, low-potency pseudoephedrine can be converted to the exceptionally potent and strongly addicting CNS stimulant methamphetamine. In addition to the negative social impact of illicit methamphetamine use, "kitchen chemists" present a public health hazard, as the explosions and fires caused when meth is "cooked" by inexperienced people using jury-rigged equipment has killed or injured many.

β$_2$-Agonists

SAR

- All β$_2$-agonists contain the phenylethylamine pharmacophore with a bulky amine substituent. A β-OH in the *R* configuration is essential for direct action (Fig 4.6).
- Short-acting β$_2$-agonists contain a salicyl alcohol or resorcinol aromatic moiety, and either an *i*-propyl or a *t*-butyl amine substituent.
- Long-acting β$_2$-agonists generally have similar ring systems and lipophilic N-aralkyl substituents.
  - N-Aralkyl substituents promote high β$_2$-selectivity (e.g., salmeterol is 50 times more β$_2$-selective than albuterol).
  - High log P results in membrane retention, delaying access to the β$_2$ active site on the extracellular membrane surface.
  - The shorter amine substituent of formoterol compared to salmeterol provides a lower log P, more facile access to the β$_2$ active site, and a faster onset of action.
  - Strong hydrophobic "tethering" of the N-aralkyl substituent to the β$_2$-receptor can result in long durations of action.
  - Formoterol's *p*-OCH$_3$ group is believed to have a receptor binding role.
- The electron withdrawing lactam of indacaterol drops the pKa to 6.7, compared to 7.9 and 10.2 for formoterol and salmeterol, respectively.
  - A higher percentage of indacaterol exists as a zwitterion, which increases drug association with the membrane-bound receptor.

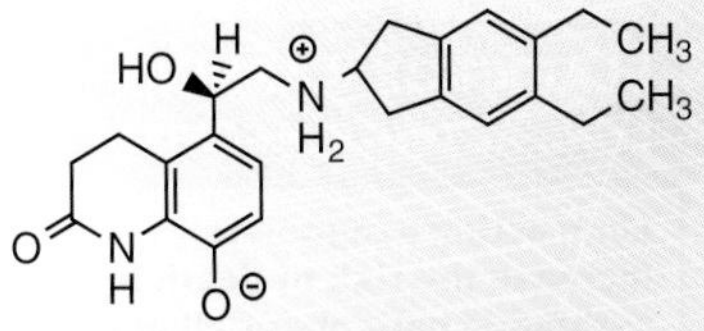

Indacaterol zwitterion

Metabolism

- β$_2$-Agonists are resistant to COMT (no catechol nucleus) and MAO (slow or inhibited dealkylation to primary amine).
- O-Dealkylation or benzylic hydroxylation occurs on vulnerable compounds (e.g., formoterol and salmeterol, respectively).
- Phase II glucuronic acid conjugation or sulfonation occurs prior to elimination.

Short Acting

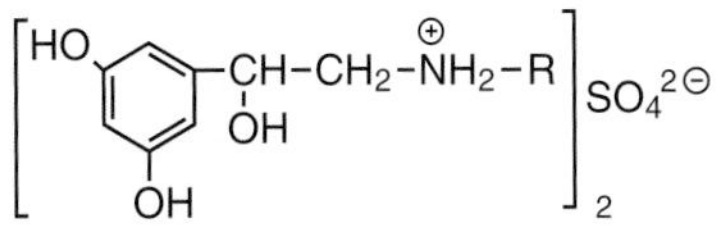

Metaproterenol sulfate (R = CH(CH$_3$)$_2$) Levalbuterol HCl (Y = CH, X = Cl$^{\ominus}$)
Terbutaline sulfate (R = C(CH$_3$)$_3$) Pirbuterol acetate (Y = N, X = CH$_3$COO$^{\ominus}$)

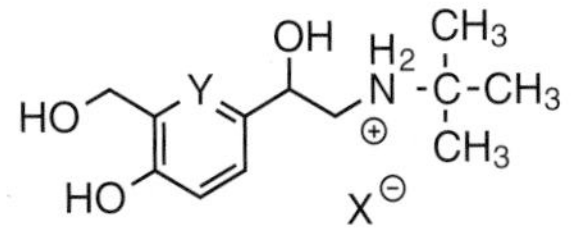

Long Acting

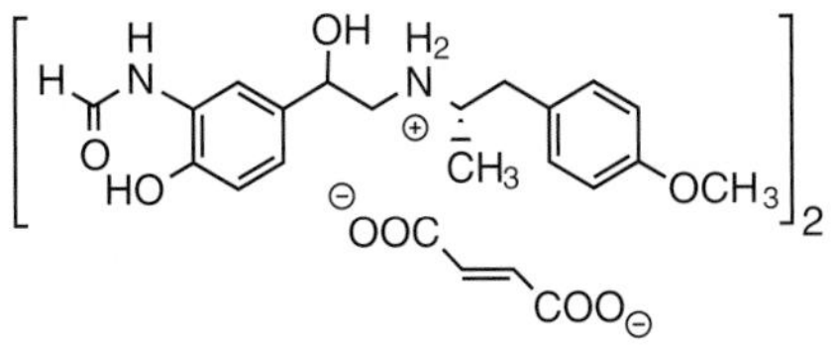

Formoterol fumerate (Log P 1.6) 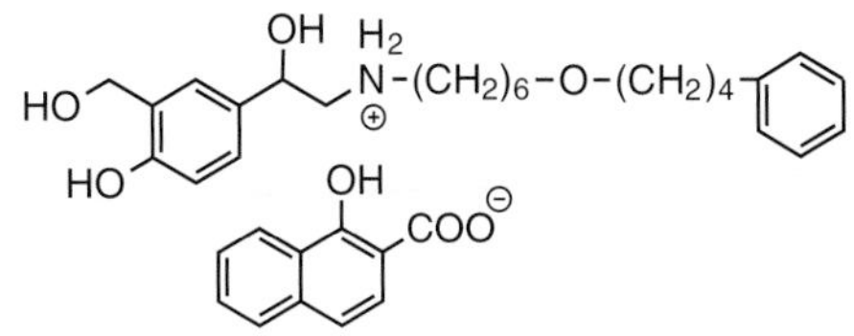 Salmeterol xinafoate (log P 3.88)

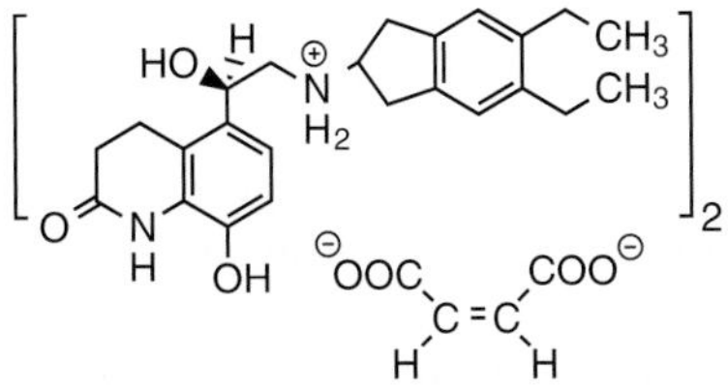

Indacaterol maleate (log P 3.88) 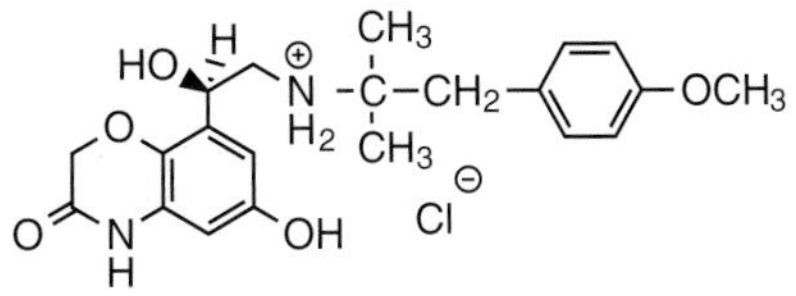 Olodaterol hydrochloride (Log P 1.88)

Figure 4.6 Direct-acting β$_2$-agonists (bronchodilators).

ADVERSE EFFECTS

Short-Acting β₂-Agonists

- Cardiovascular toxicity possible, particularly with oral or injectable preparations in high doses.
- Terbutaline: Potentially fatal cardiovascular events in pregnant women and/or the fetus (Black box warning: Do not use to halt premature labor)
- Racemic albuterol: higher airway inflammation/reactivity risk compared to levalbuterol. The proinflammatory $S(+)$ isomer causes a stronger reaction to bronchospasm triggers.

BLACK BOX WARNING
LONG-ACTING β₂-AGONISTS

Long-acting β₂-agonists increase the risk of death in asthmatic patients of all ages. The use of some agents (indacaterol, olodaterol) is restricted to COPD, and the others are used in asthmatic patients only if they are inadequately controlled with an inhaled corticosteroid. As symptoms improve, steps should be taken to eliminate the β₂-agonist from the regimen or minimize dose.

Clinical Applications

Short Acting

- Short-acting β₂-agonists are used to dilate bronchi during acute bronchospastic episodes (e.g., asthma, chronic obstructive pulmonary disease [COPD], emphysema).
- Terbutaline, metaproterenol, and racemic albuterol are orally available. Terbutaline is also available by injection.
- The remaining structures are administered only via inhalation. Bronchodilation begins within 5 minutes and lasts approximately 4 to 8 hours.

Long Acting

- Long-acting β₂-agonists are used via inhalation to control the number and severity of bronchoconstrictive episodes in patients with bronchospastic disease.
- Cotherapy with an inhaled corticosteroid is the standard in treating asthma.
- Because onset time can be up to 30 minutes, these agents cannot be used to rescue patients from acute asthma attacks.
- Salmeterol is a partial β₂-agonist with exceptionally high activity due to the ability to sequester in bronchial membranes. It has also been proposed to undergo pseudoirreversible hydrophobic bonding to the β₂-receptor.
- Formoterol is a full β₂-agonist with three-fold lower β₂-selectivity than salmeterol. Its higher potency allows for lower doses (12 μg vs. 50 μg, both administered twice daily).
- Formoterol's onset time of 3 to 5 minutes is significantly shorter than the 30-minute onset time of salmeterol. The duration is 12 hours for both agents.
- Formoterol is racemic, but the active R,R-(-) isomer is marketed as arformoterol. While approximately twice as potent as formoterol, the dosing regimen is similar (30 μg/day).
- The major therapeutic advantage of the newest long-acting β₂-agonists, indacaterol and olodaterol, is a 24-hour duration, allowing once-daily dosing (75 μg and 5 μg, respectively).
- The long duration of indacaterol and olodaterol is attributed to concentration in bronchial membranes and exceptionally high β₂ receptor affinity, respectively.

β₃-Agonist

- Mirabegron has been approved for the treatment of overactive bladder (OAB). It acts at β₃-receptors in the detrusor muscle, leading to relaxation and increased bladder capacity.
- The drug undergoes amide hydrolysis and subsequent N-acetylation.
- Unlike antimuscarinic OAB agents, mirabegron does not induce dry mouth or blurred vision.
- Mirabegron moderately inhibits CYP2D6, leading to the potential for drug–drug interactions.

Mirabegron

Major metabolite (M5)

Adrenergic Antagonists

Nonselective α-Antagonists

- Antagonists block the action of released NE and circulating Epi on postsynaptic α₁-receptors (Fig. 4.7).
- Phenoxybenzamine is the only irreversibly acting adrenergic antagonist. It spontaneously generates an electrophilic aziridinium ion that is readily attacked by α-receptor nucleophiles.
- Phenoxybenzamine is administered orally (10 mg twice daily) to patients with a benign NE/Epi-secreting tumor of the adrenal glands called pheochromocytoma.
- Predictable adverse effects include postural hypotension, nasal congestion, and reflex tachycardia.

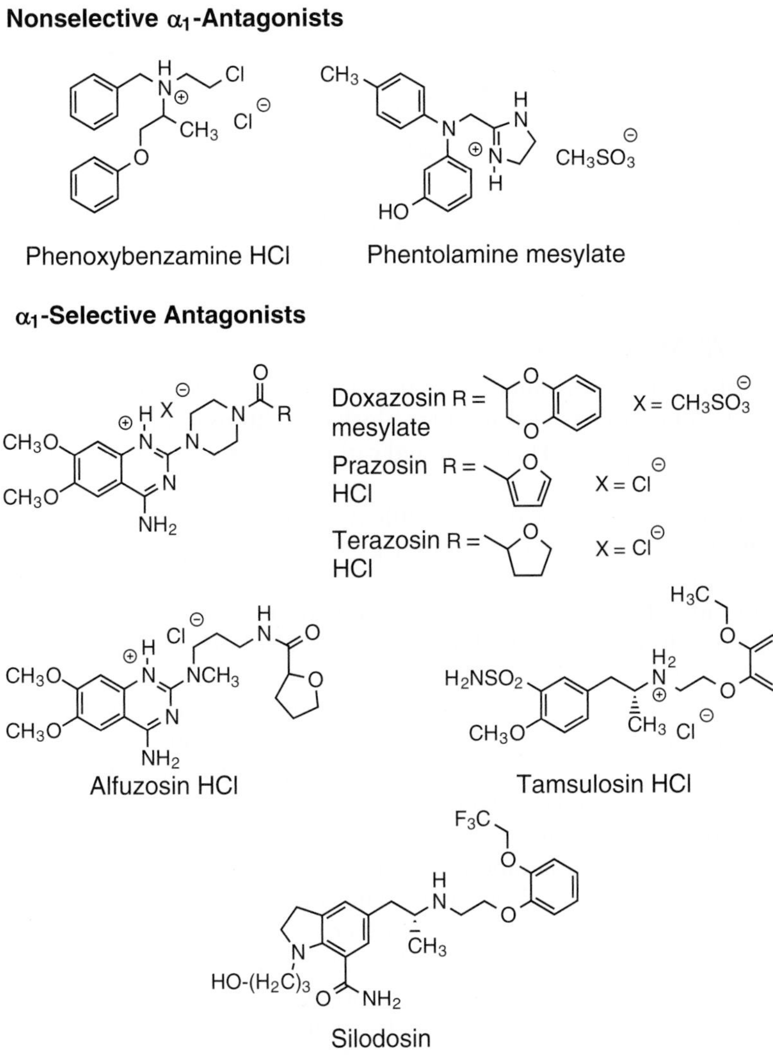

Nonselective α₁-Antagonists

Phenoxybenzamine HCl

Phentolamine mesylate

α₁-Selective Antagonists

Doxazosin R = mesylate   X = $CH_3SO_3^-$

Prazosin R = HCl   X = Cl^-

Terazosin R = HCl   X = Cl^-

Alfuzosin HCl

Tamsulosin HCl

Silodosin

Figure 4.7 α-Adrenergic antagonists in clinical use.

CHEMICAL NOTE

Aziridinium Ion

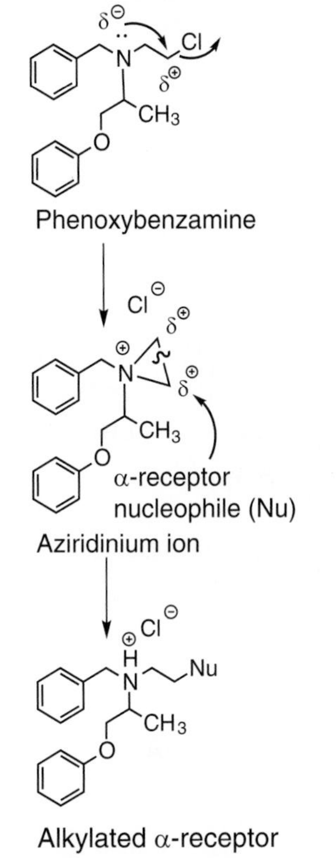

Phenoxybenzamine

Aziridinium ion

α-receptor nucleophile (Nu)

Alkylated α-receptor

Phenoxybenzamine's β-chlorethylamine moiety spontaneously cyclizes to a highly electrophilic quaternary aziridinium ion. The free base form is required, as the lone pair of electrons must attack the electrophilic β-carbon to generate the reactive quaternary species. The electron withdrawing Cl decreases amine pKa, allowing more of the active free base to predominate at cellular pH. Once generated, the aziridinium ion reacts irreversibly with α-adrenergic receptor nucleophiles.

α₁-Selective Antagonists

- Most drugs block the $α_{1A}$-receptors in the prostate and are used orally in the treatment of benign prostatic hyperplasia (BPH) (Fig. 4.7).
  - Tamsulosin and silodosin are used exclusively for this indication.
  - Prazosin is used exclusively in the treatment of hypertension.
  - Terazosin and doxazosin are marketed for both indications.
- Tamsulosin and silodosin are benzenesulfonamides. Silodosin is the more $α_{1A}$-selective agent, leading to fewer cardiovascular adverse effects.
- The other antagonists are 4-amino-6,7-dimethoxyquinazolines with related C3 side chains. Side chain chemistry impacts physicochemical properties (see Chapter 10, Figure 10.5 in *Foye's Principles of Medicinal Chemistry, Seventh Edition*, for a complete detailed discussion).

Metabolism

- Prazocin's furan ring promotes extensive metabolism leading to a shortened duration (4 to 6 hours). The tetrahydrofuran ring of the otherwise identical terazosin promotes metabolic resistance and an extended duration (>18 hours).
- Quinazoline-based $α_1$-antagonists are O-demethylated prior to excretion. Other metabolic reactions (e.g., hydrolysis) are possible.
- Benzenesulfonamide $α_1$-antagonists are also extensively metabolized. The glucuronide conjugate of silodosin is active and pervasive, and contributes to its >24-hour duration.

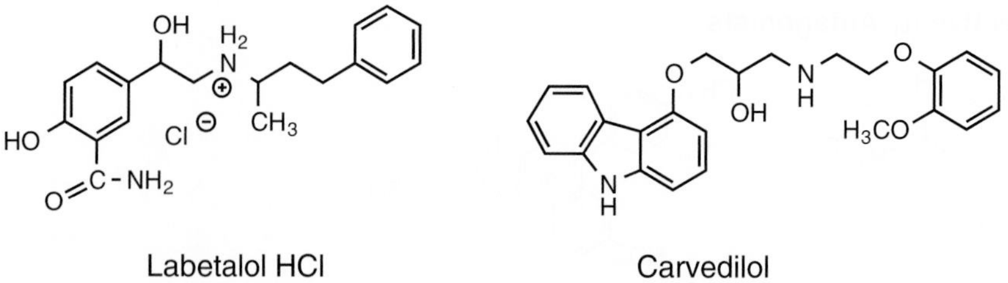

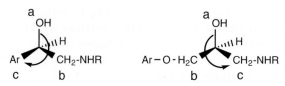

Labetalol HCl   **Carvedilol**

Figure 4.8 Nonselective α/β-adrenergic antagonists (antihypertensives).

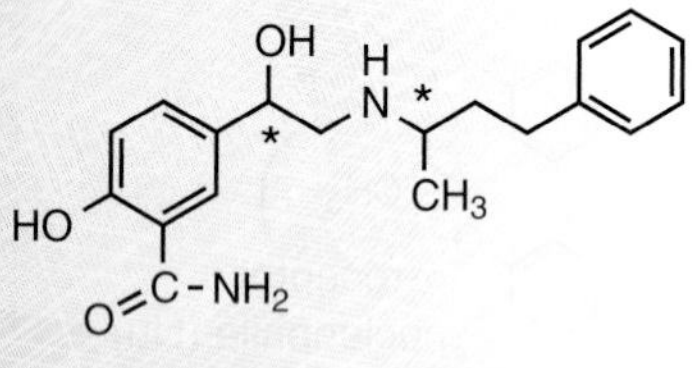

Labetalol isomers

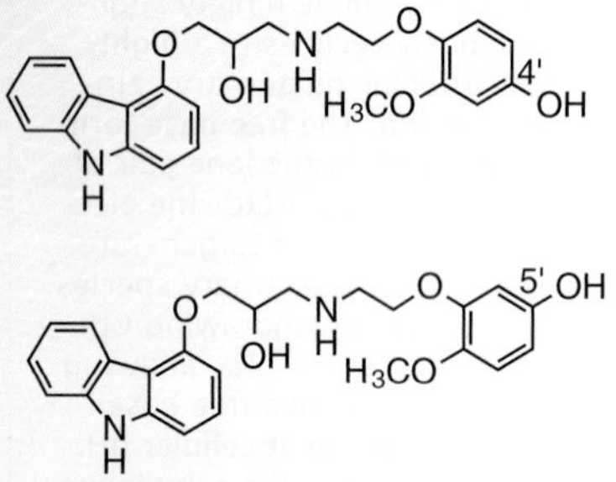

Nonselective α/β-Antagonists

- Two of labetalol's four stereoisomers are active (Fig. 4.8). The *R,R* isomer provides potent and nonselective β-receptor blockade with minimal $α_1$-antagonism. Epimerizing the CH_3-substituted carbon to *S* provides selective $α_1$-antagonism.
- $β_1$-Antagonism attenuates cardiac output, while $α_1$-antagonism blocks vasoconstriction by endogenous pressor amines (e.g., NE). The therapeutic result is decreased blood pressure.
- $β_2$-Antagonism causes bronchoconstriction, but the m-carboxamide (reminiscent of formoterol's formamide and albuterol's salicyl alcohol) provides counterbalancing $β_2$-agonism. This lowers the risk of bronchospasm.
- Labetalol is orally active (100 to 300 mg twice daily) and resistant to Phase I metabolism. Conjugation with glucuronic acid precedes excretion.
- See Chapter 10, Table 10.6 in *Foye's Principles of Medicinal Chemistry, Seventh Edition,* for a complete detailed discussion of major pharmacologic and pharmacokinetic properties of all commercially available β-antagonists.
- Carvedilol is an aryloxypropanolamine, a structural motif more commonly observed in β-antagonists. It is marketed as a racemic mixture of two stereoisomers.
  - The *R* isomer is a selective α-antagonist. The *S* enantiomer has some α-antagonist action, but nonselective β-blocking activity predominates.
  - With no inactive isomer in the formulation, carvedilol has enhanced antihypertensive potency compared to labetalol. The oral dosing regimen is 6.25 to 25 mg twice daily.
- Carvedilol can also be used in congestive heart failure. The aromatic carbazole moiety serves an antioxidant function that augments its value in heart failure.
- Carvedilol is extensively metabolized and excreted in feces as the glucuronide of active Phase I phenolic metabolites. The β-antagonist action of the 4'-hydroxy metabolite has been estimated at 13 times that of the parent drug.

Carvedilol active metabolites

Aryloxypropanolamine β-Antagonists

SAR

- The SAR of aryloxypropranolamine β-antagonists is summarized in Figure 4.9 and the compounds in clinical use are shown in Figure 4.10. These antagonists bind to the same β-receptor residues as phenylethanolamine agonists.
- The *S* configuration of the carbinol carbon is essential, binding moieties in the same 3D orientation as the *R* configuration of the agonists.

R-Phenylethanolamine agonist   *S*-Aryloxypropanolamine antagonist

R and *S* isomers of phenylethanolamines and aryloxypropanolamines

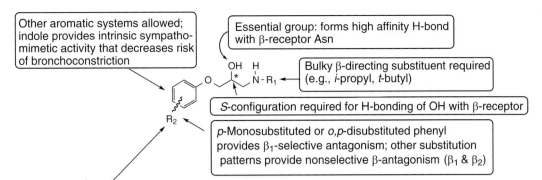

Other aromatic systems allowed; indole provides intrinsic sympatho-mimetic activity that decreases risk of bronchoconstriction

Essential group: forms high affinity H-bond with β-receptor Asn

Bulky β-directing substituent required (e.g., *i*-propyl, *t*-butyl)

S-configuration required for H-bonding of OH with β-receptor

p-Monosubstituted or *o,p*-disubstituted phenyl provides β_1-selective antagonism; other substitution patterns provide nonselective β-antagonism (β_1 & β_2)

Determines whether β-antagonist will be classified as lipophilic or hydrophilic, relative risk of central adverse effects, and metabolic vulnerability

Figure 4.9 Aryloxypro-panolamine β-adrenergic antagonist SAR.

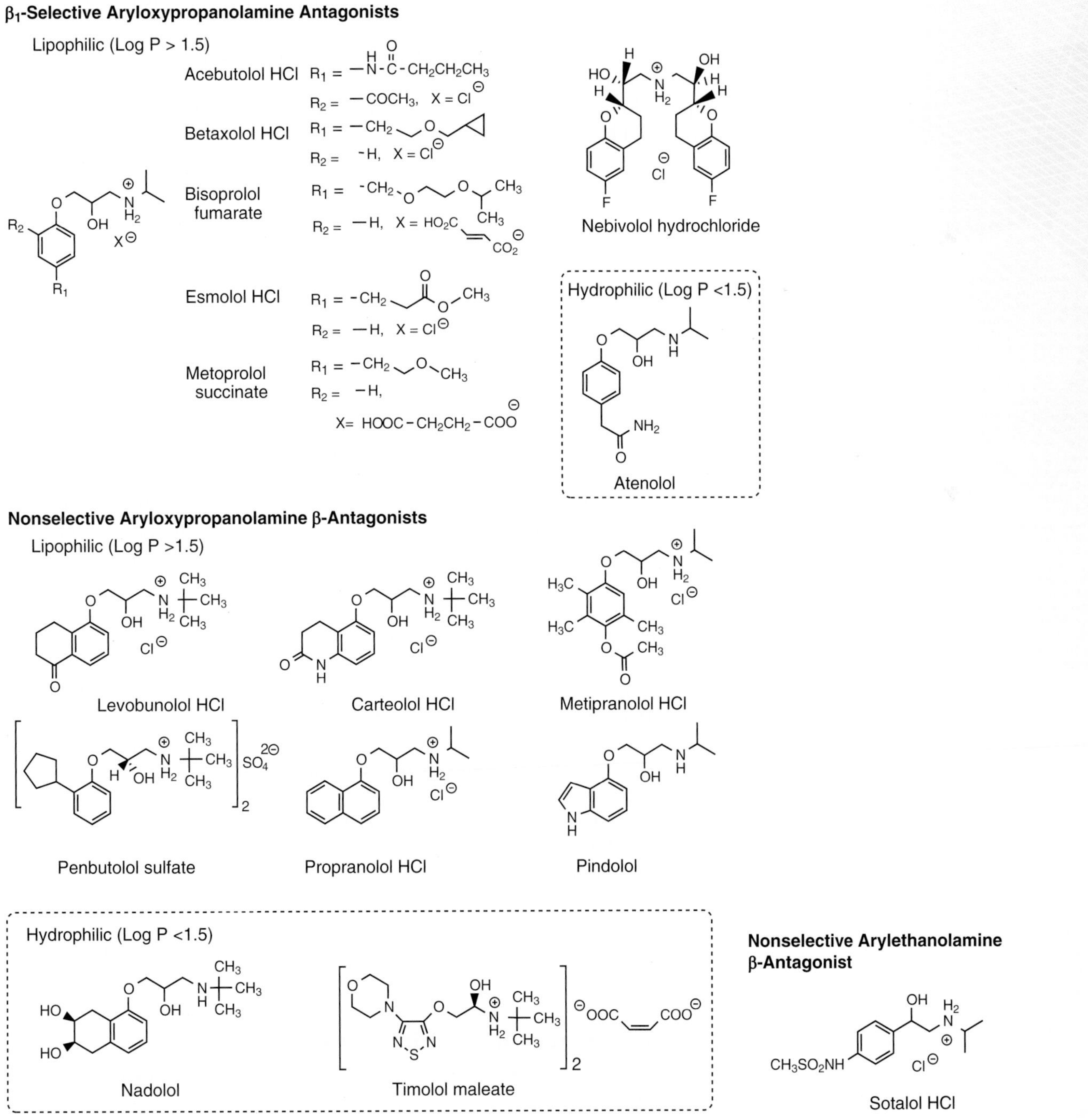

β_1-Selective Aryloxypropanolamine Antagonists

Lipophilic (Log P > 1.5)

Acebutolol HCl $R_1 =$ —N-C-$CH_2CH_2CH_3$
$R_2 =$ —$COCH_3$, X = $Cl^{\ominus}$

Betaxolol HCl $R_1 =$ —CH_2—O—◁
$R_2 =$ —H, X = $Cl^{\ominus}$

Bisoprolol fumarate $R_1 =$ —CH_2—O—O—CH(CH_3)
$R_2 =$ —H, X = HO_2C—CH=CH—$CO_2^{\ominus}$

Esmolol HCl $R_1 =$ —CH_2—C(O)—O—CH_3
$R_2 =$ —H, X = $Cl^{\ominus}$

Metoprolol succinate $R_1 =$ —CH_2—O—CH_3
$R_2 =$ —H,
X= $HOOC$—CH_2CH_2—$COO^{\ominus}$

Nebivolol hydrochloride

Hydrophilic (Log P <1.5)

Atenolol

Nonselective Aryloxypropanolamine β-Antagonists

Lipophilic (Log P >1.5)

Levobunolol HCl

Carteolol HCl

Metipranolol HCl

Penbutolol sulfate

Propranolol HCl

Pindolol

Hydrophilic (Log P <1.5)

Nadolol

Timolol maleate

Nonselective Arylethanolamine β-Antagonist

Sotalol HCl

Figure 4.10 β-Adrenergic antagonists in clinical use.

- The aryl nucleus is most commonly phenyl. Selective β_1-antagonism is achieved with *p*-monosubstituted phenyl rings. Small groups can be added at the *ortho* position without loss of selectivity.
- If indole replaces phenyl, some β-agonist activity is observed. This is known as intrinsic sympathomimetic activity.
- Polar aromatic rings (thiadiazole) or substituents (hydroxyl, acetamido, morpholino) decrease log P, minimizing central adverse effects. (see Chapter 10, Table 10.6 in *Foye's Principles of Medicinal Chemistry, Seventh Edition*, for a complete detailed discussion).
- Hydrophilic β-antagonists are primarily cleared through the renal system, while the more lipophilic blockers undergo hepatic clearance.

Metabolism

- N-Dealkylation is inhibited in *t*-butyl–substituted β-antagonists, and significantly slowed in compounds with N-*i*-propyl substituents.
- Hydrophilic β-antagonists are often resistant to hepatic metabolism (e.g., nadolol, atenolol).
- Lipophilic agents are commonly metabolized by predictable CYP-mediated reactions. For example:
  - propranolol and penbutolol undergo *p*-hydroxylation at C4′.
  - metoprolol is subject to O-dealkylation by CYP2D6.
  - betaxolol is metabolized by benzylic hydroxylation.
  - esmolol is vulnerable to ester hydrolysis.

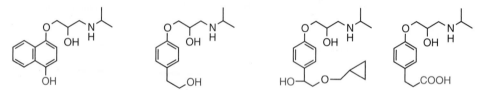

4-Hydroxypropranolol   O-Desmethylmetoprolol   α-Hydroxybetaxolol   Esmolol acid

β-Adrenergic antagonist metabolites

Clinical Applications

- Nonselective antagonists are contraindicated in patients with asthma, COPD, or emphysema. Those with intrinsic sympathomimetic activity induce less bronchoconstriction, but are not safe to use in patients at risk for bronchospasm.
- β_1-Selective antagonists have the lowest risk of inducing bronchoconstriction or bronchospasm in patients sensitive to these effects (e.g., patients with asthma).
- All aryloxypropanolamine β-antagonists except carteolol, levobunolol, and metipranolol are administered orally to reduce blood pressure. Abrupt withdrawal after chronic therapy can result in exacerbation of underlying ischemic heart disease, worsening of angina symptoms, and/or myocardial infarction. Reduce doses gradually.
- Five β-antagonists are given via ophthalmic solution for the treatment of glaucoma: betaxolol, timolol, carteolol, levobunolol, and metipranolol.
  - Polar molecules penetrate ocular tissue slowly and lipophilic molecules can be trapped at the site of action.
  - In addition to avoiding first-pass metabolism, these physicochemical properties prolong duration. Dosage regimens of one to two drops twice daily are standard.

Phenylethanolamine β-Antagonist: Sotolol

- Sotolol's lone *p*-substituent does not provide β_1-selectivity in phenylethanolamine antagonists.
- The hydrophilic methylsulfonamido substituent abolishes intrinsic activity.

- Sotolol is used to treat ventricular arrhythmias. A special formulation (sotolol AF) is marketed to treat atrial arrhythmias. The formulations have different dosing and patient safety information and are not interchangeable.
- Sotolol can induce arrhythmia. A Black Box Warning cautions that therapy initiation should occur over 3 days in a facility that can continuously monitor and, if necessary, resuscitate patients.

Chemical Significance

Adrenergic agonists can be used for disease states as seemingly innocuous as the common cold to as life threatening as cardiac arrest, and many are available to patients OTC. Adrenergic antagonists are a therapeutic mainstay in the treatment of hypertension, a disease impacting millions. Therefore, it is highly likely that patients will be self-selecting or filling prescriptions for an adrenergic agonist or antagonist at some point in their lives, many on a chronic basis.

Fortunately, the SAR of adrenergic agonists and antagonists is among the most straightforward and consistent of any class of therapeutics agents around. By looking in one or two key places on the structure, the pharmacist can readily determine appropriate indications and potential therapeutic pitfalls of the use of any adrenergic drug in specific patients. For example:

- Is your patient asking for an oral medication for relief of nasal congestion? Consider an N-CH$_3$ or N-H compound with either no aromatic substituents or a single *m*-OH to target vascular α_1-receptors. Avoid structures with bulky β-directing nitrogen substituents or a metabolically vulnerable catechol ring.
- Is the patient above predisposed to anxiety, agitation, or insomnia? Then restrict your choice to N-CH$_3$ or N-H agonists with the *m*-OH, which cannot cross the blood–brain barrier.
- Do you have a newly diagnosed asthmatic patient with high blood pressure who needs a rescue inhaler for acute life-threatening bronchospasm? Select a drug with a β_2-directing salicyl alcohol aromatic ring and *t*-butyl nitrogen substituent, as they are both fast and short acting. Avoid drugs with long, lipophilic aralkyl nitrogen substituents, as they are slower to act (sometimes fatally so).
- Are you trying to help a patient in the OTC aisle select something to help her reduce irritation-related redness in her eyes before an important job interview? Guide her towards an imidazoline-based α_1-agonist where the heterocyclic ring is joined to the phenyl ring through a methylene group. They work very well for this purpose and exhibit little systemic vasopressor effects. Avoid imidazoline agonists where bridging atom is nitrogen, as these are potent, centrally acting antihypertensive agents.
- Has your hypertensive patient with a long history of smoking developed emphysema? Be sure that any β-antagonist therapy employs an aryloxypropanolamine-based agent with a either a single *p*-substituent or an *o, p*-disubstitution pattern to avoid potentially dangerous bronchoconstriction from unwanted β_2-antagonism.
- Do you have an elderly male patient with benign prostatic hypertrophy complaining about having to take his prazocin multiple times a day? Consider switching him to a similar quinazoline structure with a nonaromatic oxygen-containing ring (e.g., tetrahydrofuran) to avoid first-pass metabolism and extend duration of action.
- Is the patient above at risk for falls from orthostatic hypotension? Consider switching him to a benzenesulfonamide that has a higher selectivity for prostatic α_1-receptors.

A sound knowledge of adrenergic SAR will always steer pharmacists in the right therapeutic direction when advising patients or consulting with prescribers. It is their unique knowledge of drug chemistry that gives pharmacists the depth of understanding needed to evaluate therapeutic alternatives competently and confidently, and counsel patients wisely.

Review Questions

1 Which of the following adrenergic agonists would be most likely to have MAO in its metabolic pathway?

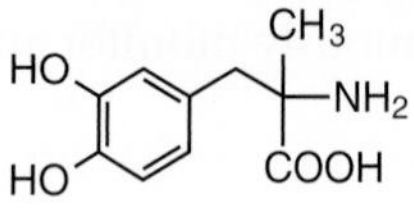
1

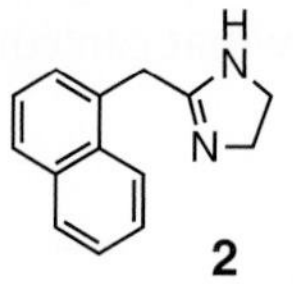
2

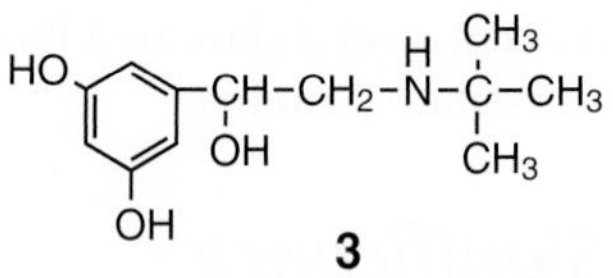
3

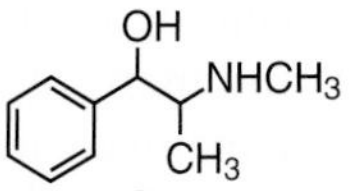
4

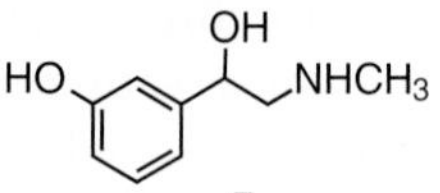
5

A. Agonist **1**
B. Agonist **2**
C. Agonist **3**
D. Agonist **4**
E. Agonist **5**

2 Which adrenergic structure, if taken orally, would put a patient at highest risk for elevated blood pressure?

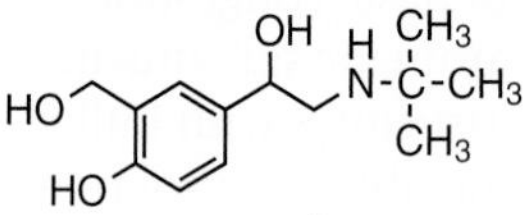
1

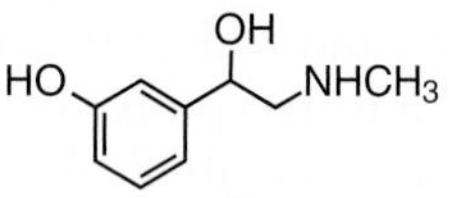
2

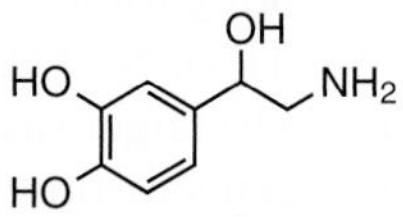
3

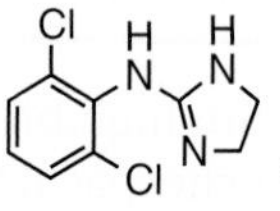
4

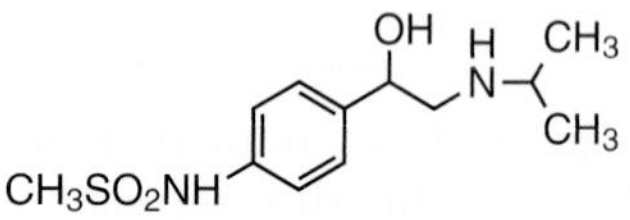
5

A. Adrenergic **1**
B. Adrenergic **2**
C. Adrenergic **3**
D. Adrenergic **4**
E. Adrenergic **5**

3 Which β-adrenergic agonist, if used via inhalation by a patient experiencing an acute, severe asthmatic episode, would be the most likely to result in a fatal outcome?

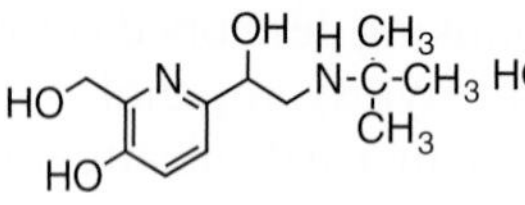
1

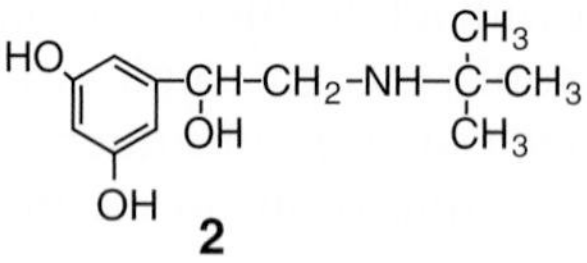
2

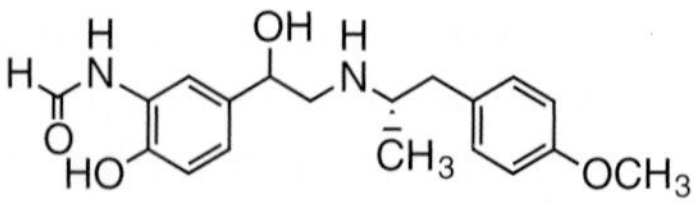
3

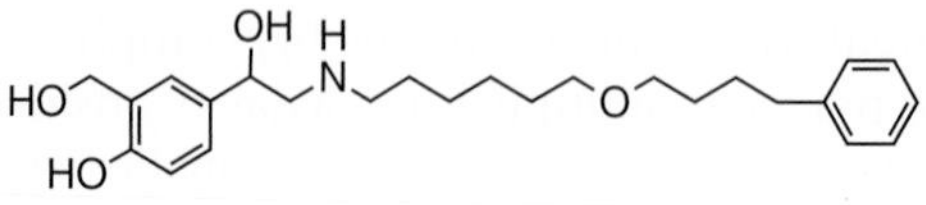
4

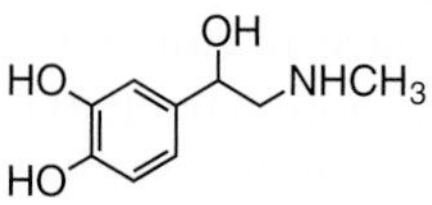
5

A. Agonist **1**
B. Agonist **2**
C. Agonist **3**
D. Agonist **4**
E. Agonist **5**

4 Which adrenergic antagonist has the proper chemistry to irreversibly inactivate α-receptors?

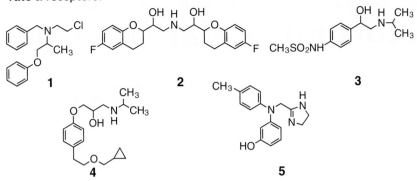

A. Antagonist **1**
B. Antagonist **2**
C. Antagonist **3**
D. Antagonist **4**
E. Antagonist **5**

5 The three adrenergic antagonists drawn below are all used in the treatment of hypertension. Which of the following represents the correct order of their relative risk in inducing bronchoconstriction in a patient with airway hypersensitivity? (> = higher risk than)

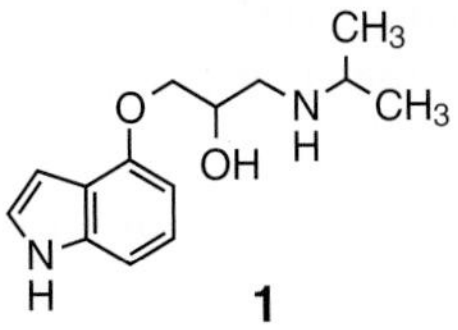

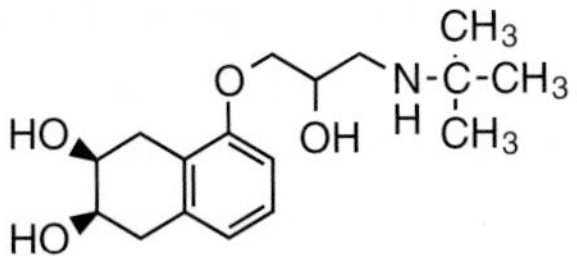

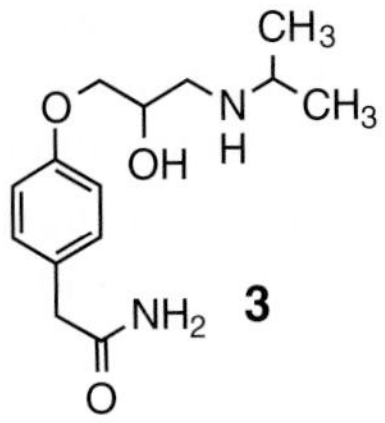

A. **1** > **2** > **3**
B. **1** > **3** > **2**
C. **2** > **1** > **3**
D. **3** > **2** > **1**
E. **3** > **1** > **2**

5

Drugs Used to Treat Neuromuscular Disorders

Learning Objectives

Upon completion of this chapter the student should be able to:

1. Describe mechanism of action of:
   a. *levo*-dihydroxyphenylalanine (L-DOPA)
   b. dopamine metabolism inhibitors
   c. dopamine receptor agonists
   d. adenosine receptor antagonists
   e. anticholinergic and antihistaminic agents used in Parkinson's Disease
   f. drugs used in the treatment of spasticity disorders.

2. Discuss applicable structure–activity relationships (SAR) of the above classes of anti-Parkinson's and antispasticity agents.

3. List physicochemical and pharmacokinetic properties that impact in vitro and in vivo stability and/or therapeutic utility of anti-Parkinson's and anti-spasticity agents.

4. Diagram metabolic pathways for all biotransformed anti-Parkinson's and antispasticity agents, identifying enzymes and noting the clinically significant metabolism and the activity, if any, of the major metabolites.

5. Apply all of the above to predict and explain therapeutic utility of anti-Parkinson's and antispasticity agents.

Introduction

Movement disorders can be categorized by their pathophysiology. Those classified as spastic involve excessive tonic stretch reflexes in skeletal muscle, resulting in spasm and accompanying pain and weakness. Spinal cord injury from disease (multiple sclerosis, cerebral palsy, stroke) or trauma leads to motor neuron hyperexcitability and the loss of reliable muscle control. The American Association of Neurological Surgeons estimates the worldwide incidence of spastic disorders at 12 million.

Parkinson's Disease (PD) is a dopamine deficiency disorder that has received significant national exposure since the 1991 diagnosis and subsequent public advocacy efforts of celebrity Michael J. Fox. Public interest was heightened with the announcement that actor/comic Robin Williams had also been diagnosed shortly before his death. The worldwide incidence of this progressively debilitating neurodegenerative disease is estimated at 4 to 6 million, with 1 million of those being US citizens. Approximately 50,000 to 60,000 cases are diagnosed annually in the United States, with Nebraska leading the nation and the world (329.3 diagnoses per 100,000). The risk of acquiring PD is highest for people living in the Midwest and Northeast, with exposure to pesticides and industrial byproducts being cited as the reason for this geographic "PD Belt." Most newly diagnosed patients are over 60, but the disease can strike in the prime of life. Muscle tremor, rigidity, and movement difficulties are hallmark symptoms.

While effective therapeutic options to treat neuromuscular disorders are somewhat limited, research is ongoing, and support in the form of robust and highly informative disease-specific websites is readily available to patients, families, and the providers who care for them.

Parkinson's Disease

Pathophysiology

- PD destroys neurons in the extrapyramidal dopaminergic pathway in the nigro-striatal region of the basal ganglia. It is most commonly acquired, and believed to be related to (1) environmental toxin exposure and (2) the normal aging process.
  - Familial history is a strong predictor of risk, but this may be related more to a shared environment than a genetic predisposition.
- Neurotoxicity results from the buildup of toxic byproducts of dopamine metabolism.
- Electrophilic quinone and semiquinone metabolites from dopamine auto-oxidation, and hydroxyl radical formed from the H_2O_2 byproduct of mono-amine oxidase-B (MAO-B) metabolism, have been implicated in neuronal destruction.
  - Cigarette smoke inhibits MAO, which may help explain the lower incidence of PD in smokers.
- Loss of dopamine neurotransmission leads first to loss of fine motor control, particularly when muscles are at rest (e.g., tremor). As the disease progresses muscle rigidity slows movement (bradykinesia) and can lead to balance and coordination difficulties. Cognition and mood impairment can also manifest.

Dopamine semiquinone, Dopamine quinone, and hydroxyl radical

Dopamine Replacement Therapy: L-DOPA

- L-DOPA is the mainstay of PD pharmacotherapy. It is administered in an effort to replace striatal dopamine lost to disease-related neuronal degeneration.
- In addition to relieving symptoms, prompt initiation of L-DOPA therapy has been shown to increase longevity.

L-DOPA

Metabolism

- L-DOPA, a prodrug, is activated via cytoplasmic decarboxylation to dopamine in the central nervous system (CNS) (Fig. 5.1). The catalyzing enzyme is aromatic L-amino acid decarboxylase (AAD), also known as DOPA decarboxylase.
- Blockade of premature DOPA decarboxylation in the periphery is essential. Dopamine generated peripherally cannot be transported to central sites of action.

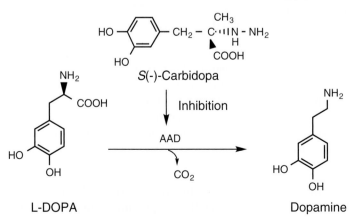

S(-)-Carbidopa

Inhibition

AAD

CO_2

L-DOPA   Dopamine

Figure 5.1 L-DOPA metabolism to dopamine.

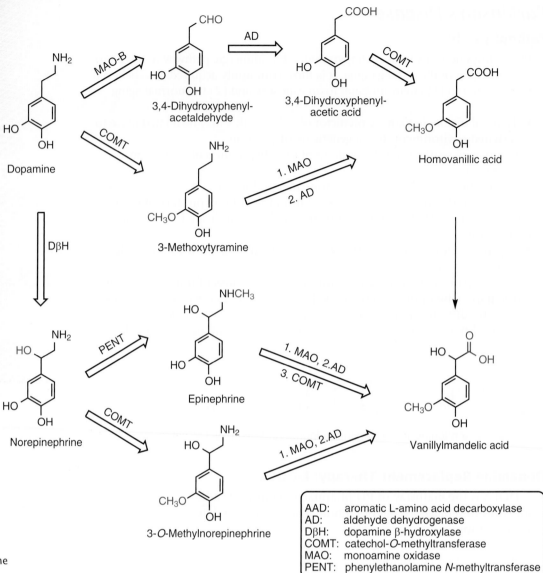

Figure 5.2 Dopamine metabolism.

- To ensure L-DOPA therapeutic effectiveness, the peripheral AAD inhibitor S(-)-carbidopa must be coadministered (Fig. 5.1).
- Without concomitant administration of S(-)-carbidopa, only about 1% of an orally administered dose of L-DOPA would reach the CNS. With S(-)-carbidopa, L-DOPA doses can be cut by up to 30-fold, decreasing toxicity risk.
- Carbidopa does not effectively cross the blood–brain barrier and does not inhibit conversion of L-DOPA to dopamine in the basal ganglia.
- Dopamine generated from L-DOPA is metabolized as shown in Figure 5.2.

Physicochemical and Pharmacokinetic Properties

- As an amino acid, L-DOPA is actively transported across the blood–brain barrier.
- Amine pKa of L-DOPA (8.72) is also lower than that of dopamine (8.93). L-DOPA can form a zwitterion that is absorbed from the intestinal tract and transported by aromatic and neutral amino acid transporters.
- L-DOPA is well absorbed by transporters upon oral administration. Bioavailability from immediate release formulations is 70% to 75%.

Clinical Applications

- L-DOPA/carbidopa is administered as immediate release, dispersible or extended release oral tablets. The ratio of L-DOPA to AAD inhibitor ranges from 4:1 to 10:1.

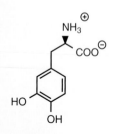

L-DOPA zwitterion

MAO Inhibitors:

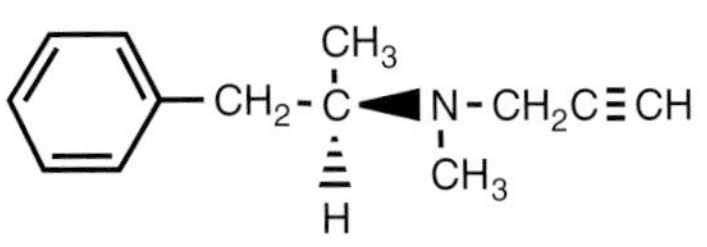

Selegiline (*R*-(-)-Deprenyl, Eldepryl, Emsam)

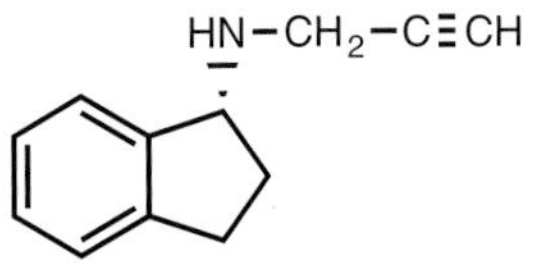

(R)-(+)Rasagiline (Azilect)

COMT Inhibitors:

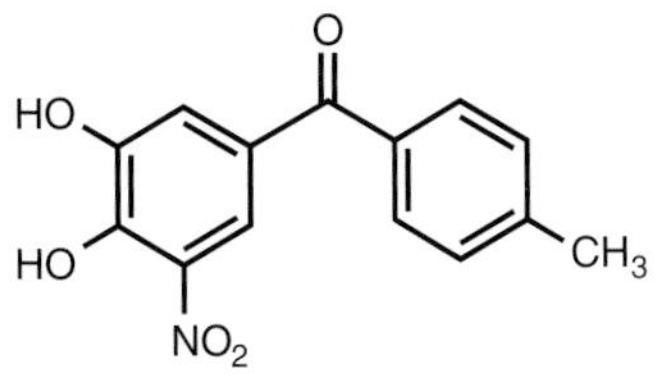

Tolcapone (Tasmar)

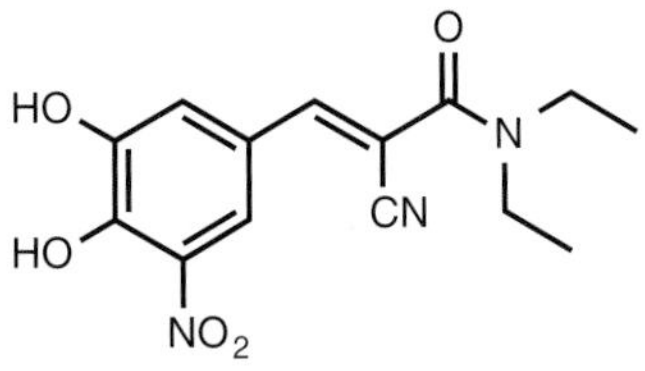

Entacapone (Comtan)

Figure 5.3 Dopamine metabolism inhibitors.

- A L-DOPA/carbidopa tablet that incorporates 200 mg of the catechol-O-methyltransferase (COMT) inhibitor entacapone (Fig. 5.3) is available.
- Patients on long-term L-DOPA therapy will experience periods of alternating drug hypersensitivity and resistance ("on–off" effect). This is attributed to increased dopamine receptor sensitivity and ongoing destruction of dopaminergic neurons, respectively.
- High protein meals or supplements retard L-DOPA transport across the gut and the blood–brain barrier by competition for amino acid active transporting proteins. L-DOPA patients should avoid high protein diets.
- High-dose pyridoxine (Vitamin B_6), the AAD cofactor, can potentially stimulate peripheral decarboxylation of L-DOPA, although the therapeutic impact is questionable if AAD is inhibited by carbidopa.
- L-DOPA interferes with assays for urinary glucose and ketones.
- Dopamine stimulates the chemoreceptor trigger zone (CTZ) in the brainstem, which induces nausea and vomiting. The lower doses of L-DOPA possible when coadministered with carbidopa decrease the risk of this adverse effect.

> **ADVERSE EFFECTS**
>
> **L-DOPA**
> - Dyskinesia
> - Nausea and vomiting
> - Psychological disturbances (mania, psychosis, impulse control problems)
> - Increased libido

> **DRUG–DRUG INTERACTIONS**
>
> **L-DOPA**
> - Typical antipsychotic agents (decreased therapeutic efficacy of both agents)
> - MAO inhibitors (hypertension)
> - Hypotensive agents (orthostatic hypotension)
> - Iron salts, including multivitamins (decreased absorption of L-DOPA)
> - Hydantoins (decreased L-DOPA therapeutic efficacy)

Dopamine Metabolism Inhibitors

MAO Inhibitors

- MAO inhibitors are sometimes used as "last resort" antidepressants, but MAO-B selective inhibitors selegiline and rasagiline have found significant utility in the treatment of PD (Fig. 5.3).
  - MAO-B is the isoform involved in dopamine metabolism.
  - Nonselective MAO inhibitors (e.g., phenelzine, tranylcypromine) can induce hypertensive crisis if coadministered with L-DOPA. They are not used in the treatment of PD.

MOA

- Irreversible inhibition of MAO-B blocks one mechanism of dopamine degradation, making more dopamine available in central nerve terminals.
- The MAO-B selective inhibitors, selegiline and rasagiline, feature an N-propargylamine (acetylene) group which is involved in the mechanism-based

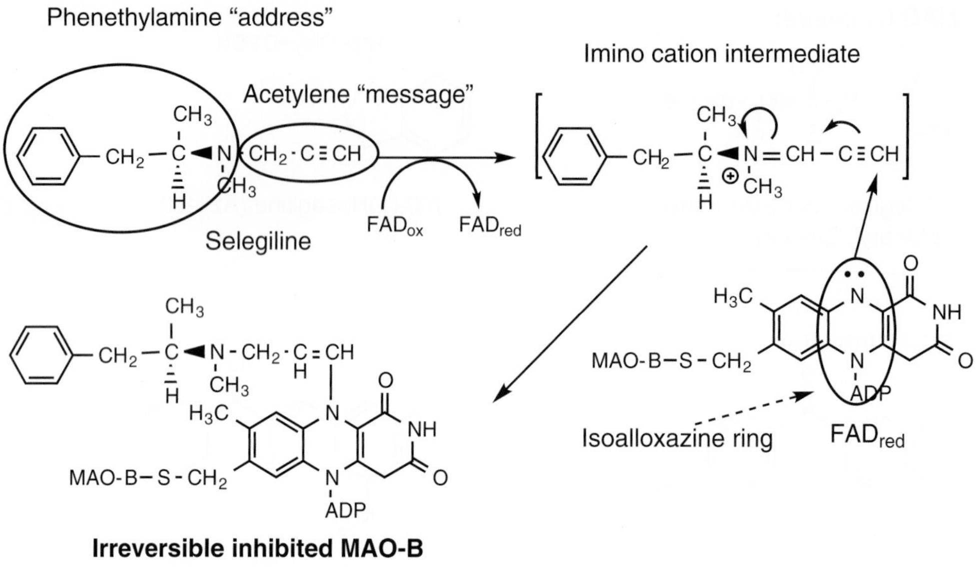

Figure 5.4 MAO-B inhibitor MOA.

irreversible inhibition of MAO-B and the formation of a stable covalent adduct (Fig. 5.4).

- FAD_{ox} catalyzes the oxidation of the amine to an iminium cation, producing an activated terminal acetylene. The acetylene moiety can be attacked by the nucleophilic N5 of the isoalloxazine ring (FAD_{red}) leading to the formation of the covalent adduct.

SAR

- Selegiline and rasagiline approximate the phenethylamine structure of endogenous MAO-B ligands. The phenethylamine moiety (phenylmethylamine for rasagiline) is the structural component that allows access to the enzyme.
- The key to irreversible binding with MAO-B lies in the acetylene moiety. The acetylene is the structural component responsible for enzyme inhibition.

Metabolism

- The metabolic pathways of selegiline and rasagiline are provided in Figure 5.5. Metabolites are renally excreted.

Physicochemical and Pharmacokinetic Properties

- The amine nitrogen of both selegiline and rasagiline retains basic character and acidic salts are marketed. Selegiline hydrochloride and rasagiline mesylate are administered orally.
- The oral bioavailability of selegiline HCl is increased three- to four-fold if taken with food. Rasagiline's oral absorption is rapid and its bioavailability is approximately 36%.

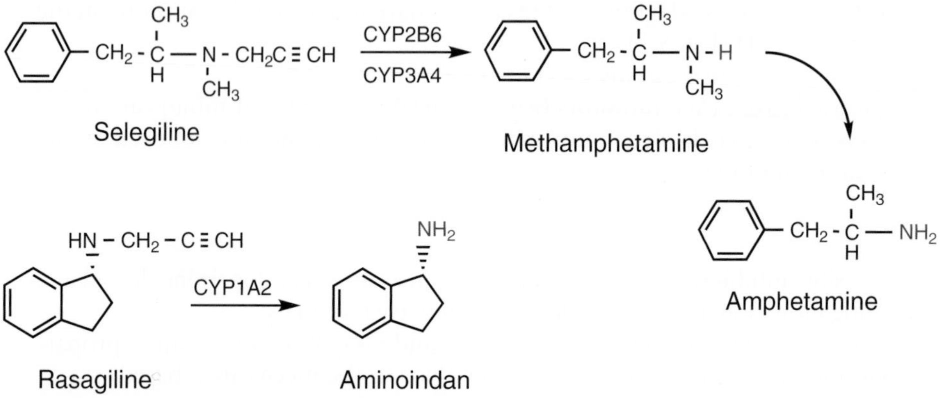

Figure 5.5 MAO-B inhibitor metabolism.

- Selegiline free base is also formulated into a 24-hour transdermal patch, which is available in multiple strengths.

COMT Inhibitors

MOA

- Tolcapone and entacapone are nitrated catechols that reversibly inhibit COMT, an enzyme that inactivates L-DOPA and dopamine (Fig. 5.3). They are termed "tight-binding inhibitors" and bind within the COMT active site cavity.
- Tolcapone inhibits COMT in the peripheral and CNS. Entacapone works more exclusively in the periphery.

SAR

- The catechol nucleus of the inhibitor permits enzyme recognition and the adjacent NO_2 is essential for the inhibition. Binding interactions between the drug's NO_2 and COMT Lys[144] and Trp[143] have been identified.
- The differing side chains of the COMT inhibitors impact enzyme affinity.

Metabolism

- Entacapone isomerizes in vivo to the *cis* isomer. Both isomers are glucuronidated prior to predominantly fecal excretion.
- Major metabolic reactions of tolcapone include glucuronidation, CYP3A4 and CYP2D6-catalyzed benzylic hydroxylation followed by cytosolic oxidation to the carboxylic acid, and COMT-methylation. Elimination is both renal and biliary.

Tolcapone metabolites

Physicochemical and Pharmacokinetic Properties

- The phenolic OH groups of both inhibitors are acidic, with pKa values lowered significantly by the strong electron withdrawing effect of the nitro moiety (pKa 5.1 to 5.7 compared to normal phenol pKa of approximately 11).
- Absorption of both drugs after oral administration is rapid and peak serum levels are obtained within 1 to 2 hours. Tolcapone's 8- to 12-hour duration is significantly longer than entacapone (2 hours) due to a low extraction ratio and a longer elimination half-life. It is dosed at half entacapone's strength (100 vs. 200 mg).
- Both drugs are marketed as tablets.

Clinical Applications

- Tolcapone's ability to work centrally increases the risk of central adverse effects compared to entacapone.
- Liver function of patients on tolcapone should be closely monitored.
- Coadministering COMT inhibitors with drugs that are CNS depressants or COMT substrates may increase the therapeutic and toxic effects of those drugs.

Dopamine Receptor Agonists

- Dopamine receptors are G-protein coupled (Fig. 5.6). A cationic amine on receptor agonists is required for anchoring at the conserved anionic Asp residue.
- Agonists used to treat PD currently target the D_2 receptor. The D_2 agonist message is expressed through inhibition of adenylyl cyclase and stimulation of phospholipase C. D_1 receptors may also be stimulated, and D_1 selective agonists are being explored as anti-Parkinson's agents.

ADVERSE EFFECTS

MAO-B Inhibitors
- Potential for hypertension
- Application site reaction (selegiline transdermal patch)
- Suicide ideation in young adults (selegiline Black Box Warning)

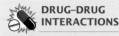

DRUG–DRUG INTERACTIONS

MAO-B Inhibitors (Many are Life-Threatening)
- Drugs metabolized by MAO
- Drugs that induce hypertension
- CYP3A4 inducers (selegiline)
- CYP1A2 inducers (rasagiline)
- Alcohol and high tyramine-content foods (hypertension)
- Serotonin reuptake inhibitors and tricyclic antidepressants (serotonin syndrome)
- Opioids (increased adverse effects of analgetics)
- Dextromethorphan (psychosis, serotonin syndrome)
- Sulfonylureas, insulin (hypoglycemia)

ADVERSE EFFECTS

COMT Inhibitors
Peripheral
- Dyskinesia
- Diarrhea, abdominal pain
- Nausea
- Hepatotoxicity (tolcapone Black Box Warning)

Central
- Hallucinations
- Headache
- Excessive and vivid dreaming
- Confusion

Table 5.1 **Dopamine Receptor Agonists Used in Parkinson's Disease**

Drug	Dosage Forms	Elimination Half-Life	Dosing Regimen	Metabolism	Notes
Bromocriptine	Tablet (0.8, 2.5 mg) and capsule (5 mg)	5–6 h	Twice daily	Proline hydroxylation (CYP3A4)	Inhibitor of prolactin secretion
Apomorphine	Sub-Q solution (10 mg/mL)	40 min	Once daily	Proposed N-demethylation, phenol conjugation (glucuronide, sulfate)	Used to treat "off" periods in advanced disease
Pramipexole	Tablets (0.15–1.25 mg), extended release tablets (0.375–3.75 mg)	8.5 h	Three times daily (immediate release); once daily (extended release)	Negligible	Renal clearance 30% lower in women
Ropinirole	Tablets (0.5–5 mg), extended release tablets (2–12 mg)	6 h	Three times daily (immediate release); once daily (extended release)	N-Dealkylation, C7 hydroxylation (both CYP1A2), deamination/oxidation to carboxylic acid, glucuronidation	C_{max} and AUC reduced 30% and 38%, respectively, in smokers
Rotigotine	Transdermal patch	1–8 mg/day	Once daily	N-Dealkylation of propyl and thienylethyl (CYP2C19 and others), phenol conjugation (glucuronide, sulfate)	Application site reactions are possible

- For best fit at dopamine receptors, agonists must assume an extended *trans* conformation between aromatic moiety and cationic amino nitrogen.

MOA

- Dopamine agonists bind to functional receptors in substantia nigra, taking the place of endogenous dopamine no longer able to be produced in degenerated nerve terminals.

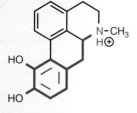

Dopamine agonist apomorphine showing dopamine in extended *trans* conformation

Specific Drugs

- Selected pharmacokinetic and therapeutic properties of currently available dopamine agonists are provided in Table 5.1 (Fig. 5.6).
- All marketed agents exert their therapeutic effect at D_2 receptors. Bromocriptine is a partial agonist.
- The dosing of agents metabolized by CYP isoforms should be evaluated when other drugs impacted by these isoforms are coadministered.

Bromocriptine
(Parlodel)

Apomorphine
(Apokyn)

S-Pramipexole
(Mixapex)

Ropinirole
(Requip)

Rotigotine
(Neupro)

Figure 5.6 Dopamine receptor agonists used in PD.

Adenosine Receptor Antagonists

MOA

- Blockade of adenosine A_2 receptors in striatum increases GABA innervation to address the loss of dopamine-mediated GABA release. Acetylcholine (ACh) release is inhibited, which better balances the central dopamine/ACh ratio.

Specific Drugs: Amantadine

- Amantadine is available in tablet, capsule, and oral syrup formulations. The elimination half-life is 17 hours and twice-daily administration is recommended.
- Metabolism is considered minimal, although the N-acetylated metabolite can account for up to 15% of the dose.
- Sudden drug discontinuation can result in significant deterioration known as "Parkinsonian crisis."
- Overdose has resulted in anticholinergic-related fatalities. While not common at recommended doses, anticholinergic action can manifest as blurred vision and arrhythmia.
- Avoid coadministration of drugs known to prolong the QT interval due to a potential additive effect.

Antimuscarinic and Antihistaminic Adjunct Therapy

MOA

- Centrally acting antimuscarinic agents attempt to restore the balance between dopamine and ACh in the CNS (Fig. 5.7). Antihistamines with a strong structural resemblance to antimuscarinics (e.g., diphenhydramine) can also block central muscarinic receptors.

SAR

- Agents useful as PD adjuncts are tertiary amines capable of penetrating the blood–brain barrier.
- Both antimuscarinics and antihistaminics must be cationic at pH 7.4 to bind to the G-protein–coupled muscarinic receptor.
- The most potent agents have two rings separated from the cationic center by the length of three to four carbon atoms. Both rings must be aromatic in antihistamines, while one ring can be saturated in antimuscarinics.
- Readers are directed to Chapters 3 and 22 of this text for additional information on antimuscarinic and antihistaminic chemistry, respectively.

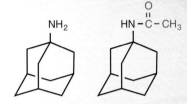

Amantadine (Symmetrel) and
N-acetylamantadine metabolite

ADVERSE EFFECTS

Dopamine Agonists
- Somnolence, including falling asleep precipitously and unexpectedly
- Orthostatic hypotension, dizziness
- Psychotic symptoms and reactions
- Nausea
- Dyskinesias

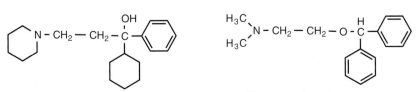

Figure 5.7 Antimuscarinics and antihistamines used as adjuncts in PD.

Table 5.2   Drugs Indicated in the Treatment of Spasticity Disorders

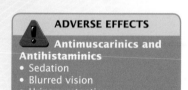

Baclofen
(Lioresal)
(*S/R*-mixture)

Dantrolene sodium
(Dantrium)

Tizanidine
(Zanaflex)

Drug	Mechanism	Elimination Half-Life	Metabolism	Dosage Form	Note
Baclofen	Central GABA$_B$ agonist	3–4 h	Negligible	Tab, cream, IT, sol	*R* isomer is active
Dantrolene	Ca^{2+} channel inhibitor skeletal muscle	9 h	Nitro reduction, subsequent N-acetylation	Cap, IV solution, IV suspension	Hepatotoxic (Black box warning)
Tizanidine	Central α$_2$-agonist, decreased release Glu and Asp	2.5 h	Aryl hydroxylation oxidation/degradation (CYP1A2)	Tab, cap	Food: Tab ↑C$_{max}$, Cap, ↓C$_{max}$
Botulinum Toxin-Type A	ACh release inhibition in motor neuron	4–6-mo duration	NA	Inj	Action occurs within 2 cm from site of inj

Tab, tablet; *IT*, intrathecal; *sol*, solution; *cap*, capsule; *IV*, intravenous; *susp*, suspension; *inj*, injection.

> **⚠ ADVERSE EFFECTS**
>
> **Antimuscarinics and Antihistaminics**
> - Sedation
> - Blurred vision
> - Urinary retention
> - Constipation
> - Cardiac arrhythmia

Clinical Applications

- Both natural antimuscarinics (e.g., atropine and scopolamine found in Belladonna alkaloids), synthetic aminoalkyl ethers (benztropine), and alcohols (trihexyphenidyl), along with anticholinergic antihistamines (diphenhydramine) can be administered in combination with L-DOPA.
- Anticholinergic adverse effects limit the therapeutic utility of these agents.

Spasticity Disorders

- Spasticity disorders are chronic conditions induced by disease or other brain trauma. The impact on skeletal muscle is painful spasms with increased tonic stretch reflex.
- Centrally active skeletal muscle relaxants can be used for a variety of indications, including the treatment of short-term muscle injury. Many of these agents have antimuscarinic properties and can be highly sedative. They are not specifically indicated in the treatment of spastic disorders.
- See Chapter 13, Table 13.2 in *Foye's Principles of Medicinal Chemistry, Seventh Edition,* for specific information on centrally active skeletal muscle relaxants.

Specific Drugs

- Baclofen, dantrolene, and tizanidine are all indicated for use in spasticity disorders. They have distinct mechanisms and sites of action, all resulting in relaxation of skeletal muscle.
- Selected pharmacokinetic and therapeutic properties of these antispastic agents are provided in Table 5.2.

> **⚠ ADVERSE EFFECTS**
>
> **Antispastic Agents**
> - Sedation
> - Weakness
> - Dizziness
> - Confusion
> - Hepatotoxicity (dantrolene, tizanidine)
> - Diarrhea (dantrolene)
> - Hypotension (tizanidine)
> - Dry mouth (tizanidine)
> - Possible compromised breathing and swallowing (botulinum toxin)

Chemical Significance

Conscientious medication therapy management by the pharmacist is critical to keep symptoms of chronic, debilitating neuromuscular disorders under control while managing adverse effects that risk further compromising quality of life. While therapeutic options are somewhat limited, the pharmacist's knowledge of drug chemistry can help minimize therapeutic problems and keep the patient as functional as possible for as long as possible. For example:

- If a Parkinson's patient who is a committed smoker requires augmentation of L-DOPA therapy, be thoughtful in the dosing of CYP1A2 substrates rasagiline (MAO-B inhibitor) and ropinirole (dopamine receptor agonist). The inactivation rate of these drugs will be increased.
- If a Parkinson's patient with a history of dependence on CNS stimulants requires MAO-B inhibitor therapy, consider rasagiline, as selegiline is metabolized to amphetamine and methamphetamine.
- Parkinson's patients at risk for cardiac arrhythmias should only be administered drugs with antimuscarinic activity with significant caution (tropine alkaloids, aminoalkylethers and alcohols, amantadine). Alternative therapies should be sought.
- If a Parkinson's patient prescribed L-DOPA and selegiline is a well-known tailgater at the local "State U." football games, he needs to be counseled to stay away from the keg and opt for burgers, as opposed to hotdogs, from the grill. Tap beer and wieners are high in tyramine.
- If a Parkinson's patient being prescribed amantadine is known to be a "fast acetylator," pay attention to the impact of that phenotype on her response to therapy.
- If a young cerebral palsy patient also suffers from childhood liver disease, try baclofen as a first approach to antispastic therapy, as it is not metabolized and is free of the overt hepatotoxicity risk posed by dantrolene sedum and tizanidine.

The importance of team-based care of patients with neuromuscular disorders, including appropriate triage with colleagues in occupational therapy and physical therapy, as well as medicine, cannot be emphasized strongly enough. Likewise, keeping up on new therapeutic approaches being researched and discussed in the literature, and informed about clinical trials patients could consider, documents the pharmacist's commitment to holistic patient-centered care for the long term.

Review Questions

1 Which PD therapy would be least likely to put patients at risk for a fall due to dizziness from orthostatic hypotension?

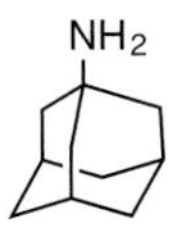

1 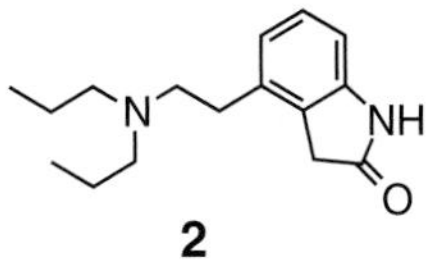   **2** 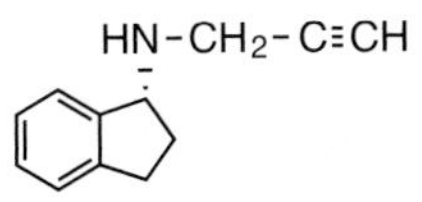   **3** 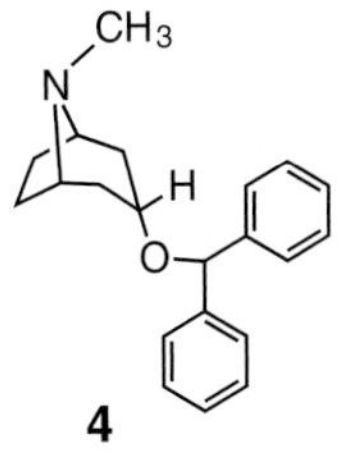   **4**

- A. Agent **1**
- B. Agent **2**
- C. Agent **3**
- D. Agent **4**

2 An adverse effect predictable from the structure of the anti-Parkinson's drug drawn below is:

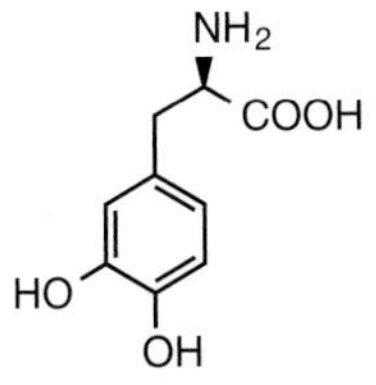

- A. Nausea
- B. Dry mouth
- C. Hepatotoxicity
- D. Muscle paralysis
- E. Diarrhea

3 What therapeutic agent must be coadministered with the anti-Parkinson's drug drawn in the previous question (drug A, below) in order for drug A to be effective?

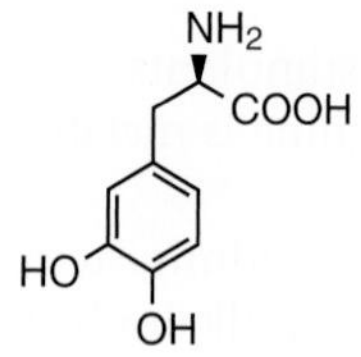
A

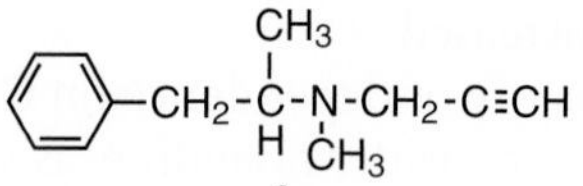
1

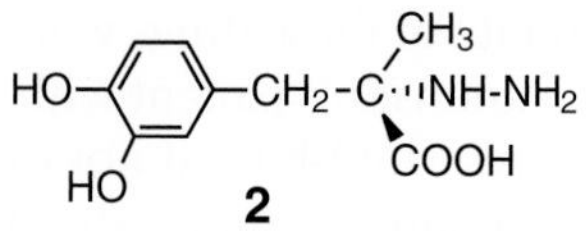
2

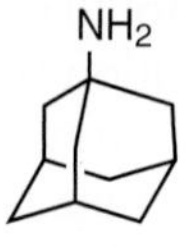
3

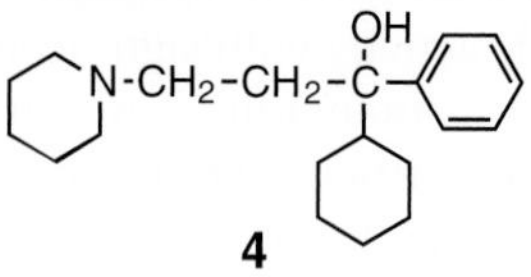
4

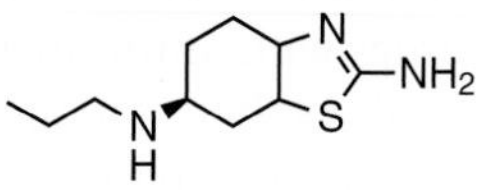
5

A. Agent **1**
B. Agent **2**
C. Agent **3**
D. Agent **4**
E. Agent **5**

4 Which structural component of tolcapone is responsible for the inhibition of COMT?

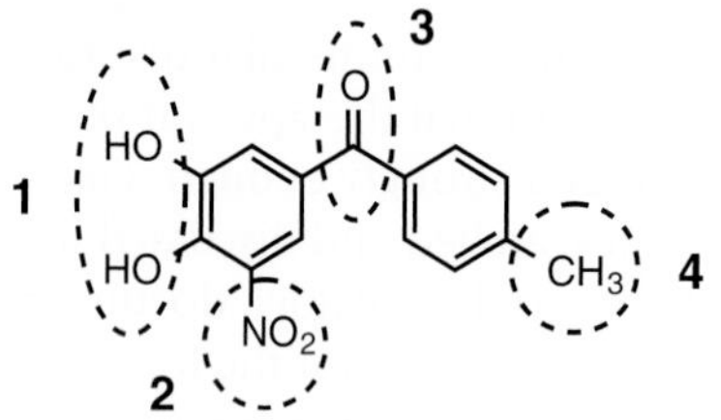

A. Component **1**
B. Component **2**
C. Component **3**
D. Component **4**

5 Which antispastic agent works directly on skeletal muscle?

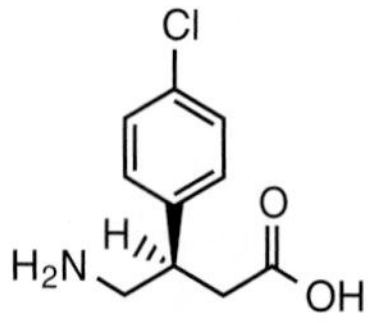
1

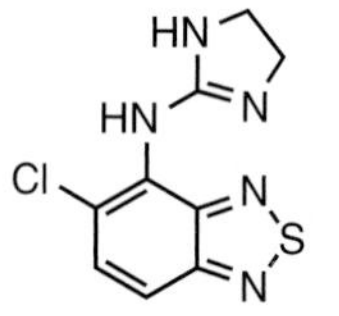
2

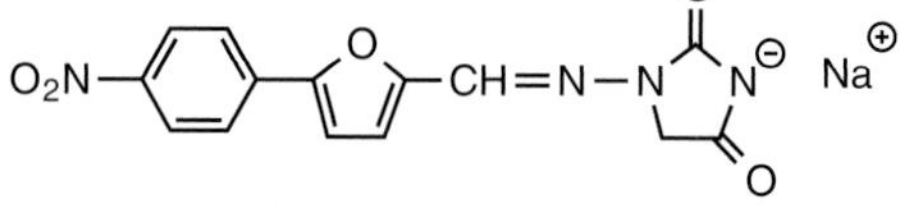
3

A. Agent **1**
B. Agent **2**
C. Agent **3**

Respiratory Agents

6

Learning Objectives

Upon completion of this chapter the student should be able to:

1. Identify or discuss key points related to the signs and symptoms of asthma and chronic obstructive pulmonary disease (COPD).

2. Describe the mechanism of action of drugs used to treat asthma and COPD:
   a. β_2-adrenergic agonists
   b. antimuscarinic agents
   c. methylxanthines
   d. adrenocorticoids
   e. mast cell stabilizers (cromolyn and leukotriene [LT] modifiers)
   f. monoclonal anti-IgE antibody
   g. phosphodiesterase (PDE) inhibitors.

3. Discuss the structure activity relationships (SAR) of the drugs used to treat asthma and COPD.

4. Identify any clinically relevant physicochemical and pharmacokinetic properties (solubility, in vivo and in vitro chemical stability, absorption, and distribution).

5. Discuss, with chemical structures, the clinically significant metabolism of each drug class.

6. Explain clinical uses and clinically relevant adverse effects, drug–drug, and drug–food interactions.

Introduction

Asthma is a common chronic, complex, airway disorder that is characterized by airflow obstruction, bronchial hyperresponsiveness, and an underlying inflammation that leads to variable degrees of symptoms, such as difficulty breathing (paroxysmal dyspnea), wheezing, and cough. The National Institutes of Health (NIH) Expert Panel Report 3: Guidelines for the Diagnosis and Management of Asthma (EPR-3) simply defines asthma as a chronic inflammatory response of the airways. The most common form of asthma is allergic asthma (allergens such as pollen, house dust mites, molds, food, and pet dander) with less common forms referred to as intrinsic asthma with no known etiology. Aside from allergens, an asthmatic attack may be precipitated by respiratory infection, strenuous exercise, polyps, drugs (aspirin, β-adrenergic antagonists), and environmental pollutants, primarily cigarette smoke.

COPD is characterized by persistent breathing difficulty that is not completely reversible and is progressive over time. It is usually the result of an abnormal inflammatory response to airborne toxic chemicals and most significantly associated with cigarette smoking. Thus, COPD is a general term and most commonly refers to chronic bronchitis and emphysema. Asthma is considered to be a disease entity unto itself and is not included in the definition of COPD. Both COPD and asthma are considered to be inflammatory diseases; however, the nature of the inflammation is different. Asthma is associated with the release of inflammatory mediators from mast cells and eosinophils, whereas chronic bronchitis is primarily associated with neutrophils and emphysema with alveoli damage. In addition, asthma is more often than not allergenic, whereas chronic bronchitis and emphysema have no allergic component. Finally, it is uncommon for asthma to be associated with smoking, whereas there is a very high incidence of both chronic bronchitis and emphysema in smokers. Patients with COPD display a variety of symptoms, ranging from chronic productive cough to severe dyspnea requiring hospitalization.

Treatment of asthma and COPD requires the use of drugs to treat both bronchoconstriction and inflammation symptoms, and approaches to treatment are directed at both of these physiological problems. Therefore, drugs that affect adrenergic or cholinergic bronchial smooth muscle tone (adrenergic agonists or anticholinergics) and drugs that inhibit the inflammatory process (corticosteroids, mast cell stabilizers, LT modifiers, IgE monoclonal antibodies, and PDE inhibitors) are used to treat and control asthma and COPD symptoms. For a more complete review of asthma and COPD pathophysiology, diagnosis and treatment Chapter 39 in *Foye's Principles of Medicinal Chemistry, Seventh Edition.*

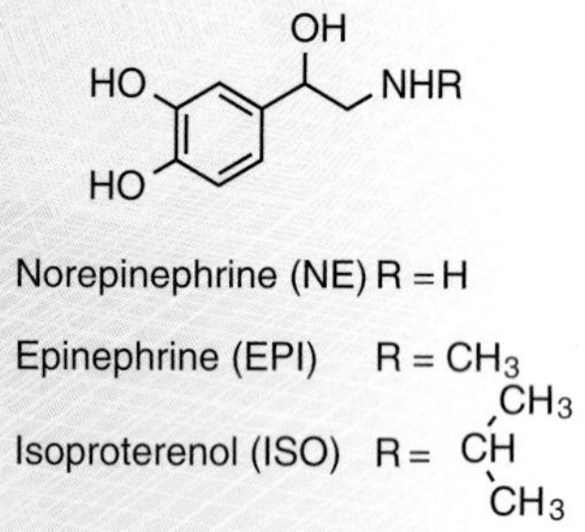

β_2-Adrenergic Agonists

- Adrenergic agonists are drugs that interact with the adrenergic receptors in the sympathetic branch of the autonomic nervous system (see Chapter 4, Chapter 10, and Chapter 39 in *Foye's Principles of Medicinal Chemistry, Seventh Edition,* for a complete detailed discussion) (Fig. 6.1).

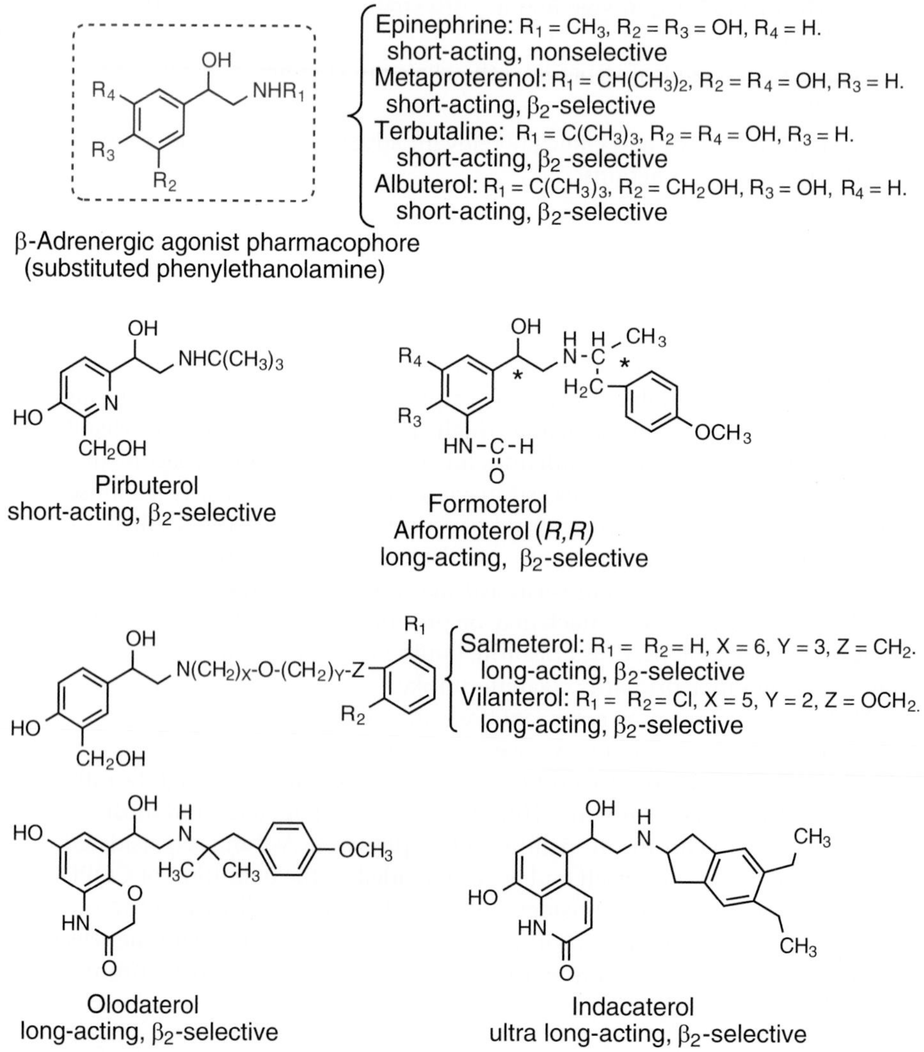

Norepinephrine (NE) R = H

Epinephrine (EPI)   R = CH_3

Isoproterenol (ISO)   R = $CH(CH_3)_2$

NE-EPI-ISO

Epinephrine: $R_1 = CH_3$, $R_2 = R_3 = OH$, $R_4 = H$. short-acting, nonselective
Metaproterenol: $R_1 = CH(CH_3)_2$, $R_2 = R_4 = OH$, $R_3 = H$. short-acting, β_2-selective
Terbutaline: $R_1 = C(CH_3)_3$, $R_2 = R_4 = OH$, $R_3 = H$. short-acting, β_2-selective
Albuterol: $R_1 = C(CH_3)_3$, $R_2 = CH_2OH$, $R_3 = OH$, $R_4 = H$. short-acting, β_2-selective

β-Adrenergic agonist pharmacophore (substituted phenylethanolamine)

Pirbuterol
short-acting, β_2-selective

Formoterol
Arformoterol (R,R)
long-acting, β_2-selective

Salmeterol: $R_1 = R_2 = H$, $X = 6$, $Y = 3$, $Z = CH_2$. long-acting, β_2-selective
Vilanterol: $R_1 = R_2 = Cl$, $X = 5$, $Y = 2$, $Z = OCH_2$. long-acting, β_2-selective

Olodaterol
long-acting, β_2-selective

Indacaterol
ultra long-acting, β_2-selective

Figure 6.1 β-Adrenergic agonist pharmacophore and the clinically important β_2-selective agonists.

- The endogenous neurotransmitters are norepinephrine (NE) and epinephrine (EPI) and the receptors are classified as either α and β.
- The adrenergic receptors are further classified into α_1, α_2, β_1, β_2 based upon their organ distribution and physiological activity.
- The β_2-adrenergic receptors occur in the lung and their stimulation results in bronchial smooth muscle dilation.

MOA

- The β_2-selective adrenergic agonists bind to the β-receptors in the lung resulting in bronchial smooth muscle relaxation and thereby helping to relieve the bronchoconstriction involved in an asthmatic attack.
- The β_2-receptors are coupled to adenylyl cyclase via specific G-stimulatory proteins (G_s). When stimulated, cyclic adenosine monophosphate (cAMP) is released, which in turn activates protein kinase A which leads to characteristic cellular responses, that is, bronchial smooth muscle dilation in the lung smooth muscle.
- Agonists bind ionically via their protonated amine to Asp^{113} in helix 3 and hydrogen bond via their phenolic hydroxyl groups to Ser^{204} and Ser^{207} in helix 5.
- Additional reinforced bonding occurs between the aromatic ring with residues Phe^{290} in helix 6 and Val^{114} in helix 3.
- N-Alkyl substituents fit into a pocket formed between aliphatic residues in helix 6 and 7.
- The β-hydroxy substituent is oriented toward residue Asn^{293} when it is in the R-configuration (Fig. 6.2).

SAR

- The fundamental pharmacophore for all β-adrenergic agonists is a substituted β-phenylethanolamine (Fig. 6.1).
- The nature and number of substituents influences both the type of action (direct or indirect) as well as specificity for the β-receptor subtypes.
- The β-hydroxy as well as an m-OH (R_2) are essential for direct-acting agonists.
- Direct-acting β-agonists bind to the receptor similar to EPI producing the β-sympathetic response.
- Both β_1- and β_2-receptor selectivity is conferred by the bulk of the N-substituent (R_1). The larger the N-substituent the greater the selectivity for the β-receptors. As a matter of fact when the N-substituent is tertiary butyl or aralkyl, all α-receptor affinity is lost and the β-receptor affinity shows preference for the β_2-receptor.

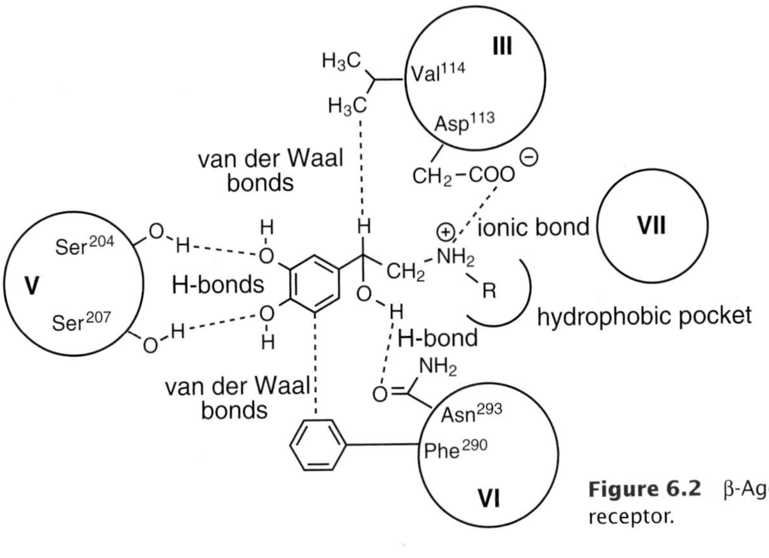

Figure 6.2 β-Agonist binding to key residues in the β-adrenergic receptor.

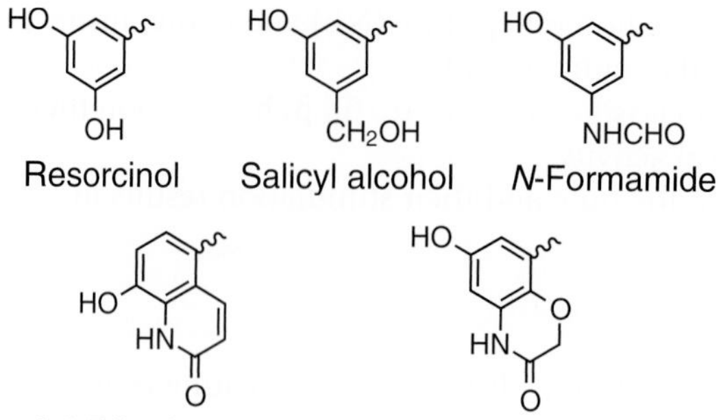

Figure 6.3 Ring configurations that contribute to β₂-receptor selectivity.

- Significant β₂-receptor selectivity occurs when there is an appropriate substitution pattern on the aromatic ring. All currently marketed adrenergic agonists are β₂-receptor selective and contain either a resorcinol ring, a salicyl alcohol moiety, an N-formamide, or aromatic ring as part of a 1,2-dihydroquinolone or 1,4-benzoxazinone ring system (Fig. 6.3).

Physicochemical and Pharmacokinetic Properties

- All β₂-adrenergic agonists have acidic and basic functional groups (phenolic hydroxyl and aliphatic amino groups). Physiologically, however, they behave as a base being predominately protonated at pH 7.4 and functioning as an ionized acid.
- β₂-Adrenergic agonists are typically administered via inhalation, that is, via metered dose inhaler (MDI), dry powder inhaler, or nebulizer (atomizes liquid drug into a mist).
- Inhalation delivers the drug to the lungs, where the highest concentrations are desired, thereby reducing the probability of side effects.
- Some β₂-adrenergic agonists are available in tablet and parenteral dosage forms. Table 6.1 compares the dosage form, route of administration, and onset and duration of action of the clinically important β₂-adrenergic agonists.
- Inhalation onset for short-acting β₂-adrenergic agonists (SABAs) is rapid (~5 minutes) and onset of the long-acting β₂-adrenergic agonists (LABAs) is similar to that of SABAs, the exception being salmeterol (20 to 60 minutes).
- Many β₂-adrenergic agonists are formulated in combination with inhaled corticoids which will be covered in the adrenocorticoid section.

Metabolism

- Adrenergic agonists with a catechol ring system are metabolized by catechol-O-methyl transferase (COMT) and those with primary amines or N-methyl are metabolized by monoamine oxidase (MAO). These enzymes occur extraneuronally in the liver, gastrointestinal tract, and the lungs.

Table 6.1   Dosage Form, Route of Administration, Onset and Duration of Action of β₂-Adrenergic Agonists

Generic Name	Dosage Form/Route of Administration	Onset (min)	Duration of Action (h)
Metaproterenol	Tablet/oral	<15	>4
	Solution/inhalation	<1	4–6
Terbutaline	Tablet/oral	1–2	4–8
	Injectable/subcutaneous (SQ)	<15	1–4
Albuterol	Aerosol/metered/inhalation	5–15	2–4
	Tablet/ extended release/oral	30	12
Pirbuterol	Aerosol/metered/inhalation	5–30	5
Formoterol	Aerosol/metered/solution/powder/inhalation	3	12
Salmeterol	Aerosol/metered/powder/inhalation	5–14	12
Olodaterol	Spray/metered/inhalation	5	24
Indacaterol	Powder/inhalation	5	24
Vilanterol	Powder/inhalation	3	22

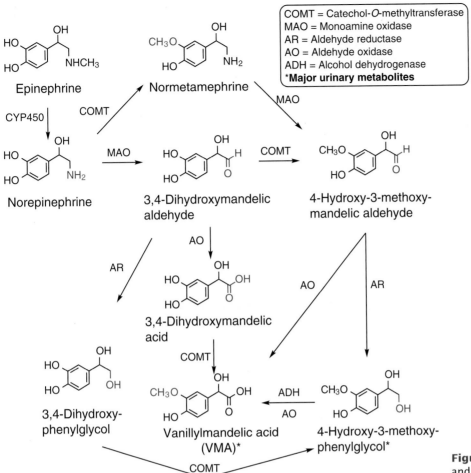

COMT = Catechol-*O*-methyltransferase
MAO = Monoamine oxidase
AR = Aldehyde reductase
AO = Aldehyde oxidase
ADH = Alcohol dehydrogenase
*Major urinary metabolites

Figure 6.4 Metabolic pathways for EPI and NE.

- Figure 6.4 depicts the metabolism of NE and EPI.
- Some clinically important β_2-adrenergic agonists are resistant to COMT due to their lack of a catechol ring system.
- Resistance to MAO depends on the size of the substituent on the nitrogen since it must be dealkylated before MAO can act to oxidatively deaminate the nitrogen (Fig. 6.5).
  - Tertiary butyl (e.g., terbutaline) groups are slowly N-dealkylated, whereas aralkyl substituents are resistant to N-dealkylation (e.g., salmeterol).

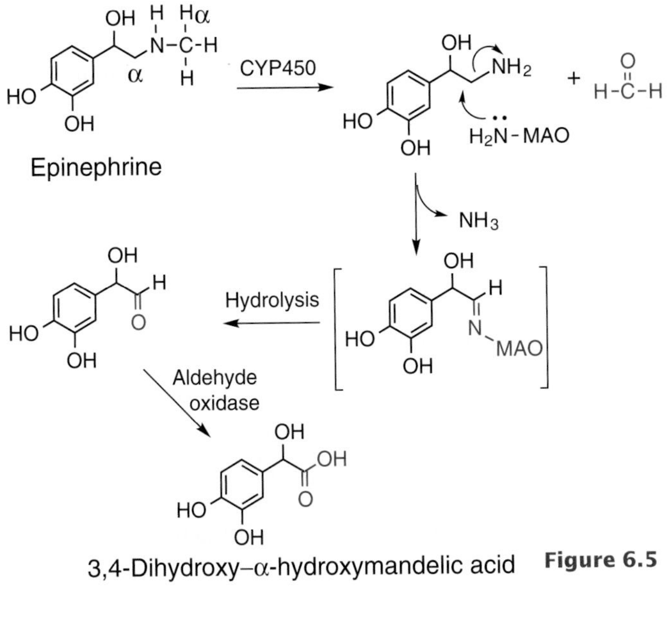

Figure 6.5 MAO mechanism of oxidative N-deamidation.

Table 6.2 Stepwise Medication Management of Asthma

All asthma patients: short-acting inhaled β_2-agonist (SABA) as needed for acute exacerbations

Severity Classification	Long-Term Control
Step 1: Intermittent asthma Persistent asthma	*Preferred:* SABA PRN
Step 2: Mild	*Preferred:* Low-dose inhaled corticosteroid (ICS) *Alternative:* Cromolyn, nedocromil, leukotriene receptor antagonist (LTRA) or theophylline
Step 3: Moderate	*Preferred:* Low-dose inhaled ICS + long-acting β_2-agonist (LABA) or medium-dose ICS *Alternative:* Low-dose ICS + either LTRA, theophylline, or zileuton
Step 4: Severe	*Preferred:* Medium-dose ICS + LABA *Alternative:* Medium-dose ICS + LTRA, theophylline, or zileuton
Step 5: Severe	*Preferred:* High-dose ICS + LABA *Consider:* Omalizumab for patients who have allergies
Step 6: Severe	*Preferred:* High-dose ICS + LABA + oral corticosteroid *Consider:* Omalizumab for patients who have allergies

⚠ ADVERSE EFFECTS

β_2-Adrenergic Agonists: Insomnia, tachycardia, tremor, sweats, agitation, hypokalemia, pulmonary edema, myocardial ischemia, and increased risk of asthma-related death.

☼ DRUG–DRUG INTERACTIONS

β_2-Adrenergic Agonists: HIV protease inhibitors (i.e., ritonavir); tricyclic antidepressants (i.e., desipramine); MAOIs (i.e., phenylzine); β-blockers (i.e., propranalol); diuretics (i.e., dihydrochlorothiazide); anticholinergics (i.e., atropine).

Clinical Application

- Therapeutic management of asthma requires the use of a quick-acting drug to relieve an acute attack as well as a drug to control symptoms over the long term.
- The quick relief medication is almost always an inhaled short-acting β_2-adrenergic agonist (i.e., albuterol).
- Long-term relief medications include LABAs, inhaled corticosteroids, LT modifiers, cromolyn sodium, and/or methylxanthines.
- FDA recommends that LABAs never be used alone because of the increased risk of asthma-related death.
- The current stepwise approach to asthma treatment is found in Table 6.2.

Antimuscarinics

- Antimuscarinics are competitive antagonists of muscarinic receptors found on the parasympathetic postganglionic nerve endings (endogenous neurotransmitter is acetylcholine).
- They are classed as muscarinic because they bind the agonist molecule muscarine (see Chapter 3 in *Foye's Principles of Medicinal Chemistry, Seventh Edition,* for a complete discussion.).

Acetylcholine and Muscarine

- Muscarinic receptors are found in smooth muscle, secretory cells, and certain central synapses.
- The muscarinic receptors are classified into M_1 to M_5 subtypes (Fig. 6.6).
- The $M_{1,3,5}$ receptors are G_q-protein coupled. The $M_{2,4}$ receptors are G_i-coupled.
- M_3 muscarinic receptors are responsible for a number of physiologic actions of the parasympathetic system, including contraction of the iris circular muscle, GI tract smooth muscle contraction, increased secretions, and bronchiole smooth muscle contraction.

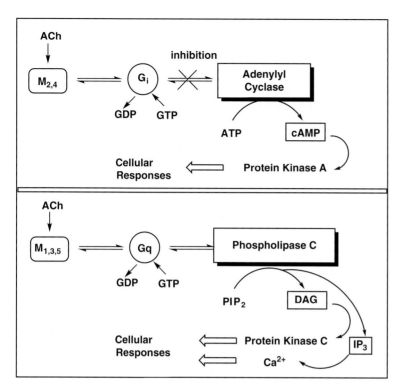

Figure 6.6 Comparison of the role of G-proteins in odd- and even-numbered muscarinic receptors (e.g., M_2, M_4 vs. M_1, M_3, M_5).

MOA

- The M_3 receptors cause bronchiole constriction and therefore counterbalance the bronchiole dilation of the β_2-adrenergic receptors in the lung, thus maintaining bronchiole tone.
- Blocking the M_3 receptor bronchiole constriction allows adrenergic bronchiole dilation to help overcome the pulmonary constriction associated with an asthmatic or COPD attack.
- Antimuscarinics bind to the M_3 receptor in a similar manner as acetylcholine (Fig. 6.7); however, they contain additional hydrophobic substituents that bind to a hydrophobic pocket in the receptor which does not allow the change in conformation needed to transfer agonist signal to the coupled G-protein.

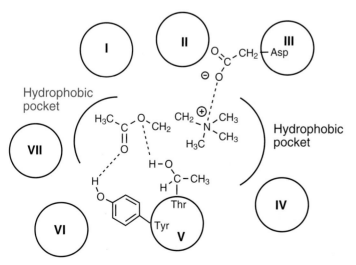

Figure 6.7 Acetylcholine binding to residues in the muscarinic receptor. Antagonists have additional hydrophobic groups that bind to the hydrophobic pocket highlighted in red.

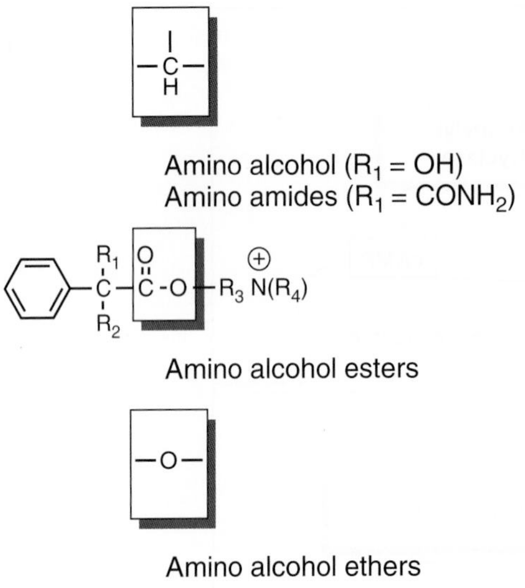

Figure 6.8 The pharmacophore for all classes of antimuscarinic agents.

SAR

- The pharmacophore for all antimuscarinic drugs is an acetylcholine analog in which the acetyl methyl group is substituted with at least one phenyl ring (Fig. 6.8).
- This pharmacophore is generally classed as an amino alcohol ester.
- The ester function can be replaced with different moieties to produce different classes, that is, amino alcohol ethers, amino alcohols, and amino amides.
- The N must be quaternary or cationic at physiological pH.
  - The primary point of receptor interaction. However, the quaternary ammonium group is not absolutely necessary for antagonism and can be replaced with t-butyl and still maintain minimal activity.
  - N-Alkyl substituents: CH₃ optimal, bulkier groups have decreased antagonism, but still yield effective compounds.
- Figure 6.9 shows the current clinically important antimuscarinic agents for the treatment of asthma and COPD as well as the naturally occurring atropine, the prototype for the amino alcohol ester class.

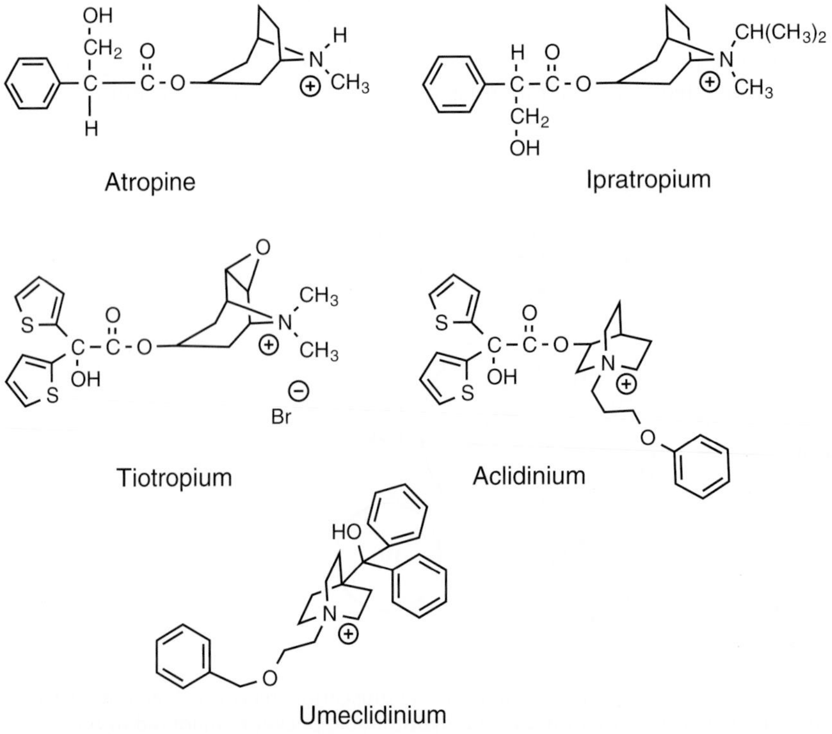

Figure 6.9 Structures of currently useful antimuscarinic agents for the treatment of asthma and COPD.

Table 6.3	Bioavailability, Onset of Action, and Duration of Action of Inhaled Antimuscarinic Agents		
Drug	Bioavailability	Onset	Duration
Ipratropium	7%	15 min	6–8 h
Tiotropium	19.5%	60 min	24 h
Aclidinium	6%	30 min	30 h
Umeclidinium	13%	15 min	24 h

Stereochemistry

- When introduced into the amino alcohol portion (see Fig. 6.9) of the molecule there is little or no difference between the *R*- and *S*-enantiomers.
- When present in the acid moiety the *R*-enantiomer is ~100 times more active than the *S*-enantiomer indicating an important binding role for the phenyl ring.

Physicochemical and Pharmacokinetic Properties

- Antimuscarinic agents for the treatment of asthma and COPD are all administered via inhalation.
- Table 6.3 gives the inhaled bioavailability, onset of action, and duration of action for these agents.

Metabolism

- Ipratropium is partially hydrolyzed by esterases to tropic acid and 1-hydroxy-8-methyl-8-isopropyl-8-azoniabicyclo[3,2,1]-octane.

2-Phenyl-3-hydroxy
propanoic acid
(Tropic acid)

1-Hydroxy-8-methyl-
8-(1-methylethyl)-8-azo-
niabicyclo[3,2,1]-octane

Ipratropium hydrolysis

- Tiotropium is 74% excreted unchanged in the urine, whereas the remainder is metabolized by both CYP3A4 and 2D6 followed by glutathione conjugation to various metabolites.
- Aclidinium is primarily metabolized by esterases to its acid and alcohol components as well as minor oxidation via CYP3A4 and 2D6 to a mono-oxygenated phenyl and the loss of one of the thienyl rings.
- Umeclidinium is a substrate for P-glycoprotein as well as CYP2D6 with O-dealkylation, hydroxylation, and O-glucuronide metabolites being formed with elimination primarily in the feces (58%) and urine (22%).

Clinical Application

- The inhaled antimuscarinic agents find their most important therapeutic application for the treatment of COPD rather than for asthma.
- Ipratropium is also available in combination with albuterol.
- Umeclidinium in combination with vilanterol has a specific indication for COPD treatment.

> **⚠ ADVERSE EFFECTS**
>
> **Inhaled Antimuscarinic Agents:** Dry mouth, blurred vision, tachycardia, urinary retention, headache, and exacerbation of narrow-angle glaucoma.

Mast Cell Degranulation Inhibitors

- Compounds synthesized to improve the bronchodilating activity of Khellin (Fig. 6.10).
- Khellin is a natural product chromone (benzopyrone) isolated from *Ammi visnaga* (Khella, Bishop's weed).

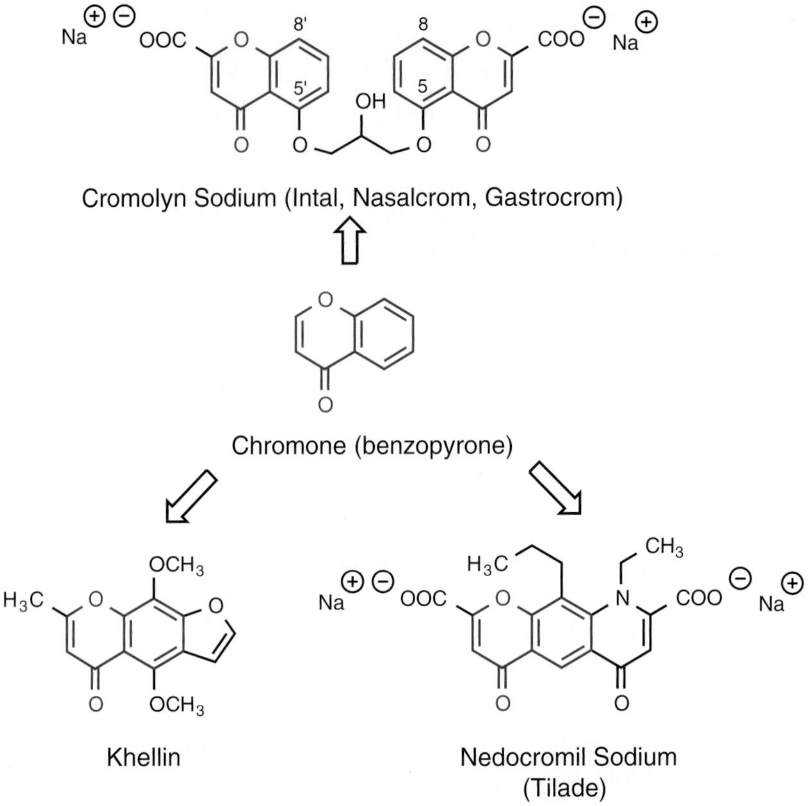

Figure 6.10 Structure relationship of the mast cell stabilizers cromolyn sodium and nedocromil sodium to benzopyrone (chromone) and Khellin.

- Khellin has vasodilation effects as well as bronchodilation activity and appears to have calcium channel blocking activity.

MOA

- Both cromolyn and nedocromil inhibit the release of the bronchial spasmogens, histamine, LTs, prostaglandins, and other inflammatory autocoids, from sensitized mast cells.
- They act on the mast cells after they have become sensitized by binding to IgE.
- They do not inhibit the binding of IgE to the mast cells or the binding of the antigen to the mast cell–IgE complex.
- They have been shown to block calcium channels in the membranes of the mast cells which may play a part in the mechanism of action.

SAR

- Cromolyn sodium
  - A bischromone where the two chromone rings must be coplanar.
  - The linking chain can be no longer than six carbons.
  - If the 5,5' linkage is changed to the 8,8' position, all activity is lost.
- Nedocromil sodium has the furan ring of Khellin replaced with a piperidinone ring and maintains same activity as cromolyn sodium.

Physicochemical and Pharmacokinetic Properties

- Cromolyn sodium is poorly absorbed from the lungs (~8%), insignificantly from the eye (~0.07%), and ~1% from the GI tract.
- Cromolyn sodium has a duration of action of <2 hours, but requires 4 weeks for onset of action.
- Eliminated intact in the urine and bile.
- Nedocromil sodium is poorly absorbed from the lungs (~3%) and 2% from the GI tract.

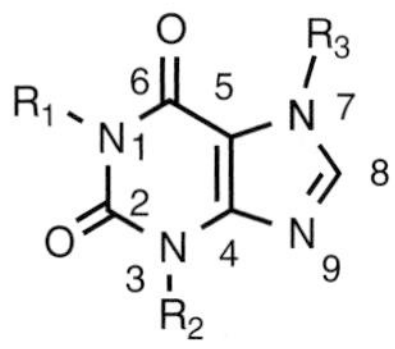

	R₁	R₂	R₃
Caffeine	CH_3	CH_3	CH_3
Theophylline	CH_3	CH_3	H
Theobromine	H	CH_3	CH_3

Figure 6.11 Structural differences between the naturally occurring methylxanthines.

- Nedocromil sodium has a duration of action of <2 hours, but with a faster onset than cromolyn (1 week).
- Eliminated unchanged in the urine and bile.

Metabolism

- Neither cromolyn sodium nor nedocromil sodium is metabolized to any great extent.

Clinical Application

- Cromolyn or nedocromil is an alternative to low-dose inhaled corticosteroids for the step 2 treatment of mild asthma (Table 6.2).

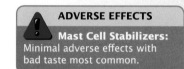

ADVERSE EFFECTS

Mast Cell Stabilizers: Minimal adverse effects with bad taste most common.

Methylxanthines

- Three major methylxanthines occur in nature (caffeine, theobromine, and theophylline) (Fig. 6.11).
- Theophylline is the clinically important agent for asthma and COPD treatment.
- They occur in coffee (~85-mg caffeine/cup), tea (~60-mg caffeine/cup), cocoa (~230-mg theobromine/cup), and are also found in cola drinks (~35-mg caffeine/12-ounce), OTC analgesics (~50-mg caffeine/dose), OTC stimulants (~100-mg caffeine/dose).

MOA

- MOA not clearly understood; however, inhibition of PDE is usually considered the primary MOA. The PDE binding sites have been identified by x-ray crystallographic studies (Fig. 6.12).
- PDE terminates the action of cAMP. cAMP starts the processes which leads to bronchodilation in smooth muscle cells.
- Other effects of methylxanthines include:
  - Nonselective adenosine receptor antagonist, antagonizing A_1, A_2, and A_3 receptors almost equally, which explains many of its cardiac effects and some of its antiasthmatic effects.

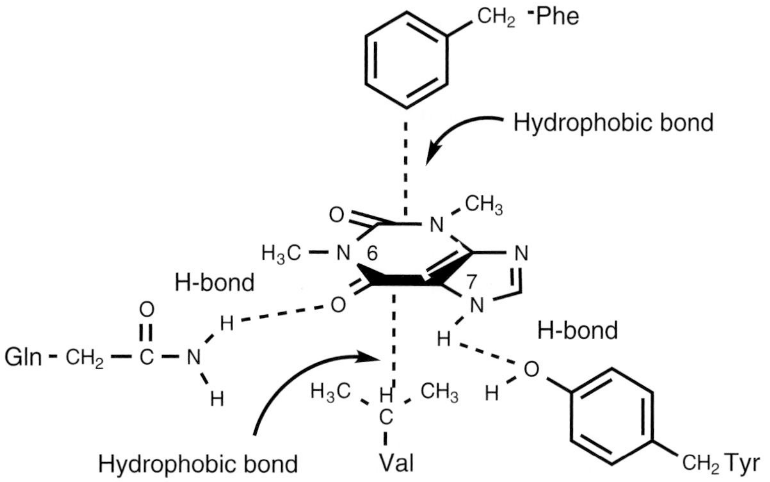

Figure 6.12 Methylxanthine binding interactions in the catalytic pocket of PDE.

- In vitro, theophylline induces apoptosis of eosinophils and neutrophils which may decrease the duration of their effects in the asthmatic.
- Theophylline has been shown to have an effect on the large conductance calcium-activated K^+ channels that are related to its bronchodilation action.
- Theophylline may reverse the clinical observations of steroid insensitivity in patients with COPD and asthmatics who are active smokers.
  - Theophylline in vitro can restore the reduced HDAC2 (histone deacetylase) activity that is induced by oxidative stress (i.e., in smokers), returning steroid responsiveness toward normal. Furthermore, theophylline has been shown to directly activate HDAC2.

SAR

- The methylxanthines contain a purine base structure (fused pyrimidine and imidazole rings).
- Theophylline is a dimethylxanthine with methyl groups at N-1 and N-3 positions (Fig. 6.11).

Physicochemical and Pharmacokinetic Properties

- Theophylline is both a weak acid and a weak base:
  - N-9 is basic and can accept a proton (pKa = 3.5).
  - N-7 is acidic and can donate its proton (pKa = 8.6).
- Theophylline acts as an acid in physiological systems.
- Theophylline has limited water solubility (1 g in 120 mL). Water solubility is increased by preparing salts:
  - With ethylenediamine (Aminophylline) – 79% theophylline.
  - With choline (Oxtriphylline)– 64% theophylline.
- Theophylline pharmacokinetics
  - Rapid and complete oral absorption
  - Peak plasma levels in 1 to 2 hours with elimination $T_{1/2}$ ~8 hours
  - ~10% excreted unchanged in the urine

Metabolism

- Theophylline is metabolized by a combination of C-8 oxidation and N-demethylations to yield methyl uric acid metabolites (Fig. 6.13).
- The major urinary metabolite is 1,3-dimethyl uric acid formed by the action of xanthine oxidase.
- Since none of the metabolites is uric acid itself, theophylline can be given safely to patients who suffer from gout.

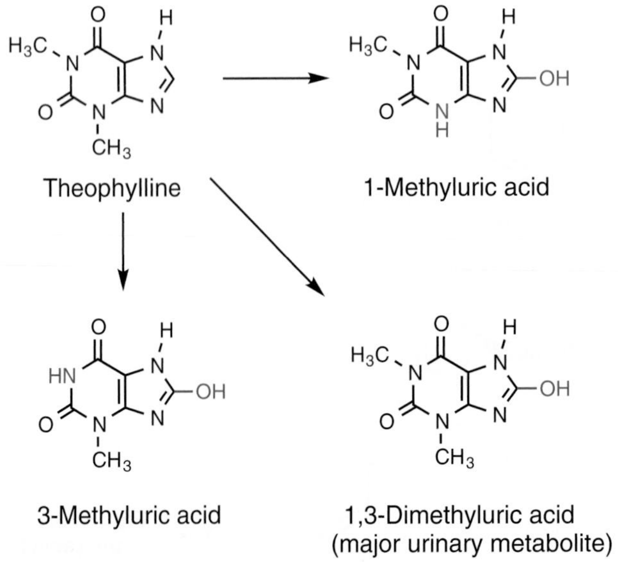

Theophylline

1-Methyluric acid

3-Methyluric acid

1,3-Dimethyluric acid
(major urinary metabolite)

Figure 6.13 Metabolism of theophylline.

Clinical Application

- Theophylline is an alternative maintenance therapy for stages 2, 3, and 4 of asthma.
- Extended-release theophylline is an option for COPD patients who do not receive adequate relief of moderate symptoms or who cannot tolerate other bronchodilators.
- Theophylline has fallen into disuse because of its narrow therapeutic window.
- Care must be taken when switching theophylline to one of its salts because of the risk of dosing outside its therapeutic window.

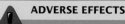

ADVERSE EFFECTS

Theophylline: Within therapeutic window: nausea, headache, tremor
Overdose: ventricular arrhythmia, convulsions, and death

DRUG–DRUG INTERACTIONS

Many coadministered drugs can increase plasma levels of theophylline: quinolone, macrolide antibiotics, nonselective β-blockers, ephedrine, calcium channel blockers, cimetidine, and oral contraceptives.

Adrenocorticoids

- The adrenocorticoids are one of the four major classes of naturally occurring steroids which include the sex hormones, the bile acids, and vitamin D.
- The adrenocorticoids are formed in the adrenal cortex and are subdivided into the glucocorticoids (glucose homeostasis) and mineralocorticoids (water and sodium retention).
- The glucocorticoids also have significant anti-inflammatory activity which makes them most important for the treatment of asthma and COPD.
- See Chapter 18 and Chapter 28 in *Foye's Principles of Medicinal Chemistry, Seventh Edition,* for a complete detailed discussion of the adrenocorticoids.

MOA

- The major role of glucocorticoids is to provide the body with levels of glucose that are compatible with life. They accomplish this by affecting endogenous levels of enzymes involved in gluconeogenesis **(use with caution in diabetic patients)**:
  - Alanine transaminase
  - Glycogen synthetase
  - Pyruvate kinase
  - Phosphoenolpyruvate carboxykinase (PEP)
  - Glucose-6-phosphate kinase
- The major role of mineralocorticoids is to maintain blood volume and regulate electrolyte balance **(avoid use in hypertensive patients)**.
- The major pharmacological use of the glucocorticoids is as anti-inflammatory agents.
  - Glucocorticoids stimulate the synthesis of **Lipocortin,** a protein that inhibits the enzyme **phospholipase A_2 (PLA$_2$)**.
  - PLA$_2$ cleaves arachidonic acid from membranes and is the first step in the arachidonic acid cascade that results in producing inflammatory prostaglandins and LTs.
  - A secondary mechanism of anti-inflammatory action involves inhibition of interleukin-1 (IL-1) which decreases the proliferation of *T*- and *B*-lymphocytes, which in turn decreases the production of interleukin-2 (IL-2) and tumor necrosis factor (TNF). As these are potent immunosuppressant agents, glucocorticoids must be used with **caution or avoided in patients with infection.**
- Antiallergic activity: Glucocorticoids inhibit the synthesis and release of histamine and other autocoids from mast cells and are important for treating symptoms of allergy, and is therefore of particular importance in the treatment of asthma.
- How do glucocorticoids affect the endogenous levels of enzymes and other proteins?
  - Glucocorticoids bind to their receptor in the cytoplasm and the steroid–receptor complex is then able to cross into the nucleus.
  - Inside the nucleus, the steroid–receptor complex binds to DNA and this usually results in the increase (and sometimes decrease) of important proteins

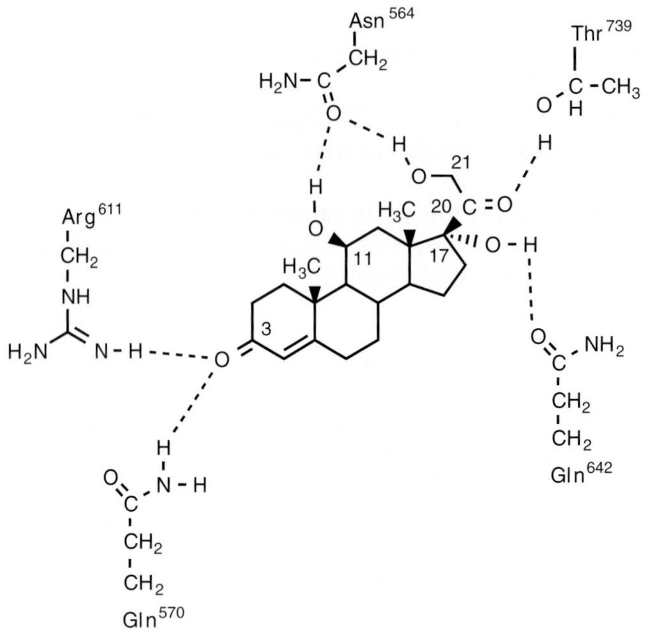

Figure 6.14 Important glucocorticoid receptor–ligand binding residues.

(see Chapter 28 and Chapter 39 in *Foye's Principles of Medicinal Chemistry, Seventh Edition,* for a complete detailed discussion).

- The glucocorticoid receptor is a soluble protein found within the cytosol of the cell and consists of 11 α-helices and 4 β-strands.
- Within the glucocorticoid receptor there is a unique binding pocket which makes it distinct from binding to the other types of natural steroids.
- Nearly every atom of the glucocorticoid contacts one or more amino acid residues via hydrogen bonding.
- Figure 6.14 depicts the five most significant receptor interactions.

SAR

- The natural glucocorticoids have both glucocorticoid activity as well as mineralocorticoid action.
- The functional groups essential for both mineralocorticoid and glucocorticoid activity include (Fig. 6.15):
  - A pregnane ring skeleton with an all *trans* backbone.
    - The substituents on the β-face are believed to interact with the glucocorticoid receptor.
    - The α-face is believed to interact with the mineralocorticoid receptor.

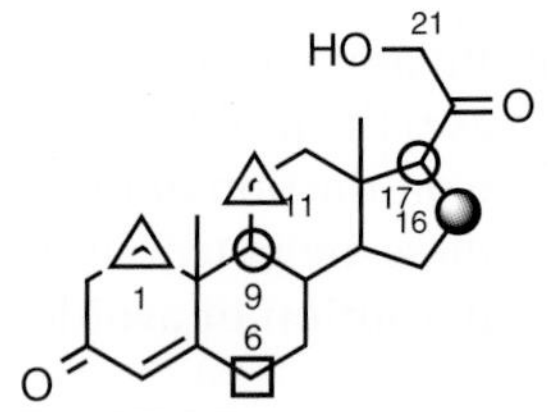

Both anti-inflammatory and mineralocorticoid activity increased (open circle)	Anti-inflammatory activity increased (triangle and square)	Mineralocorticoid activity decreased (closed circle and square)
17α-OH	1-dehydro	6α-CH₃
9α-F	6α-F	16α-& 16β-CH₃
9α-Cl	11β-OH	16α,17α-acetonide
21-OH		16α-OH

Figure 6.15 Glucocorticoid structure–activity relationships.

- A ring: an A-en-one system (Δ^4-3-one)
- A 17β-ketol side chain (C20-keto C21-hydroxy)
- A 9α-halogen
- A 16-substituent (methyl or hydroxyl)
- The C21 hydroxy must be free for both mineralocorticoid and glucocorticoid activity.
- C21 chloro and C20 fluoromethyl thioester (-SCH$_2$F) confer anti-inflammatory action when inhaled.
- Figure 6.15 depicts those SARs that will increase anti-inflammatory activity and decrease mineralocorticoid activity.

Physicochemical and Pharmacokinetic Properties

- The glucocorticoids are lipophilic despite having at least three hydroxy groups.
- The hydroxyl groups can be esterified with appropriate acids that will either enhance or decrease lipophilicity.
- The C21 hydroxy group is most accessible and therefore the easiest to esterify.
- The C21 hydroxy can also be esterified with dicarboxylic acids to monoesters or phosphoric acid making the glucocorticoid water soluble.
- C21 ester glucocorticoids are prodrugs, giving them longer durations of action.
- Both C16 and C17 hydroxy, if present, are easily esterified and can be used to form diesters as well as ketals (acetonides) which also enhance lipophilicity.
  - Diesters are prodrugs with longer duration of action.
  - Ketals are not prodrugs unless they have a C21 ester.
- Figure 6.16 gives examples of lipophilic, hydrophilic, and ketal esters of glucocorticoids.

Common Lipophilic Esters at C21

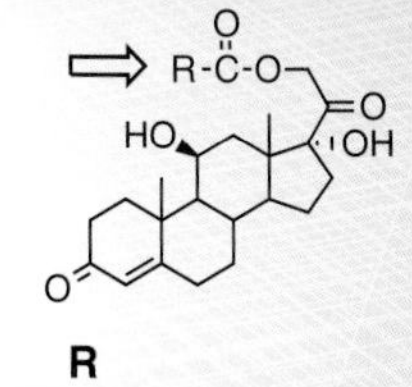

R	
—CH$_3$	Acetate
—CH$_2$·C(CH$_3$)$_2$·CH$_3$	*t*-Butylacetate
—(CH$_2$)$_2$CH$_3$	Butyrate
—C(CH$_3$)$_2$·CH$_3$	Pivalate
—(CH$_2$)$_3$CH$_3$	Valerate
—(CH$_2$)$_2$⬠	Cypionate

Diesters:

n = 1; 17α, 21-dipropionate ester (prodrug)

n = 2; buteprate or probutrate (prodrug)

16α, 21-dipropionate ester (prodrug)

Ketals:

Acetonide (active)   Hexacetonide (prodrug)

Hydrophilic esters:

R = —C(=O)-CH$_2$CH$_2$-C(=O)-O$^{\ominus}$ Na$^{\oplus}$

Sodium succinate (prodrug)

R = —P(=O)(O$^{\ominus}$Na$^{\oplus}$)-O-Na$^{\oplus}$

Sodium phosphate (prodrug)

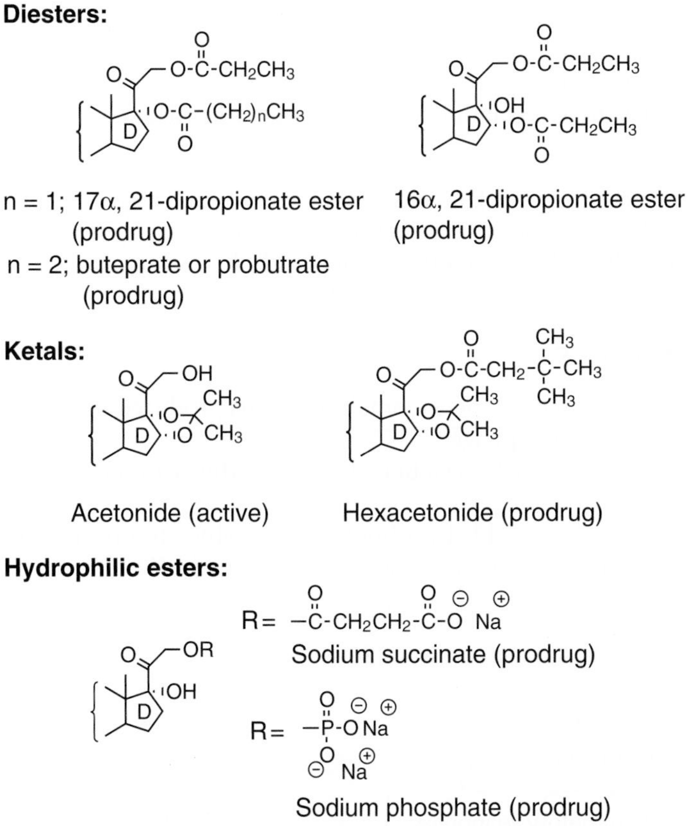

Figure 6.16 Examples of glucocorticoid lipophilic diesters and ketals and hydrophilic esters.

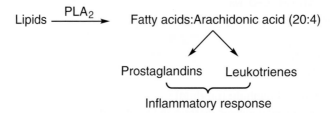

Figure 6.17 Metabolic pathways for hydrocortisone.

Metabolizing enzymes: 1 = 11β-hydroxysteroid dehydrogenase
2 = 3α-hydroxysteroid dehydrogenase
3 = 5β-reductase
4 = C17 oxidase

ADVERSE EFFECTS

Inhaled Glucocorticoid: Local: dry mouth, hoarseness, oral candidiasis, upper respiratory infections, and reflex cough.

Systemic: potential for decreased growth rate in children from low-medium doses; decrease in bone density; longterm use may lead to cataracts.

Adverse effects of oral glucocorticoid:

Fluid retention, increased appetite, weight gain, hypertension, adrenal axis suppression, diabetes, osteoporosis, glaucoma muscle weakness, and Cushing's syndrome.

DRUG–DRUG INTERACTIONS

Glucocorticoids: Rifampin, barbiturates, and other enzyme inducers reduce therapeutic levels.

Estrogens, birth control pills (BCPs), ketoconazole, and erythromycin increase plasma levels by decreasing elimination.

Isoniazid and salicylate plasma concentrations decreased by glucocorticoids.

Metabolism

- If one or more of the essential functional groups on the glucocorticoid skeleton are modified by metabolism, then activity will be lost.
- There are three major metabolic reactions that will eliminate glucocorticoid activity.
  - Ring A reductions
  - C17 oxidation
  - C11 keto-enol isomerization
- Figure 6.17 demonstrates these metabolic reactions for hydrocortisone.

Clinical Application

- Along with a rescue inhaler (β₂-agonist), inhaled low-, medium-, and high-dose glucocorticoids are the first-line treatment for the various stages of asthma and COPD.
- Table 6.4 lists the clinically important corticosteroids with their structures.

Leukotriene Modifiers

- The LTs are a group of inflammatory triene-containing lipids produced in eosinophils, mast cells, and macrophages.
- Derived from arachidonic acid (20:4) via a branch of the pathway to the prostacyclins and thromboxanes.
- Arachidonic acid is converted to the LTs by 5-lipoxygenase, which in turn is controlled by an activating protein, 5-lipoxygenase-activating protein (FLAP).

Lipids $\xrightarrow{\text{PLA}_2}$ Fatty acids:Arachidonic acid (20:4)

Prostaglandins   Leukotrienes

Inflammatory response

Table 6.4 Clinically Important Corticosteroid Structures and Indications

	Structures	Indications
Systemic Corticosteroids		
6-Methylprednisolone	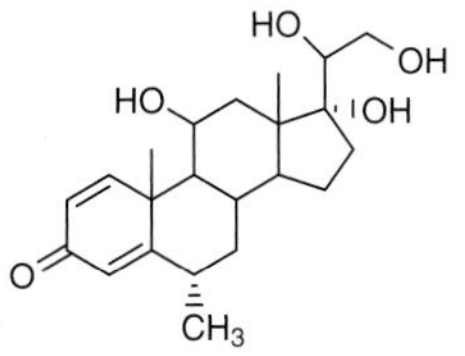	Oral administration for severe persistent asthma with LABA and high-dose inhaled corticosteroid
Prednisolone	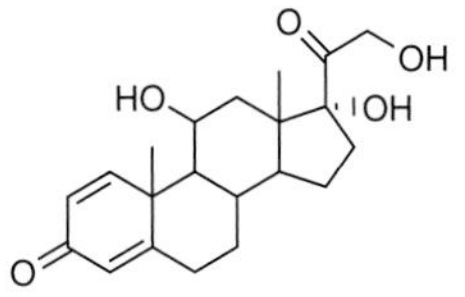	Oral administration for severe persistent asthma with LABA and high-dose inhaled corticosteroid
Prednisone	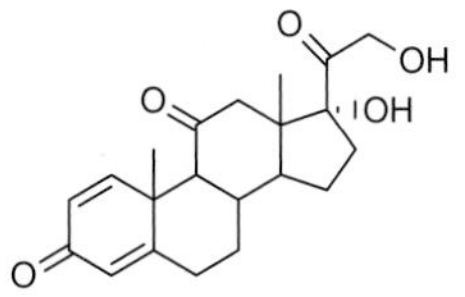	Oral administration for severe persistent asthma with LABA and high-dose inhaled corticosteroid
Inhaled Corticosteroids		
Beclomethasone	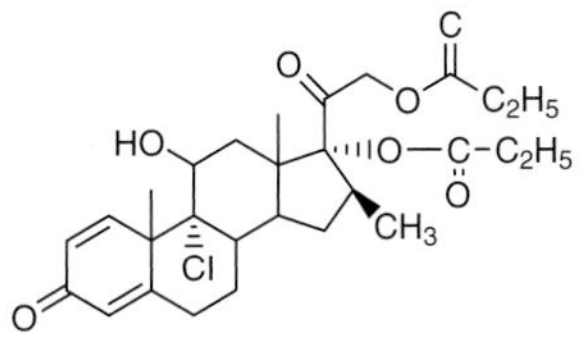	Long-term control of asthma and stage IV COPD
Budesonide	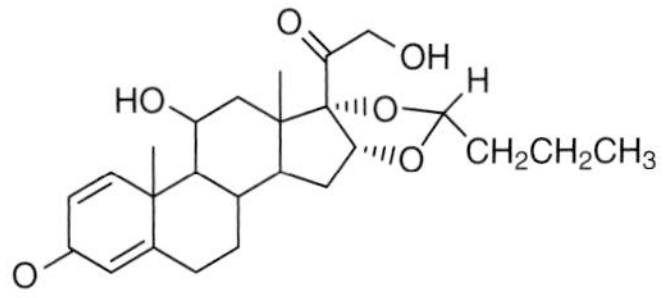	Long-term control of asthma and stage IV COPD
Flunisolide	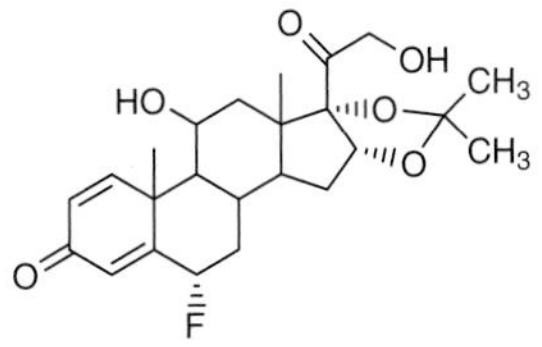	Long-term control of asthma and stage IV COPD
Fluticasone	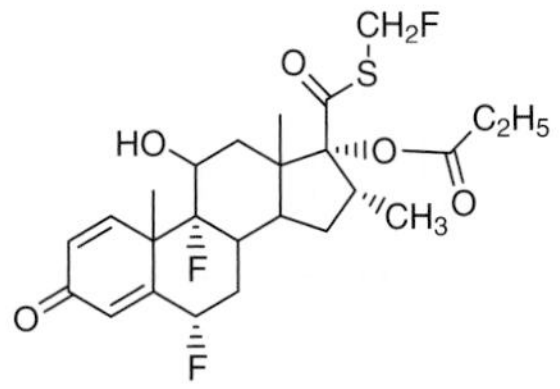	Long-term control of asthma and stage IV COPD
Mometasone	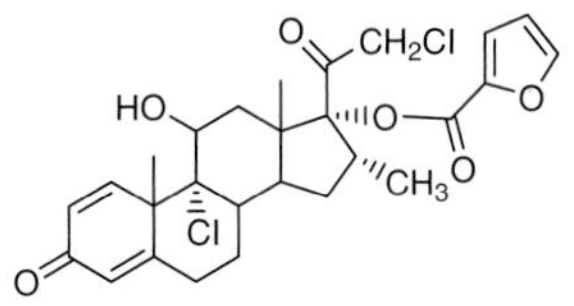	Long-term control of asthma and stage IV COPD
Ciclesonide	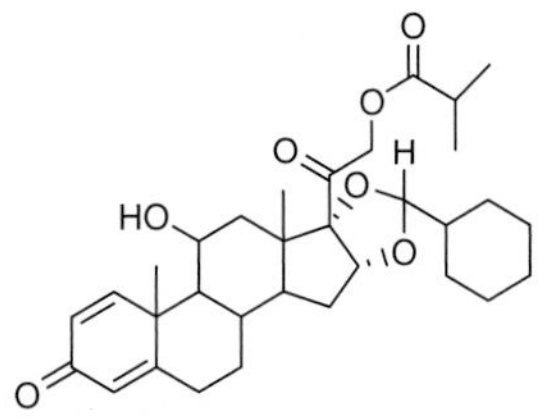	Long-term control of asthma and stage IV COPD

- The LTs are a mixture of cysteine-containing compounds referred to as LTC_4, LTD_4, and LTE_4 which were known as slow-reacting substance of anaphylaxis (SRS-A) that produce a long-lasting contraction of bronchial smooth muscle associated with asthma pathophysiology.

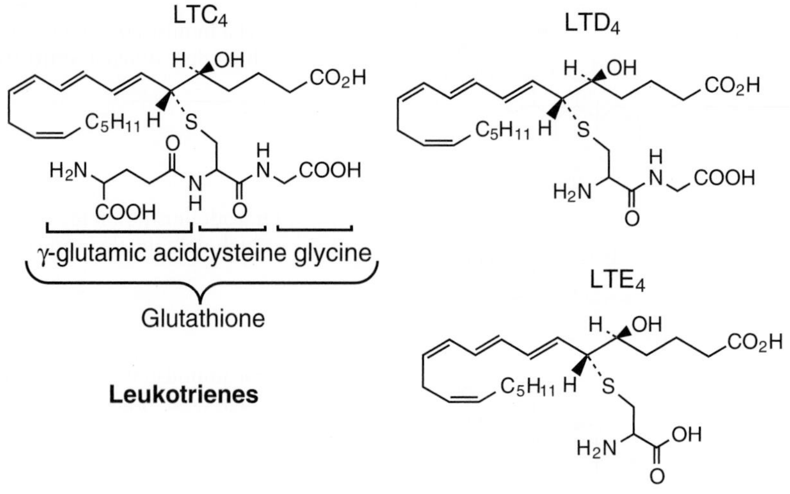

Leukotrienes

LTs

MOA

- There are two mechanisms of action for the LT modifiers:
  - LT receptor ($cysLT_1$) antagonists
  - 5-lipoxygenase inhibitors

SAR

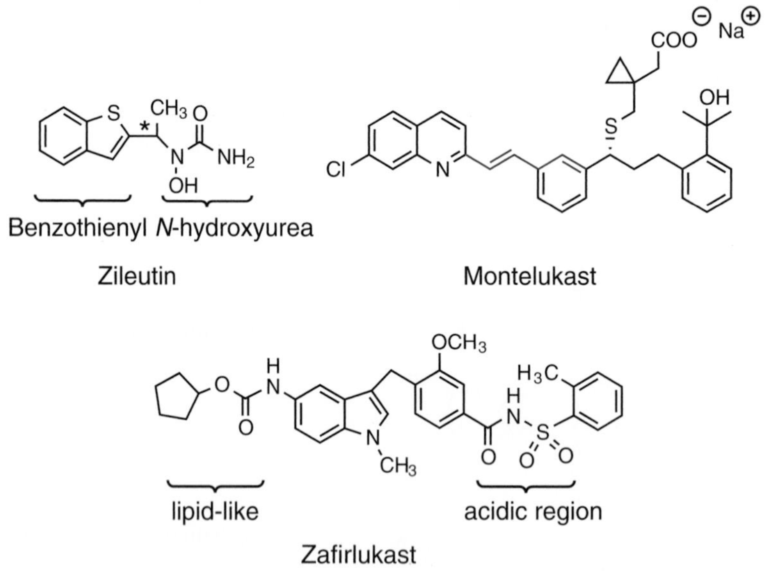

Benzothienyl *N*-hydroxyurea

Zileutin

Montelukast

Zafirlukast

LT modifiers

- LT receptor antagonists
  - Montelukast (Singulair)
    - Quinoline derivative that mimics the triene moiety of LTs.
    - The quinoline ring can be reduced and can retain activity.
    - The chloro can be replaced with a fluorine.
    - The double bond between the quinoline ring and the phenyl ring can be replaced with an ether linkage.
    - The phenyl ethyl moiety on the middle phenyl ring is not necessary for activity.

- The sulfide-acid moiety can be changed to an amide acid and can retain activity.
  - Zafirlukast
    - Acidic region can be attached to the indole *N* with decreased activity.
    - The sulfonamide moiety can be replaced with COOH with decreased activity.
    - Cyclopentyl ring can be replaced with a N-butyl or other alkyl group with decreased activity.
    - The carbamate linkage can be replaced by an amide with decreased activity.
- 5-lipoxygenase inhibitor
  - Zileuton
    - Chemical class: N-hydroxyurea.
    - The N-OH is necessary for activity.
    - The benzothienyl ring system contributes lipophilicity.
    - Stereochemistry: both enantiomers are active; however, the *S*-isomer is metabolized and eliminated more quickly.

Physicochemical and Pharmacokinetic Properties

- Zileuton is rapidly absorbed orally and 93% protein bound. It is 90% bioavailable and 95% excreted as metabolites with a $T_{1/2}$ of 2.5 hours.
- Montelukast is rapidly absorbed with 64% bioavailability and 99% protein bound in the plasma, with 86% eliminated as metabolites through the bile.
- Zafirlukast is well absorbed orally, but food decreases absorption by as much as 40%. More than 90% is excreted as metabolites in the feces.

Metabolism

- Zileuton is metabolized in the liver.
  - O-glucuronide is the major metabolite with <0.5% as inactive N-dehydroxylated or unchanged zileuton.
  - Glucuronidation is stereoselective with *S*-isomer metabolized more quickly.
- Montelukast is extensively metabolized in the liver to oxidized products.
  - CYP3A4 oxidizes the sulfur to the sulfoxide and the C21 benzylic carbon to a hydroxyl group.
  - CYP2C9 is selectively responsible for the methyl hydroxylation.
  - Glucuronidation of the carboxylic acid also occurs.
- Zafirlukast is metabolized by CYP3A4 and CYP2C9 to hydroxylated metabolites.
  - The benzylic carbon can undergo glutathione addition.
  - Carbamate hydrolysis followed by N-acetylation has also been shown to occur.

Clinical Application

- The LT modifiers are used as alternatives to the use of LABAs in addition to inhaled corticosteroids for the treatment of moderate persistent asthma (Table 6.2).

Monoclonal Anti-IgE Antibody

- In the pathophysiology of allergic asthma, IgE binds to mast cells causing their degranulation and release of asthma mediators.
- Omalizumab is a monoclonal antibody developed as an antihuman IgE antibody.

MOA

- Omalizumab complexes with free IgE forming trimers consisting of 2:1 IgE:omalizumab or a 1:2 complex of IgE:omalizumab which effectively neutralizes free IgE decreasing mast cell degranulation.

ADVERSE EFFECTS

Leukotriene Modifiers:
Zileuton: increases liver enzymes (caution in patients with hepatic impairment); headache, nausea, dyspepsia, and myalgia. Montelukast and zafirlukast: headache, dizziness, nausea, dyspepsia, and rarely, Churg–Strauss syndrome (eosinophilic vasculitis).

DRUG–DRUG INTERACTIONS

Leukotriene Modifiers:
Zileuton increases the AUC and decreases clearance of warfarin, theophylline, and propranolol.
  Montelukast serum levels are decreased by CYP3A4 inducers (e.g., phenytoin, phenobarbital, and rifampin).
  Zafirlukast inhibits both CYP3A4 and 2C9 isozymes.
- Drugs metabolized by these enzymes will have increased plasma levels.
  - 3A4: terfenadine, cyclosporine, cisapride, and dihydropyridine calcium channel blockers.
  - 2C9: warfarin, phenytoin, and carbamazepine.
- Aspirin increases zafirlukast plasma levels.
- Theophylline and terfenadine decrease AUC of zafirlukast.
- Erythromycin decreases the bioavailability of zafirlukast.

- Omalizumab also causes a downregulation of the IgE binding site on the mast cells.
- Omalizumab does not bind to IgE that is already bound to mast cells.

SAR

- Omalizumab was developed through somatic cell hybridization techniques in a murine (mouse) model with subsequent amino acid sequences added so that a humanized product resulted which only differs by 5% nonhuman amino acid residues.

Physicochemical and Pharmacokinetic Properties

- Omalizumab is 62% bioavailable after subcutaneous administration.
- Absorption is slow with peak serum levels reached in 7 to 8 days.
- Steady state serum levels reached in 14 to 29 days with multiple dosing regimens.
- Elimination of both IgE and the omalizumab complex are excreted via the bile.
- The omalizumab:IgE complex is excreted more slowly than the IgE but faster than uncomplexed omalizumab resulting in an increase of both free IgE and omalizumab:IgE complex over time.

Metabolism

- Metabolism is not known, but believed that IgE and the complex may be degraded by the hepatic reticuloendothelial system.

Clinical Application

- Primary role is for the treatment of allergic asthma.
- Approved for patients 12 years old or older with severe persistent asthma whose symptoms are inadequately controlled with inhaled glucocorticoids and who have a positive skin test for airborne allergens.
- Administration is contraindicated during acute bronchospasm and status asthmaticus.

⚠ ADVERSE EFFECTS

Omalizumab: Headache, local injection site reaction, upper respiratory tract infection, sinusitis, and pharyngitis.

BLACK BOX WARNING
ANAPHYLAXIS

Administration should be under direct medical supervision with observation for 2 hours post administration.

DRUG–DRUG INTERACTIONS

Omalizumab: Increases levels of leflunomide, natalizumab, and live vaccines.
Omalizumab levels may be increased by trastuzumab and decreased by Echinacea.

Phosphodiesterase-4 Inhibitors

- The secondary messengers, cAMP and cGMP are important regulators of intracellular signal transduction.
- PDEs are a family of enzymes that metabolize and deactivate cAMP and cGMP.
- cAMP in inflammatory cells (neutrophils, *T*-lymphocytes, and macrophages) interferes with the expression of proinflammatory mediators such as TNF-α and is therefore anti-inflammatory.
- cAMP also relaxes airway smooth muscle leading to bronchodilation.

MOA

- PDE-4 is the main cAMP metabolizing isozyme found in pulmonary tissue and inflammatory cells.
- Inhibition of PDE-4 elevates cAMP in inflammatory cells resulting in prolonging its anti-inflammatory effect.
- Inhibition of PDE-4 elevates cAMP in airway smooth muscle prolonging its bronchodilator effects.
- These two effects makes PDE-4 inhibitors ideal for the treatment of COPD and asthma.

SAR

- Roflumilast contains the dialkoxyphenyl ring making it selective as a PDE-4 inhibitor.

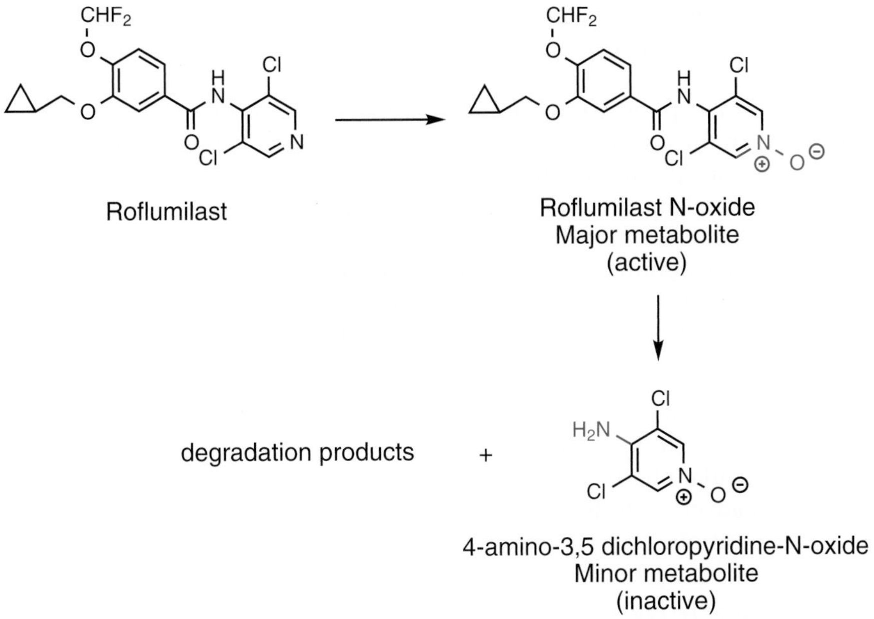

Roflumilast

Roflumilast N-oxide
Major metabolite
(active)

degradation products +

4-amino-3,5 dichloropyridine-N-oxide
Minor metabolite
(inactive)

Figure 6.18 Metabolism of roflumilast.

- The ether oxygens hydrogen-bond to glutamine in the binding pocket and the cyclopropyl and difluoromethoxy groups add additional hydrophobic interactions.
- The dichloropyridyl group forms hydrogen bonds with water which is coordinated with Mg^{2+} located in the distal end of the binding pocket.

Physicochemical and Pharmacokinetic Properties

- Well absorbed orally with $T_{1/2}$ of 10 hours.
- 80% bioavailable but food delays T_{max} by 1 hour.
- 99% protein bound and 70% excreted as metabolites via urine.

Metabolism

- Roflumilast is metabolized by CYP3A4 and CYP1A2 in the liver to its N-oxide derivative which is also a PDE-4 inhibitor (Fig. 6.18).
- Roflumilast also undergoes conjugation to roflumilast N-oxide glucuronide and metabolized to a minor metabolite 4-amino-3,5-dichloropyridine N-oxide.

Clinical Application

- Roflumilast is used to treat high-risk COPD patients with chronic bronchitis in combination with either an LA anticholinergic, LABA, or inhaled corticosteroid.
- Roflumilast did not show improvement in patients with emphysema and its clinical significance for the treatment of asthma is uncertain.

ADVERSE EFFECTS

Roflumilast: Weight loss, decreased appetite, nausea, diarrhea, dizziness, insomnia, anxiety, depression, and suicidal thoughts.

DRUG–DRUG INTERACTIONS

Roflumilast: CYP3A4 inducers decrease plasma levels (carbamazepine, phenobarbital, phenytoin).

CYP3A4 inhibitors increase plasma levels (clarithromycin, azole antifungals).

Chemical Significance

Asthma and COPD are diseases that if not controlled can lead to debilitating health problems and death. Asthma and COPD prevalence has moderated over the last 15 years and asthma-related mortality and hospitalizations have decreased. These outcomes can be related to more effective drugs and improved understanding of the pathophysiology of these conditions and the application of the SAR understanding of the drugs used to treat these diseases. For example:

- The basic pharmacophore of adrenergic agonists is a substituted β-phenylethylamine with a bulky N-substituent.

- The nature and number of substituents influences both the type of action (direct or indirect) as well as specificity for the β-receptor subtypes found in bronchial smooth muscles.
- The basic pharmacophore for the antimuscarinic drugs is an acetylcholine analog in which the acetyl methyl group is substituted with at least one phenyl ring. The pharmacophore contains an ester function which can be replaced with different functional groups to produce the various antimuscarinic classes, that is, amino alcohol ethers, amino alcohols, and amino amides.
- The basic pharmacophore for the glucocorticoid is a free 11β, a 17β, and a C21 hydroxy group for activity, a 1,2- and 4,5-diene, and 9α-halogen. The C21 ester corticosteroids are prodrugs giving them enhanced absorption and longer durations of action.

It is clear that medicinal chemistry is an integral part of a pharmacist's education that gives them insight into how the chemical structure relates to the large number of drugs in a particular pharmacologic class and how it influences a drug's pharmacokinetic and adverse effects, all contributing to the therapeutic choice of an appropriate medication.

Review Questions

1 Which of the following structures is both selective for the β2-adrenergic receptor and resistant to COMT?

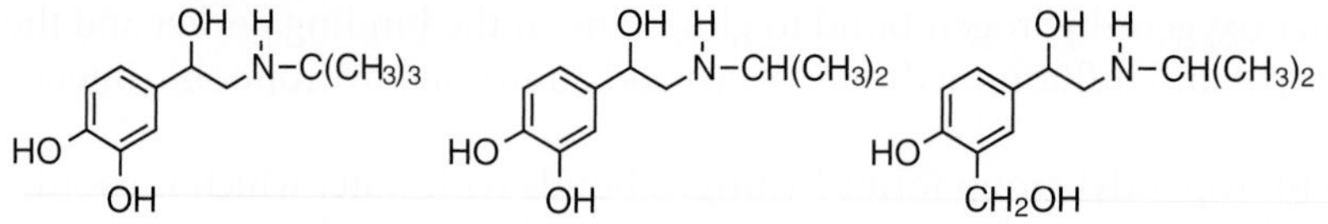

A **B** **C**

2 Ipratropium has two chiral carbons. Stereochemistry in which moiety is important for anticholinergic activity?

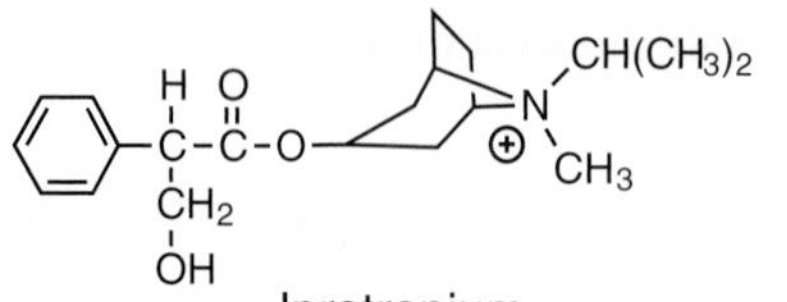

Ipratropium

A. Aminoalcohol moiety
B. Acid moiety
C. Tropane ring
D. Quaternary nitrogen
E. None of the above

3 Which of the following statements about cromolyn sodium is false?

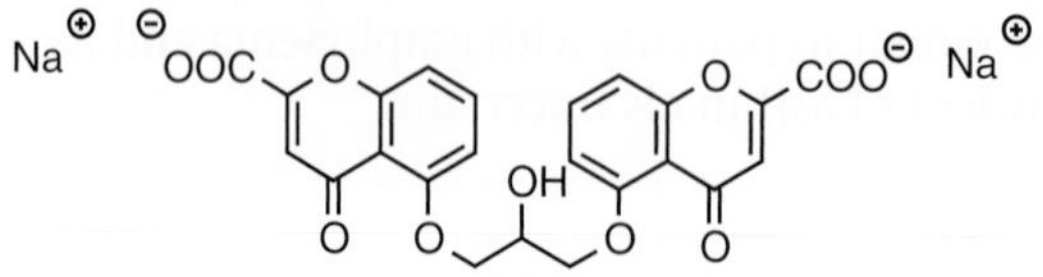

Cromolyn Sodium

A. Related to the naturally occurring chromone, Khellin.
B. A bischromone that is not metabolized.
C. The two chromones must be coplanar for activity.
D. The linking chain cannot be longer than six carbons.
E. If the 5,5' linkage is changed to the 8,8' position activity is maintained.

4 Using steroid SAR (Fig. 6.15), which of the following steroid structures would have the most anti-inflammatory activity with the least mineralocorticoid activity when administered orally?

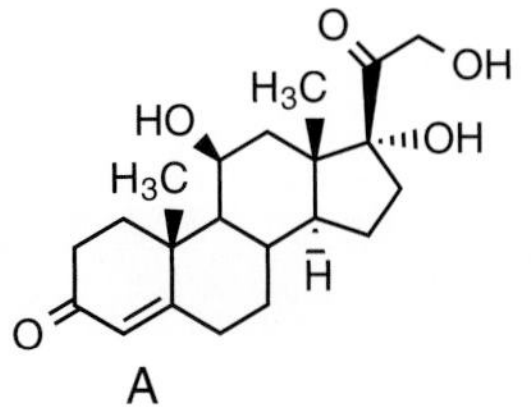

A 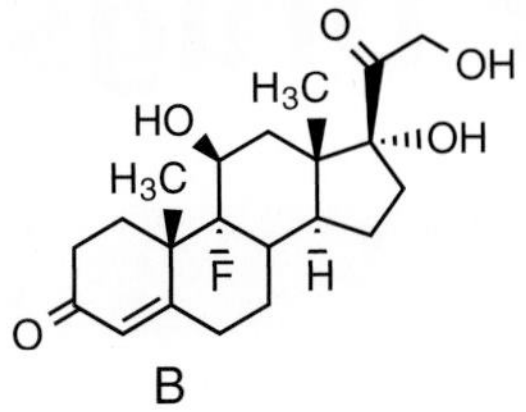   B

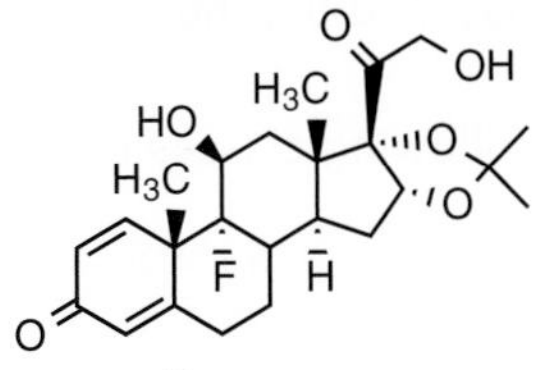

C 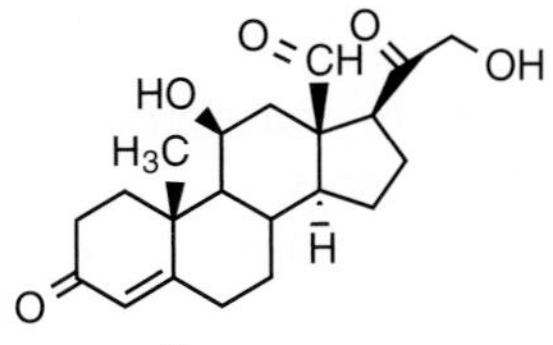   D

5 Given the structure of zileuton, is the following statement true or false?

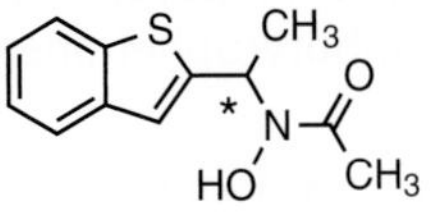

Zileuton

Both enantiomers are active; however, the S-isomer is metabolized and eliminated more quickly.
A. True
B. False

7

Antipsychotic and Anxiolytic Drugs

Learning Objectives

Upon completion of this chapter the student should be able to:

1. Discuss the hypotheses and neurotransmitters related to psychotic illnesses.

2. Identify the site and mechanism of action of the following classes of antipsychotic and anxiolytic agents:
   a. phenothiazines and thioxanthenes
   b. butyrophenones
   c. diarylazepines, benzisoxazole and benzisothiazole
   d. benzodiazepines

3. List structure–activity relationships (SARs) of drugs of the antipsychotic and antianxiety agents: chlorpromazines, thiothixenes, butyrophenones, and benzodiazepines.

4. Identify clinically significant physicochemical and pharmacokinetic properties.

5. Discuss, with chemical structures, the clinically significant metabolism of the antipsychotic and antianxiety agents and how the metabolism affects their activities.

6. Discuss the clinical application of the antipsychotic and antianxiety agents.

7. Discuss potential drug–drug interactions and the mechanism of these interactions as they relate to atypical antipsychotic agents.

Introduction

This chapter is focused on the medicinal chemistry of drugs that are used to treat psychoses and anxiety disorders. The definitive diagnostic criteria for psychiatric disorders in the United States are well described in the *Diagnostic and Statistical Manual of Mental Disorders* of the American Psychiatric Association (DSM-5). Psychoses (e.g., schizophrenia) are among the most severe mental illnesses with symptoms of delusions and sensory hallucinations while anxiety disorders (neuroses), in which the patient retains the ability to comprehend reality, involves mood changes (anxiety, panic, dysphoria) and thought (obsessions, irrational fears) and behavioral (rituals, compulsions, avoidance) dysfunction. Mood and panic disorders often include dysfunction of the autonomic nervous system (e.g., altered patterns of sleep and appetite) in addition to psychic abnormalities. In general, antipsychotic agents, which can have severe neurological and metabolic side effects, should be used to treat only the most severe mental illnesses.

Schizophrenia

Schizophrenia is defined as presenting two or more of the following characteristic symptoms during a 1-month period: delusions, hallucinations, disorganized speech, or grossly disorganized or catatonic behavior (*positive symptoms*). Schizophrenic patients may also experience *negative symptoms* (e.g., flat affect, lack of pleasure in their life, difficulty in carrying out planned activities), which are harder to recognize. Flexibility exists in the diagnostic criteria leaving room for professional psychiatric judgment. Continuous symptoms must persist for 6 months and affective disorders as well as drug and alcohol abuse must be ruled out. According to DSM parameters, the common estimate for schizophrenia incidence is about 1% of the population.

The etiology of schizophrenia remains unknown despite attempts to link the disease to a wide variety of internal and external factors. One of the most studied hypotheses involves brain dopaminergic neurotransmission and is referred to as the "dopamine hypothesis" of schizophrenia.

Dopamine Hypothesis

The dopamine hypothesis of schizophrenia suggests that the disease results from increased dopaminergic neurotransmission in key parts of the brain and that pharmacological approaches, which decrease dopaminergic neurotransmission, will alleviate psychotic symptoms. Most antipsychotic agents have activity to limit dopaminergic neurotransmission, providing some indirect evidence to support the dopamine hypothesis of schizophrenia. More specifically, the activity of many antipsychotic drugs correlate with affinity for dopamine D_2-type receptors and their extrapyramidal side effects also correlate with their dopamine D_2 affinity. The action of the drug at the D_2 receptor is complex, involving antagonism, inverse agonism, and partial agonism. And while the affinity for dopamine receptor continues to be the focus for many drugs, more recent additions to the treatment arsenal suggest a role for adrenergic and serotonergic receptor systems.

There are five different dopamine receptors, D_1, D_2 (with short and long splice variants), D_3, D_4, and D_5. All dopamine receptors are G-protein–coupled receptors (GPCRs). Because of the lack of drugs capable of distinguishing between the five subtypes, dopamine receptors are often classified as the D_1-type (D_1 and D_5) that stimulates adenylyl cyclase and D_2-type (D_2, D_3, and D_4) that inhibits adenylyl cyclase. These D_2-type receptors correlate with schizophrenia and it is on these receptors that many of the antipsychotic drugs act.

The dopamine pathways within the brain that are affected by the antipsychotic agents account not only for the beneficial effects of the antipsychotic agents, but also account for some of the significant adverse effects of the drugs. These pathways within the brain are beyond the scope of this presentation, but the biosynthesis and degradation (Fig. 7.1) and the site of action within the pre- and postsynaptic neurons of the dopaminergic pathway (Fig. 7.2) are important to understand.

Additional Neurotransmitters Involvement in Schizophrenia

In addition to the dopamine D_2-type autoreceptors, heteroreceptors, such as adenosine (A_2), histamine (H_1), and serotonin (5-$HT_{1A,2A}$ and $_{2C}$), located on or near presynaptic dopaminergic nerve terminals modulate dopamine synthesis (and release). Adrenergic (α_2) autoreceptors may also negatively modulate the release of

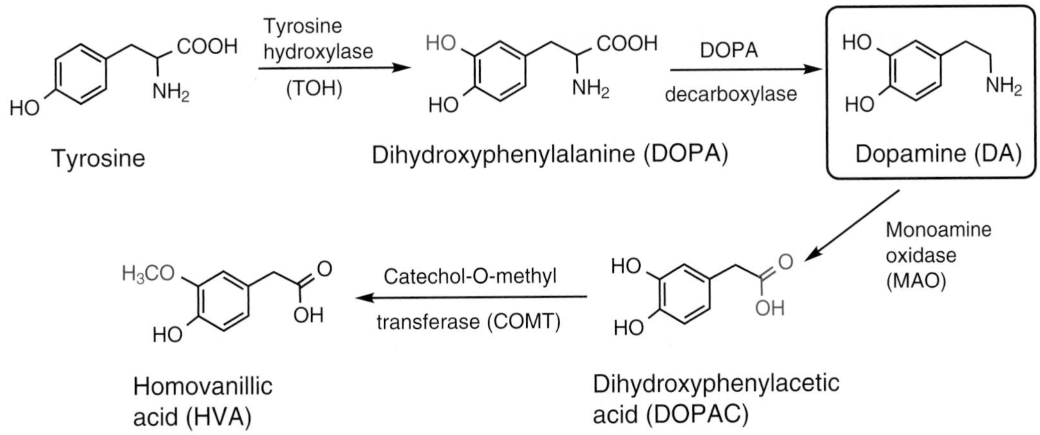

Figure 7.1 The biochemical synthesis and catabolism pathway of dopamine.

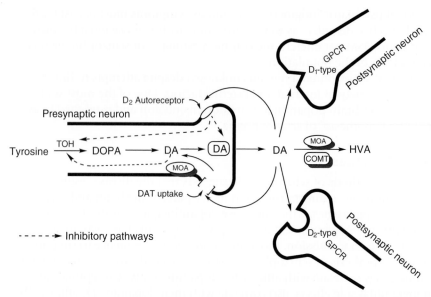

Figure 7.2 Neuronal schematic for the biosynthesis of dopamine (DA), its action on D_1- and D_2-type receptors. DA is stored in vesicles in the presynaptic neuron, released into the synaptic cleft, and in addition to its binding to D-type receptors it binds to presynaptic D_2 autoreceptors to modulate or inhibit TOH and dopamine release. DA can be transported back into presynaptic neurons via DA transporter protein (DAT), destroyed in presynaptic neurons by monoamine oxidase (MAO) bound to mitochondria or removed from the synapse via oxidized and methylated to HVA and excreted in urine.

the neurotransmitter norepinephrine (atypical antipsychotic drugs) and may interact with other neurotransmitter receptor systems (i.e., histamine, serotonin, and adrenergic) instead of (or in addition to) dopamine receptor systems.

Serotonin 5-HT_{2A} and 5-HT_{2C} GPCRs have been linked to the pathophysiology and treatment of schizophrenia. For example, activation of 5-HT_{2A} receptors mediates the psychotomimetic effects of some classes of hallucinogenic drugs, whereby activation of 5-HT_{2C} receptors leads to antipsychotic effects in various rodent models of schizophrenia. Some second generation atypical antipsychotic drugs have higher affinity for human serotonin 5-HT_2-type receptors than dopamine D_2-type receptors.

Antipsychotic Drugs

Drugs to treat schizophrenia fall into two classes: the typical antipsychotic agents, which demonstrate activity against the positive symptoms of schizophrenia (i.e., hallucinations, delusions, and disorganized speech and behavioral), commonly exhibit extrapyramidal adverse effects, and atypical antipsychotic agents that commonly exhibit reduced tendency to cause extrapyramidal adverse effects.

Typical Antipsychotic Drugs

MOA
- Site of action is direct interaction with D_2-type receptors.
- Functional activity of antipsychotic drugs spans the spectrum of inverse agonism, antagonism, and partial agonism activity. NOTE: it is not always clear as to which of these actions are involved in the clinical outcome.
- The phenothiazine antipsychotics (e.g., chlorpromazine) have appreciable affinity at the D_1-type, serotonin ($5\text{-HT}_{2A/2C/6/7}$) adrenergic (α_1- and α_2-type), cholinergic M_{1-5}, and histamine H_1 GPCRs accounting for both efficacy and numerous side effects.
- The butyrophenone-type antipsychotics have high affinity at adrenergic α_1-type, and moderate affinity at D_1-type and serotonin 5-HT_{2A}, but not 5-HT_{2C} GPCRs in addition to their action on D_2-type receptors.

SAR of Chlorpromazine and Thiothixene Antipsychotics

- Binding of a typical antipsychotic (i.e., phenothiazine) to the D_2-type dopamine receptors is thought to occur through a conformation similar to that shown by DA (*trans* α-rotamer conformation) as indicated by X-ray structures (Fig. 7.3) (Table 7.1).

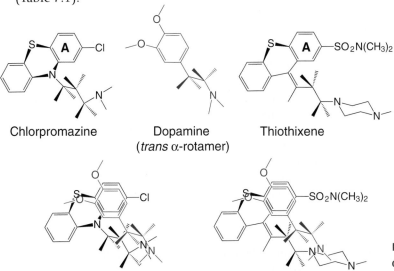

Chlorpromazine

Dopamine (*trans* α-rotamer)

Thiothixene

Overlapping of dopamine with chlorpromazine

Overlapping of dopamine with thiothixene

Figure 7.3 Representation of dopamine in a *trans* α-rotamer conformation and how it might be superimposed with specific conformations of chlorpromazine and thiothixene based upon X-ray crystallographic analysis.

Table 7.1 Phenothiazine and the Thioxanthene Derivatives Used as Antipsychotics

Generic Name	Trade Name	R_{10}	R_2	Dosage Forms
Phenothiazine Type				
Chlorpromazine hydrochloride	Generic	$(CH_2)_3N(CH_3)_2$ HCl	Cl	Tab
Thioridazine hydrochloride	Generic	$(CH_2)_2$–N(piperidine)–CH_3 · HCl	SCH_3	Tab
Perphenazine	Generic	$(CH_2)_3$–N(piperazine)N–CH_2CH_2OH	Cl	Tab
Prochlorperazine edisylate maleate	Compro Procomp	$(CH_2)_3$–N(piperazine)N–CH_3	Cl	Rectal Sup
Fluphenazine hydrochloride	Generic	$(CH_2)_3$–N(piperazine)N–CH_2CH_2OH · 2HCl	CF_3	Elixir/Tab/Inj
Fluphenazine decanoate	Generic	N–CH_2CH_2O–C(=O)–$(CH_2)_8$–CH_3	CF_3	Inj
Trifluperazine hydrochloride	Generic	$(CH_2)_3$–N(piperazine)N–CH_3 · 2HCl	CF_3	Tab
Thioxanthene Type				
Thiothixene hydrochloride	Navane Generic	(thioxanthene with $SO_2N(CH_3)_2$; H–C=(CH_2)_2·N(piperazine)N–CH_3 · HCl)		Cap

Tab, tablet; *Sup*, suppository; *Inj*, injectable; *Cap*, capsules.

Discontinued Chlorpromazine Derivatives

Several typical antipsychotic agents have been voluntarily discontinued in the United States, but are available outside of the U.S. These include the chlorpromazine derivatives: mesoridazine, thiethylperazine, and the long-acting esters: perphenazine enanthate, and fluphenazine enanthate as well as the thiothixene derivative flupenthixol and its long-acting ester flupenthixol decanoate.

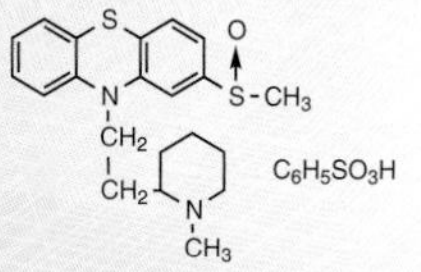

Mesoridazine mesylate

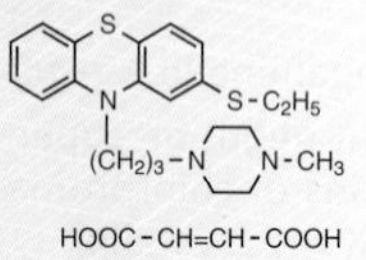

Thiethylperazine maleate

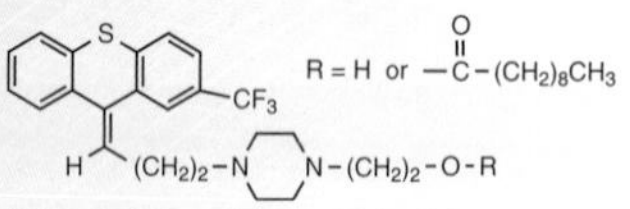

Flupenthixol/ flupenthixol decanoate

Y = O; or H and

X = F

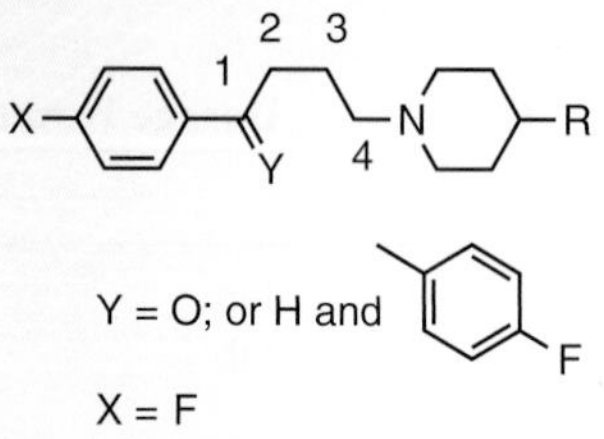

- The electronegative atom at C2 (e.g., Cl, CF_3, SCH_3) on ring "A" is essential for activity.

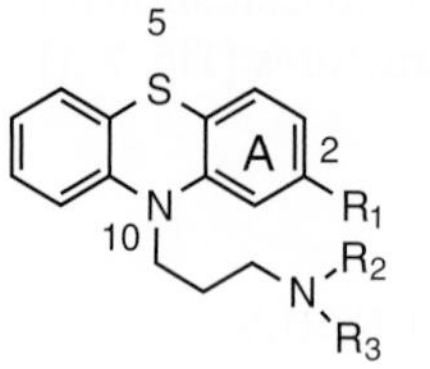

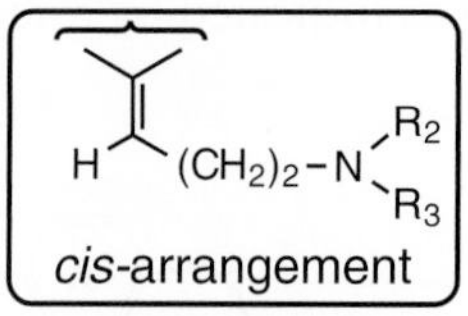

Structure B

cis-arrangement

- A side-chain amine containing three unbranched methylene groups separating the two nitrogen atoms is required.
- Replacement of N10-C11 unit with a carbon-carbon double bond as shown in Structure B gives an active compound in which the *cis* isomer exhibiting several-fold greater activity than the *trans* isomer.
- Preparation of long-chain fatty acid esters at the free hydroxyl moiety (OH, as in perphenazine, fluphenazine, and acetophenazine) results in prolonged activity of the resulting antipsychotic.

Metabolism of Chlorpromazine and Thiothixene Antipsychotics

- N-Dealkylation involving demethylation is quite common (Fig. 7.4).
- Aromatic oxidation at position C7.
- S-Oxidation to an inactive metabolite.
- Ester hydrolysis of the long-acting fluphenazine gives rise to the active fluphenazine.

SAR of Butyrophenone Antipsychotic Agents

While investigating a series of analogs of the analgesic meperidine researchers prepared substituted butyrophenones, which displayed activity resembling chlorpromazine with high affinity for D_2-type receptors. These compounds (Table 7.2) proved active as typical antipsychotic agents.

- A tertiary amino group attached at C4 of the butyrophenone skeleton is essential for antipsychotic activity; lengthening, shortening, or branching of the three-carbon propyl chain decreases antipsychotic potency.
- Variations of the basic tertiary amino group can be made without loss of potency (i.e., incorporated into a 6-membered heterocycle – piperidine, tetrahydropyridine, or piperazine) (Table 7.2).
- Replacement of the keto moiety (e.g., with the thioketone, with olefinic or phenoxy groups) or reduction of the carbonyl group decreases antipsychotic potency.
- Replacement of the keto moiety with a second 4-fluorophenyl gives a potent antipsychotic agent (pimozide, see Table 7.2).

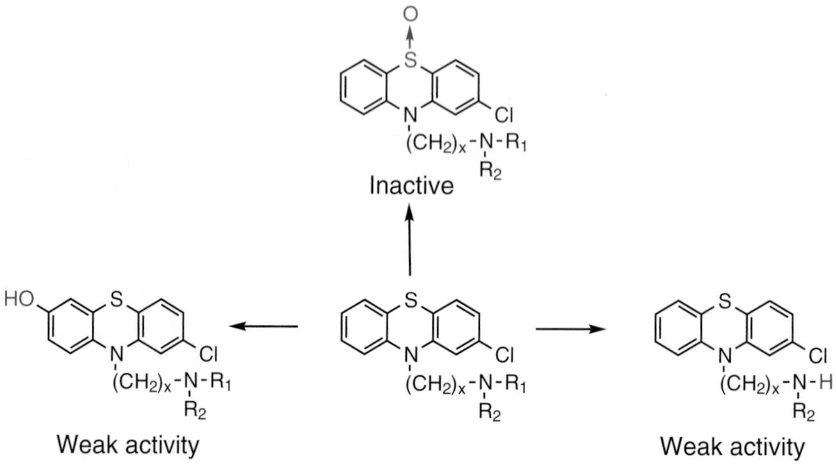

Inactive

Figure 7.4 Common metabolic pathways for phenothiazine type antipsychotic agents. N-dealkylations and 7-hydroxylation are the most common oxidative reactions catalyzed by CYP2D6 while S-oxidation metabolites have also been identified.

Weak activity

Weak activity

Table 7.2 Haloperidol and Its Derivatives Used as Antipsychotics

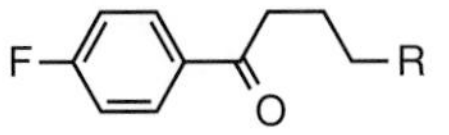

Generic Names	Trade Names	R	Dosage Forms
Haloperidol	Haldol	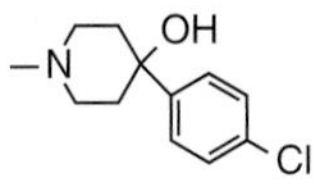	Tab/Inj
Haloperidol decanoate		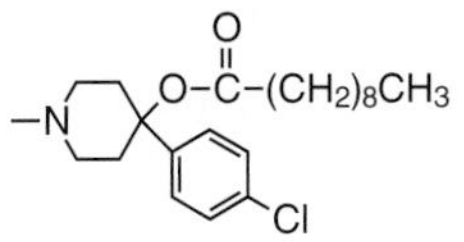	Inj
Droperidol	Inapsine	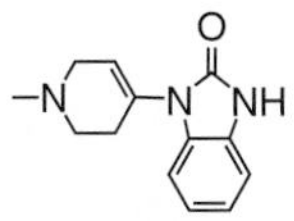	Inj
Pimozide	Orap	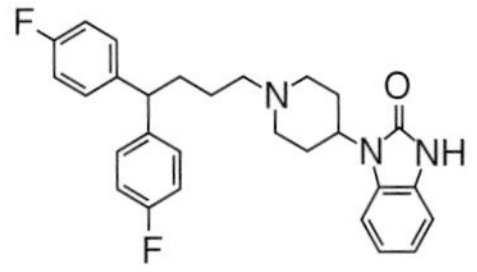	Tab

Tab, tablet; *Inj*, injectable.

- The most potent butyrophenone compounds have a fluorine substituent in the *para* position of the benzene ring.

Metabolism

- Haloperidol and its derivatives (droperidol and pimozide) appear to have common metabolic fates as shown in Figure 7.5.
- The major metabolite for pimozide involves N-dealkylation to the respective 4-substituted piperazine and a carboxylic acid as seen with haloperidol.
- CYP isoforms involved in the metabolism of pimozide are CYP3A and CYP1A2.

Atypical Antipsychotic Agents – Diarylazepine Derivatives

The atypical antipsychotic agents, sometimes are referred to as second-generation antipsychotic agents differ from typical antipsychotics by their binding to additional

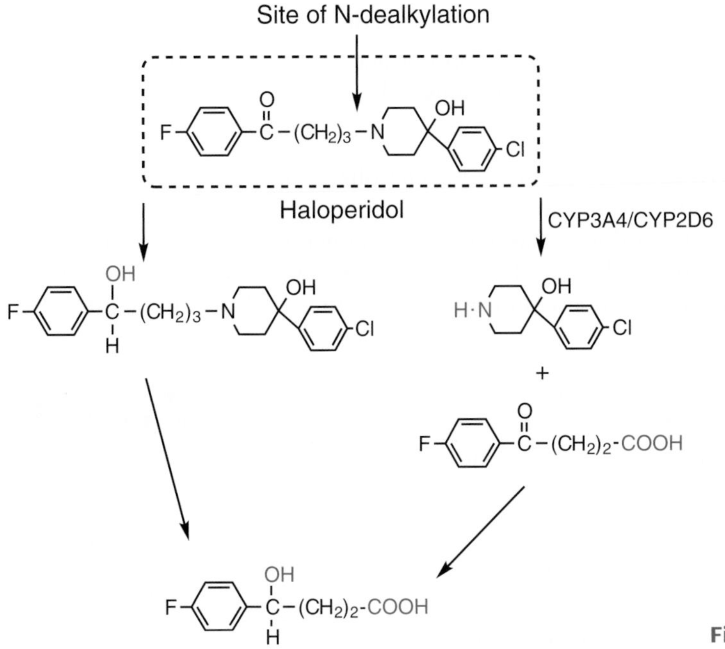

Investigative Butyrophenone Derivatives

Several typical butyrophenone antipsychotic agents have been studied in humans and may also be available outside the U.S., but have not been approved by the FDA. These include the agents spiperone and trifluperidol.

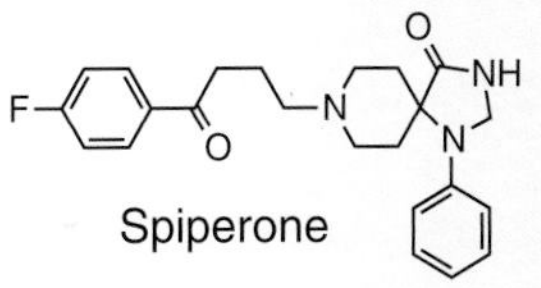

Spiperone

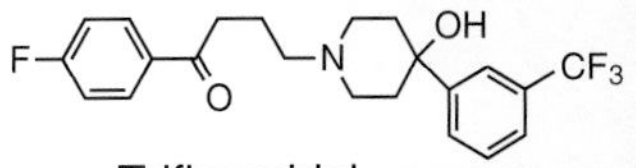

Trifluperidol

Figure 7.5 Metabolism of haloperidol.

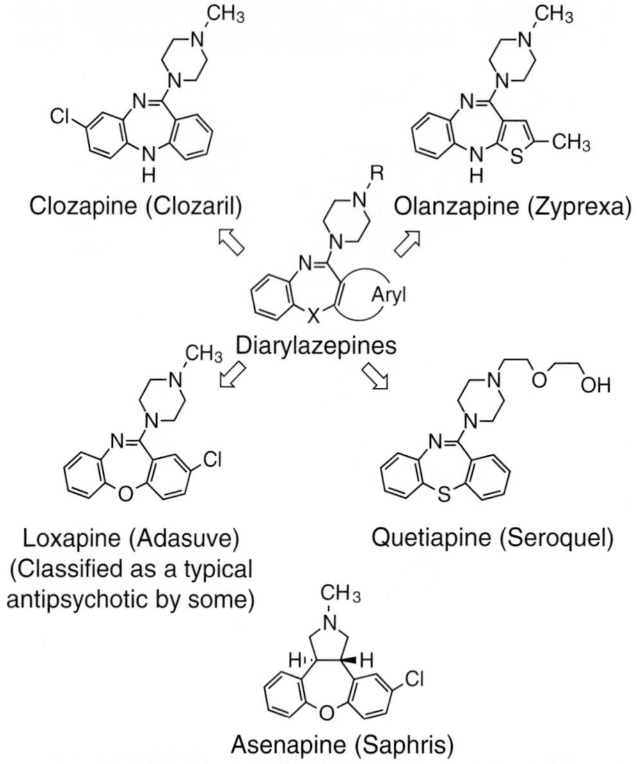

Figure 7.6 Atypical antipsychotic agents: diarylazepine derivatives.

GPCR as indicated in their mechanism of action. There are primarily two chemical classes of atypical antipsychotic agents: the diarylazepines (Fig. 7.6) and the benzisoxazole and benzisothiazoles.

MOA

- Site of action is direct interaction with D_2-type receptors as an antagonist, but to a lesser degree than typical antipsychotic agents plus partial D_2 agonist activity.
- Antagonism of 5-HT$_2$ receptors.
- Combination leads to reduced extrapyramidal adverse effects.
- Involvement of other receptors have also been suggested (i.e., 5-HT$_7$, γ-aminobutyric acid [GABA], histamine, adrenergic, and acetylcholine receptors) which may account for modulation of the negative symptoms in schizophrenics.

Physicochemical and Pharmacokinetic Properties

The diarylazepines are all orally active lipophilic agents, which exhibit similar pharmacokinetic properties as shown in Table 7.3.

Metabolism

- Diarylazepine derivatives are extensively metabolized via oxidation and conjugation reactions as indicated in Table 7.4.
- The major metabolic products of asenapine are shown in Figure 7.7, while major products of metabolism of diarylazepines are shown in Figure 7.8.
- The oxidation products of quetiapine and olanzapine are shown in Figure 7.9.

Table 7.3 Pharmacokinetic Properties of Diarylazepine Derivatives

Drug	Bioavailability (%)	Protein Binding (%)	$T_{1/2}$ (h)	Dosage Form
Asenapine	~35 (SL)	95	24	Tab/SL
Clozapine	60–70	97	9–17	Tab/Susp
Olanzapine	87	90	~33	Tab/Inj
Quetiapine	~100	83	~7	Tab

Tab, tablet; *SL*, sublingual; *Susp*, suspension; *Inj*, injection.

Table 7.4 Metabolic Reactions of Diarylazepine Derivatives

Drug	CYP Isoforms	Other Enzymes	Metabolic Process
Asenapine	1A2	UGT1A4	N-Demethylation, glucuronidation, oxidation
Clozapine	3A4	FMO	N-Demethylation, N-oxidation
Olanzapine	1A2, 2D6	FMO	N-Demethylation, glucuronidation, N-oxidation, oxidation
Quetiapine	3A4		N-Alkylation, oxidation, sulfoxidation

UGT, glucuronosyltransferase; *FMO*, flavin-containing monoxygenase.

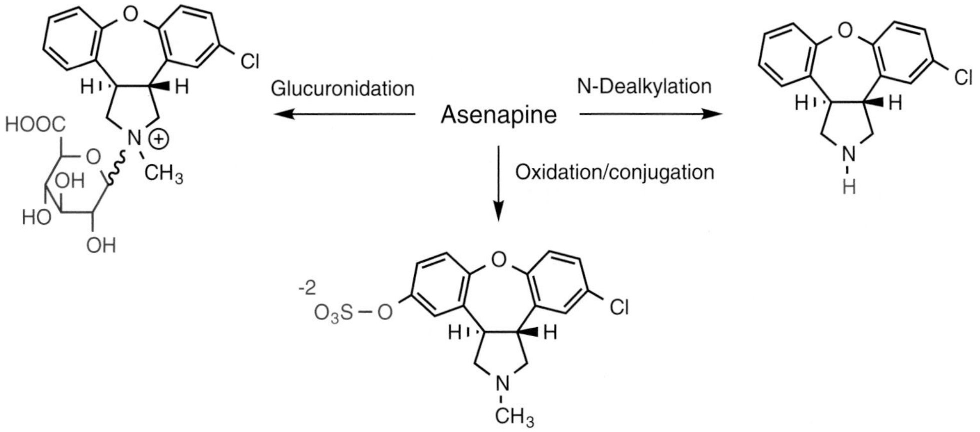

Figure 7.7 Major metabolic products for asenapine.

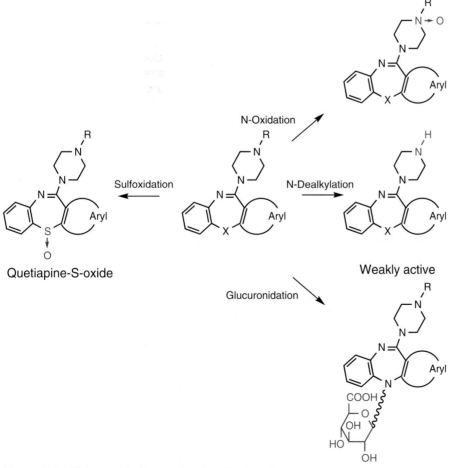

Figure 7.8 Major metabolic reactions for diarylazepines.

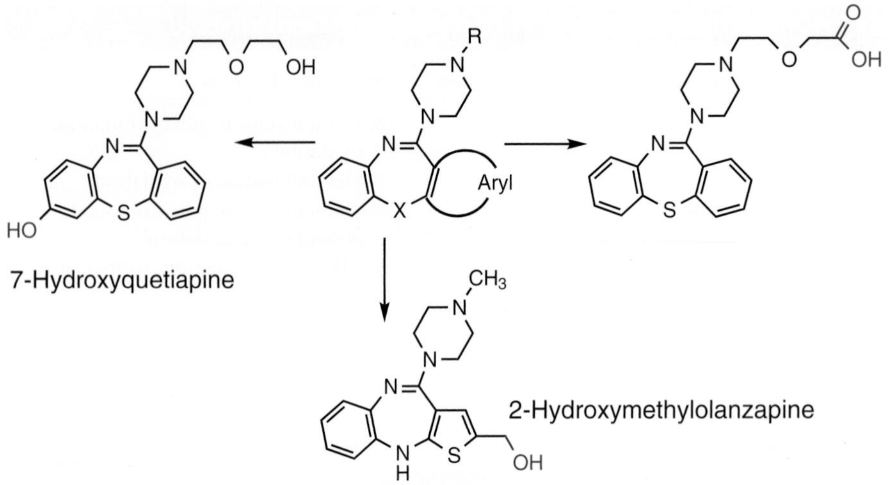

Figure 7.9 Oxidation products of quetiapine and olanzapine.

7-Hydroxyquetiapine

2-Hydroxymethylolanzapine

Atypical Antipsychotic Agents – Benzisoxazole and Benzisothiazole

MOA

- The benzisoxazole and benzisothiazole (Fig. 7.10) have a mechanism of action with similarity to the diarylazepines in that they have high affinity and antagonism at 5-HT2$_A$ receptors.
- Antagonism at D$_2$-type receptors in the striatum and cortex of the brain.

Physicochemical and Pharmacokinetic Properties

- The pharmacokinetic properties of the various benzisoxazole and benzisothiazole antipsychotic agents are shown in Table 7.5.
- Paliperidone injectable is available as an insoluble palmitate ester (Invega Sustenna) for long-term release. Risperidone solution consists of a soluble tartrate salt.
- Risperidone is available as an orally disintegrating tablet (Risperdal M-Tab).

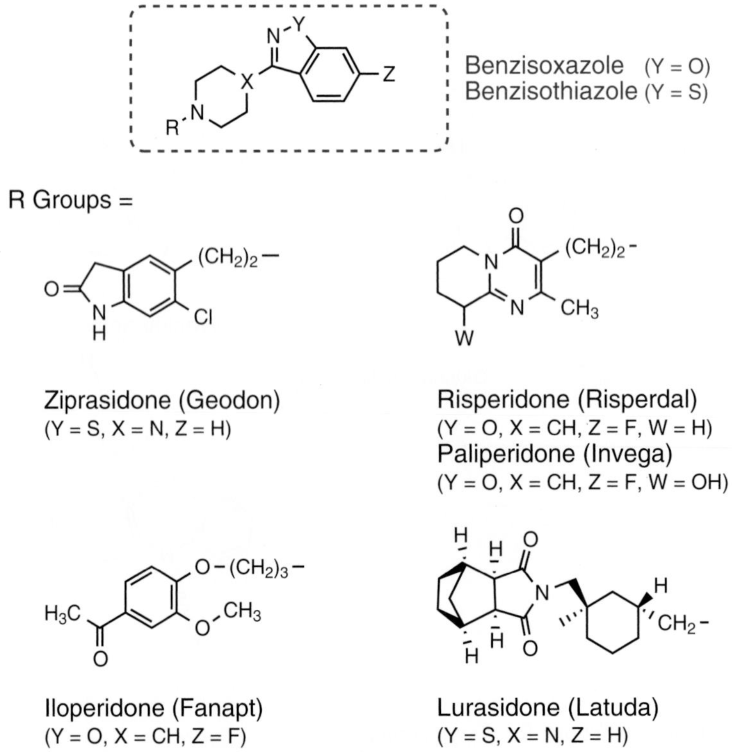

Benzisoxazole (Y = O)
Benzisothiazole (Y = S)

R Groups =

Ziprasidone (Geodon)
(Y = S, X = N, Z = H)

Risperidone (Risperdal)
(Y = O, X = CH, Z = F, W = H)
Paliperidone (Invega)
(Y = O, X = CH, Z = F, W = OH)

Iloperidone (Fanapt)
(Y = O, X = CH, Z = F)

Lurasidone (Latuda)
(Y = S, X = N, Z = H)

Figure 7.10 Benzisoxazole and benzisothiazole antipsychotic agents.

Table 7.5   Pharmacokinetic Properties of Benzisoxazole and Benzisothiazole Derivatives

Drug	Absorption/Bioavailability	Protein Binding	$T_{1/2}$ (h)	Dosage Form
Iloperidone	Well/96%	~97%	18–33	Tab
Lurasidone	9–19%	99%	18–31	Tab
Paliperidone	Rapid/28%	72%	23	Tab, Inj
Risperidone	Rapid/70–94%	90%	3–20 – poor metabolizers; 21–30 – extensive metabolizers	Tab, Sol, Inj
Ziprasidone	Well/~60%	>99%	~7	Cap, Inj, Susp

Tab, tablet; *Cap*, capsule, Sol, solution; Susp, suspension; Inj, injection.

Metabolism

- Risperidone is extensively metabolized to paliperidone as an enantiomeric mixture (+/−).
- The (+)-enantiomer is formed by CYP2D6 oxidation while the (−)-enantiomer is produced by CYP3A4 oxidation. Both enantiomers are active.
- Paliperidone itself has limited metabolism.
- Ziprasidone is extensive metabolized as shown in Figure 7.11.
- In a similar manner lurasidone undergoes N-dealkylation and S-oxidation, and the the norbornyl group is hydroxylated to an active metabolite.

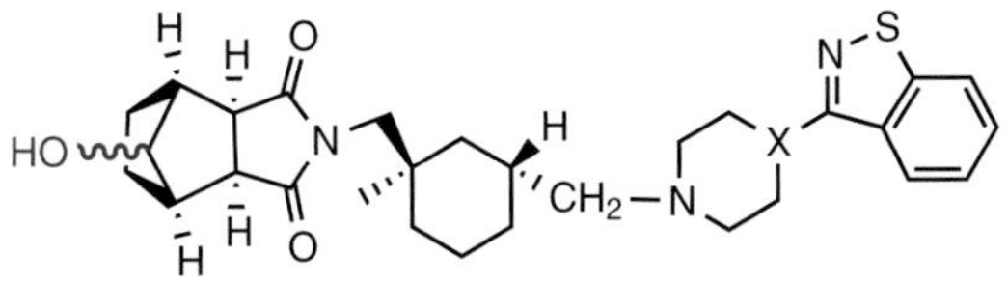

Active metabolite of lurasidone

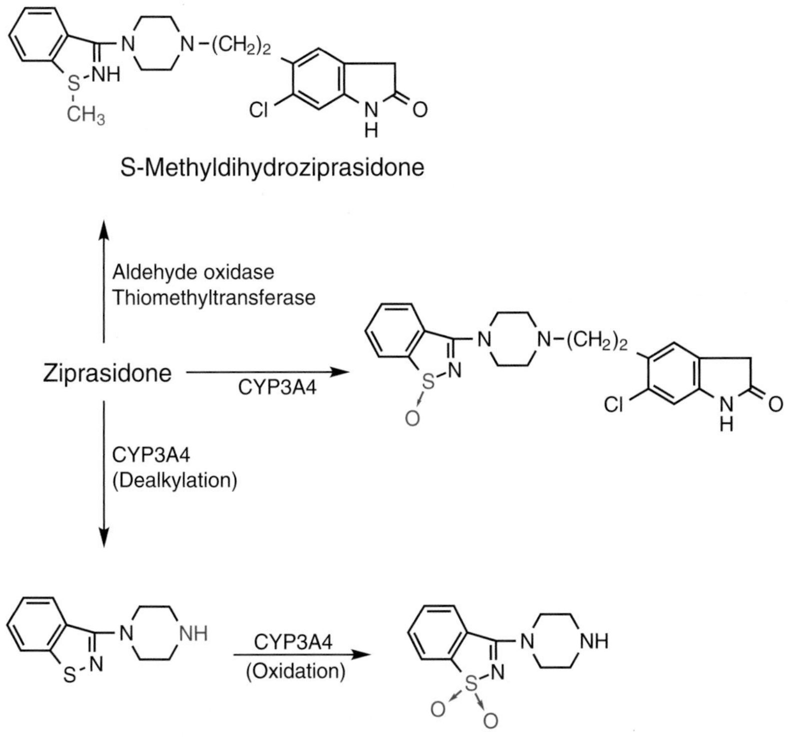

S-Methyldihydroziprasidone

Figure 7.11   Metabolism of ziprasidone.

- Iloperidone is metabolized via CYP2D6 and 3A4 to two major products. Ketone reduction to the active secondary alcohol and hydroxylation of the methyl ketone to the inactive hydroxymethylketone.

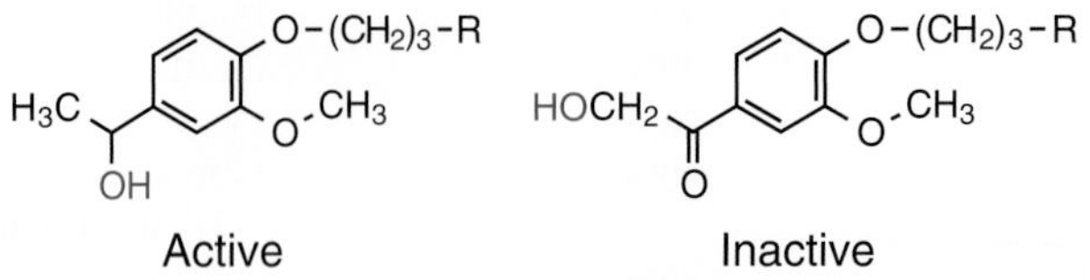

Active — Inactive

Iloperidone metabolites

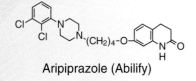

Aripiprazole (Abilify)

Atypical Antipsychotic Agent: Aripiprazole

Distinct from the two major classes of atypical antipsychotic agents is the drug aripiprazole an arylpiperazine quinolinone.

MOA

Aripiprazole has properties similar and dissimilar of those seen with other atypical antipsychotic agents including:

- high affinity with partial agonist activity at D_2-type receptors.
- high affinity for several of the serotonin GPCRs (i.e., 5-HT$_{1A}$, 5-HT$_{2A}$, 5-HT$_{2C}$, and 5-HT$_7$).
- partial agonist activity on 5-HT$_{2A}$ and 5-HT$_{2C}$ receptors.
- agonist activity at D_2 autoreceptors (negatively modulate dopamine synthesis and release).

Physicochemical and Pharmacokinetic Properties

- Aripiprazole has oral bioavailability (90%).
- Is highly bound to plasma protein (99%)
- Has a long duration of action ($t_{1/2}$ of 75 hours).

Metabolism

- Aripiprazole is a substrate for CYP3A4 and 2D6 giving rise to the metabolites shown in Figure 7.12. The active dehydroaripiprazole has a very long duration of action (~90 hours).

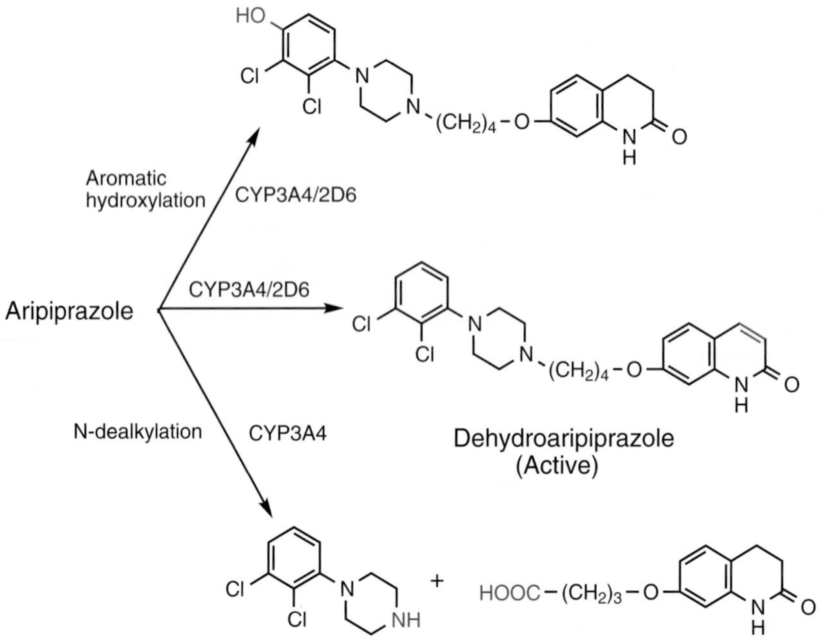

Aromatic hydroxylation / CYP3A4/2D6

Aripiprazole

CYP3A4/2D6

Dehydroaripiprazole (Active)

N-dealkylation \ CYP3A4

Figure 7.12 Metabolism of apripiprazole.

Table 7.6   Potential Drug–Drug Interactions with Atypical Antipsychotic Agents

Drug	Coadministered Drug(s)	CYP450 Isoform	Dose Adjustment of Atypical Agent
Aripiprazole	Ketoconazole	3A4-inhibitor	Reduce 1/2
	Quinidine	2D6-inhibitor	Reduce 1/2
	Carbamazepine	3A4-inducer	Increase 2 fold
Clozapine	Fluvoxamine/Ciprofloxacin	1A2-inhibitor	Reduce 1/3
Quetiapine	Ketoconazole	3A4-inhibitor	Reduce 1/6
	Phenytoin	3A4-inducer	Increase 5 fold
Iloperidone	Ketoconazole	3A4-inhibitor	Reduce 1/2
	Fluoxetine/Paroxetine	2D6-inhibitor	Reduce 1/2
Lurasidone	Ketoconazole/Rifampin	3A4-inhibitor	Contraindicated

Clinical Applications of Antipsychotic Agents

The choice of an antipsychotic agent is often dependent upon the specifics of the individual patient and may be either a typical or an atypical antipsychotic. While the adverse effects have an important role in the choice of the agent, the important over riding feature is the need to treat the positive symptoms and if possible the negative symptoms of the psychiatric behavior. Uses include:

- Delirium and dementia, while not FDA approved conditions such conditions are often treated with typical or atypical antipsychotic agents.
- Mania is commonly treated with atypical antipsychotic drugs whereas typical antipsychotic drugs are less effective.
- Major depression will commonly be treated with antipsychotic drugs combined with antidepressants.
- Schizophrenia and chronic psychotic illnesses are commonly treated over an expanded period of time with the antipsychotic agents.

Drug–Drug Interaction of Antipsychotic Agents

- The atypical antipsychotic agents can exhibit a number of significant drug–drug interactions which calls for dose adjustment (Table 7.6).

Anxiety and Anxiety Disorders

Anxiety can be considered pathological when the level of anxiety interferes with normal social or occupational functioning but should not be associated with an exogenous factor (e.g., caffeine) or a medical condition (e.g., hyperthyroidism). Examples of anxiety disorders include phobias, generalized anxiety disorder (chronic abnormally high level of anxiety), social phobia (e.g., fear of public speaking), obsessive-compulsive disorder, panic disorder with or without agoraphobia (avoidance of situations believed by the patient to precipitate panic attacks), and posttraumatic stress disorder (PTSD).

There appears to be no gross neuroanatomical lesion associated with anxiety nor is there a genetic implication to anxiety. Currently, abundant evidence exists to document the involvement of the neurotransmitters GABA with a possibly involvement of norepinephrine and serotonin, to anxiety, and thus the development of drugs that affect GABA neurotransmission.

GABA Receptors

- The major inhibitory neurotransmitter in the mammalian CNS is GABA with two major classes of GABA receptors: inotropic $GABA_A$ and metabotropic $GABA_B$ receptors.
- The $GABA_A$ receptor is a member of the gene superfamily of ligand-gated ion channels known as the "cys-loop" family (presence of a cysteine loop in the N-terminal domain).

⚠ ADVERSE EFFECTS

Antipsychotic Agents: Sedation, hypotension, sexual dysfunction are associated with histamine H_1 and adrenergic α_1/α_2 receptor binding. Antimuscarinic actions account for cardiac, ophthalmic, gastrointestinal, and genitourinary adverse effects. The extrapyramidal adverse effects (associated with antagonism of dopamine D_2-receptors) include: parkinsonian-like movement (action in the nigrostriatal pathway), acute dystonias, akathisia, bradykinesia, cogwheel rigidity, tremor, masked face, shuffling gait, and tardive dyskinesia involving involuntary, repetitive, choreiform movements of the face, eyelids, mouth (grimaces), tongue, extremities, and trunk. Metabolic and endocrine adverse effects include weight gain, hyperprolactinemia, and gynecomastia. Common dermatologic reactions (e.g., urticaria and photosensitivity) also are observed especially with the phenothiazines.

Antagonism of dopamine D_2-type receptors in the chemoreceptor trigger zone in the brainstem is responsible for beneficial antiemetic effects produced by the phenothiazines (e.g., promethazine and prochlorperazine) neuroleptics.

⚠ ADVERSE EFFECTS

Atypical Antipsychotic Agents: The atypical antipsychotic adverse effects include metabolic adverse effects such as weight gain, glucose dysregulation, and dyslipidemia. These effects are most common with clozapine and olanzapine and result from binding of the drugs to non-dopaminergic receptors (e.g., serotonin, glutamate, histamine, α-adrenergic and muscarinic receptors). An advantage of atypical antipsychotics is that they exhibit less D_2 receptor blockade.

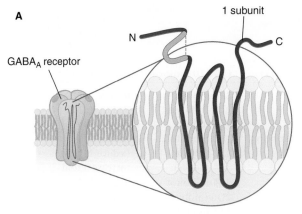

Figure 7.13 Schematic representation of the γ-aminobutyric acid A (GABA_A) receptor. The GABA_A receptors have a pentameric structure composed predominantly of α, β, and γ subunits arranged, in various proportions, around a central ion channel that conducts chloride. Each subunit has four membrane-spanning regions and a cysteine loop in the extracellular N-terminal domain (dashed line). The type and proportion of α and γ subunit composition affects affinity, pharmacologic activity, and efficacy of ligands. (Reprinted with permission from Forman SA, Chou J, Strichartz GR, Lo EH. Pharmacology of GABAergic and glutamatergic neurotransmission. In: Golan DE, ed. *Principles of Pharmacology.* Philadelphia, PA: Lippincott Williams & Wilkins; 2005:166.)

- Receptors exist as heteropentameric subunits arranged around a central ion channel (Fig. 7.13).
- The five polypeptide subunits (e.g., α_1, β_2, and γ_2 subunits) are composed of an extracellular region, four membrane-spanning α-helical cylinders, and a large intracellular cytoplasmic loop.
- The GABA_A ion channel conducts chloride, which produces an inhibitory action.
- The GABA_A extracellular N-terminal contains distinct binding sites for neuroactive drugs (e.g., barbiturates, benzodiazepines, β-carbolines, and neurosteroids).
- The benzodiazepines bind to the benzodiazepine receptor (BZR), which is found at the interphase between the α and γ subunits.
- GABA receptors with other isoforms of the α subunit (e.g., α_2–α_6) or γ component exhibit moderate or no binding sensitivity to benzodiazepines.
- The GABA_B receptors are heterodimeric GPCRs that exist as two major subtypes, GABA_{B(1)} and GABA_{B(2)}.
- These receptors produce an array of pharmacological and physiological effects including anxiolytic, anti-convulsant, anti-depressant, muscle relaxant and analgesic effects.

Drugs Used to Treat Anxiety

Benzodiazepines

The benzodiazepines are efficacious drugs for the treatment of anxiety disorders via their interaction with GABA_A receptors. A variety of benzodiazepines have been

---

Ligand-Receptor Binding

Binding of a ligand to BZR can result in agonist, partial agonists, antagonist, partial inverse agonist, or inverse agonists (see below). The model is based on the hypothesis that the BZR and GABA_A receptor exist in three oscillating conformational states: "active" or agonist, antagonist, and "inactive" or inverse agonist. The BZR agonists and partial agonists bind to and stabilize the "active" state (inducing a conformational change in the receptor complex) resulting in chloride channel opening and anticonvulsant or anxiolytic effects. The BZR inverse or partial inverse agonist binding to and stabilize the "inactive" state resulting in chloride channel remaining closed and a convulsant or anxiogenic effect. The BZR competitive antagonists bind to both states and affect no change in GABA_A receptor function or chloride conductance, but blocks access of agonists to the BZR.

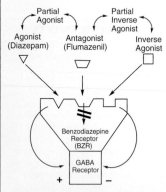

Ligand interaction with the GABA_A/benzodiazepine receptor complex leading to enhanced (**+**), blocked (**≡**), or reversed (**−**) activity of the GABA receptor.

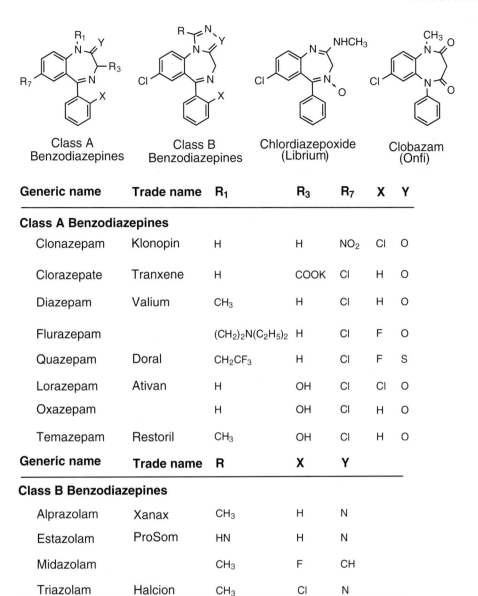

Figure 7.14 Structures of commercially available benzodiazepines.

Generic name	Trade name	R_1	R_3	R_7	X	Y
Class A Benzodiazepines						
Clonazepam	Klonopin	H	H	NO_2	Cl	O
Clorazepate	Tranxene	H	COOK	Cl	H	O
Diazepam	Valium	CH_3	H	Cl	H	O
Flurazepam		$(CH_2)_2N(C_2H_5)_2$	H	Cl	F	O
Quazepam	Doral	CH_2CF_3	H	Cl	F	S
Lorazepam	Ativan	H	OH	Cl	Cl	O
Oxazepam		H	OH	Cl	H	O
Temazepam	Restoril	CH_3	OH	Cl	H	O

Generic name	Trade name	R	X	Y
Class B Benzodiazepines				
Alprazolam	Xanax	CH_3	H	N
Estazolam	ProSom	HN	H	N
Midazolam		CH_3	F	CH
Triazolam	Halcion	CH_3	Cl	N

> **CHEMICAL NOTE**
>
> **Flumazenil:** Flumazenil is a drug with high BZR binding affinity, but does not change the conformation of the receptor. The drug therefore has antagonistic activity and is used to treat benzodiazepine overdoses. The drug is available in an injectable dosage form.
>
> Flumazenil
> (Generic)

synthesized and introduced into use (Fig. 7.14) differing in rate and extent of absorption, presence or absence of active metabolites, and degree of lipophilicity. These factors help to determine how a benzodiazepine is marketed and used.

MOA

- The benzodiazepines are BZR ligands and do not directly alter transmembrane chloride conduction to produce an anxiolytic effects.
- Benzodiazepines bind to the BZR at an allosteric site modulating the transmembrane conductance of chloride.
- Benzodiazepines increase the affinity and rate of GABA binding to $GABA_A$ receptors enhancing the action of chloride following GABA binding.
- Chloride ion hyperpolarizes the membrane potential reducing membrane firing.

SAR
Ring A

- Aromatic or heteroaromatic ring are believed to participate in π–π stacking with aromatic amino acid residues of the receptor (a heterocyclic rings generally show less pharmacological activity when compared to phenyl-substituted analogs).

- Electronegative group (e.g., halogen or nitro) substituted at the C7 markedly increases functional anxiolytic activity (Fig. 7.14).
- Substituents at C6, C8, or C9 generally decrease anxiolytic activity.

Ring B

- Bulky alkyl groups at the N1-position drastically reduce receptor affinity and in vivo activity.
- Methyl or linear N-alkyl side chains result in active drugs.
- Proton-accepting group at the C2 (i.e., O or S) is required for ligand binding to a histidine residue in the $GABA_A$ α_1 subunit.
- C3-methylene substitution with a hydroxy moiety results in comparable potency to non-hydroxylated analogs but with faster excretion.
- Neither the 4,5-double bond, nor the N4 nitrogen in ring B is required for in vivo anxiolytic activity.

Ring C

- The 5-phenyl ring C is not required for binding to the BZR in vitro, but an aromatic ring may contribute favorable hydrophobic or steric interactions to receptor binding.
- Substitution at the C4' of a 5-phenyl ring is unfavorable, but C2'-substituents are not detrimental to agonist activity.

Additional Ring

- 1,2-Annelation as in the Class B benzodiazepines (Fig. 7.14) with an additional "electron-rich" (i.e., proton acceptor) ring, such as s-triazole or imidazole, results in active benzodiazepine derivatives with high affinity for the BZR.

Stereochemistry

- 3-Methylated benzodiazepines results in an enantiomeric pair of compounds with the S-isomer possessing the anxiolytic activity.
- The S-isomer stabilizes conformation a of the B ring suggesting the importance of this conformation for binding to the BZR (NOTE: oxazepam is used as the racemic SR mixture).

Physicochemical and Pharmacokinetic Properties

- Most benzodiazepines have relatively high lipid:water partition coefficients (log P> 3) and are completely absorbed after oral administration with rapid distribution to the brain and other highly perfused organs.
- An exception to the high lipophilicity is clorazepate (Fig. 7.14), which is rapidly decarboxylated and then absorbed.
- Benzodiazepines and their metabolites bind to plasma proteins depending on lipophilicity (e.g., 70% for the more polar alprazolam to 99% for the lipophilic derivative diazepam).

Metabolism

- Hepatic microsomal oxidation results in N-dealkylation and aliphatic hydroxylation by CYP enzymes (primarily CYP3A4 and 2C19)(Fig. 7.15).
- N-Demethylation of diazepam and N-dealkylation of flurazepam are catalyzed by CYP3A4.
- Aliphatic hydroxylation occurs at C3 of N-desmethyldiazepam and on the C1 methyl and C4 methylene of the Class B benzodiazepines (R = CH_3) alprazolam, midazolam, and triazolam (Fig. 7.16).

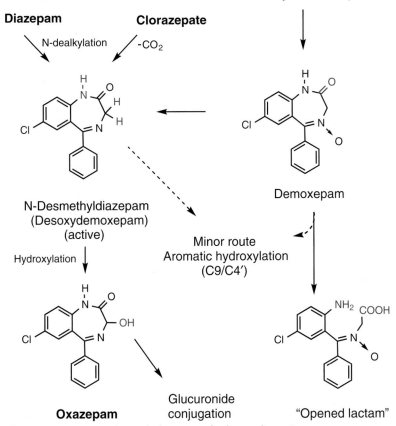

Figure 7.15 General metabolic routes for benzodiazepines.

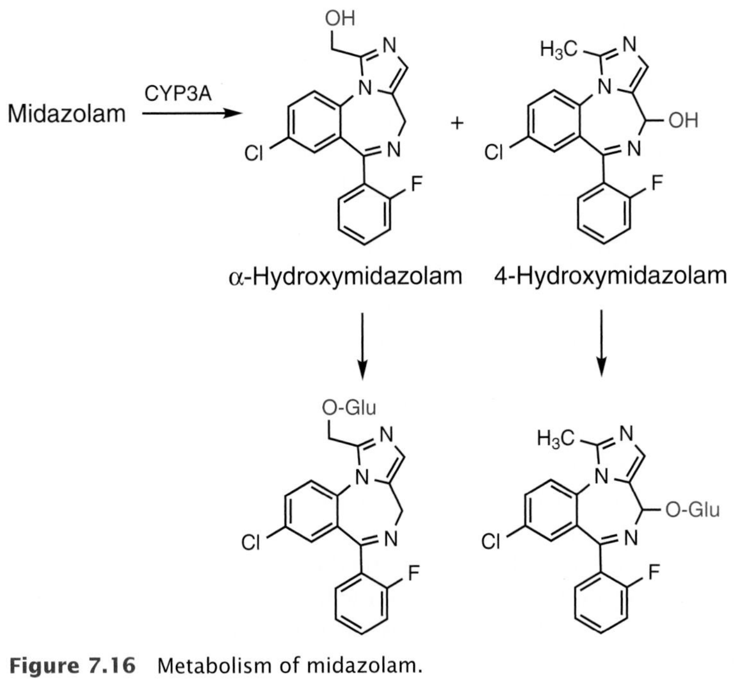

Figure 7.16 Metabolism of midazolam.

! **ADVERSE EFFECTS**

Benzodiazepines:
Drowsiness, dizziness, and loss of coordination, but may lead to dose-dependent alterations in behavior leading to bizarre, uninhibited, and confused behavior. Amnesia, tolerance, physical dependence, and abuse potential exists.

Drug Abuse:

Flunitrazepam (Rohypnol, "Roofies", "date rape drug") is a potent hypnotic with a rapid onset combined with amnesia. The drug is not approved for use in the U.S.

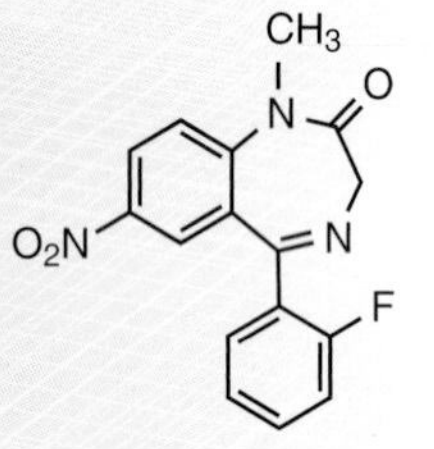

Flunitrazepam

- Conjugation by glucuronosyl transferases of the hydroxylated benzodiazepines yielding polar glucuronides that are inactive and excreted primarily in the urine.

Clinical Application

Many of the benzodiazepines are interchangeable when it comes to their clinical uses. The pharmacokinetic properties of the individual drugs may influence the choice of drug. Clinical applications include:

- Acute and chronic anxiety including panic disorders.
- Control agitation associated with alcohol withdrawal.
- Treatment of various types of seizures (anticonvulsant, see Chapter 10).
- Insomnia (See Chapter 8).
- Adjunct in the treatment of schizophrenia.
- Treatment of mania.

Non-Benzodiazepine Agonists at the Benzodiazepine Receptor

Few structural classes of non-benzodiazepine compounds have clinically relevant affinity for the BZR and show pharmacological activity in vivo. Notable examples of these classes which have led to marketed drugs include pyrazolopyrimidine (zaleplon), cyclopyrrolone (eszopiclone), and imidazopyridazine (zolpidem). These non-benzodiazepine BZR ligands show greater selectivity for $GABA_A$ receptors and are used to treat insomnia. These agents are discussed in Chapter 8 Sedative-Hypnotics.

Chemical Significance

Within this chapter the drug therapy for schizophrenia and anxiety disorders is addressed. The drugs for patients suffering from these conditions fit into a relatively small number of medicinal chemical classes: phenothiazines, butyrophenones, diarylazepine, benzisoxazole, benzisothiazole, and benzodiazepines. With a significant population of patients experiencing symptoms associated with psychoses or anxiety it is important for a pharmacist to have a working knowledge of the drug classes, their differences or similarities in mechanisms of action, physical chemical and pharmacokinetic properties, and how the body disposes of them.

For treatment of the more severe psychiatric conditions the pharmacist should recognize that all of the therapeutic agents act similarly at the molecular level through interference with dopamine transmission but that the phenothiazine and butyrophenone classes of drugs as typical antipsychotic agents are slowly being replaced with atypical agents (diarylazepine, benzisoxazole/benzisothiazole). Pharmacists serve at the interface between science and the patient and must be able to understand and explain how these newer atypical antipsychotic agents improve treatment for their patients. This is accomplished in part through an in depth understanding of the mechanisms of action. One can expect a refinement of the specific sites of action of the atypical agents and this will be translated into the selection of particular drugs for particular psychiatric conditions.

Treatment of anxiety with benzodiazepines involves a completely different neurotransmitter (GABA) and based upon pharmacokinetic properties which results in drugs with diverse onsets, durations of action, and therapeutic capabilities. Through an appreciation for chemical characteristics presented by specific functional groups within the benzodiazepines, one can explain different durations of action. An awareness and understanding of metabolic disposition is also an important characteristic explaining duration of action.

The ability of the pharmacist to call upon their medicinal chemistry background improves their ability to play an important role in the healthcare team. This becomes extremely important as new medications are introduced for the treatment of psychiatric disorders.

Review Questions

1 Which of the following accounts for the anxiolytic action of the benzodiazepines?
A. Benzodiazepines require GABA binding for activity.
B. Benzodiazepines increase potassium efflux.
C. Benzodiazepines act as GABA agonists at the GABA receptor.
D. Benzodiazepines promote depolarization.
E. Benzodiazepines act as partial agonists at the BZ receptor.

2 Given the general structure of a typical antipsychotic agent shown below and based upon the SAR of this class of drugs, which chemical would be predicted to exhibit potent antipsychotic activity?

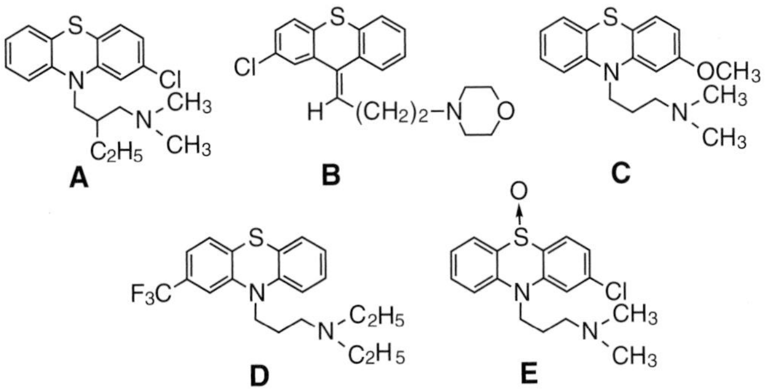

3 The major receptors involved in the action of atypical antipsychotic agents is/are?
A. Dopamine type 2 receptors/agonist
B. Dopamine type 2 receptors/antagonist
C. Combination of Dopamine type 2 and serotonin 2 antagonism
D. A balanced action at Dopamine type 2 (antagonism) and serotonin 2 agonist activity.
E. Blocking dopamine reuptake and increased release of serotonin.

4 Diarylazepines are extensively metabolized. Which of the metabolic products shown below is NOT normally found as a diarylazepine metabolite?

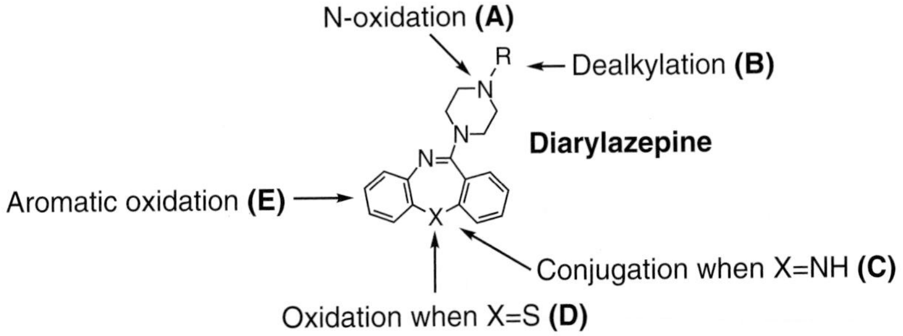

5 What is the significance of the metabolic reaction shown for risperidone?

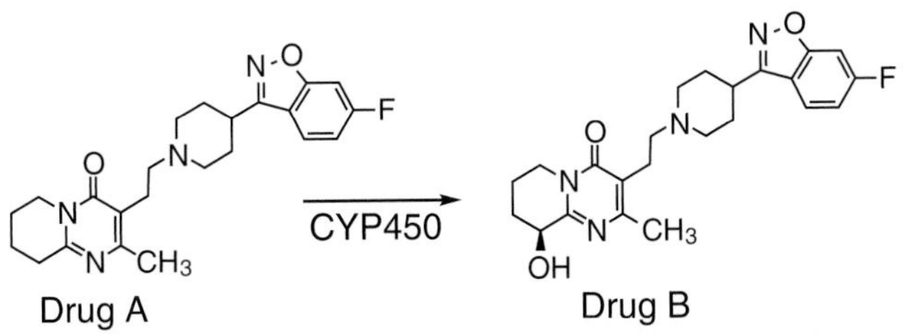

A. Accounts for the short duration of action of risperidone.
B. Accounts for the long duration of action of risperidone in extensive metabolizers in which both A and B are active drugs.
C. Accounts for the prodrug (inactive) activation of risperidone.
D. Accounts for the increase in lipophilicity that occurs upon metabolism.
E. Accounts for the inactivation of risperidone.

Sedative-Hypnotics

Learning Objectives

Upon completion of this chapter the student should be able to:

1. Explain the neurobiology of sleep.
2. Describe the mechanism of action of the different drug classes of sedative-hypnotics:
   a. Barbiturates
   b. Benzodiazepines
   c. Nonbenzodiazepine GABA$_A$ agonists
   d. Melatonin receptor agonist
3. Discuss the structure activity relationship (SAR) of the sedative-hypnotics.
4. Discuss with chemical structures the clinically significant metabolism of each drug class.
5. Identify any clinically relevant physicochemical and pharmacokinetic properties of the sedative-hypnotics.
6. Explain the clinical uses and clinically relevant adverse effects, drug–drug, and drug–food interactions of the sedative-hypnotics.

Introduction

Sedative and hypnotic drugs are used to treat sleep disorders which are generally associated with anxiety states that result in the inability to fall asleep. Such anxiety states can include times of stress and great emotional strain; chronic tension due to disease or sociologic factors, and hypertension. These drugs are capable of producing CNS depression ranging from slight sedation (awake with decreased excitability) to a state of sleep (hypnotic).

Neurobiology of Sleep–Wake Regulation

The sleep—wake rhythm is controlled by two opposing mechanisms: the homeostatic drive for sleep and the circadian promotion for wakefulness. The primary difference between the waking brain and the sleeping brain is the status of the cerebral cortex. When awake the cortex is activated and when asleep the cortex can be in either Rapid eye movement sleep (REM) or nonrapid eye movement sleep (NREM) (slow-wave sleep). In NREM sleep, the cortical blood flow is reduced and the electroencephalogram (EEG) exhibits a low activity as compared to the wake activity and the cortex is considered deactivated. In REM sleep, the EEG activity is equivalent to the wake state and one often experiences dreaming.

When awake, the cortex is activated by a complex of neurotransmitters arising from:

- The posterior hypothalamus: glutamate (Glu), histamine (Hst), and hypocretin (Hct).
- The basal forebrain: acetylcholine (ACh).
- The monoamine neuron system: norepinephrine (NE), serotonin (5HT), and dopamine (DA).

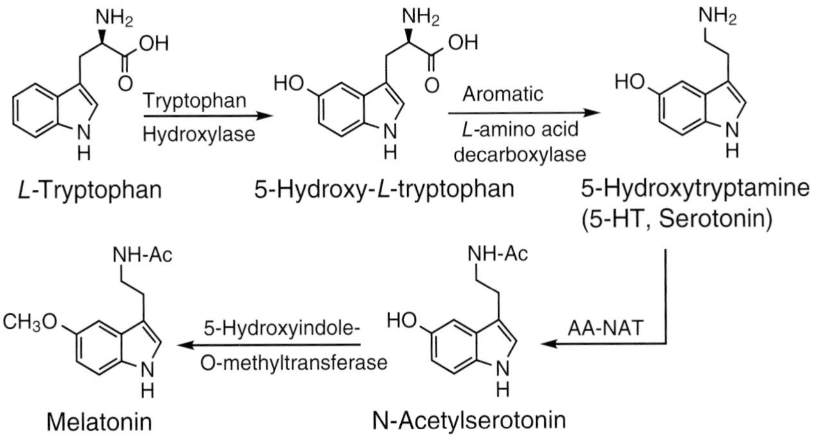

Figure 8.1 Biosynthesis of melatonin.

All of these areas receive arousal input from the ascending reticular activating system (ARAS), which itself is under the regulation of the excitatory neurotransmitter Glu and the inhibitory neurotransmitter γ-amino butyric acid (GABA).

The tuberomammillary nucleus (TMN) is found in the posterior hypothalamus and it also facilitates wakefulness and arousal by a circadian release of histamine. The suprachiasmatic nucleus (SCN) is found in the anterior hypothalamus and is involved in the induction of sleep. The SCN by receiving visual input from the retina synchronizes circadian rhythm to daylight. The SCN is under control of glutaminergic and GABAergic neurons but also works in concert with the pineal gland. Inhibitory outflow (GABAergic) from the SCN to the pineal gland is enhanced during daylight and decreased as daylight decreases. During daylight, the inhibitory GABAergic outflow inhibits the release of NE at the pineal gland. Conversely, during periods of darkness, the inhibitory GABAergic outflow from the SCN is reduced resulting in increased levels of NE at the level of the pineal gland. The NE acts on the pineal β_1 and α_{1B} adrenergic receptors to stimulate a significant synthesis and release of melatonin. Melatonin is another key regulator of circadian rhythms. In this way the SCN behaves as the brain's endogenous master clock.

Melatonin is synthesized from serotonin in the pineal gland (Fig. 8.1). Serotonin is formed from the hydroxylation of tryptophan by hydroxylation by tryptophan hydroxylase. The significant enzymes are arylalkylamine N-acetyltransferase (AA-NAT) and 5-hydroxyindole–O-methyltransferase (HIOMT). AA-NAT is activated by β_1-adrenergic production of cyclic AMP and protein kinase A phosphorylation in the pineal gland. As previously described, NE is increased during periods of darkness (hence enhanced β_1-adrenergic activity) and therefore the production of melatonin is linked to the day–night cycle.

Pharmacologic Targets for Sedative-Hypnotic Agents

There is a large number of neurotransmitter systems involved in the activation of the cerebral cortex each of which are potential targets for the development of sedative-hypnotic drugs. However, development in this area has focused on $GABA_A$ receptor agonists and drugs that modulate hypothalamic histamine or melatonin circadian systems. This chapter will therefore discuss the following drug types:

- $GABA_A$ receptor agonists.
  - Barbiturates.
  - Benzodiazepines.
  - Nonbenzodiazepine $GABA_A$ agonists.
  - Melatonin receptor agonists.
  - H_1-receptor antagonist.

Potential Pharmacologic Targets for the Development of Sedative-Hypnotic Drugs

- Adenosinergic system
- Serotonergic system
- Dopaminergic system
- Adrenergic system
- Histaminergic system
- Cholinergic system

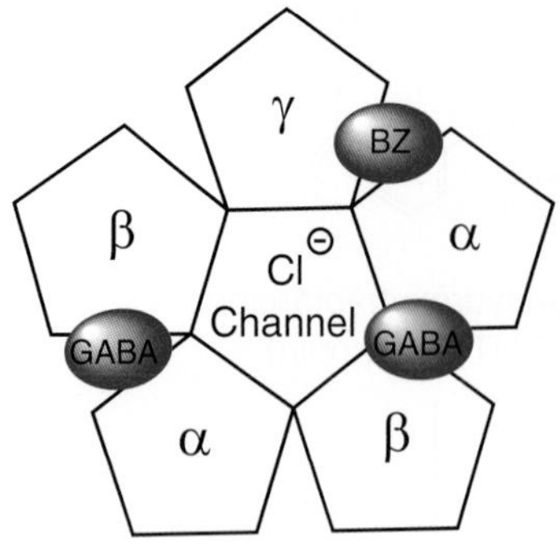

Figure 8.2 The GABA$_A$ receptor showing the pentameric subunits α, β, and γ and the distinct GABA and benzodiazepine (BZ) binding sites.

GABA$_A$ Receptor Agonists

- Drugs in this class of sedative-hypnotics include the barbiturates, the benzodiazepines, and the nonbenzodiazepine GABA$_A$ agonists.
- GABA is the major inhibitory neurotransmitter in the CNS system and is essential for balancing neuronal excitation.
- GABA$_A$ is a ligand-gated ion channel that modulates conductance of chloride ions through the cell membrane upon binding GABA.
- Activation of GABA$_A$ receptors leads to membrane hyperpolarization which increases the firing threshold potential reducing the likelihood of generating an action potential (i.e., neuronal inhibition) and leading to CNS depression.
- Therefore GABA$_A$ agonists are important drugs not only for their sedative-hypnotic effects but also for the treatment of anxiety, convulsions, and seizures, as well as for their anesthetic effects (see Chapter 12, Chapter 14, Chapter 15, and Chapter 17 in *Foye's Principles of Medicinal Chemistry, Seventh Edition*, for a complete detailed discussion).
- The GABA$_A$ receptor is a pentameric assembly of α, β, γ subunits containing separate binding sites for GABA and other receptor agonists (Fig. 8.2).

Barbiturates

The barbiturates were the agents of choice as sedative-hypnotics until they were replaced with other agents that have less tolerance, dependence, abuse potential, and lower toxicity thresholds. They have also found use as anesthetics and anticonvulsants. Currently there are five barbiturates approved by the FDA for use as sedatives-hypnotics: amobarbital, butabarbital, pentobarbital, phenobarbital, and secobarbital.

MOA

- The barbiturates bind to the GABA$_A$ receptor and enhance the binding of GABA.
- The binding site is unknown, but it is not the benzodiazepine site and it does not require specific subunits.
- Increased GABA binding prolongs the time the receptor is open which allows for increased chloride ion transport across the membrane resulting in the reversible inhibition of all excitatory neurons.

SAR

- Barbituric acid is the pharmacophore from which all barbiturates are derived (Fig. 8.3).
- Both nitrogen atoms and the hydrogen atoms on position C5 are acidic making barbituric acid inactive as a sedative-hypnotic.
- Both hydrogen atoms on C5 must be replaced for maximal activity.
- Increasing the chain length of alkyl groups at C5 increases activity up to 6 carbons; after that activity decreases because they become too lipophilic and are therefore too insoluble to be poorly absorbed.
- Branched, cyclic, or unsaturated groups at C5 produce barbiturates with a shorter onset of action and decreased duration of action.
- The duration of action correlates to the lipophilicity (log P) contributed by a combination of alkyl and aryl substituents at C5:

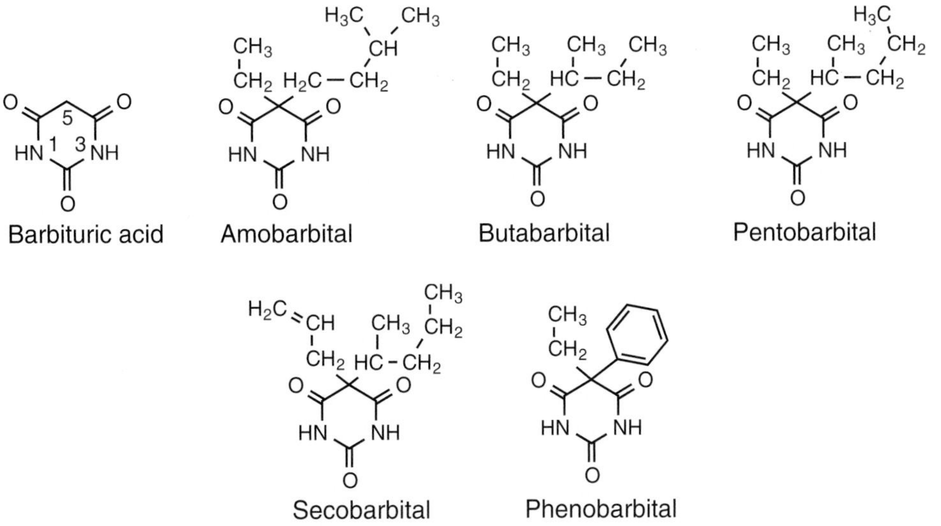

Figure 8.3 Barbituric Acid Pharmacophore and the barbiturates approved for sedative-hypnotic use.

- ethyl, pentyl, or propenyl – short duration of action – log P = 2.10 to 2.36 (e.g., secobarbital).
- ethyl, butyl, or isopentyl – intermediate duration of action – log P = 1.60 to 2.07 (e.g., butabarbital).
- ethyl, phenyl – long duration of action – log P = 1.46 (e.g., phenobarbital).
- C5 is a chiral carbon and as such the L-isomers are more active than the D-isomers indicating stereoselective binding to the $GABA_A$ receptor.
- Alkyl substituents at N1 or N3 lead to increased lipophilicity due to the decrease in the nitrogen acidity which results in the drugs being more union-ized at pH 7.
- Replacement of oxygen with sulfur results in the thiobarbiturates which are ultra-short acting. These derivatives are not used as sedative-hypnotics but rather as anesthetics (see Chapter 16 in *Foye's Principles of Medicinal Chemistry, Seventh Edition*, for a complete detailed discussion).

Physicochemical and Pharmacokinetic Properties

- The barbiturates form salts with bases at N1 or N3.
- The salts are rapidly and completely absorbed orally whereas the free acids exhibit a much slower rate of absorption because of being very insoluble in gastric juices.
- Table 8.1 gives the pharmacokinetic properties of the barbiturates approved for sedative-hypnotic use.
- They are classified according to their duration of action:
  - Short-acting: 3 to 4 hours.
  - Intermediate-acting: 6 to 8 hours.
  - Long-acting: 10 to 16 hours.
- The duration of action is the direct result of the redistribution of the drugs from the brain into other body sites, primarily adipose tissue. However, metabolism may also contribute to the loss of activity.

Table 8.1 Pharmacokinetic Parameters of Sedative-Hypnotic Barbiturates

Barbiturate	Log P	Onset Time (min)	Duration of Action (h)	Classification
Pentobarbital	2.10	10–15	3–4	Short-acting
Secobarbital	2.36	10–15	3–4	Short-acting
Amobarbital	2.07	45–60	6–8	Intermediate-acting
Butabarbital	1.60	45–60	6–8	Intermediate-acting
Phenobarbital	1.46	30–60	10–16	Long-acting

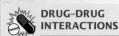

DRUG–DRUG INTERACTIONS

Barbiturates: Phenobarbital might require dose adjustments of the following when co-administered with: aripiprazole, phenytoin, furosemide, levothyroxine, acetaminophen, and ranitidine.

⚠ **ADVERSE EFFECTS**

Barbiturates: CNS: drowsiness, confusion, impaired judgment and fine motor skills, tolerance, and dependence. Respiratory and Cardiovascular: respiratory depression, bradycardia, and orthostatic hypotension. Other: nausea, diarrhea, headache, hepatotoxicity, and megaloblastic anemia (chronic use).

Definitions

Agonist: A ligand that binds to a receptor and causes a pharmacological response.

Antagonist: A ligand that binds to the receptor and blocks the agonist from acting.

Inverse Agonist: A ligand that binds to a receptor and causes an action opposite that of the agonist.

Metabolism

- The barbiturates are metabolized in the liver leading to loss of activity (Fig. 8.4).
- Loss of activity is due to a decrease in lipophilic character and hence inability of metabolites to enter the brain.
- Barbiturate metabolic pathway includes oxidation of the C5 substituents to alcohols and subsequent oxidation to ketones or carboxylic acids.
- The oxidation metabolites are excreted as O-glucuronide or O-sulfate conjugates.
- It should be noted that phenobarbital is a potent inducer of CYP450 isozymes (2C and 3A4) and as such it can influence the pharmacokinetics of co-administered drugs metabolized by those isoforms.

Clinical Application

- Barbiturates were used as anxiolytics and sedative-hypnotics but have fallen into disuse for these indications as a result of the availability of the safer benzodiazepines.
- Long-acting barbiturates are still used as anti-epileptics (see Chapter 17 in *Foye's Principles of Medicinal Chemistry, Seventh Edition,* for a complete detailed discussion).
- Ultra-short-acting barbiturates are used for the induction of general anesthesia and as general anesthetics for short surgical procedures.

Benzodiazepines

The benzodiazepines were introduced into clinical use in 1960. Since this time they have been the drug of choice for the treatment of sleep disorders as well as other anxiety-related diseases (see Chapter 14 in *Foye's Principles of Medicinal Chemistry, Seventh Edition,* for a complete detailed discussion). They are relatively safe and effective and have virtually replaced the barbiturates as sedative-hypnotic agents. There are currently five benzodiazepines approved by the FDA for use as sedative-hypnotics: estazolam, flurazepam, quazepam, temazepam, and triazolam.

MOA

- The benzodiazepines modulate the function of the $GABA_A$ receptors leading to neuronal hyperpolarization and CNS depressant effects.
- The benzodiazepine binding site is at the α and γ subunits of the $GABA_A$ receptor.
- Binding of benzodiazepines causes an increased affinity and rate of GABA binding to the $GABA_A$ receptor.
- Increasing the affinity and rate of GABA binding correlates to an increase in the frequency of the open-channel state allowing for an increased opportunity for Cl^- transport.
- The benzodiazepines increase the frequency of the open-channel state whereas the barbiturates increase the duration of the open-channel state.
- Benzodiazepines have no effect in the absence of GABA whereas the barbiturates do.

SAR

- The basic pharmacophore for binding to the $GABA_A$ receptor is 5-phenyl-1,2-benzodiazepin-2-one (Fig. 8.5).
- Minor modification to the pharmacophore can shift intrinsic activity from agonist to antagonist to inverse agonist.

Ring A

- Must be aromatic for optimal activity. Necessary for π–π stacking with aromatic amino acids at the binding site.
- An electronegative substituent (halogen, nitro) at C7 enhances sedative-hypnotic effect.

Ring B

- Requires a proton-accepting group at C2 (a carbonyl or sulfonyl) for interaction with proton-donating amino acids in the receptor (usually histidine).

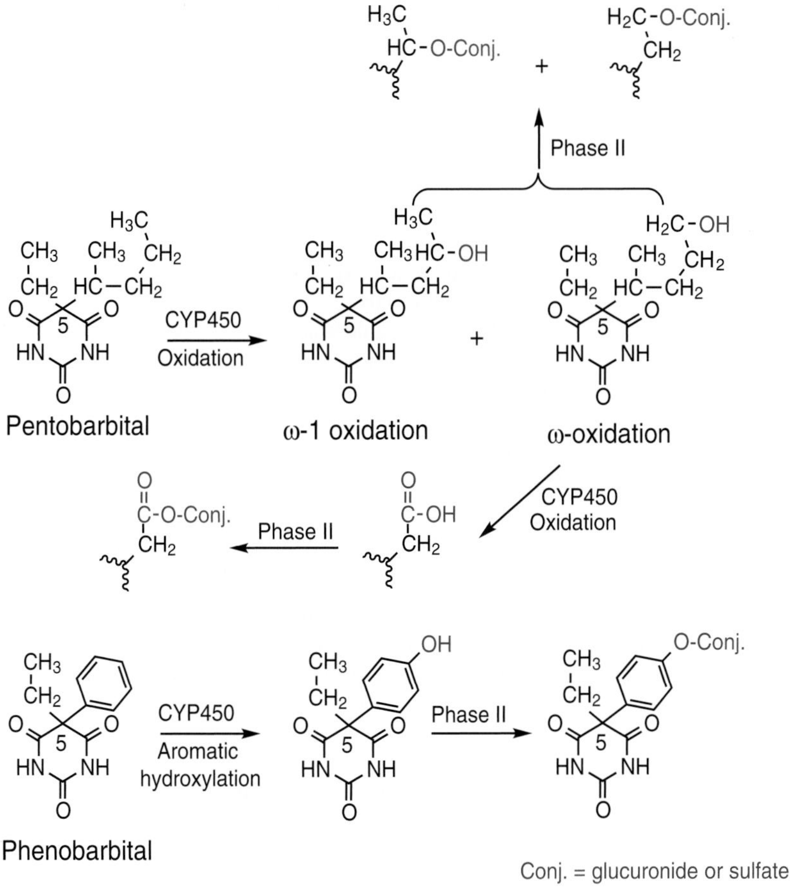

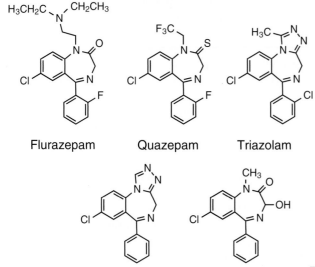

Figure 8.4 Metabolism of pentobarbital and phenobarbital showing the major routes of metabolism.

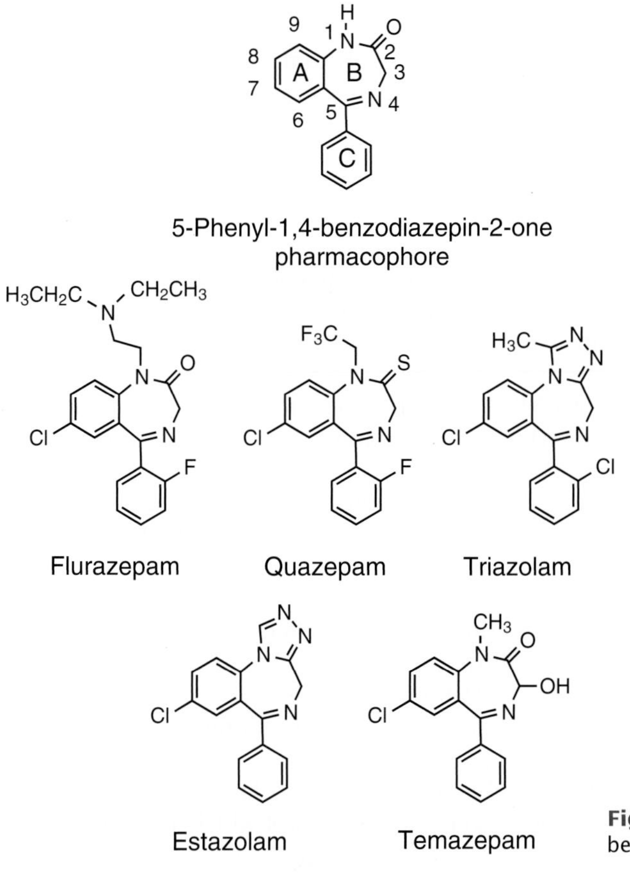

Conj. = glucuronide or sulfate

5-Phenyl-1,4-benzodiazepin-2-one pharmacophore

Flurazepam

Quazepam

Triazolam

Estazolam

Temazepam

Figure 8.5 The benzodiazepine pharmacophore and the benzodiazepines approved for sedative-hypnotic use.

Table 8.2   Pharmacokinetic Parameters of Benzodiazepines Approved as Sedative-Hypnotics

Benzodiazepine	Log P	Time to Peak Conc. (h)	Parent Elimination Half-Life (h)	Major Metabolite ($t_{1/2}$, h)	CYP450 Isoform
Flurazepam	2.35	0.5–1.0	~2	N_1-desalkyl (47–100, active) N_1-hydroxyethyl (2–4, active)	3A4
Quazepam	4.03	~2.0	39	2-Oxo (40, active) N_1-desalkyl (73, active)	3A4/2C9
Estazolam	3.51	0.5–6.0	10–24	4'- hydroxy; 4-hydroxy; and 1-Oxo (all inactive)	3A4
Triazolam	2.42	<2.0	1.5–5.5	α-hydroxy (50%–100% active) 4-hydroxy (inactive)	3A4
Temazepam	2.39	1.2–1.6	0.4–0.6	O-glucuronide	—

- Optimal activity occurs when the carbonyl at C2 is co-planar with the aromatic Ring A.
- When the substituent at C3 is hydroxyl, there is equivalent activity; however, it will be excreted faster.
- N-1 amide can be unsubstituted or substituted with alkyl, alkylamino or fused into an electron rich triazole ring system which contributes to high affinity for the receptor. The triazole derivatives prevent oxidative metabolism that prolongs the duration of action.

Ring C

- Not required for binding, however, contributes to hydrophobic and/or steric interactions at the receptor.
- Can be unsubstituted or substituted with halogen in the ortho (2') position.
- Substituents at the para (4') position are inactive.

Physicochemical and Pharmacokinetic Properties

- Table 8.2 summarizes the pharmacokinetic characteristics of the benzodiazepines approved as sedative-hypnotics.
- All are readily absorbed orally and distribute easily to the brain.
- The parent elimination half-life closely corresponds to its log P.
- Duration of action can be prolonged by the formation of active metabolites. The triazolo derivatives inhibit the formation of active metabolites, and therefore have shorter durations of action.

Metabolism

- The duration of action and side-effect potential is influenced by their metabolism.
- All benzodiazepines used as sedative-hypnotics are primarily metabolized by CYP3A4 oxidations which are usually followed by glucuronidation and subsequent elimination.
- Flurazepam is metabolized by CYP3A4 to two active metabolites (N1-desalkyl and N1-hydroxyethylflurazepam) with elimination $t_{1/2}$ lives of 47 to 100 and 2 to 4 hours, respectively (Fig 8.6).
- Quazepam is metabolized by CYP3A4 to 2-oxoquazepam and N1-desalkyl-2-oxoquazepam, both of which are active with elimination $t_{1/2}$ of 40 to 73 hours (Fig. 8.6).
- The major active metabolites of both flurazepam and quazepam with long elimination half-lives lead to excessive residual hypnotic effects including daytime drowsiness, over-sedation, and cognitive confusion.
- Incorporation of a 1,4-triazole ring into the benzodiazepine structure prevents oxidative metabolism that leads to active metabolites resulting in metabolites with a shorter duration of action.
- Estazolam is metabolized by CYP3A4 to 4'-hydroxyestazolam (major), 2-oxoestazolam (minor), and 4-hydroxyestazolam (minor) all of which are inactive (Fig. 8.7).

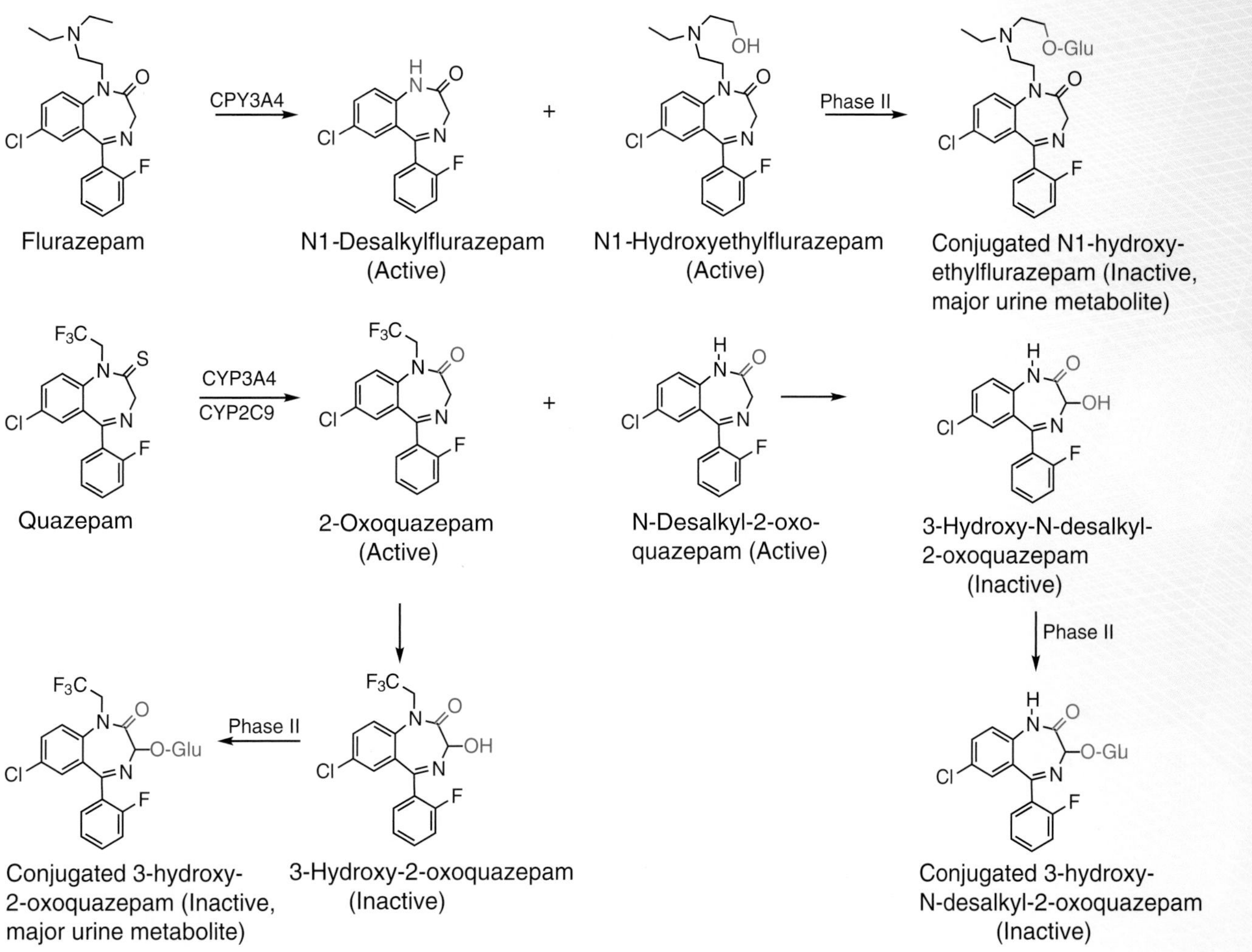

Figure 8.6 Flurazepam and quazepam metabolism showing formation of their long-acting active metabolites.

- Triazolam has a relatively short duration of action due to its metabolism by CYP3A4 to α-hydroxytriazolam and 4-hydroxytriazolam both of which are eliminated as glucuronides (Fig. 8.8).
- The α-hydroxytriazolam retains 50% to 100% of the activity of triazolam, but since it occurs in the plasma as the glucuronide it does not contribute very much to their clinical effects.
- Temazepam already has a 3-hydroxy substituent which quickly gets glucuronidated and excreted giving temazepam a short elimination $t_{1/2}$ of about 2 hours (Fig. 8.8).

Clinical Application

- The primary clinical use of the benzodiazepines as sedative/hypnotics is to achieve restorative sleep with ease of falling asleep.
- Restorative sleep will decrease the likelihood of secondary complications to sleep loss:
  - Induction of depression, anxiety, substance abuse, and other psychiatric disorders.
  - Decreased ability to concentrate, memory loss.
  - Decreased productivity.
  - Increased irritability.
- Use of benzodiazepines in the elderly requires drugs without active metabolites (e.g., temazepam) or those metabolized by phase II only.
- Use should be limited to 2 to 3 nights in a row followed by one or more treatment-free nights. This will avoid the development of dependence and tolerance.

ADVERSE EFFECTS

Benzodiazepines: May cause daytime drowsiness, mental cloudiness, anterograde amnesia, and rebound insomnia (triazolam). Use with caution with other CNS depressants; when patients must perform complex activities requiring mental alertness; in breastfeeding since they enter breast milk.

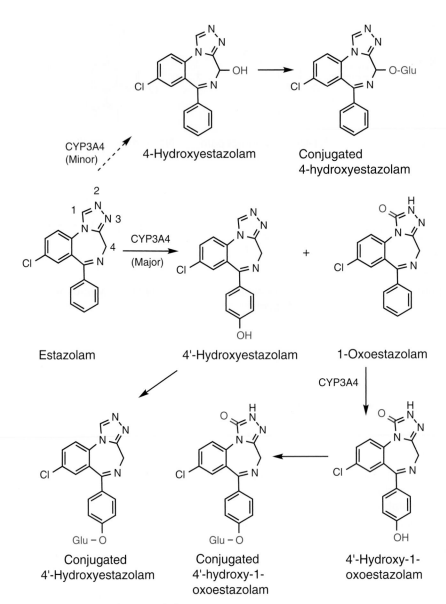

Figure 8.7 Estazolam metabolism.

Nonbenzodiazepine GABA~A~ Agonists

The discovery that there are several GABA$_A$ receptor subtypes (α_1, α_2, α_3, and α_5) and that the α_1-subtype is related most closely to benzodiazepine-induced sedation and hypnosis has led to the search for subtype-selective chemicals that would yield appropriate therapeutic outcomes. Currently there are three structurally distinct α_1-subtype selective nonbenzodiazepines: zolpidem, eszopiclone, and zaleplon.

Zolpidem, Eszopiclone, and Zaleplon

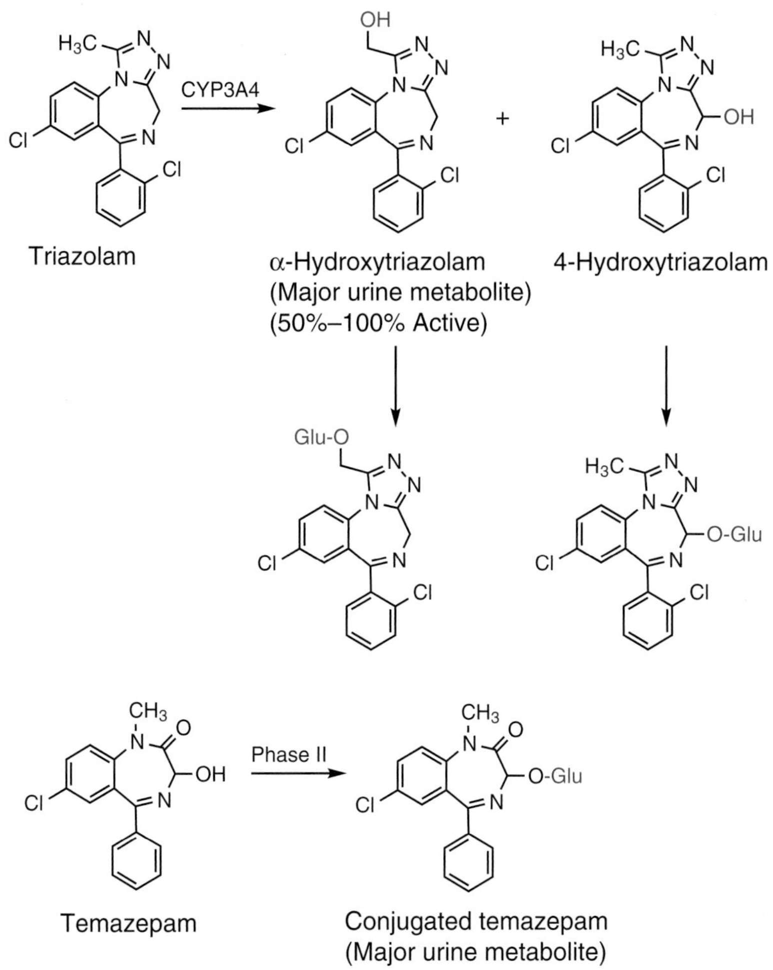

Figure 8.8 Metabolism of Triazolam and temazepam.

Zolpidem (Table 8.3)

MOA

- Highly selective for the α_1-GABA$_A$ subtype giving zolpidem weaker anxiolytic, anti-convulsant, and muscle relaxant effects compared to classical benzodiazepines.
- It has a five- and ten-fold greater affinity for the α_1 versus α_2 and α_3 subtypes, respectively, with no appreciable affinity for the α_4 and α_5 subtypes.
- Hypnotic effects are similar to temazepam and triazolam with decreased pharmacologic side effects.

SAR

- Zolpidem is an imidazopyridine with selectivity for the α_1 subtype.

Table 8.3 Pharmacokinetic Parameters of Nonbenzodiazepines Approved for Sedative-Hypnotic Use

Drugs	Log P	Time to Peak Conc. (h)	Parent Elimination Half-Life (mean h)	Major Metabolites ($t_{1/2}$, h)	Predominant CYP Isoform(s)
Zolpidem	2.31	Immediate-release $1.6^a/1.6^b$ Control-release 1.5^c Oral spray 0.25	Immediate-release $2.5^a/2.6^b$ Control-release 2.6^c	None active	3A4 (major) 2C9, 1A2, 2D6 (minor)
Eszopiclone	−0.34	1	6.5	N-oxide (inactive) N-desmethyl (active, lower affinity)	1A2, 3A4
Zaleplon	1.23	1	1	None active	3A4 (minor)

[a]5 mg tablet;
[b]10 mg tablet;
[c]12.5 mg tablet

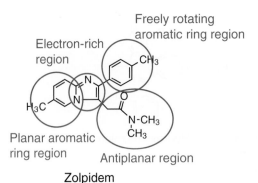

Figure 8.9 Zolpidem structure activity regions.

- Structure–activity relationship studies revealed three activity regions necessary for binding to the GABA$_A$ receptor (Fig. 8.9).

An Electron-Rich Planar Aromatic Ring Region
- Substitution with electronegative groups (e.g., chloro) decreases selective affinity for the α_1 subtype.
- Imidazole ring is necessary for selectivity since conversion of either of the imidazole nitrogens to hydrogen bond donors leads to loss of α_1 subtype selectivity.

Freely Rotating Aromatic Ring
- Substitution with electronegative groups (e.g., chloro) decreases selective affinity for the α_1 subtype.

Antiplanar Region
- The carbonyl group in this region can hydrogen bond with key residues in loop C of the α_1 subunit (Ser^{204} and Thr^{206}/Gly^{207}) and loop F of the γ_2 subunit (Arg^{194}) that forms the key binding site complex ($\alpha_1\gamma_2$).
- Bulky N-alkyl substituents decrease binding to the α_1 subtype.

Physicochemical and Pharmacokinetic Properties
- Time to peak concentration and bioavailability are related to the three available formulations:
  - Immediate release: 1.6 hours to peak concentration; 70% bioavailable.
  - Controlled release: 1.6 hours to peak concentration with sustained plasma levels over 8 hours; 70% bioavailable.
  - Oral spray: 15 minutes to peak concentration; 70% bioavailable.
- Elimination $t_{1/2}$ is 2.6 hours for all three formulations.
- Food interferes with absorption lowering peak concentrations by 30%.

Metabolism
- Zolpidem undergoes metabolic oxidations by CYP3A4 (61%), CYP1A2 (14%), and CYP2D6 (<3%) (Fig. 8.10).
- Major excretory metabolites result from hydroxylation of the p-tolyl methyl group (M III) followed by alcohol dehydrogenase oxidation to the corresponding carboxylic acid (M I).
- Other oxidative metabolites:
  - Hydroxylation of the p-methyl group on the imidazopyridine ring (M IV) which can be further oxidized to the carboxylic acid (M II).
  - Hydroxylation of the imidazopyridine ring (M X).
  - Hydroxylation of the amide nitrogen methyl group (M XI).

Eszopiclone (Table 8.3)

MOA
- Eszopiclone is not subtype selective; however, it binds with higher affinity to the α_1 subtype with appreciable affinity for the α_3 receptor subtype.

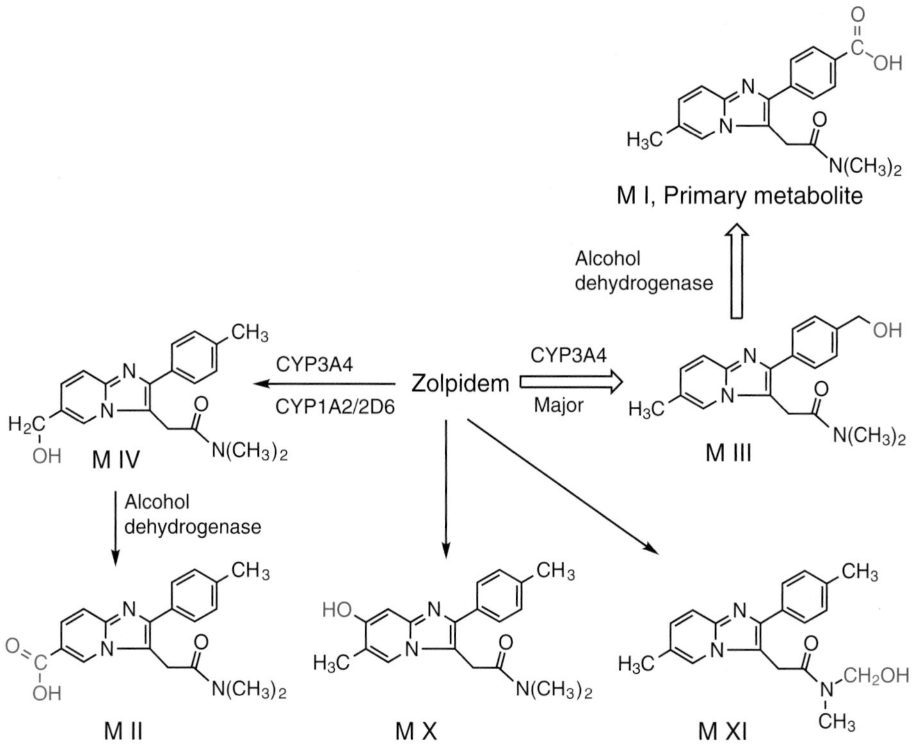

Figure 8.10 Zolpidem metabolism.

- Its affinity for the α_1 subtype is lower than that for zolpidem and zaleplon.

SAR

- Eszopiclone is a pyrrolopyrazine cyclopyrrolone which is active as the (*S*)-enantiomer. The (*R*)-enantiomer has 50-fold less binding affinity for the GABA$_A$ receptors.
- This subclass of nonbenzodiazepines binds to a distinct allosteric site within the benzodiazepine binding site (Arg[144], Tyr[209], Tyr[159]) and its binding is not affected by GABA or barbiturates.
- Molecular model binding studies revealed that eszopiclone's binding to the GABA$_A$ receptor is different than that of zolpidem which may account for the lack of its α_1 subtype selectivity.

Physicochemical and Pharmacokinetic Properties

- Rapidly absorbed orally with peak plasma levels reached within 1 hour.
- Elimination $t_{1/2}$ is 6.5 hours which is the longest of this class of sedative/hypnotics.
- Food interferes with absorption lowering time to peak by 1 hour as well as peak concentration by 20%.

Metabolism

- Eszopiclone is metabolized by both CYP3A4 and CYP1A2 (Fig. 8.11).
- CYP3A4 N-dealkylates the piperazine ring to yield (*S*)-N-desmethylzopiclone which exhibits weak sedative activity.
- CYP1A2 oxidizes the nitrogen of the pyrrolopyrazine ring to form (*S*)-N-oxidezopiclone which is inactive.
- Less than 7% of eszopiclone is excreted unchanged.

Zaleplon (Table 8.3)

MOA

- Selective for the α_1 subtype similar to zolpidem but with one-third to one-half the potency.
- Also able to potentiate the GABA effects at α_5 subtype, however, with more than 15-fold less potentiation than at the α_1 subtype.

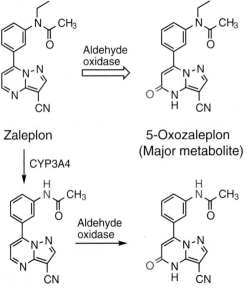

Figure 8.11 Eszopiclone metabolism.

(S)-N-Desmethylzopiclone
(low affinity)

(S)-N-Oxidezopiclone
(inactive)

SAR

- Zaleplon is a pyrazolopyrimidine with much the same pharmacologic profile as zolpidem.
- Very little SAR studies have been done; however, there is some evidence that the cyano group contributes significantly to the α_1 subtype selectivity.

Physicochemical and Pharmacokinetic Properties

- Rapid onset (<1 hour) and relatively faster rate of elimination (1 hour).
- Completely absorbed orally but undergoes first-pass metabolism resulting in an effective bioavailability of 30%.
- Food delays time to peak plasma concentrations (3 hours) and decreases peak plasma concentrations by 35%.

Metabolism

- Zaleplon undergoes significant first-pass metabolism with less than 0.1% recovered unchanged in the urine.
- Primary metabolite is formed by the action of aldehyde oxidase to yield 5-oxozaleplon which is inactive (Fig. 8.12).

Zaleplon

Aldehyde oxidase

5-Oxozaleplon
(Major metabolite)

CYP3A4

Aldehyde oxidase

Figure 8.12 Zaleplon metabolism. N-Desethylzaleplon

5-Oxozaleplon

- 5-oxozaleplon can be eliminated as such or further oxidized and excreted as a glucuronide conjugate.
- A minor metabolite is formed by CYP3A4 N-dealkylation to yield N-desethylzaleplon which is also inactive and may be oxidized further by aldehyde oxidase to 5-oxo-desethylzaleplon.

Clinical Applications for Nonbenzodiazepine GABA_A Agonists

- The nonbenzodiazepine GABA_A agonists have the same clinical application as the benzodiazepines; however, they are believed to lead to fewer adverse effects because they have decreased anticonvulsant, muscle relaxant, and anxiolytic activity.
- They are said to decrease the time it takes to fall asleep and to increase total sleep time.
- Zaleplon improves the time it takes to fall asleep but does not increase total sleep time.
- All require a dose adjustment in elderly patients and patients with impaired liver function.
- Required by the FDA to be dispensed with a medication guide warning patients that they may do things during the night that they do not remember doing (amnesic sleep-related behaviors) such as sleep walking, sleep driving, sleep eating, sleep cooking, sleep talking, and sleep sex.

> **⚠ ADVERSE EFFECTS**
>
> **Nonbenzodiazepines:** CYP3A4 inhibitors and inducers may decrease or increase activity respectively. Valerian and St. John's wort and kava kava may increase CNS depression. Use with caution with other CNS depressants. Not recommended for women who are breastfeeding.
> Required by the FDA to be dispensed with a medication guide that warns patients that they may do things during the night that they do not remember doing (amnesic sleep-related behaviors) such as sleep walking, sleep driving, sleep eating, sleep cooking, sleep talking, and sleep sex.

Melatonin Receptor Agonists

Melatonin is a neurohormone that is synthesized in the pineal gland from serotonin. Its synthesis is concurrent with sleep and circadian rhythms (Fig. 8.13). Melatonin is a poor chemotherapeutic agent due to its hydrophilicity, poor absorption, and ubiquitous effects on sleep and circadian rhythms. This has prompted efforts to develop melatonin receptor agonists. To date, only ramelteon is approved as a sedative/hypnotic and tasimelteon is approved for non-24 hour sleep–wake disorder (usually occurring with blind individuals but can also occur with sighted individuals).

MOA

- Ramelteon, similar to melatonin, binds to the melatonin receptors MT_1 (MT_1R) and MT_2 (MT_2R) found in the SCN and pineal gland. It does not bind to the MT_3 receptor which is not found in the CNS.
- Binding as an agonist to the MT_1R inhibits SCN neuron firing promoting sleep.
- Binding as an agonist to the MT_2R affects circadian rhythm settings related to the CNS clock.
- Ramelteon binds to both MT_1R and MT_2R with an 8- to 10-fold higher affinity for MT_1R and a 6-fold greater affinity for MT_1R than melatonin.

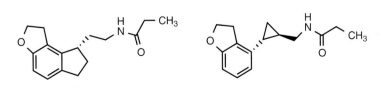

(S)-Ramelteon   Tasimelteon

Figure 8.13 Structures of melatonin and the melatonin receptor agonists (S)-ramelteon and tasimelteon.

SAR

- Requires either an indole (melatonin) or indane (ramelteon) aromatic ring system for π–π stacking within the receptor.
- The aromatic ring system is also necessary for optimal distance between the 3-amide and the 5-methoxy side chain responsible for binding and functional activity.
- The 3-amide side chain binds to Ser[110] and Ser[114] in transmembrane helix III (TMH III) of MT$_1$R and Asn[175] in TMH IV in MT$_2$R.
- The 5-methoxy oxygen interacts with His[195] and His[208] in TMH VI of both MT$_1$R and MT$_2$R.
- In addition, Val[192], located ~ one turn above His-195 interacts with the methyl component of the methoxy group.
- Binding is stereoselective with the (S)-enantiomer having a 500-fold greater binding affinity than the (R)-enantiomer.

Physicochemical and Pharmacokinetic Properties

- Ramelteon is rapidly absorbed orally with peak plasma concentrations reached in 45 minutes.
- Ramelteon undergoes significant first-pass metabolism; therefore, the absolute oral bioavailability is only ~2%.
- Ramelteon has a short elimination $t_{1/2}$ from 1 to 2.5 hours.
- Food interferes with absorption delaying time to peak plasma concentration by 45 minutes and decreasing plasma concentrations by 22%.

Metabolism

- Ramelteon undergoes oxidative metabolism by CYP1A2 (49%), CYP2C19 (42%), and CYP3A4 (8.6%) (Fig. 8.14).
- Major metabolite is the hydroxylated propionamide (M II) which is active with an elimination $t_{1/2}$ of 2 to 5 hours and may contribute to the overall sedative/hypnotic effect of ramelteon.

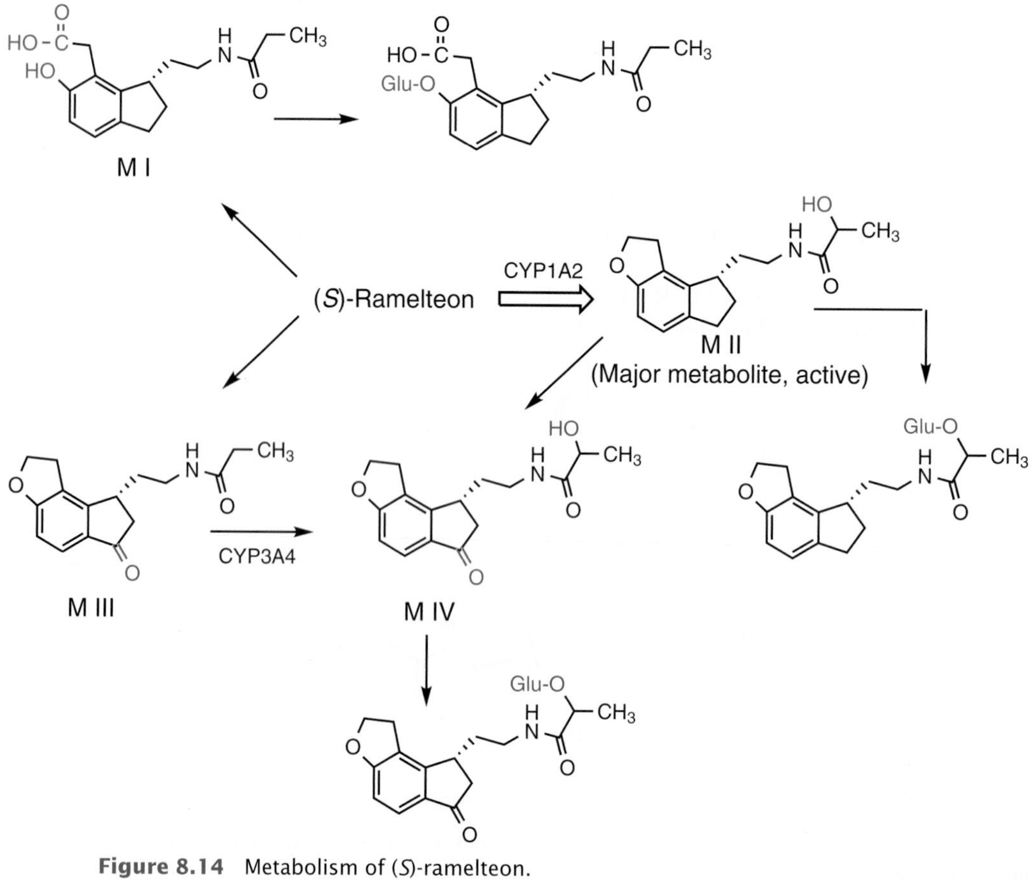

Figure 8.14 Metabolism of (S)-ramelteon.

- Also oxidized to ring-opened M I metabolite and the keto-metabolite M III, both of which are inactive.
- Biotransformation of M II and M III occurs via sequential oxidation to M IV.
- All three hydroxylated metabolites are eliminated as O-glucuronide conjugates in the urine.

Clinical Application

- The melatonin receptor agonists are primarily indicated for the treatment of patients who have difficulty falling asleep.
- May be useful in increasing duration of sleep.
- Should be used with caution in patients with severe hepatic impairment.
- Tasimelteon is currently approved for non-24 hour sleep–wake disorder.

Histamine H$_1$ Receptor Antagonists

Diphenhydramine and Doxylamine

- The first-generation ethanolamine ether H$_1$-receptor antagonists (antihistamines) cross the blood–brain barrier and have sedative side effects.
- Examples of these antihistamines are diphenhydramine and doxylamine.
- Their MOA is that they block the release of histamine in the TMN of the posterior hypothalamus where its release causes cerebral cortex arousal.
- These agents are marginally effective and may actually delay the onset and duration of REM sleep.
- OTC availability allow for self-medication for insomnia.
- They, however, have a number of side effects:
  - They embody the pharmacophore for anticholinergic action and therefore cause dry mouth, urinary retention, and blurred vision.
  - Patients can experience next-day drowsiness and impaired cognition.
  - Tolerance develops on long-term use.
- See Chapter 32 in *Foye's Principles of Medicinal Chemistry, Seventh Edition*, for a complete detailed discussion of these agents.

Orexin-A and Orexin-B Antagonist

- Orexin-A protein is composed of 33 amino acids with two intrachain disulfide links while Orexin-B protein is composed of 28 amino acids. Both proteins are derived from prepro-orexin which is produced in the hypothalamus.
- The orexin receptors (OX$_1$R and OX$_2$R, also known as HCRT$_1$ and HCRT$_2$) are G protein-coupled receptors and found primarily in the neurons of the brain.
- Orexin-A binds with high affinity to OX$_1$R while orexin-B binds with low affinity.
- Both orexin-A and -B bind with equal affinity to OX$_2$R.
- The two receptors have 64% amino acid homology.

MOA

- Stimulation of the OXRs by orexin-A and -B results in increased wakefulness and the orexin neurons are active during wakefulness and fall silent during non-REM and REM sleep.
- It is postulated that dual antagonism of orexin-A and -B at OX$_1$R and OX$_2$R promotes sleep.

ADVERSE EFFECTS

Melatonin Agonists: Common adverse effects: drowsiness, tiredness, dizziness, and next day drowsiness which could impair motor function. Serious adverse effects: severe allergic reactions (tongue and throat swelling), amnesic sleep-related behaviors (sleep walking), worsening of depression, nightmares, and suicidal ideation.

DRUG–DRUG INTERACTIONS

Melatonin agonists: CYP inhibitors: fluvoxamine (CYP1A2 inhibitor; do not co-administer), ketoconazole (CYP3A4 inhibitor; use with caution), Fluconazole (CYP2C9 inhibitor; use with caution). CYP inducers: rifampin may decrease plasma concentration and effective therapeutic outcomes.

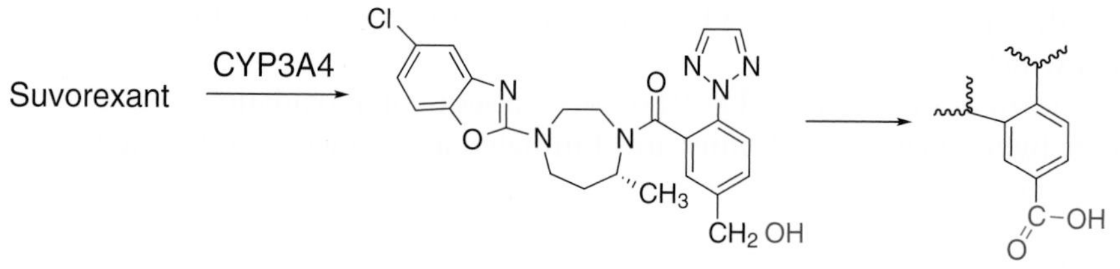

Figure 8.15 Metabolism of Suvorexant.

⚠ ADVERSE EFFECTS

Suvorexant: May cause sleepiness during the day, cloudy thinking, acting strangely and confused, sleepwalking, or performing other acts when asleep (eating, talking).

SAR
- Currently there is only one clinically useful orexin-A and orexin-B antagonist—suvorexant.
- Suvorexant is a diazepanyl compound substituted on its nitrogens with a benzoxazole and triazylphenylmethanone.

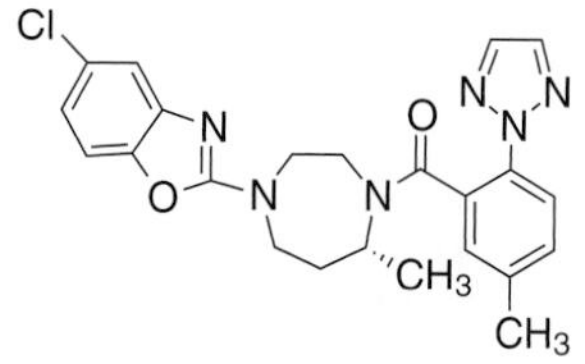

Suvorexant

Physicochemical and Pharmacokinetic Properties
- Suvorexant is rapidly absorbed following oral administration with an oral bioavailability of 82%.
- Suvorexant is highly bound to plasma protein (~99%) and is excreted primarily in the feces with an elimination $t_{1/2}$ of 12.2 hours.
- It is metabolized primarily by CYP3A4 to a hydroxylated metabolite which is further metabolized to a benzoic acid derivative (Fig. 8.15).

Clinical Application
- Suvorexant has been approved as a sleep aid for use in the treatment of insomnia for individuals who have trouble with onset and maintenance of sleep.

Chemical Significance

Understanding the neurobiology of sleep and the role of GABA chloride channels in cerebral cortex arousal has allowed for the development of the different classes of drugs used as sedative-hypnotics. The appropriate choice of a sedative hypnotic relies on the patient's specific need. If a patient has difficulty falling asleep or if staying asleep is the problem, knowing the metabolism and physicochemical and pharmacokinetic properties of the different classes of sedative-hypnotics can allow the pharmacist to contribute to the discussion of the appropriate drug within a class to be recommended.

Medicinal chemistry provides the basis for understanding the mechanism of action as well as the pharmacokinetic parameters needed for the appropriate therapeutic outcome. Pharmacists are the only members of the healthcare team with specific education in the area of medicinal chemistry which gives them a unique perspective when it comes to making the appropriate drug recommendation.

Review Questions

1 Which of the following structures represent the basic pharmacophore for the barbiturate sedative/hypnotics?

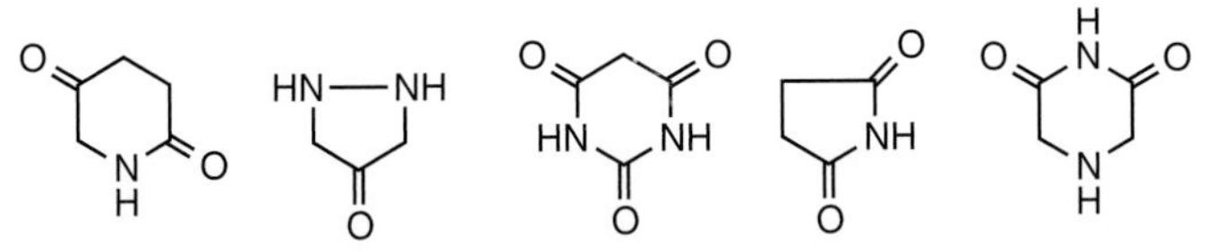

A   B   C   D   E

2 Which of the following benzodiazepines would you expect to have metabolites that contribute to its long duration of action?

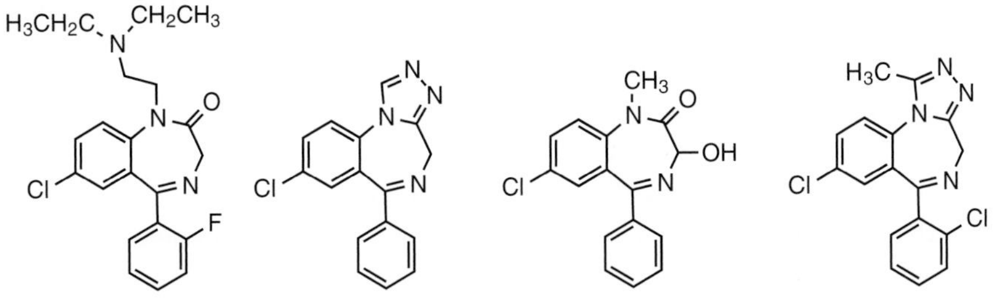

A   B   C   D

3 Which of the following statements is correct?
A. The benzodiazepines have no effect in the absence of GABA whereas the barbiturates do have an effect in the absence of GABA.
B. The barbiturates have no effect in the absence of GABA whereas the benzodiazepines do have an effect in the absence of GABA.
C. Neither benzodiazepines nor barbiturates have an effect in the absence of GABA.
D. Both benzodiazepines and barbiturates have an effect in the absence of GABA.

4 Which of the circled areas on zolpidem, if substituted with an electron-rich moiety, would cause loss of selectivity for the α_1-subtype GABA$_A$ receptor?

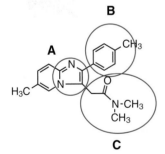

5 The three-position amide group on ramelteon interacts with which of the following amino acid residues in the MT$_1$ receptor?
A. Asn^{175}
B. His^{208}
C. Ser^{110}
D. Tyr^{226}
E. Val^{192}

9 Anesthetic Agents: General and Local Anesthetics

Learning Objectives

Upon completion of this chapter the student should be able to:

1. Define general anesthesia and distinguish general anesthetics from local anesthetics.
2. Describe the mechanism of action of the different types of general and local anesthetics:
   a. Volatile general anesthetics
   b. Intravenous general anesthetics
   c. Cocaine (Benzoic acid derivatives) local anesthetics
   d. Isogramine (Anilide derivatives) local anesthetics
3. Discuss the structure activity relationship (SAR) of the general and local anesthetics.
4. Identify any clinically relevant physicochemical and/or pharmacokinetic properties of the general and local anesthetics.
5. Discuss with structures the significant metabolism of the general and local anesthetics.
6. Explain the clinical uses, important adverse effects, and drug–drug interactions of the general and local anesthetics.

Introduction

Anesthesia is defined as the loss of sensation with or without loss of consciousness. There are a large number of drugs with many different chemical structures that can be used as anesthetic agents. They include the classic general and local anesthetics as well as CNS depressants such as analgesics, sedative-hypnotics, anticonvulsants, and skeletal muscle relaxants. Fundamentally, all produce anesthesia by interfering with conduction in sensory neurons and motor neurons. The anesthetics are useful in clinical practice to facilitate surgical and other medical and dental procedures. This chapter will discuss those agents classically referred to as "general" and "local" anesthetics.

General Anesthetics

The history of general anesthesia began in the middle of the 19th century with the introduction of nitrous oxide (N_2O) by Horace Wells for dental procedures; diethyl ether by William Morton for surgery, and chloroform by James Young Simpson for surgery. With the exception of N_2O, diethyl ether and chloroform are no longer used because of chloroform's toxicity and diethyl ether's flammability and tendency to form explosive peroxides. The ideal general anesthetic is characterized by a loss of all sensations along with analgesia, skeletal muscle relaxation, and nontoxicity. Classically, there are four stages for anesthesia: analgesia (stage 1), delirium (stage 2), surgical anesthesia (stage 3), and respiratory paralysis (stage 4). Stages 1 and 2 are referred to as the induction period and stage 3 is, as its name implies, the stage when surgery can commence. Stage 4 represents toxic overdosing and is never reached during surgical procedures. However, many anesthetic agents in use today do not follow the classical four stages and therefore anesthesiologists monitor reflexes, blood pressure, and respiratory rate to effectively maintain an appropriate depth of surgical

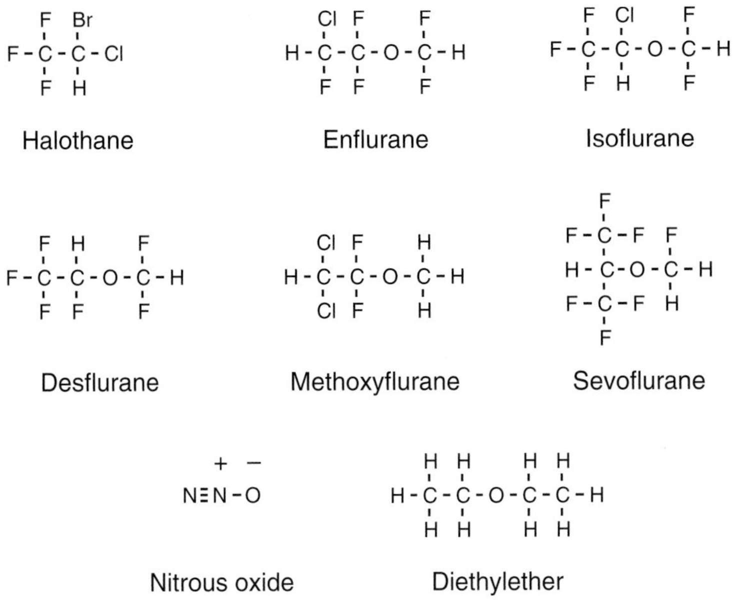

Figure 9.1 Volatile general anesthetics.

anesthesia without producing toxicity. Modern general anesthetics are classified as volatile and intravenous anesthetic agents.

Volatile General Anesthetics: Halogenated Hydrocarbons and Ethers

- The clinically important useful volatile anesthetics are shown in Figure 9.1.
- Of those depicted, methoxyflurane is not used because of its propensity to cause renal toxicity.
- Toxic degradation products can form by reaction of the volatile anesthetic with soda lime which is used as a CO_2 absorber during anesthesia procedures:
  - Halothane forms 2-Br-2-Cl-1,1,-difluoroethylene.
  - Sevoflurane forms 2-(fluoromethoxy)-1,1,3,3,3,-pentafluoro-1-propene (Compound A) which is nephrotoxic.
  - Desflurane, isoflurane, and enflurane form carbon monoxide.
  - These toxic products do not generally cause clinical problems since exposure to these anesthetics is of relatively short duration.

MOA

There are two major theories of how the volatile general anesthetics produce anesthesia.

- The Meyer–Overton Theory – the potency of a substance as an anesthetic is directly related to its lipid solubility or oil/partition coefficient and not its structure. The dissolved anesthetic nonspecifically disrupts the physical properties of the lipid membrane bilayer which in turn alters the excitable cell membrane ion-channel protein inducing anesthesia.
  - Also known as the out-dated "unitary theory of anesthesia".
  - Compounds with high lipid solubility required lower concentrations to produce anesthesia, that is, lower MAC (minimum alveolar concentration; Table 9.1).
- The Ion-Channel and Protein Receptor Hypothesis – Anesthetics interact with transmembrane ligand-gated ion channels such as the inhibitory GABA_A (chloride-selective ion channels) receptors and excitatory nicotinic acetylcholine receptors (Na+ ion channels), which are respectively potentiated and inhibited by general anesthetics by either allosterically modulating the opening of ion channels (e.g., Cl−, K+) or modulate activity by directly interacting with the ion channel (e.g., Na+).
- Many other hypotheses continue to be postulated (see Chapter 16 in *Foye's Principles of Medicinal Chemistry, Seventh Edition*, for a complete detailed discussion).

Stereochemical Aspects of Volatile Anesthetics

Halothane, isoflurane, desflurane, and enflurane each have an asymmetric carbon and therefore can exist as (+) and (−)-enantiomers. They are clinically used as racemates, however the (+)-enantiomer of isoflurane has been shown to be at least 50% more potent as the (−)-enantiomer in the rat. This may indicate, at least for isoflurane, a more complex mechanism with probable involvement of protein–anesthetic interactions.

Table 9.1   Partition Coefficients and MACs of Volatile Anesthetics

| Anesthetics | Partition Coefficients at 37°C | | | MAC (vol. %)[a] |
	Oil/Gas	Blood/Gas	Without N$_2$O	With N$_2$O (%)
Methoxyflurane	970	12	0.16	0.07 (56)
Halothane	224	2.3	0.77	0.29 (66)
Enflurane	98.5	1.19	91.7	0.60 (70)
Isoflurane	90.8	1.4	1.15	0.50 (70)
Sevoflurane (2)	53.4	0.60	1.71	0.66 (64)
Desflurane (3)	16.7	0.42	6.0	2.83 (60)
Nitrous oxide	1.4	0.47	104	—

[a]*MAC*, minimum alveolar concentration, expressed as volume %, required to produce immobility at 50% of middle-aged humans.

SAR

- The volatile anesthetics are structurally nonspecific, that is, their action is not related to chemical structure and their action requires relatively high doses. They vary in structure but have the same activity and slight changes in structure has no effect on activity.

Physicochemical and Pharmacokinetic Properties

- Reaching and maintaining the anesthetic state using volatile anesthetics depends on the concentration (partial pressure) in the brain (areas not yet identified).
- To reach the brain the inhaled anesthetic must saturate the lung alveoli, the blood, and other tissues (adipose) as well as the brain.
- Since each compartment is in equilibrium, saturation is usually expressed as the minimum alveolar concentration (MAC).
- The MAC is defined as the concentration in volume % required to produce immobility in 50% of adult patients. A MAC of 1.3 will cause immobility in 99% of patients.
- The MAC is equivalent to the potency of an individual anesthetic agent such that the lower the MAC, the more potent is the anesthetic (Table 9.1).
- Most clinically used volatile anesthetics are low boiling point liquids at room temperature and administered, either alone or in combination with another anesthetic, by bubbling oxygen through the liquid and the resulting gas mixture inhaled by the patient. Table 9.2 gives the boiling points and the chemical stability of the volatile anesthetics.

Table 9.2   Physicochemical Properties of Clinically Important Volatile Anesthetics

Generic Name and Structure	Boiling Point (°C)	Chemically Stable[a]
Desflurane $CHF_2\text{-}O\text{-}CHF\text{-}CF_3$	23.5	Yes
Enflurane $CHF_2\text{-}O\text{-}CF_2\text{-}CHFCl$	56.5	Yes
Halothane $CF_3\text{-}CHBrCl$	50.2	No
Isoflurane $CHF_2\text{-}O\text{-}CHCl\text{-}CF_3$	48.5	Yes
Methoxyflurane $CH_3\text{-}O\text{-}CF_2\text{-}CHCl_2$	104.7	No
Sevoflurane $(CF_3)_2CH\text{-}O\text{-}CH_2F$	58.5	No
Nitrous Oxide N_2O	−8.0	Yes

[a]Indicates stability to soda lime, ultraviolet light, and common metals.

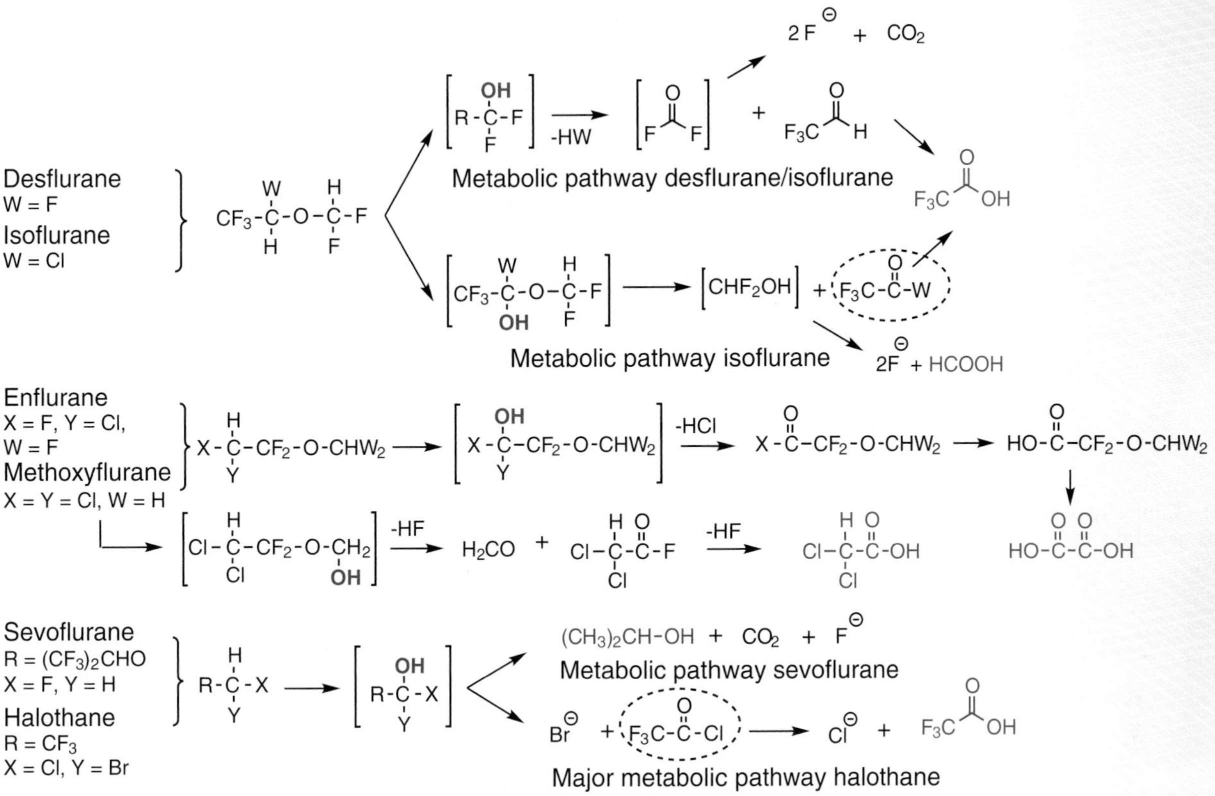

$$2F^{\ominus} + CO_2$$

Desflurane
W = F

Isoflurane
W = Cl

$$CF_3-\overset{\overset{W}{|}}{\underset{\underset{H}{|}}{C}}-O-\overset{\overset{H}{|}}{\underset{\underset{F}{|}}{C}}-F$$

$$\left[R-\overset{\overset{OH}{|}}{\underset{\underset{F}{|}}{C}}-F\right] \xrightarrow{-HW} \left[\overset{O}{\underset{F}{\|}}_{F}\right] + \overset{O}{\underset{H}{\|}}_{F_3C}$$

Metabolic pathway desflurane/isoflurane

$$F_3C-\overset{O}{\underset{}{\|}}-OH$$

$$\left[CF_3-\overset{\overset{W}{|}}{\underset{\underset{OH}{|}}{C}}-O-\overset{\overset{H}{|}}{\underset{\underset{F}{|}}{C}}-F\right] \longrightarrow \left[CHF_2OH\right] + \left(F_3C-\overset{O}{\underset{}{\|}}-W\right)$$

Metabolic pathway isoflurane

$$2F^{\ominus} + HCOOH$$

Enflurane
X = F, Y = Cl, W = F

Methoxyflurane
X = Y = Cl, W = H

$$X-\overset{\overset{H}{|}}{\underset{\underset{Y}{|}}{C}}-CF_2-O-CHW_2 \longrightarrow \left[X-\overset{\overset{OH}{|}}{\underset{\underset{Y}{|}}{C}}-CF_2-O-CHW_2\right] \xrightarrow{-HCl} X-\overset{O}{\underset{}{\|}}-CF_2-O-CHW_2 \longrightarrow HO-\overset{O}{\underset{}{\|}}-CF_2-O-CHW_2$$

$$\left[Cl-\overset{\overset{H}{|}}{\underset{\underset{Cl}{|}}{C}}-CF_2-O-\overset{}{\underset{\underset{OH}{|}}{CH_2}}\right] \xrightarrow{-HF} H_2CO + Cl-\overset{\overset{H}{|}}{\underset{\underset{Cl}{|}}{\overset{O}{\underset{}{C}}}}-F \xrightarrow{-HF} Cl-\overset{\overset{H}{|}}{\underset{\underset{Cl}{|}}{\overset{O}{\underset{}{C}}}}-OH \qquad HO-\overset{O}{\underset{}{\|}}-\overset{O}{\underset{}{\|}}-OH$$

Sevoflurane
R = (CF$_3$)$_2$CHO
X = F, Y = H

Halothane
R = CF$_3$
X = Cl, Y = Br

$$R-\overset{\overset{H}{|}}{\underset{\underset{Y}{|}}{C}}-X \longrightarrow \left[R-\overset{\overset{OH}{|}}{\underset{\underset{Y}{|}}{C}}-X\right]$$

$$(CH_3)_2CH-OH + CO_2 + F^{\ominus}$$

Metabolic pathway sevoflurane

$$Br^{\ominus} + \left(F_3C-\overset{O}{\underset{}{\|}}-Cl\right) \longrightarrow Cl^{\ominus} + F_3C-\overset{O}{\underset{}{\|}}-OH$$

Major metabolic pathway halothane

Figure 9.2 Metabolites of the fluorinated anesthetics.

Metabolism

- The fluorinated anesthetics are metabolized in the liver to acid products, including trifluoroacetic acid, oxalic acid, fluoromethoxydifluoroacetic acid, and dichloroacetic acid (Fig. 9.2).
- They are metabolized to varying degrees by CYP2E1: halothane (20%); enflurane (2%); isoflurane (0.2%), desflurane (0.02%), sevoflurane (3%), and methoxyflurane (50%).
- Although not usually a clinical problem, there is a minimal potential for hepatotoxicity associated with trifluoroacetyl halide.

Clinical Application

- Volatile anesthetics are used clinically to keep the patient in the stage 3 surgical state during surgery.
- During surgery, heart rhythm, blood pressure, blood oxygen, body temperature, and consciousness, and during certain surgeries (e.g., neurosurgery, orthopedic) where neurological functions are monitored.

> **⚠ ADVERSE EFFECTS**
>
> **Volatile Anesthetics:** Hepatotoxicity, malignant hyperthermia, and nephrotoxicity are rare since exposure is limited.
> Cardiac arrhythmia liability (except N$_2$O) during Stage 2 due to catecholamine release.
> Spontaneous abortions, congenital birth defects, and increased incidence of cancers in chronically exposed medical personnel.

Factors That May Alter MAC		
Increase MAC:	Increased catecholamine levels in CNS Hypernatremia Hyperthermia	
Decrease MAC:	Decreased catecholamine levels in CNS Alcohol ingestion Clonidine Lidocaine Lithium Opioids	Hyponatremia Hypotension Hypothermia Hypoxia Increased age Pregnancy
No effect on MAC:	Plasma potassium Gender Hypertension Duration of anesthesia	

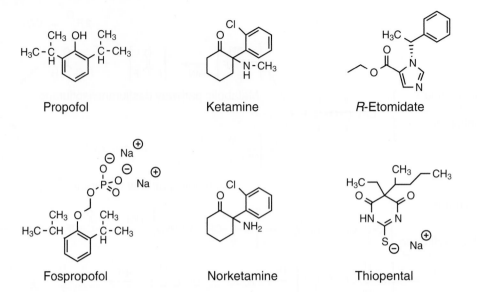

Figure 9.3 Clinically useful intravenous general anesthetics.

- It is common practice to co-administer a halogenated anesthetic with N_2O at the same time since this technique allows for reduction of the adverse reactions from either one alone (Table 9.1).

Intravenous General Anesthetics

General anesthetics administered intravenously are found in Figure 9.3. They are commonly used to induce a rapid, short-acting anesthesia which is then maintained with volatile anesthetics.

MOA

Propofol, R-Etomidate, and Thiopental

- Act by enhancing GABAergic neurotransmission by interacting at the $GABA_A$ receptors but not at the benzodiazepine site.

Ketamine

- Acts as an antagonist within the cationic channel of the NMDA receptor complex (N-methyl-D-aspartate).
- It prevents the flow of cations through the channel causing deactivation of neurons.
- May also interact with the opioid, muscarinic, and serotonin receptors.

SAR

- The intravenous general anesthetics are individual chemicals with no common pharmacophore.
- Propofol is a 1,5-diisopropyl phenol; ketamine is 2-(2-chlorophenyl)-2-methyl-amino–cyclohexanone.
- R-Etomidate is (+)-ethyl-1-(α-methylbenzyl)imidazole-5-carboxylate.

Physicochemical and Pharmacokinetic Properties

Propofol

- Poor water solubility therefore dispensed as an emulsion.
- Onset is 30 to 60 seconds and duration of action 5 to 10 minutes.
- Ninety-eight percent bound to plasma proteins.
- Fospropofol is the phosphate ester of propofol which is the water soluble pro-drug for propofol.
- Fospropofol is hydrolyzed to propofol by plasma alkaline phosphatase and therefore has a longer onset (4 to 10 minutes) and duration of action as compared to propofol.

Ketamine

- Rapid onset with a duration of action of 10 to 25 minutes.
- Oral bioavailability is <16% with elimination $t_{1/2}$ of 2 to 3 hours.
- Metabolized to norketamine which has significant anesthetic activity contributing to the duration of action.

R-Etomidate

- A very lipophilic weak base (PC = 2,000; pKa = 4.5).
- Administered as a 35% solution in propylene glycol.
- Rapid onset with extremely short duration of action of <3 minutes and elimination $t_{1/2}$ of 5 to 6 hours due to redistribution to adipose and other body tissues.

Thiopental

- Ultrashort-acting thiobarbiturate with onset of 30 to 40 seconds.
- Eighty percent plasma protein bound and elimination $t_{1/2}$ 3 to 8 hours.
- The barbiturates are covered in Chapter 8 in *Foye's Principles of Medicinal Chemistry, Seventh Edition.*

Metabolism

Propofol

- Rapidly metabolized to O-glucuronide and O-sulfate conjugates with <0.3% excreted unchanged.
- Fospropofol is hydrolyzed to propofol in the plasma by alkaline phosphatase.

Ketamine

- Metabolized to hydroxynorketamine glucuronide conjugate (inactive) as well as norketamine by the action of CYP3A4 and CYP2B6 which retains activity.
- Less than 4% eliminated unchanged.

R-Etomidate

- Hydrolyzed by hepatic esterases to the inactive carboxylic acid which is excreted in both the urine and bile.

Thiopental

- Oxidatively metabolized to pentobarbital which is further metabolized by CYP450 isozymes to alcohol and acid metabolites.
- The acid metabolites are excreted as the glucuronide conjugate through the kidney (see Chapter 8 in *Foye's Principles of Medicinal Chemistry, Seventh Edition*).

Clinical Application

- Propofol, *R*-etomidate, and thiopental are useful for the induction of anesthesia which is then maintained with volatile anesthetics.
- Propofol has a special application for outpatient surgical procedures; for example, colonoscopy, dermatology, tubal ligations, and laparoscopy.
- Ketamine is recommended for diagnostic and surgical procedures that do not require skeletal muscle relaxation.
- *R*-Etomidate is useful for anesthesia in hemodynamically unstable patients prone to hypotension; for example, hypovolemia, coronary artery disease, or cardiomyopathies.

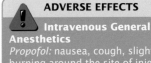

ADVERSE EFFECTS

Intravenous General Anesthetics

Propofol: nausea, cough, slight burning around the site of injection, severe allergic reaction, confusion, agitation, anxiety, and muscle pain.

Ketamine: can cause "dissociative" anesthesia with the patient appearing cataleptic with open and oscillating eyes. Can also produce analgesia and amnesia with the patient appearing awake but incapable of communication.

R-Etomidate: can cause myoclonic jerks and inhibit adrenal steroid synthesis.

Local Anesthetics

Local anesthetics are drugs that produce a state of local anesthesia when applied topically or locally by infusion. They are useful for the temporary relief of pain in dentistry and minor surgery. They also find use when applied topically to relieve pain and itching due to minor burns, insect bites, allergic responses, and hemorrhoids. Unlike the general anesthetics the local anesthetics produce a local anesthesia without the loss of consciousness or impairment of central

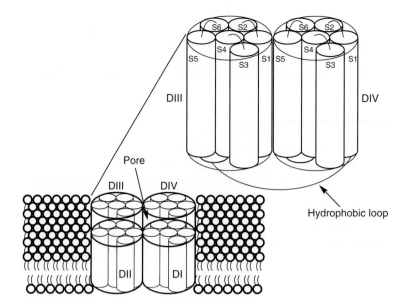

Figure 9.4 A schematic representation of the α-subunit of the sodium channel and the pore-forming unit.

cardiorespiratory functions. The ideal local anesthetic would produce reversible blockade, be selective for sensory neurons only, and have a rapid onset and short duration of action. In addition, the ideal local anesthetic would be free of adverse effects and systemic toxicity.

MOA

- The local anesthetics act by decreasing the excitability of sensory nerve cells without effecting the resting potential.
- They reversibly bind to the voltage-gated Na^+ channel blocking Na^+ transport and therefore nerve conduction (Fig. 9.4).
- Current evidence suggests that the local anesthetics bind to hydrophobic amino acid residues near the center and intracellular end of the S6 segment in the DIV domain.

SAR

- The local anesthetics are derived from either cocaine or isogramine (Fig. 9.5).
- Cocaine was first used as a topical anesthetic in ophthalmology and as a spinal anesthetic, but its acute toxicity and addiction liability prompted synthesis of analogs without those adverse effects.
- Cocaine-derived local anesthetics are benzoic acid derivatives (Fig. 9.5).

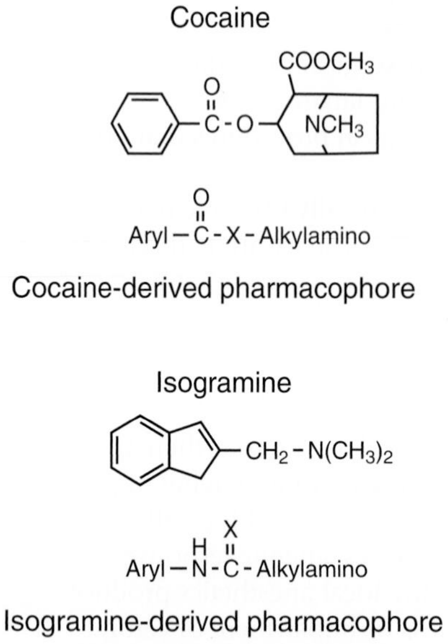

Cocaine

Aryl—C—X—Alkylamino

Cocaine-derived pharmacophore

Isogramine

—CH₂—N(CH₃)₂

Aryl—N—C—Alkylamino

Figure 9.5 Local anesthetic pharmacophores based on cocaine and isogramine.

Isogramine-derived pharmacophore

- Isogramine is a synthetic indole alkaloid modeled on gramine which was isolated from barley.
- Isogramine-derived local anesthetics are anilide derivatives.
- It should be noted that local anesthetics that contain 2 "i's" in their names are all anilides (e.g., L*I*DOCA*I*NE).

SAR of Cocaine Analogs (Benzoic acid derivatives; Fig. 9.5 and Table 9.3)
Aryl Group

- The aromatic ring must be directly attached to the carbonyl for maximum activity but can be conjugated to it which decreases the activity.
- Alkoxy, amino, and alkylamino groups in the ortho or para positions enhance activity because they increase the electron density of the carbonyl oxygen (Table 9.3).

The X Bridge

- Can be C, O, N, or S with activity S>O>C>N.
- When X = N the local anesthetic is an amide which is resistant to hydrolysis and therefore has a longer duration of action.

Aminoalkyl Group

- Not absolutely necessary for activity but is used to form water soluble salts (see benzocaine; Table 9.3).
- Most contain a tertiary nitrogen which is important for salt formation and contributing to the correct pKa for optimal activity.
- The length of the alkyl group primarily influences the lipid solubility of the unionized species.

SAR of Isogramine Analogs (Anilides; Fig. 9.5 and Table 9.4)
Aryl Group

- The aromatic ring is attached directly to a nitrogen which is then attached to an sp^2 hybridized carbon.
- Methyl groups in the 2,6 positions increase the duration of action by hindering amide hydrolysis.

Table 9.3 Cocaine Analog Local Anesthetics (Benzoic Acid Derivatives)

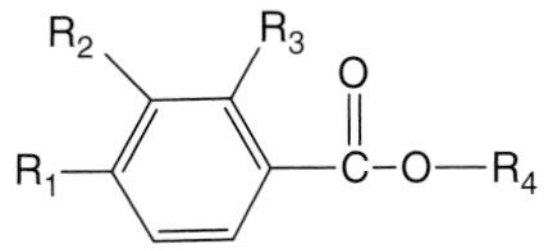

Name	R_1	R_2	R_3	R_4
Cocaine	H	H	H	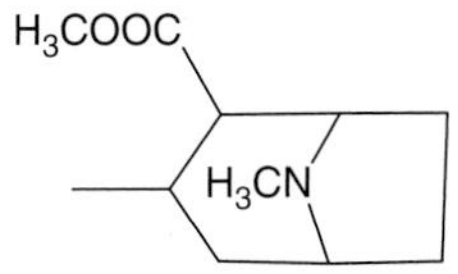 H_3COOC ... H_3CN
Benoxinate	NH_2-	C_4H_9O-	H	$-CH_2CH_2N(C_2H_5)_2$
Benzocaine	NH_2-	H	H	$-C_2H_5$
Butamben	NH_2-	H	H	$-C_4H_9$
Chloroprocaine	NH_2-	H	Cl	$-CH_2CH_2N(C_2H_5)_2$
Procaine	NH_2-	H	H	$-CH_2CH_2N(C_2H_5)_2$
Proparacaine	C_3H_7O-	NH_2-	H	$-CH_2CH_2N(C_2H_5)_2$
Propoxycaine	NH_2-	H	C_3H_7O-	$-CH_2CH_2N(C_2H_5)_2$
Tetracaine	C_4H_9NH-	H	H	$-CH_2CH_2N(CH_3)_2$

The X Atom

- Most clinically important local anesthetics are those when X is an oxygen; however, the X can be an N or a C and retain activity (e.g., dibucaine, Table 9.4).

Aminoalkyl Group

- Similar function as with the cocaine analogs, that is, functions for salt formation and contributes to lipophilicity.

Physicochemical and Pharmacokinetic Properties

- The majority of the useful local anesthetics have a pKa value between 7.0 and 9.0 (Table 9.5); therefore, the percent ionized at physiologic pH will vary among these compounds. A balance must be struck between the lipophilic (unionized)

Table 9.4   Isogramine Analog Local Anesthetics (Anilides)

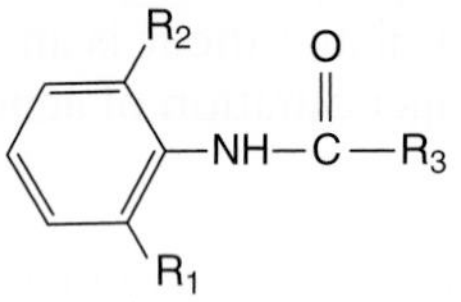

Name	R$_1$	R$_2$	R$_3$
Bupivacaine	CH$_3$—	CH$_3$—	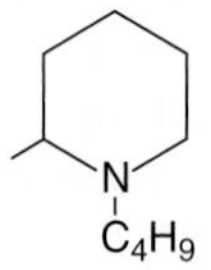
Etidocaine	CH$_3$—	CH$_3$—	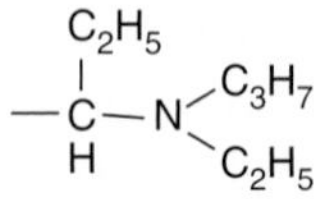
Lidocaine	CH$_3$—	CH$_3$—	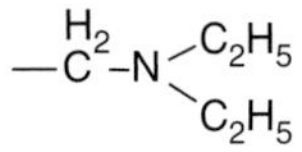
Mepivacaine	CH$_3$—	CH$_3$—	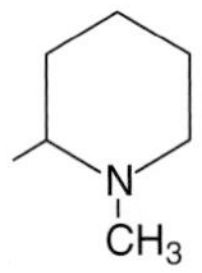
Prilocaine	H	CH$_3$—	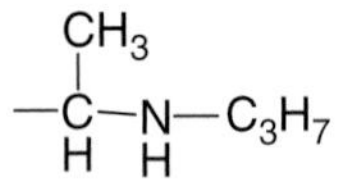
Ropivacaine	CH$_3$—	CH$_3$—	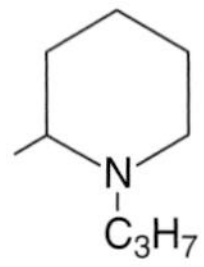
Articaine	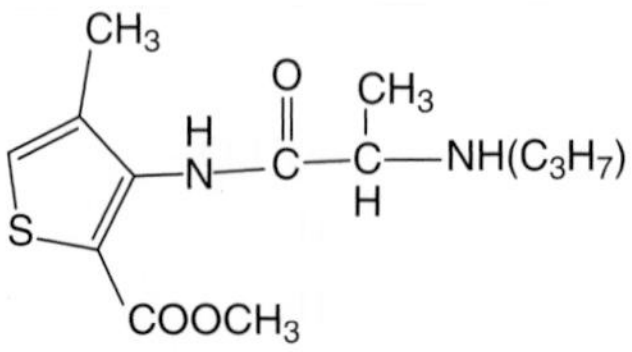		
Dibucaine	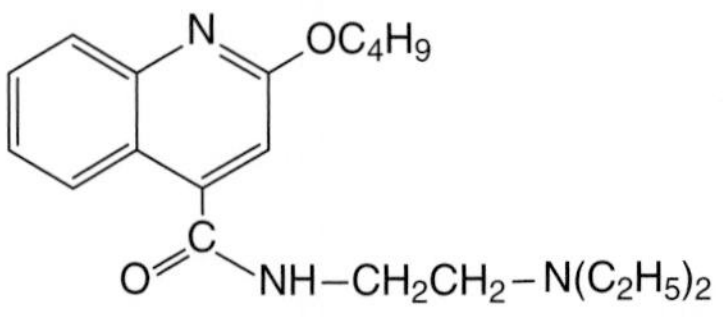		

Table 9.5 Pharmacokinetics and pKa's of Some Clinically Useful Local Anesthetics

Name	Onset (min)	Duration (hr)	pKa
Articaine	1–6	1	7.8
Benoxinate	≤15	2–3	9.0
Benzocaine	0.5	≤1	2.8
Bupivacaine	5	2–4	2.1
Chloroprocaine	6–12	0.5	9.0
Etidocaine	3–5	5–10	7.7
Lidocaine	<2	0.5–1	7.8
Mepivacaine	3–5	0.75–1.5	7.6
Prilocaine	<2	≥1	7.9
Procaine	2–5	0.25–1	8.8
Ropivacaine	11–26	1.7–3.2	8.2
Tetracaine	≤15	2–3	8.4

and hydrophilic (ionized) forms for maximal activity and binding to the Na^+ channels.

- For optimal activity, the unionized form is necessary for diffusion through and across the neuronal membrane and then the re-protonated (ionized) form binds to the Na^+ channel (Fig. 9.6). Therefore, the current theory for local anesthetics postulates two pathways, lipophilic and hydrophilic, by which they reach their binding sites. It should be noted that they do not bind to either the tetrodotoxin (TTX) or the saxitoxin (STX) binding site.
- Stereochemistry plays only a minor role in local anesthetic interaction and activity with the Na^+ channels; however, stereochemistry does have an important effect on toxicity. For example, levobupivacaine (S-(−)-bupivacaine) has lower cardiotoxicity than the racemic mixture and the R-(+)-enantiomer.

Metabolism

- The benzoic acid ester-type local anesthetics are rapidly hydrolyzed by pseudo-cholinesterase in the plasma as well as other tissue cholinesterases, and therefore are metabolized not only in the plasma but in the kidneys, liver, and at the site of administration (Fig. 9.7).
- The anilide-type local anesthetics are metabolized primarily in the liver by CYP1A2 isoenzymes; for example, lidocaine is metabolized to a metabolite that exhibits CNS toxicity (monoethylglycinexylidide; Fig 9.7).

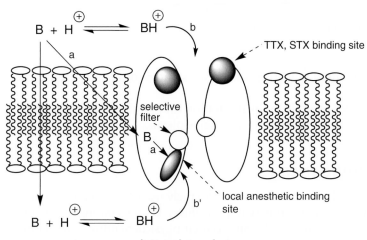

Figure 9.6 Model of a sodium channel depicting a hydrophilic pathway (denoted by b and b') and a hydrophobic pathway (denoted by a) by which local anesthetics may reach their binding sites.

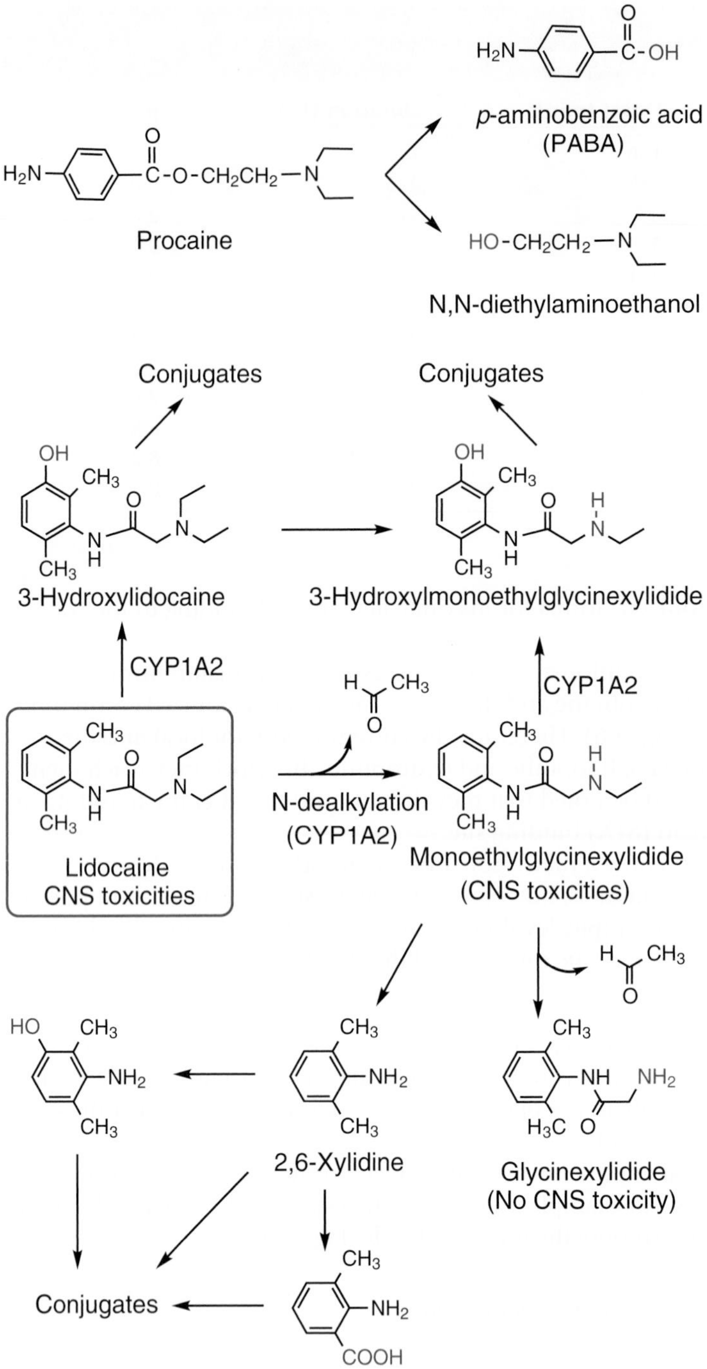

Figure 9.7 Metabolism of procaine and lidocaine.

ADVERSE EFFECTS

Local Anesthetics

Cardiovascular effects: myocardial depression, hypotension, decreased cardiac output, heart block, bradycardia, ventricular arrhythmia, and cardiac arrest all associated with high plasma concentrations.

CNS effects: excitation, depression, dizziness, tinnitus, drowsiness, disorientation, muscle twitching, seizures, and respiratory arrest.

DRUG–DRUG INTERACTIONS

Local Anesthetics

Benzoic acid esters: cholinesterase inhibitors prolong duration of action; atropine-like anticholinergics prolong duration of action by competing for cholinesterases.

Anilide local anesthetics: increased or decreased duration of action when coadministered with enzyme inducers or inhibitors (e.g., cimetidine and barbiturates).

Clinical Application

- Local anesthetics are used in many primary care settings for the short-term relief of regional pain.
- They are administered topically or by infiltration into joints, in the epidural or intrathecal space, along nerve roots or in a nerve plexus.

Chemical Significance

General anesthetics and local anesthetics are extremely useful for both major and minor surgeries as well as many outpatient procedures. Understanding the chemistry of both classes of anesthetics as it relates to their physicochemical and pharmacokinetic properties gives the clinician insight into establishing optimal application and use. The ideal general anesthetic has yet to be designed; however, understanding the mechanism of action at the chemical level gives insight that helps to develop newer agents. In the community, the pharmacist is often asked to recommend topical anesthetics for a variety of regional pain complaints. The medicinal chemistry of the local anesthetics gives the pharmacist special knowledge into the appropriate choice for a specific complaint, from dental pain to insect bites to minor burns, that renders optimal patient outcomes.

Review Questions

1 All of the following are characteristics of an ideal general anesthetic except:
A. Causes loss of all sensations
B. Produces analgesia
C. Produces skeletal muscle relaxation
D. Is nontoxic
E. Slow onset and withdrawal

2 All of the following intravenous general anesthetics work by interacting with the GABAergic neurons except:
A. Propofol
B. Ketamine
C. Thiopental
D. *R*-Etomidate
E. None of the above

3 Which of the following local anesthetics can be considered an analog of isogramine?

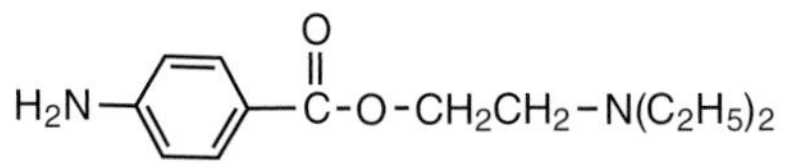

A 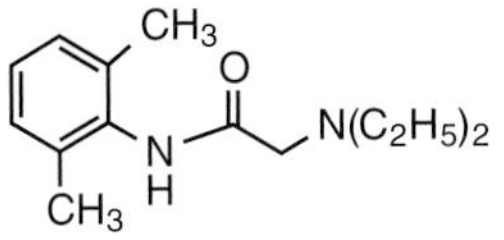 **B** 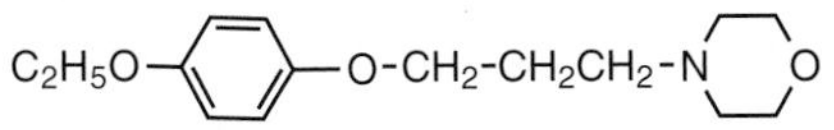 **C** 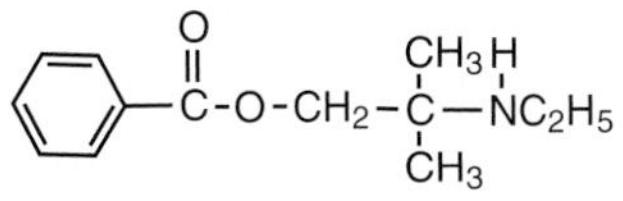 **D** 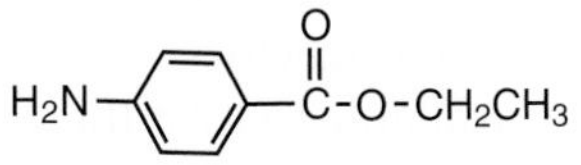 **E**

4 Which of the local anesthetics in Question 3 would you expect to be metabolized by CYP1A2?

5 Which of the following statements regarding stereochemistry and local anesthetics is correct?
A. Stereochemistry plays a major role in the interaction with Na^+ channels, but has minimal effects on toxicity.
B. Stereochemistry plays a minor role in the interaction with Na^+ channels, but has an important effect on toxicity.
C. Stereochemistry plays a major role in both the interaction with Na^+ channels and the development of toxicity.
D. Stereochemistry is relatively unimportant in terms of interaction with Na^+ channels and the development of toxicity.

10

Antiseizure Agents

Learning Objectives

Upon completion of this chapter the student should be able to:

1. Identify the various types of epilepsy and their predominant symptoms.
2. Describe the mechanism of action of:
   a. Hydantoins
   b. Succinimides
   c. Barbiturates
   d. Benzodiazepines
   e. Miscellaneous adjunct antiseizure agents
3. Discuss applicable structure activity relationship (SAR) of the above classes of antiseizure agents.
4. List physicochemical and pharmacokinetic properties that impact in vitro stability and/or therapeutic utility of antiseizure agents.
5. Diagram metabolic pathways for all biotransformed antiseizure drugs, identifying enzymes and noting clinically significant metabolism and the activity, if any, of the major metabolites.
6. Apply all of above to predict and explain therapeutic utility of antiseizure agents.

Introduction

According to the Epilepsy Foundation, seizure disorders affect over 50 million people worldwide. Over 3 million of these individuals are US citizens. The disease knows no age limits; it impacts infants, elders, and all age groups in between. Epilepsy claims as many lives as breast cancer and approximately 500 new US cases are diagnosed daily, 30% of them in children. The toll of seizure risk on patient quality of life is enormous, and can result in significant threats to personal safety, independence, and longevity.

A variety of medications are available to treat the various types of epilepsy, but seizure control is often not complete with monotherapy. Many patients require combination therapy to adequately manage symptoms without use-limiting adverse effects. Hydantoins and their oxygen (oxazolidinedione) and carbon (succinimide) analogs, barbiturates, and benzodiazepines have been used in epilepsy for decades. However, as the intricate neurophysiology of the disease in its various forms has become better understood, newer agents have been designed that target specific ion channels and central excitatory (glutamate-related) or inhibitory (GABA) receptors. These second-generation agents have proven valuable as adjuncts for the more effective management of complex and/or persistent seizure disorders.

Classification of Epilepsy

- The various types of epilepsy are categorized in Table 10.1.
- Focused scarring of brain tissue post-injury is thought to underlie partial seizures. Generalized seizures may result from diffuse injury, or from metabolic or genetic-related dysfunctions.
- A prolonged seizure, or repetitive seizures, of any type is called status epilepticus. Grand mal is the most common type of seizure involved.

Table 10.1 Seizure Types

Partial	Neuronal Discharge	Symptoms	Impaired Consciousness?
Simple	Temporal (most common) or frontal lobe foci	Contralateral; psychic, autonomic, sensory (temporal lobe foci); motor (frontal lobe foci)	No
Complex	Similar to simple	Often bilateral; can include repetitive involuntary movements	Yes, including preseizure and postseizure dulling of consciousness
Progressive	Initially similar to simple	Step-wise or rapid progression to generalized clonic–tonic seizure	Yes during generalized stage; partial stage symptoms perceived as aura in step-wise progression

Generalized			
Absence (petit mal)	Bilateral gray matter involvement	Loss of awareness, blank staring, multiple daily occurrences; complex or atypical types may include clonic and autonomic symptoms	Brief, with postseizure amnesia
Myoclonic	Same as absence	Generalized or focused muscle jerking	No, due to short seizure duration
Tonic	Same as absence	Excessive extensor muscle tone with subsequent falling	Possible, but seizure duration is brief
Atonic	Same as absence	Muscle tone loss with possible falling	Possible
Clonic	Same as absence	Muscle tone loss, tonic contractions, asymmetric jerking	Yes
Tonic–clonic (grand mal)	Same as absence	Complete tonic response with falling, bilateral symmetric jerking	Yes, can be prolonged

MOA

- Seizure disorders are of central origin, and involve spontaneous and excessive neuronal depolarization (firing).
- Genetic mutations in Na^+, K^+, Ca^{2+}, and/or Cl^- ion channels, as well as in the structure and function of inhibitory (GABA) and excitatory (NMDA, AMPA, KA) receptors, are associated with idiopathic generalized seizures.
- Antiseizure drugs act by altering ion channel or receptor function to promote synaptic inhibition or attenuate synaptic excitation (Table 10.2).

Table 10.2 Antiseizure Drug Mechanisms of Action

Mechanism	Antiseizure Drugs
Stabilization of inactive (closed) Na^+ channels	Primary: carbamazepine, lamotrigine, phenytoin Secondary: barbiturates benzodiazepines, lacosamide, oxcarbazepine, rufinamide, topiramate, valproate, zonisamide
Stabilization of active (open) K^+ ion channels	Ezogabine
Stabilization of inactive (closed) HVA Ca^{2+} ion channels	Gabapentin
T-type Ca^{2+} ion channel blockade	Ethosuximide, valproate, zonisamide
Stabilization of active (open) Cl^- channel of the $GABA_A$ receptor	Barbiturates, benzodiazepines, topiramate
NMDA receptor blockade	Felbamate
AMPA/KA receptor blockade	Barbiturates, topiramate
GABA reuptake transporter blockade	Tiagabine, valproate

HVA, high voltage activated; *NMDA*, N-methyl-D-aspartate; *AMPA*, α-amino-3-hydroxy-5-methyl-4-isoxazole propionate; *KA*, kainate; *GABA*, γ-aminobutyric acid

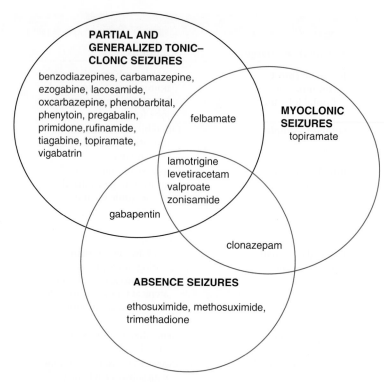

Figure 10.1 Antiseizure drugs and their indications.

- Most currently available antiseizure drugs, and the types of epilepsy they treat, are identified in Figure 10.1.

Ureides

- Phenobarbital, hydantoins, oxazolidinediones, and succinimides are all chemically classified as cyclic ureides (Fig. 10.2).

SAR

- Hydantoins and barbiturates are imidazolidinedione and hexahydropyrimidine-trione structures, respectively (Fig. 10.3).

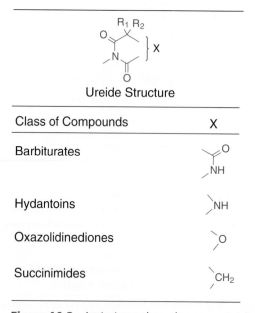

Ureide Structure

Class of Compounds	X
Barbiturates	$\begin{array}{c} O \\ NH \end{array}$
Hydantoins	NH
Oxazolidinediones	O
Succinimides	CH_2

Figure 10.2 Antiseizure drug classes containing the ureide structure.

Hydantoins:

Phenytoin
(Dilantin)

Fosphenytoin
(Cerebyx)

Succinimide: **Barbiturate:**

Ethosuximide
(Zarontin)

Phenobarbital

Figure 10.3 Most commonly used ureide antiseizure drugs.

- Activity against generalized seizures is optimal with at least one C5-phenyl substituent.
- Among hydantoins, a 5,5-diphenyl substitution pattern (e.g., phenytoin) provides the highest activity.
- Substitution at hydantoin N3 with small alkyl groups (e.g., methyl, ethyl) decreases potency, alters seizure specificity, and increases the potential for adverse effects, including sedation, severe rash, and blood dyscrasias (e.g., mephenytoin, ethotoin).
  - The impact of a similar substitution on barbiturates is less dramatic (e.g., mephobarbital).

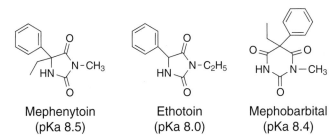

Mephenytoin
(pKa 8.5)

Ethotoin
(pKa 8.0)

Mephobarbital
(pKa 8.4)

- Replacement of hydantoin N3 with oxygen (oxazolidinedione) or methylene (succinimide) decreases utility in generalized tonic–clonic (grand mal) and complex partial seizures and increases activity against absence (petit mal) seizures.
  - Toxic effects of oxazolidinediones, including potentially fatal blood dyscrasias, limit therapeutic utility.
  - Ethosuximide's relatively low toxicity makes it preferred in the treatment of absence seizures.

Selected Ureide Antiseizure Drugs (Fig. 10.3)

Phenytoin

Physicochemical and Pharmacokinetic Properties

- Free phenytoin contains an acidic secondary imide (pKa 8.33). The sodium salt of phenytoin is poorly water soluble but absorbs CO_2, prompting the formation of the water insoluble free acid over time.
  - Mephenytoin and ethotoin are tertiary imides, but the strongly electron withdrawing hydantoin ring gives acidic character to the normally neutral secondary amides.

- Immediate and extended release capsules are available. Peak serum concentrations are achieved within 3 and 12 hours, respectively.
- Oral bioavailability of sodium phenytoin ranges from 70% to 100% based on the manufacturer. Once the patient is stabilized on a specific medication, careful monitoring of serum levels and/or therapeutic response is required if formulations are changed.
- Phenytoin exhibits nonlinear kinetics. The elimination half-life in adults is two to three times that of children (12 to 36 hours vs. 5 to 14 hours) because of an age-related decrease in inactivating metabolism.

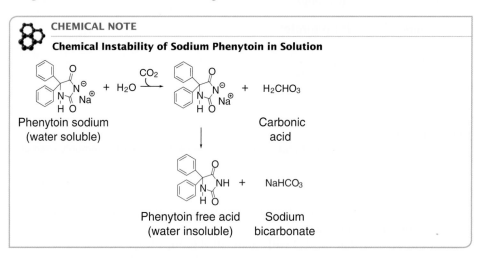

CHEMICAL NOTE

Chemical Instability of Sodium Phenytoin in Solution

Phenytoin sodium (water soluble) + H_2O $\xrightarrow{CO_2}$ + H_2CHO_3 Carbonic acid

Phenytoin free acid (water insoluble) + NaHCO3 Sodium bicarbonate

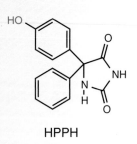

HPPH

Metabolism

- Approximately 60% to 70% of a dose of phenytoin is metabolized by CYP2C9-catalyzed aromatic hydroxylation to 5-hydroxyphenyl-5-phenylhydantoin (HPPH). Both glucuronide and sulfate conjugates are found in urine.
- Phenytoin induces CYP3A4 and glucuronosyltransferase (UGT) enzymes, leading to a significant risk of drug–drug interactions.
  - Dosage adjustments must be considered when phenytoin is used in combination with antiseizure drugs metabolized by these enzymes (Table 10.3).

Clinical Applications

- Phenytoin can be used as monotherapy for grand mal and complex partial seizures. It is also used to reverse convulsive status epilepticus.

Table 10.3   Examples of Metabolism-Based Drug–Drug Interactions Between Antiseizure Drugs

Antiseizure Drug	Increases Serum Levels of	Decreases Serum Levels of
Felbamate	Phenytoin, valproate	Carbamazepine
Lamotrigine	Carbamazepine epoxide	
Carbamazepine	Phenytoin	Clonazepam, felbamate, lamotrigine, oxcarbazepine, perampanel, pregabalin, rufinamide, tiagabine, topiramate, valproate, zonisamide
Oxcarbazepine	Phenobarbital, phenytoin	Lamotrigine
Phenobarbital		Carbamazepine, clonazepam, felbamate, lamotrigine, oxcarbazepine, perampanel, pregabalin, rufinamide, tiagabine, topiramate, valproate, zonisamide
Phenytoin		Carbamazepine, clonazepam, felbamate, lamotrigine, oxcarbazepine, perampanel, pregabalin, rufinamide, tiagabine, topiramate, valproate, zonisamide
Rufinamide	Phenobarbital, phenytoin	Carbamazepine, lamotrigine
Topiramate	Phenytoin	
Valproic acid	Carbamazepine, lamotrigine, phenobarbital, phenytoin, rufinamide	Oxcarbazepine, zonisamide
Vigabatrin	Carbamazepine	Phenytoin

- Phenytoin increases petit mal seizure frequency and is contraindicated in this seizure type.
- Phenytoin sodium injection is incompatible in D5W and with coadministered drugs that are acidic salts or which result in an admixture pH <11.
- Significant interactions between many antiseizure drugs, including phenytoin, occur from competition for serum protein.

Fosphenytoin

- Fosphenytoin is the disodium phosphate ester of phenytoin. As a highly water soluble prodrug, it is superior to the physically unstable sodium phenytoin for parenteral administration.
- Drug absorption after IM administration, and plasma phosphatase hydrolysis to phenytoin post-administration, is rapid.
- Fosphenytoin can be used for all phenytoin indications, including status epilepticus. Its adverse effect profile mimics that of phenytoin, including the risk for severe bradycardia.

Phenobarbital

Physicochemical and Pharmacokinetic Properties

- Phenobarbital's secondary imide groups are weakly acidic, allowing for the formation of a water soluble sodium salt for parenteral administration.
- Kinetics are linear. Oral absorption of the free acid is slow, but close to complete. Steady state serum concentrations may take up to 21 hours to achieve.

Metabolism

- Like phenytoin, phenobarbital is metabolized by aromatic hydroxylation and excreted as glucuronide or sulfate conjugates. It is also a potent inducer of CYP3A4 and UGT, leading to many drug–drug interactions (Table 10.3).

Clinical Applications

- Phenobarbital is preferred for the treatment of generalized seizures in infants ≤2 months. It can be used to treat convulsive status epilepticus.
- The sedative and cognitive adverse effects can be particularly problematic in school-aged children and the elderly.

Ethosuximide

Physicochemical and Pharmacokinetic Properties

- Ethosuximide free acid is marketed in capsule and oral solution form.
- Maximum serum concentrations are achieved within 3 hours. The half-life is 20 to 40 hours in adults and approximately 30 hours in children.

Metabolism

- While approximately 20% of a dose is excreted unchanged, ethosuximide undergoes inactivating ω-1 hydroxylation of the 2-ethyl moiety by CYP3A4 and CYP2E1 to form diastereomeric alcohols.
- 3-Hydroxyethosuximide has also been identified as a major metabolite in humans.

1-Hydroxyethosuximide diastereomers   3-Hydroxyethosuximide

Ethosuximide metabolites

> ⚠️ **ADVERSE EFFECTS**
>
> **Ureides and Iminostilbenes**
>
> *Phenytoin and fosphenytoin*
> - CNS toxicity (nystagmus, ataxia, sedation)
> - Reversible gingival hyperplasia
> - Bradycardia
> - Hypersensitivity reactions (rare, but serious)
>
> *Phenobarbital*
> - Sedation
> - Cognitive impairment
>
> *Ethosuximide*
> - Sedation
> - Gastrointestinal distress
>
> *Carbamazepine*
> - Anticholinergic effects (drowsiness, dry mouth, blurred vision)
> - Decreased serum 25-hydroxyvitamin D (calciferol)

Clinical Applications
- Ethosuximide is the preferred drug for absence seizures.
- In cases of absence status epilepticus, it is administered after a parenterally administered benzodiazepine (diazepam, midazolam, lorazepam).

Iminostilbenes

Carbamazepine and Analogs

SAR
- Carbamazepine and its active analogs are dibenzazepine-5-carboxamides.

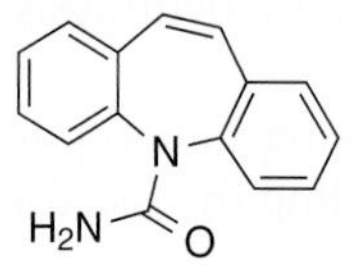

Carbamazepine
(Tegretol)

- Substitution is only allowed at azepine C10. The 10-keto analog is marketed as oxcarbazepine. While less potent, its mechanism is similar to the parent drug.
  - The reduced 10-hydroxycarbazepine metabolite (licarbazepine) is believed responsible for most of oxcarbazepine's activity.
- The *S*-acetate ester of 10-hydroxycarbazepine is marketed as eslicarbazepine.
  - Esterase-catalyzed hydrolysis to licarbazepine is believed to be responsible for its therapeutic effect.

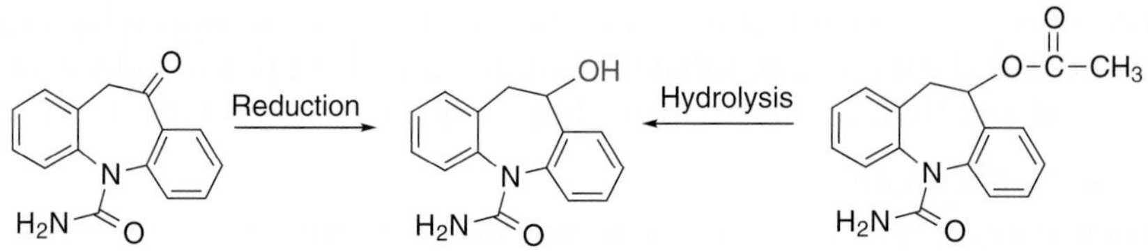

Oxcarbazepine
(Oxtellar) →Reduction→ 10-hydroxycarbazepine
(Licarbazepine, active metabolite) ←Hydrolysis← Eslicarbazepine acetate
(Aptiom)

Physicochemical and Pharmacokinetic Properties
- Oral absorption of carbamazepine is prolonged. Peak serum concentrations are attained within 5 and 12 hours for immediate and extended release formulations, respectively.
- The normal carbamazepine elimination half-life is 12 to 17 hours.
- The oral absorption of oxcarbazepine and eslicarbazepine is complete. The active 10-hydroxy metabolite of both drugs (licarbazepine) has a 9-hour half-life.

Metabolism
- CYP3A4 catalyzes the formation of the active 10,11-epoxide metabolite of carbamazepine. Prior to predominantly renal elimination, the epoxide is hydrolyzed by epoxide hydrase to an inactive *trans* diol.
- Carbamazepine induces its own CYP3A4-catalyzed metabolism, leading to a prolonged elimination half-life of up to 29 hours with chronic use.
  - UGT is also induced, leading to a high risk of drug–drug interactions. Oxcarbazepine also induces CYP3A4 and UGT.
- CYP is not significantly involved in the metabolism of oxcarbazepine or eslicarbazepine, so neither induces its own biotransformation.
- Hypersensitivity reactions are possible with carbamazepine and attributed to the formation of electrophilic 2-hydroxyiminostilbene and quinoneimine metabolites.

Carbamazepine-10,11-epoxide
(active metabolite)

Carbamazepine-10,11-*trans* diol
(inactive metabolite)

2-Hydroxyiminostilbene
(reactive metabolite)

Carbamazepine iminoquinone
(reactive metabolite)

Carbamazepine metabolites

Clinical Applications

- Along with phenytoin, carbamazepine is considered one of the safest antiseizure drug for the treatment of grand mal, partial, or mixed seizure disorders. It is ineffective in treating absence seizures.
- Oxcarbazepine and eslicarbazepine use is restricted to partial seizure disorder.
- Taking carbamazepine with food can decrease gastrointestinal adverse effects.
- Vitamin D supplements may be considered to counteract the decreased serum calciferol levels induced by carbamazepine and its analogs.
- Despite the lack of 2-hydroxy and iminoquinone metabolites, oxcarbazepine and eslicarbazepine can exhibit hypersensitivity cross-reactivity with carbamazepine.

> **Rare, but Serious, Carbamazepine Toxicities**
>
> While rare, the potential for serious hypersensitivity, dermatologic reactions (including Stevens–Johnson syndrome), hepatotoxicity, and blood dyscrasias with carbamazepine demand periodic blood and liver function tests. Carbamazepine carries a Black Box warning for dermatologic reactions and blood dyscrasias. Patients of Asian ancestry who carry the HLA-B*1502 allele are at highest risk for potentially fatal Stevens–Johnson syndrome.

Benzodiazepines (Fig. 10.4, Table 10.4)

SAR

- All benzodiazepines have an electron withdrawing group at C7.
- Imidazolo analogs have an abbreviated duration of action.
- Most benzodiazepines with a 3-OH have short durations. With a half-life of up to 72 hours, lorazepam is an exception to this SAR rule.
- Substitution at C3 with an ionizable COOH does not detract from activity and provides another site for water soluble salt formation (e.g., clorazepate dipotassium).
- Substitution on the C5-phenyl ring with an *o*-halogen increases potency and sedative potential, particularly in long-acting compounds.

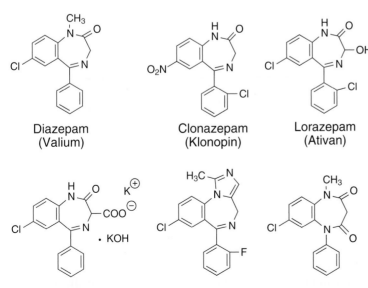

Diazepam
(Valium)

Clonazepam
(Klonopin)

Lorazepam
(Ativan)

Clorazepate dipotassium
(Tranxene)

Midazolam

Clobazam
(Onfi)

Figure 10.4 Most commonly used benzodiazepine antiseizure drugs.

Table 10.4 Benzodiazepines Used in Seizure Disorders

Benzodiazepine	Indications	Half-Life (hrs.)	Metabolism
Diazepam	All seizure types except absence, status epilepticus	36–46	N-demethylation to nordiazepam[a] (CYP3A4, 2C19), C3-hydroxylation[a] (CYP3A4)
Clorazepate	Partial	Rapidly decarboxylated to nordiazepam	C3-hydroxylation[a] (CYP3A4)
Clonazepam	Atonic, myoclonic, absence (third line)	30–40	Nitro reduction, aromatic hydroxylation (CYP3A4)
Lorazepam	Status epilepticus	16	C3-glucuronic acid conjugation
Midazolam	Status epilepticus	4–6	α-methyl hydroxylation[a] (CYP3A4)
Clobazam	Lennox–Gastaut associated seizures	36–42	N-demethylation[a] (CYP3A4), 4'-hydroxylation of norclobazam (CYP2C19)

[a]active metabolite

- Modest structural modifications of the diazepine ring system are permitted (e.g., clobazam).

Physicochemical Properties and Pharmacokinetics

- The benzodiazepine imino nitrogen is a weak base (e.g., pKa diazepam = 3.3). Water soluble hydrochloride salts are available for parenteral administration.
- Midazolam undergoes pH-dependent opening of the diazepine ring.
  - At pH <4.0, the ring opens to provide a more highly basic primary amine that enhances water solubility and absorption after IM administration compared to other benzodiazepines. IM solutions are buffered to pH 3.5.
  - At pH >4.0 (e.g., physiologic pH), the ring closes to regenerate the less basic imino nitrogen. Lipophilicity and distribution to central sites of action increase.
- Benzodiazepines are lipophilic and readily absorbed upon oral administration. With the exception of midazolam, absorption via the IM route is slower.
- Onset of action is in the 1- to 4-hour range for all benzodiazepines, but duration ranges from 3 to 4 hours for midazolam to 72 hours for lorazepam.

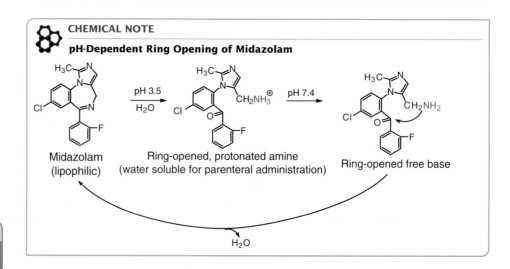

CHEMICAL NOTE

pH-Dependent Ring Opening of Midazolam

Midazolam (lipophilic)

Ring-opened, protonated amine (water soluble for parenteral administration)

Ring-opened free base

ADVERSE EFFECTS

Benzodiazepine Antiseizure Agents
- Sedation
- Ataxia, dizziness
- Behavior changes (clonazepam)
- Hypotension (lorazepam)
- Respiratory depression/arrest (midazolam Black Box Warning)

Specific Drugs

- Indications, half-lives, and major metabolic pathways of anticonvulsant benzodiazepines are provided in Table 10.4.
- Diazepam, lorazepam, and midazolam can be administered parenterally to halt seizures in status epilepticus. Midazolam's onset of action is faster if administered IM in an acidic (pH 3.5) vehicle. Lorazepam's duration of action is the longest.

- Diazepam is available as a rectal gel for control of acute, repetitive seizures in patients stabilized on other antiseizure drugs.
- Clorazepate is rapidly decarboxylated in the acidic environment of the stomach to nordiazepam, the major active metabolite of diazepam. Since this conversion is nonenzymatic, clorazepate is not classified as a prodrug.
- Clonazepam can precipitate grand mal seizures in patients with mixed seizure disorder, and can induce absence seizures when coadministered with valproic acid.
- Lorazepam's lack of CYP-mediated metabolism minimizes drug–drug interactions in comparison with other antiseizure benzodiazepines.

Miscellaneous Adjunct Antiseizure Drugs

- Several "one only" antiseizure drugs have been developed to fill the therapeutic gaps left by the more commonly administered "first-line" agents.
- Important chemical and therapeutic properties of these agents are provided in Table 10.5.

GENERAL DRUG–DRUG INTERACTIONS

Benzodiazepines, Barbiturates, Phenytoin, and Carbamazepine
- CYP3A4 inducers, inhibitors, or substrates (all but lorazepam)
- CYP2C19 inducers, inhibitors, or substrates (phenytoin, diazepam, clobazam, carbamazepine)
- CYP2C9 inducers, inhibitors, or substrates (phenytoin, carbamazepine)
- UGT substrates (phenytoin, phenobarbital, carbamazepine)
- Alcohol and other CNS depressants

Table 10.5 Miscellaneous Adjunct Antiseizure Agents

Antiseizure Drug	Indication(s)	Primary Transmitter/ Ion Channel Impacted	Major Metabolism
Ezogabine (Potiqua)	Partial	K$^+$	N-glucuronidation
Felbamate (Felbatol)	Partial	NMDA	Ester hydrolysis, aldehyde oxidase oxidation
Gabapentin (Neurotin)	Partial	Ca^{2+}	Minimal
Lacosamide (Vimpat)	Partial	Na$^+$	O-demethylation (CYP2C19)
Lamotrigine (Lamictal)	Partial, grand mal	Na$^+$	N-glucuronidation
S-(−)-Levetiracetam (Keppra)	Partial, myoclonic, grand mal	Synaptic vesicle 2A (SV2A) proteins	Amide hydrolysis

(Continued on following page)

Table 10.5 Miscellaneous Adjunct Antiseizure Agents (*Continued*)

Antiseizure Drug	Indication(s)	Primary Transmitter/ Ion Channel Impacted	Major Metabolism
Perampanel (Fycompa) 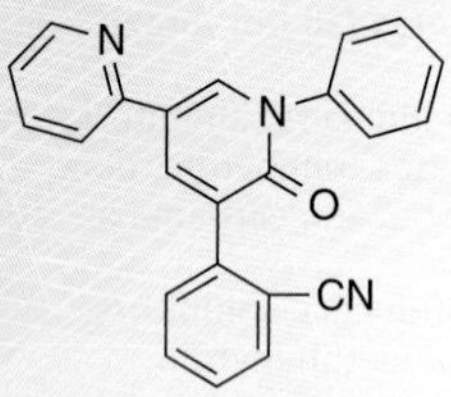	Partial	AMPA	Hydroxylation at multiple sites on phenyl and 2-pyridyl rings (CYP3A4)
Pregabalin (Lyrica) 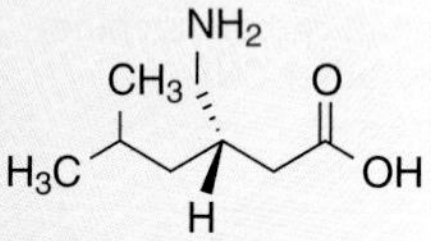	Partial	Ca^{2+}	Minimal
Rufinamide (Banzel) 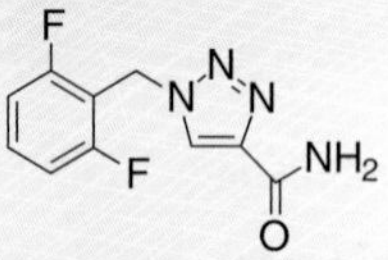	Lennox–Gastaut seizures	Na^+	Hydrolysis
Tiagabine (Gabitril) 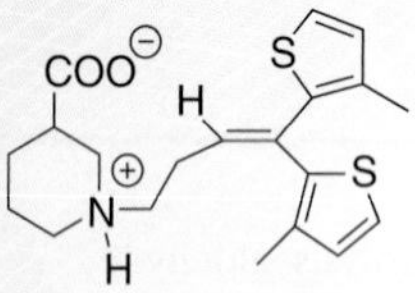	Partial	GABA	Oxidation at thiophene C5 (CYP3A4)
Topiramate (Topamax) 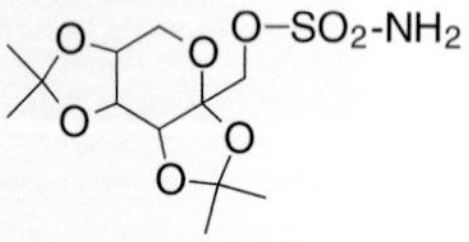	Partial, grand mal, Lennox–Gastaut seizures	Na^+ Cl^-, AMPA/KA	Hydrolysis of the 2,3-isopropylidine to the 2,3-diol, hydroxylation at C10, sulfamylester cleavage
Valproic acid (Depakene, sodium salt Depakote) 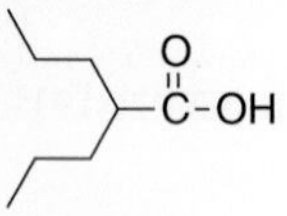	Absence, mixed seizures involving absence	Ca^{2+}, GABA	ω-oxidation, β-oxidation, acyl glucuronidation
Vigabatrin (Sabril) 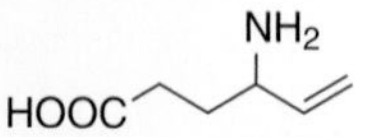	Complex partial (refractory)	GABA	Minimal
Zonisamide (Zonegran) 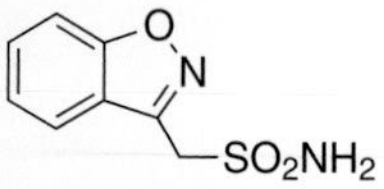	Partial	Na^+, Ca^{2+}	N-acetylation, ring hydrolysis, reduction (CYP3A4)

Clinical Applications

- Black Box Warnings are associated with several adjunct antiseizure drugs (Table 10.5). Some toxicities are associated with reactive metabolites (e.g., felbamate, valproate).
- Some agents can induce rebound seizures if withdrawn too quickly (e.g., rufinamide, vigabatrin). In the absence of life-threatening toxicity, gradual dose tapering is recommended.

- While rare, pregabalin can induce myopathy, including potentially fatal rhabdomyolysis. Creatinine kinase levels should be monitored.
- Antacids decrease the oral absorption of gabapentin.
- Antiseizure drugs that are extensively metabolized, or which influence levels and/or activity of metabolizing isoforms, must be used with caution in the elderly and in patients with hepatic dysfunction.

> **⚠ ADVERSE EFFECTS**
>
> *Ezogabine* – Black Box Warning: retinopathy with potential vision loss; urinary retention, syncope, nephritis.
>
> *Felbamate* – Black Box Warnings: aplastic anemia, hepatotoxicity.
>
> *Gabapentin* – sedation, ataxia.
>
> *Lacosamide* – dizziness, diplopia, possible PR interval prolongation.
>
> *Lamotrigine* – Black Box Warning: dermatologic toxicity, including Stevens–Johnson syndrome; aseptic meningitis, syncope, diplopia, sedation.
>
> *Levetiracetam* – sedation, syncope, ataxia.
>
> *Perampanel* – Black Box Warning: sedation, syncope, ataxia, psychosis.
>
> *Pregabalin* – sedation, ataxia, behavioral effects, potentially severe constipation, dependence syndrome.
>
> *Rufinamide* – shortened QT interval, sedation, syncope, ataxia.
>
> *Tiagabine* – sedation, syncope, abnormal thinking, depression, psychosis (adverse effects more common than with other adjunct antiseizure drugs).
>
> *Topiramate* – sedation, syncope, altered cognition, memory and speech, induction of acute open angle glaucoma and/or kidney stones.
>
> *Valproic acid* – Black Box Warning: anorexia, nausea, potentially fatal hepatotoxicity (believed due to the 4-ene metabolite).
>
>
>
> *Vigabatrin* – Black Box Warning: retinopathy, irreversible visual field disturbances.
>
> *Zonisamide* – sedation, syncope, cognitive impairment, kidney stones, sulfonamide-induced allergic response.

Chemical Significance

If not appropriately managed, antiseizure drug use can be fraught with opportunities for toxicity and patient harm. While the choice of agent may not be as unrestricted as in some disease states, the pharmacist's knowledge of drug chemistry can play an important role in assuring effective therapy. For example:

- If a patient in status epilepticus requires the fastest possible therapy, consider midazolam since its labile diazepine ring system allows for greater water solubility and permits more rapid absorption from IM sites than other benzodiazepines.
- If a patient suffering from absence seizures is prescribed a ureide, make sure that a succinimide (not a hydantoin or oxazolidindione) is selected.
- If a patient has had dosing problems with intravenously administered phenytoin, consider use of the more highly solution-stable phenytoin prodrug, fosphenytoin.
- If a patient stabilized on carbamazepine for complex partial seizures requires adjunct therapy, evaluate whether an agent resistant to CYP3A4 or 2C9 metabolism is feasible.
- If a patient on potentially hepatotoxic drug therapy (e.g., acetaminophen, diclofenac, dasatinib, all of which can generate electrophilic quinoneimines) requires antiseizure therapy, avoid carbamazepine if possible, as it can also generate a hepatotoxic quinoneimine metabolite.

Helping people get the most out of life despite potentially disabling health challenges is the key goal of medication therapy management. Working collaboratively with other health care professionals, the pharmacist's understanding of antiseizure drug action can help patients with this chronic disorder live their lives with a greater sense of personal safety, freedom, and overall quality.

Review Questions

1 Which benzodiazepine antiseizure drug would be the most water soluble at acidic pHs?

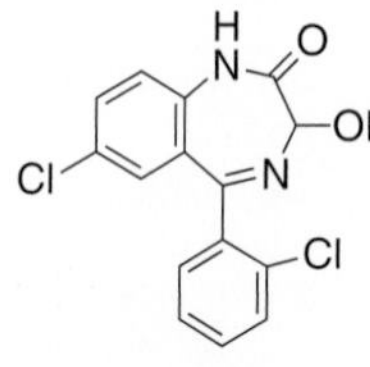

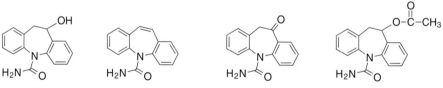

1 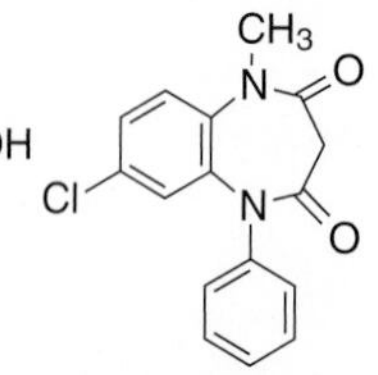   2 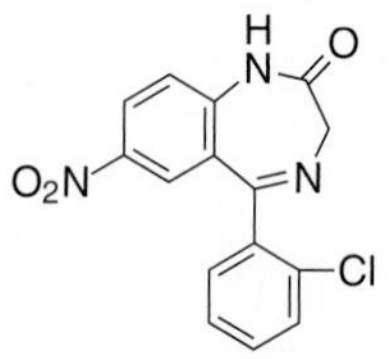   3

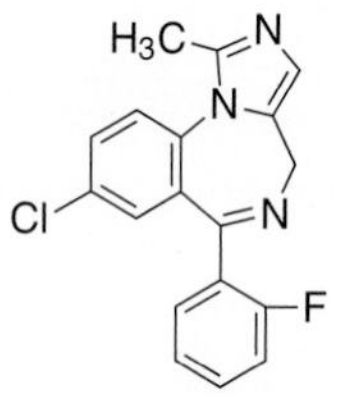

4 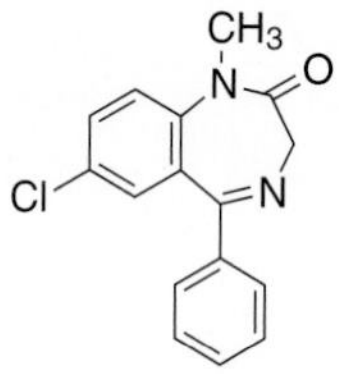   5

A. Antiseizure drug **1**
B. Antiseizure drug **2**
C. Antiseizure drug **3**
D. Antiseizure drug **4**
E. Antiseizure drug **5**

2 The iminostilbene metabolite drawn below is responsible for the therapeutic activity of: (select all correct answers)

Iminostilbene metabolite   Carbamazepine   Oxcarbazepine   Eslicarbazepine acetate

A. Carbamazepine
B. Oxcarbazepine
C. Eslicarbazepine

3 Which ureide would be the most appropriate to treat an adolescent with absence seizures?

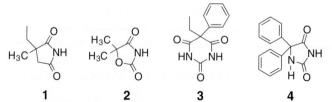

1   2   3   4

A. Antiseizure drug **1**
B. Antiseizure drug **2**
C. Antiseizure drug **3**
D. Antiseizure drug **4**

4 Which antiseizure drug drawn below is a true prodrug?

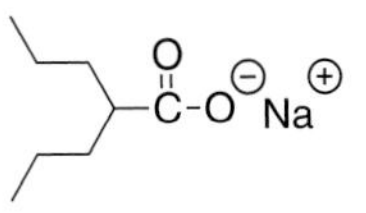

1 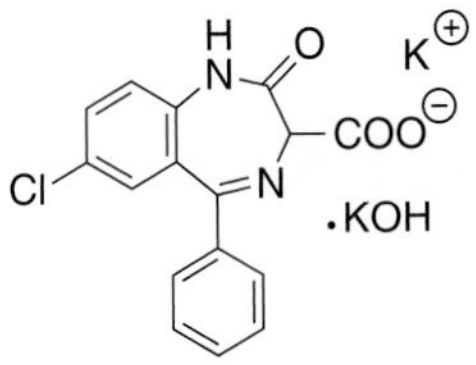 **2** 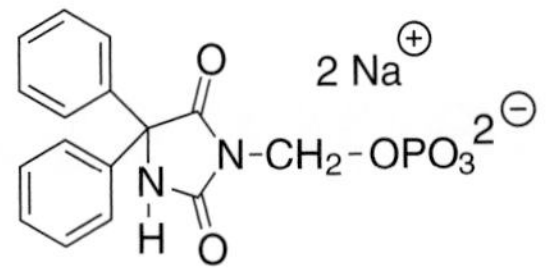 **3**

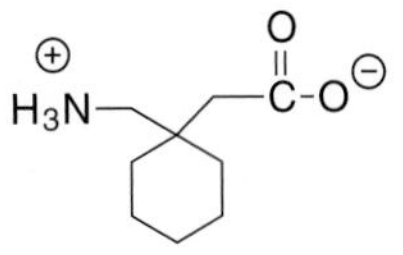

4 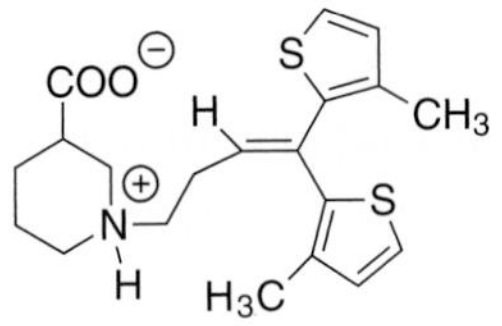 **5**

A. Antiseizure drug **1**
B. Antiseizure drug **2**
C. Antiseizure drug **3**
D. Antiseizure drug **4**
E. Antiseizure drug **5**

5 Which antiseizure drug metabolites are hepatotoxic electrophiles? (select all correct answers)

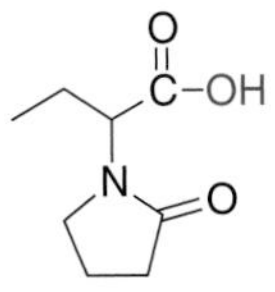

1
(Levetiracetam metabolite)

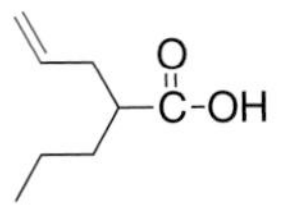

2
(Valproic acid metabolite)

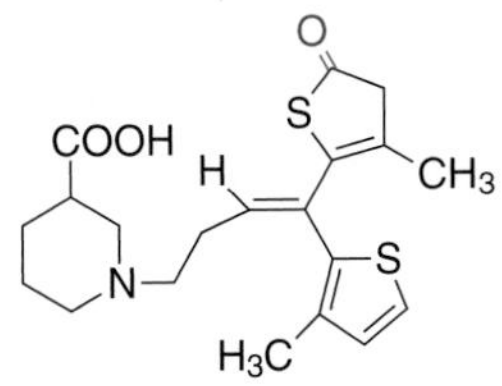

3
(Tiagabine metabolite)

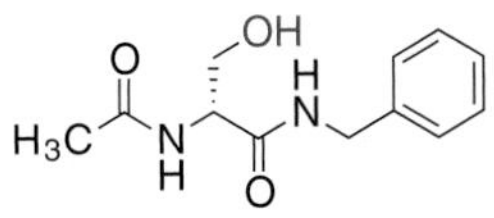

4
(Lacosamide metabolite)

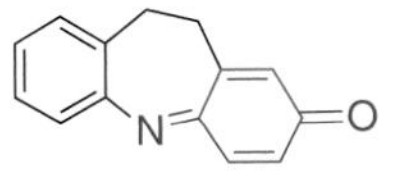

5
(Carbamazepine metabolite)

A. Antiseizure drug **1**
B. Antiseizure drug **2**
C. Antiseizure drug **3**
D. Antiseizure drug **4**
E. Antiseizure drug **5**

11

Antidepressants

Learning Objectives

Upon completion of this chapter the student should be able to:

1. Discuss the hypotheses associated with the biologic basis of depression.
2. Identify the site and mechanism of action of the following classes of antidepressants:
   a. reuptake inhibitors (i.e., SNRIs, SSRIs, NSRIs, NDRI, SARI)
   b. α_2-adrenergic/selective serotonin antagonist
   c. monoamine oxidase inhibitors
   d. mood stabilizers
3. List structure–activity relationships (SARs) of antidepressant drugs: for SNRIs and SSRIs.
4. Identify clinically significant physiochemical and pharmacokinetic properties.
5. Discuss, with chemical structures, the clinically significant metabolism of the antidepressant agents and how the metabolism affects their activity.
6. Discuss the clinical application of the antidepressant drugs.
7. Discuss potential drug–drug interactions and the mechanism of these interactions especially as they relate to SNRIs, SSRIs, NSRIs, and SARIs.

Introduction

A person experiencing depression feels worthlessness, helplessness, and hopelessness with an overwhelming sense of despair. Depression can immobilize an individual affecting the way one eats and sleeps, the way one feels about oneself, and the way one thinks about things. Without treatment, symptoms can last for weeks, months, or years. On the other hand depression is one of the most treatable mental illnesses with 80% of those seeking treatment showing improvement. Depression affects more than 19 million Americans each year ranking in the top five diseases in the United States. At least 50% of those with major depression will suffer one or more repeated episodes of depression during their adult lifetime. Depression affects at least 1 in 50 children under the age of 12 years, mostly girls during puberty, suggestive of hormonal implications in the disease. About half of all cases of depression go unrecognized and untreated, and approximately 29% of those with severe depression (bipolar disorder) will attempt suicide sometime in their life. Depression and the resulting suicide is the 11th leading cause of death in the United States. Depression in the elderly (17% to 35%) often goes undiagnosed and untreated, causing needless suffering for the family and the individual who could otherwise live a fruitful life.

Biologic Basis of Depression

Current theories regarding the causes of depression support the role of the neurotransmitters serotonin (5-hydroxytryptamine, 5-HT) and norepinephrine (NE) in depression and their interrelationships with each other and with dopamine (DA) (Fig. 11.1). Although the precise nature of the depression is not fully understood at the level of the chemistry in the brain, several theories have been proposed to explain the role of NE and 5-HT as causative factors in depression.

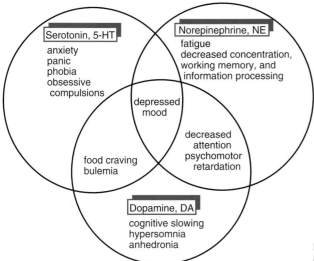

Figure 11.1 Neurotransmitter deficiency syndromes and their interactions.

In addition, there is evidence for hypercholinergic mechanisms in depression involving a cholinergic imbalance. Specifically, a role for a central nicotinic cholinergic (nACh) pathway is suggested. While most clinically prescribed antidepressants target NE and 5-HT neurotransmitter systems, some also act as potent noncompetitive antagonists of nACh receptors at clinically effective doses for treatment of major depression.

Monoamine Hypothesis

- The monoamine hypothesis proposes that depression results from a deficiency in 5-HT and/or NE with a possible involvement of DA although its role may be dependent upon the initial effects of 5-HT and NE.
- The clinical efficacy/toxicity produced by antidepressants is thought to be due to a combination of inhibitory actions on targets involving reuptake transporters for the neurotransmitters: 5-HT, NE, DA, acetylcholine receptors (nAChRs), histamine H_1 receptors, and α_1-adrenergic receptors.

Receptor Sensitivity Hypothesis

- This hypothesis proposes that it is not simply the level of NE or 5-HT in the synapse that matters but, rather, the sensitivity of the postsynaptic receptors to these neurotransmitters.
- With depression postsynaptic receptors become hypersensitive to NE and 5-HT because of their depletion in the synaptic cleft.
- Low level of NE and 5-HT at their respective receptors can lead to changes in the receptors leading to increasing receptor sensitivity (hypersensitivity) and an increase in receptor numbers on the neuronal cell membrane, events that may correlate with the start of depression.
- Relief from the symptoms of depression comes from a normalization of receptor sensitivity by increasing the concentration of NE and 5-HT in the synaptic cleft.
- Antidepressants increase the concentration of NE and/or 5-HT in the synaptic cleft causing the postsynaptic neuron to compensate by decreasing receptor sensitivity (desensitization) and the number of receptor sites (downregulation).

Permissive Hypothesis

- The permissive hypothesis emphasizes the importance of the balance between 5-HT and NE in regulating mood, not the absolute levels of these neurotransmitters or their receptors.

Hormonal Hypothesis

- The hormonal hypothesis suggests that changes in the hypothalamus–pituitary–adrenal axis (HPA) (hyperactivity) can influence the levels of 5-HT, NE, and ACh release and function.
- This hypothesis suggests that nAChRs play important roles in mediating stress-related/depression-inducing effects of ACh stimulating the HPA through activation of nAChRs.
- Antidepressants may reduce symptoms of depression through blockade of nAChRs involved with stress-induced activation of the HPA.

Antidepressants in Psychotherapy

- Treatment of depression is based upon distinct mechanisms of actions of the antidepressants with the drugs falling into specific classes based upon these mechanisms (Table 11.1).

Table 11.1   Antidepressant Classes of Drugs with Generic and Trade Names

Drug Classes	Abbreviation	Drug Examples	Trade Names
Reuptake Inhibitors:			
Selective Norepinephrine Reuptake Inhibitor	SNRIs	Amoxapine	Generic
		Atomoxetine	Strattera
		Desipramine	Norpramin
		Maprotiline	Generic
		Nortriptyline	Aventyl, Pamelor
		Protriptyline	Vivactil
Selective Serotonin Reuptake Inhibitor	SSRIs	Citalopram	Celexa
		Escitalopram	Lexapro
		Fluoxetine	Prozac, Sarafem
		Fluvoxamine	Luvox, Selfemra
		Paroxetine	Paxil, Brisdelle, Pexeva
		Sertraline	Zoloft
Norepinephrine and Serotonin Reuptake Inhibitor	NSRIs	Amitriptyline	Generic
		Clomipramine	Anafranil
		Desvenlafaxine	Pristiq, Khedezla
		Doxepin	Silenar
		Duloxetine	Cymbalta
		Imipramine	Tofranil
		Milnacipran	Fetzima, Savella
		Trimipramine	Surmontil
		Venlafaxine	Generic
Norepinephrine and Dopamine Reuptake Inhibitor	NDRI	Bupropion	Forfivo, Wellbutrin, Zyban
Serotonin-2 Antagonists/ Serotonin Reuptake Inhibitor	SARI	Trazodone	Oleptro
		Vilazodone	Viibryd
		Aripiprazole	Abilify
		Vortioxetine	Brintellix
α_2-Adrenergic/Selective Serotonin Antagonist	NaSSA	Mirtazapine	Remeron
Monoamine Oxidase Inhibitors	MAOIs	Phenelzine	Nardil
		Tranylcypromine	Parnate
		Moclobemide	Not available in US
Mood Stabilizers		Lithium carbonate	Lithobid
		Valproic acid	Depakene
		Carbamazepine	Tegretol, Teril, Epitol, etc.

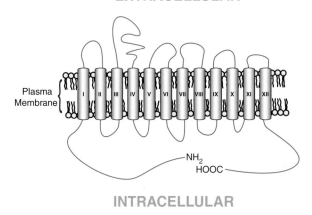

EXTRACELLULAR

Plasma Membrane

I II III IV V VI VII VIII IX X XI XII

NH$_2$
HOOC

INTRACELLULAR

Figure 11.2 The transporter proteins of human dopamine (DAT), norepinephrine (NET), and serotonin (SERT) all consist of 12 transmembrane domains (TMs) connected by intracellular and extracellular loops. There is considerable similarity between these transporter proteins consisting of 620 amino acids in DAT, 617 amino acids in NET, and 630 amino acids in SERT. Monoamine affinity appears to be associated with the extracellular loop between TM1–TM3 and TM8–TM12.

Monoamine Signaling

- The intensity and duration of monoamine signaling at synapses involves release of the neurotransmitter from nerve terminals followed by the rapid reuptake via high-affinity plasma membrane transporter proteins.
- The neurotransmitter's action is terminated by binding to a transporter protein in the extracellular fluid of the synapse transporting the monoamine across the presynaptic plasma membrane back into the intracellular fluid.
- Transporter proteins are specific to their respective neurotransmitter: serotonin reuptake transporter (SERT), NE reuptake transporter (NET), and DA reuptake transporter (DAT) (Figs. 11.2 and 11.3).

Mechanisms of Action of Reuptake Inhibitor Antidepressants

- The transporter proteins are targets for five classes of antidepressant drugs (Table 11.1).
- The binding of the monoamine to the respective transporter protein within the perisynaptical neurons leads to a fast passive diffusion of the transporter across the presynaptic membrane terminating its actions.
  - SNRIs selectively block NET.

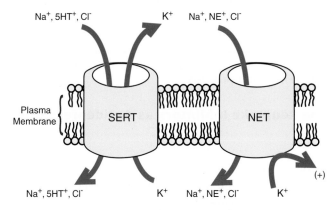

EXTRACELLULAR

Na$^+$, 5HT$^+$, Cl$^-$ K$^+$ Na$^+$, NE$^+$, Cl$^-$

Plasma Membrane

SERT   NET

(+)

Na$^+$, 5HT$^+$, Cl$^-$ K$^+$ Na$^+$, NE$^+$, Cl$^-$ K$^+$

INTRACELLULAR

Figure 11.3 Model of the NET and SERT and the ion-coupled NE and 5-HT reuptake. Reuptake of 5-HT is dependent on the cotransport of Na$^+$ and Cl$^-$ and countertransport of K$^+$. Reuptake of NE is dependent on the cotransport of Na$^+$ and Cl$^-$ with intracellular K$^+$ stimulation but without K$^+$ efflux. DAT is not shown, but it has a similar need for cotransport ions for binding of the DA to membrane-bound Na$^+$/K$^+$ ATPase with two Na$^+$ and one Cl$^-$ ions.

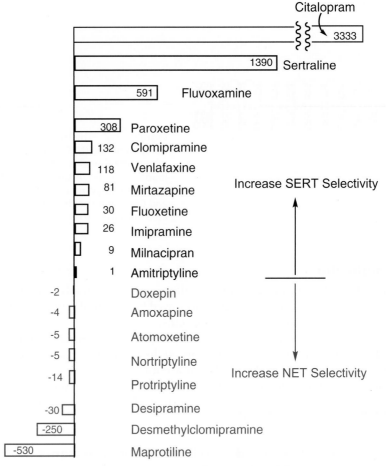

Figure 11.4 In vitro selectivity ratios for reuptake inhibitors.

- SSRIs selectively block SERT (NOTE: the degree of selectivity is based upon protein binding ratios of the inhibitors for SERT versus NET (Fig. 11.4 [Ratio > 100 suggest clinical selectivity])).
- TCAs and NSRIs block the reuptake transporters for both NE and 5-HT.
- SARIs block SERT in addition to action at postsynaptic sites.
- The drug bupropion as a DNRI acts at DAT sites and NET sites (Fig. 11.5).
- It is generally agreed that increased levels of NE and 5-HT (and presumably DA) in the synaptic cleft accounts for the antidepressant action of the reuptake inhibitors.
- Antidepressants may also contribute to relief of depression by decreasing the expression of their respective transporter proteins.

Selective Norepinephrine Reuptake Inhibitors as Antidepressants

SAR

- The SNRIs with one exception are TCAs (Fig. 11.6).
- The secondary amine functionality is essential for SNRI selectivity (i.e., nortriptyline vs. amitriptyline – see discussion below and Fig. 11.11).
- The side chain, attached to any one of the atoms in the central seven-membered ring, or within the open chain of atomoxetine must be three carbon atoms (either saturated or unsaturated) terminating with an amine.
- The tricyclic ring system is important for its lipophilicity, with little significance on selectivity for inhibiting NET or SERT, but does appear important for DA transporter inhibition.

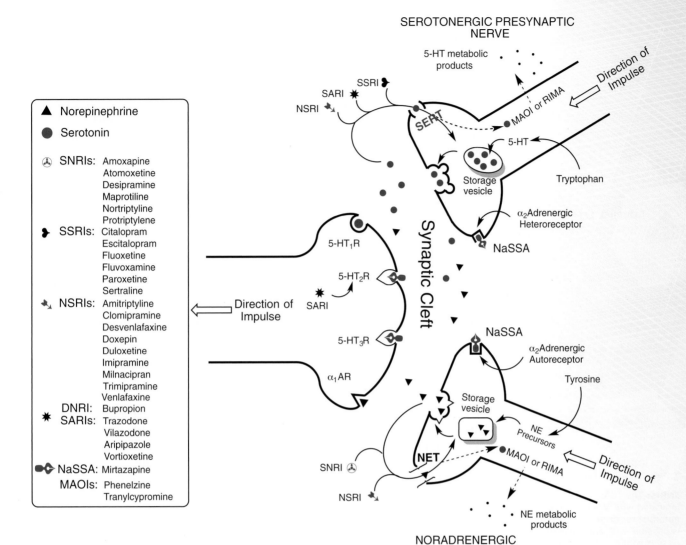

Figure 11.5 Site of action of antidepressants.

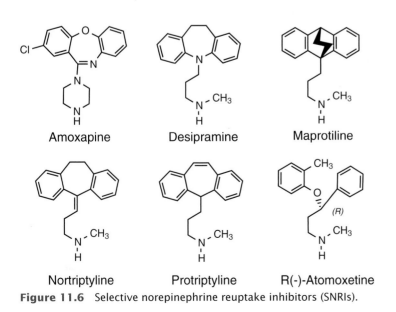

Figure 11.6 Selective norepinephrine reuptake inhibitors (SNRIs).

- Substituting a halogen (i.e., chlorine; amoxapine or clomipramine) onto the aromatic ring enhances preferential affinity for NET or SERT.
- Branching the propyl side chain (e.g., 2-methyl group) significantly reduces affinity for both the SERT and NET.
- Atomoxetine, is the only open chain SNRI approved for antidepressant use in the United States. Structurally similar drugs (SSRIs: fluoxetine, paroxetine) are marketed but are not SNRI selective. The selectivity for NET over SERT is the result of 2-aryl substitution (2-methyl). 4-Aryl substitution gives rise to SERT selective agents.

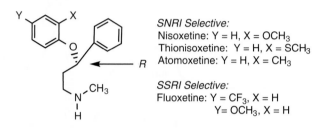

SNRI Selective:
Nisoxetine: Y = H, X = OCH₃
Thionisoxetine: Y = H, X = SCH₃
Atomoxetine: Y = H, X = CH₃

SSRI Selective:
Fluoxetine: Y = CF₃, X = H
Y= OCH₃, X = H

⚠ ADVERSE EFFECTS

SNRIs: The family of TCAs has many undesirable adverse effects associated with blocking muscarinic receptors (anticholinergic), α_1-adrenergic receptors, H_1-receptors (antihistamine) and sodium channels. The adverse effects include dry mouth, constipation, and bladder complaints (i.e., weak urine stream, painful urination, and difficulty emptying bladder). Adverse effects reported for atomoxetine (Non-TCA) include dry mouth, nausea, decreased appetite, and insomnia.

DRUG–DRUG INTERACTIONS

SNRIs: The TCAs are contraindicated with MAOIs and the potential for drug interactions is possible with anticholinergic or sympathomimetic drugs. A 2-week washout of the TCA should be allowed before dosing with an MAOI. The high protein binding of secondary TCAs could lead to possible drug–drug interactions with other highly bound drugs with narrow therapeutic indices.

- The *R* enantiomer of atomoxetine is significantly more active than the *S* isomer.

Physicochemical and Pharmacokinetic Properties

- The secondary TCAs are rapidly and well absorbed following oral administration.
- Although the pharmacokinetics are similar within the secondary and tertiary amine groups, they are different between the two groups (Table 11.2; also see Table 11.4 for NSRIs).
- The secondary amine TCAs have relatively high bioavailability.
- Serum plasma levels are reached with 1 to 2 days; antidepressant action is not reached for 2 to 3 weeks.

Metabolism

- N-Demethylation to a primary amine is a major route of metabolism and inactivation for the SNRIs. CYP2D6 is the usual cytochrome isoform responsible for this metabolism.
- Aromatic hydroxylation gives active metabolites in the case of amoxapine (Table 11.2) and an inactive 2-hydroxydesipramine.
- Aliphatic hydroxylation at C10 of nortriptyline by CYP2D6 leads to an active metabolite (Table 11.2).
- Atomoxetine undergoes N-demethylation and aryl hydroxylation (Fig. 11.7).

Table 11.2 Pharmacokinetics of the Selective Norepinephrine Reuptake Inhibitors (SNRIs)

Parameters	Amoxapine	Atomoxetine	Desipramine	Maprotiline	Nortriptyline	Protriptyline
Oral bioavailability (%)	ND	>70%	60–70	66–75	32–79	77–93
Protein binding (%)	92	>97	91	88	92	92
Elimination half-life (h)	8 (8–30)	4 (3–6) (EM)ᵃ 17–21 (PM)	30 (12–30)	43 (27–58)	30 (18–44)	67–89
Major active metabolites	8-OH, 7-OH	4-hydroxy	None	N-Desmethyl	10-E-hydroxy-	None
Excretion (%)	Urine 60 Feces 7–18	Urine Feces	Urine Feces	Urine 65 Feces 30	Urine 40 Feces minor	Urine 50 Feces minor
Plasma half-life (h)	8	4.3	14–62	21–52	18–93	54–198

h, hours; *ND*, not determined
ᵃ*EM*, extensive metabolizer, *PM*, poor metabolizer

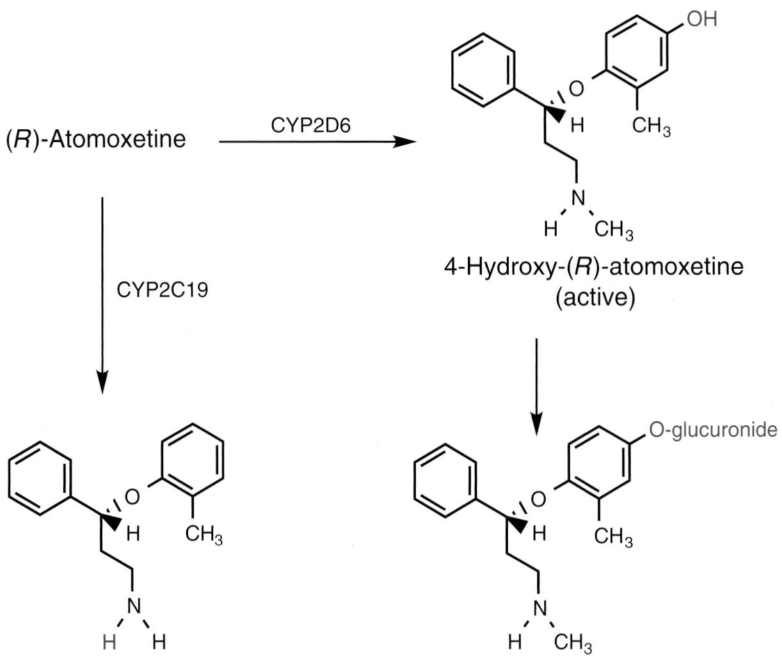

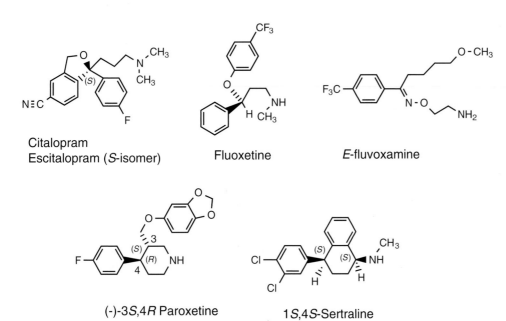

Figure 11.7 Metabolism of (R)-atomoxetine.

Selective Serotonin Reuptake Inhibitors as Antidepressants
(Table 11.1 and Fig. 11.8)

MOA

- The monoamine hypothesis of depression suggests that low levels of 5-HT in neural synapse of the serotonergic pathway may lead to depression.
- The SSRIs preferentially act to inhibit SERT (minimal or no affinity on NE/DA transporters, Fig. 11.5) with high and selective affinity for SERT (Fig. 11.5).
- The 5-HT in the synaptic cleft ultimately causes a downregulation of pre- and postsynaptic receptors, a reduction in the amount of 5-HT produced in the CNS and a reduction in the number of SERTs expressed.
- These compensatory responses at receptors and transporters are thought to produce the antidepressant effects of SSRIs.

Citalopram
Escitalopram (S-isomer)

Fluoxetine

E-fluvoxamine

(-)-3S,4R Paroxetine

1S,4S-Sertraline

Figure 11.8 Selective serotonin reuptake inhibitors (SSRIs).

SAR

- Stereochemistry of the SSRIs significantly affects selectivity for SERT:
  - Escitalopram (*S* isomer) >> selectivity over (±)-citalopram.
  - (±)-Fluoxetine the *R* isomer > potency than *S* isomer (*S* isomer has >> affinity for SERT than the *R* isomer), but the N-desmethyl metabolite (norfluoxetine) the *S* isomer is a stronger inhibitor of SERT.
  - *E* isomer of fluvoxamine is essential for inhibition of SERT.
  - (-)-3*S*,4*R*-Paroxetine is the marketed and most active SERT inhibitor.
  - 1*S*,4*S*-Sertraline is the more active SERT inhibitor over the remaining three isomers.
- Electronegative 4-substitution (i.e., CF_3 in fluoxetine and fluvoxamine, F in paroxetine, 3,4-diCl in sertraline) is necessary for selectivity and affinity for SERT.
- 2,4- or 3,4-substitution of phenoxy in fluoxetine leads to loss of selectivity.
- N-methylation or F-replacement with H- or CH_3- leads to loss of activity in paroxetine.

Physicochemical and Pharmacokinetic Properties

- The SSRIs are well absorbed orally (Table 11.3).
- The SSRIs are highly lipophilic and are highly plasma protein bound.
- Current SSRIs tend to be characterized by high volumes of distribution, which results in relatively long plasma concentration (typically 4 to 8 hours).
- Only sertraline and citalopram exhibit linear pharmacokinetics.

> ### ⚠ ADVERSE EFFECTS
>
> **SSRIs:** The adverse effects observed for the SSRIs include nausea, diarrhea, anxiety, agitation, insomnia, and sexual dysfunction. Sexual dysfunction is reported in both men and women (e.g., decreased libido, anorgasmia, ejaculatory incompetence, ejaculatory retardation, or inability to obtain or maintain an erection). Pharmacological similarities among the SSRIs suggest that the effects on sexual function should be similar for each SSRI and the effects may be dose-related. Orgasm difficulties and impotence occurred more frequently with paroxetine as compared with sertraline and fluoxetine.

> ### 💊 DRUG–DRUG INTERACTIONS
>
> **SSRIs:** The most serious drug–drug interaction for the SSRIs is their potential to produce the "serotonin syndrome," which develops within hours or days following the addition of another serotonergic agent. Symptoms of 5-HT syndrome include agitation, diaphoresis, diarrhea, fever, hyperreflexia, incoordination, confusion, myoclonus, shivering, or tremor. This syndrome is most common with MAOIs and SSRIs and necessitates a washout ranging from 2 to 5 weeks depending on the plasma half-life of the SSRI. The drug interaction between TCAs and SSRIs is importance because of the potential for the development of toxic TCA concentrations and 5-HT syndrome.
>
> Drugs such as paroxetine and fluoxetine which are metabolized by CYP2D6 have the greatest potential for drug–drug interaction with other drugs which are metabolized principally by CYP2D6 and have a narrow therapeutic index.
>
> The SSRIs are highly protein bound and may affect the pharmacodynamic effect of other protein-bound drugs with narrow therapeutic indices (e.g., warfarin). The changes appear to be clinically significant, however, only for fluoxetine, fluvoxamine, and paroxetine. Close monitoring of prothrombin time and international normalized ratio may be necessary.

Metabolism

- The SSRIs are extensively metabolized by CYP isoforms (Table 11.3).

Table 11.3 Pharmacokinetics of the Selective Serotonin Reuptake Inhibitors (SSRIs)

Parameters	Citalopram (Escitalopram)	Fluoxetine	Fluvoxamine	Paroxetine	Sertraline
Oral bioavailability (%)	80 (51–93)	70	>50	50	20–36
Protein binding (%)	~56 (70–80)	95	77	95	96–98
Elimination half-life (h)	36 (27–32) Elderly ~48	50	15–20	22	24 (19–37)
Major active metabolites	N-desmethyl	N-desmethyl	None	None	N-desmethyl
Major cytochrome isoform	2C19, 2D6, and 3A4	2D6	2D6	2D6	3A4
Excretion (%)	Feces 80–90 Urine <5	Urine 25–50	Urine ~40	Urine ~50	Urine 51–60 Feces 24–32
Plasma half-life (h)	36 (23–75)	24–106 (norfluoxetine 7–15 days)	7–63	24	22–35
Steady-state concentration in days	7	~4 wks	10	7–14	2–3 wks 7–10 elderly

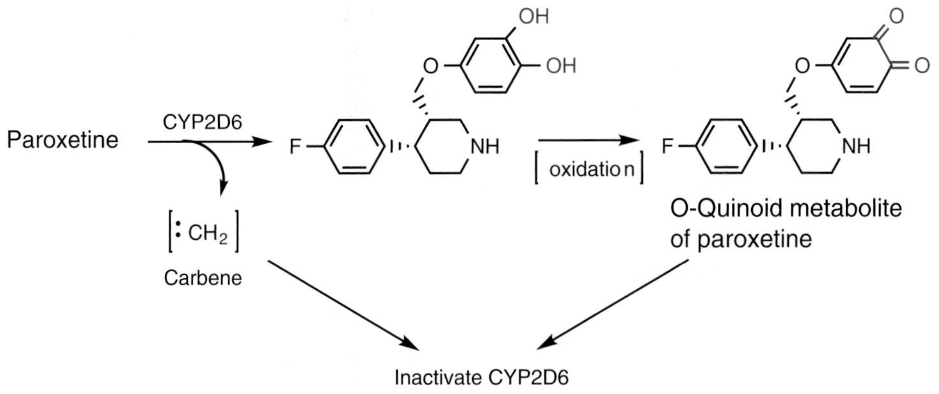

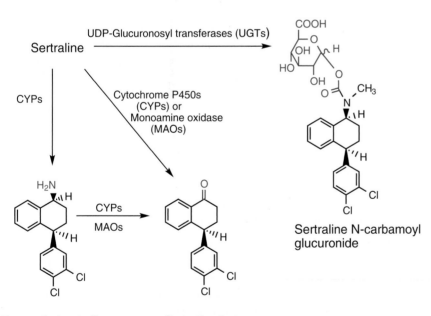

Figure 11.9 Metabolism of paroxetine and inactivation of CYP2D6.

- The N-methyl SSRIs undergo N-demethylation to active metabolites, which are then excreted in urine and feces (Table 11.3).
- The half-lives are variable (Table 11.3) and dependent on the presence and plasma concentration of an active metabolite (fluoxetine and its active metabolite norfluoxetine both have long half-lives resulting in an extended washout period before switching from fluoxetine to another SSRI or MAOI).
- Paroxetine metabolism is shown in Figure 11.9. Formation of *o*-quinone or carbene may account for inhibition of CYP2D6.
- Sertraline is extensively metabolized (Fig. 11.10).

Figure 11.10 Metabolic products formed from sertraline metabolism.

Norepinephrine and Serotonin Reuptake Inhibitors (NSRIs) as Antidepressants

The clinical outcome of administering tertiary amine TCA NSRIs ultimately results in dual inhibition of 5-HT and NE presynaptic reuptake. Several of these agents also bind to other neuroreceptors, resulting in a narrow therapeutic window with significant adverse effects. Clinical studies suggest that dual-acting inhibitors of 5-HT and NE reuptake may be more beneficial than selective inhibitors in managing depression and has led to the development of a second class of NSRIs, the nontricyclics (Fig. 11.11).

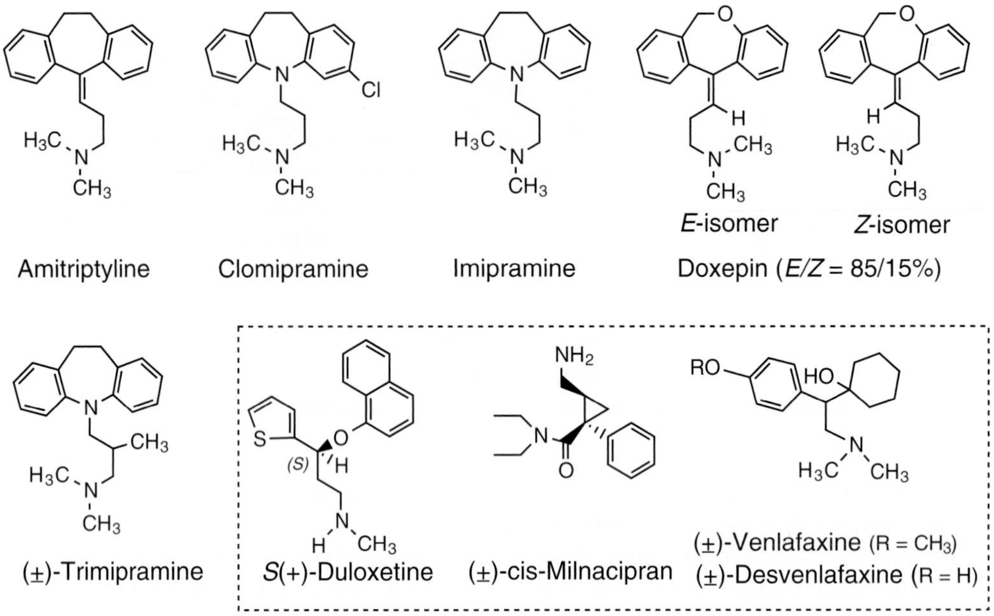

Figure 11.11 Norepineph-rine and Serotonin Reuptake Inhibitors (NSRIs) made up of tertiary amine tricyclic antide-pressants (TCA) and noncyclic agents.

MOA

- The NSRI antidepressant drugs (Table 11.1) block both NET and SERT as a result of their affinity for these transporter proteins (low NE:5-HT potency ratio) (Figs. 11.4 and 11.5).
- The rapid N-demethylation of the tertiary amine gives the secondary TCAs (see SNRIs above) and thus it is difficult to assign the degree of selectivity to the tertiary amine TCAs (e.g., amitriptyline to nortriptyline and imipramine to desipramine).
- Clomipramine and imipramine preferentially inhibit 5-HT reuptake over NE, but their N-desmethyl metabolites inhibit NE reuptake at the presynaptic neuro-nal membrane.
- The oxygen replacement of a methylene group in the two-carbon bridge leads to an E-Z mixture (85:15) in doxepin which is used as such clinically (NOTE: the E-isomer inhibits NET while the Z-isomer is selective for SERT).
- N-demethylated doxepin is also a selective inhibitor of NET.
- The nontricyclic NSRIs do not differ significantly from their tricyclic counter-parts and result in inhibition of both SERT and NET.

Physicochemical and Pharmacokinetic Properties of NSRIs

- The pharmacokinetic properties of the TCA NSRIs is presented in Table 11.4. Unique properties are listed below:
  - Amitriptyline with its extended conjugation exhibits sensitivity to photo-oxidation; thus solutions must be protected from light.

Table 11.4 Pharmacokinetics of the Tricyclic Norepinephrine and Serotonin Reuptake Inhibitors (NSRIs)

Parameters	Amitriptyline	Clomipramine	Doxepin	Imipramine	Trimipramine
Oral bioavailability (%)	31–61	~50	13–45	29–77	44
Protein binding (%)	95	96–98	~80	89–95	95
Elimination half-life (h)	10–26	32 (19–37)	11–16	11–25	9–11
Major active metabolites	N-desmethyl	N-desmethyl	N-desmethyl	N-desmethyl	N-desmethyl
Cytochrome P450 major isoform	2D6	2D6	2D6 and 2C19	2D6 and 2C19	2D6 and 2C19
Excretion (%)	Urine 25–50	Urine 51–60	Urine ~50	Urine ~40	Feces 80–90
	Feces minor	Feces 24–32	Feces minor	Feces minor	Urine <5
Plasma half-life (h)	21 (10–46)	34 (22–84)	8–24	18 (9–24)	9

Table 11.5 Pharmacokinetics of Nontricyclic Norepinephrine and Serotonin Reuptake Inhibitors

Parameters	Venlafaxine	O-Desvenlafaxine	Milnacipran	Duloxetine
Oral bioavailability (%)	~15	80	90	>70%
Protein binding (%)	25–30	30	15–30	>90
Elimination half-life (h)	4 (2–7)	11	~8	11–16
Major active metabolites	ODV	None	None	4-hydroxy
Cytochrome P450 major isoform	2D6	3A4	None	2D6/1A2
Excretion (%)	Urine 87	Urine 45	Urine 90	Urine >70
	Feces 2		Feces minor	Feces 15
Plasma half-life (h)	5	~11	~8	12 (8–17)
Time to steady state conc.	3 d	3 d	32–48 h	3–7 d

- Imipramine is available as both a hydrochloride and pamoate salts. The insoluble pamoate allows for one daily dosing.
- The bioavailability of trimipramine is strongly influenced by whether or not the patient is a poor metabolizer (without CYP2D6) or ultrarapid metabolizer (two and three active genes of CYP2D6) leading to bioavailability of 44% versus 12%, respectively, and varying clinical effectiveness.
- The pharmacokinetic data for the nontricyclic NSRIs is shown in Table 11.5.

Metabolism
- As indicated in Table 11.4, N-demethylation of the tertiary amine TCAs leads to the secondary amines, which are NET inhibitors.
- Clomipramine undergoes extensive metabolism (Fig. 11.12).
- The E,Z-isomers of doxepin exhibit oxidative N-demethylation by different CYP isoforms: E-doxepin by CYP2D6 and CYP2C19; Z-doxepin primarily by CYP2C19. Following demethylation, the ratio of the two isomers is 1:1 with the Z-isomer being a stronger inhibitor of SERT and NET than its E-counterpart.
- The two diastereomers of trimipramine are oxidized by different CYP isoforms (Fig. 11.13).
- Venlafaxine is metabolized (~70%) by CYP2D6 to the O-desvenlafaxine (ODV) which in turn is relatively stable to further metabolism.
- Duloxetine undergoes extensive oxidative metabolism (e.g., 4-hydroxy, 4,6-dihydroxy, 5,6-dihydroxy) to inactive products (Fig. 11.14).

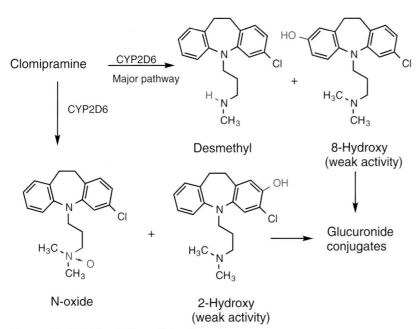

Figure 11.12 Metabolism of clomipramine.

ADVERSE EFFECTS

Tertiary Amine TCA NSRIs: Because of their potent and multiple pharmacodynamic effects at histamine H_1, muscarinic, and α_1-adrenergic receptors, the tertiary TCAs exhibit greater anticholinergic, antihistaminic, and α_1-antiadrenergic adverse effects than the secondary TCAs. Increased cardiotoxicity or frequency of seizures is higher for the tertiary TCAs, because they are potent inhibitors of sodium channels, leading to changes in nerve conduction. Cardiotoxicity occurs at plasma concentrations approximately 5 to 10 times higher than therapeutic blood levels. These concentrations can occur in individuals who take an overdose of the TCA or who are slow metabolizers.

DRUG–DRUG INTERACTIONS

NSRIs: Two concerns for drug–drug interactions should be considered. Since the NSRIs are bound strongly to plasma proteins the potential exist for drug–drug interactions with other highly bound drugs, which could lead to drug displacement. This is especially true for drugs with a narrow therapeutic index. Secondly as previously indicated with secondary amine TCAs, drug–drug interactions are possible, especially with CYP2D6 inhibitor or inducers.

ADVERSE EFFECTS

Nontricyclic NSRIs: The most common adverse effects include nausea, headaches, anxiety, insomnia, drowsiness, and loss of appetite. Some patients may experience dry mouth and constipation. Seizures have been reported for patients on venlafaxine. Antidepressants in general carry a risk for suicidal thinking and behavior especially in children.

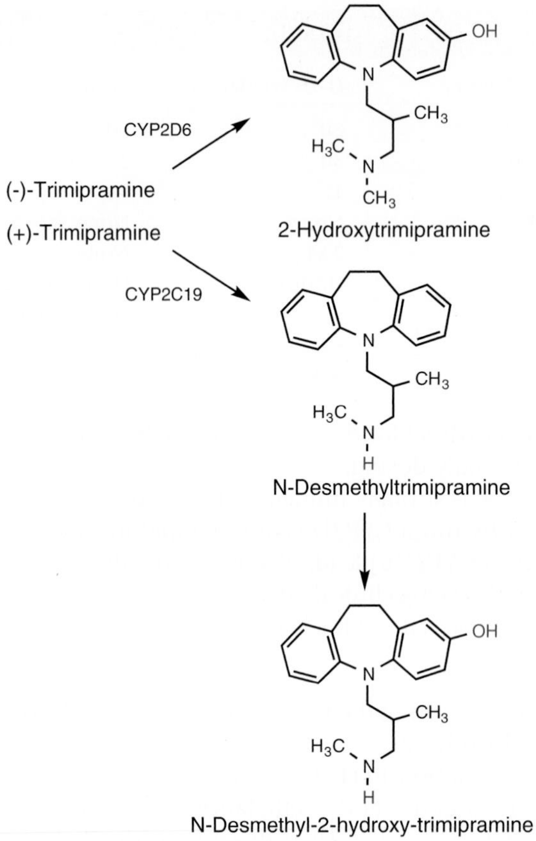

Figure 11.13 Preferential metabolism of (±)-trimipramine.

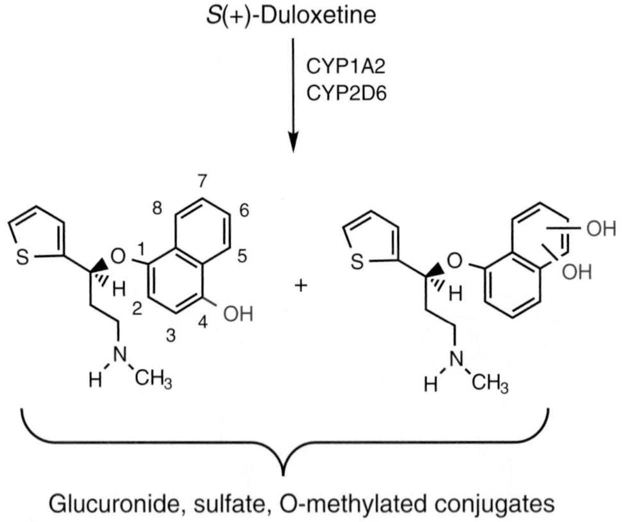

Figure 11.14 Metabolism of duloxetine.

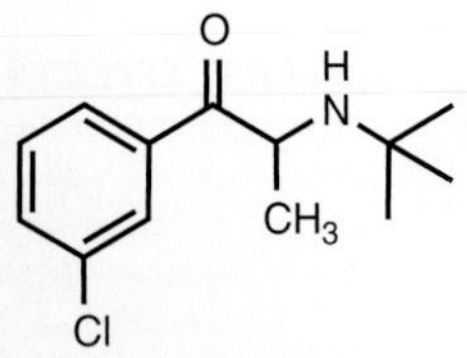

Glucuronide, sulfate, O-methylated conjugates

Norepinephrine and Dopamine Reuptake Inhibitor (NDRI)

A somewhat unique compound, bupropion, has a mix of activity at monoamine receptors resulting in clinical utility in the treatment of depression and smoking cessation.

Bupropion

MOA

- Bupropion appears to be a selective inhibitor of DA reuptake at the DA presynaptic neuronal membrane (DAT) (major effect) and at NET (Table 11.1).
- In addition, the drug induces the release of DA and NE from presynaptic receptors.

- Complicating these mechanisms, bupropion's principal metabolites contribute to its antidepressant action (see the discussion of pharmacokinetics and metabolism below).
- Bupropion is also a noncompetitive antagonist of nicotine at nAChRs (i.e., noncompetitive blocking of $\alpha_3\beta_4$, $\alpha_4\beta_2$, and α_7 neuronal nAChRs).

Physicochemical and Pharmacokinetic Properties
- Bupropion is absorbed from the GI tract, with a low oral bioavailability (5%–20%).
- The drug is bound to plasma proteins (80%) and has a half-life of 10 to 17 hours.
- The active metabolites (see below) have half-lives of 21 to 26 hours with elimination primarily occurring in the urine.

Metabolism
- Alkyl hydroxylation catalyzed by CYP2B6 leads to racemic hemiketals and reduction giving the *threo* and *erythro* isomers with antidepressant activity (Fig. 11.15).
- Further metabolism to inactive metabolites (3-chlorobenzoic acid and conjugates) occurs prior to excretion in the urine.

> **⚠ ADVERSE EFFECTS**
>
> **Bupropion:** The most serious adverse effect associated with bupropion is epileptic seizures. The seizure potential appears to be dose dependent and more prominent with immediate release preparations. Daily doses below 450 mg are recommended (normal dose is 300 mg/d). Other adverse effects include insomnia and headaches.

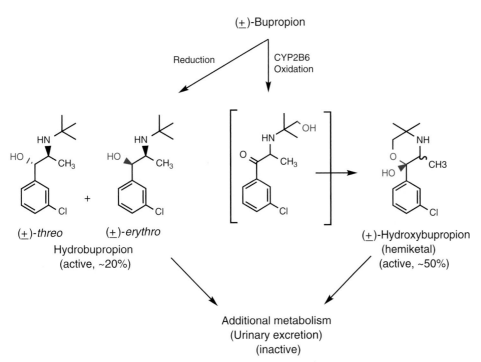

Figure 11.15 Bupropion metabolism.

> **Bupropion for Smoking Cessations**
>
> Bupropion reduces the discomfort and craving associated with withdrawal symptoms following smoking cessation. The current presumed mechanism of action of bupropion involves modulation of dopaminergic and noradrenergic systems as a result of its inhibition of DAT and NET. As nicotine CNS concentrations drop with smoking cessation the firing rates of noradrenergic neurons increase while bupropion and its active metabolites reduce the firing rates of these noradrenergic neurons. Bupropion's ability to block the nicotinic receptors also may prevent relapse by attenuating the reinforcing properties of nicotine.

Serotonin-2 Antagonists/Serotonin Reuptake Inhibitors

These drugs may also be referred to as serotonin receptor modulators (SRMs) and are chemically related in that they are all phenylpiperazine derivatives and include trazodone, nefazodone, vilazodone and the atypical antipsychotic agent

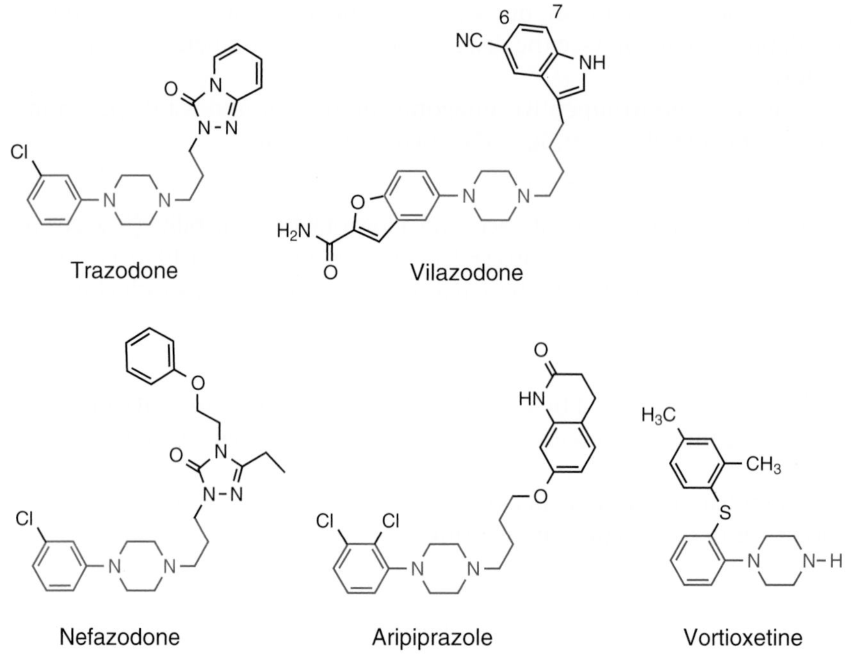

Figure 11.16 Serotonin-2 antagonists/serotonin reuptake inhibitors (SARI).

aripiprazole, and vortioxetine (Fig. 11.16). Nefazodone is not commonly used today due to severe hepatotoxicity.

MOA

- Although the exact mechanism of action is unknown, these drugs act as selective inhibitors at SERT (Fig. 11.5).
- The SARIs potentiate the postsynaptic effects of 5-HT via several mechanisms:
  - Trazodone is an antagonist at 5-HT$_{2A}$ receptors (Fig. 11.5).
  - Vilazodone is a partial agonist at 5-HT$_{1A}$ receptors.
  - Aripiprazole is an antagonist at 5-HT$_{2A/7}$, a partial agonist at 5-HT$_{1A/2C}$, and a partial agonist at D$_2$ receptors (atypical antipsychotic, Chapter 7).
  - Vortioxetine is an antagonist at 5-HT$_3$, 5-HT$_7$, and 5-HT$_{1D}$, a partial 5-agonist at HT$_{1B}$, an agonist at 5-HT$_{1A}$ receptor while blocking reuptake at SERT.
- In the case of trazodone, its mechanism is complicated by the presence of its metabolite, *m*-chlorophenylpiperazine, which is a 5-HT$_{2C}$ antagonist (see metabolism section).
- At therapeutic dosages these drugs do not appear to affect the reuptake of DA or NE within the CNS.
- Long-term administration of trazodone may lead to postsynaptic serotonergic receptor modification and influence antidepressant action.

Physicochemical and Pharmacokinetic Properties

The pharmacokinetic properties of the SARIs are shown in Table 11.6.

Table 11.6 Pharmacokinetics of Serotonin-2/Serotonin Reuptake Inhibitors (SARIs)

Parameters	Trazodone	Vilazodone	Aripiprazole	Vortioxetine
Oral bioavailability (%)	65 (60–70)	72 (with food)	90	75
Protein binding (%)	90–95	96–99	~100	96
Elimination half-life (h)	7 (4–9)	5–6 days	75	57
Major active metabolites	*m*-chlorophenyl-piperazine	None	Dehydroaripiprazole	Several
Cytochrome major isoform	3A4	3A4	3A4/2D6	Several
Excretion (%)	Urine 70–75	Urine 20	Urine 27	Urine 59
	Feces 25–30	Feces 65	Feces 60	Feces 26
Plasma half-life (h)	6 (4–8)	~25	3–5	—
Steady-state concentration in days	3–7	3	14	—

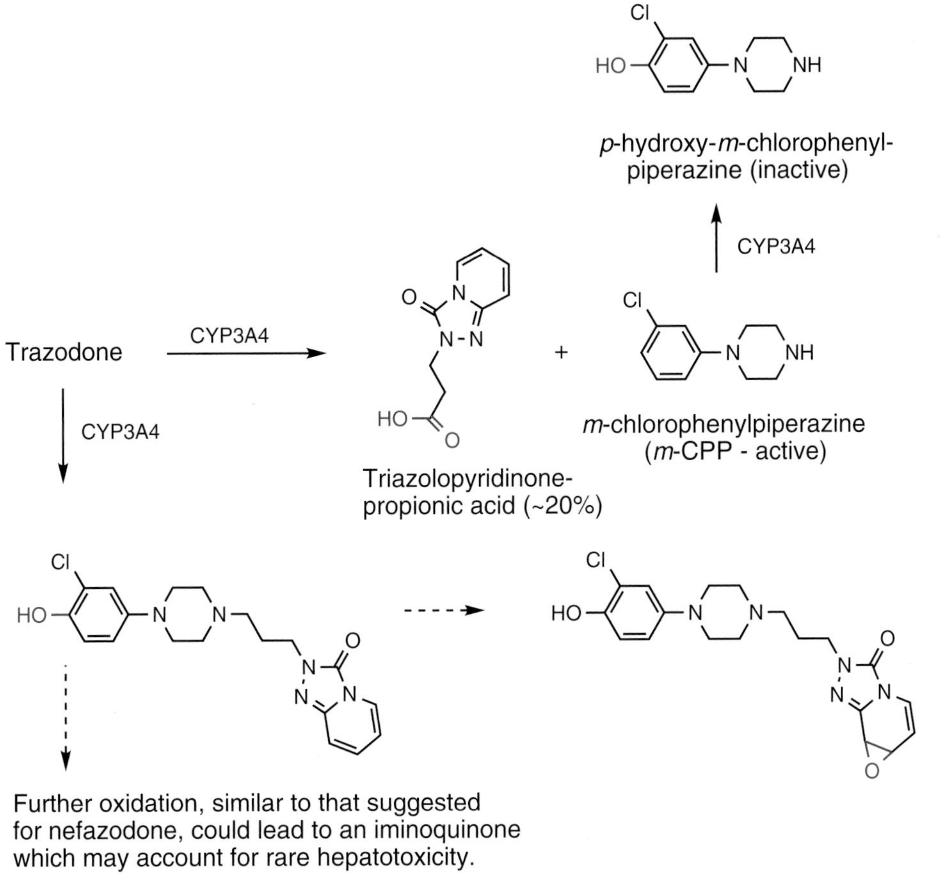

Figure 11.17 Metabolism of trazodone.

Metabolism

- The SARIs are extensively metabolized primarily by CYP3A4 giving rise to both active and inactive metabolites.
- Trazodone undergoes both N-dealkylation and aromatic hydroxylation (Fig. 11.17). While *m*-CPP is active by virtue of its affinity for multiple 5-HT receptors as well as SERT its action may lead to depressive and anxiogenic effects.
- Aripiprazole undergoes a similar metabolism (catalyzed by CYP2D6/3A4) to trazodone, with N-dealkylation, aromatic hydroxylation, and dehydroaripiprazole which possess antidepressant activity. The dichlorophenylpiperazine is not reported as being an active metabolite.

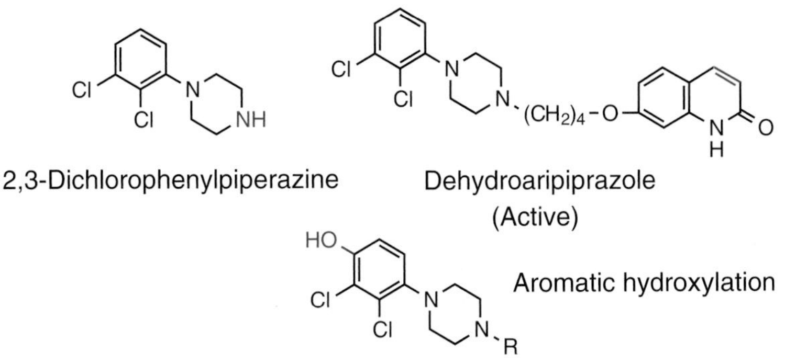

2,3-Dichlorophenylpiperazine   Dehydroaripiprazole (Active)

Aromatic hydroxylation

- Vilazodone's major metabolites result from CYP3A4 oxidation of the alkyl chain and aryl oxidation, and carboxylesterase hydrolysis of the amide moiety (Fig. 11.18). None of these metabolites has significant biologic activity.
- Vilazodone is reported to be a strong inhibitor of CYP2C8.

ADVERSE EFFECTS

SARIs: The warning for antidepressants in general is that in some individuals (people 25 or younger) these drugs can worsen depression leading to suicidal thoughts and attempts. Common adverse effects include nausea, vomiting, diarrhea, insomnia, drowsiness, dizziness, blurred vision, dry mouth, headache, or changes in sexual dysfunction.

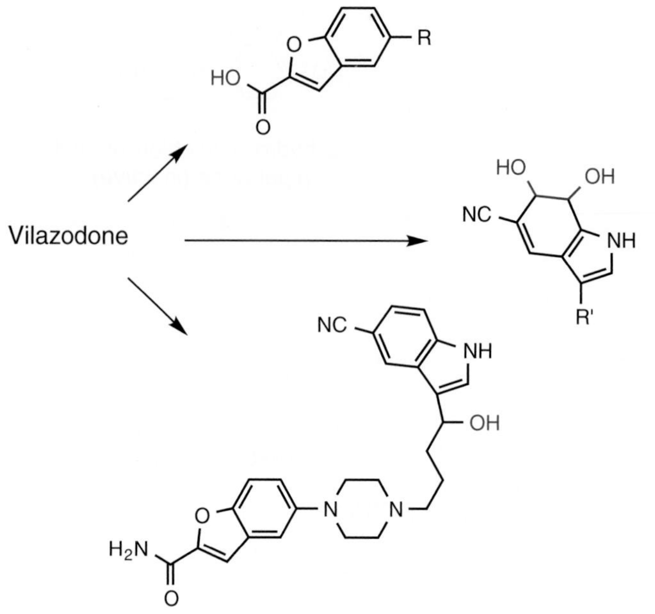

Figure 11.18 Metabolism of vilazodone.

DRUG–DRUG INTERACTIONS

SARIs: Drug–drug interactions between the SARIs and CYP3A4 inhibitors or inducers should be considered as possibilities. In the case of vilazodone if administered with ketoconazole increased plasma levels of vilazodone will occur and the dose of vilazodone should not exceed 40 mg/day. Carbamazepine will reduce the plasma levels of vilazodone and the dose should be monitored. Vilazodone is a potent inhibitor of CYP2C8 and can lead to drug–drug interaction if combined with drugs that are metabolized by CYP2C8. Finally, aripiprazole combinations with CYP2D6 inhibitors or inducers have not been reported to cause clinically significant interactions, but one should be aware of the possibility of interactions with drugs that are substrates for CYP2D6.

- Vortioxetine is a substrate for several CYP450 isoforms as well as alcohol dehydrogenase. The four major metabolites are all inactive and appear as their conjugates in the urine and feces (Fig. 11.19).

Clinical Applications of Reuptake Inhibitors

- The SSRIs, SNRIs, and NSRIs have proven to be effective for a broad range of depressive illnesses, dysthymia, several anxiety disorders, and bulimia.

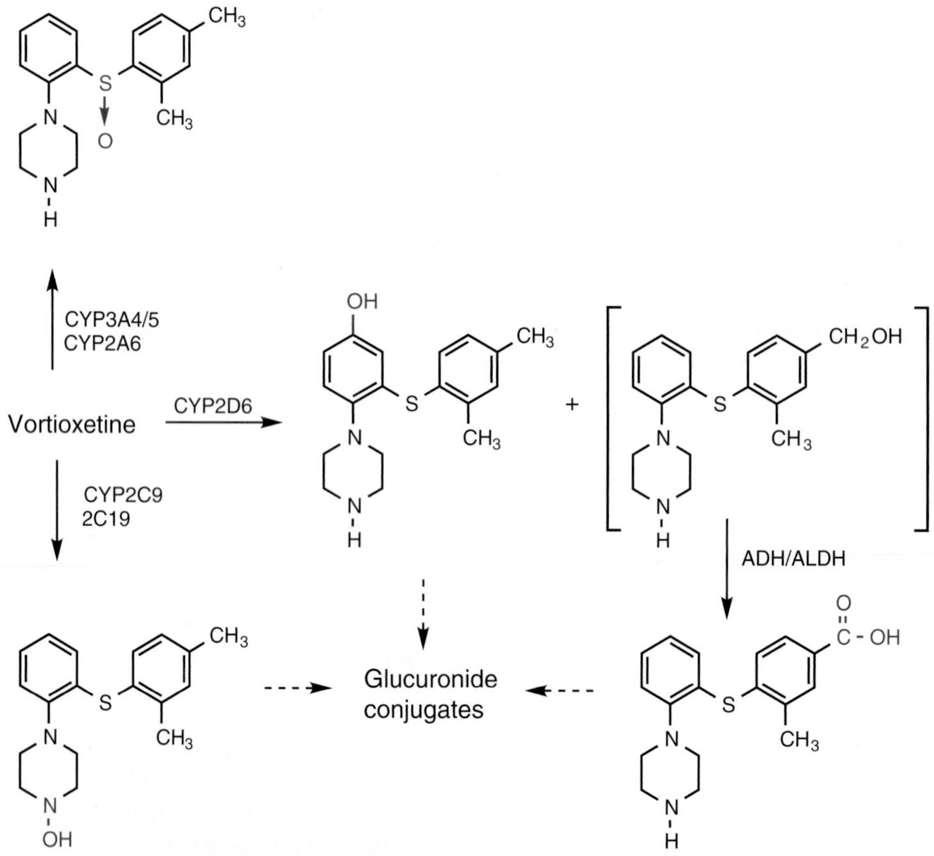

Figure 11.19 Metabolism of vortioxetine.

- The SSRIs are first-line treatments for panic disorder, obsessive–compulsive disorder, social phobia, posttraumatic stress disorder, and bulimia. They also may be the best medications for treatment of dysthymia and generalized anxiety disorder.
- Despite pharmacological differences in their mechanisms of actions, the general view has been that all antidepressants are of equal efficacy.
- Within the last few years questions have been raised about whether dual acting antidepressants outperform single mechanism of action drugs.
- Antidepressants are often prescribed and the dosage adjusted before the most effective antidepressant or combination of antidepressants is found.
- Improvements may be seen during the first few weeks of treatment with an antidepressant, but in most cases the drug must be taken regularly for 3 to 4 weeks (or longer) before the full therapeutic effect occurs.
- Patient compliance can become an issue, with the patient often stopping medication as soon as he or she feels better. It is important to continue the medication for at least 4 to 9 months (or longer) to prevent a recurrence of the depression.
- Antidepressants alter the brain chemistry; therefore, they must be stopped gradually to give the brain time to adjust. Patients who are prescribed antidepressants should be informed that discontinuation/withdrawal symptoms may occur on stopping, missing doses, or on reducing the dose of the drug. These symptoms are usually mild and self-limiting, but may be severe, particularly if the drug is stopped abruptly.

α_2-Noradrenergic/Selective Serotonin Antidepressants (NaSSA)

A single representative drug of this class of antidepressants, mirtazapine, has a unique and complex mechanism of action, which is described below. Mirtazapine may be considered another arylpiperazine, but since it distinguishes itself mechanistically from this class it is referred to as a tetracyclic antidepressant.

S,R-Mirtazapine

MOA

- Mirtazapine acts as an antagonist at presynaptic α_2-adrenergic autoreceptors and α_2-adrenergic heteroreceptors in noradrenergic and serotonergic presynaptic nerves, respectively (Fig. 11.5).
- By blocking these receptors the drug increases the levels of both NE and 5-HT in the synaptic cleft leading to increased activity at α_1-adrenergic receptors (α-AR) and 5-HT$_1$ receptors (5-HT$_1$R).
- Mirtazapine also selectively acts as an antagonist at 5-HT$_2$ and 5-HT$_3$ receptors (5-HT$_2$R, 5-HT$_3$R).
- *S*-Mirtazapine is reported to be responsible for the antagonism at the autoreceptors and 5-HT$_2$ receptors while the *R*-isomer is responsible for the antagonism at heteroreceptors and 5-HT$_3$ receptors.
- Mirtazapine also exhibits moderate anticholinergic activity at muscarinic sites and is a potent histamine 1 (H$_1$) blocking agents.

Physicochemical and Pharmacokinetic Properties

- Mirtazapine is absorbed via the oral route with a bioavailability of ~50%.
- The drug is highly bound to plasma proteins (85%).

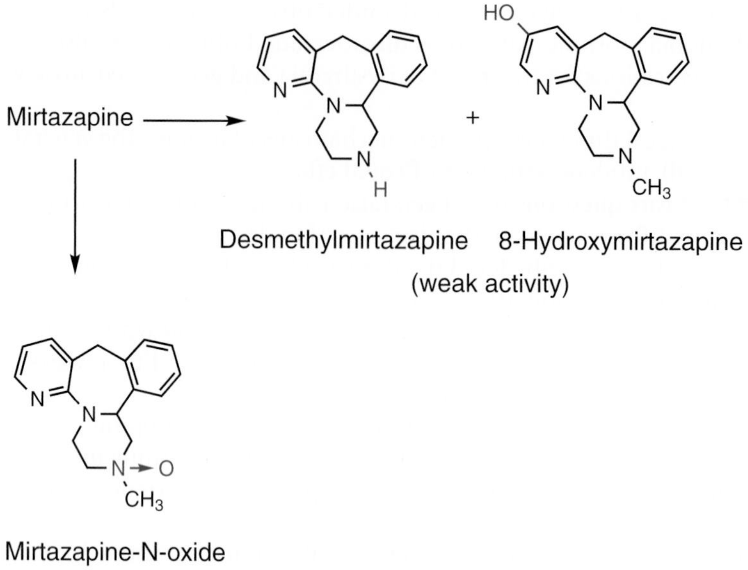

Mirtazapine ⟶ Desmethylmirtazapine   8-Hydroxymirtazapine

(weak activity)

Mirtazapine-N-oxide

Figure 11.20 Metabolism of mirtazapine by CYP3A4/2D6.

- Mean elimination half-life of 21.5 (13.1 to 33.6) hours, with the $t_{1/2}$ of the *R*-isomer is twice that of the *S*-isomer.
- In addition, the $t_{1/2}$ for women is significantly longer than for men (may not be clinically important).

Metabolism

- Mirtazapine undergoes N-demethylation and N-oxidation (CYP3A4) and aromatic hydroxylation (CYP2D6 – major isoform) (Fig. 11.20).

Clinical Applications

Although mirtazapine has a mechanism of action quite different from the monoamine reuptake inhibitors the drug is approved for use in the treatment of major depressive episodes similar to the utility of the reuptake inhibitors. Much of what was said about the clinical application of reuptake inhibitors applies to mirtazapine.

⚠ **ADVERSE EFFECTS**

Mirtazapine: Mirtazapines major adverse effects are drowsiness and sedation, dry mouth, weight gain associated with the H_1 blocking action and, orthostatic hypotension related to the α_1-adrenergic blocking action. Additional adverse effects appear less commonly.

Monoamine Oxidase Inhibitors

MAOIs represent a group of drugs, which act by altering the level of various monoamines within and outside of the CNS. Such agents have been found to possess antidepressant activity.

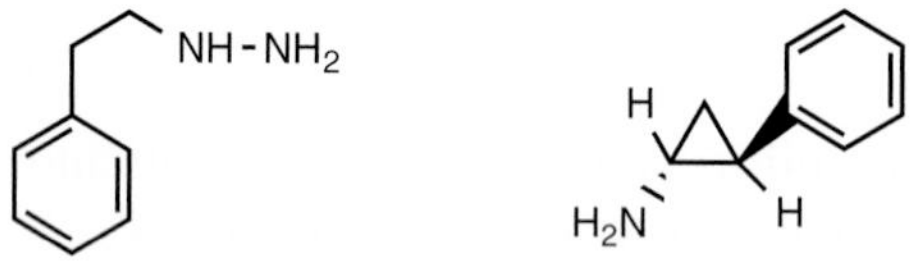

Phenelzine   (±)-Tranylcypromine

MOA

- MAO is an enzyme found mainly in nerve tissue, in the liver, and lungs where it catalyzes the oxidative deamination of various amines including epinephrine, NE, DA, and 5-HT.
- Two isoforms of MAO exist (MAO-A and MAO-B) with different substrate specificity and tissue distribution. 5-HT is a substrate for MAO-A while epinephrine, NE, and DA are substrates for both MAO-A and MAO-B.
- The MAOIs available in the United States are nonselective, irreversible inhibitors of MAO and demonstrate antidepressant effects by inhibiting 5-HT and NE metabolism in presynaptic neurons within the CNS thus increasing the levels of these neurotransmitters (Table 11.1 and Fig. 11.5).

- Pharmacological effects of MAOIs are cumulative with a latency of a few days to several months before onset and are persistent for up to 3 weeks following discontinuance of therapy.

Physicochemical and Pharmacokinetic Properties

- Both phenelzine and tranylcypromine are well absorbed following oral administration.
- Because these agents are irreversible inhibitors of MAO, blood levels are not particularly important pharmacokinetic parameters.
- Tranylcypromine is utilized as its enantiomeric mixture, although the enantiomers do exhibit different pharmacokinetic properties.

Metabolism

- The metabolic fate of phenelzine is oxidation to phenylacetic acid, 4-hydroxy-phenylacetic acid, and N-acetylation.
- Tranylcypromine is also N-acetylated and undergoes aromatic hydroxylation.

CHEMICAL NOTE

Moclobemide

Moclobemide

The drug moclobemide is available outside of the United States and acts as a selective reversible MAO inhibitor (RIMA). The drug is selective for MAO-A (80%) with lesser inhibitory action on MAO-B (20%). The result being that potentially life-threatening hypertensive crises related to tyramine containing foods is greatly reduced (tyramine can displace the RIMA from MAO-A resulting in tyramine's metabolism thus preventing its buildup).

Clinical Applications

- The MAOIs are indicated in patients with atypical (exogenous) depression and in some patients who are unresponsive to other antidepressant therapy.
- They are rarely drugs of first choice.
- Unlabeled uses for MAOIs have included bulimia, treatment of cocaine addiction (phenelzine), night terrors, posttraumatic stress disorder; some migraines resistant to other therapies, seasonal affective disorder (30 mg/d), and treatment of some panic disorders.

Mood Stabilizers

Manic depression or bipolar affective disorder, is a prevalent mental disorder with a global impact. Mood stabilizers have acute and long-term effects and are prophylactic for manic or depressive disorders. Three drugs – lithium ion, valproic acid (as its semisodium salt), and carbamazepine – are cited as having value as mood stabilizers exhibiting significant effects on mania and depression. The biochemical basis for mood stabilizer therapies or the molecular origins of bipolar disorder is unknown, but all three drugs have been implemented in inositol trisphosphate signaling.

Lithium carbonate
(Lithobid, generic)

Valproate semisodium
Divalproex
(Depakote, generic)

Carbamazepine
(Various trade names
and generic)

> **ADVERSE EFFECTS**
>
> **MAOIs:** A wide variety of adverse effects can be expected with these agents including but not limited to: dizziness, blurry vision, dry mouth, headaches, sedation, somnolence, insomnia, GI disturbance (nausea, vomiting, diarrhea, constipation), and sexual dysfunction. In addition, various foods should be avoided due to the potential for hypertensive crisis due to tyramine content and potential effects associated with this compound (i.e., cheeses – cheddar, camembert, sour cream; spirits – chianti and champagne; fish – pickled herring and anchovies; meats – chicken livers and salami; fruits – raisins and overripe bananas; vegetable products – yeast extracts, soy sauce, and avocado; and some OTC products – cold decongestants and sympathomimetics).

Lithium Ion

MOA

- Lithium therapy is believed to be effective because of its ability to reduce signal transduction through the phosphatidylinositol signaling pathway (Fig. 11.21).
- The second messengers, diacylglycerol (DAG) and inositol 1,4,5-triphosphate, are produced from membrane-bound phosphatidylinositol-4,5-bisphosphate and phosphatidylinositol-specific phospholipase C.
- Termination of activity of the second messengers results from hydrolysis by inositol monophosphatases (IMP) to inactive inositol.
- To recharge the signaling pathway, inositol is recycled back to phosphatidylinositol bisphosphate.
- Uncompetitive inhibition of inositol phosphatases in the signaling pathway by lithium ion depletes the pool of inositol reducing the enzymatic formation of the second messengers.
- Lithium ion restores the balance of signaling pathways in critical regions of the brain.
- Inositol cannot cross the blood–brain barrier and lithium ion exerts its effect when the lithium ion concentration is at saturation conditions.
- Valproic acid and carbamazepine also lead to decrease in inositol through other mechanisms.

Physicochemical and Pharmacokinetic Properties

- Lithium is rapidly and completely absorbed within 6 to 8 hours.
- Lithium is not protein bound.
- The elimination half-life varies by age (elderly patients: 39 hours, average adult: 24 hours, adolescent: 18 hours).
- Steady-state serum concentration is reached in 4 days with a maintenance plasma concentration of lithium ion of 0.5 to 1.2 mEq/L.

Clinical Applications

- Lithium has been the treatment of choice for bipolar disorder, because it can be effective in smoothing out the mood swings common to this condition.

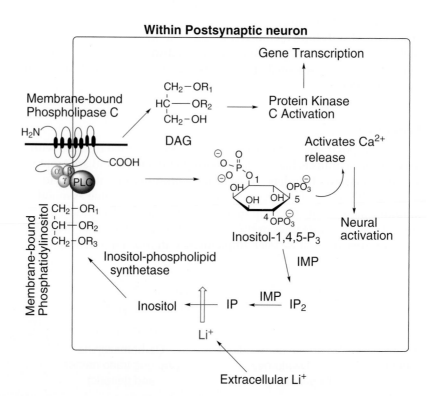

Figure 11.21 Intracellular signaling pathway and site of action of lithium ion. Phospholipase C (PLC); Diacetylglycerol (DAG); Inositol monophosphatase (IMO); Inositol diphosphate (IP$_2$), and inositol phosphate (IP).

- Divalproex and carbamazepine are also used to treat bipolar disorders as are a number of antipsychotic agents: aripiprazole, asenapine, quetiapine, risperidone, and ziprasidone.
- The level of lithium ion in a patient's blood level must be closely monitored (with a low blood level of lithium the patient's symptoms will not be relieved, but a high blood level can lead to a toxic reaction).
- Serum concentrations of lithium ion need to be markedly reduced in the elderly population – and particularly so in the very old and frail.

ADVERSE EFFECTS

Lithium: Common adverse effects of lithium include tremors, increased urination, weight gain, hypothyroidism, and potential kidney failure especially in patients with kidney disease. Other side effects of lithium include, increased white blood cell count, skin rashes, and potential for birth defects.

Chemical Significance

The economic cost for depressive illnesses in the United States is estimated to be $53 billion per year (2005), but the cost in human suffering cannot be estimated. In 2009, antidepressant drugs (primarily NE and NSRIs and SSRIs) ranked in the top 50 drugs dispensed, third in total prescriptions written, and third in total dollar prescription sales, at approximately $12.5 billion (~5% of total prescription drug sales); as a result the pharmacists have and can expect to see increasing numbers of patients entering their practice on a daily basis. With these patients will come more and more questions about the actions of these drugs.

While the antidepressants are commonly classified by their mechanism of action, the drugs also fall into specific chemical classes of which the tricyclic antidepressants represent nearly a third of the prescribed antidepressants. The metabolic processing of the antidepressants shows a commonality involving oxidative demethylation with more than half of the drugs undergoing this chemical modification. Thus, a working knowledge of basic chemical structures can easily explain how the body handles many of the antidepressants. Often the metabolic products retain antidepressant activity, which in some cases changes the mechanism of action or explains duration of drug action. The metabolic catalyst for the oxidative demethylation involves specific cytochrome oxidases, which can explain the potential for drug–drug interactions. Thus, medicinal chemistry application to the antidepressants explains various clinically significant outcomes associated with these drugs supporting the pharmacist in his/her daily interactions with patients and other health care professionals.

Review Questions

1 Which of the following comparisons is true of desipramine versus imipramine?

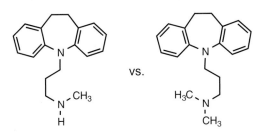

vs.

Desipramine   Imipramine

A. The two drugs are basically the same in their selectivity and action.
B. Desipramine is an SSRI while imipramine is an SNRI.
C. Desipramine is a selective inhibitor of NET while imipramine is a nonselective SERT and NET inhibitor.
D. Desipramine is a prodrug while imipramine is active without metabolic transformation.
E. Both drugs increase reuptake of 5-HT.

2 What is known about the activity of doxepin?

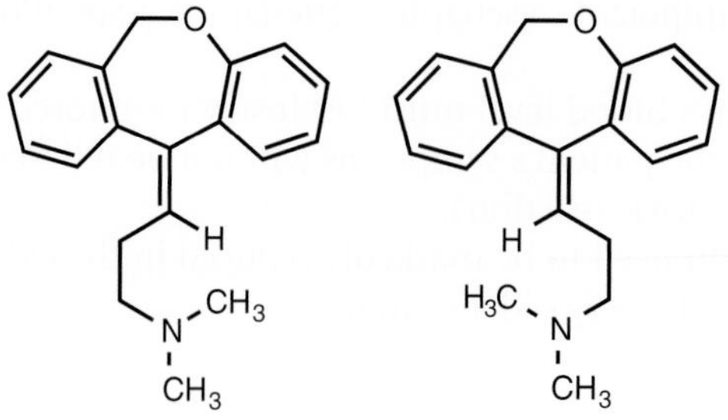

Doxepin

A. The active isomer is the *Z* isomer and the *E* isomer is an impurity.
B. The active isomer is the *E* isomer and the *Z* isomer is an impurity.
C. Both *E* and *Z* isomers are active, but because they block the same receptor they tend to work against each other.
D. Neither isomer is active, doxepin is a prodrug, and both isomers are converted to the same active drug.
E. Both isomers are active but at different reuptake receptors.

3 The basis for the mechanism of action of mirtazapine is that:
A. It blocks SERT, NET, and DAT receptors.
B. It blocks 5-HT metabolism and thus increases the postsynaptic concentration of 5-HT.
C. It blocks selective 5-HT receptors and increases the presence of both 5-HT and NE neurotransporters.
D. It blocks MAO-B.
E. The drug is a prodrug that requires metabolism before effecting the concentration of monoamines.

4 Which of the structures would be the predicted metabolite of venlafaxine?

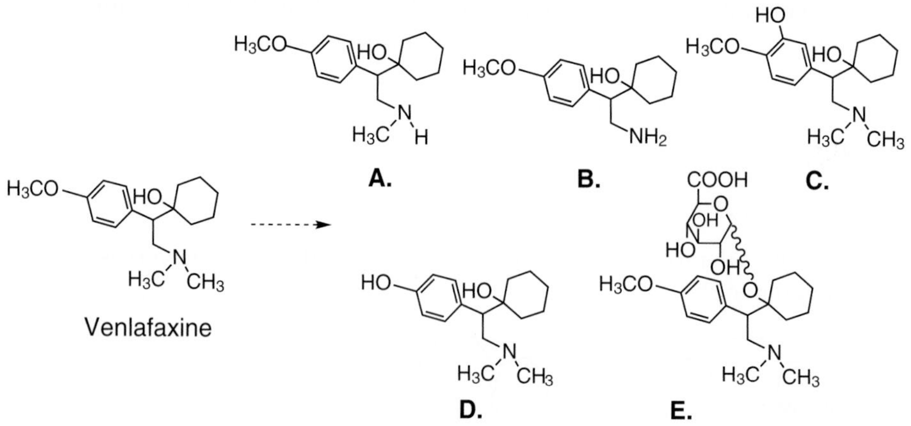

Venlafaxine

5 Which of the following statements is correct for the drugs shown?

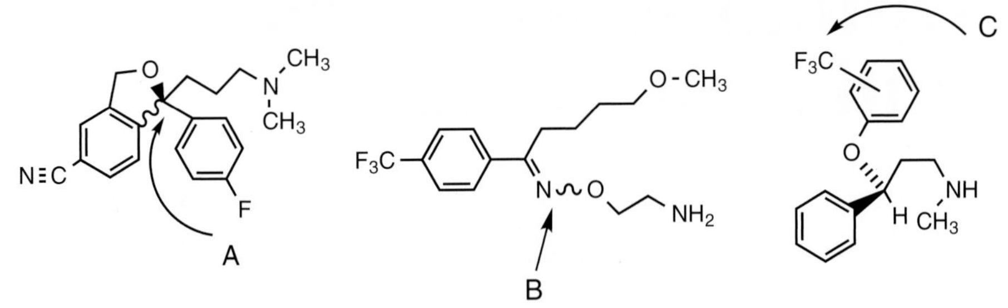

Citalopram

Fluvoxamine

Fluoxetine

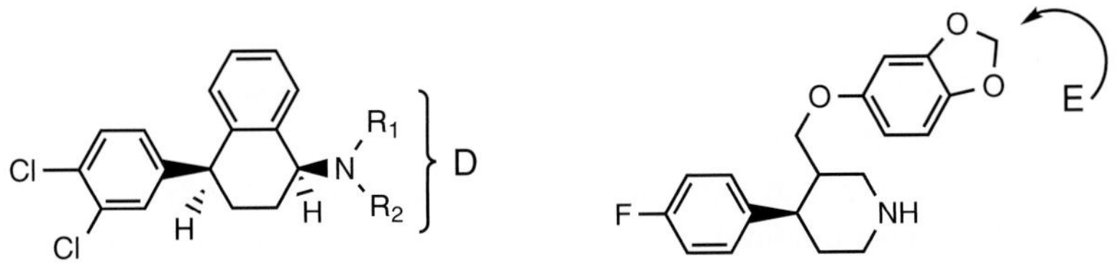

Sertraline

Paroxetine

A. Stereochemistry at site A is extremely important in promoting antidepressant effects.

B. Stereochemistry at site B is unimportant.

C. Substitution at any of the open position, C, will improve activity.

D. Primary, secondary, or tertiary amines at D all exhibit equal activity as an SSRI.

E. Paroxetine is a prodrug which requires removal of the methylene (E) for activation.

12

Opioid Analgesics and Antagonists

Learning Objectives

Upon completion of this chapter the student should be able to:

1 Describe the mechanism of action of:
  a. Multicyclic opioid analgesics (morphines, morphinans, benzazocines, oripavines)
  b. Flexible opioid analgesics (anilinopiperidines, meperidine, methadone)
  c. Dual-acting opioid analgesics (tramadol, tapentadol)
  d. Opioid antagonists

2 Discuss applicable structure activity relationships (SAR) (including key aspects of receptor binding) of the above classes of opioid analgesics and antagonists.

3 Identify clinically relevant physicochemical and pharmacokinetic properties that impact the in vitro and in vivo stability and/or therapeutic utility of opioid analgesics and antagonists.

4 Diagram metabolic pathways for all biotransformed opioid analgesics and antagonists, identifying enzymes and noting the clinically significant metabolism and the activity, if any, of the major metabolites.

5 Apply all of above to predict and explain the therapeutic utility of opioid analgesics and antagonists.

Introduction

Pain that is ongoing and inescapable can be one of the most demoralizing and debilitating experiences of the human condition. Among the most powerful pain-relieving molecules known are alkaloids found in the latex of the opium poppy (*Papaver somniferum*) and their synthetic analogs. Referred to as opioid analgesics, these molecules stimulate one of two opioid receptors, μ or κ, to dull the perception of pain by interfering with the neurotransmission of nociceptive or, less commonly, neuropathic stimuli.

The euphoria induced by μ-opioid agonists gives these analgesics the potential for misuse and abuse. While patients using opioids for the treatment of pain are much less likely to become pathologically dependent on these drugs than patients who do not need them for pain management, addiction potential is still viewed as a major use-limiting side effect by some health care providers, and patients can be undermedicated because of it. The important role of the pharmacist in clinically competent pain management cannot be overestimated.

This chapter will focus on drugs used in the treatment of chronic and/or intractable pain, and the antagonists used to reverse adverse effects or support addiction recovery. The reader is referred to Chapter 20 in *Foye's Principles of Medicinal Chemistry, Seventh Edition,* for an in-depth discussion of nociceptive pathways, endogenous opioid peptides, additional opioid receptor subtypes, and the biochemically based phenomena of opioid tolerance, dependence, and addiction.

Opioid Receptors Mediating Analgesia

Mu (μ) Receptors

- The prototypical μ agonist is (−)-morphine.
- The μ receptor is coupled to G_i and G_o proteins. Analgesia is mediated through a decrease in intracellular cAMP and Ca^{2+}.

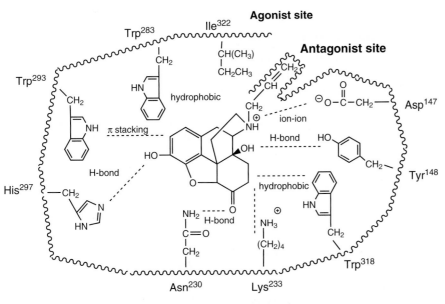

Mu receptor-oxymorphone interactions

Mu receptor-naloxone interactions

Figure 12.1 Mu-opioid receptor.

- Stimulation of μ receptors increases dopamine release in the nucleus accumbens, leading to euphoria that can drive addictive behaviors. Stimulation of μ receptors in the gut leads to chronic constipation that can be use-limiting.
- The residues important for binding opioid agonists and antagonists have been identified. Figure 12.1 depicts a potent agonist (oxymorphone) and antagonist (naloxone) binding at the μ receptor. Similar interactions are known to occur at κ receptors (Table 12.1).

Table 12.1 Corresponding μ and κ Receptor Binding Residues

μ-Receptor Residue	κ-Receptor Residue	Opioid-Binding Functional Group
Asp[147]	Asp[138]	cationic amine
His[297]	His[291]	C3-phenol
Asn[230]	Leu	C6-OH or keto
Tyr[148]	Glu[297]	C14-OH

⚠️ **ADVERSE EFFECTS**

Multicyclic Opioid Analgesics

μ agonists
- Sedation
- Constipation (no tolerance)
- Respiratory depression (cause of death in overdose)
- Nausea
- Miosis or "pinpoint pupils" (no tolerance)
- Histamine release (from 6-OH agonists)
- Increased intracranial pressure
- Dependence liability (no tolerance)

κ agonists
- Sedation
- Nausea
- Increased cardiac workload (pentazocine and butorphanol only)
- Dysphoria (primarily pentazocine and butorphanol)
- Risk of pain exacerbation in some male patients
- Increased intracranial pressure
- Precipitation of withdrawal symptoms in μ agonist–dependent patients

Opioid antagonists
- Precipitation of severe withdrawal symptoms when used in opioid-dependent patients

Pinpoint Pupils

Pinpoint pupils can be a life-saving action of opioids. Overdosed patients who are unconscious will always have this diagnostic sign of opioid use, signaling their immediate need for an antagonist to reverse life-threatening respiratory depression.

Kappa (κ) Receptors

- The prototypical κ agonist is (+)-ethylketocyclazocine.
- Like the μ receptor, the κ receptor is G_i and G_o coupled.
- Unlike the μ receptor, κ receptor stimulation inhibits dopamine release in the nucleus accumbens, so euphoria/addiction potential are not use-limiting issues. Some patients experience dysphoria from κ agonists and cannot take these analgesics.
- When used chronically, patients can still become dependent on κ agonists. This underscores the difference between dependence and addiction.
- Kappa receptors are found in the heart, and an enhanced cardiac workload is observed with the use of some (but not all) κ agonists.

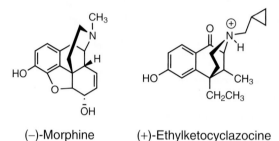

(–)-Morphine   (+)-Ethylketocyclazocine

Opioid Receptor–Mediated Actions

- Opioids exhibit one of three structure-dependent activities.
  - Pure agonist (full or partial) analgesics act at the μ receptor. They produce potent analgesia along with the adverse effects typically associated with morphine use.
  - Most agonist/antagonist analgesics produce potent analgesia at κ receptors and strong μ antagonism. Buprenorphine is the exception, with dose-dependent agonist and antagonist action elicited at μ receptors.
  - Pure antagonists block the action of opioids at all receptor subtypes, but have their greatest affinity at μ receptors.

⬡ **CHEMICAL NOTE**

Cardiovascular Toxicity of Codeine: An increased risk of serious or fatal cardiovascular toxicity has been noted in patients administered codeine or oxycodone for 6 months or longer. An increased risk for all-cause mortality has been noted when the drugs are used for only 1 month. Hydrocodone does not appear to carry the same cardiovascular and all-cause mortality risk of the other two 3-methoxy analgesics.

Codeine   Oxycodone   Hydrocodone

Multicyclic Opioids

SAR

- Multicyclic opioids are most commonly pentacyclic (morphines). Only one clinically available analgesic is found in each of the bicyclic (oripavines), tetracyclic (morphinans), and tricyclic (benzazocines) structural classes.
- The pharmacophore found in all opioid ligands is 4-phenylpiperidine.

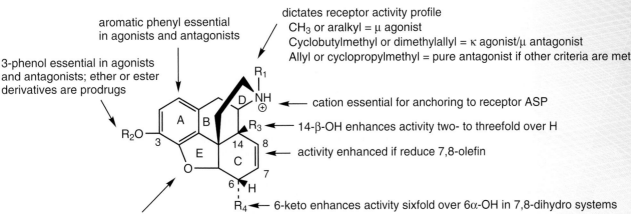

Figure 12.2 Multicyclic opioid SAR.

- A general overview of multicyclic opioid SAR is provided in Figure 12.2.

Morphine with 4-phenylpiperidine
pharmacophore highlighted

Nitrogen Substituent (R₁)

- A cationic amine is essential. Tertiary amines provide the most potent opioid activity.
- N-CH₃ and N-aralkyl (e.g., phenylethyl, thienylethyl) substituted opioids are μ agonists.
  - Aralkyl substituents provide greater CNS distribution, hydrophobic receptor binding, and analgesic potency.
- N-allyl and N-cyclopropylmethyl substituted opioids are μ antagonists.
  - The three carbon chain length and partial negative (δ^-) charge increase affinity for the μ-antagonist site.
  - Pure antagonists have one of these two amine substituents. They block agonist action at all opioid receptor subtypes.
- N-cyclobutylmethyl and N-dimethylallyl substituted opioids are agonists at κ receptors and antagonists at μ receptors.
  - These substituents bind with high affinity to the agonist site of κ receptors.
  - The bulkier (dimethylallyl) or less strained (cyclobutylmethyl) substituents do not fit or bind as well as allyl or cyclopropylmethyl within the sterically restricted μ-antagonist site. Antagonist potency decreases.

C3 Substituent (R₂)

- A phenolic OH is essential for H-bonding to the μ receptor.
- Masking the 3-phenol through ether or ester formation generates a prodrug. Metabolic conversion to the phenol through O-dealkylation or hydrolysis, respectively, is required for activity.

C14 Substituent (R₃)

- Multicyclic opioids have a hydrogen or an OH group at C14β.

- A 14β-OH adds polarity that decreases CNS distribution but significantly increases receptor affinity. Analgesic potency increases.
- Opioids with a 14β-OH have no clinically useful antitussive (cough suppressant) activity.
- A 14β-OH group is a required component of pure opioid antagonists.

C6 Substituent (R₄)

- Pentacyclic opioids have an OH group or keto at C6.
- In 7,8-dihydro opioids, the 6-keto has a higher activity due to more efficacious H-bonding with the μ receptor. The opposite is true for 7,8-dehydro opioids.
- The 6α-OH group is believed, at least in part, responsible for histamine release associated with codeine and morphine. This can result in severe itching, nausea, rash, and, in some cases, profound hypotension.
  - Codeine is not administered parenterally for this reason.
- A 7,8-dihydro-6-keto substitution pattern in a pentacyclic opioid is a required component of pure opioid antagonists.
- Tetracyclic opioids are unsubstituted at C6.
- Tricyclic opioids have no C ring. There is no position equating to the C6 of pentacyclic and tetracyclic opioids.

Dihydrofuran (E Ring)

- Removing the dihydrofuran ring to form tetracyclic (morphinan) opioids results in increased molecular lipophilicity and flexibility. This provides increased CNS distribution, receptor affinity and analgesic potency.

Stereochemistry

- Pentacyclic opioids have five asymmetric centers. The $5R,6S,9R,13S,14R$ stereochemistry provides the most active levorotatory isomer.
- Dextrorotatory multicyclic opioids have no analgesic activity but, along with the *levo* isomers, are effective cough suppressants. They act on specific receptors found in cough centers of the medulla oblongata.
- The aromatic A ring is held in an axial orientation with respect to the piperidine (D) ring, forcing a π stacking interaction with the μ receptor Trp[293].
  - Trp[293] normally binds the N-terminal Tyr residue of enkephalin pentapeptides. The fact that Tyr is a phenolic amino acid may help explain the strict requirement for a phenolic OH on multicyclic opioids.

Physicochemical Properties

- Log P values of multicyclic opioids range from 4.98 (buprenorphine) to 0.89 (morphine).
- Lipophilicity can impact the extent of oral bioavailability, hepatic penetration on first pass, and CNS distribution.
- A summary of the impact of common functional groups on CNS distribution, μ-receptor affinity, and analgesic potency is provided in Table 12.2.
- Most multicyclic opioids provide durations of analgesic action of 4 to 6 hours.

Table 12.2 Impact of Common Multicyclic Opioid Agonist Functional Groups

Functional Group	Impact on CNS Distribution	Impact on μ-Receptor Affinity	Impact on Analgesic Potency
3-Phenol	Decrease	Increase	Increase
3-OCH₃	Increase	Decrease	Decrease
14-OH	Decrease	Increase	Increase
6-OH (compared to H)	Decrease	Slight increase	Decrease
7,8-dihydro-6-keto (compared to H)	Decrease	Increase	Increase
N-CH₂CH₂-C₆H₅	Increase	Increase	Increase

- After metabolic biotransformation, opioids are excreted primarily through the kidneys. Some fecal excretion also occurs.
- Selected pharmacokinetic parameters of commercially available multicyclic opioids are provided in Chapter 20, Table 20.2 in *Foye's Principles of Medicinal Chemistry, Seventh Edition*.

Metabolism

- Multicyclic opioids undergo predictable Phase I and Phase II reactions. Common to most are N-dealkylation and glucuronic acid conjugation or sulfonation of the phenol. These biotransformations strongly attenuate activity or inactivate the drug, respectively.
- The 3-methoxy ethers on codeine-based analgesics must O-dealkylate to be active. CYP2D6 catalyzes this reaction.
  - Poor and intermediate CYP2D6 metabolizing phenotypes are fairly common in some ethnic groups. Patients with poor metabolizing (PM) CYP2D6 alleles receive little or no analgesic benefit from codeine-based analgesics.
  - Ultrafast (UM) CYP2D6 metabolizers are also known. These patients rapidly convert the 3-methoxy group to the active phenol, putting them at risk for serious and potentially fatal adverse effects.
- The 6α-OH found on morphine and codeine undergoes Phase II glucuronic acid conjugation, catalyzed by UGT2B7. Glucuronidation at C6 produces a conjugate that is many times more active than the parent drug.
- The metabolic reactions common to multicyclic opioids are illustrated by the biotransformation pathway of codeine (Fig. 12.3).

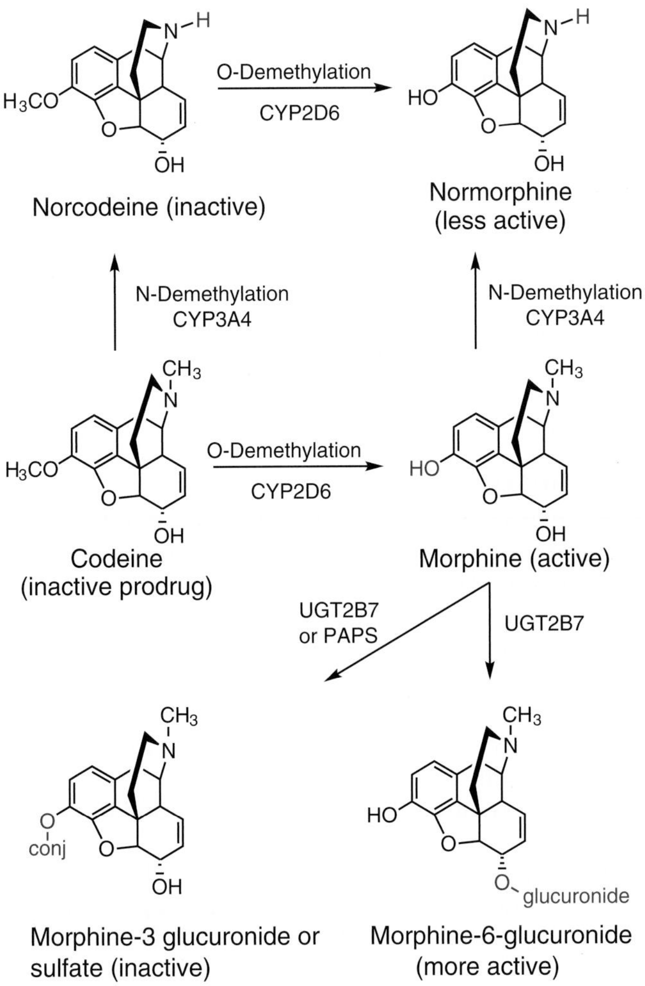

Common Metabolic Reactions of Multicyclic Opioids

- N-dealkylation (significantly decreases activity)
- C3-glucuronidation (abolishes activity)
- C3-sulfonation (abolishes activity)
- C6-glucuronidation (enhances activity)
- O-dealkylation of 3-methoxy ethers in codeine congeners (initiates activity)
- Hydrolysis of 3,6-diacetoxy esters in heroin (initiates activity)

CHEMICAL NOTE

Metabolism Matters: The importance of knowing the CYP2D6 phenotype of patients taking codeine-based analgesics cannot be overestimated. There are reports in the literature of deaths related to codeine dosing without regard to metabolizing phenotype. This includes infants with UM 2D6 activity being nursed by mothers given standard or high doses of codeine-based analgesics.

Figure 12.3 Codeine metabolism.

Table 12.3 Full μ-Agonist Analgesics

Morphine sulfate (R = H)
(Duramorph, Kadian)
Codeine sulfate (R = CH₃)

Hydromorphone hydrochloride
(R = H, X = H) (Dilaudid)
Oxymorphone hydrochloride
R = H, X = OH) (Opana)
Oxycodone hydrochloride
(R = CH₃, X = OH) (Roxicodone)

Hydrocodone bitartrate
(Zohydro)

Levorphanol tartrate
(Levo-Dromeran)

μ Agonist Analgesic	Approximate Potency Relative to Morphine	Dosage Forms Available
Morphine	Equipotent	Inj, cap, tab, ext rel tab/cap, oral sol, supp, epi susp
Hydromorphone	6 x	Inj, tab, ext rel abuse deterrant tab[a], oral liquid, supp
Oxymorphone	12 x	Inj, tab, ext rel tab, ext rel abuse deterrant tab[b]
Codeine	0.083 x	Tab, oral sol
Hydrocodone	0.5 x	Cap ext. release
Oxycodone	Equipotent	Tab, abuse deterent tab[b,c], ext rel abuse deterrant tab[b,d], cap, oral sol, oral concentrate
Levorphanol	4–8 x	Tab

inj, injection; *cap*, capsule; *ext rel*, extended release, *sol*, solution; *supp*, suppository; *epi*, epidural; *susp*, suspension
[a]Naltrexone incorporated into dosage form; [b]Crush resistant; [c]Formulation with naloxone approved by FDA in 2013; [d]Resistant to dissolution; forms viscous gel that is difficult to inject.

BLACK BOX WARNING

Morphine – Risk of severe adverse reaction from epidural administration. Patients must be observed for 24 hours.

Hydromorphone – For use in opioid-tolerant patients only. High abuse potential.

Oxymorphone – For use in opioid-tolerant patients only. High abuse potential. Alcohol use can increase risk of fatal respiratory depression.

Codeine – Risk of fatal respiratory depression when used in children who are ultrafast 2D6 metabolizers.

Hydrocodone – High addiction potential and risk of fatal respiratory depression. Alcohol use increases risk.

Oxycodone – High addiction potential and risk of fatal respiratory depression. Caution related to dosing/dispensing errors.

Specific Drugs: Opioid Agonists

- The relative analgesic potency of μ agonists follows logically from an interpretation of SAR and an analysis of the relative importance of distribution versus receptor affinity to potency (Table 12.3).
- The relative analgesic potency of the three κ agonists/μ antagonists vary widely (Table 12.4). Relative lipophilicity may, in part, explain potency differences between butorphanol and nalbuphine.
- Pentacyclic nalbuphine appears to put patients at lowest risk for use-limiting dysphoria and cardiovascular toxicity. It is one of two opioids that decrease myocardial oxygen demand (morphine is the other).
- Buprenorphine, a partial μ agonist, is unique from both structural and pharmacologic perspectives (Fig. 12.4).
  - Despite the N-cyclopropylmethyl substituent normally associated with opioid antagonism, buprenorphine is used clinically as an analgesic. Its potency is 20 to 50 times morphine. It is the only partial μ agonist available to treat pain.
  - Mu antagonism can occur in doses higher than those recommended for analgesia. Buprenorphine is not used clinically as an opioid antagonist.

Table 12.4 κ-Agonist/μ-Antagonist Analgesics

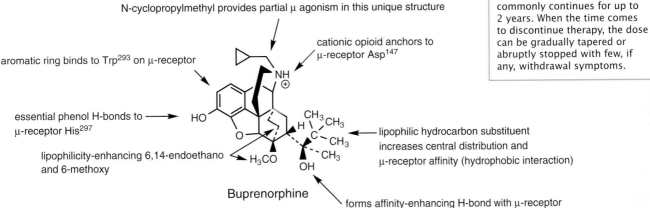

Nalbuphine hydrochloride
(Nubain)

Butorphanol tartrate
(Stadol)

Pentazocine lactate
(Talwin)

Nalorphine
(reference antagonist)

κ Agonist/μ Antagonist	Log P (calc)	Approx. Agonist Potency Relative to Morphine	Approx. Antagonist Potency Compared to Nalorphine	Dosage Forms
Nalbuphine	2.0	0.5–1 x	0.25 x	In
Butorphanol	3.6	5 x	Equipotent	Inj, nasal spray
Pentazocine (racemic)	4.4	0.25 x	0.02 x	Inj

Inj, injection

- Receptor affinity of buprenorphine is high enough to be termed "pseudoirreversible." With a 37-hour half-life, it can be administered once daily.
- Because of the exceptionally high receptor affinity there are few, if any, withdrawal symptoms associated with buprenorphine use, even with abrupt discontinuation.
- Buprenorphine is useful in opioid addiction recovery. It is cross tolerant with commonly misused/abused μ agonists but does not elicit dependence-promoting euphoria due to the partial agonist action.
- Sublingual tablets and transdermal patches are available for this indication. Sublingual naloxone-containing abuse deterrent formulations are also available.

CHEMICAL NOTE

Buprenorphine in Opioid Addiction Recovery: Buprenorphine precipitates withdrawal in μ agonist–dependent individuals and should only be administered to patients who have been opioid free for 7 to 10 days. Alternatively, patients can be allowed to enter the first phase of withdrawal and then administered buprenorphine to suppress symptoms. Once daily sublingual doses of 12 to 16 mg (4 to 24 mg in naloxone-containing formulations) are standard for maintenance therapy, which commonly continues for up to 2 years. When the time comes to discontinue therapy, the dose can be gradually tapered or abruptly stopped with few, if any, withdrawal symptoms.

N-cyclopropylmethyl provides partial μ agonism in this unique structure

aromatic ring binds to Trp[293] on μ-receptor

cationic opioid anchors to μ-receptor Asp[147]

essential phenol H-bonds to μ-receptor His[297]

lipophilicity-enhancing 6,14-endoethano and 6-methoxy

lipophilic hydrocarbon substituent increases central distribution and μ-receptor affinity (hydrophobic interaction)

Buprenorphine

forms affinity-enhancing H-bond with μ-receptor

Figure 12.4 SAR of buprenorphine.

Table 12.5 Opioid Antagonists

Naloxone hydrochloride
(Evzio)

Naltrexone hydrochloride
(R = H, X = Cl) (ReVia, Vivitrol)
Methylnaltrexone bromide
(R = CH₃ , X = Br) (Relistor)

Opioid Antagonist	Therapeutic Use	Dosage Forms
Naloxone	Reversal of life-threatening overdose	Inj, including autoinject formulation
Naltrexone	Adjunct in opioid addiction recovery	Tab, susp for IM inj
Methylnaltrexone	Relief of opioid-induced constipation	SC inj

inj, injection; *tab*, tablet; *susp*, suspension; *IM*, intramuscular; *SC*, subcutaneous

Specific Drugs: Opioid Antagonists

- Three pure opioid antagonists are marketed (Table 12.5).
- Naloxone and naltrexone (tertiary amines) cross the blood–brain barrier. They displace the μ agonist from central receptors and precipitate withdrawal in opioid agonist–dependent patients.
- Tertiary opioid antagonists also reverse life-threatening respiratory depression and CNS depressions (unconsciousness) in overdosed patients. Naloxone is used exclusively for this purpose.
  - Naloxone is inactivated in the gut, primarily by C3 glucuronidation. It is orally ineffective and administered by injection.
  - The 1- to 2-hour half-life limits the severity and duration of withdrawal symptoms.
- Naltrexone is more metabolically stable and orally active than naloxone. It is also more lipophilic and more potent.
  - Naltrexone is used in addiction recovery. It blocks μ agonist–induced euphoria in patients who relapse. It is also valuable in alcohol addiction recovery.
  - While orally active, naltrexone is routinely given IM. The suspension contains a polymer that allows for the slow release of active drug over 4 weeks, increasing convenience and adherence. IM administration also avoids first pass inactivation.
  - Naltrexone's primary active metabolite is 6β-naltrexol.
- Methylnaltrexone bromide, a quaternary amine, is administered subcutaneously to reverse opioid-induced constipation in patients on chronic opioid therapy.
  - Methylnaltrexone bromide is dosed based on patient weight. It is usually given every other day, but may be given once daily if needed.
  - Since there is no CNS distribution, it can be used in opioid-dependent patients without risk of precipitating withdrawal.

CHEMICAL NOTE

Naloxone Autoinject Delivery System: In April, 2014, the FDA approved a hand-held naloxone autoinject delivery system (Evzio) for use by family and/or friends of an individual who has experienced a life-threatening opioid overdose. The injector surface is placed on the patient's body and the user pushes the device to deliver the antagonist subcutaneously before calling 911 for emergency assistance (http://www.fda.gov/NewsEvents/Newsroom/PressAnnouncements/ucm391465.htm).

CHEMICAL NOTE

Nalmefene: The One That Got Away: The opioid antagonist nalmefene was once marketed as Revex for IV use in reversing life-threatening opioid overdose. It is identical in structure to naltrexone with the exception of an exocyclic methylene moiety at C6 in place of the 6-keto. Nalmefene is orally active and has 11-hour half-life due to an exceptionally high "pseudoirreversible" affinity for μ receptors. The resultant prolonged withdrawal may have led to nalmefene's discontinuation from the U.S. market in 2008.

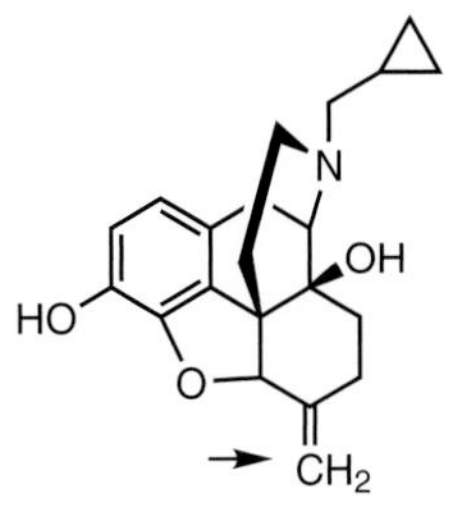

Nalmefene 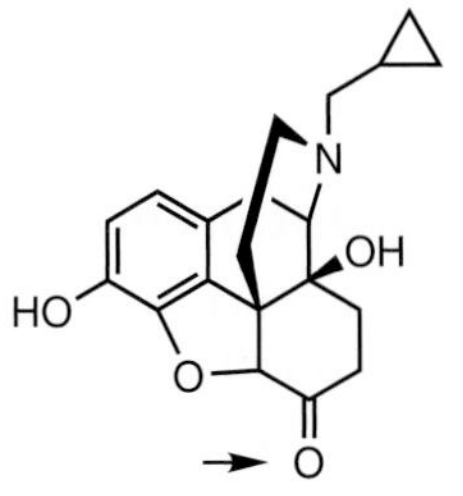   Naltrexone

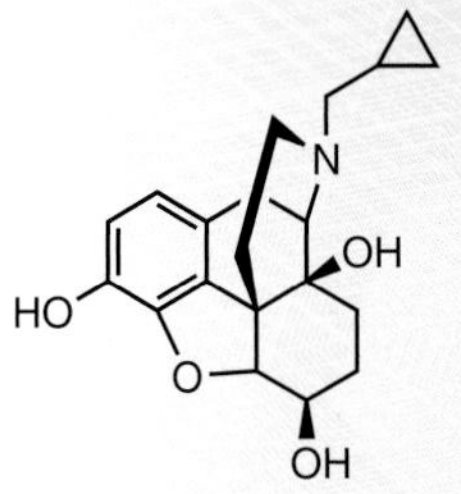

6β-naltrexol

Flexible Opioids

- Flexible opioids are not conformationally restrained. They can orient their aromatic ring equatorial to the piperidine ring, which permits binding of the former at Trp^{283} of the μ receptor.
- Trp^{238} normally binds with Phe of the enkephalins. Since Phe is nonphenolic, a phenolic OH group is neither required nor desired in flexible opioids.
- All flexible opioids are pure agonists and bind selectively to μ receptors. (See Fig. 12.5, with fentanyl as ligand). The 4-phenylpiperidine pharmacophore can be found or envisioned in the flexible opioids (Fig. 12.6).

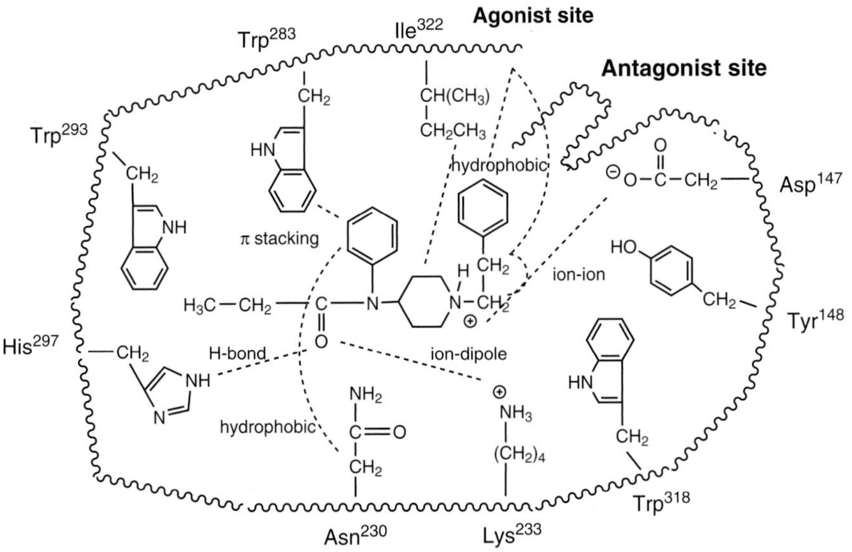

Mu receptor-fentanyl interactions

Figure 12.5 Mu receptor-fentanyl binding interactions.

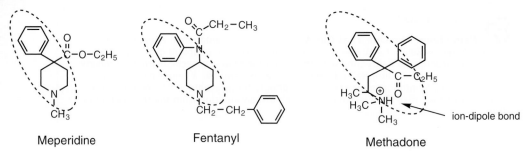

Figure 12.6 4-Phenylpiperidine pharmacophore in flexible opioids.

Anilinopiperidine Analgesics (Fentanyls)

- Four anilinopiperidine analgesics, known collectively as the "fentanyls," are currently marketed. Select physicochemical properties, relative potencies, and available dosage forms are provided in Table 12.6.
- All anilinopiperidine analgesics are orally inactive due to high metabolic vulnerability. Fentanyl undergoes CYP3A4-catalyzed N-dealkylation in the liver. Remifentanil is rapidly inactivated in the bloodstream via hydrolysis.

Physicochemical and Pharmacologic Properties

- The relative potencies and durations of anilinopiperidines are influenced by both Log P and pKa.
  - The higher Log P for sufentanil as compared to fentanyl explains its superior potency and shorter duration (Table 12.6). The terminal administration half-lives of sufentanil and fentanyl are 2.7 versus 3.6 hours, respectively.

Table 12.6 Anilinopiperidine (Fentanyl) Analgesics

Fentanyl citrate
(Sublimaze)

Sufentanil citrate
(Sufenta)

Alfentanil hydrochloride
(Alfenta)

Remifentanil hydrochloride
(Ultiva)

Fentanyl	Log P	pKa	Approx. Potency Relative to Morphine	Therapeutic Use	Dosage Forms
Fentanyl	8.4	2.86	80 x	Analg, adjunct to anesthesia	Inj, transdermal patch, sublingual tab and liq, eff sublingual tab, buccal lollipop and film, nasal spray (REMS)[a]
Sufentanil	8.5	3.45	600–800 x	Epi and surgical analg, adjunct to anesthesia	Inj[b]
Alfentanil	6.5	1.26	25 x	Analg, as adjunct to anesthesia	Inj
Remifentanil	7.1	1.25	400–500 x	Analg, as adjunct to anesthesia	Inj

[a]Risk Evaluation and Mitigation Strategy. Prescribers and pharmacies must be enrolled in a REMS program.
[b]A sufentanil transdermal patch is in clinical trials. *Analg,* analgesia; *epi,* epidural; *Inj,* injection; *liq,* liquid; *eff,* effervescent; *tab,* tablet

- The lower pKa of alfentanil as compared to sufentanil and fentanyl is due to the electron withdrawing tetrazol-5-one ring. It results in a shorter duration and a lower potency.
  - The percentage of unionized alfentanil in the bloodstream is 89% as compared to approximately 9% for sufentanil and fentanyl. CNS distribution and onset of action is faster. Its terminal elimination half-life drops to 1.3 to 1.6 hours.
  - Fewer cationic molecules are available for anchoring to the μ-receptor Asp[147]. Affinity and potency decrease.
- Remifentanil has a pKa between that of alfentanil and the other anilinopiperidines.
  - Remifentanil's unionized:ionized (u:i) ratio at pH 7.4 is 2:1, as compared to 8:1 for alfentanil. Remifentanil's higher log P facilitates greater blood–brain barrier penetration.
  - As the % ionized of remifentanil at the receptor is higher than alfentanil (33% vs. 11%), potency increases.
  - Despite a very short duration (3 to 8 minutes) and rapid inactivating hydrolysis in the bloodstream (elimination half-life of 10 to 21 minutes), remifentanil delivers exceptionally high analgesic potency. It is used in maintaining and recovering from anesthesia.

Specific Drugs

- Alfentanil and sufentanil are currently available only by IV injection for analgesic support during anesthesia. Sufentanil is also used via the epidural route during labor and vaginal delivery.
- Fentanyl can be used for the treatment of pain from many sources, most commonly breakthrough cancer pain. In addition to IV administration, several transdermal and transmucosal dosage forms merit mention.
  - Transdermal patch: Fentanyl free base is formulated into transdermal patches designed to release drug over a 72-hour period. Strengths ranging from 12 to 100 μg/h are available.
    - To avoid inadvertent overdose, patients should rotate patch placement and avoid exposure to heat (e.g., heating pads, hot baths).
    - Patches should be disposed of carefully and safely. Individuals (including children) have died as a result of exposure to inappropriately discarded patches.
  - Buccal lollipop and soluble film: Both of these dosage forms contain fentanyl citrate.
    - The ionized:unionized ratio of fentanyl at normal salivary pH 6.4 is 100:1. Only unionized molecules are absorbed across buccal membranes.
    - Approximately 25% of the dose is absorbed into the bloodstream while the remaining 75% is swallowed and destroyed in the gut and liver.
  - Effervescent buccal tablets: This formulation contains fentanyl citrate, citric acid, sodium bicarbonate, and sodium carbonate.
    - Citric acid lowers salivary pH to 5.0, generating soluble protonated fentanyl and allowing tablet dissolution.
    - Sodium bicarbonate and sodium carbonate react with citric acid to yield H_2O, CO_2 and sodium citrate.
    - As the patient exhales CO_2, citric acid is consumed and salivary pH rises to 9 to 10, favoring the formation of unionized, absorbable fentanyl.
    - Drug absorption from the effervescent tablet is approximately twice that of the buccal lollipop formulation.

Meperidine

- Meperidine has all anticipated μ-agonist actions except constipation and cough suppression (Table 12.7). It can cause profound hypotension secondary to histamine release.

BLACK BOX WARNING

Fentanyl – Persistent and potentially fatal respiratory depression. Caution against use in opioid-naïve patients and coadministration with CYP3A4 inhibitors.

Remifentanil carboxylic acid
(inactive metabolite)

BLACK BOX WARNING

Methadone – Potentially fatal respiratory depression and arrhythmias.

Table 12.7   Meperidine and Methadone

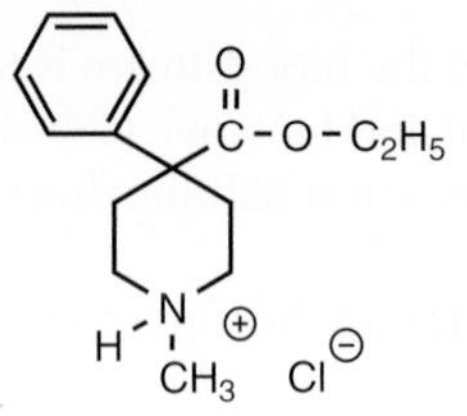

Meperidine hydrochloride
(Demerol)

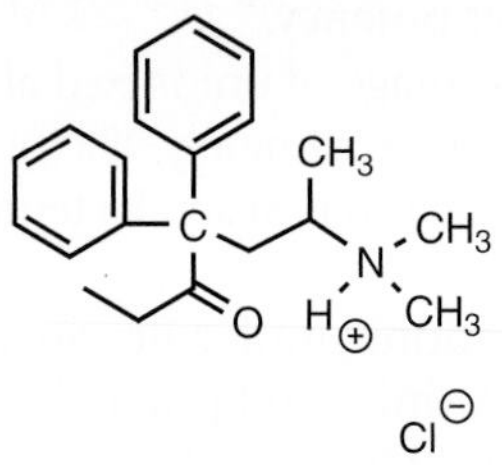

Methadone hydrochloride
(Dolophine)

Opioid	Elimination Half-life	Approx. Potency Relative to Morphine	Therapeutic Use	Dosage Forms
Meperidine	3–8 hours	0.1 x	Analgesia, adjunct to anesthesia	Inj, tab
Methadone	Up to 60 hours	Equipotent parenterally; 0.5 x orally	Analgesia, opioid addiction recovery	Inj, oral tab[a], sol, conc

inj, injection; *tab,* tablet; *conc,* concentrate
[a]Risk Evaluation and Mitigation Strategy. Prescribers of oral methadone must be enrolled in a REMS program.

- Meperidine's oral bioavailability is about 40% to 60%. It is available in tablet form, as well as a solution for injection. Absorption from IM injection sites is variable.
- Meperidine carries a neurotoxicity risk (seizures) attributed to the N-demethylated (nor) metabolite. The CYP3A4-generated nor metabolite has half the analgesic action of the parent drug and a greatly extended half-life of 14 to 21 hours, which increases toxicity risk.

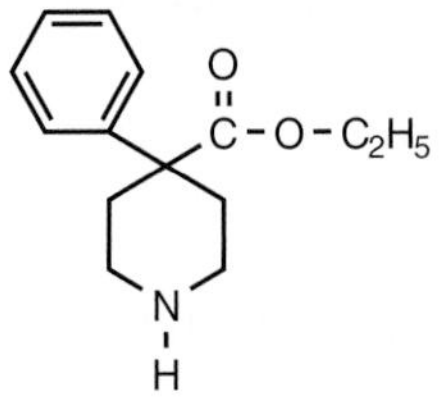

Normeperidine
(neurotoxic metabolite)

Methadone

- Methadone's most widely recognized use is in addiction recovery (Table 12.7). It has a low physical dependence and its long duration permits once daily dosing, which enhances patient convenience and adherence.
- Methadone is marketed as the racemic mixture, but the *R*(−) isomer has all the analgesic activity. The *S*(+) isomer blocks N-methyl-D-aspartate (NMDA) receptors, which can ease neuropathic pain but predispose patients to cardiotoxic adverse effects.
- Methadone's prolonged duration is due to the generation of several active metabolites (Fig. 12.7).
- Methadone N-dealkylation is CYP mediated.
  - CYP2C19 N-dealkylates the *R*(−) isomer. Inhibition or competition for this isoform can result in potentially fatal respiratory and CNS depression.
  - CYP2B6 N-dealkylates the *S*(+) isomer. Limiting access to this isoform can result in prolongation of the QT interval and potentially fatal arrhythmia.

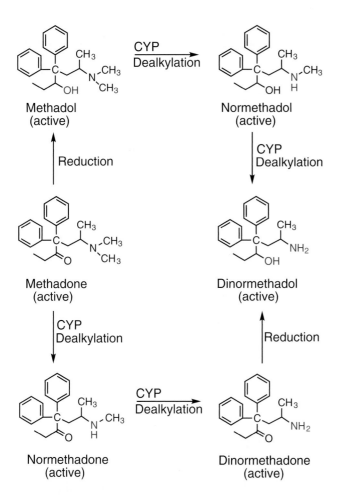

Figure 12.7 Methadone metabolism.

- CYP3A4 N-dealkylates both methadone isomers with apparently equal ease. Inhibiting or competing for CYP3A4 can increase risk for all methadone toxicities.
- Methadone's pharmacokinetic profile is pH sensitive.
  - Methadone is excreted readily in acidic urine and reabsorbed from basic urine. Basifying urine can increase elimination half-life from 19 to 42 hours.
  - Methadone absorption from the oral solution formulation is increased to 85% at salivary pH 8.5 (normal salivary pH = 6.4).

Dual-Acting Opioids

- Dual-acting opioids act by stimulating μ receptors and inhibiting the reuptake of central serotonin (5-HT) and/or norepinephrine (NE) (Table 12.8). They do not contain the 4-phenylpiperidine pharmacophore found in multicyclic and flexible opioids.
- 5-HT and NE reuptake is inhibited by (+) and (−)-tramadol, respectively. Elevated 5-HT levels can lead to suicidal thoughts, so tramadol should not be given to suicidal patients.
- Tramadol must O-dealkylate (CYP2D6) before it can bind to μ receptors. Analgesic action is time dependent.
- Tapentadol inhibits only NE reuptake. It is an active μ agonist in parent drug form so analgesic action is consistent over time.
  - Despite predicted vulnerability to N-dealkylation, only 10% to 13% of a dose undergoes this reaction. CYP2C9 and 2C19 are the catalyzing enzymes.

⚠ ADVERSE EFFECTS

Dual-Acting Opioids

Tramadol
- Serotonin syndrome (dextro isomer)
- Suicide ideation (dextro isomer)
- Tachycardia (levo isomer)
- Hypertension (levo isomer)

Tapentadol
- Respiratory depression (potentially fatal with extended release formulation)
- Constipation
- Hypertension (when coadministered with vasopressors)

Table 12.8 Tramadol and Tapentadol

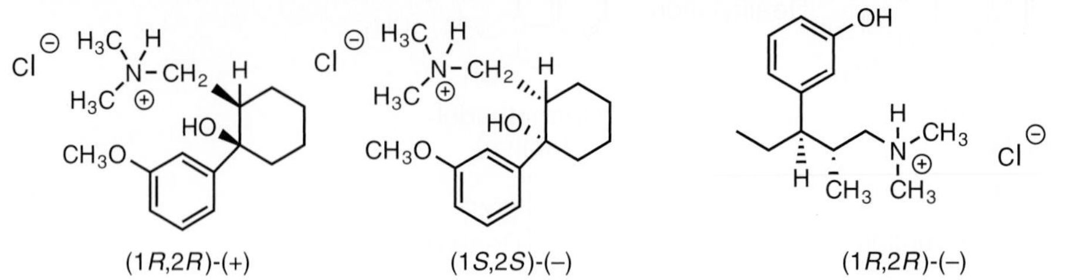

	(1R,2R)-(+)	(1S,2S)-(−)	(1R,2R)-(−)
	Tramadol hydrochloride (Ultram)		Tapentadol hydrochloride (Nucynta)

Analgesic	Neurotransmitter Reuptake Inhibition Selectivity	Appr. μ Agonist Potency Relative to Morphine	Dosage Forms
(1R,2R)-(+)-Tramadol	Serotonin	0.03 x (racemic mixture)	Tab, dispersible tab, ext rel tab, ext rel cap, topical cream
(1S,2S)-(−)-Tramadol	Norepinephrine		
(1R,2R)-(−)-Tapentadol	Norepinephrine	0.33–0.5 x	Tab, ext rel tab

Tab, tablet; *ext rel,* extended release; *cap,* capsule

Opioid-Based Antidiarrheals

- Modifying meperidine with cyano or amide substituents that inhibit CNS distribution has yielded three peripherally acting antidiarrheal medications (Fig. 12.8).
- Difenoxin is the active metabolite of diphenoxylate and has five times the therapeutic activity. Both compounds are formulated with atropine to discourage abuse.
  - Difenoxin's zwitterionic structure makes it more resistant to blood–brain barrier penetration than the esterified diphenoxylate.
  - Difenoxin is excreted as conjugates of inactive hydroxylated metabolites.
- Loperamide is excluded from the CNS by P-gp efflux. It is available OTC.
  - Loperamide is N-dealkylated by CYP3A4 and CYP2C8 to an inactive desmethyl metabolite. Only 40% of a dose reaches the systemic circulation after oral administration.

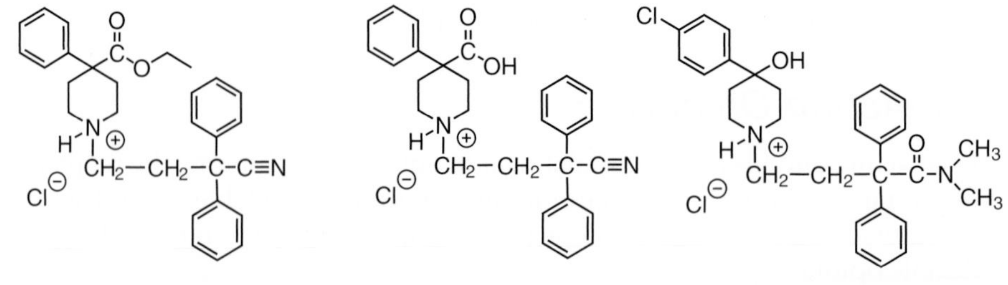

Diphenoxylate hydrochloride (Lomotil) Difenoxin hydrochloride (Motofen) Loperamide hydrochloride (Imodium)

Figure 12.8 Opioid-based antidiarrheals.

Treatment of Neuropathic Pain

- Neuropathic pain is insidious and difficult to treat. It is caused by injury to peripheral nerves as a result of disease or trauma, although central causes are also known.

- First-line agents in the treatment of neuropathic pain include antidepressants, calcium channel blockers, and local anesthetics.
- Mu agonist analgesics tramadol, methadone, morphine, and oxycodone have been relegated to second-line status, in part because of their adverse effect profile. S(+)-methadone inhibits NMDA receptors and has shown benefit in attenuating neuropathic pain.

Chemical Significance

Anyone who has ever been in severe pain or watched loved ones suffer knows the critical importance of high-quality care related to pain management. Pharmacists are on the front lines of providing that level of competent and compassionate care, as they are well-versed in both the therapeutic and adverse activity profiles of opioid analgesics from their understanding of opioid SAR. By analyzing opioid structure against patient needs, the pharmacist can guide prescribers in making wise therapeutic choices and help chronic pain patients live happier, safer, and more productive lives. For example:

- Is a patient on codeine for control of mild pain complaining of nausea and problematic itching? Consider recommending a switch to a μ agonist that does not have the 6α-OH group that has been associated with histamine release.
- Is a patient on hydrocodone complaining that he is experiencing no pain relief from the medication? It is possible that he is a poor CYP2D6 metabolizer, and will need to be switched from a 3-methoxy opioid to one containing the active 3-phenol.
- Has a chronic pain patient new to your community and stabilized on an N-methyl opioid (e.g., hydromorphone) presented you with a prescription from her new provider for an N-cyclobutylmethyl analgesic (e.g., butorphanol)? Phone the physician to explain that this new prescription, while providing potent analgesia, will precipitate withdrawal symptoms in this patient due to its μ-antagonist action.
- Does another patient with a new prescription for butorphanol complain of feeling highly distressed and uneasy for no apparent reason? She might be experiencing κ agonist–related dysphoria. A change to the pentacyclic κ agonist (nalbuphine) might resolve those symptoms. If not, consider an N-methyl (μ agonist) opioid of equivalent potency.
- Is a hospitalized cancer patient stabilized on fentanyl transdermal patches suffering from chronic and severe constipation? Make sure that the antagonist used to reverse that serious adverse effect has a quaternary nitrogen so that it cannot penetrate the blood–brain barrier and precipitate a withdrawal episode.
- Is a severely burned patient who is medicated with the fentanyl lollipop before painful dressing changes now being switched to the effervescent buccal tablet? Cut the dose in half (at least) to account for the approximately twofold increase in drug absorption that will occur due to the impact of the formulation on the amount of unionized fentanyl that will be generated.
- Is a patient who has developed an addiction to oxycodone that he no longer needs for pain control seeking your help to recover? Share with him the relative benefits and risks of methadone and buprenorphine in helping him successfully complete a recovery program. Tell him too that, if he and his physician elect to go with buprenorphine therapy, you will be able to dispense the medication and provide essential support while he is on the road to recovery.

No one should have to suffer from chronic, intractable pain when drugs as powerful as opioids are available. Pain management is truly a team effort, with prescriber, pharmacist, and patient all working collaboratively around goals established to maximize quality of life. While opioid analgesics have serious adverse effects, the pharmacist's in-depth understanding of their chemistry allows wise therapeutic decisions to be made so that pain can be controlled and distressing adverse reactions kept to a minimum.

Review Questions

1 Which of the following opioid structures would precipitate withdrawal in a μ agonist–dependent patient?

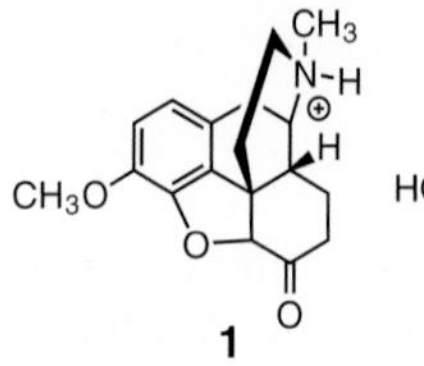
1

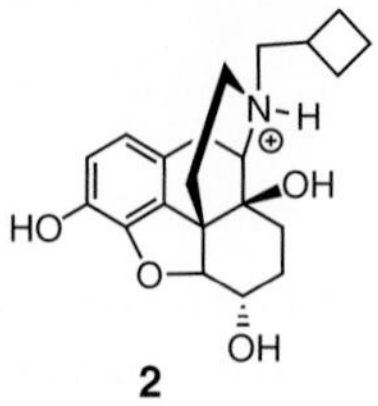
2

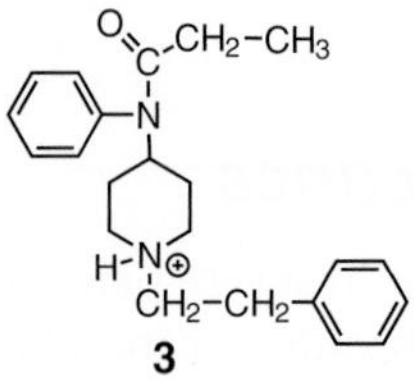
3

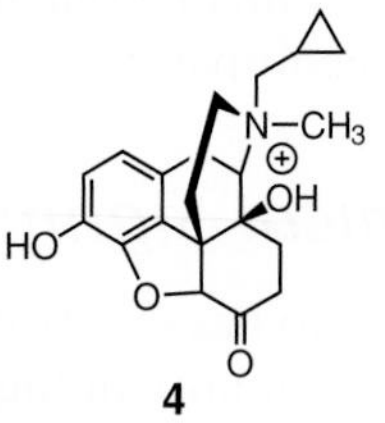
4

A. Opioid **1**
B. Opioid **2**
C. Opioid **3**
D. Opioid **4**
E. Opioids **2** and **4**

2 Which opioid agonist would have the highest μ-receptor affinity?

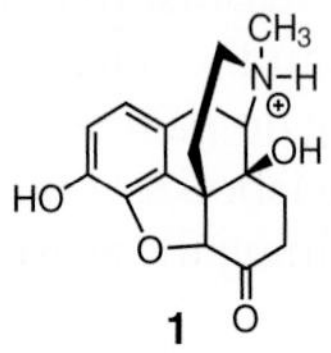
1

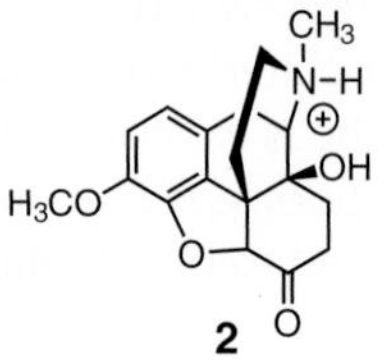
2

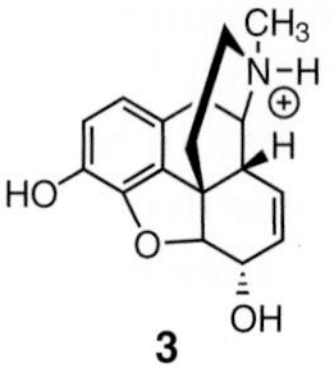
3

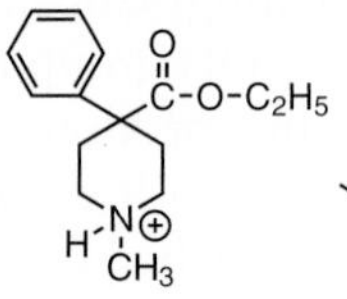
4

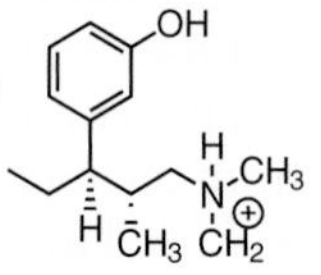
5

A. Opioid **1**
B. Opioid **2**
C. Opioid **3**
D. Opioid **4**
E. Opioid **5**

3 Which opioid analgesic would be totally ineffective in a patient who did not express CYP2D6?

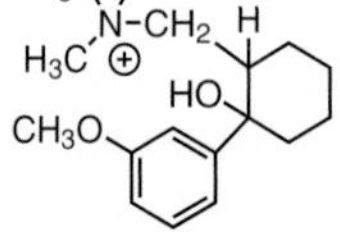
1

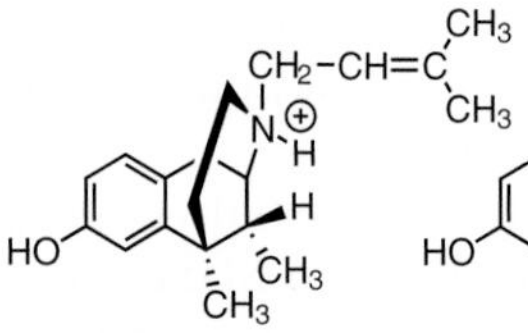
2

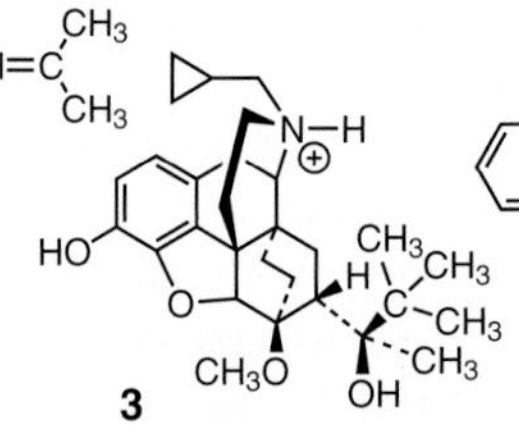
3

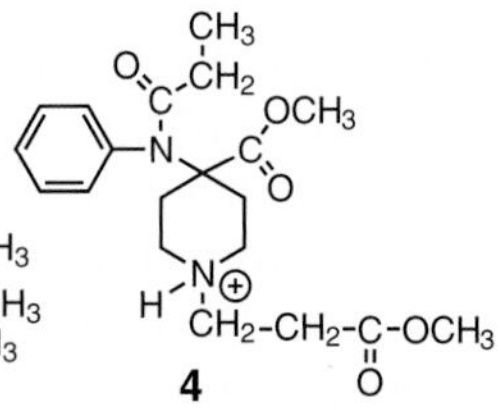
4

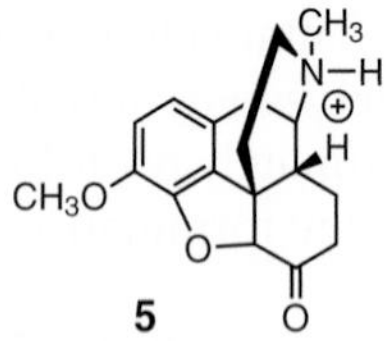
5

A. Opioid **1**
B. Opioid **2**
C. Opioid **3**
D. Opioid **4**
E. Opioid **5**

4 What is the role of citric acid in promoting the analgetic potency of fentanyl effervescent tablets?

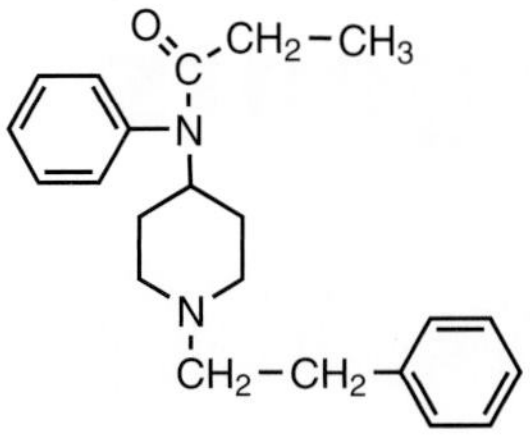

Fentanyl

A. It improves palatability of the fentanyl buccal tablet so it can be used in children
B. It increases salivary pH
C. It promotes the formation of fentanyl free base needed for absorption
D. It promotes dissolution of fentanyl into saliva
E. It protects fentanyl from prehepatic metabolism, which increases oral activity

5 The pKa of alfentanil is approximately 2 units below that of fentanyl and sufentanil. As a result:

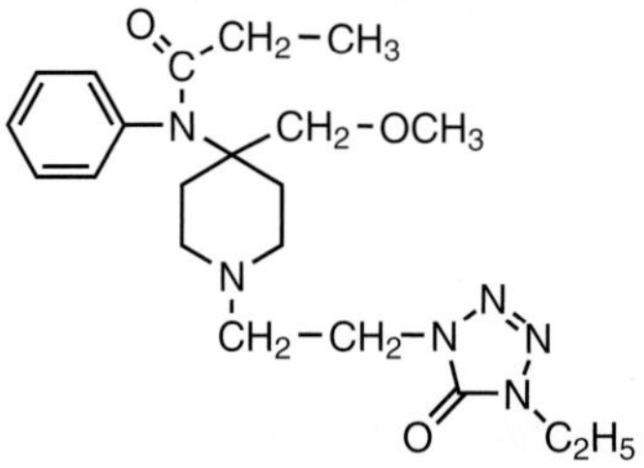

Alfentanil

A. Alfentanil is more strongly ionized in the bloodstream
B. Alfentanil binds with higher affinity to the µ receptor
C. Alfentanil more readily penetrates the blood–brain barrier
D. Alfentanil has a longer duration of action
E. Alfentanil has a higher analgetic potency

13

Cardiovascular Agents

Learning Objectives

Upon completion of this chapter the student should be able to:

1. Identify the mechanism of action of the following classes of cardiovascular drugs.

   Drugs used for the treatment of hypertension
   a. diuretics – the various classes and their sites of action
   b. peripheral acting sympatholytics – β- and α_1-adrenergic receptor blockers
   c. central acting sympatholytics
   d. calcium channel blockers – for hypertensives, for ischemic heart disease and angina
   e. adrenergic blocking agents

   Drugs used for the treatment of cardiac arrhythmia
   f. antiarrhythmic drugs – classes 1A, 1B, 1C, II, III, IV
   g. vasodilators
   h. nitrodilators

   Drugs used to treat heart failure
   i. cardiac glycosides
   j. phosphodiesterase inhibitors
   k. β-Adrenergic receptor agonists
   l. hyperpolarization-activated cyclic nucleotide-gated channel inhibitor

   Drugs used for the treatment of ischemic heart disease
   m. organic nitrates
   n. late sodium current inhibitor

   Drugs used to treat pulmonary hypertension
   o. endothelin receptor antagonists
   p. prostanoids

2. List structure activity relationships (SAR) for the following classes of drugs:
   a. thiazide and thiazide-like diuretics
   b. mineralocorticoid receptor antagonists
   c. calcium channel blockers (1,4-DHPs)
   d. cardiac glycosides

3. Identify clinically significant physiochemical and pharmacokinetic properties of cardiovascular drugs.

4. Discuss, with chemical structures, the clinically significant metabolism of the cardiovascular drugs and how the metabolism affects their activity.

5. Discuss the clinical application of the cardiovascular drugs.

6. Identify drug–drug interactions associated with various classes of cardiovascular drugs.

Introduction

According to the Centers for Disease Control and Prevention, each year more than 385,000 Americans die of heart disease and an estimated $109 billion is spent on heart disease–related services, medications, and lost productivity. In the broadest sense cardiovascular agents may include a wide variety of therapeutic modalities. Within this chapter the following subject will be presented: hypertension, cardiac arrhythmia, heart failure, ischemic heart disease (IHD) (coronary artery disease [CAD], angina), along with the drugs used to treat these conditions. In subsequent chapters drugs that affect the renin–angiotensin–aldosterone (RAA) system as well as those used to treat hyperlipidemia and coagulation disorders will be discussed.

Drugs Used for the Treatment of Hypertension

Hypertension is a major risk factor in heart disease, stroke, end-stage renal failure, and peripheral vascular disease. Hypertension affects nearly one in every three American adults. Most hypertensive patients (~90%) have *essential hypertension,* which is hypertension with no identifiable underlying cause. Disease symptoms and organ damage caused by hypertension is slow developing, but lowering blood pressure can reduce cardiovascular morbidity and mortality rates and slow overall mortality. Most hypertensive individuals require multiple antihypertensive drugs (e.g., diuretics, β- and α_1-adrenergic blockers, centrally acting sympatholytics, calcium channel blockers, vasodilators, angiotensin-converting enzyme inhibitors [ACEIs] and angiotensin II receptor blockers [ARBs]) to reduce and maintain blood pressure within acceptable ranges.

Classification of Hypertension in Adults		
	Blood Pressure (mm Hg)	
Classes	**Systolic**	**Diastolic**
Normal	<120	and <80
Prehypertension	120–139	or 80–89
Stage 1	140–159	or 90–99
Stage 2	≥160	or ≥100

Diuretics

Diuretics are chemicals that increase the rate of urine formation. By increasing the urine flow rate, diuretic usage leads to increased excretion of electrolytes (especially sodium and chloride ions) and water from the body without affecting other components of the blood. These pharmacologic properties have led to the use of diuretics in the treatment of **edematous conditions** (e.g., congestive heart failure [CHF], nephrotic syndrome, and chronic liver disease) and in the management of **hypertension**. The primary target organ for diuretics is the kidney or more specifically the nephron (Fig. 13.1). Each component of the nephron contributes to the functions of the kidney and thus the potential targets for various classes of diuretic agents. For a comprehensive discussion of the normal physiology of the nephron, the processing of blood filtration and urine formation/regulation, and the general therapeutic approaches for which diuretics are administered see Chapter 22 in *Foye's Principles of Medicinal Chemistry, Seventh Edition.*

Osmotic Diuretics

Mannitol

MOA

- Mannitol is a compound that is freely filtered through the Bowman's capsule into the renal tubules.

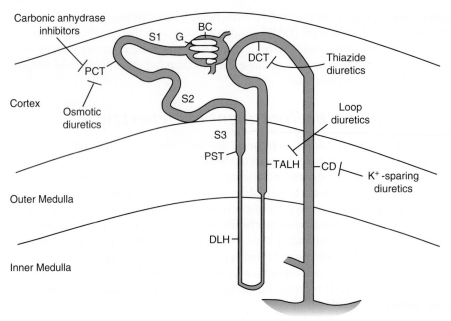

Figure 13.1 Sites of action of the diuretics in the nephron. The various portions of the nephron consist of the Bowman's capsule, BC; glomerulus, G; proximal convoluted tubule, PCT with its various segments, S1, S2, S3; proximal straight tubule, PST; descending limb of the loop of Henle, DLH; thick ascending limb of the loop of Henle, TALH; distal convoluted tubule, DCT; and the collecting duct, CD.

Drug Note

Previously several additional osmotic diuretics were available but most have been discontinued. These include glycerin, isosorbide, and urea. These products and mannitol have a very limited use primarily for treatment of acute renal failure.

- It is a non-reabsorbable solute which increases intraluminal osmotic pressure, causing water to pass from the body into the tubule increasing the volume of urine and excretion of water (PCT in Fig. 13.1).

Physicochemical and Pharmacokinetic Properties

- Mannitol is water soluble and administered IV in solutions of 5% to 50%.
- Large doses are required for effectiveness (50 to 200 g).
- The drug is excreted unchanged.

Carbonic Anhydrase Inhibitors

MOA

- Carbonic anhydrase inhibitors block the formation of carbonic acid within the proximal convoluted tubule (S2, Fig. 13.1) and distal tubular cells to limit the number of hydrogen ions available for exchange with sodium (Fig. 13.2). This promotes sodium/water excretion.

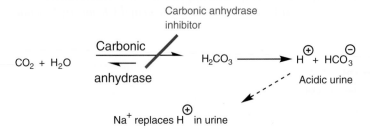

- Diuretic response is observed when >99% of the carbonic anhydrase is inhibited.
- The acidic unsubstituted sulfonamides bind strongly to carbonic anhydrase with the unsubstituted sulfonamide acting as an isostere for carbonic acid.
- Due to inhibition of Na^+/H^+ exchange, an increase in urine pH occurs.

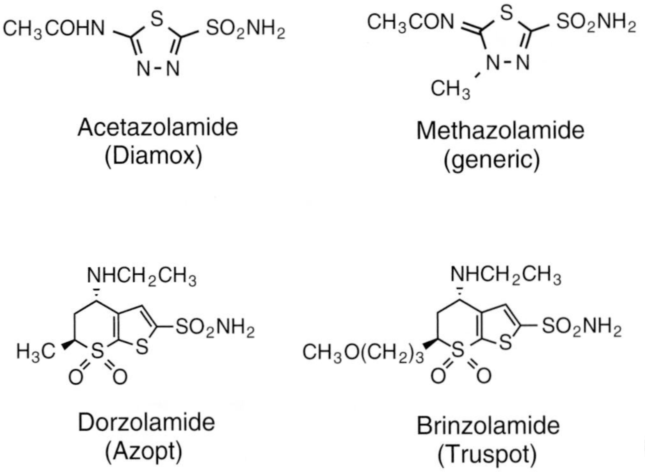

Acetazolamide
(Diamox)

Methazolamide
(generic)

Dorzolamide
(Azopt)

Brinzolamide
(Truspot)

Figure 13.2 Carbonic anhydrase inhibitors.

Clinical Applications

- Carbonic anhydrase inhibitors have limited diuretic use (prolonged use causes blood acidosis and ineffectiveness of these drugs).
- These drugs are commonly used to treatment of glaucoma (inhibition of ocular carbonic anhydrase reduce aqueous humor formation and intraocular pressure is reduced).

> ⚠️ **ADVERSE EFFECTS**
>
> **Carbonic Anhydrase Inhibitors:** may produce gastrointestinal (e.g., nausea, vomiting), nervous system (e.g., sedation, headache), and renal effects (kidney stones).

Thiazide and Thiazide-Like Diuretics

MOA

- The thiazide and thiazide-like diuretics act by a similar mechanism of action (Fig. 13.3).
- These diuretics are actively secreted into the filtrate in the PCT and are carried to the DCT (Fig. 13.1).

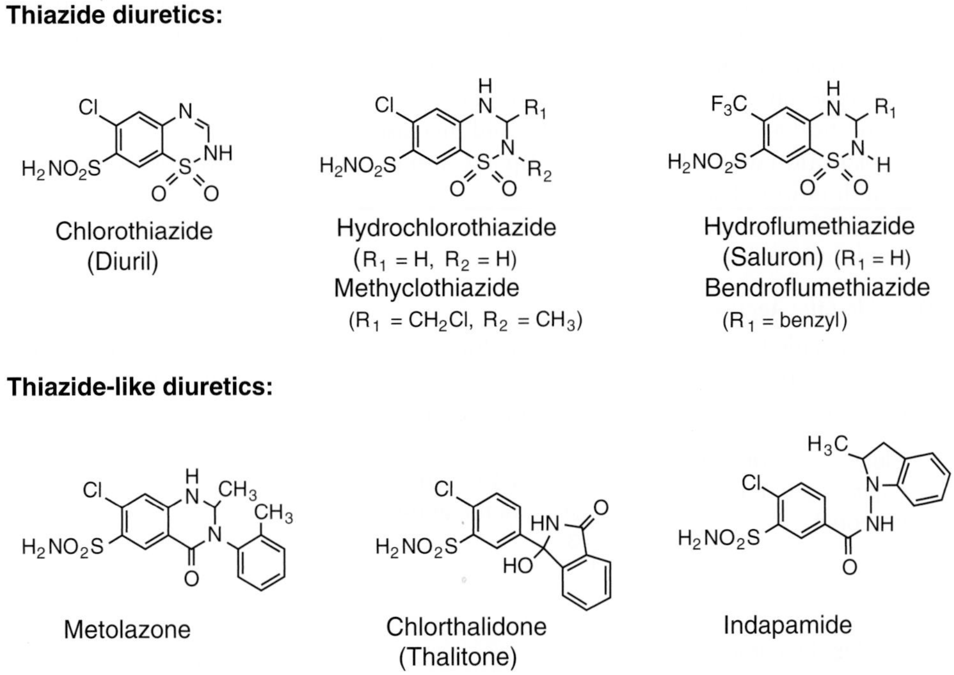

Thiazide diuretics:

Chlorothiazide
(Diuril)

Hydrochlorothiazide
$(R_1 = H, R_2 = H)$
Methyclothiazide
$(R_1 = CH_2Cl, R_2 = CH_3)$

Hydroflumethiazide
(Saluron) $(R_1 = H)$
Bendroflumethiazide
$(R_1 = benzyl)$

Thiazide-like diuretics:

Metolazone

Chlorthalidone
(Thalitone)

Indapamide

Figure 13.3 Thiazide and thiazide-like diuretics.

Tubular Filtrate

Blood

DCT cell

Na$^+$ Cl$^-$ Symporter

Na$^+$ → Na$^+$

Cl$^-$ → Cl$^-$

Thiazide diuretics

Na$^+$ → Na$^+$

Na$^+$/K$^+$ ATPase

K$^+$ ← K$^+$

K$^+$ ← K$^+$

Cl$^-$ → Cl$^-$

Apical cell membrane

Basolateral cell membrane

Figure 13.4 Thiazide diuretics act on the cells of the distal convoluted tubule (DCT) preventing Na$^+$/Cl$^-$ absorption into the cell which ultimately results in Na$^+$/Cl$^-$ reabsorption into the blood stream.

- Normally Na$^+$ is reabsorbed into the DCT cell and then pumped out of the DCT cell and into the bloodstream by the Na$^+$/K$^+$ ATPase (Na$^+$/K$^+$ pump) with Cl$^-$ leaving through chloride channels (Fig. 13.4).
- The thiazides/thiazide-like agents compete for the Cl$^-$ binding site of the Na$^+$/Cl$^-$ symporter (a member of the SLC12 cotransporter family) in DCT cells inhibiting the absorption of Na$^+$ and Cl$^-$ thus increasing their excretion.
- Additionally thiazides and thiazide-like agents are weak inhibitors of carbonic anhydrase, and increase HCO$_3^-$ excretion and an increase in urine pH.
- The thiazides also inhibit the reabsorption of K$^+$/Mg^{2+} increasing their excretion and decrease Ca^{2+} excretion.

SAR

- It should be noted that the initial difference between the thiazide and thiazide-like diuretics is the replacement of a "sulfonamide" by an "amide."

Thiazides

- An electron-withdrawing group is necessary at C6 (e.g., Cl, CF$_3$) for diuretic activity.
- The trifluoromethyl-substituted diuretics are more lipid soluble and have a longer duration of action than their chloro-substituted analogs.
- The sulfonamide group at C7 is essential.
- Saturation at C3-C4 results in a 10-fold increase in activity.
- Substitution with a lipophilic group at C3 markedly increases diuretic potency and duration of action.
- Alkyl substitution on the N2 position decreases polarity increasing the duration of diuretic action.

Thiazide

Thiazide-Like Agents

- The thiazide-like agents exhibit similar functionality, which is important for activity.
- Note that the thiazide-like agents all have an unsubstituted sulfonamide and an electron-withdrawing chloride group present.

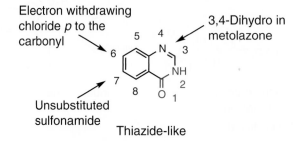

Electron withdrawing chloride *p* to the carbonyl

3,4-Dihydro in metolazone

Unsubstituted sulfonamide

Thiazide-like

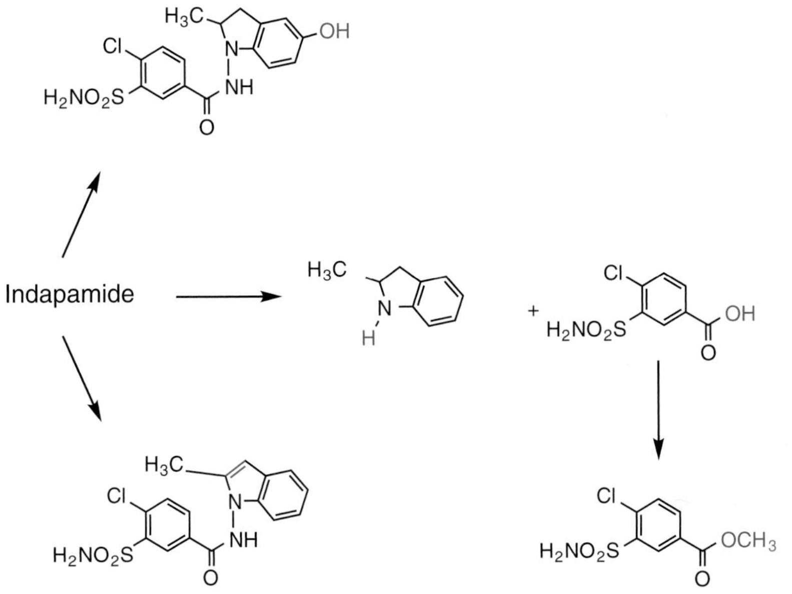

Figure 13.5 Metabolism of indapamide.

Physicochemical and Pharmacokinetic Properties

- The thiazide and thiazide-like diuretics all have a weakly acidic sulfonamide group.
- Water-soluble sodium salts can be prepared leading to products that can be used for intravenous administration.
- The thiazides are excreted in the urine unchanged with the exception of hydroflumethiazide (metabolite: 2,4-disulfamyl-5-trifluoromethylaniline).
- Thiazide-like diuretics also do not undergo significant metabolism except for indapamide, which is extensive metabolized (Fig. 13.5).
- Some pharmacokinetic properties are presented in Table 13.1.

Clinical Applications

- The thiazide and thiazide-like diuretics are indicated for the treatment of edema associated with heart, liver, and renal diseases.
- These agents are also indicated for the treatment of hypertension either alone (initial therapy) or in combination with other antihypertensive drugs.
- These agents are usually administered when the glomerular filtration rate is >30 to 40 mL/min (GFR), although both metolazone and indapamide are effective in patients with a GFR <40 mL/min.

> **⚠ ADVERSE EFFECTS**
>
> **Thiazide Diuretics:** may induce a number of adverse effects including: hypersensitivity reactions, gastric irritation, nausea, and electrolyte imbalances (e.g., hypo: natremia, kalemia, magnesemia, chloremic alkalosis and hyper: calcemia, uricemia). Thiazide-like diuretics indapamide and chlorthalidone: dryness of mouth, thirst, irregular heartbeat, mood changes, muscle cramps, and fatigue while metolazone's adverse effects appear minor and include: dizziness, and headaches.

Table 13.1 Pharmacokinetic Properties of Thiazide and Thiazide-Like Diuretics

Generic Name	Protein Binding	Bioavailability (%)	Peak Plasma (h)	Half-Life (h)	Duration of Effect (h)
Thiazide Diuretics					
Chlorothiazide	~40%	<25	4	1–2	12–16
Hydrochlorothiazide	68%	>80	4–6	5–15	12–16
Methyclothiazide	75%	~93	6	—	>24
Hydroflumethiazide	74%	—	3–4	6–14	18–24
Thiazide-Like Diuretics					
Metolazone	50–70%[a]	<65	8–12	14	12–24
Chlorthalidone	~98%[a]	>90	2	35–50	48–72
Indapamide	80%[a]	>90	2–3	14–18	8 wks

h, hours
[a]Bound to erythrocytes

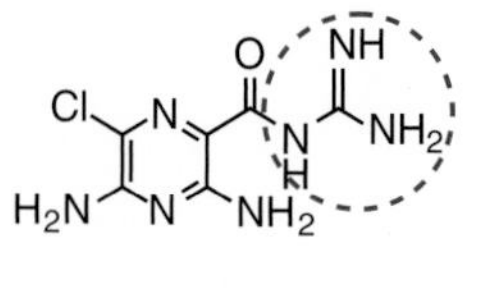

206 —— CARDIOVASCULAR AGENTS

Guanidino group

Figure 13.6 Potassium-sparing diuretics.

Amiloride
(Midamor)

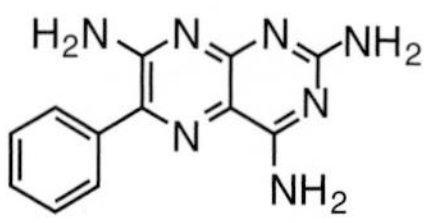

Triamterene
(Dyrenium)

Potassium-Sparing Diuretics

MOA

- Potassium-sparing diuretics exert a diuretic effect by blocking the Na^+ channel in the late DCT and CD (Figs. 13.1 and 13.6).
- The two drugs are weak organic bases and inhibit Na^+ channel by binding to negatively charged regions within the channel.
- The binding is pH dependent with the stronger base, amiloride (pKa = 8.7) being ~100-fold more active than triamterene (pKa = 6.2).
- Na^+ channel inhibitors which increase Na^+ and Cl^- excretion inhibit K^+ secretion.

Physicochemical and Pharmacokinetic Properties

- The guanidino group in amiloride (Fig. 13.6) is a strong base.
- Amiloride exhibits variable bioavailability (30% to 90%) while triamterene is well absorbed with moderate bioavailability (~52%).
- The duration of action is ~24 hours.
- Amiloride is excreted nearly equally in feces and urine unchanged, triamterene is a substrate for CYP1A2 with extensive metabolism resulting in 4'-hydroxylation of the aromatic ring followed by sulfate conjugation.

Clinical Applications

- Both drugs are used in combination with a thiazide or a loop diuretic in the treatment of edema or hypertension.
- The major advantage of these drugs is the overall reduced loss of K^+ in the combination products.

ADVERSE EFFECTS

Potassium-Sparing Diuretics: Hyperkalemia is a serious and potentially fatal adverse effect of the potassium-sparing diuretics. These drugs are contraindicated in patients with hyperkalemia or may be at risk of hyperkalemia (e.g., renal failure, patients receiving angiotensin-converting enzyme inhibitors, or on potassium supplements).

Mineralocorticoid Receptor Antagonists

The adrenal cortex secretes the potent mineralocorticoid aldosterone (Fig. 13.7), which promotes Na^+, Cl^-, and water retention while excreting K^+ and H^+. Aldosterone exerts its biologic effects through binding to the mineralocorticoid receptor (MR), a nuclear transcription factor.

MOA

- The MR is located in the late DCT and CD of the nephron.
- Aldosterone~MR complex translocates to the nucleus, binds to DNA response elements, and initiates expression of gene products leading to protein synthesis.
- MR antagonists (Fig. 13.7) competitively inhibit aldosterone binding to MR, interfering with NaCl transporter protein synthesis, and reducing Na^+, Cl^-, and water reabsorption (K^+ sparing activity).
- MR antagonists require the presence of a γ-lactone ring at C17 and a substituent on C7 (the unhindered C7 position is required for aldosterone binding to MR).

Physicochemical and Pharmacokinetic Properties

- Both spironolactone and eplerenone are readily absorbed with good bioavailability (~70%).
- Spirolactone and eplerenone are bound to plasma protein, >90% and ~50%, respectively. Some biliary recycling may occur.

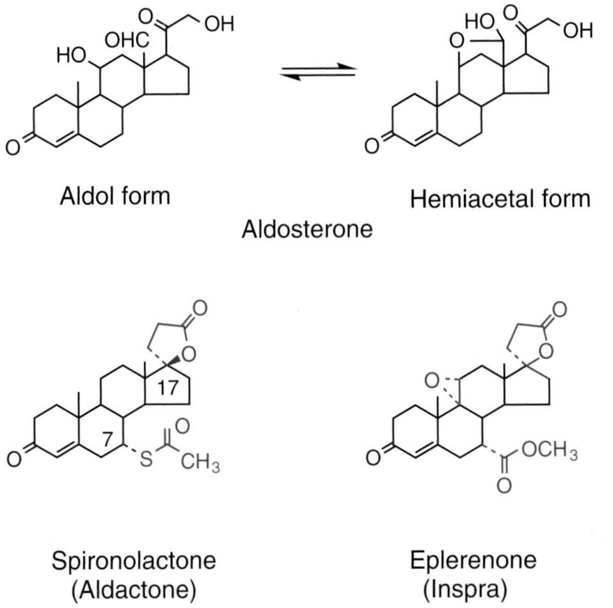

Aldol form   Hemiacetal form
Aldosterone

Spironolactone
(Aldactone)

Eplerenone
(Inspra)

Figure 13.7 Mineralocorticoid aldosterone and the mineralocorticoid antagonists.

- Spironolactone has a short half-life and may actually act as a prodrug.

Metabolism

- Both drugs are extensively metabolized (Fig. 13.8) and excreted in the urine and feces.
- Eplerenone was developed with the hope of reducing the formation of the spironolactone metabolites.

High-Ceiling or Loop Diuretics

MOA

- The site of action of the loop diuretics is believed to be on the thick ascending limb of the loop of Henle (TALH, Fig. 13.1) thus referred to as loop diuretics (Fig. 13.9).

ADVERSE EFFECTS

Spironolactone: exhibits a number of significant adverse effects including: hyperkalemia, hypersensitivity reactions, gastrointestinal disturbances, and peptic ulcer. Sexual adverse effects (i.e., gynecomastia, decreased libido, and impotence) related to nonselective binding of spironolactone to the androgen (AR), glucocorticoid (GR), and progesterone receptors (PR). While eplerenone may also lead to potential toxic hyperkalemia most other adverse effects are minor since eplerenone exhibits specificity for MR without action on other steroid receptors.

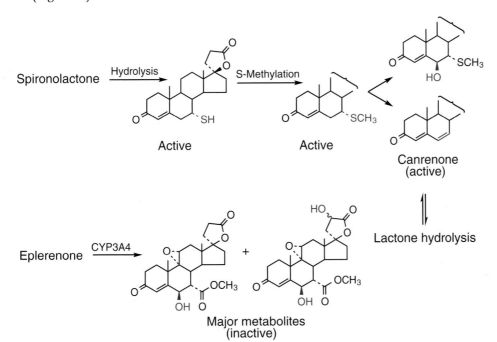

Figure 13.8 Metabolism of spironolactone and eplerenone.

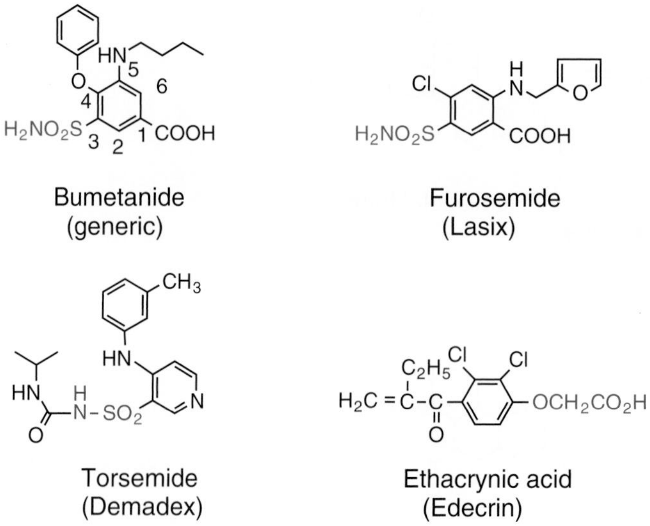

Figure 13.9 The high-ceiling or loop diuretics.

- These drugs inhibit the luminal $Na^+/K^+/2Cl^-$ symporter a protein located in the cell apical membrane – a member of the SLC12 cotransporter family.
- The result of inhibition leads to significant Na^+ and Cl^- loss along with Ca^{2+} and Mg^{2+} loss.

Physicochemical and Pharmacokinetic Properties

- Loop diuretics are characterized by a fast onset and short duration of action (see Table 13.2).
- These drugs are excreted primarily via the urine.

Metabolism

- The loop diuretics undergo various metabolic reactions including: γ–hydroxylation of the butyl chain of bumetanide (primarily unchanged drug); glutathione conjugate at the α position of the α,β double bond of ethacrynic acid and glucuronidation of furosemide.
- Torsemide is prone to CYP2C9 oxidation of the methyl tolyl group to a carboxylic acid.

Clinical Applications

- The primary use of the loop diuretics is for treatment of acute pulmonary edema.
- Unlike thiazide diuretics the loop diuretics, also referred to as high-ceiling diuretics, do not exhibit a ceiling effect and therefore are quite valuable in treatment of edema.
- These drugs are prescribed for treatment of hypertension, but their short duration of action limits their use to some extent.

> **⚠ ADVERSE EFFECTS**
>
> **Loop Diuretics:** may cause hypokalemia an important adverse effect that can be prevented or treated with potassium supplements or coadministration of potassium-sparing diuretics. Hypersensitivity reactions also are possible with sulfonamide-based drugs, and cross-reactivity with other sulfonamide-containing drugs is possible. Other adverse effects may include: headache, easy bruising, hearing problems. Ethacrynic acid being a nonsulfonamide-containing drug has a different set of adverse effects including: thirst, nausea, easy bruising, muscle pain, and hearing problems.

Table 13.2 Pharmacokinetic Properties of Loop Diuretics

Generic Name	Protein Binding	Bioavailability (%)	Peak Plasma (h)	Half-Life (h)	Duration of Effect (h)
Bumetanide	95%	~80	<2	1–1.5	5–6
Ethacrynic acid	>98%	~100	2	0.5–1	6–8
Furosemide	91–97%	~60	4–5	0.5–4	6–8
Torsemide	97–99%	~80	1–2	0.8–4	6–8

h, hours

Peripheral Acting Sympatholytics

The discussion of peripheral acting sympatholytics will focus on β-blockers (β-antagonists) and α₁-blockers (α₁-antagonists), which are beneficial for the treatment of hypertension. There are additional utilities for β-blockers (e.g., glaucoma) and the reader is directed to Chapter 4 for this discussion. For a discussion of regulation of vascular tone and smooth muscle contraction and relaxation the reader is directed to Chapter 24 in *Foye's Principles of Medicinal Chemistry, Seventh Edition*.

β-Adrenergic Receptor Blockers

MOA

- β-Adrenergic receptor blockers bind to β-adrenoceptors blocking the binding of epinephrine to these receptors.
- Some of these drugs when bound to the β-adrenoceptors partially activate the receptor as β-agonists with intrinsic sympathomimetic activity (ISA).
- First-generation β-blockers are nonselective, blocking both β₁- and β₂-adrenoceptors (Fig. 13.10).
- Second-generation β-blockers are more cardioselective with selectivity for β₁-adrenoceptors.
- Third-generation β-blockers (mixed α₁/β₁-adrenergic blockers) possess vasodilator actions through blockade of vascular α-adrenoceptors

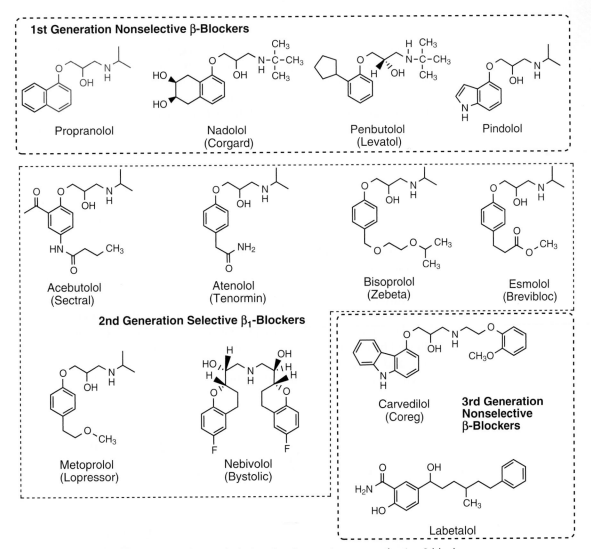

Figure 13.10 Peripheral Acting Sympatholytics: 1st Generation – nonselective β-blockers, 2nd Generation – β₁-selective blockers, 3rd Generation – α₁/β₁-adrenergic blockers.

Table 13.3   Pharmacokinetic Properties of Antihypertensive β-Adrenergic Blocking Agents

Drug	Adrenergic Receptor Blocked	Lipophilicity log D - pH$_{7.4}$	Oral Bioavailability (%)	Half-Life (h)	Protein Binding (%)	Metabolism/ Excretion
Acebutolol	β$_1$	−0.9	20–60	3–4	26	Hepatic/renal
Atenolol	β$_1$	−2.01	50–60	6–9	5–16	Unchanged/renal
Bisoprolol	β$_1$	+2.4	80	9–12	30	Unchanged/renal
Esmolol	β$_1$	+1.7	—	0.15	55	Red blood/cells/ renal
Metoprolol	β$_1$	−0.24	40–50	3–7	12	Hepatic/renal
Nebivolol	β$_1$	+2.34	12–96	12–19	98	Hepatic/renal
Nadolol	β$_1$ β$_2$	−1.43	30–50	20–24	30	Unchanged/renal
Penbutolol	β$_1$ β$_2$	+2.06	>90	5	80–98	Hepatic/renal
Pindolol	β$_1$ β$_2$	−0.36	>90	3–4	40	Hepatic/renal
Propranolol	β$_1$ β$_2$	+1.41	30	3–5	90	Hepatic/renal
Labetalol	β$_1$ β$_2$ α$_1$	+1.08	30–40	5.5–8	50	Conjugation/ renal
Carvedilol	β$_1$ β$_2$ α$_1$	+3.53	25–35	7–10	98	Hepatic/bile

h, hours.

SAR

- For a discussion of SAR of β-adrenergic receptor blockers, see Chapter 4 and Figure 4.7.

Physicochemical and Pharmacokinetic Properties

- The lipophilicity varies with the particular β-blocker as shown in Table 13.3. Lipophilic drugs such as propranolol create the potential for CNS adverse effects while hydrophilic drugs such as atenolol and nadolol have reduced CNS effects.
- The β-adrenergic receptor blockers are all readily absorbed following oral administration and have good bioavailability (Table 13.3).
- Several of the drugs are excreted unchanged, usually in the urine, while the other undergo extensive metabolism.

Metabolism

- Several of the β-adrenergic receptor blockers are substrates for CYP2D6 (e.g., metoprolol, nebivolol, propranolol, carvedilol), which raises the potential for significant difference in metabolic rates between poor and extensive metabolizers, which can effect drug half-life.
- Metoprolol's major metabolite is O-demethylation by CYP2D6 along with N-dealkylation and α-hydroxylation.
- Nebivolol's major metabolite results from O-glucuronidation but CYP-catalyzed aromatic hydroxylation is also reported.
- CYP2D6 oxidations of propranolol was reported earlier in the chapter (4-hydroxylation and N-dealkylation) while aromatic hydroxylation at multiple sites followed by glucuronidation is significant in carvedilol.
- Hydrolysis and acetylation of acebutolol results in the formation of diacetolol a product with activity equivalent to the administered drug (Fig. 13.11).

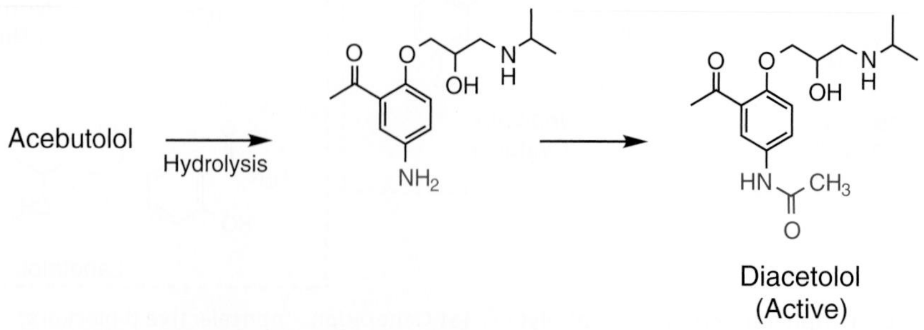

Figure 13.11   Metabolism of acebutolol.

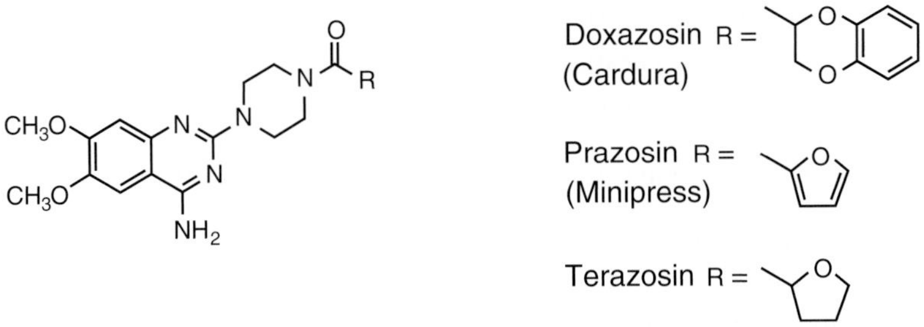

Figure 13.12 α_1-Selective adrenergic blockers used in treatment of hypertension.

- Pindolol is extensively metabolized to a hydroxyl derivative, but the catalyst is not reported and esmolol undergoes hydrolysis as reported earlier.

Clinical Applications

- β-Adrenergic receptor blockers are recommended for treatment of heart failure and CAD in combination with an ACE inhibitor or angiotensin receptor blocker (ARB).
- Their effectiveness may be associated with inhibition of release of renin, which is regulated by β_1-adrenoceptors in the kidney (see Chapter 14).
- β-Blockers may also be used as monotherapy in the treatment of angina, arrhythmias, mitral valve prolapse, myocardial infarction, migraine headaches, performance anxiety, excessive sympathetic tone or "thyroid storm" in hyperthyroidism.

α_1-Adrenergic Blockers

MOA

- α_1-Adrenoceptor antagonists (Fig. 13.12) are selective competitive blocker of α_1-adrenoceptors inhibiting the binding of NE causing vasodilation of both arteries and veins (Note: α_1-adrenoceptors are the predominant α-receptor located on vascular smooth muscle [VSM]) (Fig. 13.12).
- α_1-Adrenoceptors are linked to G_q-proteins that activate smooth muscle contraction through the IP_3 signal transduction pathway. These drugs block the effect of sympathetic nerves on blood vessels.
- The vasodilator effect is more pronounced in the arterial resistance vessels especially during elevated sympathetic activity (e.g., during stress).

Physicochemical and Pharmacokinetic Properties

- Pharmacokinetic properties are presented in Table 13.4.
- Doxazosin and prazosin are extensively metabolized and primarily excreted in the feces. Terazosin is eliminated in both the urine and feces and with minimal first-pass metabolism.

Metabolism

- Doxazosin and prazosin undergo O-demethylation (6 and 7 demethylation) followed by conjugation. Doxazosin also undergoes 6' and 7'-hydroxylation on the benzodioxan ring.
- Terazosin is hydrolyzed by an amidase enzyme resulting in an active metabolite.

Ophthalmic Preparation:

Several nonselective β-adrenergic blockers are used exclusively as ophthalmic preparations, but have a structural similarity to the prototype β-blockers propranolol.

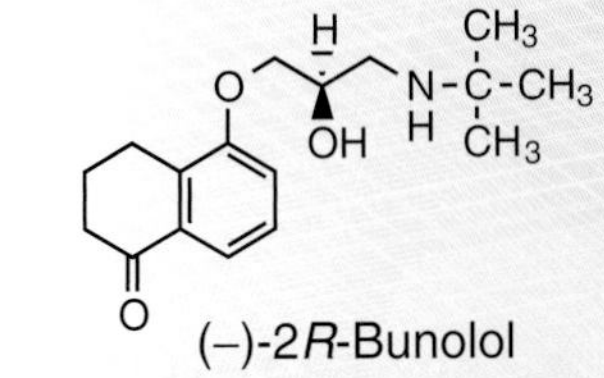

(−)-2R-Bunolol
(Akbeta, Betagen)

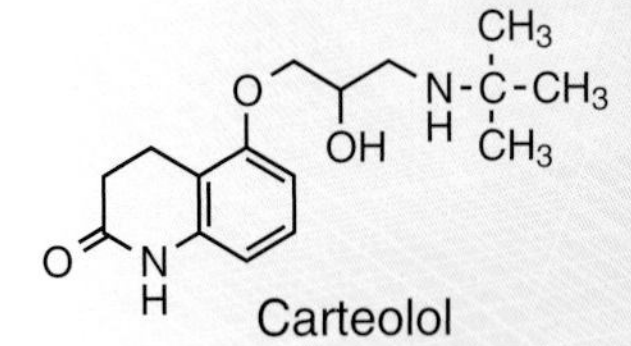

Carteolol
(Ocupress)

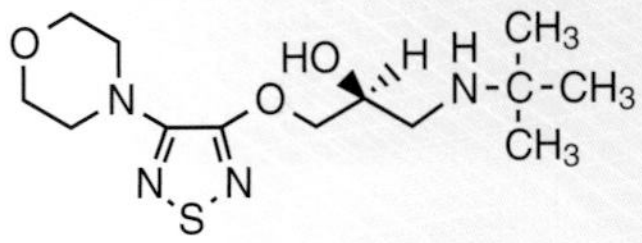

S(−)Timolol
(Timoptic)

ADVERSE EFFECTS

β_1-**Adrenergic Blockers:** Common adverse effects for the β_1-adrenergic blockers include decreased exercise tolerance, cold extremities, depression, sleep disturbance, and impotence. The lipid-soluble β-blockers, but not hydrophilic drugs, have been associated with more CNS effects such as dizziness, confusion, or depression. At high doses, these cardioselective agents can adversely affect asthma, peripheral vascular disease, and diabetes.

Table 13.4 Pharmacokinetic Properties of α_1-Adrenergic Blockers

Generic Name	Protein Binding	Bioavailability	Peak Plasma (h)	Half-Life (h)	Duration of Effect (h)
Doxazosin	98.3%	~65%	<2	~20	36
Prazosin	92–97%	50–70%	2	~3	7–10
Terazosin	90–94%	>90%	1–2	~12	>18

h, hours.

Mixed α/β-Adrenergic Receptor Blockers

- The mixed α/β-receptor blockers carvedilol and labetalol are third-generation nonselective blockers shown in Figure 13.10 and their pharmacokinetic properties are presented in Table 13.3.
- The mechanism of action, as the title implies, involves blocking both types of adrenergic receptors.
- Vasodilation occurs via α₁-blockade while the β-blocker helps avoid reflex tachycardia.
- Both drugs are administered as racemates: S-carvedilol has α- and β-blocking activity while the R-isomer exhibits α₁-blocking activity; labetalol has four stereoisomers with the R,R-isomer being the active isomer.

Clinical Applications

- α₁-Blockers and mixed-acting blockers are effective agents for management of hypertension.
- And while they have been shown to be as effective as other major classes of antihypertensives in lowering blood pressure these drugs are not first-line drugs.
- Selection of mixed α/β-blockers are recommended for management of more severe hypertension (e.g., stage 2) and may be effective for both essential and renal hypertension.
- α₁-Adrenoceptor antagonists are also indicated for the treatment of benign prostatic hyperplasia (BPH). (See Chapter 18 in *Foye's Principles of Medicinal Chemistry, Seventh Edition,* for a complete detailed discussion.)

> **ADVERSE EFFECTS**
>
> **α₁-Adrenoceptor Blockade:** The most common adverse effects are related directly to α₁-adrenoceptor blockade include: dizziness, orthostatic hypotension (loss of reflex vasoconstriction on standing), nasal congestion (dilation of nasal mucosal arterioles), headache, and reflex tachycardia. Fluid retention is also a problem that can be rectified by the addition of a diuretic.

Central Acting Sympatholytics

The sympathetic adrenergic nervous system plays a major role in the regulation of arterial pressure. Sympatholytic drugs block the sympathetic adrenergic system on three different levels: (1) peripheral sympatholytic drugs (e.g., α-adrenoceptor and β-adrenoceptor antagonists discussed above); (2) ganglionic blockers that block impulse transmission at the sympathetic ganglia; (3) centrally acting sympatholytic drugs block sympathetic activity within the brain.

MOA

- Centrally acting sympatholytics (Fig. 13.13) block sympathetic activity by binding to and activating centrally inhibiting α₂-adrenoceptors (in the brainstem). More specifically the guanidine-containing sympatholytics act at α₂ₐ-adrenoceptors.
- Binding reduces sympathetic output to the vasculature decreasing sympathetic vascular tone leading to vasodilation and reduced systemic vascular resistance thus decreasing arterial pressure.

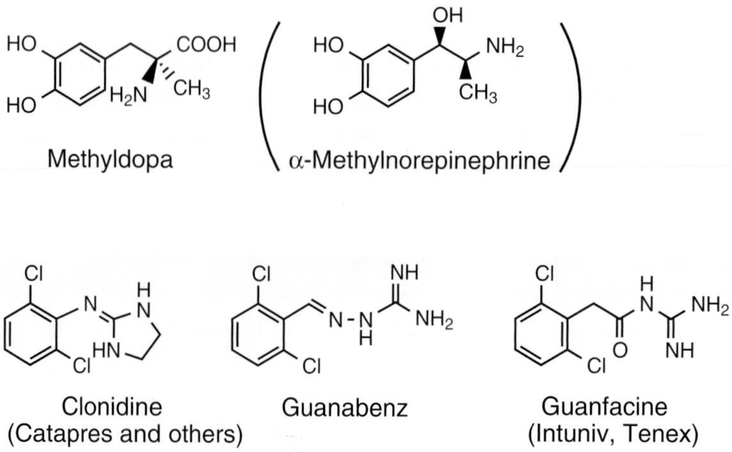

Figure 13.13 Centrally acting sympatholytics.

Table 13.5 Pharmacokinetic Properties of Central Acting Sympatholytics

Generic Name	Protein Binding	Bioavailability	Peak Plasma (h)	Half-Life (h)	Duration of Effect (h)
Methyldopa	<15	20–50%	variable	~2	24–48
Clonidine	20–40%	>90%	2–4	6–20	~8
Clonidine is well absorbed percutaneously from a transdermal patch (release 50–70%)					
Guanabenz	90%	70–80%	2–7	4–14	~12
Guanfacine	64%	>80%	1–4	14–23	~36

h, hours.

- There is some evidence that the action of the guanidine-containing sympatholytics may involve binding to imidazoline receptors I_1 and I_2 (see Chapter 24 in *Foye's Principles of Medicinal Chemistry, Seventh Edition,* for a complete detailed discussion).

Physicochemical and Pharmacokinetic Properties

- Methyldopa is a prodrug, which is converted into α-methylnorepinephrine following transport into the CNS via aromatic amino acid transporter (see metabolism below).
- The drug is unstable to air, alkaline pH and light.
- Methyldopa is available for parenteral use as a hydrochloride salt of the ethyl ester.
- The pharmacokinetic properties of the centrally acting sympatholytics are shown in Table 13.5.

Metabolism

- Metabolism of methyldopa is extensive and occurs in the GI tract and centrally with the majority of metabolites appearing in the urine (Fig. 13.14).
- Metabolism of the guanidine-containing sympatholytics is shown in Figure 13.15.
- The metabolites are inactive and primarily appear in the urine. Approximately 24% to 37% of guanfacine is excreted unchanged.

Clinical Applications

- Methyldopa may be used in the management of hypertension during pregnancy and also for the management of pregnancy-induced hypertension (i.e., preeclampsia).

DRUG–DRUG INTERACTIONS

The hypotensive actions of the α_2-adrenergic agonists can be additive with, or potentiated by the action of CNS depressants (e.g., opiates, barbiturates/sedatives, anesthetics, or alcohol). Tricyclic antidepressants (i.e., imipramine and desipramine) may inhibit the hypotensive effect of α_2-adrenergic agonists, and increase in blood pressure. Sudden withdrawal of clonidine, guanabenz, and guanfacine can result in an excess of circulating catecholamines. Because α_2-adrenergic agonists can produce bradycardia the possibility of additive effects should be considered combined with hypotensive drugs or cardiac glycosides.

ADVERSE EFFECTS

Methyldopa: Adverse effects are common with methyldopa and include: drowsiness, decrease in mental acuity, lapses of memory, and difficulty in performing simple calculations. Nightmares, mental depression, orthostatic hypotension, and symptoms of cerebrovascular insufficiency can also occur. Positive direct antiglobulin (Coombs') test results have been reported in 10% to 20% of patients receiving methyldopa. Drowsiness, tiredness, dizziness, weakness, bradycardia, headache, and dry mouth are common adverse effects for patients receiving clonidine, guanabenz, and guanfacine.

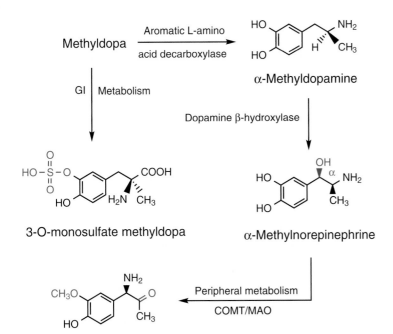

Figure 13.14 Metabolism of methyldopa.

Clonidine

Guanabenz

O-glucuronide
conjugate

O-sulfate
conjugate

Guanfacine

O-glucuronide
conjugate

mercapturic
conjugate

O-sulfate
conjugate

Figure 13.15 Metabolic products formed from centrally acting sympatholytics.

- Clonidine, guanabenz, and guanfacine are generally reserved for patients who fail to respond to therapy with a stage 1 drug (e.g., diuretics, calcium channel blockers, ACE inhibitors, or angiotensin II blockers).
- These drugs can be used in combination with diuretics and stage 1 hypotensive agents.

Calcium Channel Blockers – L-Type Channels

The calcium (Ca^{2+}) channel blockers are prescribed for several disease states, the most important of which is hypertension. Additionally, several Ca^{2+} channel blockers have as an indication the prevention and treatment of IHD or angina. The effectiveness of the Ca^{2+} channel blockers in treatment of hypertension is associated with selectivity for VSM while action on myocardial tissue as well as VSM opens their use in treatment of IHD. The latter drugs will be discussed at the end of this section. The antihypertensive Ca^{2+} channel blockers are members of the 1,4-dihydropyridines (1,4-DHPs) class of drugs as shown in Figure 13.16.

MOA

- Ca^{2+} channel blockers exert their effects by binding to specific receptor sites located within the central α_1 subunit of L-type slow channels (e.g., potential-dependent channels).
- Potential-dependent channels can exist in one of three conformations: a resting state; an open state, which allows the Ca^{2+} to enter; and an inactive state, which is refractory to further depolarization.
- Ca^{2+} channel blockers preferentially bind to Ca^{2+} channel in either the open or inactive state reducing Ca^{2+} flux.
- The 1,4-DHPs are primarily vasodilators although an increased heart rate may be seen resulting from a reflex mechanism tied to the vasodilation.

SAR

The following structural features are important for activity of the 1,4-DHPs:

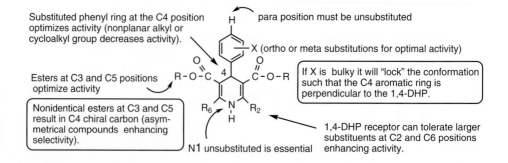

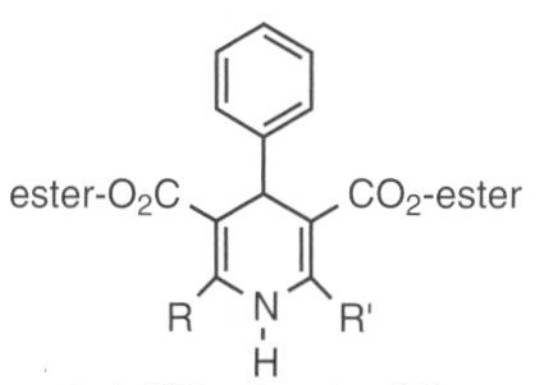

1,4-Dihydrophyridine pharmacophore

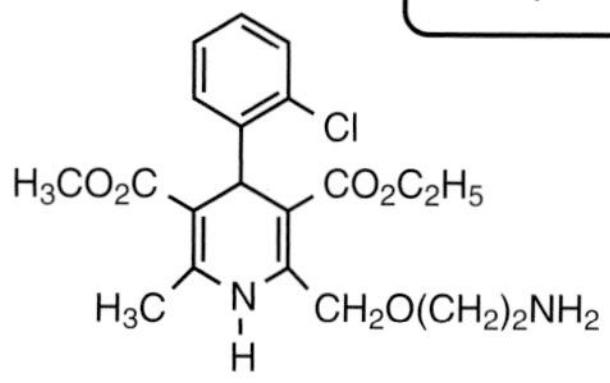

Amlodipine (Norvasc, Lotrel)

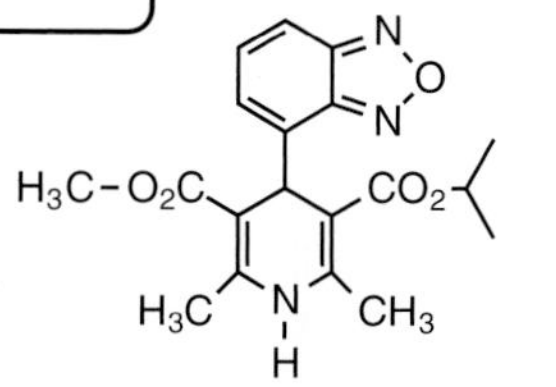

Isradipine

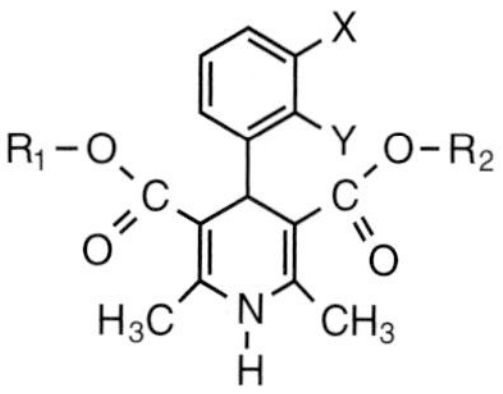

Clevidipine: $R_1 = CH_3$; $R_2 = CH_2CH_2O_2CC_3H_7$
(Cleviprex) X = Y = Cl

Felodipine: $R_1 = CH_3$; $R_2 = CH_2CH_3$
(Plendil) X = Y = Cl

Nicardipine: $R_1 = CH_3$; $R_2 = CH_2CH_2N(CH_3)(CH_2{-}C_6H_5)$
(Cardene) X = NO_2; Y = H

Nifedipine: $R_1 = R_2 = CH_3$
(Adalat, Procardia) X = NO_2; Y = H

Nimodipine: $R_1 = CH(CH_3)_2$; $R_2 = CH_2CH_2OCH_3$
(Nymalize) X = NO_2; Y = H

Nisoldipine: $R_1 = CH_3$; $R_2 = CH_2CH(CH_3)_2$
(Sular) X = H; Y = NO_2

Figure 13.16 Ca^{2+} channel blockers indicated for hypertension.

Physicochemical and Pharmacokinetic Properties

- The N1 nitrogen of the 1,4-DHPs is part of a conjugated carbamate and as a result is not basic.
- The 1,4-DHPs are all marketed as their racemic mixtures.
- The 1,4-DHPs possess good lipid solubility and excellent oral absorption, however rapid first-pass metabolism results in variable oral bioavailability depending on the extent of metabolism (Table 13.6).
- Nicardipine and clevidipine are available as parenteral preparations. Parenteral nicardipine is incompatible with IV sodium bicarbonate resulting in the precipitation of the calcium channel blocker.

Table 13.6 Pharmacokinetic Properties of Calcium Channel Blockers

Drug	Oral Bioavailability (%)	Effect of Food on Absorption	Protein Binding (%)	T_{max} (h)	Elimination Half-Life (h)	Route of Elimination (Renal/Fecal %)
Amlodipine	64–90	None	93–97	6–12	35–50	60/20–25
Clevidipine	NA	NA	>99	2–4 min	0.15	63–74/7–22
Felodipine	10–25	Increase	>99	2.5–5	11–16	70%/10
Isradipine	15–24	Reduced rate	95	7–18 (CR)	8	60–65/25–30
Nicardipine	35	Reduced	>95%	0.5–2 (IR) 1–4 (SR)	2–4	60/35
Nifedipine	45–70 86 (SR)	None	92–98%	0.5 (IR) 6 (SR)	2–5 (IR) 7 (SR)	60–80/15 (Biliary/Fecal)
Nimodipine	13	Reduced	>95%	1	8–9	Renal only
Nisoldipine	5	High-fat meal increases IR, lowers overall	>99%	6–12	7–12	70–75/6–12

CR, controlled release product; *h*, hours; *IR*, immediate release product; *SR*, sustained release product; T_{max}, time to maximum blood concentration

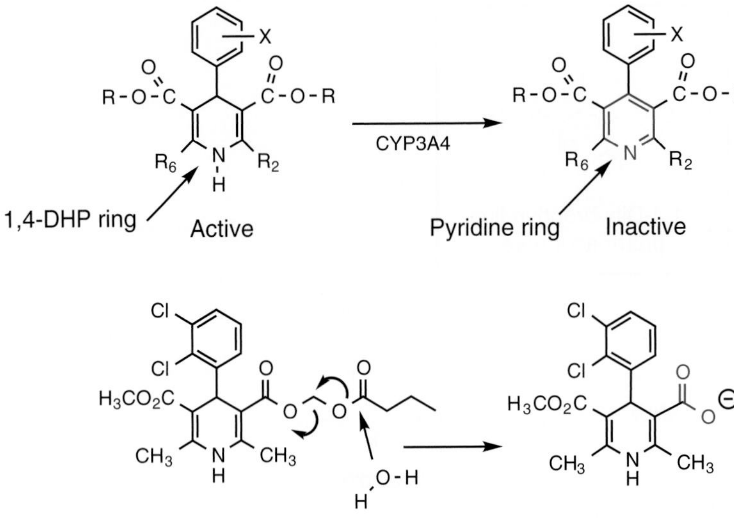

Clevidipine

Inactive metabolite

Figure 13.17 Metabolic reactions of 1,4-DHP Ca^{2+} channel blockers.

Metabolism

- With the exception of clevidipine, the 1,4-DHPs are oxidatively metabolized to inactive pyridine analogues (Fig. 13.17) followed by additional transformations (e.g., ester hydrolysis and conjugation).
- Nisoldipine also is subject to hydroxylation of the isobutyl ester producing a weakly active metabolite (~10% of the activity of the parent compound).
- Clevidipine, an ultra-short duration acting drug, undergoes hydrolytic inactivation upon IV infusion, but is used in patients with either renal or hepatic dysfunction.

Calcium Channel Blockers – Treatment of Ischemic Heart Disease or Angina

One 1,4-DHP (nifedipine), a benzothiazepine (diltiazem) and an arylalkylamine (verapamil) represent Ca^{2+} channel blockers used in the treatment of angina (Fig. 13.18).

Physicochemical and Pharmacokinetic Properties

- Nifedipine, diltiazem, and verapamil are lipophilic.

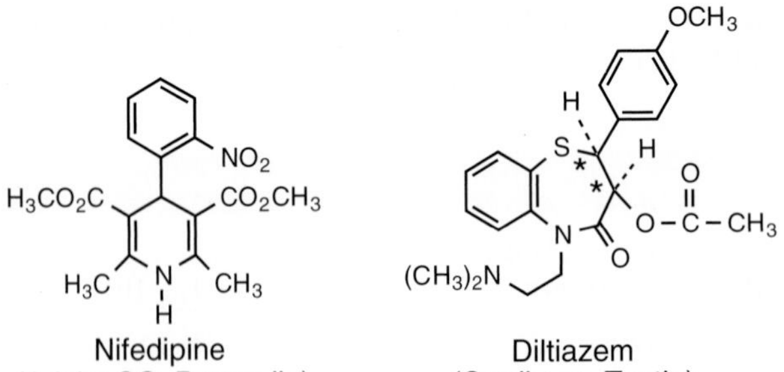

Nifedipine
(Adalat CC, Procardia)

Diltiazem
(Cardizem, Taztia)

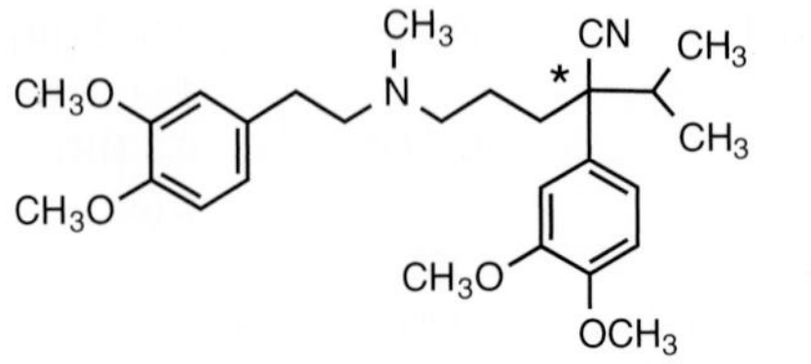

Verapamil
(Calan, Verelan)

Figure 13.18 Ca^{2+} channel blockers indicated for angina.

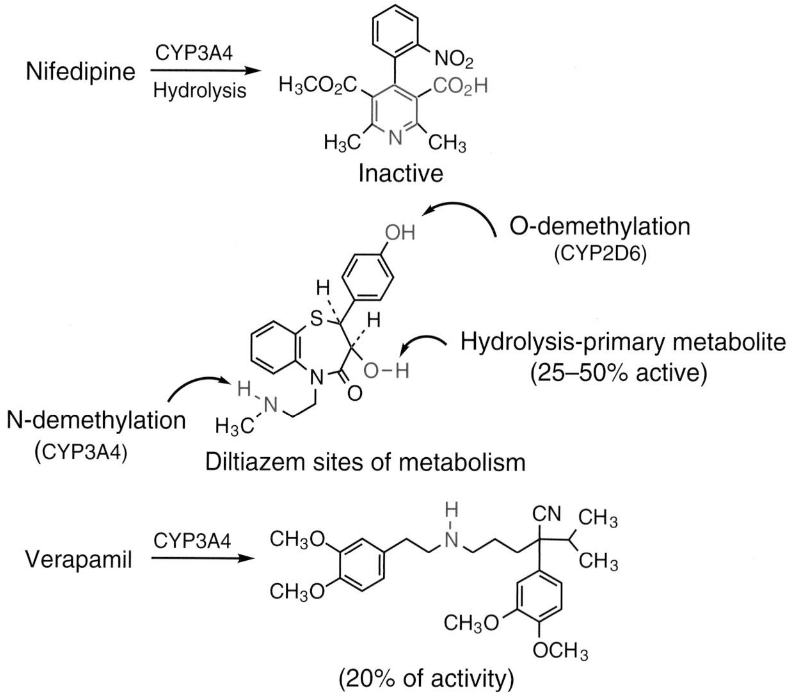

Figure 13.19 Metabolism of the Ca^{2+} channel blocker nifedipine, diltiazem, and verapamil.

- Although diltiazem and verapamil are asymmetric they are used as a racemic mixture, $S(-)$-verapamil is more potent than the $R(+)$ isomer. The *cis* stereochemistry of diltiazem is essential for activity.
- These drugs are rapidly and completely absorbed following oral administration, but have moderate bioavailability (nifedipine: 45% to 70%, diltiazem: 40% to 67%, verapamil: 20% to 35%).
- These drugs are highly bound to plasma protein (70% to 98%).

Metabolism

- Nifedipine, diltiazem, and verapamil are extensively metabolized to inactive or metabolites of reduced activity (Fig. 13.19).
- CYP3A4 and CYP2D6 are commonly associated with drug metabolism.

Clinical Applications

- The 1,4-DHP Ca^{2+} channel blockers, with the exceptions of nimodipine (treatment of subarachnoid hemorrhage), are approved for the treatment of hypertension.
- Both verapamil and diltiazem are used clinically in the management of angina, hypertension (as sustained release formulations), and cardiac arrhythmia, whereas the nifedipine is used more frequently as an antianginal and antihypertensive agents.
- Drug–drug interactions are common with the Ca^{2+} channel blockers (Table 13.7).

ADVERSE EFFECTS

Ca^{2+} Blockers: The adverse effects of Ca^{2+} blockers in general include: edema, flushing, hypotension, nasal congestion, palpitations, chest pain, tachycardia, headache, fatigue, dizziness, rash, nausea, abdominal pain, constipation, diarrhea, vomiting, shortness of breath, weakness, bradycardia, and AV block. Serious adverse effects of verapamil are uncommon although some patients may experience CHF and pulmonary edema. Verapamil has been reported to cause constipation. Patients on diltiazem usually will experience mild adverse effects.

DRUG–DRUG AND DRUG–FOOD INTERACTIONS

In addition to the drug–drug interactions, an interesting drug–food interaction occurs with the 1,4-DHPs and grapefruit juice. Coadministration of 1,4-DHPs with grapefruit juice produces an increase in systemic concentration of the 1,4-DHPs.

Table 13.7 Selective Drug–Drug Interactions with Ca^{2+} Channel Blockers

Ca^{2+} Blocker	Drug	Result
All	β-Blockers	Increased cardiodepressant effects
All	H_2 antagonists	Increased bioavailability of 1,4-DHPs
Felodipine	Carbamazepine	Decreased felodipine levels
Isradipine	Diclofenac	Decreased hypertensive effect
Felodipine, nifedipine, nisoldipine	Digoxin	Increased digoxin levels
Felodipine and potentially other 1,4-DHPs	Erythromycin	Increased felodipine levels and toxicity
Isradipine	Lovastatin	Decreased effects of lovastatin
Felodipine, nifedipine	Phenobarbital	Decreased bioavailability of 1,4-DHPs

Adrenergic Blocking Agents

Reserpine

Metyrosine
(Demser)

MOA

- Reserpine acts to deplete norepinephrine and dopamine from adrenergic storage vesicles by inhibiting the vascular catecholamine transporter (VMAT2).
- Catecholamines leaving the vesicles are destroyed by MAO and COMT.
- Adrenergic transmission is inhibited and sympathetic tone decreased and vasodilation occurs.
- Metyrosine is a competitive inhibitor of tyrosine hydroxylase inhibiting DOPA synthesis and depleting adrenergic stores.

Clinical Applications

- Neither reserpine nor metyrosine find significant use today.
- Reserpine can be used to treat hypertension, but adverse effects and the lack of availability the last several years has limited its use (there is only a single supplier).
- Metyrosine is used primarily for treatment of phenochromocytoma, but its adverse effects are also significant.

Drugs Used for the Treatment of Cardiac Arrhythmia

Arrhythmias are electrical disturbances that interfere with the ability of the heart to pump blood, and may underlie the onset of angina or heart failure. Severe arrhythmias can cause ventricular fibrillation and sudden death. Arrhythmia occurs when the heart's natural pacemaker develops an abnormal rate or rhythm, when the normal conduction path is interrupted, or when another part of the heart takes over as pacemaker. The rhythm of the heart normally is determined by a pacemaker site called the SA node, with spontaneous generation of action potentials at a rate of 100 to 110 action potentials ("beats") per minute. This intrinsic rhythm is strongly influenced by the vagus nerve, overcoming the sympathetic system at rest. This "vagal tone" brings the resting heart rate down to a normal sinus rhythm of 60 to 100 beats per minute (sinus rates <60 are termed "sinus bradycardia" and sinus rates >100 are termed "sinus tachycardia"). An in-depth discussion of normal physiology of cardiac contractions, cause of arrhythmias, and classification of antiarrhythmic drugs can be found in Chapter 21 in *Foye's Principles of Medicinal Chemistry, Seventh Edition*. It should be noted that the commonly used antiarrhythmic drug classifications may be an over simplification since these drugs often have overlapping actions. The drug classifications are shown in Table 13.8.

Table 13.8	**Classification of Antiarrhythmic Drugs**	
Class	**Drugs**	**Site of Action**
IA	Disopyramide, Procainamide, Quinidine	Atrial and ventricular tissue
IB	Lidocaine, Mexiletine (Phenytoin)	Ventricular tissue
IC	Flecainide, Propafenone	Ventricular tissue
II	Propranolol, Sotalol, Esmolol	SA/AV node
III	Amiodarone, Dronedarone, Dofetilide, Ibutilide, Sotalol	Atrial and ventricular tissue
IV	Diltiazem, Verapamil	SA/AV node

Class IA antiarrhythmic agents:

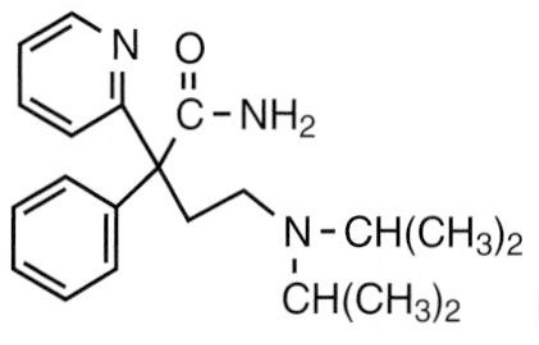

Disopyramide
(Norpace)

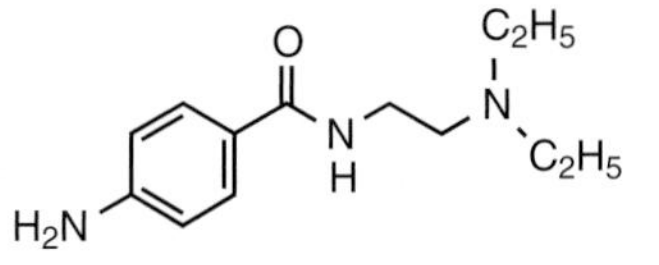

Procainamide

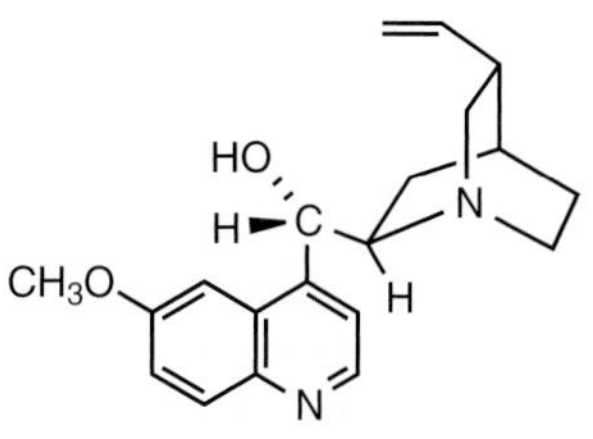

Quinidine

Class IB antiarrhythmic agents:

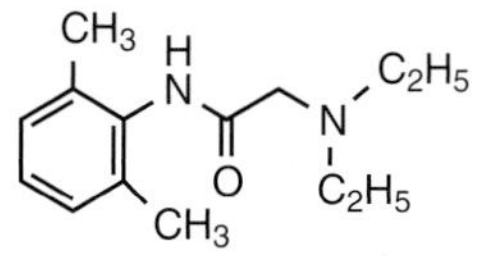

Lidocaine
(Xylocaine)

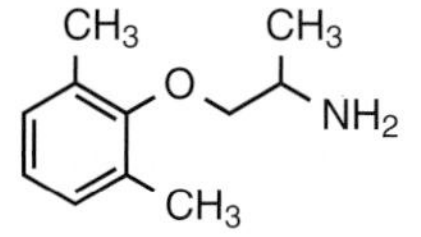

Mexiletine

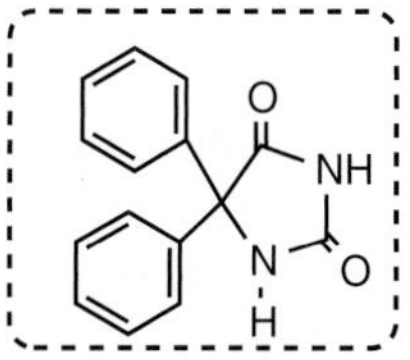

Phenytoin
(not approved by
the FDA)

Class IC antiarrhythmic agents:

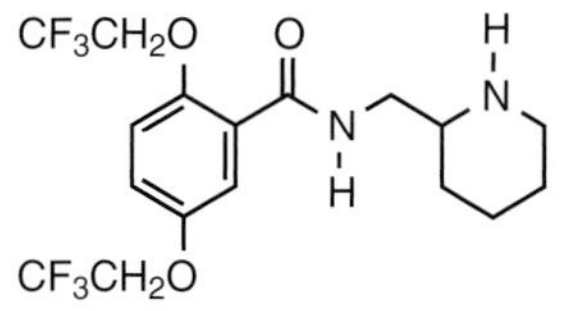

Flecainide

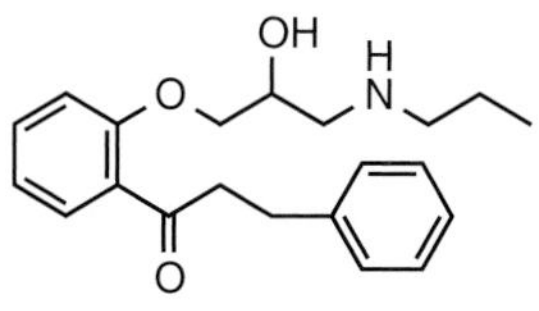

Propafenone
(Rythmol)

Figure 13.20 Class IA, IB, and IC antiarrhythmic agents.

Class IA Antiarrhythmic Drugs

MOA

- Class IA antiarrhythmics act on nerve and myocardial membranes to slow conduction by blocking fast Na^+ channels (moderate action), inhibiting phase 0 of the action potential (Figs. 13.20 and 13.21).
- Myocardial membranes show the greatest sensitivity leading to a decrease in the maximal rate of depolarization without changing the resting potential.
- Class IA drugs increase the threshold of excitability, increase the effective refractory period, decrease conduction velocity, and decrease spontaneous diastolic depolarization in pacemaker cells.

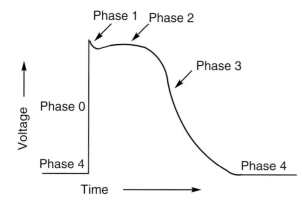

Figure 13.21 Action potential for a single cardiac cell over time and represents the changes in voltage. Phase 0: Depolarization with Na^+ channels stimulated to open with Na^+ entering the cell. Phase 1: Initial stage of repolarization. Phase 2: Plateau stage with repolarization and a slow influx of Ca^{2+} with extension of refractory period. Phase 3: Late stage of repolarization. When repolarization is complete a new stimulus can occur. Phase 4: Repolarization is complete (quiescent phase).

Physicochemical and Pharmacokinetic Properties

- *R*-quinidine is the cardioactive isomer.
- The IA drugs are used as their water-soluble salts: disopyramide phosphate; procainamide hydrochloride; quinidine sulfate or gluconate.
- All three of the class IA drugs are readily absorbed (~80% to 95%) following oral administration.
- A significant amount of the class IA drugs are excreted unchanged.
- Quinidine is highly bound to plasma protein (85%) as is disopyramide (82%) while its N-dealkyl metabolite is moderately bound to plasma protein (22% to 35%).
- Quinidine is a substrate for P-gp and inhibits digoxin secretion by blocking digoxin's efflux.

Metabolism

- Disopyramide and quinidine are substrates for CYP3A4: Disopyramide undergoes N-dealkylation (active); quinidine undergoes O-demethylation and vinyl hydroxylation (hydroxyethyl), both metabolites exhibit weak activity.
- Procainamide undergoes N4 acetylation (active) and amide hydrolysis (inactive).

⚠ **ADVERSE EFFECTS**

IA Drugs: Procainamide-induced lupus syndrome is common in prolonged therapy especially in "slow acetylators" (reversed when drug is removed). Because procainamide and disopyramide have proarrhythmic effects they are not recommended for lesser arrhythmias or asymptomatic patients. The common adverse effects of IA drugs are tachycardia, dry mouth, urinary retention, blurred vision, and constipation, as well as diarrhea, nausea, headache, and dizziness (anticholinergic effects). Quinidine enhances digitalis toxicity, especially if hypokalemia is present. Quinidine can precipitate Torsades de pointes (especially in patients with long QT syndrome), a ventricular tachyarrhythmia caused by after depolarizations.

Class IB Antiarrhythmic Drugs

MOA

- Class IB antiarrhythmics block open cardiac Na^+ channel proteins (weak blockers on Phase 0 – Fig. 13.21) (Fig. 13.20).
- Possibly causing conformational changes in the Na^+ channel protein leading to a nonconducting conformation.
- Shorten repolarization decreasing the duration of the action potential.
- Exhibits selectivity for the ventricular tissue.
- It should be noted that mexiletine is an analog of lidocaine and while phenytoin has antiarrhythmic activity it is not approved for this use in the United States.

Physicochemical and Pharmacokinetic Properties

- Lidocaine is effective as an antiarrhythmic only when given via the intravenous route.
- Lidocaine is ineffective orally because of rapid first-pass metabolism, but mexiletine is orally active with a slower rate of metabolism.
- Elimination half-life of lidocaine is 1.5 to 2 hours while that of mexiletine is 10 to 12 hours.
- Both drugs are bound to plasma protein (50% to 70%).

Metabolism

- Both drugs are metabolized in the liver: lidocaine is a substrate for CYP3A4 while mexiletine is a substrate for CYP2D6 (Fig. 13.22).
- Lidocaine is rapidly metabolized as indicated by the $t_{1/2}$ of 1.5 to 2 hours.
- Mexiletine was developed with the goal of reducing the rapid hydrolysis seen with lidocaine.

⚠ **ADVERSE EFFECTS**

Lidocaine: The adverse effects of lidocaine include emetic and convulsant properties that predominantly involve the central nervous system and heart. The central nervous system effects can begin with dizziness and paresthesia and, in severe cases, ultimately lead to epileptic seizures.

Mexiletine: The adverse effects of mexiletine are quite mild and consist of gastric disturbance, lightheadedness, tremor, and coordination difficulties. Some of these effects can be reduced by taking the drug with food or an antacid.

Class IC Antiarrhythmic Drugs

MOA

- Class IC antiarrhythmic drugs blocks Na^+ channel proteins (Phase 0) of the heart slowing conduction of electrical impulse (stabilizing action on myocardial cells) (Fig. 13.20).
- These drugs exhibit selectivity for cells with a high conductance rate.
- The effective refractory period is prolonged.

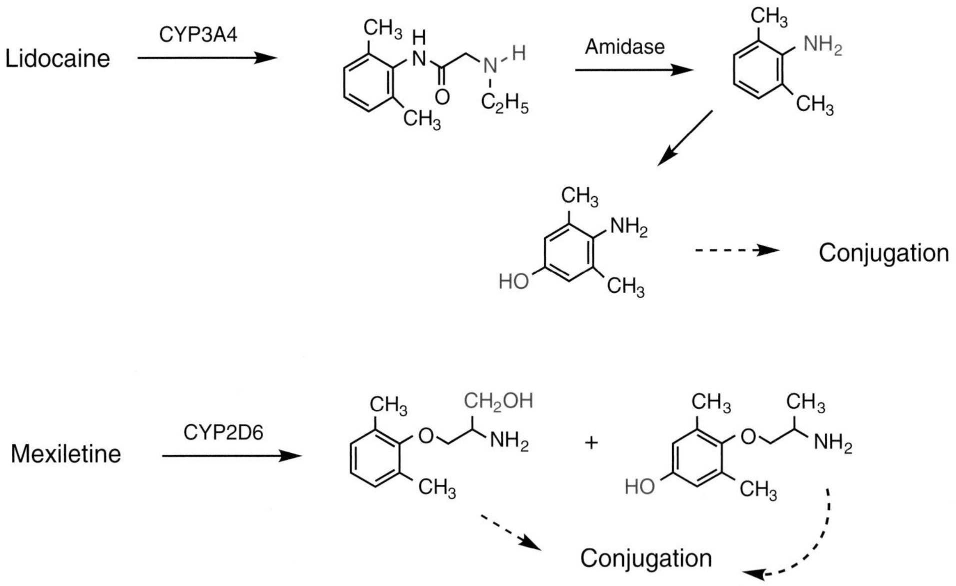

Figure 13.22 Metabolism of lidocaine and mexiletine.

- Also have β-adrenergic blocking activity.
- Their greatest effect is on the ventricular myocardium.

Physicochemical and Pharmacokinetic Properties

- Both class IC drugs are well absorbed following oral administration, but because of propafenone's first-pass metabolism the drug exhibits a low bioavailability.
- Protein binding: flecainide ~40%; propafenone ~97%.
- Flecainide has a long plasma half-life (14 to 20 hours) while propafenone has a short half-life (2 to 10 hours).
- Since both drugs are substrates for CYP2D6 blood levels may differ in fast or slow metabolizers.
- Propafenone is administered as the *S,R* mixture with both enantiomers displaying similar antiarrhythmic activity.

Metabolism

- Extensive metabolism of both drugs results in decreased activity (Fig. 13.23).

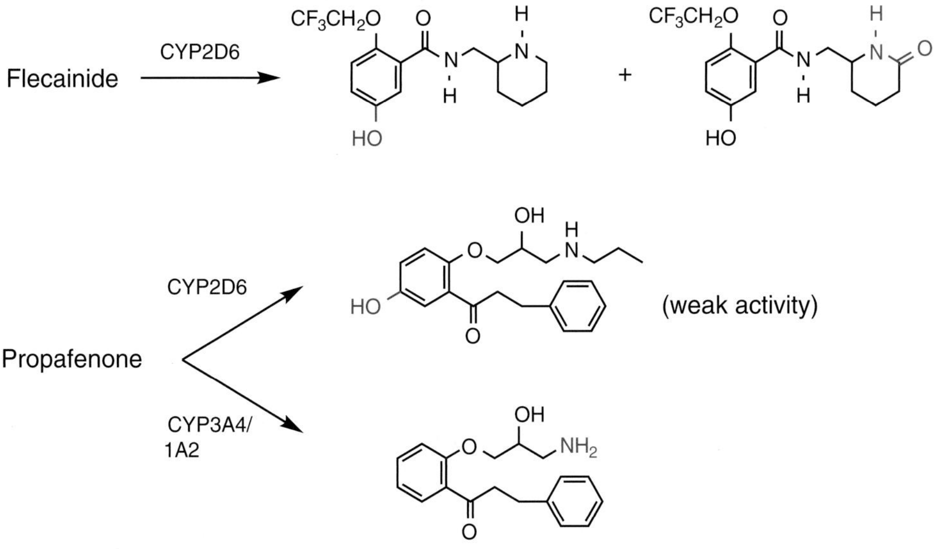

> **ADVERSE EFFECTS**
>
> **Class IC Antiarrhythmic Drugs:** may increase the chances of fatal and nonfatal cardiac arrest especially in postmyocardial infarct patients. Other than this the drugs have minor adverse effects which include: dizziness, visual disturbances, dyspnea, and headache (flecainide) and changes in taste perception, nausea and vomiting, dizziness, constipation, and headache (profaenone).

Figure 13.23 Metabolism of flecainide and propafenone.

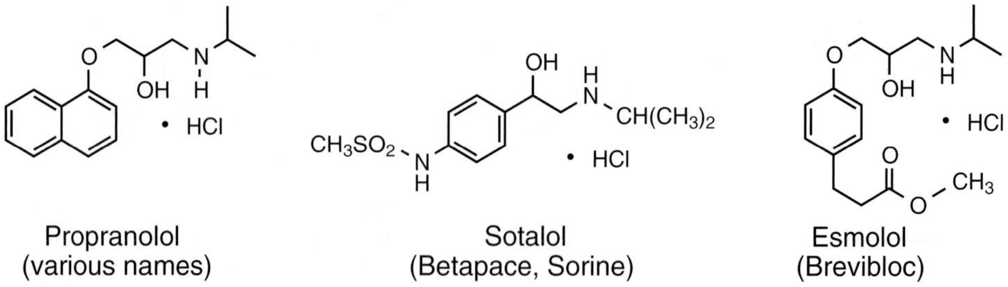

Figure 13.24 Class II antiarrhythmic drugs.

Class II Antiarrhythmic Drugs

MOA

- The class II antiarrhythmic drugs are β-adrenergic blockers (Fig. 13.24). Both propranolol and sotalol are nonspecific while esmolol is a selective β₁-adrenergic blocker.
- By blocking sympathetic stimulation of cardiac tissue cyclic adenosine monophosphate (cAMP) levels are reduced which in turn results in decreased calcium influx and decreased force and rate of cardiac contraction.
- Sotalol is considered a class II agent because of its β-adrenergic effects but is also considered a class III agent since it additionally blocks K⁺ channels (see Class III Antiarrhythmic Drugs).

Physicochemical and Pharmacokinetic Properties

- The class II antiarrhythmic drugs are water soluble as their hydrochloride salts and are administered as enantiomeric mixtures.
- Based upon their therapeutic indication esmolol and propranolol are administered via intravenous infusion while sotalol is administered orally as a tablet or syrup.
- Esmolol is rapidly hydrolyzed to an inactive carboxylate with a half-life of ~9 minutes.
- Sotalol is nearly completely absorbed (oral bioavailability 90% to 100%) following oral administration, is not protein bound and is excreted unchanged.
- Propranolol experiences rapid first-pass metabolism (e.g., 4-hydroxylation, N-dealkylation, and direct O-glucuronidation), low bioavailability (30% to 40%), and is highly bound to plasma protein.

> **⚠ ADVERSE EFFECTS**
>
> **Class II Antiarrhythmic Drugs:** In general the adverse effects with the class II antiarrhythmic drugs are mild with the exception of sotalol which causes a low incidence (4%) of Torsades de Pointes (polymorphic ventricular tachycardia). This drug has also been reported to be proarrhythmic leading to ventricular tachycardia or ventricular fibrillation which is potentially fatal.

Class III Antiarrhythmic Drugs

MOA

- The class III antiarrhythmic drugs block K⁺ channels (blocking outward K⁺ current) leading to a prolonged action potential (Fig. 13.25).
- Most of the class III drugs act through phase 3 of the action potential (Fig. 13.21).
- Several of the class III drugs have overlapping mechanisms affecting other part of the action potential: Amiodarone – Na⁺ channel blocker, inhibits β-receptors, weak Ca²⁺ channel blocker; dronedarone – similar to amiodarone plus inhibits α-adrenergic receptors; ibutilide – produces influx of Na⁺ in slow Na⁺ channels (this may be its major action); sotalol – nonselective β-blocker (see β-blockers above).

Physicochemical and Pharmacokinetic Properties

- Some pharmacokinetic properties are shown in Table 13.9.
- Amiodarone is lipophilic and insoluble in water.
- This drug has structural similarity to the thyroid hormones accounting for some of its adverse effects.
- The variable oral bioavailability of amiodarone and dronedarone can be improved by taking with food.

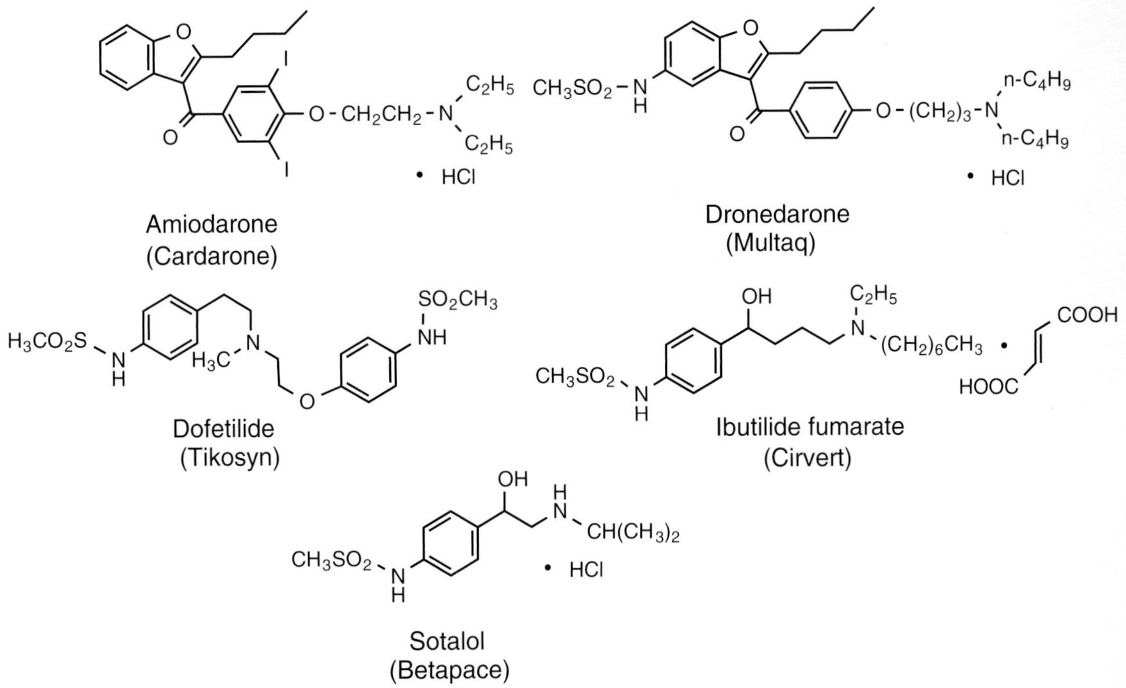

Figure 13.25 Class III antiarrhythmic drugs.

Metabolism

- Amiodarone, dronedarone, and dofetilide are all substrates for CYP3A4 with N-dealkylation being a major metabolite (Fig. 13.26).
- Inhibitors of CYP3A4 can significantly increase blood levels of dronedarone.
- Ibutilide undergoes extensive metabolic oxidation, but it does not involve CYP3A4 or 2D6.
- Sotalol is excreted without significant metabolism.

> ⚠️ **ADVERSE EFFECTS**
>
> **Amiodarone:** has the potential for severe adverse effects including pulmonary toxicity, hepatic dysfunction, neuromuscular symptoms (e.g., peripheral neuropathy or proximal muscle weakness), photosensitivity, hypo- or hyperthyroidism (associated with structural similarity to thyroid hormones), and QT prolongation. 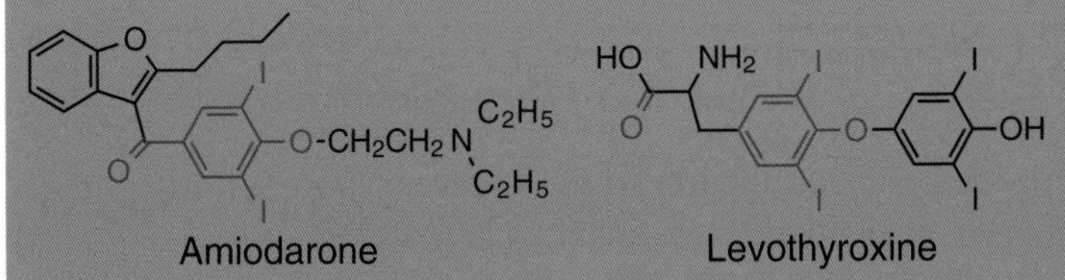
>
>
> Amiodarone   Levothyroxine
>
> Additionally, adverse effects include nausea, vomiting, constipation, weight loss, fatigue, heart block, and heart failure. Dronedarone also has been associated with QT prolongation, heart failure, liver injury, and pulmonary toxicity. As with sotalol reported above, Torsades de Pointes may be seen with ibutilide and dofetilide.

Table 13.9 Pharmacokinetics Properties of Class III Antiarrhythmic Drugs

Drug	Amiodarone	Dofetilide	Dronedarone	Ibutilide	Sotalol
Oral bioavailability (%)	22–86	96–100	4–15	—	90–100
Route of administration	Oral	Oral	Oral	Injection	Oral
Protein binding (%)	~96	60–70	>98	~40	0
Elimination half-life (h)	58 d (ave)	—	30	2–12	10–20
Excretion (%)...feces	>90		86	~19	
...urine		80	4	~82	80–90

h, hours

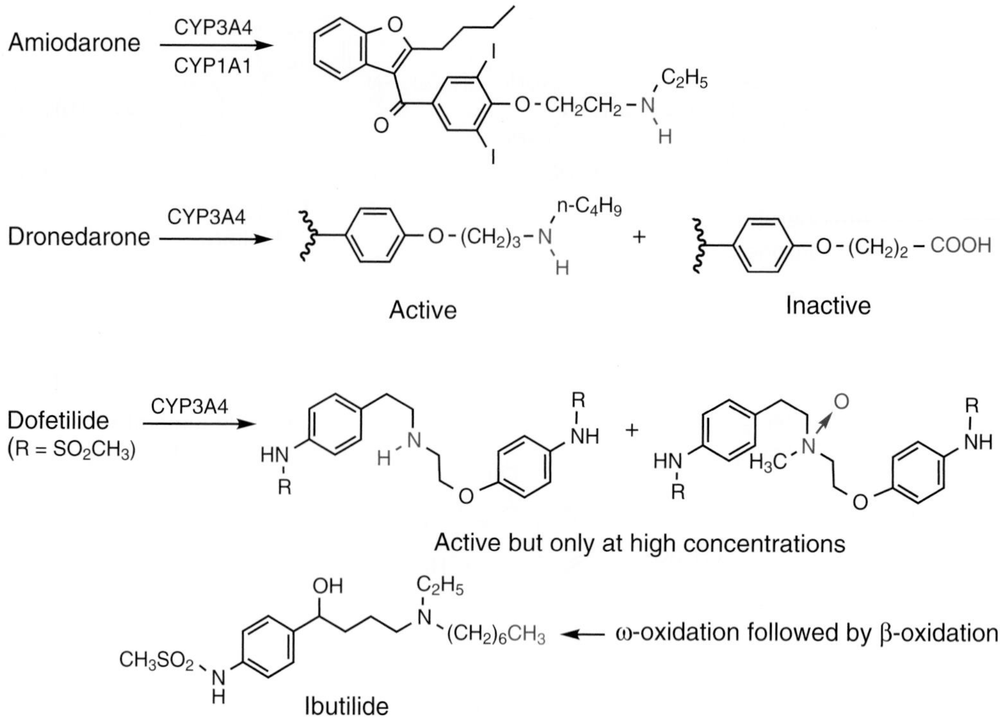

Figure 13.26 Metabolic reactions common to class III antiarrhythmic drugs.

Class IV Antiarrhythmic Drugs

Class IV antiarrhythmic drugs consist of the calcium channel blockers diltiazem and verapamil which have already been discussed (Fig. 13.18 and Table 13.8). These drugs selectively block the slow inward current carried by calcium. The slow inward current in cardiac cells has been shown to be of importance for the normal action potential in SA node cells. It has also been suggested that this inward current is involved in the genesis of certain types of cardiac arrhythmias. Administration of a class IV drug causes a prolongation of the refractory period in the AV node and the atria, a decrease in AV conduction, and a decrease in spontaneous diastolic depolarization. These effects block conduction of premature impulses at the AV node and, thus, are very effective in treating supraventricular arrhythmias.

Clinical Application of Antiarrhythmic Drugs
The clinical intent is:

- to restore normal rhythm and conduction so as to prevent more serious and lethal arrhythmias from occurring.
- decrease or increase conduction velocity.
- alter the excitability of cardiac cells by changing the duration of the effective refractory period and to suppress abnormal automaticity.
- For specific conditions see Table 13.10.

Table 13.10 Clinical Application of Antiarrhythmic Drugs

Specific Condition	Antiarrhythmic Drugs
Sinus tachycardia	Class II, IV
Atrial fibrillation/flutter	Class IA, IC, II, III, IV
Paroxysmal supraventricular tachycardia	Class IA, IC, II, III, IV
Ventricular tachycardia	Class I, II, III
Premature ventricular complexes	Class II, IV
Digitalis toxicity	Class IB

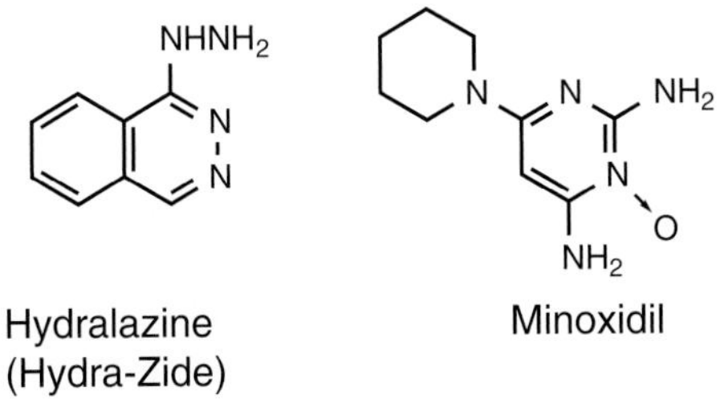

Hydralazine
(Hydra-Zide)

Minoxidil

Figure 13.27 Antihypertensive vasodilators.

Vasodilators

Hydralazine, minoxidil, and diazoxide are all classified as vasodilators and all share a common site of action, the VSM, and overlapping mechanisms of action (Fig. 13.27).

MOA

- Hydralazine and minoxidil vasodilatory action appears to be associated with opening of ATP-sensitive K^+ channels, increased efflux K^+ from cells with hyperpolarization of VSM and closure of the voltage-gated Ca^{2+} channels.
- Reduced intracellular Ca^{2+} leads to vasodilation.
- Hydralazine also inhibits the second messenger, IP_3 – induced release of Ca^{2+} from the smooth muscle storage sites and reduced contraction of smooth muscle cells.

Physicochemical and Pharmacokinetic Properties

- Hydralazine and minoxidil are readily absorbed following oral administration.
- Hydralazine exhibits low bioavailability due to metabolism in the GI lining.
- Minoxidil is a prodrug requiring oxidative metabolism to the active N-O-sulfate (see Fig. 13.28).
- Plasma binding for hydralazine is 85% while minoxidil is not bound to plasma protein.

Metabolism

- The metabolic products formed from the vasodilators are shown in Figure 13.28. With the exception of minoxidil N-O sulfate all metabolites are inactive.

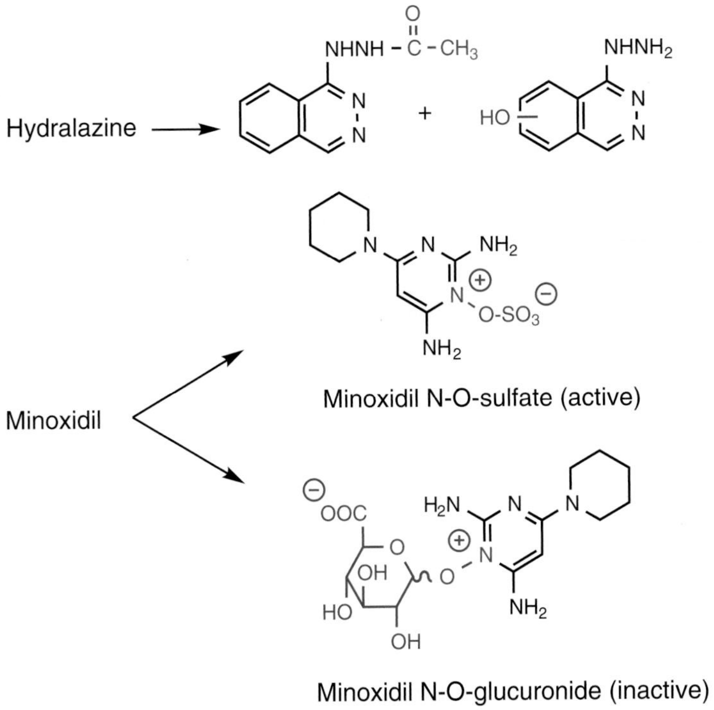

Minoxidil N-O-sulfate (active)

Minoxidil N-O-glucuronide (inactive)

Figure 13.28 Metabolic reactions of vasodilators.

ADVERSE EFFECTS

Hydralazine and Minoxidil: Common adverse effects of hydralazine and minoxidil include headaches, flushing, and tachycardia. Reflex cardiac stimulation can precipitate angina in patients with coronary artery disease. Drug-induced lupus syndrome has been seen with hydralazine while minoxidil may cause hypertrichosis, a thickening and enhanced pigmentation of body hair.

ADVERSE EFFECTS

Sodium Nitroprusside: Profound hypotension and the accumulation of cyanide and thiocyanate may be seen with sodium nitroprusside. Cyanide can also bind to hemoglobin producing methemoglobinemia while thiocyanate can accumulate in the blood of patients with impaired renal function. Thiocyanate is mildly neurotoxic at low serum concentrations and life-threatening at high concentrations.

Additional Source of NO

Sodium nitroprusside is an inorganic source of nitric oxide radical (Fig. 13.32) which is used intravenously for its vasodilatory effects. The drug is used for treatment of severe congestive heart failure and acute hypertensive crises.

- It should be noted that the acetylation of hydralazine is genetically influenced with slow metabolizers exhibiting higher plasma levels than fast acetylators.

Clinical Applications

- Neither hydralazine nor minoxidil are first-line agents used in the treatment of hypertension.
- Both are reserved for the treatment of severe hypertension or where other drugs have proven ineffective. This is due in part to their significant adverse effects.
- Minoxidil is used topically to stimulate regrowth of hair in men (male-pattern alopecia, hereditary alopecia, or common male baldness) and in women to treat hair loss.

Nitrodilator

Sodium nitroprusside
(Nitropress)

A single agent, sodium nitroprusside, with limited therapeutic utility is a source of nitric oxide (NO) and belongs to the class of drugs referred to as NO-releasing drugs. The mechanism of action of nitroprusside is discussed under organic nitrates (Drugs used for the treatment of ischemic heart disease, p. 229). The release of NO from sodium nitroprusside through the action of glutathione in erythrocytes and tissue forming a S-nitrosothiol. This occurs with the spontaneous release of NO.

Clinical Applications

- Intravenous infusion of sodium nitroprusside is used to treat hypertensive crises in emergency situations.
- Sodium nitroprusside can also be used to manage acute CHF.

Drugs Used to Treat Heart Failure

Heart failure or CHF can be described as the inability of the heart to pump blood effectively at a rate that meets the needs of metabolizing tissues. This is the direct result of reduced contractility of the cardiac muscles, especially those of the ventricles, which causes a decrease in cardiac output, increasing the blood volume of the heart (hence the term "congested"). A group of drugs known as the cardiac glycosides were found to reverse most of these symptoms and complications. Today only a single cardiac glycoside, digoxin, is available for use in the United States.

Cardiac Glycosides

Digoxin

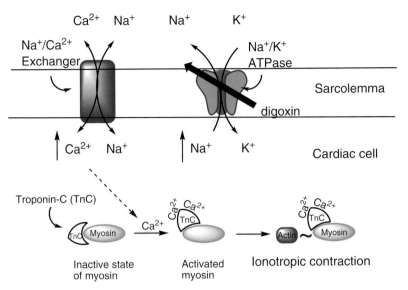

Figure 13.29 MOA of digoxin involves inhibition of Na^+/K^+ ATPase leading to increased intracellular Na^+ and Ca^{2+}, Ca^{2+} catalyzed change in shape of troponin C, exposure of actin binding sites, and myocardial muscle contraction.

MOA

- The cardiac glycoside, digoxin, is an inhibitor of Na^+/K^+ ATPase located in the sarcolemma layer of the cardiac muscle (Fig. 13.29).
- By blocking Na^+/K^+ ATPase Na^+ ion concentration increases within the cell which in turn leads to an increase in intracellular Ca^{2+} through the Na^+/Ca^{2+} exchanger protein which becomes a less effective at exporting of Ca^{2+}.
- Calcium binding to troponin C activates myosin by exposure of binding sites for actin leading to ionotropic muscle contraction.

SAR

Essential to the activity of digoxin is:

- Rings A-B and C-D must be *cis* fused for binding to Na^+/K^+ ATPase. Digoxin is a U-shaped molecule.
- C14 unsubstituted hydroxyl is necessary for cardiotonic activity.
- The α,β-unsaturated 17-lactone is required for binding.

Physicochemical and Pharmacokinetic Properties

- C12 hydroxy on the aglycone and the C3 triglucose glycoside decrease lipophilicity as compared with from other cardiac glycosides resulting in a reduced half-life (1 to 2 days).
- Digoxin is well absorbed following oral administration (70% to 85%).
- Moderate protein binding (25% to 30%).
- Substrate for P-glycoprotein (P-gp) efflux in intestinal lining and kidney with potential for drug–drug interaction at P-gp.
- Excreted primarily unchanged (50% to 70%) via the kidney with less than 13% metabolism.

Phosphodiesterase Inhibitors

Positive inotropic action on the heart can also be initiated by increasing the levels of cAMP. In theory this can be done through the actions of a β-adrenergic agonist stimulating cAMP synthesis or by inhibiting the hydrolytic breakdown of cAMP with phosphodiesterase 3 (PDE3) inhibitors including those shown in Figure 13.30.

MOA

- Cardiac contraction involves a G-protein signal transduction pathway (involves cAMP and phosphokinase A) in which increased intracellular cAMP increases intracellular Ca^{2+} stimulating cardiac muscle contraction.

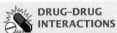

DRUG–DRUG INTERACTIONS

Competitively binding of digoxin and quinidine to P-gp reduces digoxin secretion by as much as 60%. Verapamil inhibits P-gp efflux of digoxin increasing blood levels while rifampin induces intestinal P-gp expression decreasing digoxin blood levels. Drugs that interfere with the absorption of oral digoxin include: laxatives, cholestyramine, antacids, especially magnesium trisilicate, and antidiarrheal adsorbent suspensions. Potassium-depleting diuretics, such as thiazides, can increase the possibility of digitalis toxicity because of additive hypokalemia.

ADVERSE EFFECTS

Digoxin: Extreme toxicity including ventricular tachycardia, AV block, and ventricular fibrillation has been reported with digoxin. Additionally salivation, anorexia, nausea, vomiting, headache, drowsiness, weakness, faintness, visual changes (halo around lights), confusion have been reported. The severe toxicity is associated with inhibition of the Na^+/K^+-ATPase pump and increased intracellular levels of Ca^{2+}. Hypokalemia which can be induced by coadministration of thiazide diuretics or glucocorticoids may initiate a toxic response. In treating any cardiac glycoside-induced toxicity, it is important to discontinue administration of the drug in addition to administering a potassium salt (e.g., potassium chloride).

PDE 3 Inhibitors:

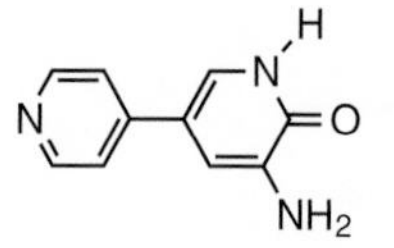

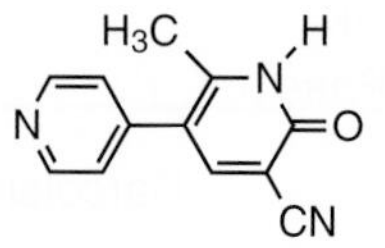

Inamrinone lactate   Milrinone lactate

β-Adrenergic Agonist:

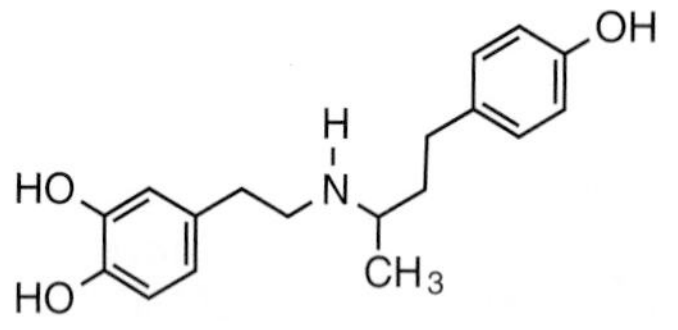

Dobutamine hydrochloride

Figure 13.30 Positive inotropic agents.

- Intracellular Ca^{2+} binds to troponin C increasing cardiac contraction through the actin~myosin complex (Fig. 13.29).
- PDE3 inhibitors also promote vasodilation via relaxation of VSM.

Physicochemical and Pharmacokinetic Properties

- Both milrinone and inamrinone are administered IV as their water-soluble lactate salt.
- Milrinone is chemically incompatible with furosemide leading to precipitation.
- Milrinone exhibits greater selectivity for PDE3 than inamrinone.
- Milrinone has a short plasma half-life (30 to 60 minutes), elimination half-life of 2.3 hours, is 70% bound to plasma protein, and is eliminated primarily unchanged (83%) in the urine.

> ⚠ **ADVERSE EFFECTS**
>
> **Inamrinone:** reported to cause gastrointestinal disturbances, thrombocytopenia, and impairment of the liver function with a severe adverse effect of ventricular arrhythmias, which may be life-threatening. Adverse effects of milrinone tend to be less severe than inamrinone although arrhythmias may still be possible. Additional adverse effects include headaches and hypotension. Long-term use of either drug can result in an increased mortality in CHF patients.

β-Adrenergic Receptor Agonists

MOA of a β-Adrenergic Agonist

- Dobutamine (Fig. 13.30) is a β-adrenergic agonist, which stimulates the myocardium primarily through β_1-adrenergic receptors.
- The β_1-adrenergic receptors act through a G-protein signal transduction leading to an increased intracellular cAMP.
- As with PDE3, increased cAMP results in an increase in intracellular Ca^{2+} and myocardial contraction (Fig. 13.29).

Physicochemical and Pharmacokinetic Properties

- The water-soluble dobutamine hydrochloride is formulated for intravenous administration (IV).
- The catecholamine structure of dobutamine increases the potential for air oxidation (sodium bisulfite is added to the IV preparation for stabilization).
- The drug is a substrate for COMT and O-sulfate conjugation with elimination in the urine.

Hyperpolarization-Activated Cyclic Nucleotide-Gated Channel (HCN) Inhibitor

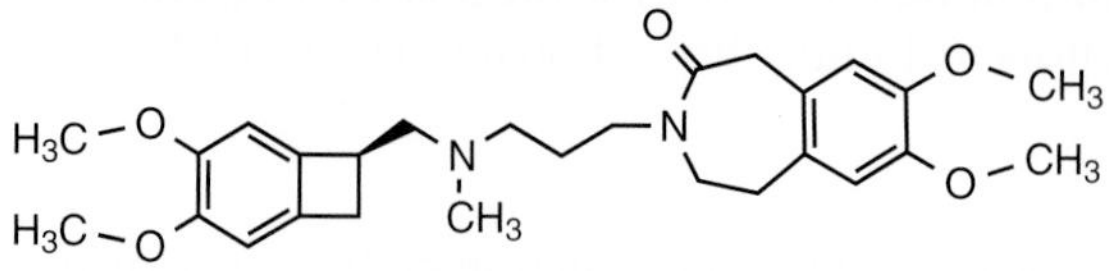

Ivabradine
(Corlanor)

MOA of Ivabradine

- Ivabradine selectively blocks the "funny" (I_f) current/channel from the sinoatrial node (terminology was associated with unusual properties of this current compared to other known current systems).
- The I_f current is carried by Na^+/K^+ ions across the sarcolemma.
- I_f current is activated by intracellular cAMP carried by hyperpolarization-activated cyclic ion channels.
- Blocking open I_f current/channels results in unbound cAMP/I_f, leading to a closed channel, lower heart rate, and decreased myocardial oxygen demand.

Physicochemical and Pharmacokinetic Properties

- Nearly complete oral absorption with a bioavailability of ~40%.
- Food may slow absorption and increase plasma exposure.
- Elimination $t_{1/2}$ is ~2 hours.
- Moderate plasma protein binding (50% to 70%).
- The active N-desmethyl metabolite is bound at 60% to 80%.
- Excreted primarily via the kidney.

Metabolism

- Major metabolite is the CYP3A4 oxidative N-desmethyl metabolite.
- Additional metabolites consist of O-desmethyl metabolite and aromatic oxidation.

Clinical Applications

- The primary clinical use for digoxin is in the treatment of CHF.
- Digoxin is also used in cases of atrial flutter or fibrillation and paroxysmal atrial tachycardia.
- Both milrinone and inamrinone are used only for short-term treatment of CHF.
- The PDE3 inhibitors increase inotropy while the vasodilatory actions reduce ventricular wall stress and oxygen demands.
- Milrinone is preferred over inamrinone due to its shorter half-life and reduced toxicity.
- Dobutamine is used for short-term treatment of advanced CHF in patients with systolic dysfunction in which immediate improvement in myocardial contraction and increased cardiac output is intended.
- Ivabradine is indicated to reduce hospitalization from worsening heart failure in patients with stable, symptomatic chronic heart failure with left ventricular ejection fraction ≤35% with resting heart rate of ≥70 beats per minute. Due to the drugs selectivity it does not have the common adverse effects associated with β–blockers.

> **ADVERSE EFFECTS**
>
> **Ivabradine:** Adverse effects were minimal with ivabradine and consisted of bradycardia, hypertension, atrial fibrillation, and temporary vision disturbance (flashes of light). Ivabradine is contraindicated with strong CYP3A4 inhibitors and CYP3A4 agonists, may also effect the plasma concentration of the drug.

Drugs Used for the Treatment of Ischemic Heart Disease (Coronary Artery Disease and Angina)

Ischemic heart disease (IHD) also referred to as coronary artery disease (CAD) is a common type of heart disease associated with a buildup of plaque on the inner walls of the arteries often leading to heart attacks. Symptoms associated with CAD include angina pectoris and decreased tolerance to exercise. With a decreased lumen size of the coronary artery there is a reduced supply of blood and oxygen to the heart, and the heart is said to be "ischemic" (oxygen deficient). Angina is characterized by a sudden, severe pain originating in the chest, often radiating to the left shoulder and down the left arm provoked by stress, food, exercise, and emotional factors (stable angina). Unstable angina is more difficult to treat and the coronary spasm occurs when the person is at rest.

Organic Nitrates

The organic nitrates are esters of an organic alcohol and nitric acid (Fig. 13.31). As such these compounds are not particularly stable as indicated below.

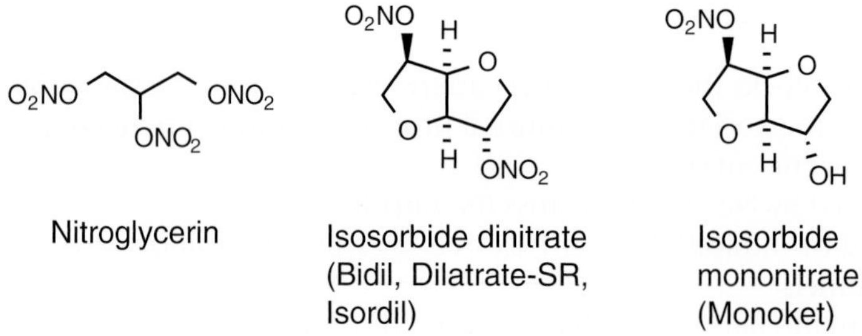

Figure 13.31 Organic nitrates.

Nitroglycerin

Isosorbide dinitrate
(Bidil, Dilatrate-SR,
Isordil)

Isosorbide
mononitrate
(Monoket)

MOA

- Nitric oxide (NO), naturally produced by vascular endothelial cells, causes relaxation of VSM.
- Organic nitrates mimic the actions of endogenous NO by releasing NO (Fig. 13.32).
- NO is released via a thionitrite-complex (e.g., sulfhydryl group of cysteine within the aldehyde reductase) followed by a nonenzymatic decomposition to NO.
- NO leads to increases in cyclic guanosine monophosphate (cGMP) by stimulating guanylyl cyclase elevating levels of cGMP.
- cGMP blocks Ca^{2+} influx into the cell blocking vascular contractions (Fig. 13.29).
- Organic nitrates may also increase intraplatelet cGMP concentrations, thereby inhibiting platelet aggregation.

Physicochemical and Pharmacokinetic Properties

- The organic nitrates are unstable to moisture leading to ester hydrolysis.
- The nitrates are lipophilic and low boiling and as such can be used by inhalation.
- The nitrates are administered by infusion, sublingual, sustained release tablets, capsules, oral spray, transdermal patch, or as ointments.
- Sublingual administration leads to rapid absorption (1 to 5 minutes) with a rapid onset and short duration (~30 minutes) for nitroglycerin, while isosorbide dinitrate has a longer duration of action (4 to 6 hours).
- Metabolism is catalyzed by mitochondrial aldehyde dehydrogenase by removal of the nitrate (isosorbide dinitrate to isosorbide mononitrates [2- or 5-nitrate]) with reduced activity and nitroglycerin to 1,2- or 1,3-dinitrate (greatly reduced activity <10%).

Clinical Applications

- The organic nitrates are indicated for relief of attacks or acute prophylaxis of angina pectoris.
- The nitrates are used in both stable and unstable angina.
- The nitrates have some use in myocardial infarct.

> **ADVERSE EFFECTS**
>
> **Nitrates:** The development of nitrate tolerance and clinical rebound should be considered with long-term nitrate use. Nitrate tolerance may develop within 1 to 2 days, resulting in decreased angina control. The tolerance is associated with depletion of thiols needed for activation of the nitroglycerin. Headache, tachycardia, hypotension (potential for falling), nausea, and vomiting are additional adverse effects. Drug–drug interaction leading to more profound hypotension may occur with erectile dysfunction drugs (PDE 5 inhibitors), β-blockers, calcium channel blockers, alcohol, and other vasodilators drugs.

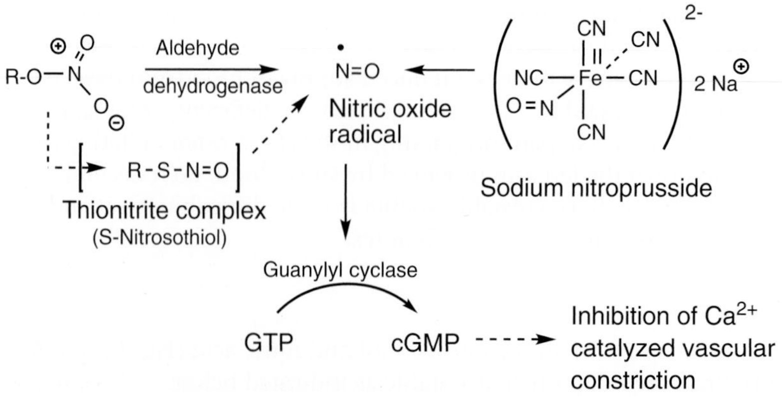

Figure 13.32 Mechanism of action of nitric oxide precursors leading to vasodilation.

Late Sodium Current Inhibitor – Ranolazine

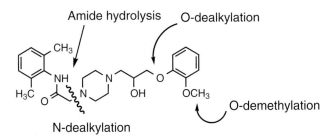

Ranolazine
(Ranexa)

MOA

- Ranolazine is thought to alter the intracellular Na^+ level, which affects the Na^+-dependent Ca^{2+} channels during myocardial ischemia.
- The drug selectively inhibits late Na^+ current (I_{Na}) decreasing intracellular Na^+ and Na^+ overload.
- Ranolazine attenuates the abnormalities of ventricular repolarization and contractility thus contributes to the cardioprotective effects.

Physicochemical and Pharmacokinetic Properties

- Ranolazine exhibits good oral bioavailability (76%) from extended-release tablets.
- The drug is excreted primarily in the urine (~75%).
- Ranolazine is an inhibitor of P-gp and is extensively metabolized primarily by CYP3A4 to a large number of inactive metabolites.

	ADVERSE EFFECTS

Ranolazine: Common adverse effects reported for ranolazine include dizziness, nausea, constipation, headache, swelling in hands, ankles, and feet and irregular heartbeats. Additionally, ranolazine is reported to prolong the QT interval limiting its use and it is contraindicated with strong inhibitor or inducers of CYP3A.

Ranolazine - sites of metabolism

Clinical Applications

- Ranolazine is approved for use in treatment of angina.
- Ranolazine when combined with a calcium channel blocker, β-adrenoceptor antagonist, or a nitrate in patients who are unresponsive to other antianginals is successful without increasing myocardial workload.

Drugs Used to Treat Pulmonary Hypertension

Five major types of pulmonary hypertension are known (e.g., arterial, venous, hypoxic, thromboembolic, miscellaneous) of which pulmonary arterial hypertension (PAH) represents one of these hypertensive conditions. Drugs used in the treatment of PAH will be addressed below.

Endothelin Receptor Antagonists

PAH is defined as a group of diseases characterized by a progressive increase of pulmonary vascular resistance, leading to right ventricular failure and death. PAH is a disease in which three factors have been associated with the pulmonary vascular resistance: vasoconstriction, remodeling of the pulmonary vessel wall, and thrombosis. Related to the genesis of these effects are an impaired production of vasoactive mediators (e.g., NO, prostacyclin) as well as an overexpression of vasoconstrictors such as endothelin-1 (ET-1). An appreciation of the role of these

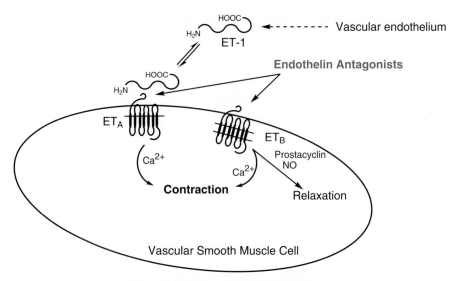

Figure 13.33 Action of ET-1 on vascular smooth muscle cells.

factors suggests logical pharmacologic targets and thus the development of ET-1 receptor antagonists. For a more in-depth discussion of PAH and its pharmacologic treatment the reader is referred to Chapter 24 in *Foye's Principles of Medicinal Chemistry, Seventh Edition*.

Physiologic Action of ET-1

- ET-1 is a 21–amino acid peptide that is produced by the vascular endothelium, which upon binding to VSM endothelin receptors ET_A and ET_B produces very strong vasoconstriction through G_q protein receptors (Fig. 13.33) causing sarcoplasmic reticulum release of Ca^{2+}.
- ET_A is located predominantly on smooth muscle cells, whereas ET_B is found on both endothelial and smooth muscle cells.
- Activation of ET_A leads to vasoconstriction resulting through increase in cytosolic Ca^{2+} levels, sustained vasoconstriction, and proliferation of vascular smooth muscle cells.
- Activation of ET_B on VSM cells leads to vasoconstriction, but also stimulates formation of prostacyclin from arachidonic acid and NO through endothelium nitric oxide synthase (prostacyclin and NO are potent vasodilators) leading to transient vasodilation.

MOA

- Bosentan is an ET_A and ET_B antagonist while ambrisentan is an ET_A selective antagonist (77× >selectivity for ET_A as compared to ET_B) (Fig. 13.34).
- Some studies suggest that in the pulmonary hypertensive state, blockade of both ET_A and ET_B is necessary.

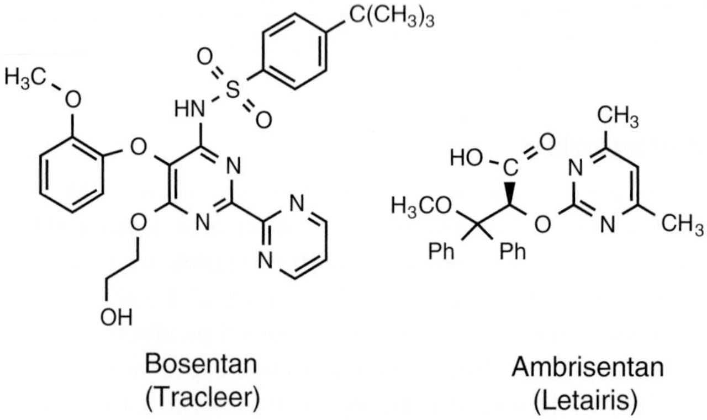

Figure 13.34 Endothelin receptor antagonists.

Bosentan
(Tracleer)

Ambrisentan
(Letairis)

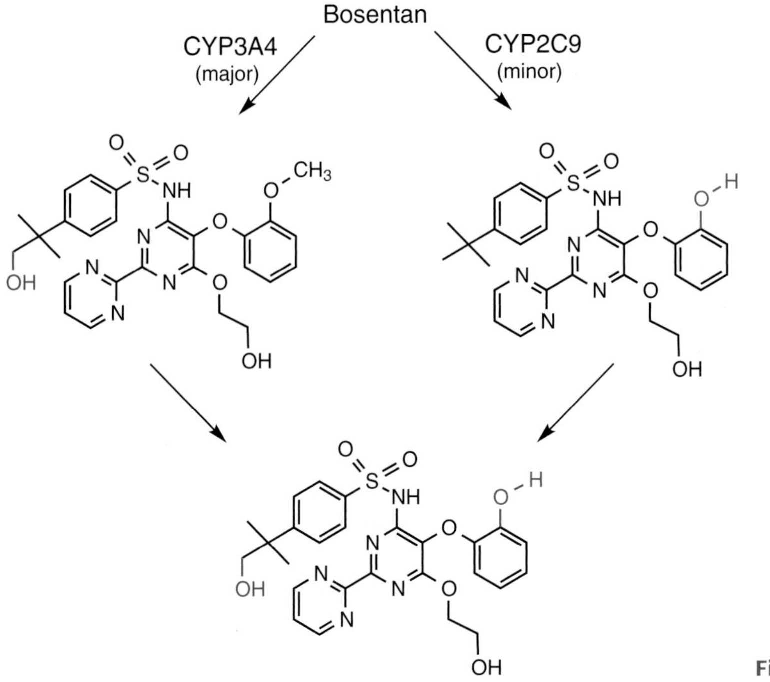

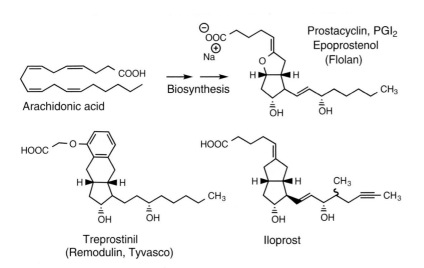

Bosentan

CYP3A4
(major)

CYP2C9
(minor)

Figure 13.35 Metabolism of bosentan.

Physicochemical and Pharmacokinetic Properties

- Both ET-1 inhibitors are highly bound to plasma protein (~98%) with bioavailabilities of 50% for bosentan and 80% for ambrisentan.
- Half-life of 5.4 hours and 15 hours for bosentan and ambrisentan, respectively.

Metabolism

- Bosentan is a substrate for CYP450 as shown in Figure 13.35.
- Inhibitors of CYP3A4 may increase bosentan plasma levels and bosentan is an inducer of CYP2C9 reducing clotting time if the drug is combined with warfarin.
- Ambrisentan's primary metabolites involve UGT (1A9, 2B7, 1A3) leading to acyl glucuronide. The drug is also a substrate for P-glycoprotein.

Prostanoids

Prostacyclin (PGI$_2$) a potent vasodilator, antithrombotic with antiproliferative properties is produced in the vascular endothelium from the precursor arachidonic acid (Fig. 13.36). Prostacyclin and its analogs (prostanoids) are effectively utilized in the treatment of pulmonary hypertension.

> **ADVERSE EFFECTS**
>
> **Bosentan and Ambrisentan:** Adverse effects of bosentan include hypotension, headache, flushing, increased liver aminotransferases (potential liver injury), leg edema, and anemia. Patients with hypersensitivity to sulfonamides should be excluded. Bosentan and ambrisentan are teratogenic in animal models and therefore can cause birth defects and is contraindicated in pregnancy. Ambrisentan has several severe warnings associated with its use: hepatotoxicity, pulmonary and peripheral edema, a marked decrease in hemoglobin, and reduced sperm count.

Arachidonic acid

Biosynthesis

Prostacyclin, PGI$_2$
Epoprostenol
(Flolan)

Treprostinil
(Remodulin, Tyvasco)

Iloprost

Figure 13.36 Prostanoid analogs.

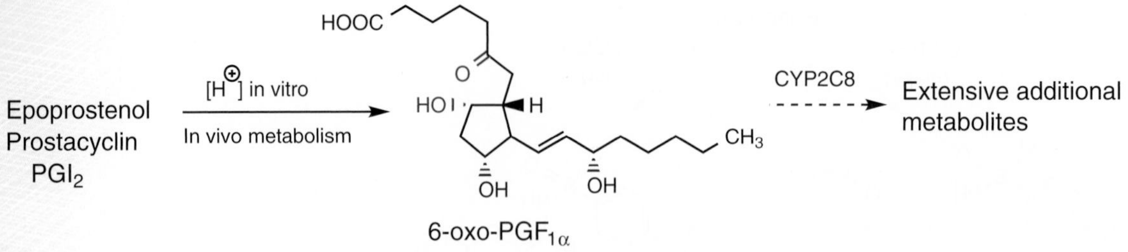

Figure 13.37 In vitro and in vivo instability of PGI_2.

MOA

- Platelet and endothelial cell G-protein–coupled receptors are activated by PGI_2 resulting in adenylyl cyclase production of cAMP.
- The cAMP goes on to reduce platelet aggregation and produce smooth muscle relaxation/vasodilation through activation of protein kinase A (PKA).
- The synthetic prostanoids, treprostinil and iloprost, mimic the effects on the PGI_2 receptor.
- It should be noted that all three prostanoids structurally quite similar.

Physicochemical and Pharmacokinetic Properties

- Epoprostenol sodium salt (pH 10), is unstable leading to acid-catalyzed hydrolysis (Fig. 13.37).
- The drug must be reconstituted shortly prior to use and administered via continuous IV infusion.
- Treprostinil, a synthetic derivative of PGI_2, is available as a stable sodium salt used via injection (e.g., IV or SC), or in solution for inhalation. The latter is packaged in foil pouches, which protects the drug from light.
- Iloprost is used by inhalation or IV therapy.
- The half-life is ~10 minutes for epoprostenol, 85 and 34 minutes for treprostinil, SC or IV respectively, and 20 to 30 minutes for iloprost.

Metabolism

- The prostanoids are extensively metabolized to inactive metabolites. Structures of the metabolites are not available.
- Several of the metabolites of epoprostenol retain the 6-oxo structure (Fig. 13.37).

Clinical Applications

- The only indication for ET-1 inhibitors is for the treatment of PAH to improve exercise ability and prevent worsening of the condition.
- A similar indication is listed for the prostanoids. Outcomes of treatment with ET-1 or prostanoids are an increase in walk distance.
- Other medications that may be prescribed for PAH include: anticoagulants, calcium channel blockers, and diuretics.

> **⚠ ADVERSE EFFECTS**
>
> **Prostanoids:** Common adverse effects of the prostanoids include nausea, vomiting, headache, hypotension, and flushing. In addition, epoprostenol may cause chest pain, anxiety, bradycardia, and tachycardia. Treprostinil has been reported to cause jaw pain and localized pain at the delivery site under the skin. This pain has been reported as slight to severe irritation. In addition to the above common reactions patients on iloprost report a high incidence of cough.

Chemical Significance

Cardiovascular diseases cover a broad range of conditions and therefore it is not surprising that medications to treat these conditions also represent a myriad of chemical entities from inorganic nitric oxide to families of analogs such as the β-blockers and calcium channel blockers. Recognizing the chemical nucleus of aryloxypropylamines or dihydropyridines can immediately allow the pharmacist to categorize new chemical entities into specific classes with their corresponding mechanisms of action, physicochemical and pharmacokinetic properties, potential metabolic traits, and ultimately therapeutic scopes. This may also allow one to evaluate advantages over traditionally prescribed agents as to onset, duration, and potential adverse effects.

A strong background in functional group-related metabolism can be quite helpful in recognizing or predicting metabolic reactions seen with individual drugs presented in this chapter. In many cases this allows the practitioner the ability to understand the metabolism and the consequence of metabolism rather than memorizing these individual reactions. In other cases such as the thiazide and thiazide-like diuretics one can easily see the similarities between these classes through the presence of specific required functional groups.

Review Questions

1 Which of the following structures represents an in vitro instability product of dobutamine?

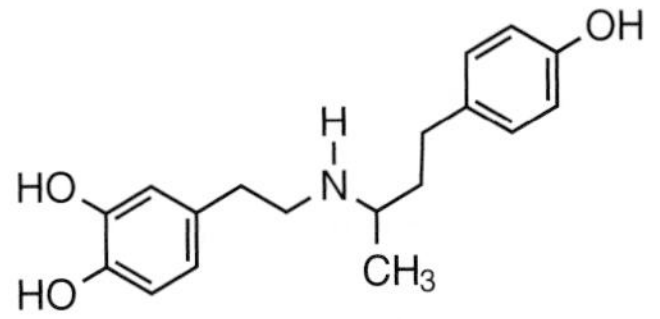

Dobutamine

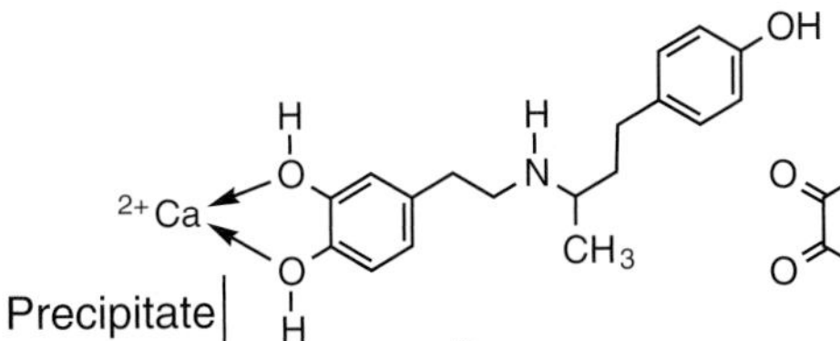

A. 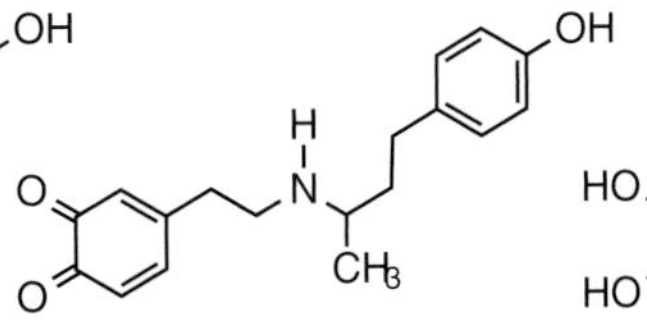 B. 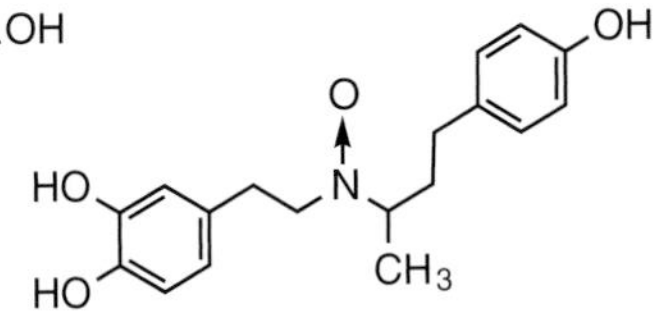 C.

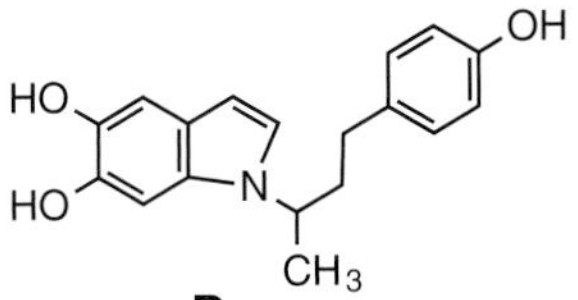

D. 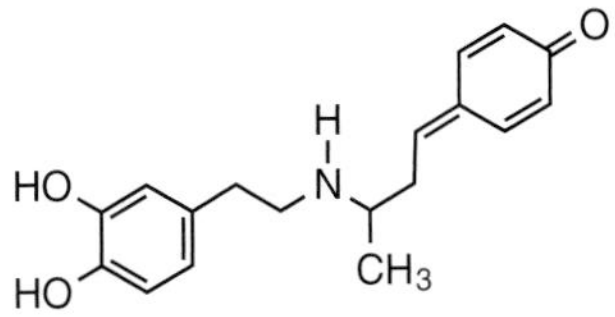 E.

2 Isosorbide dinitrate is a prodrug. What is the active form of isosorbide dinitrate?
A. Isosorbide mononitrate
B. Sodium nitrate
C. Nitrous oxide anion
D. Nitric oxide radical
E. Calcium nitrate

3 The Class I antiarrhythmics operate through a common mechanism of action which involves
A. blocking fast Na^+ channels.
B. decrease the levels of cAMP which in turn decreases Ca^{2+} flux.
C. block K^+ channels.
D. directly block cellular influx of Ca^{2+}.
E. inhibit Na^+/K^+ ATPase channels.

4 Which statement is correct as it relates to the SAR of the calcium channel blockers as depicted in the structure shown below?

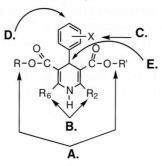

A. R and R′ must be the same for maximum activity.

B. R_2 and R_6 must be small groups for biological activity.

C. X must be a group in the *ortho* or *meta* positions to create an appropriate conformation.

D. An electron-withdrawing group must be located at this position for activity.

E. The *R* isomer is an inhibitor while the *S* isomer is an agonist.

5 Which of the following structure represents the active form of minoxidil?

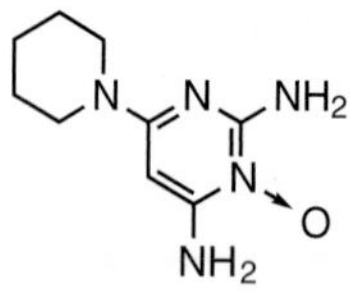

A. (Minoxidil) 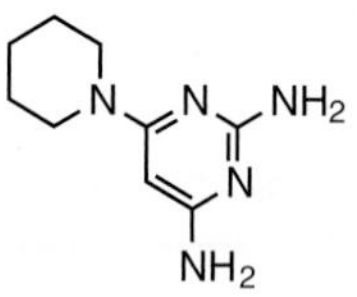   B. 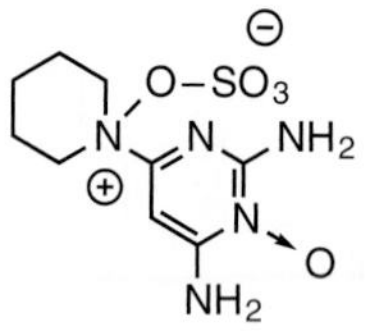   C.

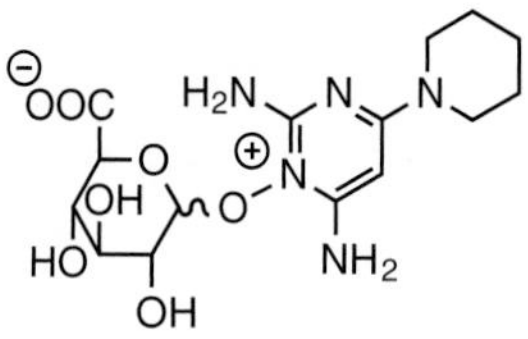

D. 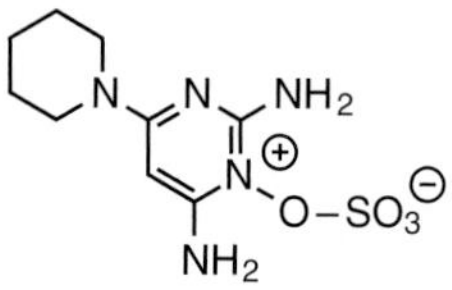   E.

Agents Affecting the Renin–Angiotensin Pathway

14

Learning Objectives

Upon completion of this chapter the student should be able to:

1. Describe the mechanism of action of drugs affecting the renin–angiotensin pathway:
   a. Angiotensin-converting enzyme inhibitors (ACEIs)
   b. Angiotensin II receptor blockers (ARBs)
   c. Renin inhibitors

2. Discuss applicable structure activity relationship (SAR) (including key aspects of receptor binding) of the ACEIs, ARBs, and renin inhibitors.

3. Identify any clinically relevant physicochemical properties, solubility, in vivo and in vitro chemical stability, absorption and distribution, and pharmacokinetic properties that impact the in vitro and in vivo stability and/or therapeutic utility of ACEIs, ARBs, and renin inhibitors.

4. Discuss and diagram metabolic pathways for all biotransformed ACEIs, ARBs and renin inhibitors identifying enzymes and note the clinically significant metabolism and the activity, if any, of the major metabolites.

5. Explain clinical uses and clinically relevant drug–drug, drug–food interactions.

6. Apply all of above to predict and explain therapeutic utility of ACEIs, ARBs, and renin inhibitors.

Introduction

The renin–angiotensin pathway is an important biochemical pathway for maintaining cardiovascular homeostasis including blood volume, arterial blood pressure, and electrolyte balance. Over production of renin and especially angiotensin II can result in hypertension and/or heart failure. Excess angiotensin II contributes to high blood pressure and heart failure through both fast and slow pressor responses including direct vasoconstriction, increasing proximal tubule sodium reabsorption, stimulating release of aldosterone, hypertrophy, and remodeling of vascular and cardiac cells.

The renin–angiotensin pathway consists of two main enzymes: renin and angiotensin-converting enzyme (ACE) whose purpose is to release angiotensin II from endogenous angiotensinogen (Fig 14.1). Angiotensinogen is an α_2-globulin found in the plasma and continually synthesized and secreted by the liver. Renin is an aspartyl protease that cleaves the leucine–valine bond in angiotensinogen to form angiotensin I. Angiotensin I is converted to angiotensin II by the action of ACE which is a zinc-containing dipeptidyl carboxypeptidase (Fig. 14.2). Renin is substrate-specific and the rate limiting step in the formation of angiotensin II. ACE is relatively nonspecific and only requires a tripeptide sequence as substrate. The nonspecific nature of ACE gives it additional proteolytic activity. ACE degrades bradykinin which is a vasodilator, a bronchoconstrictor, stimulates natriuresis, stimulates prostaglandin synthesis, and increases vascular permeability.

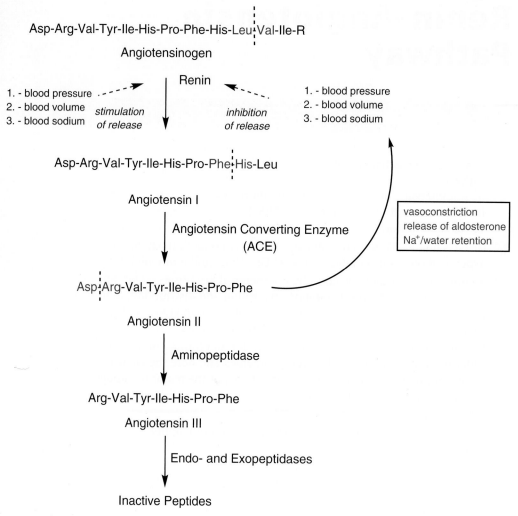

Figure 14.1 The renin–angiotensin pathway showing the formation of the angiotensins from angiotensinogen.

Since the renin–angiotensin pathway produces a large number of effects on the cardiovascular system, it stands to reason that compounds that can inhibit the effects of renin or ACE, as well as blocking the binding of angiotensin II to its receptors, would be important drugs for the treatment of hypertension and other cardiovascular diseases.

Angiotensin-Converting Enzyme Inhibitors

There are currently 10 angiotensin-converting enzyme inhibitors (ACEIs) approved for therapeutic use in the United States. They can be subclassified into three groups based on their chemical structures: sulfhydryl-containing, dicarboxylate-containing, and phosphonate-containing inhibitors.

MOA

- ACEIs block the conversion of angiotensin I to angiotensin II by interacting with three major sites in the catalytic pocket (Fig. 14.2):
  - a cationic site for binding to the anionic carboxylate terminus of the peptide substrate.
  - a zinc atom located close to the labile peptide bond and serves to stabilize the negatively charged tetrahedral intermediate.
  - a hydrophobic pocket that provides some specificity for the C-terminal aromatic or nonpolar residue.

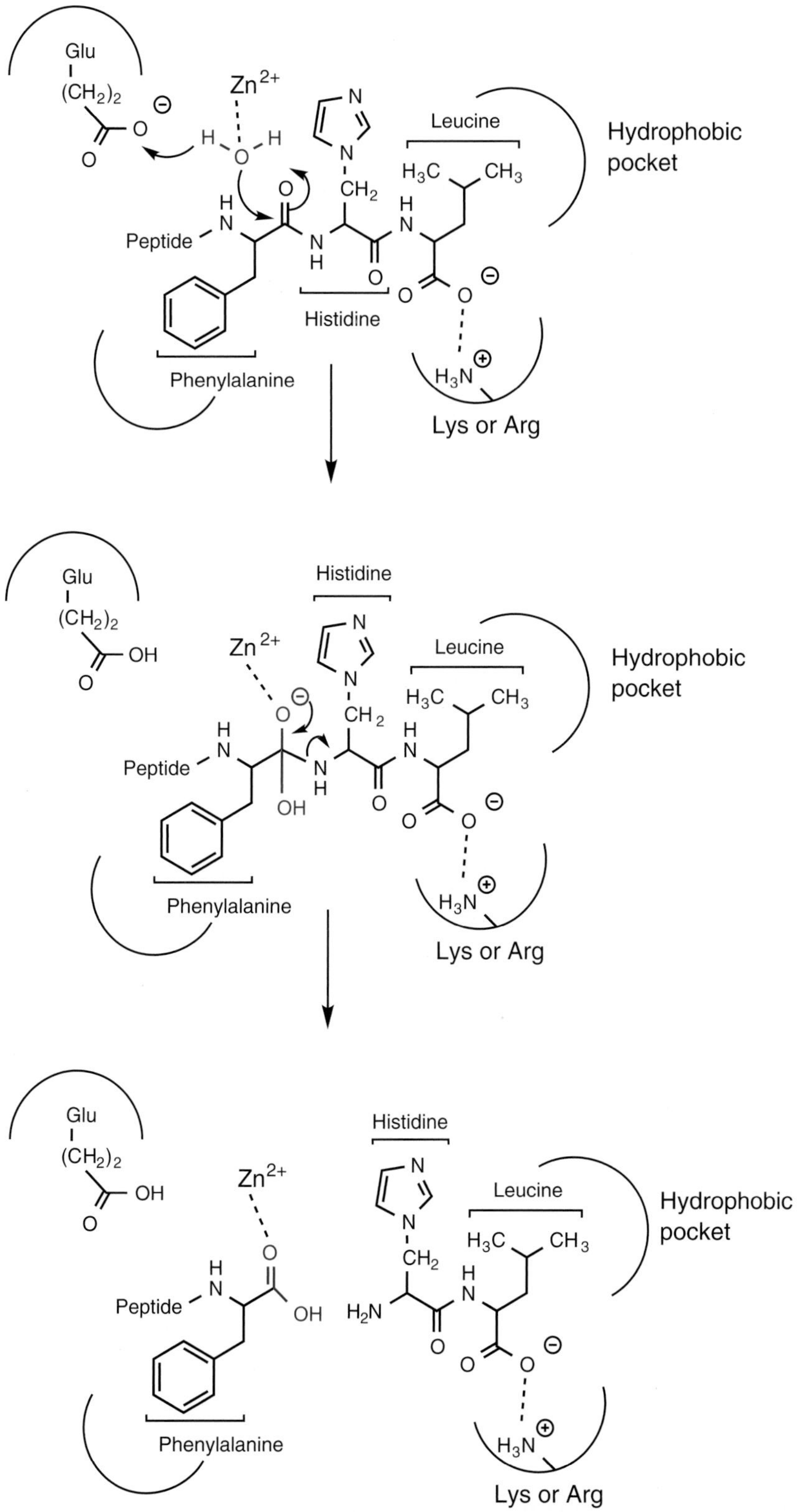

Figure 14.2 The binding of angiotensin I to key residues in ACE showing the formation of the tetrahedral transition state and the role of water and Glu in the hydrolysis process.

- In addition, ACEIs mimic the tetrahedral transition state of peptide hydrolysis (Fig. 14.2) and are resistant to hydrolysis (Fig. 14.3). Since ACE is a relatively nonspecific dipeptidyl carboxypeptidase, ACEIs also inhibit the metabolism of bradykinin which leads to a number of physiological effects:
  - vasodilation potentiating the hypotensive action of the ACEIs
  - bronchoconstriction is manifested by a dry cough

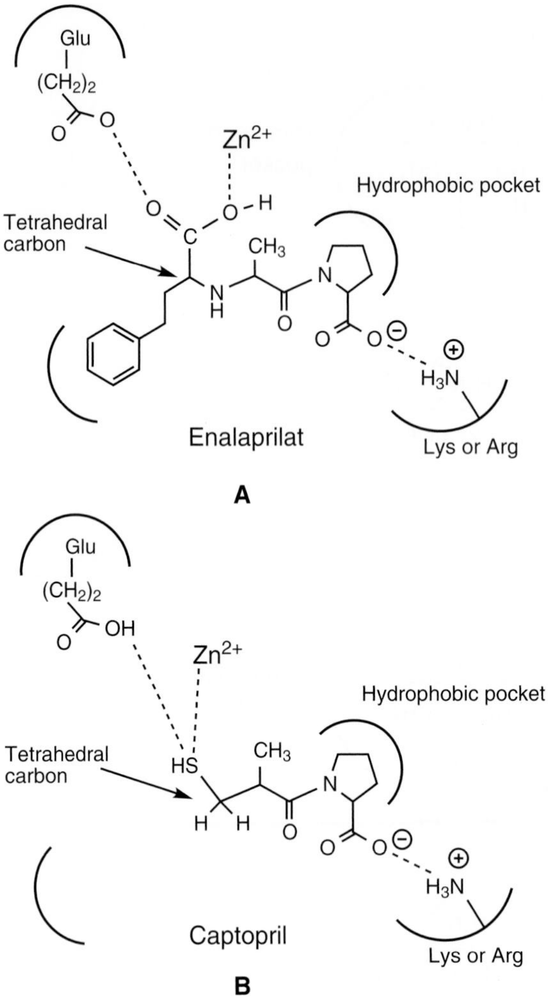

Figure 14.3 Examples of inhibitor binding to ACE demonstrating the tetrahedral transition state. **A:** Enalaprilat. **B:** Captopril.

- increased prostaglandin synthesis contributing to vasodilation, vascular permeability, and the production of some types of pain and inflammation.

SAR

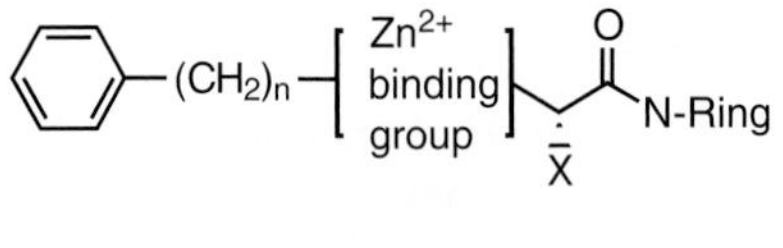

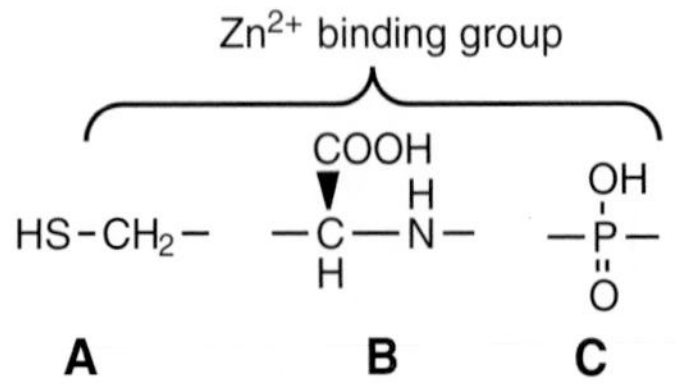

- The N-ring portion must contain a carboxylic acid for binding to the cationic site in ACE.
- Large heterocyclic rings in the N-ring portion increase potency and alter pharmacokinetics.
- The zinc binding group can be either (A) sulfhydryl, (B) carboxylate, or (C) phosphinate.
- -SH group shows superior binding to Zn^{2+}.
- -SH containing compounds produce a high incidence of skin rashes and a loss of taste. They can also form disulfides and dimers which may shorten the duration of action.

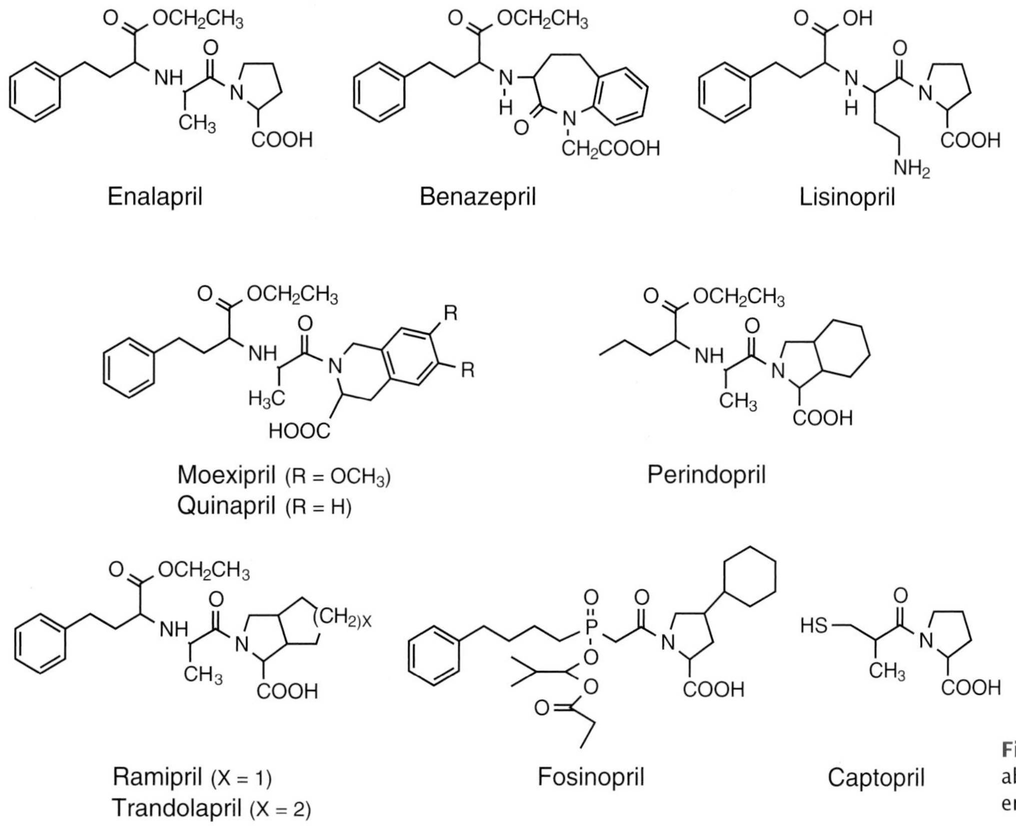

Enalapril   Benazepril   Lisinopril

Moexipril (R = OCH₃)
Quinapril (R = H)   Perindopril

Ramipril (X = 1)
Trandolapril (X = 2)   Fosinopril   Captopril

Figure 14.4 Currently available angiotensin-converting enzyme inhibitors.

- Binding of Zn^{2+} through either a carboxylate or phosphinate (O = P-OH) mimics the peptide hydrolysis transition state.
- X is usually a methyl group. Stereochemistry mimicking the stereochemistry of L-amino acid is optimal.
- Esterification of the carboxylate or phosphinate produces orally bioavailable prodrugs. Nomenclature is designated as …pril (ester prodrug) versus …prilat (active drug).
- Figure 14.4 shows all the currently available ACEIs.

Physicochemical and Pharmacokinetic Properties

- All ACEIs are amphoteric except captopril and fosinopril which are acidic.
- The carboxylic acid attached to the N-ring has a pKa range from 2.5 to 3.5 and is ionized at physiological pH.
- The second carboxylic acid in the dicarboxylate series is ionized in the active form but unionized when in the prodrug (ester) form.
- Table 14.1 gives the pharmacokinetic parameters of the ACEIs.
  - All possess good lipid solubility with the exception of captopril, enalaprilat, and lisinopril.
  - Lisinopril exhibits good oral bioavailability even though it is the most hydrophilic, because in the duodenum, lisinopril most likely exists as a di-zwitterion in which the ionized groups are internally bound to one another.

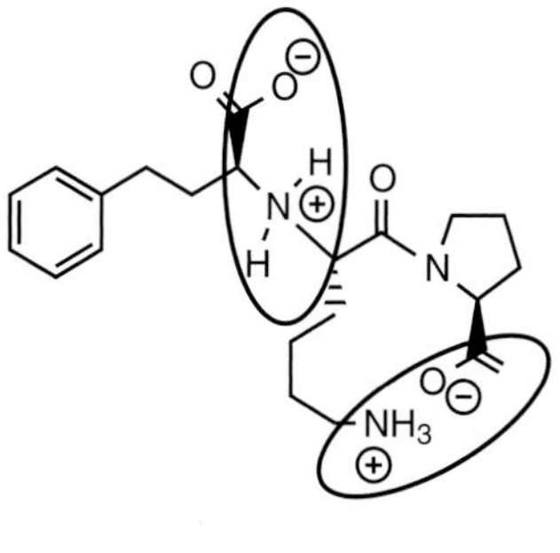

> **ADVERSE EFFECTS**
>
> **ACEIs:** Hypotension, hyperkalemia (due to inhibition of aldosterone release), dry cough, rash (captopril), taste disturbances (captopril), headache, dizziness, fatigue, nausea, vomiting, diarrhea, acute renal failure, neutropenia, proteinuria, and angioedema.
>
> Dry cough is by far most prevalent and is likely the result of the increase in the combination of bradykinin and prostaglandins.

Table 14.1 Pharmacokinetic Parameters of ACE Inhibitors

Drug	Calculated Log P	Oral Bioavailability (%)	Effect of Food on Absorption	Active Metabolite	Protein Binding (%)	Onset of Action (h)	Duration of Action (h)	Major Route(s) of Elimination
Benazepril	5.50	37	Slows absorption	Benazeprilat	>95	1	24	Renal (primary) Biliary (secondary)
Captopril	0.27	60–75	Reduced	NA	25–30	0.25–0.50	6–12	Renal
Enalapril	2.43	60	None	Enalaprilat	50–60	1	24	Renal/Fecal
Enalaprilat	1.54	NA	NA	NA	—	0.25	6	Renal
Fosinopril	6.09	36%	Slows absorption	Fosinoprilat	95	1	24	Renal (50%) Hepatic (50%)
Lisinopril	1.19	25–30%	None	NA	25	1	24	Renal
Moexipril	4.06	13%	Reduced	Moexiprilat	50	1	24	Fecal (primary) Renal (secondary)
Perindopril	3.36	65–95%	Reduced	Perindoprilat	60–80	1	24	Renal
Quinapril	4.32	60%	Reduced	Quinaprilat	97	1	24	Renal
Ramipril	3.41	50–60%	Slows absorption	Ramiprilat	73	1–2	24	Renal (60%) Fecal (40%)
Trandolapril	3.97	70%	Slows absorption	Trandolaprilat	80	0.5–1.0	24	Fecal (primary) Renal (secondary)

NA, not applicable, data not available.

DRUG–DRUG INTERACTIONS

ACEIs: Captopril/Allopurinol – Increased risk of hypersensitivity; Captopril/iron salts – Reduction of captopril levels unless administration is separated by at least 2 hours; Captopril/probenecid – Decreased clearance and increased blood levels of captopril; Enalapril/rifampin – Decreased pharmacological effects of enalapril; Quinapril/Tetracycline – Decreased absorption of tetracycline (may be due to high magnesium content of quinapril tablets); All ACEIs and Antacids – Decreased bioavailability of ACE inhibitor (more likely with captopril and fosinopril); Capsaicin – Exacerbation of cough; Digoxin – Either increased or decreased plasma digoxin levels; Diuretics – Potential excessive reduction in blood pressure. The effects of loop diuretics may be reduced; K^+ preps or K^+ sparing diuretics – Elevated serum potassium levels; Lithium – Increased serum lithium levels; NSAIDs – Decreased hypotensive effects; Phenothiazines – Increased pharmacological effects of ACE inhibitor.

Metabolism

- Lisinopril and enalaprilat are excreted unchanged, although all others undergo some type of metabolism (Fig 14.5). All dicarboxylate and phosphinate prodrugs undergo hydrolysis to active forms via hepatic esterases.
- Captopril forms dimers or conjugation with cysteine.

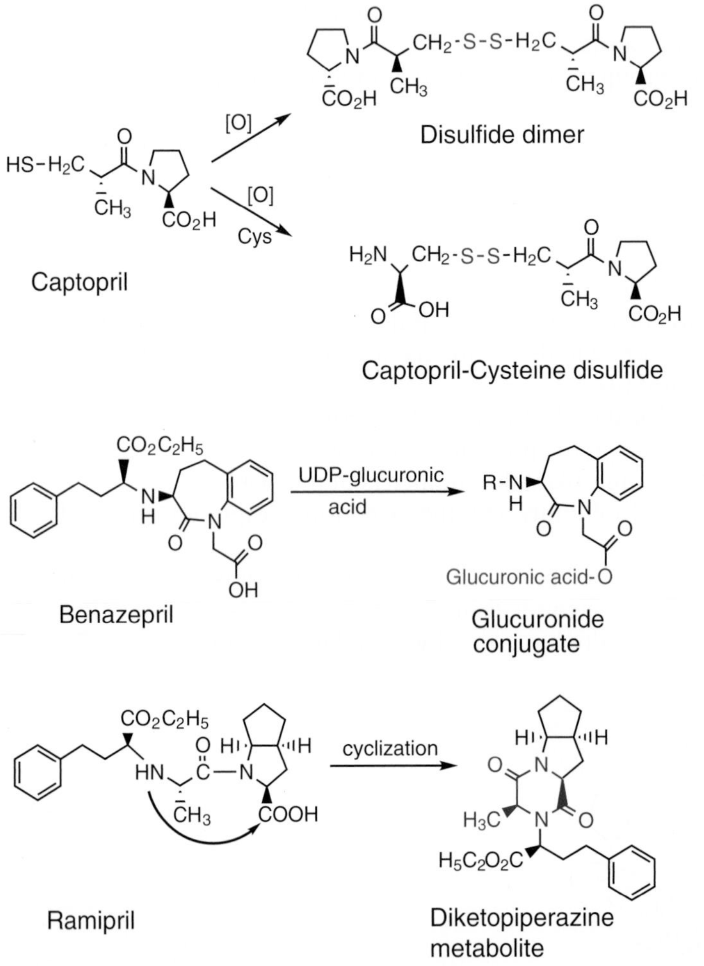

Figure 14.5 Examples of metabolic routes of ACEIs.

- Benazepril, fosinopril, quinapril, and ramipril undergo glucuronide conjugation of carboxylic acid.
- Moexipril, perindopril, and ramipril can undergo intramolecular cyclization to produce diketopiperazines.

Clinical Applications

- ACEIs are approved primarily for the treatment of hypertension and heart failure.
- Some are indicated also for left ventricular dysfunction postmyocardial infarction (MI) (captopril, enalapril, trandolapril).
- ACEIs are reported to slow the progression of diabetic nephropathy and are therefore recommended to be used in hypertensive patients with diabetes.
- Some unlabeled uses for the ACEIs include, hypertensive crisis, stoke prevention, migraine prophylaxis, nondiabetic nephropathy, and chronic kidney disease.

Angiotensin II Receptor Blockers

Angiotensin II produces its biological effects by interacting with two receptor subtypes – AT_1 and AT_2. The AT_1 receptors are located in brain, neuronal, vascular, renal, hepatic, adrenal, and myocardial tissues. AT_2 receptors are found in the brain, CNS, and myocardial and renal tissue.

MOA

- AT_1 effects include vasoconstriction, aldosterone and vasopressin secretion, sodium reabsorption, catechol release, and left ventricular remodeling. AT_2 interaction results in cardioprotective effects, for example, vasodilation, release of nitric oxide (NO), regression of hypertrophy, and apoptosis.
- All currently available ARBs are 10,000 times more selective for AT_1 subtype and are competitive antagonists.
- It is more desirable to have a selective AT_1 blocker since blocking AT_2 would reduce or eliminate its apparent cardioprotective effects.

SAR

- Initial research focused on peptide analogs of Angiotensin II.
- The prototypic compound that was developed was saralasin, an octapeptide in which the Asp^1 and Phe^8 residues of angiotensin II were replaced by Sar (sarcosine, N-methylglycine) and Ala, respectively (Fig 14.6).
- Saralasin however lacks oral bioavailability and expresses partial agonist action.
- This lead to the development of peptide mimetics to circumvent these peptide-based drug drawbacks.
- 1-benzyl-5-acetyl-imidazoles mimic the key amino acid residues (Tyr^4, Ile^5, His^6, Phe^8) crucial for blocking the action of angiotensin II.
- The SAR of the ARBs is based on the 1-benzyl-5-acetyl-imidazole pharmacophore.
  - R can be an H, a Cl or varies as seen in Figure 14.7.
  - $-CH_2$-COOH can be replaced with an H-bonding moiety such as:
    - Hydroxymethyl or a ketone, a part of a benzimidazole.
  - R_2 can be -COOH or an o-substituted phenyl.
    - o-substituents can be either a -COOH or a tetrazole ring.
  - The n-butyl chain can be either an ethyl ether or an n-propyl group.
- Figure 14.8 shows the structures of all currently available ARBs.

Combination Products with ACEIs
ACEI with hydrochlorothiazide: benazepril, captopril, enalapril, fosinopril, lisinopril, moexipril, and quinapril
ACEI with calcium channel blockers: benazepril/amlodipine, enalapril/diltiazem, enalapril/felodipine, trandolapril/verapamil

Peptidomimetic Design
Peptidomimetics are designed using a three-step rational approach: (1) identification of the amino acid residues that are crucial for action, (2) determination of the spatial arrangement of these residues, and (3) use of a template to mount the key functional groups in their proper conformation. 1-benzyl-5-acetyl imidazoles fulfill these three requirements as a peptidomimetic of angiotensin II.

Asp^1–Arg—Val—Tyr^4 Ile^5—His^6—Pro—Phe^8
Angiotensin II

Sar—Arg—Val—Tyr—Ile—His—Pro—Ala
Saralasin

Figure 14.6 Comparison of angiotensin II to saralasin.

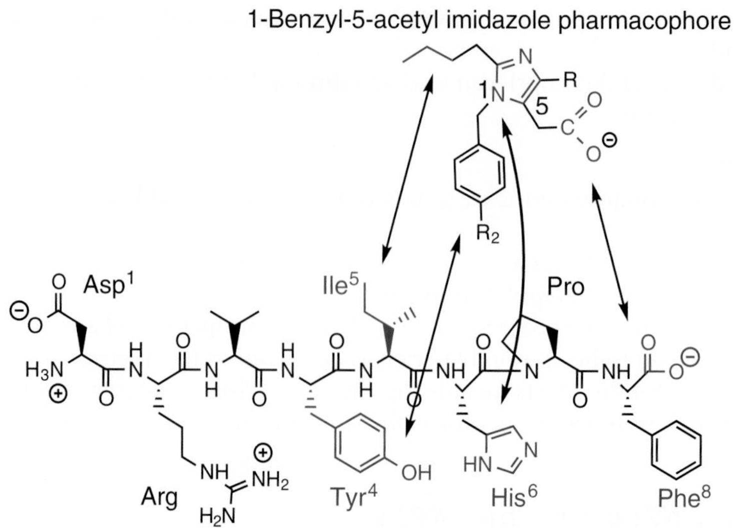

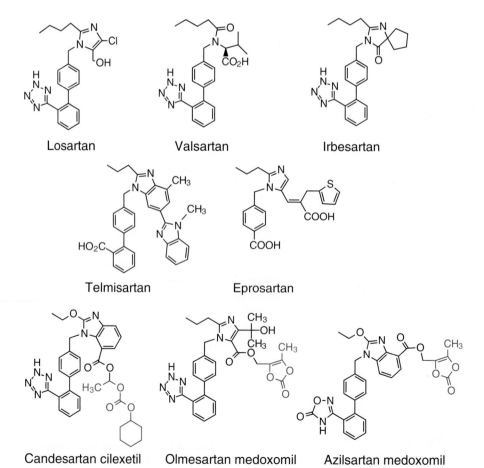

1-Benzyl-5-acetyl imidazole pharmacophore

Figure 14.7 Comparison of the peptidomimetic 1-benzyl-5-acetyl-imidazole with angiotensin II. (Adapted with permission from Timmermans PB, Wong PC, Chui AT, et al. Angiotensin II receptors and angiotensin II receptor antagonists. *Pharmacol Rev.* 1993;45:205–251. Copyright © 1993 Pharmacological Reviews Online by American Society for Pharmacology and Experimental Therapeutics.)

ADVERSE EFFECTS

ARBs: Headache, dizziness, fatigue, hypotension, hyperkalemia (due to inhibition of aldosterone release), dyspepsia, diarrhea, abdominal pain, upper respiratory tract infection, myalgia, back pain, pharyngitis, and rhinitis. Notably absent are the dry cough and angioedema seen with ACEIs.

Physicochemical and Pharmacokinetic Properties

- All ARBs are acidic due to the tetrazole ring (pKa = 6) and carboxylate groups (pKa = 3–4).
- The tetrazole and carboxylic acid groups are ionized at physiologic pH.
- The tetrazole ring containing ARBs have greater binding than those without the tetrazole ring and are more lipophilic.
- Table 14.2 gives the pharmacokinetic properties of the ARBs.
- All have low but adequate oral bioavailability (15% to 33%).
  - Exceptions are irbesartan (60% to 80%) and telmisartan (42% to 58%).
- All can be taken with meals.

Losartan   Valsartan   Irbesartan

Telmisartan   Eprosartan

Figure 14.8 Structures of currently available ARBs. The highlighted portions of candesartan cilexetil, olmesartan medoxomil, and azilsartan medoxomil are hydrolyzed by esterases to produce their respective active carboxylate metabolites.

Candesartan cilexetil   Olmesartan medoxomil   Azilsartan medoxomil

Table 14.2 Pharmacokinetic Parameters of Angiotensin II Receptor Blockers

Drug	Oral Bioavailability (%)	Active Metabolite	Protein Binding (%)	Time to Peak Plasma Concentration (h)	Elimination Half-Life (h)	Major Route(s) of Elimination
Azilsartan Medoxomil	60	Azilsartan	99	1.5–3.0	11	Fecal (55%) Renal (42%)
Candesartan Cilexetil	15	Candesartan	99	3–4	9	Fecal (67%) Renal (33%)
Eprosartan	15	None	98	1–2	5–9	Fecal (90%) Renal (10%)
Irbesartan	60–80	None	90	1.5–2	11–15	Fecal (80%) Renal (20%)
Losartan	33	EXP-3174	98.7 (losartan) 99.8 (EXP-3174)	1 (losartan) 3–4 (EXP-3174)	1.5–2 (losartan) 6–9 (EXP-3174)	Fecal (60%) Renal (35%)
Olmesartan Medoxomil	26	Olmesartan	99	1.5–3	10–15	Fecal (35–50%) Renal (50–65%)
Telmisartan	42–58	None	100	5	24	Fecal (97%)
Valsartan	25	None	95	2–4	6	Fecal (83%) Renal (13%)

h, hours

- All are highly protein bound.
- All have elimination half lives that allow once or twice a day dosing.
- All are primarily eliminated via the fecal route (exception olmesartan).
- None require dosing adjustment in renal impairment.
- Losartan and telmisartan are the only two that require dose reductions in patients with impaired hepatic function.

Metabolism

- Losartan is metabolized by CYP2C9 and 3A4 to an active metabolite (EXP-3174; Fig 14.9) and both contribute to total hypertensive effects, that is, losartan is not a prodrug.
- Candesartan and olmesartan are hydrolyzed to their active metabolites in the intestinal cell wall.
- All other ARBs are eliminated primarily unchanged (80%). However irbesartan, telmisartan, and eprosartan are metabolized to inactive carboxylic glucuronides.

Clinical Applications

- The ARBs are approved for treatment of hypertension either alone or in combination with ACEIs, diuretics, β-blockers, and calcium channel blockers.
- Unlike ACEIs they do not affect bradykinin-induced vasodilation and bronchoconstriction.

DRUG–DRUG INTERACTIONS

ARBs: Hyperkalemia when coadministered with potassium salts, K-sparing diuretics or drospirenone. NSAIDs alter effect by inhibiting the vasodilatory prostaglandins. Telmisartan increases digoxin levels. Losartan and its active metabolite levels decreased by rifampin.

Combination Products with ARBs

ARBs with hydrochlorothiazide: candesartan, eprosartan, irbesartan, losartan, olmesartan, telmisartan, and valsartan

ARBs with amlodipine: olmesartan, telmisartan, and valsartan

ARBs with hydrochlorothiazide and amlodipine: olmesartan and valsartan

ARBs with renin inhibitor: valsartan/aliskiren

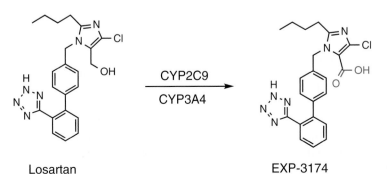

Figure 14.9 The metabolic conversion of losartan to EXP-3174 by cytochrome P450 isozymes.

- Other indications for ARBs.
  - Irbesartan and losartan are approved for treatment of nephropathy in type 2 diabetics.
  - Losartan is approved for stroke prevention in patients with left ventricular hypertrophy.
  - Candesartan and valsartan are approved for treatment of heart failure.
  - Telmisartan is approved to reduce the risk of MI and stroke.

Renin Inhibitors

- Developed as peptide mimetics to the octapeptide sequence (Pro^7-Phe^8-His^9-Leu^{10}-Val^{11}-Ile^{12}-His^{13}-Asn^{14}) of angiotensinogen recognized by renin.
- Currently, aliskiren is the only renin inhibitor available in the market.
- Aliskiren is based on an amino-amide nonpeptide template.

Nonpeptide Template   Aliskiren

MOA

- Aliskiren directly inhibits the formation of both angiotensin I and angiotensin II.
- Renin hydrolysis of angiotensinogen is the rate limiting step in this pathway and is regulated by hemodynamic, neurogenic, and humoral signals.
- Unlike ACEIs and ARBs, aliskiren does not cause a compensatory increase of renin in the plasma.

SAR

- Since there are currently no other renin inhibitors marketed there is no SAR to note.
- Aliskiren contains four chiral centers and is marketed as the pure 2S, 4S, 5S, 7S enantiomer.

Physicochemical and Pharmacokinetic Properties

- Aliskiren is a basic compound and marketed as the hemifumarate salt.
- Aliskiren has a log P of 4.32 for the unionized form and the salt form is highly water soluble.
- Oral bioavailability is approximately 2.5% with peak plasma levels reached in 1 to 3 hours.
- High fat meals decrease absorption.
- Elimination is primarily via the hepatobiliary tract.
- No dose adjustment necessary with renal or hepatic impairment but should be avoided in patients with severe renal impairment.

Metabolism

- Approximately 90% of aliskiren is excreted unchanged.
- Ten percent metabolized to the O-demethylated alcohol derivative and a carboxylic acid derivative via CYP3A4.

Clinical Application

- Aliskiren is approved for the treatment of hypertension as either monotherapy or in combination with hydrochlorothiazide, amlodipine, or valsartan.

ADVERSE EFFECTS

Renin Inhibitors: Diarrhea, abdominal pain, dyspepsia, gastroesophageal reflux, and rash.

DRUG–DRUG INTERACTIONS

Renin Inhibitors: Hyperkalemia when coadministered with potassium salts, K-sparing diuretics. Irbesartan decreases both plasma levels and efficacy. Plasma levels increased when coadministered with potent P-glycoprotein inhibitors such as atorvastatin, cyclosporine, and ketoconazole. Aliskiren decreases the maximum plasma concentration of furosemide.

Chemical Significance

The development of drugs that inhibit the renin–angiotensin pathway have significantly improved the treatment of hypertension and heart failure. Identification of the key residues for the binding of substrates to the proteolytic enzymes (renin and ACE) as well as the key residues in angiotensin II receptors (AT_1 and AT_2) has allowed medicinal chemists to design renin inhibitors, ACE inhibitors, and ARBs for the treatment of hypertension and other cardiovascular diseases. Renin inhibitors and ARBs are peptide mimics that resemble the natural peptide substrates yet have no peptide bonds and therefore have better bioavailability and are stable to proteolytic degradation. For example, ARBs are modeled after 1-benzyl-5-acetyl-imidazoles which contain moieties that mimic Tyr^4, Ile^5, His^6, and Phe^8 that are the key amino acid residues present in angiotensin II for binding to the AT_1 receptor. In addition knowing the name of an ACE inhibitor ends in ...prilat (i.e., enalaprilat) refers to the active free carboxylate form of the drug that is necessary for binding to the zinc atom in the active site of the enzyme.

The pharmacist is the only member of the health care team that has the knowledge of those chemical facts necessary for contributing to the discussion of the appropriate drug choice for use in a given therapeutic scenario. Understanding the role of the chemical structure not only for the mechanism of action but also in the effect that chemical structure has on absorption, distribution, metabolism, and elimination, arms the pharmacist with special knowledge that can be applied for the benefit of the patient and enhancing their quality of life.

Review Questions

1 Which of the following ACEIs forms dimers that may shorten the duration of action?
A. Enalapril
B. Captopril
C. Moexipril
D. Ramipril
E. Fosinopril

2 Which of the following groups found on ACEIs binds to the zinc ion in ACE?
A. A carboxyl on the nitrogen ring
B. A phenyl ring
C. A benzimidazole group
D. A sulfhydryl group
E. An alkyl group adjacent to the nitrogen ring

3 Which of the following ARBs has the greatest oral bioavailability?
A. Candesartan
B. Olmesartan
C. Irbesartan
D. Losartan
E. Eprosartan

4 Which of the following groups found on ARBs is acidic?
A. Biphenyl rings
B. Imidazole ring
C. N-butyl moiety
D. Hydroxyl methyl
E. Tetrazole ring

5 Which of the following statements about aliskiren is true?
A. 90% metabolized to O-demethylated product
B. Primarily excreted via the kidney
C. High fat meals decrease absorption
D. It is an acidic drug
E. Salt form is highly insoluble

15

Antihyperlipidemics

Learning Objectives

Upon completion of this chapter the student should be able to:

1. Distinguish between the terms dyslipidemia, hyperlipidemia, and hyperlipoproteinemia.
2. Describe mechanism of action of:
   a. Bile acid sequestrants (BAS)
   b. Niacin
   c. Ezetimibe
   d. Phenoxyisobutyric acid derivatives (fibrates)
   e. Hydroxymethylglutaryl- (HMG-) CoA reductase inhibitors (statins)
   f. Orphan drugs used in the treatment of homozygous familial hyper-cholesterolemia (HoFH).
3. Discuss applicable structure–activity relationships (SAR) (including key aspects of receptor binding) of the above classes of antihyperlipidemic agents.
4. List physicochemical and pharmacokinetic properties that impact in vitro stability and/or therapeutic utility of antihyperlipidemic agents.
5. Diagram metabolic pathways for all biotransformed drugs, identifying enzymes, and noting the activity, if any, of major metabolites.
6. Apply all of the above to predict and explain therapeutic utility.

Introduction

Cholesterol serves many life-sustaining functions. For example, the biosynthesis of corticosteroids, sex steroids, and cell membranes depends on the presence of this polycyclic structure. However, high levels of cholesterol, along with the lipoproteins that transport it and its esters through the bloodstream, lead to atherosclerosis, a predisposing factor in the development of coronary artery disease/coronary heart disease (CAD/CHD). Likewise, an excess of serum triglycerides leads to negative cardiovascular consequences and can induce pancreatitis. Millions of individuals are at risk for these potentially fatal pathologies.

The positive benefit of a low fat diet and regular exercise on maintaining healthy plasma lipid and lipoprotein levels is well known. But, since lipids are produced endogenously as well as acquired exogenously through food consumption, heredity sometimes wins out over even the healthiest lifestyle. Drugs that positively impact serum levels of lipids and lipoproteins can serve as the therapeutic lifeline for patients with moderate or severe aberrations in serum cholesterol, triglycerides, and/or lipoprotein levels.

Dyslipidemia

- Dyslipidemia: aberrations in the level of serum lipids and/or lipoproteins.
  - Can lead to negative cardiovascular events, specifically, atherosclerosis and CHD.
  - Primary dyslipidemias result from genetic predisposition.
  - Secondary dyslipidemias result from pathologic conditions or lifestyle choices.

- Hyperlipidemia: elevation of serum cholesterol, cholesterol esters, triglycerides, and/or phospholipids.
  - Increases risk of CHD.
  - Hypertriglyceridemia increases risk of pancreatitis.
- Hyperlipoproteinemia: elevation of the lipoproteins that transport lipids through the bloodstream.
  - Involves elevated low-density lipoproteins (LDLs) or very low-density lipoproteins (VLDLs) and/or decreased high-density lipoproteins (HDLs).
- Therapeutic approaches to the treatment of hyperlipidemia and hyperlipoproteinemia include:
  - inhibiting intestinal reabsorption of bile acids (BAS).
  - inhibiting triglyceride biosynthesis and VLDL formation (niacin).
  - inhibiting intestinal absorption of dietary cholesterol (ezetimibe).
  - stimulating serum triglyceride cleavage and clearance (fibrates).
  - inhibiting de novo cholesterol biosynthesis (HMG-CoA reductase inhibitors).

Cholesterol and Bile Salts (Fig. 15.1)

- The rate-limiting step in cholesterol biosynthesis is the stereospecific conversion of 3-hydroxy-3-methylglutaryl-CoA to R(−) mevalonic acid.
  - The catalyzing enzyme is HMG-CoA reductase (HMGR).
- Cholesterol is the synthetic starting point for corticosteroids, sex steroids, and bile acids.
  - The anionic conjugate base of a bile acid is called a bile salt.
- Bile acids promote the intestinal absorption of lipids and fat-soluble vitamins.

Triglycerides

- Triglycerides are long-chain fatty acid esters of glycerol.
  - Enzymes involved in triglycerides biosynthesis include phosphatidic acid phosphatase (PAP), monoacylglycerol acyl transferase (MAGAT), and diacylglycerol acyltransferase (DAGAT).
- Triglycerides in normal concentrations are stored in adipose tissue.
  - When metabolic energy is needed, stored triglycerides are hydrolyzed by lipases to release free fatty acids (FFAs).
- Triglycerides are transported in serum solubilized by VLDL.

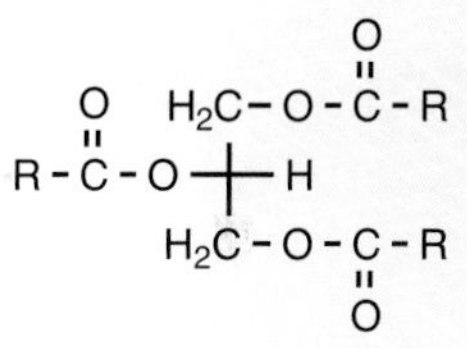

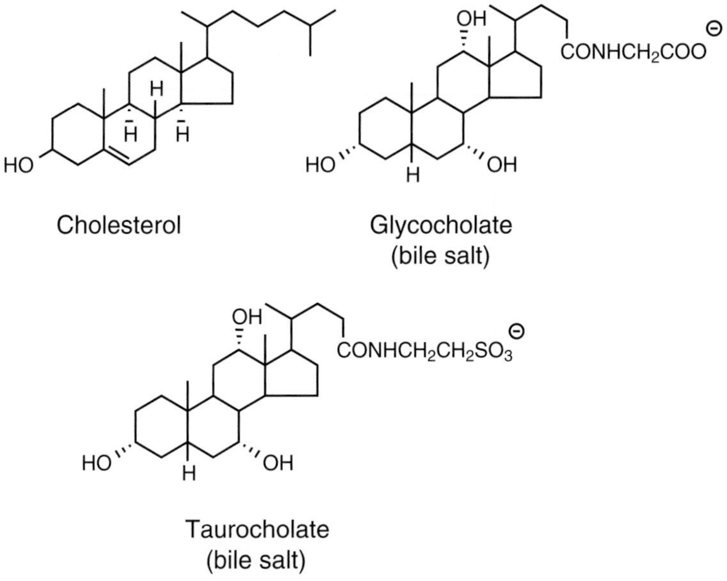

Figure 15.1 Cholesterol bile salts.

Bile Acid Sequestrants (BAS)

- Cholestyramine, colestipol, and colesevelam are used to treat hypercholesterolemia (Fig. 15.2).

Administration of BAS

- Available as dry resins (powder or granules) that are commonly administered as thick slurries by mixing with a noncarbonated beverage.
- Patients often find this formulation distasteful, which impacts adherence.
- Can sprinkle drug on foods such as applesauce or other pulpy fruits, or cereals.
- Colestipol and colesevelam are also marketed in tablet form. Must be swallowed whole with plenty of water or other liquid.
- Cholestyramine and colesevelam are administered with meals. Colestipol administration is not restricted with regard to meals.

MOA

- BAS are nonabsorbable anionic exchange resins that trade chloride anions bound to strongly cationic centers for intestinal bile salts glycocholate and taurocholate.
- Bile salts have higher affinity for the resin's cationic amines than chloride anion.
  - Bile salts are held to (sequestered by) the resin through strong ion–ion bonds.
  - Bound bile acids are excreted in the feces rather than being returned to the liver.
- Loss of hepatic return of bile acids stimulates 7α-hydroxylase-mediated oxidation of hepatic cholesterol and an increase in LDL clearance.
- De novo cholesterol biosynthesis is stimulated, but cannot overcome cholesterol loss from oxidation.
- The net result is a decrease in total serum cholesterol and LDL.
  - Triglyceride and VLDL levels may rise transiently or, in patients with hypertriglyceridemia, persistently.

SAR

- BAS contain permanently or potentially cationic amines that strongly bind intestinal glycocholic and taurocholic acids.
- Cholestyramine and colesevelam are quaternary amines and exhibit pH independent action.

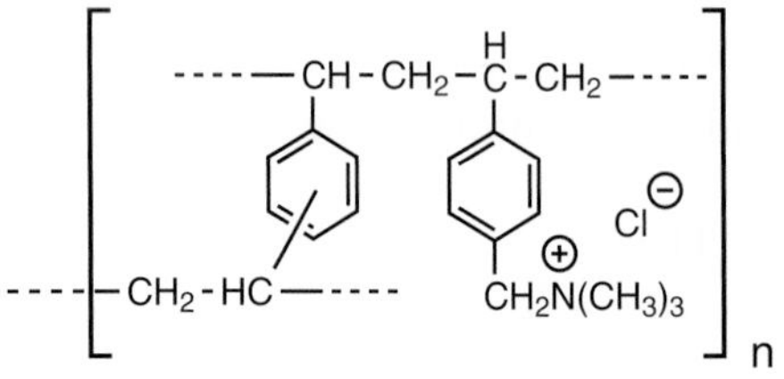

Cholestyramine (Questran)

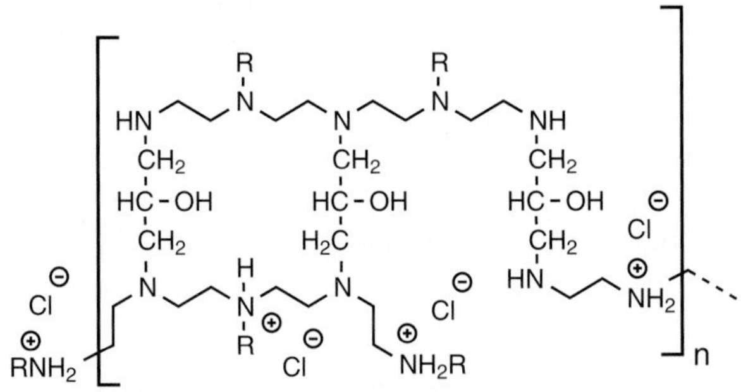

Colestipol (Colestid)
(R = H or epichlorohydrin cross link)

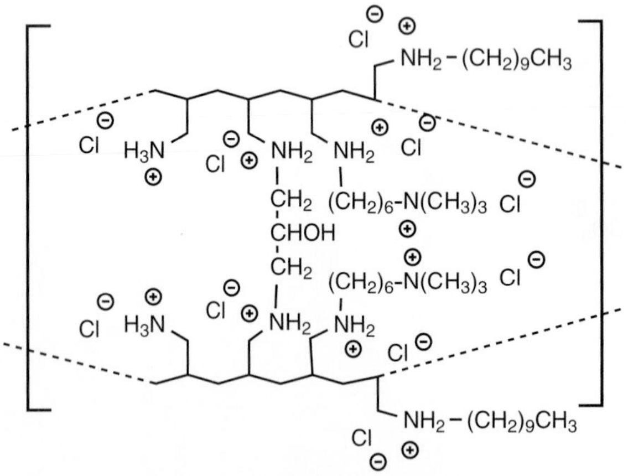

Colesevelam (Welchol)

Figure 15.2 Bile acid sequestrants.

- Colestipol's secondary and tertiary amines must protonate in the intestine to sequester bile salts. The pH dependent action lowers anion exchange capacity.
- Colesevelam's unique polymer structure allows for greater bile salt selectivity. This leads to fewer drug–drug interactions.

Physicochemical and Pharmacokinetic Properties

- BAS are water-insoluble resins. They are not absorbed across gastrointestinal membranes.
- BAS are metabolically inert and, along with their irreversibly bound bile salts, are excreted in feces.
- The normal transit time of BAS within the gastrointestinal tract is 4 to 6 hours.

Clinical Applications

- BAS are administered once or twice daily.
  - Cotreatment with niacin or statins requires careful attention to administration timing.
  - The statin in BAS–statin cotherapy blocks the cholesterol biosynthesis surge induced by the fecal loss of bile acids.
- BAS tablets are large and should not be used in patients with swallowing disorders.
- Therapeutic benefit is realized within 1 week (decreased LDL) to 1 month (decreased cholesterol).

Niacin (Nicotinic Acid)

Nicotinic acid (niacin)

MOA

- Niacin stimulates the GPR109A receptor found in adipocytes, spleen, and macrophage.
  - An ion–ion bond with a cationic receptor Arg is important to activity.
  - Lipolysis of stored triglycerides is inhibited, resulting in decreased production of triglycerides, FFA, VLDL, and LDL (Table 15.1). The FFA decrease is transient.
- Niacin also lowers serum triglycerides by inhibiting DAGAT2.
  - Acylation of diglycerides to triglycerides is blocked.
- Niacin inhibits receptor-mediated uptake of HDL, resulting in increased serum HDL.

SAR

- Niacin must be anionic to be an effective antihyperlipidemic.
  - The carboxylic acid is essential. Nonionizable amides (e.g., nicotinamide) are inactive.
- Essentially, any change made on the niacin structure results in inactivation.

Table 15.1	Effect of Niacin (>1.5 g/day) on Plasma Concentrations of Lipids/Lipoproteins
VLDL	Decrease 25–40%
LDL-cholesterol	Decrease 6–22%
HDL-cholesterol	Increase 18–35%
Total cholesterol	Decrease 4–16%
Triglycerides	Decrease 21–44%

DRUG–DRUG INTERACTIONS

BAS
- Drugs anionic at intestinal pH (e.g., warfarin, thyroid hormone, phenytoin) and some nonanionic drugs that are coadministered with a BAS can be sequestered by the resin and will not be absorbed.
  - Give potentially interacting drugs 1 hour before or 4 to 6 hours after the BAS unless clinical evidence exists to document safe and effective concurrent administration.
- Colesevelam's unique polymer structure allows for the coadministration of some anionic drugs, including antihyperlipidemic statins.
- BAS can absorb dietary vitamin K, leading to possible hemorrhage.

ADVERSE EFFECTS

BAS
- Bloating, abdominal discomfort
- Potentially severe constipation or bowel obstruction
- Aggravation of pre-existing hemorrhoids
- Gallstones (cholelithiasis)
- Pancreatitis
- Hyperprothrombinemia and bleeding

CHEMICAL NOTE

Cholestyramine in Diaper Rash: Cholestyramine can be compounded into ointments for the topical treatment of diaper rash. The chemical rationale for this therapeutic use is that the resin will sequester the highly irritating bile acids that are excreted in stool and held against the diapered skin. In some cases, resolution of even severe rash and perianal irritation can be totally resolved in a matter of days.

DRUG–DRUG INTERACTIONS

Niacin

- Can be sequestered if coadministered with BAS. Follow dose separation guidelines.
- Coadministration with statins can slightly enhance the risk of myotoxicity, including potentially fatal rhabdomyolysis.
- Coadministration with anticoagulants can prolong bleeding time to unsafe levels.
- Coadministration with vasodilating antihypertensive agents (calcium channel blockers, adrenergic blockers, nitroglycerin) worsens niacin-induced flushing, itching, and headache.

ADVERSE EFFECTS

Niacin

- Cutaneous vasodilation leading to flushing, itching, and headache. Less common with extended release formulations.
- Gastrointestinal disturbances (nausea, diarrhea, flatulence). Taking with nonspicy foods or cold beverages can decrease GI distress.
- Hepatotoxicity (high dose):
  - Extended release formulations contraindicated in patients with active liver disease.
  - Liver damage worsened by alcohol consumption.

Figure 15.3 Niacin metabolism.

Physicochemical and Pharmacokinetic Properties

- The carboxylic acid (pKa 4.76) and pyridine nitrogen (pKa 2.0) make niacin amphoteric. It exists predominantly as the active anion at pH 7.4.
- Absorption of niacin from the gut is rapid. The short 1-hour half-life necessitates frequent dosing of the immediate release formulation.
- Extended release formulations prolong the duration to 8 to 10 hours and lengthen the time to peak plasma concentration from 0.75 to 4 to 5 hours.
  - Niacin dosage forms are not interchangeable.

Metabolism

- Niacin administered as an antihyperlipidemic is excreted in the urine unchanged or conjugated with glycine (nicotinuric acid) (Fig. 15.3).

Clinical Applications

- When used as an antihyperlipidemic, niacin is dosed up to 6 g/day. Niacin administered as vitamin B3 is dosed at 13 to 20 mg/day.
- Niacin induces cutaneous vasodilation when given in multigram doses.
  - GPR109A stimulation activates phospholipase A₂.
  - Prostaglandin D_2 (PGD₂) is responsible for adverse effects.
- Approaches to combating vasodilation include pretreatment with OTC nonsteroidal anti-inflammatory drugs (NSAIDs) such as aspirin or indomethacin.
  - NSAIDs inhibit cyclooxygenase and block conversion of arachidonic acid released by phospholipase A₂ to PGD₂.
- Other tactics to minimize vasodilation-related adverse effects include:
  - Bedtime administration.
  - Titrating the dose upward over 1 to 4 months.
  - Letting normal tolerance take effect (3 to 6 weeks).

Ezetimibe

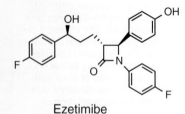

Ezetimibe

MOA

- Ezetimibe selectively blocks a cholesterol-active transporting protein at the intestinal brush border.
  - Inhibition of dietary cholesterol absorption increases serum LDL clearance and decreases total serum cholesterol.
- A cholesterol biosynthesis surge occurs, but the net result is a decrease in serum LDL.

SAR

- The 1,4-diaryl-β-lactam structure is important to activity.
- Phenolic and alcoholic hydroxyls keep ezetimibe localized in the small intestine.
- *p*-Fluoro groups block intestinal CYP-mediated aromatic hydroxylation, prolonging duration of action.

Physicochemical and Pharmacokinetic Properties

- Ezetimibe's phenolic hydroxyl (pKa 9.72) is predominantly unionized at intestinal pH.
- Oral absorption is rapid and food independent.

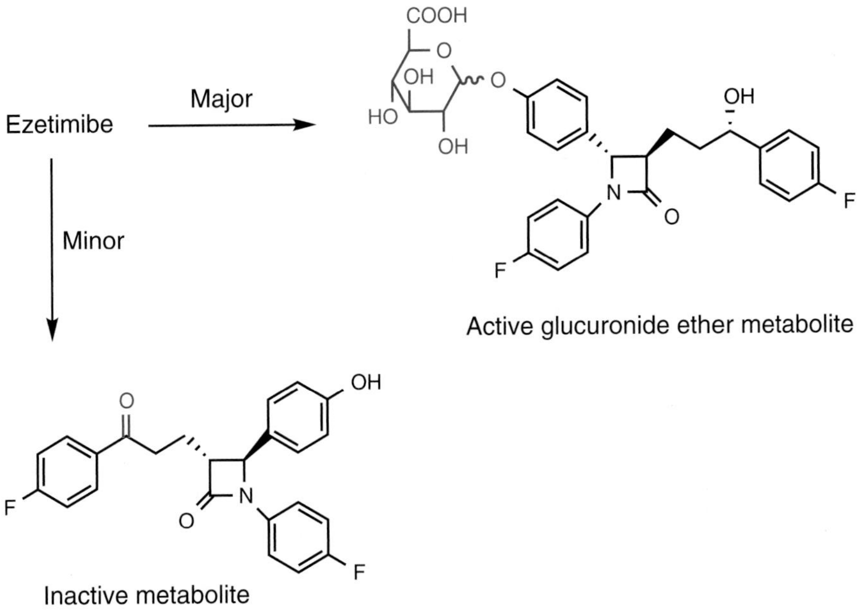

Figure 15.4 Ezetimibe metabolism.

- Approximately 60% of an administered dose is absorbed.
- Ezetimibe and its active metabolite are highly protein bound and have half-lives approaching 22 hours.

Metabolism

- The glucuronide metabolite of the *p*-phenol is active (Fig. 15.4).
  - The metabolite is formed in both gut and liver.
  - The fraction generated in hepatocytes is actively returned to the intestine via the bile.
  - Approximately 9% of the dose is excreted in urine as the glucuronide conjugate.

Clinical Applications

- Ezetimibe is marketed alone and in combination with the statin prodrug simvastatin.
- The pure compound is generally well tolerated, although it is not advised for use in patients with moderate to severe hepatic dysfunction.

> **DRUG–DRUG INTERACTIONS**
>
> **Ezetimibe**
> - BAS (ezetimibe sequestration. Space doses properly.)
> - Gemfibrozil (increased risk of cholelithiasis. Does not apply to fenofibrate.)
> - Aluminum- and magnesium-containing antacids (decreased ezetimibe plasma concentration)
> - Cimetidine (increased ezetimibe plasma concentration)
> - Cyclosporine (increases in maximum and total plasma concentrations of ezetimibe as high as 290% and 240%, respectively)

Phenoxyisobutyric Acids (Fibrates)

- The common name of fibrate comes from the chemical name of the active anionic form; <u>ph</u>en<u>o</u>xy<u>iso</u>butyrate (phibrate = fibrate) (Fig. 15.5).

MOA

- Fibrates activate peroxisome proliferator-activated receptor alpha (PPARα), a hepatic nuclear protein that regulates genes controlling fatty acid metabolism.

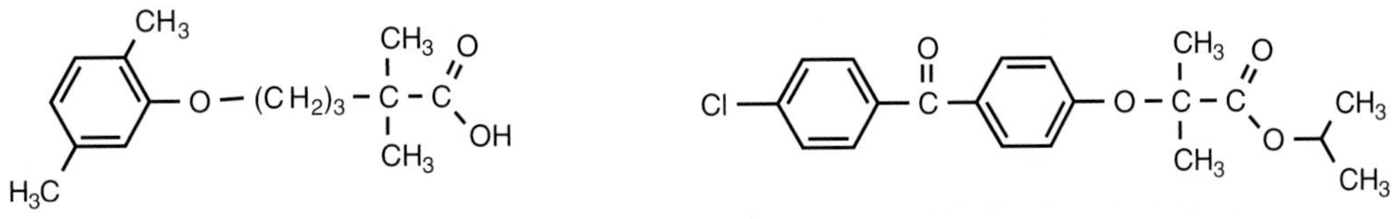

Gemfibrozil (Lopid)

Fenofibrate (Tricor and others)

Figure 15.5 Fibrate antihyperlipidemics.

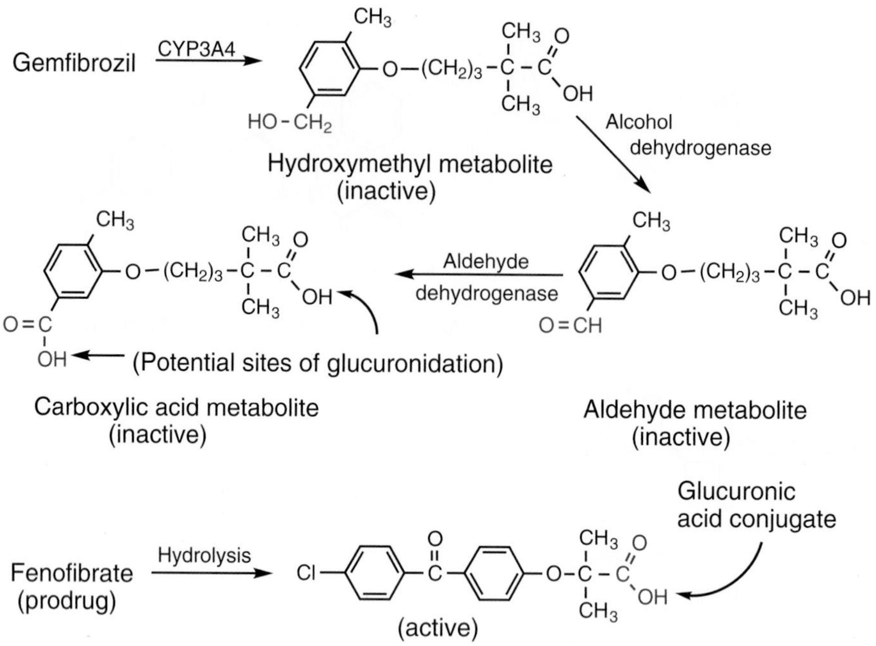

Figure 15.6 Fibrate metabolism.

- PPARα stimulation enhances lipoprotein lipase expression. This results in:
  - Triglyceride cleavage from VLDL, which facilitates receptor-mediated clearance.
  - FFA oxidation.
  - Inhibition of triglyceride synthesis.
- Fibrates decrease serum triglyceride and VLDL levels. HDL levels increase.
- Fibrates facilitate cholesterol removal from liver. Biliary cholesterol levels that exceed solubility can result.

Metabolism

- Gemfibrozil's aromatic methyl groups are rapidly oxidized by CYP3A4-catalyzed benzylic hydroxylation (Fig. 15.6). The primary alcohol metabolite is inactive.
  - The 1.5-hour elimination half-life requires twice-a-day dosing.
  - Cytosolic oxidation to the carboxylic acid is followed by conjugation to the glucuronide ester.
  - Interactions with CYP3A4 substrates, inducers, or inhibitors are possible, but toxicity is usually insignificant.
- Fenofibrate must hydrolyze to liberate the active fenofibric acid anion.
  - Steric hindrance from the branched alkyl moieties slows prodrug activation.
- Fenofibrate and fenofibric acid are resistant to oxidation. The 22-hour elimination half-life allows once-daily dosing.
- Both fibrates can be excreted as glucuronide esters of the isobutyric acid moiety.
- Gemfibrozil competes with all statins but fluvastatin for a specific glucuronidating isoform, increasing the risk of serious toxicity from both.

SAR

- The pharmacophore for fibrate antihyperlipidemics is phenoxyisobutyric acid. SAR is summarized in Figure 15.7.
- Fibrates anchor to PPARα through an ion–dipole bond with Tyr.[464]
  - Fibric acid pKa is approximately 3.5. The fibrate anion predominates at pH 7.4.
- Fibrate esters must hydrolyze to release the active anion.
- PPARα is flexible. A spacer of up to three carbons between isobutyrate and aryloxy groups is permitted.
  - Spacer groups augment molecular lipophilicity and promote gastrointestinal and hepatic membrane penetration.

Variable substitution pattern impacts lipophilicity (oral absorption, hepatic extraction) and metabolic vulnerability (duration of action, dosing regimen, risk of drug–drug interactions)

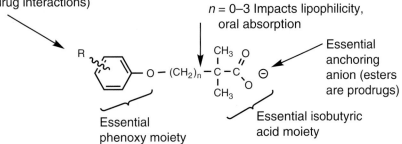

Figure 15.7 Fibrate pharmacophore and SAR.

Physicochemical and Pharmacokinetic Properties

- Fibrates are highly lipophilic (Table 15.2). They distribute passively out of the gut and into the liver.
- Although strongly anionic at pH 7.4, gemfibrozil is lipophilic due to the *n*-propyl spacer and two aromatic CH_3 groups.
- Despite the lack of a hydrocarbon spacer, fenofibrate has a higher log *P* than gemfibrozil due to:
  - The carbon-rich nonionizable isopropylcarboxylate ester.
  - The second phenyl ring and its *p*-chloro substituent.
- Gemfibrozil must be given with meals.
  - Administration 30 minutes prior to meals results in an elevated C_{max} and AUC.
- Fenofibrate's dependency on food is formulation specific. Absorption efficiency is increased if administered with meals.
- Fenofibrate's need for bioactivation delays peak serum concentration of active drug to 4 to 8 hours, compared to 1 to 2 hours for gemfibrozil.
- Fibrates are excreted predominantly by the kidney (60% to 90%) with 5% to 25% excreted in feces. Fecal excretion is more likely with fenofibrate.

Clinical Applications

- Fibrates are well tolerated and effective in lowering serum triglyceride and VLDL levels.
- Fibrates are ineffective in treating Fredrickson's type I hypertriglyceridemia (elevated chylomicron levels).
- With appropriate precautions/restrictions, fibrates can be used in combination with other antihyperlipidemics in complex dyslipidemias.
- Fibrates can sometimes induce liver function test abnormalities.
- Fibrates are contraindicated (gemfibrozil) or used with caution (fenofibrate) in severe renal dysfunction. Fenofibrate, but not gemfibrozil, is generally well tolerated in patients with mild to moderate impairment.
- If gemfibrozil–statin cotherapy is warranted, fluvastatin should be considered since it does not compete for the glucuronidating isoform used by gemfibrozil.

DRUG–DRUG INTERACTIONS

Fibrate
- Oral anticoagulants (increased International Normalized Ratio and risk of hemorrhage)
- Statins (primarily gemfibrozil: inhibition of hepatic OATP1B1, increased risk of myopathy)
- BAS (inhibition of fibrate intestinal absorption, if coadministered)
- Ezetimibe (gemfibrozil: increased risk of cholelithiasis)
- Antidiabetic agents (increased risk of hypoglycemia)

Fibrates and Vision Preservation

Dyslipidemia is a significant risk factor for diabetic retinopathy.
- Fenofibrate has shown value in preventing/halting progression of blinding macular edema in diabetic patients taking statins.
- Fenofibrate/simvastatin combination therapy has been shown to reduce diabetic retinopathy progression by 40% compared to simvastatin monotherapy.
- Patients on fenofibrate have a lower risk for retinopathy-related laser treatment and nontraumatic limb amputation compared to placebo.

Table 15.2   Pharmacokinetic Parameters of Fibrates

Drug	Calculated Log P	Oral Bioavailability (%)	Active Metabolite	Protein Binding (%)	Time to Peak Concentration (h)	Elimination Half-Life (h)	Major Route of Elimination
Fenofibrate	5.24	60–90	Fenofibric acid	99	4–8	20–22	Renal (60–90%) Fecal (5–25%)
Gemfibrozil	3.9	>90	None	99	1–2	1.5	Renal (70%) Fecal (6%)

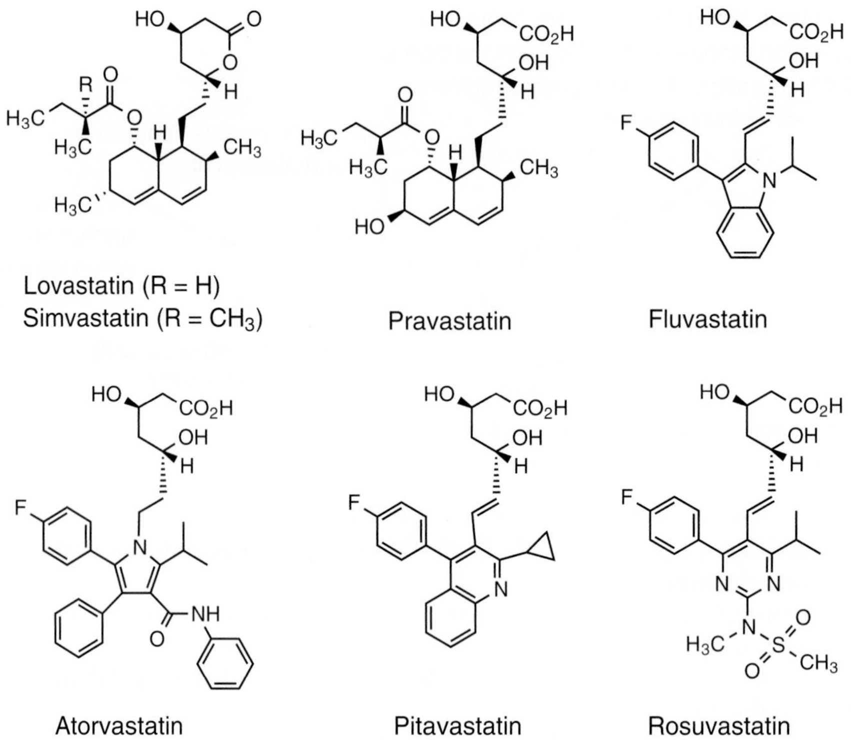

Lovastatin (R = H)
Simvastatin (R = CH₃)

Pravastatin

Fluvastatin

Atorvastatin

Pitavastatin

Rosuvastatin

Figure 15.8 Statin antihyperlipidemics.

HMG-CoA Reductase (HMGR) Inhibitors

- HMGR inhibitors are commonly known as statins (Fig. 15.8).
- Five active and two prodrug statins are currently marketed to lower LDL and total serum cholesterol levels.

MOA

- Statins are competitive inhibitors of HMGR, the rate-limiting enzyme in cholesterol biosynthesis.
- Statins successfully compete with the endogenous 3-hydroxy-3-methylglutaryl-CoA (HMG-CoA) substrate for access to the HMGR active site.
- The dihydroxyheptan(en)oic acid segment of the statins mimics the chemistry of HMG and binds in a similar fashion to HMGR.
  - An ion–ion bond between anionic statins and cationic Lys[735] is the anchoring interaction.
- The remainder of the statin structure (ring component) binds to hydrophobic and polar residues in a flexible receptor pocket (Fig. 15.9).
  - Affinity is enhanced 1000- to 10,000-fold compared to HMG-CoA, ensuring effective enzyme inhibition.
- LDL reuptake receptor expression is augmented, leading to increased LDL clearance.
- The two most potent statins (rosuvastatin, atorvastatin) also lower serum triglycerides.

SAR

Dihydroxyheptanoic acid moiety

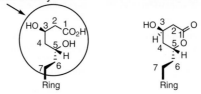

Statin (active intact)   Statin (lactone prodrug)

Statin pharmacophores

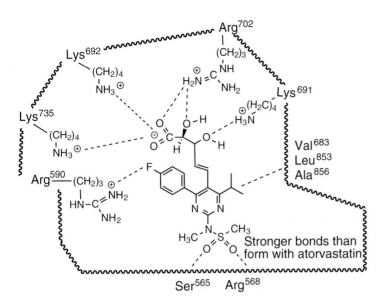

Figure 15.9 Rosuvastatin–HMGR binding.

- Statins must be anionic to anchor to HMGR Lys[735]. The dihydroxyheptan(en)oic acid segment is essential.
- Hydroxyls at chiral C3 and C5 have important interactions at HMGR and must have the proper absolute configuration.
  - C3 requires the *R* configuration.
  - Optimal configuration at C5 depends on C6–C7 saturation status. Dihydroxyheptanoic acid statins have *5R* stereochemistry and dihydroxyheptenoic acids have the *5S* configuration.
- The ring component of statins is of two general types:
  - Naturally occurring statins have a 2′,6′-dimethylhexahydronaphthylene ring system substituted with a methylbutyrate ester at C8′.
    - Addition of an α-CH₃ to the methylbutyrate group (lovastatin vs. simvastatin) increases activity two-fold.
  - Synthetic statins have heteroaromatic ring systems. Isopropyl (or cyclopropyl) and *p*-fluorophenyl substituents contribute to receptor affinity.
- Statins with polar functional groups positioned to bind to Arg[568] and Ser[565] (rosuvastatin, atorvastatin) show significant increases in affinity and potency. Statins with polar functional groups forced to interact with lipophilic HMGR residues (pravastatin) show decreased affinity and potency.

Metabolism

- Prodrug statins must hydrolyze to generate the essential carboxylate anion (Fig. 15.10).
- CYP3A4 and 2C9 metabolize some statins.
  - With one exception (pravastatin), saturated dihydroxyheptanoic acid statins are metabolized by CYP3A4.
  - Unsaturated dihydroxyheptenoic acids are metabolized by CYP2C9.
  - Pravastatin is CYP resistant. It is oxidized/decomposed in gut and liver to a less active major metabolite (Fig. 15.11).

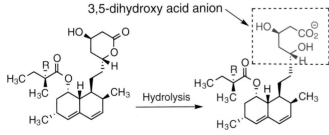

Lovastatin (R = H)
Simvastatin (R = CH₃)

Figure 15.10 Activation of statin prodrugs.

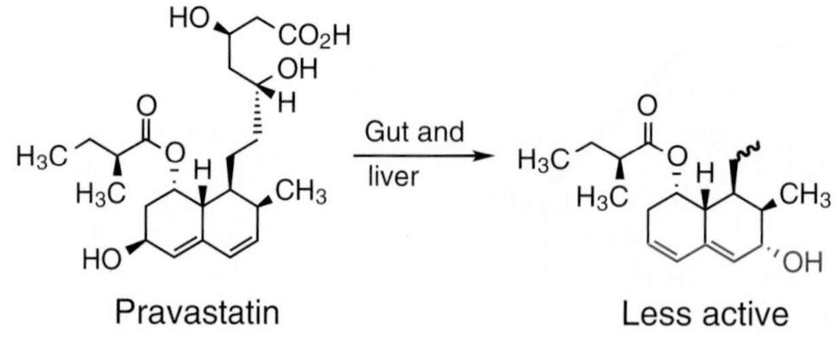

Figure 15.11 Pravastatin metabolism and acid-catalyzed degradation.

- CYP3A4: (Fig. 15.12).
  - Catalyzes inactivating alicyclic hydroxylation and ω-1 hydroxylation of lovastatin and simvastatin acids.
  - Catalyzes the aromatic hydroxylation of atorvastatin.
    - These two metabolites are equally active with the parent drug.
- CYP2C9: (Fig. 15.13).
  - Catalyzes the N-demethylation of rosuvastatin.
    - The metabolite retains activity but is formed in low amounts (5% to 10% of the dose).
  - Catalyzes aromatic hydroxylation and N-dealkylation of fluvastatin.
  - Catalyzes the inactivating aromatic hydroxylation of pitavastatin.
    - Pitavastatin toxicity can increase if coadministered with CYP3A4 inhibitors.
    - Competition for organic anion transporting protein (OATP) 1B1 may be responsible.
- All statins can be glucuronidated at either the original or the hydrolysis generated COOH prior to biliary (predominant) or urinary excretion.
  - Pitavastatin glucuronide undergoes a reversible lactonization prior to excretion. The lactone can hydrolyze to regenerate active pitavastatin.
  - Reversible lactonization has been demonstrated with atorvastatin and simvastatin, and is presumed to occur to some extent with all statins.

Physicochemical and Pharmacokinetic Properties
- Selected physicochemical and pharmacokinetic characteristics of marketed statins are provided in Table 15.3.

Lipophilic/Hydrophilic Classification
- Statins are classified as lipophilic or hydrophilic based on the chemistry of the ring component of the structure.

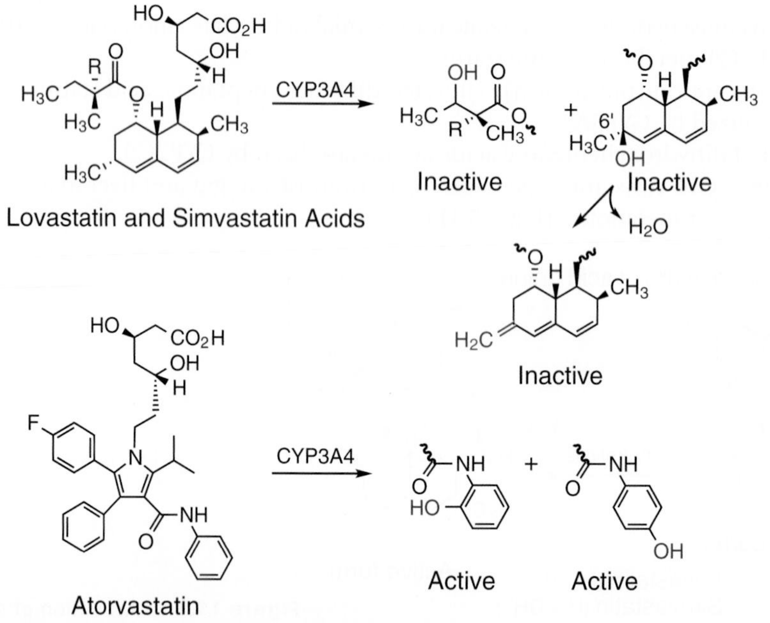

Figure 15.12 CYP3A4-catalyzed statin metabolism.

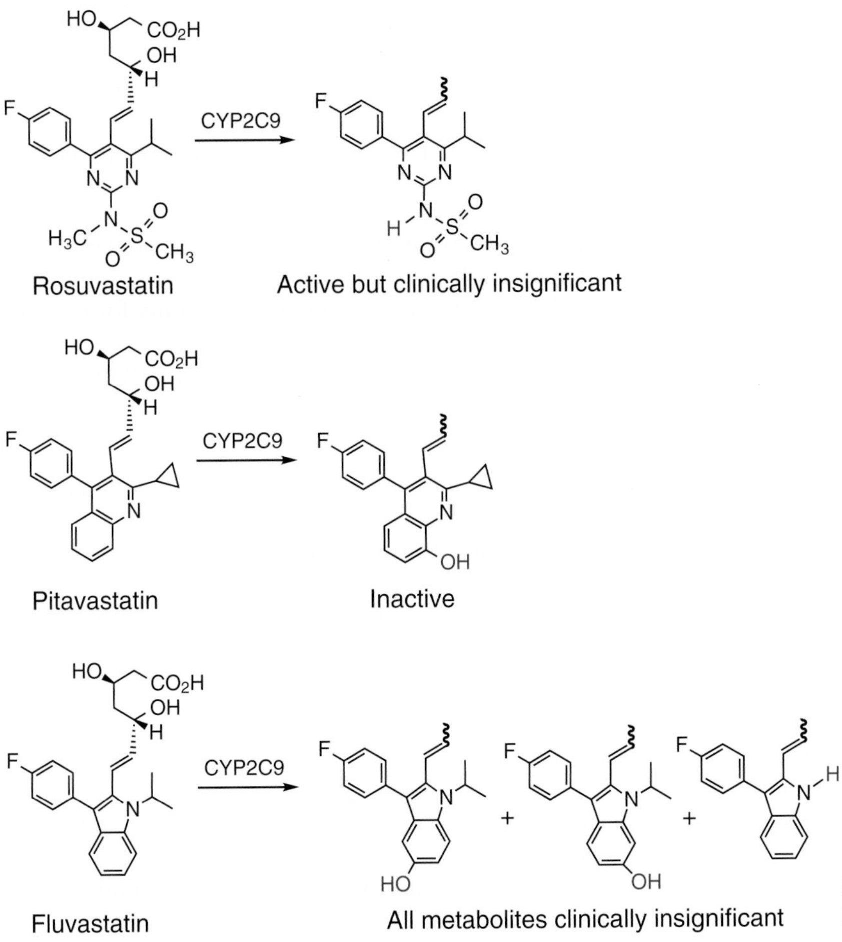

Figure 15.13 CYP2C9-catalyzed statin metabolism.

- Lipophilic statins (log $P > 3$) include fluvastatin, pitavastatin, atorvastatin, lovastatin, and simvastatin. Carbon-rich ring systems override the impact of any polar substituent.
- Hydrophilic statins (log $P < 1.5$) include pravastatin and rosuvastatin. They are carbon-poor and contain a polar functional group.
- Lipophilic, but not hydrophilic, statins are absorbed across gastrointestinal, hepatic, and muscle cell membranes primarily by passive diffusion.

Table 15.3 Pharmacokinetic Parameters of HMG–CoA Reductase Inhibitors

Drug	Calculated Log P[a]	Oral Bioavailability(%)	Active Metabolite(s)	Protein Binding (%)	Time to Peak Concentration (h)	Elimination Half-Life (h)	Major Route(s) of Elimination
Atorvastatin	4.13	12–14	Ortho- and para-hydroxylated	98	1–2	14–19	Biliary/fecal (>90) Renal (<2%)
Fluvastatin	3.62	20–30	None	98	0.5–1	1	Biliary/fecal (95%) Renal (5%)
Lovastatin	4.07 (4.04)	5	3,5-Dihydroxy acid	>95	2	3–4	Fecal (83%) Renal (10%)
Pravastatin	1.44 (0.5)	17	None	43–55	1–1.5	2–3	Fecal (70%) Renal (20%)
Pitavastatin	3.45	51	None	>99	1	12	Fecal (79%) Renal (15%)
Rosuvastatin	0.42	20	N-desmethyl	88	3–5	19–20	Fecal (90%) Renal (10%)
Simvastatin	4.42 (4.2)	5	3,5-Dihydroxy acid	95	4	3	Fecal (60%) Renal (13%)

[a]Calculated using CLOG program.

⚠ **ADVERSE EFFECTS**

Statins
- Myotoxicity (myalgia, myopathy, rhabdomyolysis)
- Hepatotoxicity (possible elevations in transaminase and creatinine phosphokinase)
- GI distress (usually mild)

SLCO1B1 Polymorphism and Statin Toxicity

A genetic variation in the SLCO1B1 gene that encodes the plasma membrane localization of the hepatic OATP1B1 transporter has been associated with an increased risk of statin-related myotoxicity. The SLCO1B1*5 allele allows higher serum concentrations of active statin because fewer hepatic transporters are available to ferry them into liver. Patients are at highest risk if administered high doses of a statin that can be actively transported into muscle by OATP2B1.

Rosuvastatin chelation

OATP Affinity

- Prodrug statins are unionized and have no affinity for OATP.
- Statins that are anionic at intestinal pH can be actively transported across gastrointestinal membranes by OATP2B1.
- Hydrophilic statins are actively transported across hepatic membranes predominantly by OATP1B1.
  - Lipophilic statins utilize this carrier system to some extent.
- OATP2B1 on muscle cell membranes actively transport some anionic statins into myocytes. Muscle toxicity can result.
  - Atorvastatin, fluvastatin, and rosuvastatin have the highest affinity for muscle cell OATP2B1.
  - Pitavastatin and pravastatin bind to OATP2B1 in intestine (pH 6.0), but not on muscle (pH 7.4).

Hepatoselectivity

- Hydrophilic statins (pravastatin, rosuvastatin) are considered among the most hepatoselective.
  - Hydrophilic statins enter hepatocytes almost exclusively by one-way OATP1B1 carrier-mediated transport.
  - Hydrophilic statins cannot exit hepatocytes by passive diffusion and are trapped at the site of action.
- Fluvastatin is also highly hepatoselective.

Clinical Applications

- Liver enzymes must be checked at 6 and 12 weeks after beginning therapy, and every 6 months thereafter.
- Unexplained muscle tenderness should be reported to the pharmacist and/or physician.
- Statins with short half-lives (fluvastatin, lovastatin, simvastatin) must be taken at bedtime to be active at early morning peak cholesterol biosynthesis times.
- Statins with active metabolites (atorvastatin) or which are resistant to CYP-mediated inactivation (rosuvastatin, pitavastatin, pravastatin) can be taken at any time of day.
- Lovastatin is taken with food. All other statins (including simvastatin) are food-independent.
- Fluvastatin serum levels increase dramatically if alcohol is consumed within 2 hours of dosing.
- Pravastatin bioavailability increases in the presence of antacids, H_2 antagonists, or proton pump inhibitors due to inhibition of acid-catalyzed degradation.
- Rosuvastatin bioavailability decreases in the presence of antacids due to chelation of divalent and trivalent metal ions.

Orphan Drugs for the Treatment of Homozygous Familial Hypercholesterolemia

- HoFH is characterized by high levels of LDL secondary to a genetic mutation that results in a complete lack of LDL receptors.
  - Serum LDL levels soar to between 250 and 500 mg/dL.
  - Traditional LDL-lowering agents (e.g., statins) are ineffective.

Individual Drugs

Lomitapide

Lomitapide (Juxtapid)

- Lomitapide inhibits microsomal triglyceride transfer protein (MTTP) which catalyzes the transfer of triglycerides to apolipoprotein B (apoB). Production of hepatic VLDL, LDL, and chylomicrons is halted.
- Lomitapide is given orally and can induce gastrointestinal distress and hepatic steatosis.
- Lomitapide is inactivated by CYP3A4, and the parent drug and metabolites are excreted in feces and urine.
- Lomitapide is contraindicated in patients with moderate to severe liver disease or who are taking other drugs that are potent 3A4 inhibitors.

Mipomersen Sodium

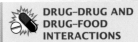

DRUG–DRUG AND DRUG–FOOD INTERACTIONS

Statins

- CYP3A4 inhibitors or substrates (e.g., protease inhibitors, calcium channel blockers, azole antifungal agents, amiodarone, grapefruit juice: increased myotoxicity risk for CYP3A4 vulnerable statins)
- CYP2C9 inhibitors or substrates (e.g., warfarin, phenytoin, diclofenac, rifampin: increased toxicity risk for CYP2C9 vulnerable statins)
- Gemfibrozil (all but fluvastatin: increased risk of myotoxicity from both drugs)
- Alcohol (increased risk of hepatotoxicity)
- Antacids (increased therapeutic effectiveness of pravastatin, decreased therapeutic effectiveness of rosuvastatin)
- Pectin/Oatbran (decreased intestinal absorption of statins)

Pleiotropic Statins

Statins are pleiotropic drugs that have multiple beneficial effects. Statins decrease levels of C-reactive protein (CRP), a major biomarker of inflammation, and they have shown promise in the treatment of inflammation-based diseases, including CAD and vascular dysfunction secondary to diabetic insulin resistance. Lipophilic statins may have a positive impact on bone density, and statin–NSAID cotherapy has demonstrated a chemoprotective effect for some (but not all) cancers.

Mipomersen sodium

- Mipomersen is an antisense inhibitor of apoB synthesis. It binds to the coding region of apoB/mRNA and promotes its degradation.
  - Circulating apoB levels are cut in half, and VLDL and LDL concentrations drop significantly.
- As an oligonucleotide, mipomersen must be given by subcutaneous or intravenous injection.
- The 2′-O-2-methoxy ethyl moieties inhibit nuclease action and prolong the duration of action.
- The risk of serious hepatotoxicity mandates that both patients and providers document practices that promote safe drug use.

Chemical Significance

CAD impacts over 15 million people in the United States alone, and many of these individuals are being treated with antihyperlipidemic medications. The therapeutic guidelines released in November 2013 by the American College of Cardiology and the American Heart Association have the potential to increase the number of patients taking cholesterol-lowering drugs (particularly statins), and this has significant implications for pharmacists. In addition, advertisements for drugs that lower serum cholesterol or triglycerides appear regularly on television, and patients are coming to their health care providers with more questions about these therapies than ever before. As the most accessible health care provider, pharmacists will be fielding many of these questions, and they are in the best position to advise both consumers and their health care team colleagues on the risk:benefit ratio of specific drugs for individual patients.

While the structures of the drugs in the various antihyperlipidemic classes appear very similar to the untrained eye, pharmacists' knowledge of medicinal chemistry allows them to recognize both therapeutic and adverse actions that are unique to select subsets of these agents. For example:

- Does the statin have a double bond in the dihydroxyheptan(en)oic acid portion of the structure? Then it can be taken safely with grapefruit juice and drugs that inhibit or are metabolized by CYP3A4.
- Is there a sulfonamide present on the ring component of the statin structure? Then it will chelate metal ions and shouldn't be taken with antacids.
- Does the fibrate contain an aromatic methyl group? Then it will need to be taken twice a day, rather than once daily.
- Does the fibrate have an unesterified isobutyric acid moiety? Then you cannot take it with a statin without increasing the risk of serious muscle toxicity.
- Does the BAS contain only quaternary amines? Then it will have a higher capacity for binding intestinal bile acids than those that have only secondary and/or tertiary amines.
- Is niacin causing intolerable itching or distressful headaches? The pharmacist can explain to the patient why premedicating with aspirin (but not Tylenol) can help.

The time and effort it takes to learn to interpret antihyperlipidemic structures and uncover the wealth of clinically relevant information embedded therein distinguishes the pharmacist as the true drug expert of the health care team.

Review Questions

1 Which BAS exhibits pH dependent action?

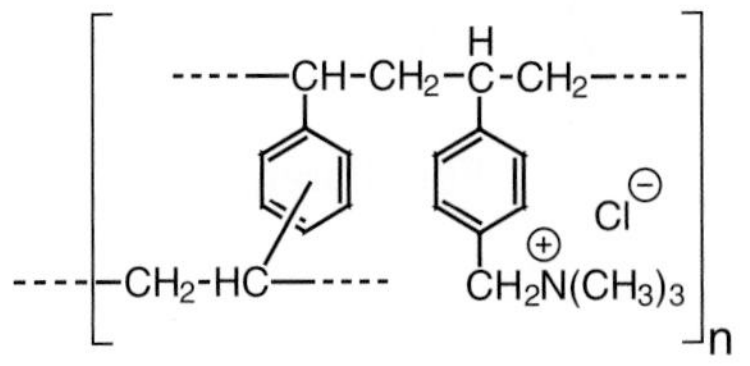

Cholestyramine

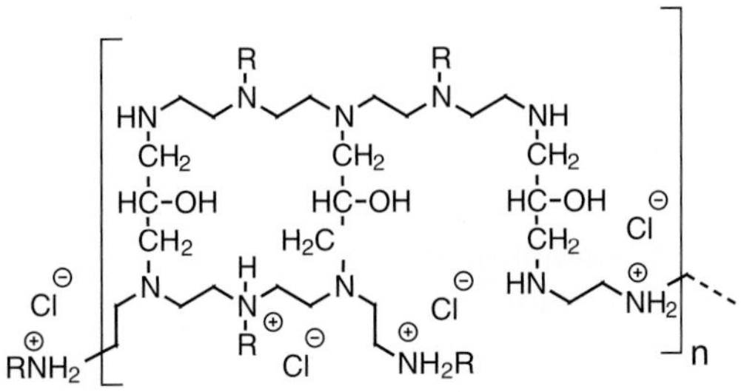

Colestipol

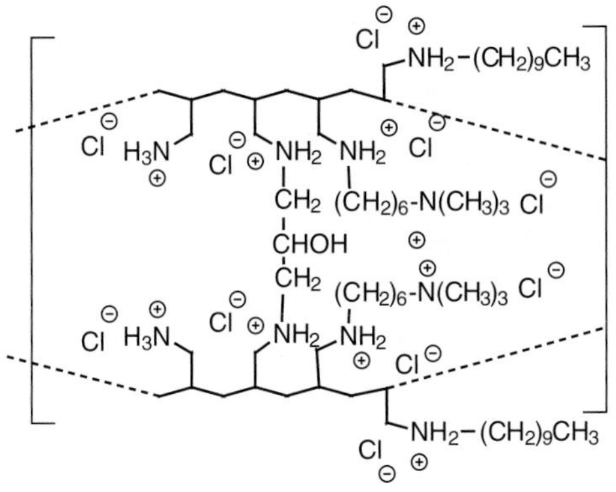

Colesevelam

A. Cholestyramine
B. Colestipol
C. Colesevelam

2 What OTC product can be used to avert the facial flushing and pruritus associated with niacin use?
A. Chlorpheniramine (an antihistamine)
B. Benzocaine (a local anesthetic)
C. Pseudoephedrine (a vasoconstrictor)
D. Aspirin (a nonsteroidal anti-inflammatory agent)
E. Omeprazole (a proton pump inhibitor)

3 Which ezetimibe structural component helps prolong duration of action?

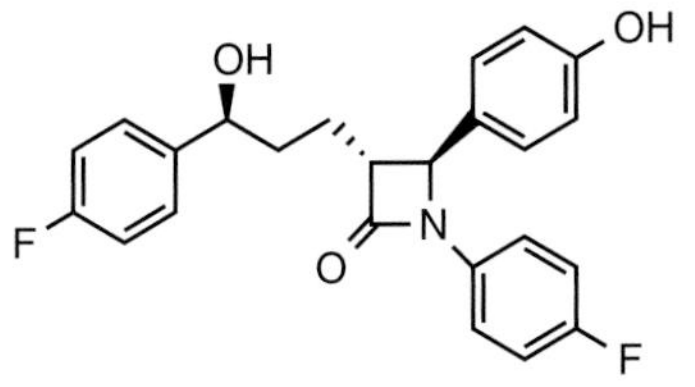

Ezetimibe

A. The *p*-fluorine atoms
B. The phenolic OH
C. The alcoholic OH
D. The β-lactam ring
E. The propyl chain connecting the β-lactam and one *p*-fluorophenyl moiety

4 Which fibrate antihyperlipidemic has a duration of action long enough to be administered once daily?

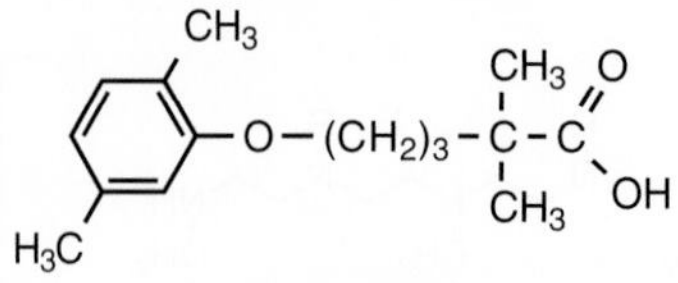

Gemfibrozil

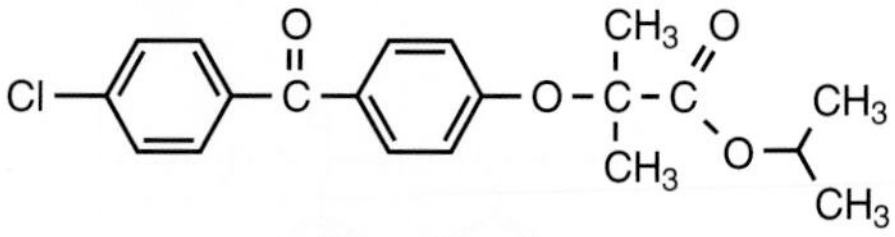

Fenofibrate

A. Gemfibrozil
B. Fenofibrate
C. Both fibrates
D. Neither fibrate

5 Which statin shows diminished bioavailability in patients taking antacids?

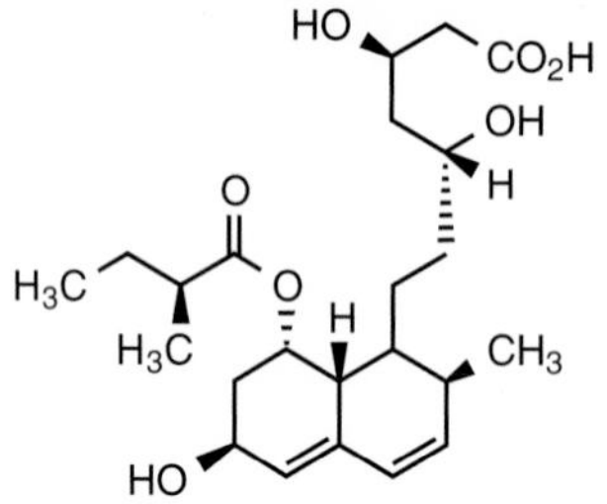

1

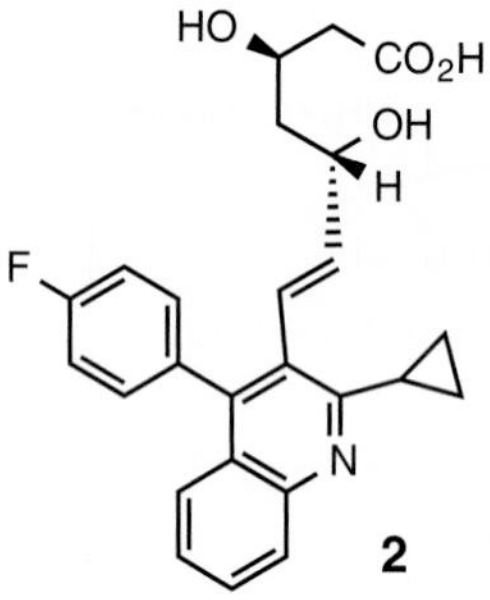

2

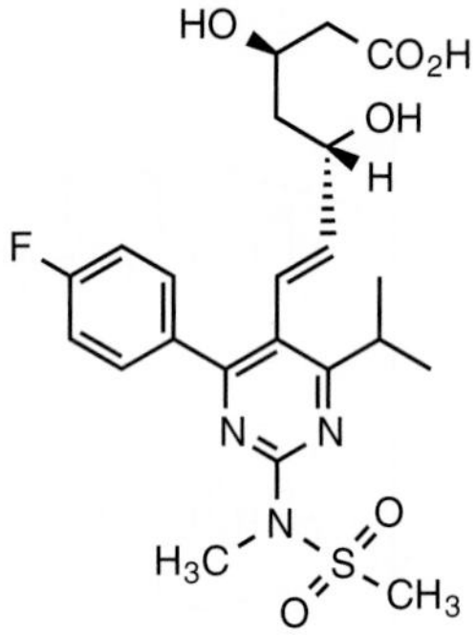

3

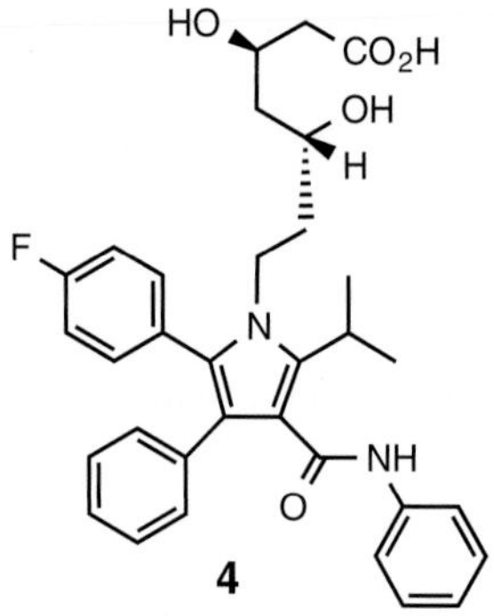

4

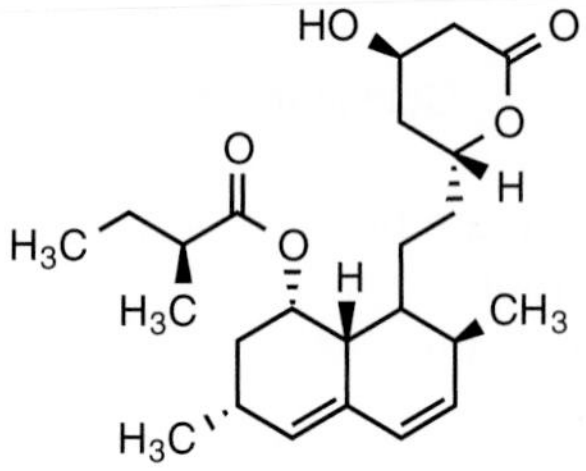

5

A. Statin **1**
B. Statin **2**
C. Statin **3**
D. Statin **4**
E. Statin **5**

Anticoagulant, Antiplatelet, and Fibrinolysis Agents

16

Learning Objectives

Upon completion of this chapter the student should be able to:

1 Identify the sites in the coagulation cascade where anticoagulants and thrombolytics produce their effects.

2 Identify the sites on or within the platelet where antiplatelet drugs produce their effect.

3 Describe the mechanisms of action (MOA) of the anticoagulants, thrombolytics, and antiplatelet drugs which include:
   a. coumarin-derived anticoagulants
   b. heparin-based anticoagulants
   c. direct thrombin inhibitors (DTIs)
   d. direct factor Xa (FXa) inhibitors
   e. Antiplatelet drugs: COX-1, phosphodiesterase, and P2Y purinergic receptor inhibitors, as well as protease-activated receptor-1 (PAR-1) and glycoprotein (GP) IIb/IIIa receptor antagonists
   f. thrombolytic drugs

4 List structure activity relationships (SARs) of key drug classes where presented.

5 Identify clinically significant physiochemical and pharmacokinetic properties.

6 Discuss, with chemical structures, the clinically significant metabolism of discussed drugs and how the metabolism affects their activity.

7 Identify the site and mechanism of action of the coagulants.

8 Discuss the clinical applications of the various classes of anticoagulants used in the treatment of diseases associated with coagulation conditions.

Introduction

There is a fine line between free-flowing blood and the development of a clot due to vascular injury. The balance between blood coagulation and dissolving of the clot is controlled by a complex physiological system. Imbalance in this system can lead to hemorrhage or thrombosis either of which can result in death. During specific disease states therapeutic agents may become necessary to insure a state of homeostasis through inhibition of clotting (antithrombotic, anticoagulant, antiplatelet, or fibrinolytic drugs), or dissolving of clots (thrombolytic drugs). This chapter will deal with drugs in these various groups as well as a short discussion of coagulants.

The major components of a blood clot consist of fibrin, aggregated platelets, nonnucleated cell fragments, and trapped red blood cells. The diseases associated with clots often involve major vasculature, heart, brain, and lungs, and consist of venous thromboembolism (VTE), myocardial infarction (MI), valvular heart disease, unstable angina, atrial fibrillation (AF), and pulmonary embolism (PE). In fact, in Western society, thrombotic conditions are a major cause of morbidity and mortality leading to heart attacks and strokes.

Biochemical Mechanism of Blood Coagulation

- Tissue factor (TF) is a globular GP, which initiates the coagulation process within the extrinsic pathway as shown in Figure 16.1.
- Other coagulation factors are serine proteases which exist in inactive forms, indicated by Roman numerals, and the active form, indicated by Roman numerals with a lower case *a*.
- TF is expressed on the surface of macrophages and is associated with thrombogenesis in the arteries.
- TF binds to factor VII to form the TF-VIIa complex, which results in activation of factors IX (arising from the intrinsic pathway) and X leading to IXa and Xa.
- Factor V/Va is a key protein involved in blood coagulation and serves as a cofactor in the prothrombinase complex. This complex consists of factors Xa, Va, prothrombin, and free Ca^{2+} ions on a phospholipid surface. The net effect of the prothrombinase complex is the conversion of prothrombin (factor II, FII) to thrombin (factor IIa, FIIa) (Fig. 16.1).
- Ca^{2+} is an essential component of many of these activation steps (Fig. 16.1).
- Factor Xa catalyzes the conversion of prothrombin to thrombin, a reaction in which Ca^{2+} is chelated to γ-carboxyglutamate. The role of chelation involves

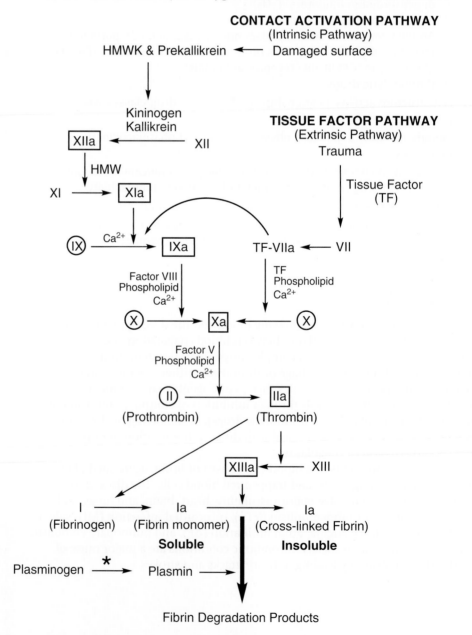

Figure 16.1 The coagulation cascade. *Circled factors* are those inhibited by warfarin-like drugs while *boxed factors* are affected by heparin. The *star* indicates the site of action of thrombolytic drugs such as streptokinase and urokinase.

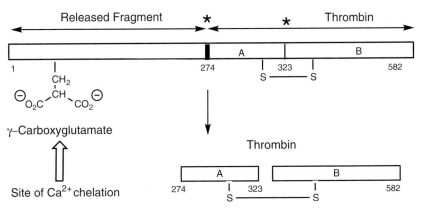

Figure 16.2 Structure of prothrombin. Thrombin is liberated through the cleavage of the Arg^{274}-Thr^{275} and Arg^{323}-Ile^{324} peptide bonds (indicated with *stars*). The γ-carboxyglutamate residues are in the released N-terminal portion of prothrombin and are not part of thrombin. The A and B chains of thrombin are joined by a disulfide bond.

Measure of Extrinsic Pathway Coagulation Time

Prothrombin time (PT) and international normalized ratio (INR) are measures of the extrinsic pathway of coagulation and are used in combination with anticoagulation drugs. PT measures factors I, V, VII, and X. PT is usually 12 to 13 seconds while INR is 0.8 to 1.2.

Measure of Intrinsic Pathway Coagulation Time

Activated partial thromboplastin time (aPTT) is used in conjunction with heparin therapy which deactivates factors II and X. aPTT is usually in the range of 23 to 35 seconds.

conformational changes in the affected protein promoting binding to a phospholipid membrane of platelets.

- Cleavage occurs at two locations in prothrombin resulting in thrombin fragments held together by a disulfide bond (Fig. 16.2).
- Thrombin's function is twofold: (1) catalyze the conversion of fibrinogen to soluble fibrin (fibrin monomer) and (2) catalyze the conversion of factor XIII to XIIIa (a transamidase) which converts soluble fibrin to cross-linked insoluble fibrin (Fig. 16.3).
- The cross-linking involves amide formation (isopeptide bond) between a side chain amine of a lysine and a side chain carboxyl of a glutamine from separate soluble fibrins.

Anticoagulants

Coumarin Derivatives

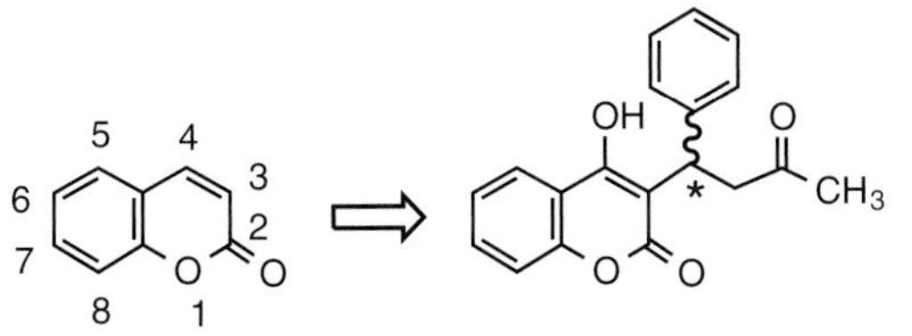

Coumarin

Warfarin (Coumadin, Jantoven, generic)

MOA

- Warfarin blocks the interconversion of vitamin K and vitamin K 2,3-epoxide (Fig. 16.4), specifically preventing the reduction of vitamin K 2,3-epoxide to vitamin K quinone, the rate limiting step in the recycling of vitamin K.

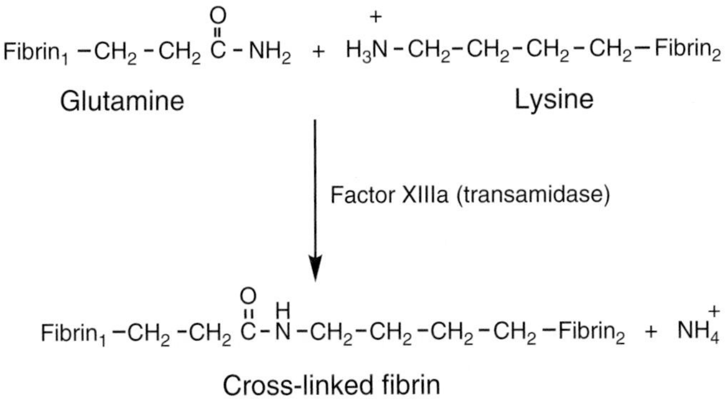

Fibrin$_1$ −CH$_2$ - CH$_2$ $\overset{\overset{O}{\|}}{C}$ - NH$_2$ + H$_3$N$^+$ − CH$_2$−CH$_2$−CH$_2$− CH$_2$−Fibrin$_2$

Glutamine   Lysine

Factor XIIIa (transamidase)

Fibrin$_1$ −CH$_2$ -CH$_2$ $\overset{\overset{O}{\|}}{C}$-$\overset{\overset{H}{|}}{N}$−CH$_2$−CH$_2$−CH$_2$− CH$_2$−Fibrin$_2$ + NH$_4^+$

Cross-linked fibrin

Figure 16.3 Cross-linking of soluble fibrin monomers (factor Ia) through the activity of factor XIIIa, a transamidase enzyme.

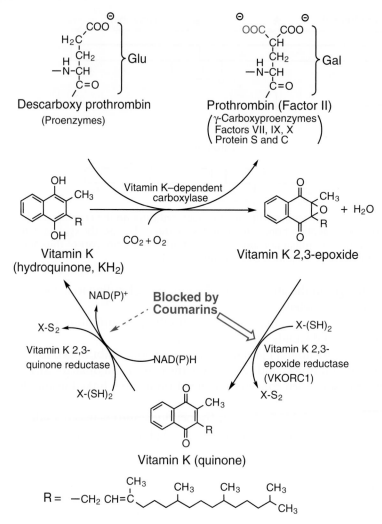

Figure 16.4 Redox cycling of vitamin K in the activation of blood clotting which involves conversion of glutamate residues to γ-carboxyglutamates.

- Vitamin K is essential for the posttranslation carboxylation of glutamic acid to γ-carboxyglutamic acid.
- γ-carboxyglutamic acid is needed for the activation factors II, VII, IX, and X (Fig. 16.2).
- The onset of action of warfarin is delayed (3 to 5 days) until clotting factors already present are cleared.

SAR

- The structural features important for activity of the coumarin's are shown in Figure 16.5.

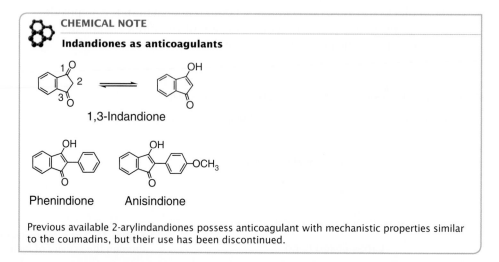

CHEMICAL NOTE

Indandiones as anticoagulants

1,3-Indandione

Phenindione   Anisindione

Previous available 2-arylindandiones possess anticoagulant with mechanistic properties similar to the coumadins, but their use has been discontinued.

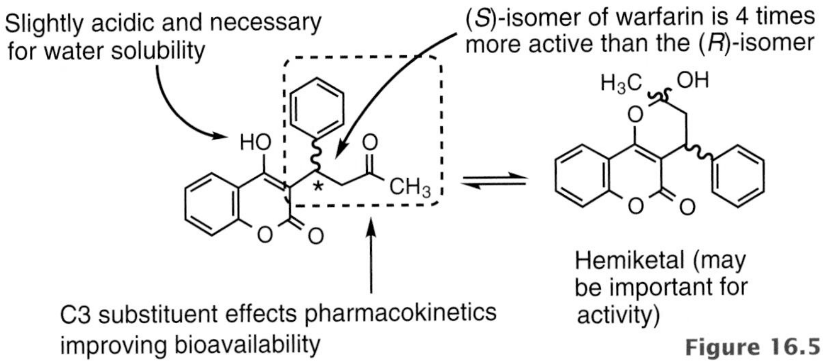

Slightly acidic and necessary for water solubility

(S)-isomer of warfarin is 4 times more active than the (R)-isomer

C3 substituent effects pharmacokinetics improving bioavailability

Hemiketal (may be important for activity)

Figure 16.5 SAR features for coumarin activity.

Physicochemical and Pharmacokinetic Properties

- Warfarin, a coumarin, is slightly acidic forming a water-soluble sodium salt at C4 enol.
- The product is used as a racemic mixture although the S-isomer is the more active form (4X R).
- Warfarin is nearly 100% absorbed via the oral route and bound to plasma protein (95% to 99%).
- Drug–drug interaction results from displacement of plasma protein–bound warfarin as well as via enhancement or inhibition of CYP2C9 metabolism of warfarin (see Table 26.2 in *Foye's Principles of Medicinal Chemistry, Seventh Edition*, for an extensive list of drugs affecting warfarin therapy).
- Unbound drug is responsible for biologic activity.
- Relatively long half-life (warfarin 0.62 to 2.5 days).
- Available in tablet and injectable dosage forms.

Metabolism

- Warfarin is extensively (100%) metabolized as shown in Figure 16.6.
- S-isomer is primarily metabolized via CYP2C9 to inactive products.
- R-isomer is primarily metabolized via CYP3A4, 1A2, 2C19.

> **ADVERSE EFFECTS**
>
> **Warfarin:** The major adverse effect is bleeding which can lead to death. It is important to monitor PT. PT is a measure of blood coagulation time and the activity of vitamin K–dependent clotting factors (II, VII, IX, and X). Normally, PT is 10 to 13 seconds. Clotting time may also be reported in INRs. For warfarin, the INR is 2 to 3. In patients with prosthetic heart valves the INR is 2.5 to 3.5. Various foods containing vitamin K can reduce warfarin's effectiveness if consumed in large quantities (e.g., kale, spinach, Brussels sprouts, collard greens, green tea, parsley). Caution should be taken with certain drinks which could increase bleeding (i.e., cranberry juice, alcohol).

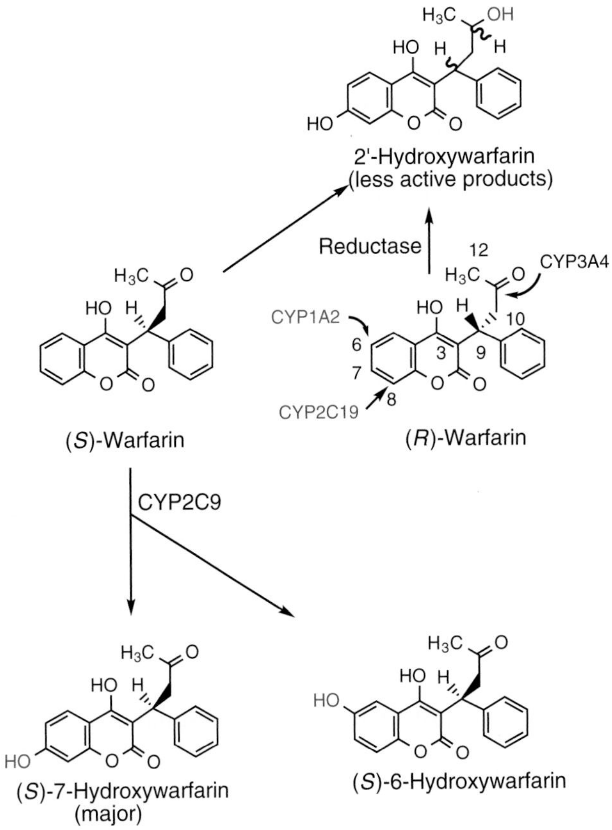

Figure 16.6 Metabolism of warfarin.

Heparin-Based Anticoagulants

- Heparin is a natural-occurring polysulfated polysaccharide found in the vascular endothelium.
- Various forms of heparin-based products exist including unfractionated heparin (UFH), low molecular weight heparin (LMWH), and synthetic pentasaccharide fondaparinux.
- The heparin-based drugs exist as polysulfate ions at physiological pH and may appear as various salts (e.g., Na^+, Ca^{2+}, Li^+).

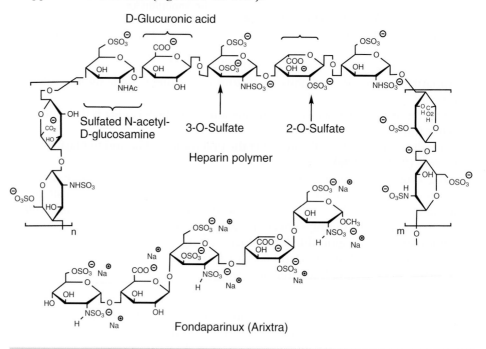

MOA

- Heparin-based products cause a conformational change in antithrombin III (AT or ATIII) peptide through ion–ion bond (COO^- and SO_3^- of heparin to the NH_3^+ of lysine and arginine in AT) (Fig. 16.7).
- The conformationally modified AT exhibits an accelerated binding to thrombin and factor Xa. The AT~heparin complex binds to two sites within thrombin in a 1:1 ratio.
- Heparin is then released from the AT~thrombin/AT~factor Xa complex leaving an inactive form of thrombin and/or factor Xa (heparin is a catalyst).
- LMWHs exhibit selectivity toward factor Xa with less binding to thrombin while fondaparinux does not bind to thrombin but only inactivates Xa (Fig. 16.7).

Physicochemical and Pharmacokinetic Properties

- The heparin-based anticoagulants for human use are all water-soluble sodium salts administered parenterally (IV, SC).
- These drugs are unstable if administered orally.
- The three heparin-based anticoagulants are shown in Table 16.1.
- UFH has poor bioavailability (SC) and low selectivity—dose is monitored via aPTT.
- LMWH demonstrates high bioavailability (~90%) and high selectivity.
- Fondaparinux has high bioavailability (100% SC) and binds only to Xa.
- Monitoring of patients on UFH is essential due to potential for bleeding. Monitoring is less important for patients on LWMH or fondaparinux.

Metabolism

- UFH and LMWH are metabolized via depolymerization and desulfation in the liver with renal excretion.
- Fondaparinux is excreted unchanged in the urine.

ADVERSE EFFECTS

Heparin: Bleeding is the most common adverse effect with heparin. This may be associated with binding of heparin to platelet factor 4 (PF4) which can lead to thrombocytopenia (HIT). Since heparin deactivates factors II and X, monitoring measures activated partial thromboplastin time (aPTT). aPTT assay uses a surface activator (i.e., elegiac acid, kaolin, silica) to activate the intrinsic pathway. When the activator comes in contact with citrated plasma, Ca^{2+}, and phospholipid clot formation begins. Normal plasma clots in 25 to 45 seconds while the aPTT in heparin patients is usually 70 to 140 seconds.

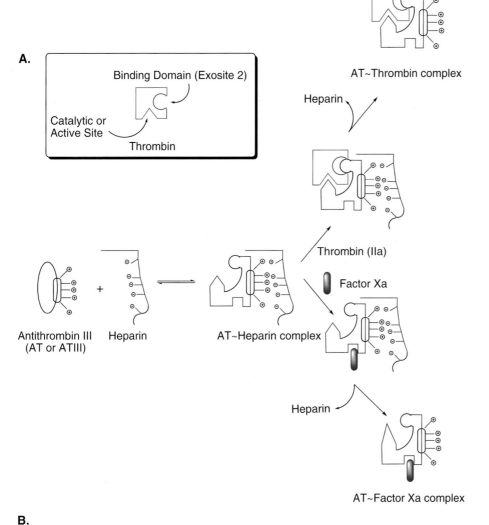

Figure 16.7 Schematic representation of catalytic role of the heparin-based drugs in promoting antithrombin factor Xa and antithrombin–thrombin complexes (**A**) and antithrombin–fondaparinux complex (**B**).

Table 16.1	**Properties of Heparin and Heparin-Based Drugs**			

Drug	Trade Name	Dosing	Molecular Weight (daltons)	Binding Ratio
Unfractionated Heparin (UFH)				
Heparin	Generic	bid, tid	5–30	1:1
Low–Molecular-Weight Heparin (LMWH)				
Dalteparin	Fragmin	qd	3–8	2.2:1
Enoxaparin	Lovenox Generic	qd, bid	3.5–5.5	2.7–3.9:1
Pentasaccharide				
Fondaparinux	Arixtra	qd	1.728	Xa only (1:1)

bid, twice daily; *tid*, three times daily; *qd*, once daily

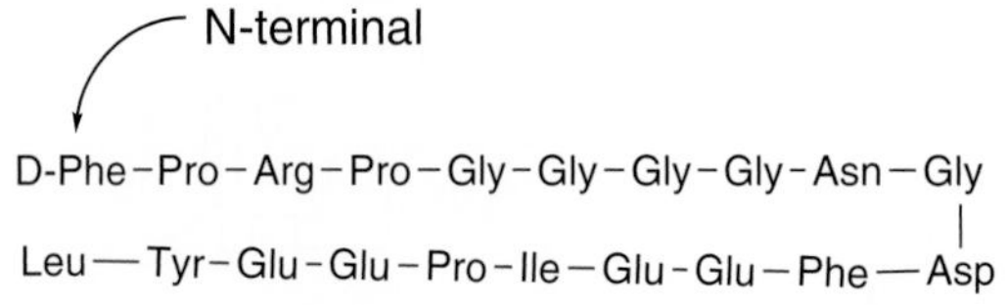

N-terminal

D-Phe—Pro—Arg—Pro—Gly—Gly—Gly—Gly—Asn—Gly

Leu—Tyr—Glu—Glu—Pro—Ile—Glu—Glu—Phe—Asp

Bivalirudin (Angiomax)

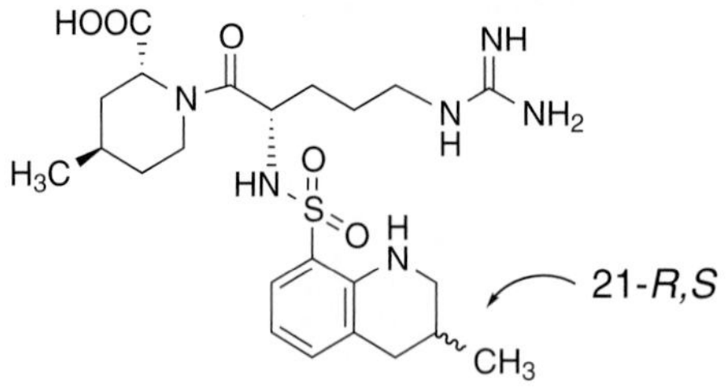

21-R,S

Argatroban (Acova)

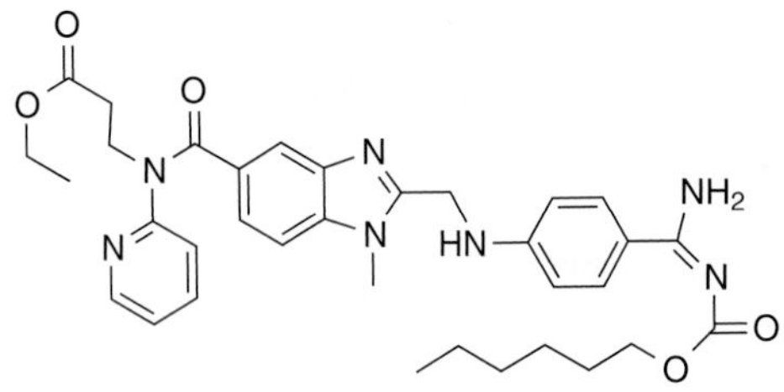

Dabigatran etexilate (Pradaxa)

Figure 16.8 Structures of the DTIs.

⚠ **ADVERSE EFFECTS**

DTIs: The most common concerns cited are associated with life-threatening hemorrhage, major bleeding, and other bleeding. In most instances these effects are less than those seen with warfarin or heparin-based drugs. Bleeding with desirudin appears to be minor, while with angiomax and argatroban bleeding may be somewhat more common. With dabigatran, major bleeding is similar to that reported with warfarin and GI bleeding is greater with dabigatran than with warfarin.

Direct Thrombin Inhibitors (DTIs)

- Original lead was hirudin isolated from *Hirudo medicinalis* (medical leech) (Fig. 16.8).
- The natural hirudin and its recombinant products lipirudin and desirudin are proteins of 65 amino acids.
- Desirudin replaces the N-terminal amino acids of lipirudin (leucine-1, threonine-2) with two valines (lipirudin sales have been discontinued).
- Bivalirudin is a 20-amino acid derivative of hirudin.
- Argatroban is a peptidomimetic analog of the natural product hirudin.
- Dabigatran etexilate is a nonpeptidomimetic orally active DTI.

MOA

- The DTIs are bound to thrombin either through the exosite 1 and the active site (catalytic site) of thrombin (bivalent binding—desirudin and divalirudin) or bind only to the active site (univalent binding—argatroban and dabigatran) (Fig. 16.9).
- More detailed studies suggest that the four N-terminal amino acids of bivalirudin bind to the active site with the 12 carboxy terminal amino acids binding to the exosite 1 (Fig. 16.10).
- The DTIs are usually capable of binding to both free and fibrin-bound thrombins.
- Divalirudin, argatroban, and dabigatran are reversible DTIs while desirudin is an irreversible DTI.
- Dabigatran etexilate is a prodrug; see metabolism below.

Physicochemical and Pharmacokinetic Properties

- Pharmacokinetic data is presented in Table 16.2.
- The DTIs have rapid onsets of action.
- 21-S-Argatroban is nearly twice as active as its *R*-isomer.

Exosite 1
Exosite 2
Catalytic or Active Site
Thrombin
Bivalirudin complex
Argatroban/Dabigatran complex
+
+
+

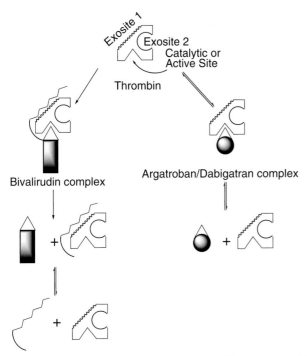

Figure 16.9 Schematic representation of the reaction of DTIs (bivalirudin, argatroban, or dabigatran) with thrombin.

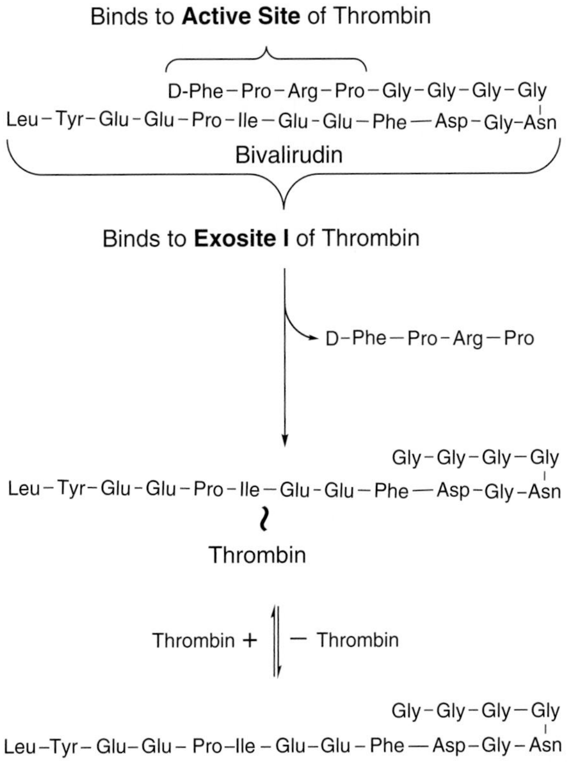

Binds to **Active Site** of Thrombin
D-Phe–Pro–Arg–Pro–Gly-Gly-Gly-Gly
Leu–Tyr–Glu-Glu–Pro-Ile–Glu-Glu–Phe—Asp-Gly-Asn
Bivalirudin
Binds to **Exosite I** of Thrombin
D-Phe–Pro-Arg–Pro
Gly-Gly-Gly–Gly
Leu–Tyr-Glu-Glu–Pro-Ile–Glu-Glu–Phe—Asp-Gly-Asn
Thrombin
Thrombin +
— Thrombin
Gly-Gly-Gly–Gly
Leu–Tyr–Glu–Glu– Pro–Ile – Glu–Glu – Phe—Asp-Gly-Asn

Figure 16.10 Bivalirudin binding sites to thrombin and release from thrombin.

Table 16.2 Direct Thrombin Inhibitors (DTIs)

Drugs	Route of Administration	Site of Binding	Half-Life (min)	Route of Excretion	Protein Binding
Desiruden	SC	CS, exosite I	120–180	Kidney	0
Bivalirudin	IV	CS, exosite I	25	Kidney	0
Argatroban	IV	CS	39–51	Hepatobiliary	54%
Dabigatran etexilate	PO	CS	~900	Kidney/bile	35%

CS, catalytic site; *IV*, intravenous; *PO*, orally; *SC*, subcutaneous

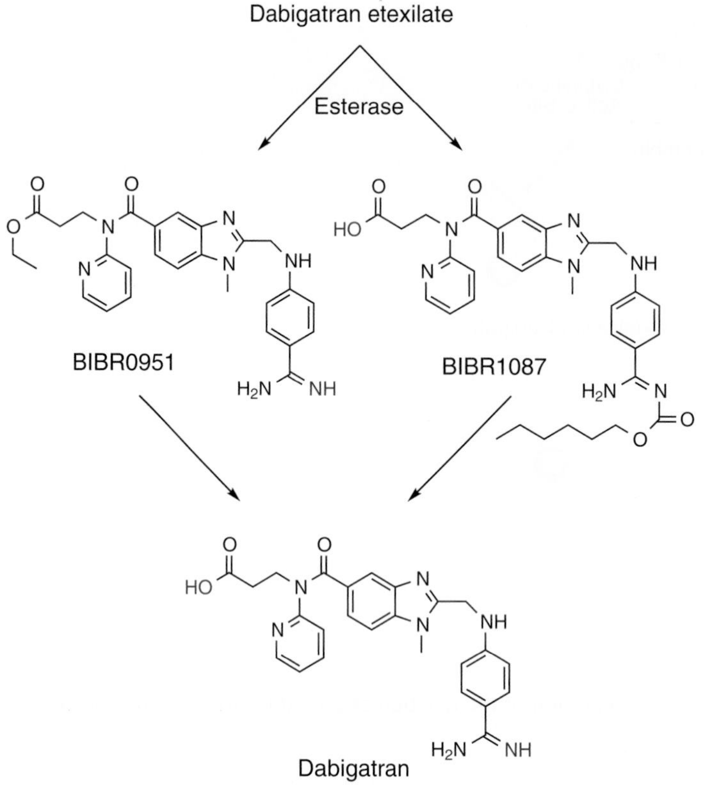

Dabigatran etexilate

Figure 16.11 Metabolic activation of dabigatran etexilate to dabigatran.

- Dabigatran is not absorbed orally while the prodrug is orally active but requires an acidic environment.
- Dabigatran etexilate capsule is coated with tartaric acid to promote an acidic media, although the drug has poor bioavailability (~6.5%).

Metabolism

- Desiruden degrades nearly 50% via proteolytic hydrolysis prior to excretion.
- Bivalirudin undergoes proteolytic hydrolysis (~30% recovered unchanged).
- Dabigatran etexilate, as previous mentioned is a prodrug, which is activated via esterase (Fig. 16.11).
- Argatroban is metabolized via CYP3A4/5 oxidation (Fig. 16.12).

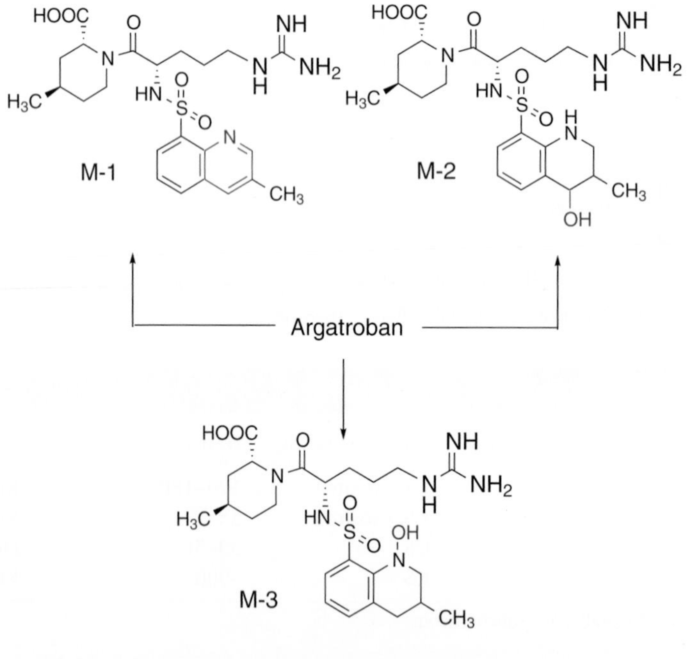

Figure 16.12 Metabolism of argatroban, a partially active product (M-1), and inactive products (M-2, M-3).

Direct Factor Xa Inhibitors

A second successful approach to the prevention of clot formation has been through the introduction of orally active direct inhibitors of factor Xa (FXa). Recently approved therapies consist of apixaban, rivaroxaban, and edoxaban.

Apixaban (Eliquis)
(a carboxamide)

Rivaroxaban (Xarelto)
(an oxazolidine)

Edoxaban
(Savaysa)

MOA

- FXa inhibitors apixaban, rivaroxaban, and edoxaban are capable of inhibiting both free (prothrombinase) as well as clot-bound FXa.
- The prothrombinase complex consists of FXa, factor Va, prothrombin, and Ca^{2+}, on a phospholipid surface.
- The FXa inhibitors are highly specific inhibiting at a single site, the convergent step of the intrinsic and extrinsic pathways (Fig. 16.1).
- The inhibition of thrombin formation also blocks the activation of platelets while not interfering with existing thrombin levels, thus improving safety.
- X-ray crystal studies of rivaroxaban bound to human FXa indicates hydrogen bonding binding between Gly^{219} and the carbonyl of the oxazolidinone, hydrogen bonding to the amide hydrogen of the chlorothiophene carboxamide, and hydrophobic bonding to two pockets (S1 and S4) within the FXa (Fig. 16.13).

> **ADVERSE EFFECTS**
>
> **Direct Factor Xa Inhibitors:** Black box warnings are listed for both apixaban and rivaroxaban that premature discontinuance without use of adequate alternatives can place nonvalvular AF patients at an increased risk of thrombotic events including stroke. With rivaroxaban an increase of hematomas following spinal puncture has also been reported. In addition, rivaroxaban use in patients with renal impairment may be problematic. Avoidance of combined use of rivaroxaban with P-gp and strong CYP3A4 inhibitors and inducers has been recommended.

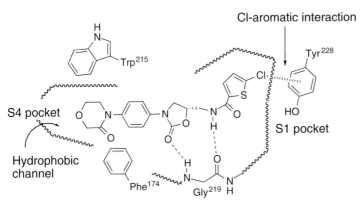

Figure 16.13 Binding sites of rivaroxaban to human FXa and potential binding sites in apixaban and edoxaban.

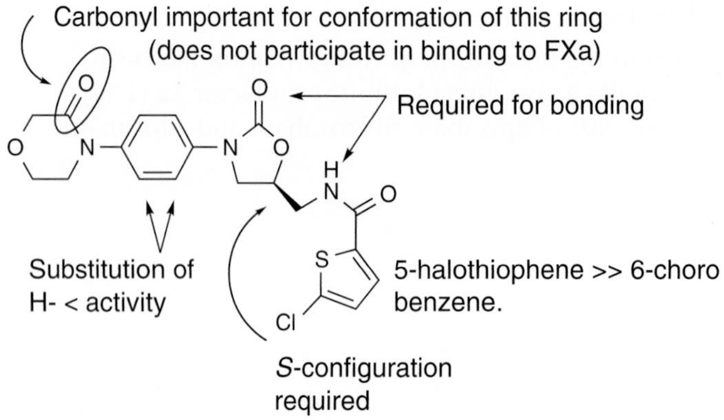

Figure 16.14 SAR requirements for the direct factor Xa inhibitor rivaroxaban and related analogs.

- Similar structural regions can be identified in apixaban and edoxaban (Fig. 16.13).

SAR—Rivaroxaban

- Lipophilic chlorothiophene versus substituted benzenes or furans decrease aqueous solubility and improve plasma binding.
- Morpholinone greatly improves activity versus morpholine, piperazine, or pyrolidinone.
- Additional structural requirements for activity are shown in Figure 16.14.

Physicochemical and Pharmacokinetic Properties

- The direct FXa inhibitors are reversible, highly selective, and orally active drugs.
- Exhibiting rapid absorption and high binding to plasma protein (Table 16.3).

Metabolism

- A large number of metabolites are reported for apixaban and rivaroxaban, none of which are active.
- Oxidative metabolism involves CYP3A4/5 and, to a minor extent, CYP2J2.
- Hydrolytic cleavage also has been reported for all three drugs and represents the major metabolites for edoxaban (Fig. 16.15).
- The extent of metabolism appears to be low in humans for these drugs (<50% of apixaban, <33% of rivaroxaban, and <30% of edoxaban).
- Studies suggest little or no possibility for drug–drug interactions associated with apixaban metabolism while CYP inhibitors can produce clinically significant increases in rivaroxaban exposure.
- Edoxaban is a substrate for P-glycoprotein (P-gp) efflux in the intestine with drug–drug interactions requiring dose adjustment if edoxaban is used in combination with a strong P-gp inhibitor (e.g., verapamil, quinidine).

Clinical Application of Anticoagulants

- The four classes of anticoagulants (vitamin K antagonists, heparin-based anticoagulants, DTIs, and direct FXa inhibitors) often have overlapping treatment indications, which include: prevention/treatment of VTE, deep vein thrombosis (DVT), and PE.
- Indications for vitamin K antagonists: prevent the progression or recurrence of acute DVT or PE following initial course of heparin, prevent recurrent coronary

Table 16.3 Pharmacokinetic Properties of Direct FXa Inhibitors

Generic Name	Protein Binding	Bioavailability	Half-Life (h)	Elimination Route
Apixaban	87%	50%	12	Rental/fecal
Rivaroxaban	92–95%	66–100%	5–9	Renal/hepatic
Edoxaban	40–59%	62%	5.8–10.7	Renal/hepatic

h, hours

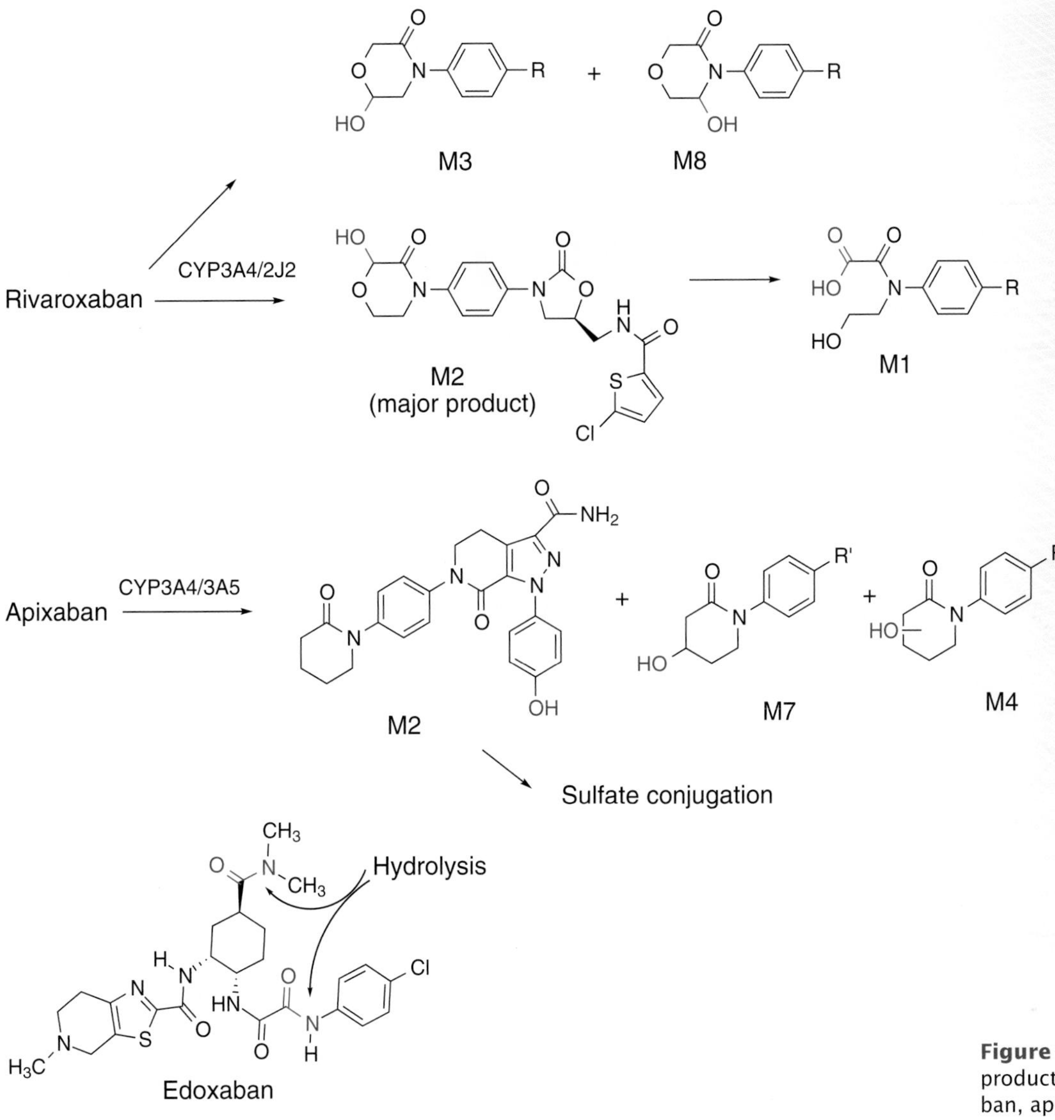

Figure 16.15 Major metabolic products formed from rivaroxaban, apixaban, and edoxaban.

ischemia in patients with acute MI and systemic embolization in patients with prosthetic heart valves or chronic AF, reduce the risk of clotting in patients with AF, patients with prosthetic heart valves, or following a heart attack.

- Indications for heparin-based anticoagulants: manage patients with unstable angina or acute MI, used during coronary balloon angioplasty, and treat acute coronary syndrome, AF.
- Indications for DTIs include prevention of stroke and blood clots in patients with AF, treatment of VTE in patients undergoing hip or knee replacement surgery, prophylaxis of DVT in patients undergoing hip replacement surgery, reduce the risk of stroke and systemic embolism in patients with nonvalvular AF, and treat patients undergoing coronary angioplasty or cardiopulmonary bypass surgery.
- Indications for direct FXa inhibitors include reduction of risk of stroke and systemic embolism in patients with nonvalvular AF, treatment of VTE and stroke prevention in patients with AF, treatment of PE, and prevent recurrent coronary infarctions.

Antiplatelet Drugs

Platelets are intimately involved in thrombus formation and the use of antiplatelet drugs represents another approach to prevention of blood coagulation. Platelet aggregation can be activated by vascular wall defect and the release of von Willebrand

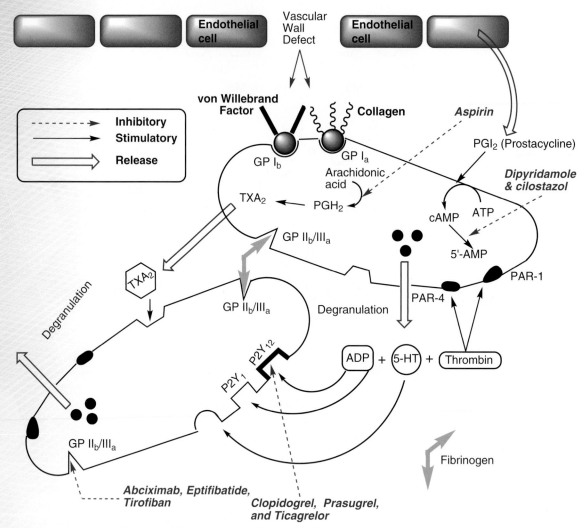

Figure 16.16 Scheme describing platelet activation as it relates to blood clot formation. The thrombus is formed at the site of a damaged wall in the vasculature. Normal endothelial cells in vascular wall provide prostacyclin, which stimulates the conversion of adenosine triphosphate (ATP) to cAMP, preventing platelet aggregation. In injury, GP receptors bind substances such as von Willebrand factor and collagen, activating the platelet. The GP IIb/IIIa receptors cross-link platelets via fibrinogen binding. As the platelet degranulates, additional aggregating substances including TXA$_2$, serotonin (5-HT), and ADP are released. ADP by binding to P2Y$_1$ and P2Y$_{12}$ promote and sustain platelet aggregation, respectively. Also shown, in italics, drugs and sites of inhibition of platelet aggregation.

factor and collagen, which bind to GP receptors (i.e., GP Ia and GP Ib). In addition, aggregation can be stimulated through secondary chemical messengers such as thromboxane A$_2$ (TXA$_2$), serotonin (5-HT), thrombin, and adenosine diphosphate (ADP) which recruits additional platelets, thus amplifying the aggregation process. The role of each of these substances and their site of action are outlined in Figure 16.16.

COX-1 Inhibitors

Aspirin

MOA

- Aspirin is an irreversible inhibitor of platelet COX-1 enzyme near the active site of cyclooxygenase, thus inhibiting TXA$_2$ synthesis (Fig. 16.16).
- Acetylation occurs on a serine residue deactivating the platelet for the life of this platelet (7 to 10 days) (Fig. 16.17).
- Aspirin has other antithrombotic actions also, including enhanced fibrinolysis.

COOH

Aspirin

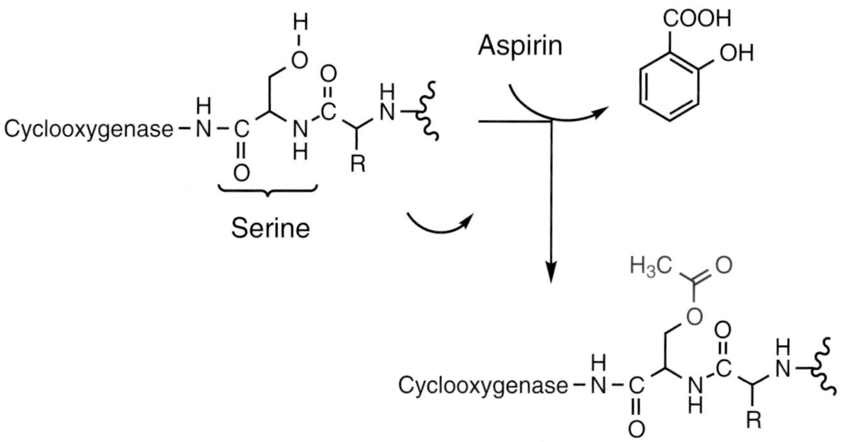

Figure 16.17 Irreversible acetylation of Ser[530] in cyclooxygenase-1 (COX-1) by aspirin.

Physicochemical and Pharmacokinetic Properties

- Aspirin is unstable in vitro and in vivo: hydrolysis of the acetyl group by water and plasma cholinesterase, respectively.
- Dose for aspirin's antiplatelet effect is 50 to 100 mg/day (below analgesic effects).
- Rapidly absorbed following oral administration.

Phosphodiesterase Inhibitors

Of the various phosphodiesterases, phosphodiesterase 3 (PDE3) specifically catalyzes the hydrolysis of cyclic adenosine monophosphate (cAMP) to 5′-AMP which is a potent inducer of platelet aggregation.

Dipyridamole and Cilostazol

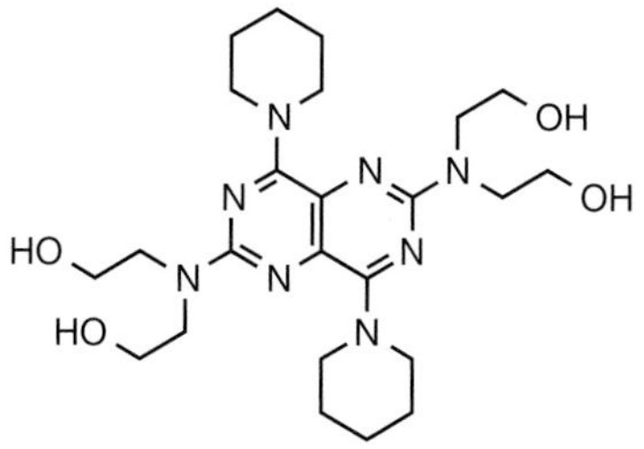

Dipyridamole
(Aggrenox, Persantine)

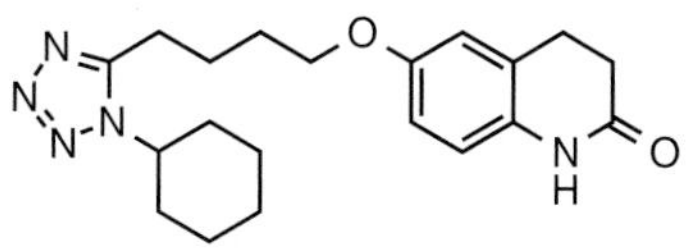

Cilostazol
(Pletal)

> **ADVERSE EFFECTS**
>
> **Cilostazol:** Cilostazol has a black box warning suggesting a decrease in survival in patients with class III/IV congestive heart failure and is therefore contraindicated with patients with congestive heart failure. The drug has also been associated with severe headaches, painful urination, chest pain, and shortness of breath.

MOA

- Both drugs inhibit PDE3, thus increasing the level of cAMP in platelets (Fig. 16.16).
- Cilostazol exhibits selectivity for the PDE3 A isoform (PDE3A) above that shown by dipyridamole.
- Both drugs inhibit adenosine reuptake, thus increasing adenosine binding to adenosine A_2 receptors stimulating cAMP synthesis (Fig. 16.18).

Physicochemical and Pharmacokinetic Properties

- Cilostazol is rapidly absorbed following oral administration, especially if taken with high-fat meals.

Metabolism

- Dipyridamole is metabolized primarily by glucuronidation at one of the alcohol groups.

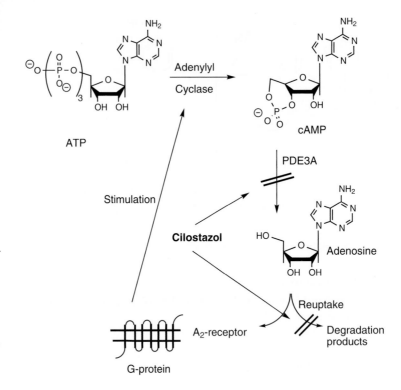

Figure 16.18 Cilostazol's mechanism of action: 1). blocks PDE3A directly increasing cellular levels of cAMP; 2). inhibits reuptake of adenosine increasing adenosine binding to adenosine 2 receptors (A_2-receptor a G-protein coupled receptor) which stimulates adenylyl cyclase indirectly increasing cAMP cellular levels.

- Cilostazol is extensively metabolized (11 metabolites are known). The two major metabolites exhibit activity and are shown in Figure 16.19.
- Coadministered inhibitors of CYP3A or CYP2C19 may require reduced dosing of cilostazol. The drug's exposure increases significantly when administered with grapefruit juice.

Clinical Application

- Dipyridamole is used in a fix combination with aspirin (Aggrenox) to treat thrombosis.
- Also used in combination with warfarin in treating patients with prosthetic heart valves.
- Cilostazol is approved for treating intermittent claudication.
- Off-label use of cilostazol includes prevention of secondary strokes and intracranial atherosclerosis.

P2Y Purinergic Receptor Inhibitors

As shown in Figure 16.16, ADP binds to the P2Y purinergic receptors ($P2Y_1$ and $P2Y_{12}$) on the platelet surface activating this G protein–coupled receptor. Activation of $P2Y_1$ induces changes in the platelet shape, mobilization of intracellular Ca^{2+},

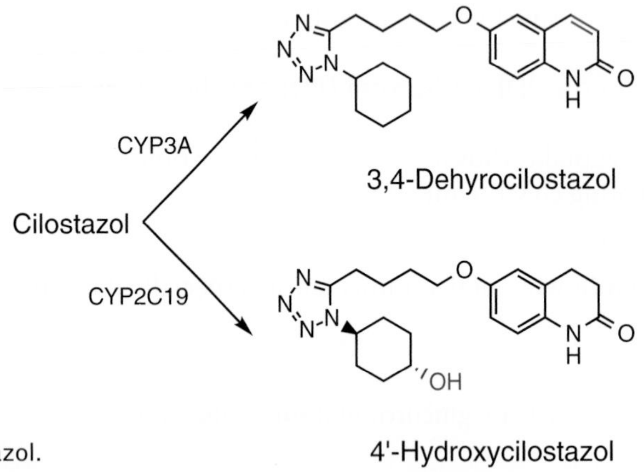

CYP3A

3,4-Dehyrocilostazol

Cilostazol

CYP2C19

Figure 16.19 Metabolism of cilostazol.

4'-Hydroxycilostazol

and initiating aggregation. Binding to $P2Y_{12}$ sustains platelet aggregation by inhibiting adenylyl cyclase resulting in a decrease in cAMP.

Ticlopidine, Clopidogrel, and Prasugrel

Ticlopidine

(*S*)-Clopidogrel
(Plavix)

Prasugrel
(Effient)

MOA

- All three drugs are prodrugs which following metabolic activation are irreversible inhibitors of $P2Y_{12}$ (Fig. 16.20).
- Metabolic activation of ticlopidine and clopidogrel require CYP oxidation while activation of prasugrel requires simple hydrolysis.

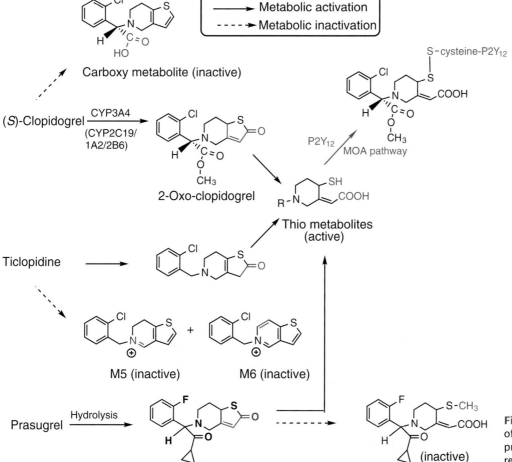

Figure 16.20 Metabolism of clopidogrel, ticlopidine and prasugrel. The thio metabolites react irreversibly with $P2Y_{12}$.

- The thio metabolites bind irreversibly with a cysteine (Cys^{17} or Cys^{270}), which is near the ADP binding site in $P2Y_{12}$, thus blocking ADP binding.
- Inactivation of the platelet occurs in vivo and lasts for 7 to 10 days and will occur irrespective of which of the two P2Y receptors is inhibited.

Physicochemical and Pharmacokinetic Properties

- The P2Y purinergic receptor inhibitors are rapidly absorbed following oral administration.
- Prasugrel is highly bound to plasma protein (~98%).
- These drugs are eliminated via urine and feces.

Metabolism

- In addition to the activating metabolism each of the inhibitors undergoes inactivation via metabolism (Fig. 16.20).
- Clopidogrel is hydrolyzed to the inactive carboxylic acid metabolite, a reaction that appears to be more significant than the inactivation of prasugrel.
- Ticlopidine is metabolized to metabolites M5 and M6, which have been associated with toxic side effects (life-threatening neutropenia).
- Prasugrel is methylated to an inactive metabolite.
- The FDA has a boxed warning of a potential for reduced effectiveness among poor metabolizers (CYP2C19) using clopidogrel.

Ticagrelor

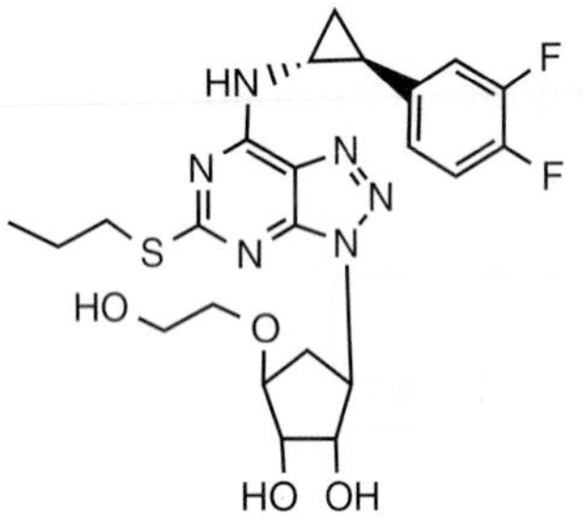

Ticagrelor (Brilinta) 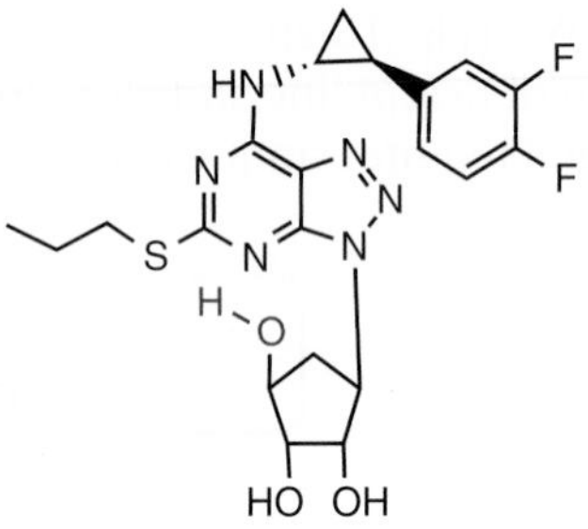 Ticagrelor major metabolite

MOA

- Ticagrelor and its O-deethylation metabolite are reversible inhibitors of $P2Y_{12}$.
- The binding of ticagrelor occurs at a site distinct from that of the irreversible inhibitor of $P2Y_{12}$ and ADP (allosteric site), and acts as a noncompetitive antagonist (Fig. 16.21).
- The binding of ticagrelor locks the $P2Y_{12}$ conformation in an inactive state, which is not able to bind to the G-protein complex, and although ADP can still bind the signaling is inhibited.

Physicochemical and Pharmacokinetic Properties

- Ticagrelor is rapidly absorbed following oral administration and the deethylation metabolite appears in the plasma shortly after administration of the drug.
- The deethylated metabolite represents 30% to 40% of the concentration of ticagrelor.
- Total recovery of drug and metabolites is high with most being recovered from the bile and feces with the remainder found in urine (~2:1 ratio).
- Ticagrelor is highly bound to plasma protein (>99.7%).

Metabolism

- Ticagrelor is extensively metabolized to 10 or more products with the O-deethylated metabolite being the only active metabolite.

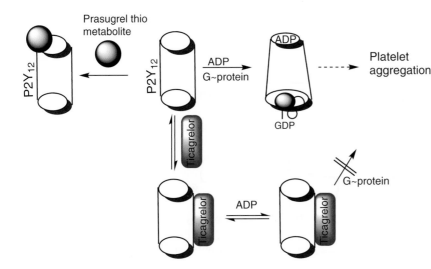

Figure 16.21 Scheme describing platelet aggregation resulting from ADP binding to the $P2Y_{12}$ receptor, conformational changes in the receptor, and G-protein coupling. This process can be inhibited irreversibly by the thio metabolite of prasugrel binding near the ADP binding site or reversibly by ticagrelor binding to a remote site resulting in blocking conformational changes to the receptor.

- Ticagrelor is metabolized by CYP3A4/5 although potential drug–drug interactions are of low likelihood since ticagrelor direct acting.
- Ticagrelor as well as many of its metabolites appear as glucuronide metabolites.

Clinical Application of P2Y Purinergic Receptor Inhibitors
- Reduction of MI and stroke in patients with a history of recent MI or stoke.
- Treatment of peripheral arterial disease.
- Acute coronary syndromes in patients with non–ST-segment elevation (commonly used in combination with aspirin).

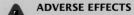

ADVERSE EFFECTS

P2Y Purinergic Receptor Inhibitors: Major adverse effects of ticlopidine, clopidogrel, prasugrel, and ticagrelor are bleeding which could lead to fatal/life-threatening outcomes. Additional effects include dyspnea, headache and cough, dizziness, and nausea. Ticagrelor also may produce hypovolemic shock or severe hypotension. This is also seen, but to lesser effect with the other agents.

Protease-Activated Receptor-1 Antagonist

In spite of the effectiveness of aspirin, clopidogrel, and other newer $P2Y_{12}$ antagonists for the treatment and prevention of atherothrombotic events in patients with acute coronary syndrome, MI, and other existing peripheral arterial diseases, the risk of bleeding events, potential drug resistance, and other factors contributing to the interpatient variability remains a therapeutic challenge in these patients. A new antiplatelet drug has been approved which acts at a site different from previously approved drugs.

Vorapaxar (Zontivity)

MOA
- α-thrombin can initiate platelet aggregation by its catalytic action at the protease-activated receptor (PAR) (Fig. 16.16).
- Four PARs (PARs 1 to 4) have been identified, with PAR-1 and PAR-4 being present on the surface of human platelets and involved in the process of α-thrombin–initiated platelet aggregation.
- These receptors are glycoprotein coupled receptors (GPCRs) in which thrombin, a serine protease, binds to PAR-1 promoting cleavage between Arg^{41} and Ser^{42} of the N-terminus exomembrane 7-transmembrane spanning GPCR. The N-terminal serine then binds to a site on loop II of the GPCR to initiate platelet aggregation (Fig. 16.22).
- Vorapaxar is a highly selective competitive inhibitor/antagonist of thrombin binding to PAR-1 and although the binding is reversible, a long half-life suggests nearly irreversible binding.

Vorapaxar (Zontivity)

Physicochemical and Pharmacokinetic Properties
- Vorapaxar is rapidly absorbed following oral administration with an oral bioavailability of nearly 100%.
- The drug is highly bound to plasma proteins (>99%).

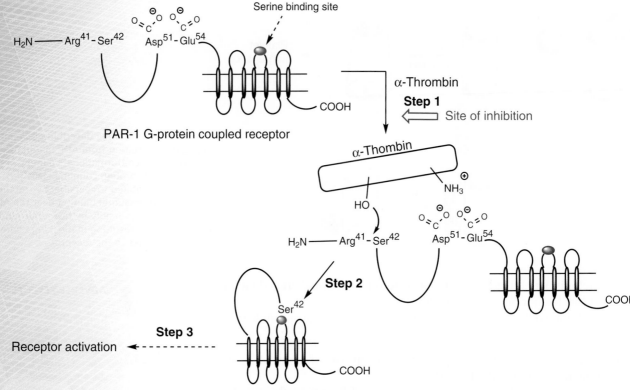

Figure 16.22   Activation of platelet aggregation via thrombin. The thrombin binds to the PAR-1 receptor as shown in step 1 through ionic binding involving a charge portion of amino acids 51 to 56 in the PAR-1 while the serine protease portion of thrombin catalyzes the hydrolysis at arginine[41]/serine[42]. The tethered serine[42] then binds to a portion of exomembrane loop II of the GPCR shown in step 2 which initiates platelet aggregation as shown in step 3. Vorapaxar is a reversible competitive inhibitor of thrombin receptor–activation peptide, thus selectively blocking platelet aggregation.

- The drug has a long half-life (159 to 311 hours) suggesting that while it is a reversible inhibitor of PAR-1, its release from PAR-1 is quite slow.
- The drug is eliminated in the feces (58%) and urine (25%).
- CYP3A4 inhibitors and inducers significantly affect plasma levels of vorapaxar and this may require dose modification in the presence of such drugs.

Metabolism

- Vorapaxar is metabolized by CYP3A4 to the primary metabolite (M19) with minor amounts of M20 (Fig. 16.23).

Clinical Application

- Vorapaxar has been approved for use in reducing the risk of MI, stroke, or cardiovascular death in patients with a previous MI or peripheral artery disease.
- Vorapaxar may find use in combination with other platelet aggregation inhibitors such as aspirin or clopidogrel for optimal management of various cardiovascular events.

> ⚠ **ADVERSE EFFECTS**
>
> **Vorapaxar:** Similar to the P2Y purinergic inhibitors, vorapaxar has a black box warning of increased risk of bleeding including ICH and therefore potential for fatal bleeding. The drug should not be used in patients with a history of stroke, transient ischemic attack, or ICH. The drug has only recently been approved by the FDA and therefore additional adverse effects may appear over time.

Glycoprotein IIb/IIIa Receptor Antagonists

The platelet surface is covered with inactive GP IIb/IIIa receptors (Fig. 16.16). These receptors are dimeric glucoproteins, referred to as platelet–surface integrin ($\alpha_{IIb}\beta_3$). The GP IIb/IIIa receptors exist in an inactive conformation, but conformational change can be initiated by various platelet agonists including thrombin, collagen, and TXA$_2$. Activation of the receptor results in cross-linking of platelets through bonding to fibrinogen and von Willebrand factor, thus mediating aggregation. Chemically diverse antagonists are capable of bonding to GP IIb/IIIa receptors to block the platelet aggregation.

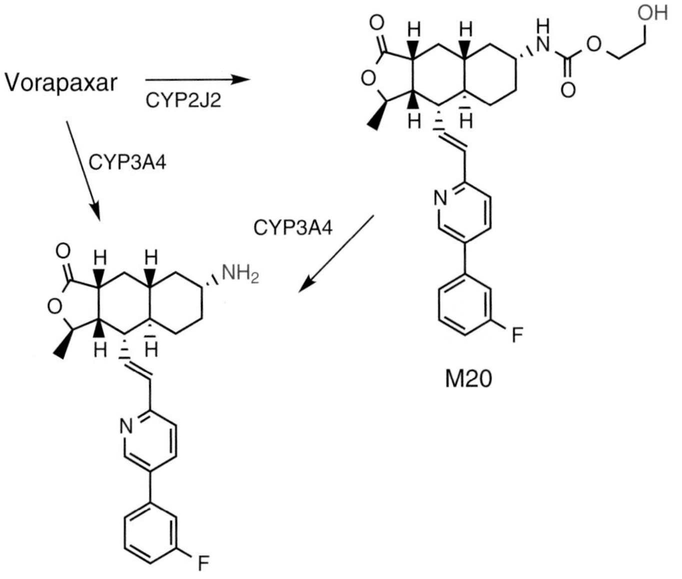

Figure 16.23 Metabolism of vorapaxar.

Primary product
(M19)

M20

Abciximab (ReoPro)

- Abciximab is a chimeric human/mouse antibody (7E3 Fab).
- The drug is a monoclonal antibody directed as an antagonist against the $\alpha_{IIb}\beta_3$ receptor (integrin receptor).
- The drug is nonspecific and binds to additional receptors.
- As a GP the drug is administered IV and has a very short half-life (~10 to 30 min), but may remain bound to platelets for days (see Table 16.4 for pharmacokinetic properties).
- Used in patients undergoing percutaneous coronary intervention (i.e., angioplasty).
- Major adverse effects are associated with long-term binding to platelets: hemorrhage, thrombocytopenia.

Eptifibatide and Tirofiban

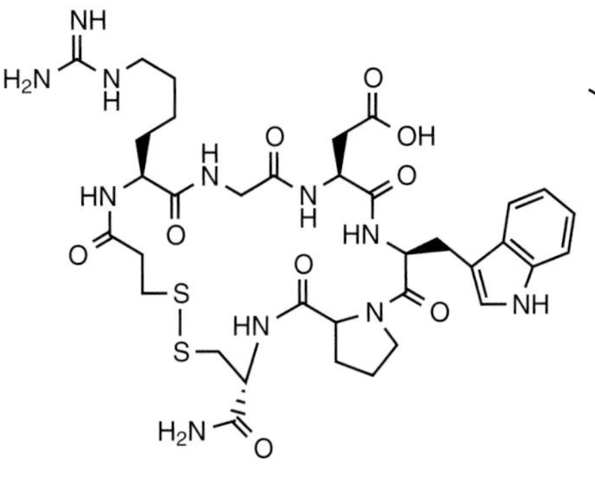

Tirofiban
(Aggrastat)

Eptifibatide
(Integrilin)

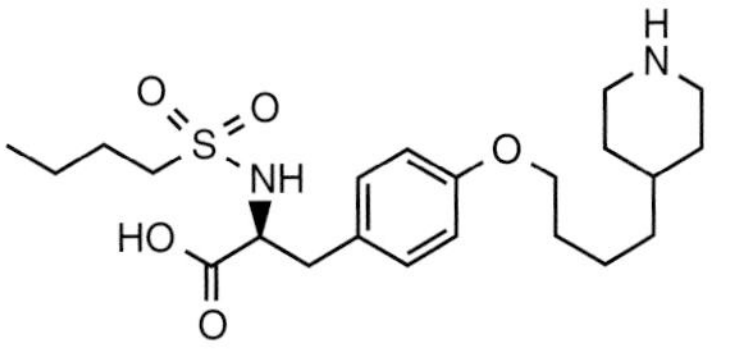

Table 16.4	**Pharmacokinetic Properties of the GP IIb/IIIa Receptor Antagonists**				
Drug	**Route of Administration**	**Molecular Weight (daltons)**	**Dissociation Constant (nmol/L)**	**Plasma Half-Life (h)**	**Protein Binding (%)**
Abciximab	IV	47,615	5	72	
Eptifibatide	IV	800	120	4	25
Tirofibin	IV	495	15	3–4	65

IV, intravenous; *h*, hours

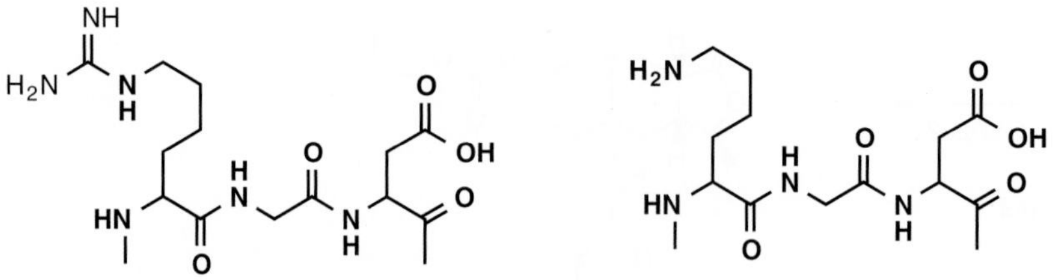

RGD
(Arginine-Glycine-Aspartic acid)

KGD
(Lysine-Glycine-Aspartic acid)

Figure 16.24 RGD sequence of amino acids found in snake venom and von Willebrand factor and KGD sequence of amino acids important to synthetic antiplatelet drugs.

- Snake venom contains a family of proteins (disintegrins), which are potent inhibitors of platelet aggregation and integrin-dependent cell adhesion.
- Within the snake venom, a sequence of amino acids consisting of arginine-glycine-aspartic acid (RGD) is essential for binding to the $\alpha_{IIb}\beta_3$ receptor. This sequence is also present in von Willebrand factor and fibronectin.
- Synthetic derivatives make use of the lysine-glycine-aspartic acid (KGD) sequence, which is also capable of binding to the $\alpha_{IIb}\beta_3$ receptor (Fig. 16.24).

MOA

- Eptifibatide has the KGD/RGD template as part of the cyclic heptapeptide structure and as a result is highly specific for the receptor.

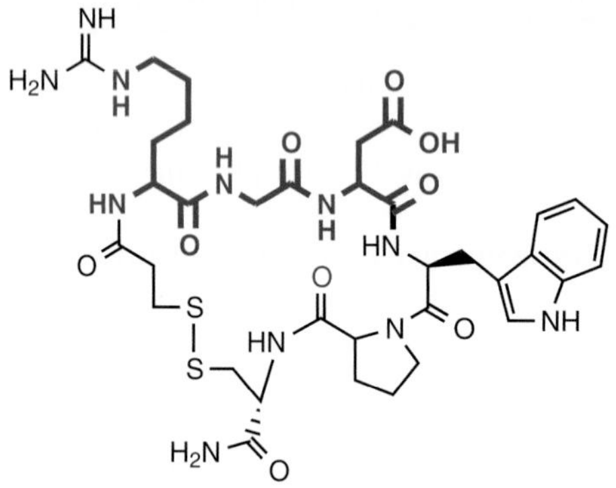

Eptifibatide (red indicates KGD portion of drug)

- Tirofiban (a member of the "fibans") is a nonpeptide peptidomimetic with a carboxylic acid and basic amine separated by a distance similar to that seen in the RDG sequence of amino acids.
- Both drugs are reversible antagonists of the GP IIb/IIIa receptors.

Physicochemical and Pharmacokinetic Properties

- The short half-life for eptifibatide and tirofiban results in a reduced risk of bleeding.
- Neither drug appears to exhibit significant metabolism.
- Additional pharmacokinetic properties are shown in Table 16.4.

Clinical Application

- The GP IIb/IIIa receptor antagonists (e.g., abciximab, eptifibatide, and tirofiban) are intended to treat ischemic complications in patients undergoing percutaneous coronary intervention or those with unstable angina.
- Abciximab is used as an adjunct with aspirin and heparin while eptifibatide may be used in combination with aspirin, clopidogrel and heparins.
- Eptifibatide and tirofiban may also be used to reduce the risk of acute cardiac ischemic in patients being treated for MI.

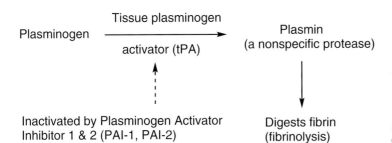

Figure 16.25 Scheme describing role of tPA in the digestion of clots.

Thrombolytic Drugs

Acute MI or stroke requires the digestion of insoluble fibrin clots. The common standard of treatment calls for the use of thrombolytic drugs for these and other conditions associated with clot formation. These drugs act through their catalytic role as tissue plasminogen activators (tPAs) in generating plasmin as shown in Figure 16.25. The action of tPA is modulated by plasminogen activator inhibitors (PAIs) which reduce plasmin levels.

⚠️ **ADVERSE EFFECTS**

Thrombolytic Drugs: The thrombolytic drug exhibit a common adverse effect of bleeding which may occur in the GI tract but may also involve ICH. Allergic reactions have also been reported for these drugs. The allergic reactions with streptokinase is related to circulating antistreptococcal antibodies. These antibodies are present in individuals who have had a streptococcal infections sometime in their past. The allergic reactions include rash and fever, but could lead to anaphylaxis.

MOA

- All of the thrombolytic drugs (Fig. 16.26) act by increasing the conversion of plasminogen to plasmin. They differ by their source (bacterial versus unmodified or modified human [tPA]) and their specificity.
- All are serine protease peptides, which increase the production of plasmin.
- All thrombolytic drugs form various complexes:
  - Streptokinase forms 1:1 complex with plasminogen, which then converts unbound plasminogen to plasmin.
  - Alteplase and reteplase specifically binds to plasminogen–fibrin clot complex in thrombus, leading to clot lysis. Reteplase has reduced affinity for fibrin, but a prolonged half-life.
  - Tenecteplase has high specificity for fibrin as well as low binding to PAI-1.

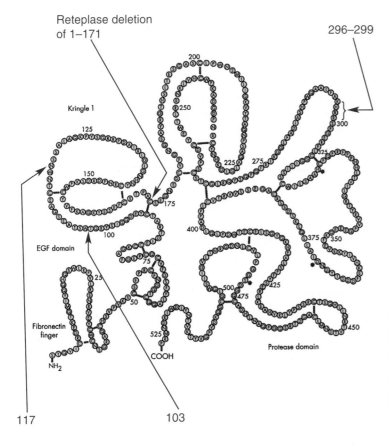

Figure 16.26 Schematic diagram of alteplase, reteplase (removal of amino acids 1 to 172), and tenecteplase in which threonine[103] (T) is replaced with lysine, asparagine[117] (N) is replaced with glutamine, and lysine (K)-histidine-arginine-arginine at positions 296 to 299 are replaced with four alanines. (Adapted with permission from Nordt TK, Bode C. Thrombolysis: newer thrombolytic agents and their role in clinical medicine. *Heart.* 2003;89:1358–1362.)

- Tenecteplase stimulates fibrin formation and inhibits PAI-1 to produce thrombolytic activity.

Clinical Application

- The thrombolytics are commonly used for treatment of MI, massive PE, and acute ischemic stroke.

Specific Antithrombotic Drugs

Streptokinase (First-Generation Thrombolytic)

- Streptokinase is a fibrin nonspecific drug (Table 16.5).
- May also catalyze breakdown of factors V and VII.

Alteplase (Second-Generation Thrombolytic)

- Alteplase is unmodified tPA (Fig. 16.26) and at low dose degrades only fibrin, but at therapeutic doses also activates plasminogen which can lead to hemorrhage.
- Alteplase is very specific for plasminogen bound to fibrin with a preference for older clots.

Reteplase (Third-Generation Thrombolytic)

- Structurally modified tPA in which amino acids 1 to 172 have been removed from natural tPA (Fig. 16.26).
- After removal of these amino acids the half-life improves (Table 16.5).

Tenecteplase (Third-Generation Thrombolytic)

- Structurally modified tPA in which the amino acids at 103, 117, and 296 to 299 are replaced with six new amino acids (Fig. 16.26).
- The replacement of threonine (T^{103}) with asparagine^{103} leads to a new glycosylation site, which enlarges the molecular size and reduces the elimination rate.
- The asparagine^{117} (N)-glycosylate increased elimination, but replacement with glutamine^{117} reduces elimination.
- Replacement of lysine (K)-histidine-arginine-arginine at 296 to 299 with four alanines, which taken together with other changes increases binding to PAI-1 by 80-fold.
- The amino acid replacements are done through point mutations.

Coagulants

Excessive bleeding or hemorrhage results from inadequate coagulation and thus the need for coagulating agents. Several of the coagulating agents act at the same sites as the anticoagulants (Fig. 16.1).

Vitamin K

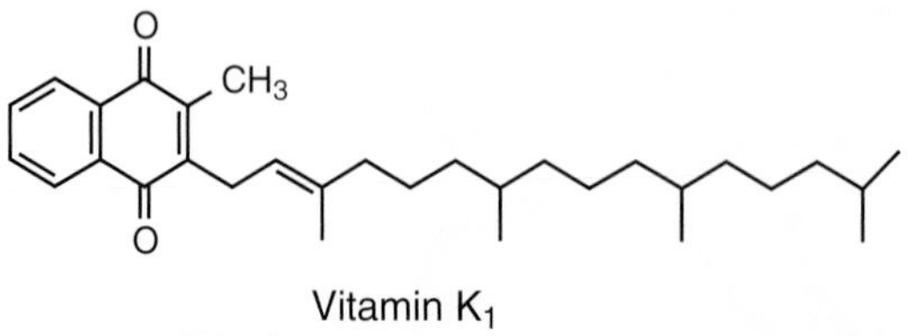

Vitamin K$_1$
(Mephyton, phytonadione)

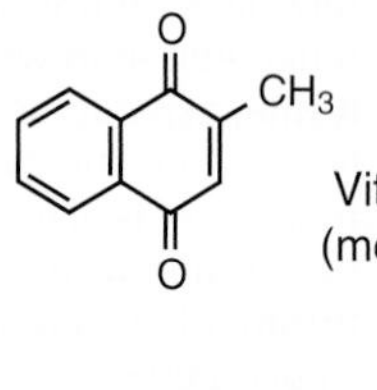

Vitamin K$_3$
(menadione)

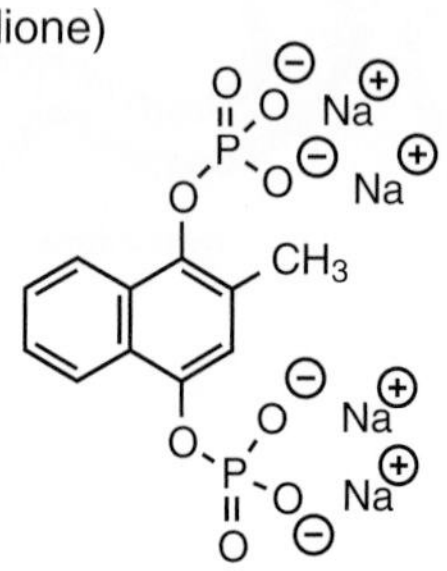

Vitamin K$_4$
(menadione sodium diphosphate)

Thrombolytics	Trade Names	Amino Acids	Source	Half-Life (min)
Streptokinase	Streptase	414	β-Hemolytic streptococci	~30
Alteplase	Activase	527	Recombinant DNA tPA	3–4
Reteplase	Retavase	357	Modified human tPA	14–18
Tenecteplase	TNKase	527	Modified human tPA	~17

Table 16.5 **General Properties of Thrombolytic Drugs**

- Vitamin K_1, the fat soluble form of vitamin K, is useful for the treatment of bleeding associated with warfarin-induced bleeding.
- Vitamin K_1 requires the presence of bile for its absorption.
- Vitamin K_3 and Vitamin K_4 are water-soluble forms of vitamin K and are absorbed passively from the intestine.
- Vitamin K_3 and Vitamin K_4 have potential toxicities in neonates.
- Vitamin K_1 is available in an oral as well as injectable dosage forms, but all forms have short half-lives requiring divided dosing over a period of time.
- Vitamin K products have a delayed onset (up to 24 hours) and this delay can increase the amount of hemorrhage.

Protamine Sulfate

- Protamine sulfate is an arginine-rich protein, which can compete with heparin for antithrombin III (AT) (Fig. 16.7).
- Protamine sulfate can actually reverse the binding of AT~heparin complex and tie up heparin (heparin antagonist).
- Protamine sulfate can interact with fibrinogen and platelets causing anticoagulant effects.

Antifibrinolytic Agents

Bleeding associated with the use of fibrinolytic drugs as well as prevention of rebleeding in ICHs and as adjunctive therapy in hemophilia suggests the need for antifibrinolytic agents.

Tranexamic and ε-Aminocaproic Acids

H_2N ⬡ COOH $H_2N - CH_2-CH_2-CH_2-CH_2-CH_2-COOH$

Tranexamic acid
(Cyklokapron, Lysteda)

ε-Aminocaproic acid
(Amicar)

- Plasmin binds to fibrin through lysine-binding sites to promote the fibrinolysis.
- Tranexamic and ε-aminocaproic acids have a strong affinity for the five lysine-binding sites on plasminogen and are thus able to effectively compete for plasminogen, preventing the availability of plasminogen for complexation with fibrin or conversion to plasmin (Fig. 16.27).

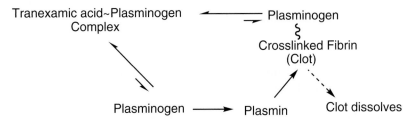

Figure 16.27 The high affinity of tranexamic acid/ε-aminocaproic acid for plasminogen decreases the availability of plasminogen for conversion to plasmin and prevents clot dissolution.

- Tranexamic and ε-aminocaproic acids are readily absorbed orally or can be administered IV.
- These drugs exhibit little metabolism and undergo renal excretion.

Clinical Application

- The antifibrinolytic agents are used to reduce bleeding during surgery (heart and liver surgery).
- They are also used for short-term treatment of patients with hemophilia to reduce hemorrhage.

Thrombopoietin Receptor Agonists

Life-threatening bleeding associated with thrombocytopenia is the hallmark of chronic idiopathic thrombocytopenia purpura (ITP). This condition involves antibody-mediated platelet destruction leading to platelet counts of <30,000 platelets/μL. Thrombopoietin receptor agonists appear to be beneficial in treating ITP.

Eltrombopag

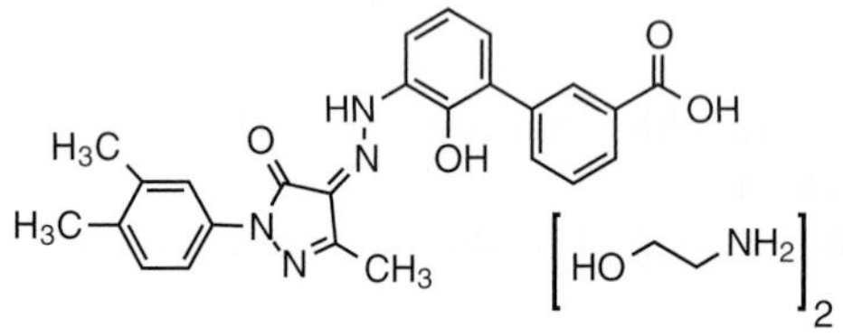

Eltrombopag
(Promacta)

- Eltrombopag is a small-molecule agonist which binds to human thrombopoietin (TPO) on the surface of platelets (megakaryocytes and their precursors). This binding triggers gene expression and the production of megakaryocytes, ultimately leading to the release of platelets.
- The drug acts through Janus Kinase/Signal Transducer and Activator Transcription (JAK/STAT) pathway.
- As a nonprotein, eltrombopag is not immunogenic.
- Eltrombopag is orally active, rapidly absorbed, and highly bound to plasma protein (>99%).
- Eltrombopag's absorption is adversely affected by polyvalent cations in antacids.
- Minimal metabolism occurs in the liver by CYP isoforms followed by glucuronidation.

Romiplostim (Nplate)

- Romiplostim is a fusion protein (Fc-peptide) analog of thrombopoietin with a molecular weight of 59 kDa.
- Romiplostim, as does eltrombopag, acts as an agonist at receptors that stimulate the release of platelets through the JAK/STAT pathway.
- The drug is used subcutaneously with a slow and prolonged peak serum concentration and a half-life of several days.

Clinical Application

- Both eltrombopag and romiplostim are beneficial for the treatment of ITP.
- Administration leads to significant improvement in platelet count in patients with ITP.
- Both drugs reduced the degree of bleeding in IPT patients.
- Both drugs were well tolerated during clinical studies.

Chemical Significance

Blood flows through arteries and veins as a continuous process with a delicate balance between flow and clotting. A clot forming too readily is known as a hypercoagulation state and involves blood proteins and platelets. Hypercoagulation can be dangerous and is associated with any number of conditions including trauma or surgery, pregnancy, prolonged bed rest or immobility (lengthy airplane travel), heart attack, congestive heart failure or stroke, obesity, and various diseases including cancer, HIV/AIDS, and inflammatory bowel disease. Today, with increased hip and knee replacement, more and more elderly are prone to hypercoagulation. VTE, which encompasses DVT, affects more than 2 million people annually and this can progress to PE, which is estimated to lead to 200,000 deaths annually. The use of anticoagulants and antiplatelet and fibrinolytic agents for the various hypercoagulation conditions is fairly common today.

The applications of medicinal chemistry to the clinical situations involving hypercoagulation are many and can be utilized in daily conversations with health professionals, patients, and caretakers.

- A chemical evaluation of the structures of the heparin-based anticoagulants immediately suggests high water solubility (ionic compounds) with little likelihood of oral activity.
- The DTIs bivalirudin, a protein, and argatroban, a peptidomimetic, suggest nonoral administration, while the etexilate addition to dabigatran leads to an oral prodrug product.
- The irreversible antiplatelet action of aspirin is well understood by viewing the chemical reaction of aspirin with the COX-1 enzyme, an action not associated with other COX inhibitor anti-inflammatory drugs.
- The metabolic activation of ticlopidine, clopidogrel, and prasugrel immediately identifies the difference between prasugrel and ticlopidine/clopidogrel as it relates to potential drug–drug interactions involving CYP for ticlopidine/clopidogrel, but not with prasugrel. In addition, these differences account for a boxed warning found for ticlopidine/clopidogrel.
- The identification of the thrombolytics as peptides immediately dictates the dosage form and route of administration since such drugs must bypass the GI tract to be effective.

The above examples represent practical applications of medicinal chemistry to address everyday problems encountered in pharmacy practice. A pharmacist utilizing his medicinal chemistry background is enabled to converse more effectively with the patient and health care professional to explain complicated chemically based problems.

Review Questions

1 The mechanism of action of coumadin involves _____.
  A. Blocking reduction of vitamin K epoxide to vitamin K.
  B. Direct inhibition of vitamin K–catalyzed carboxylation of glutamic acid to γ-carboxyglutamate.
  C. Inhibition of the catalytic cleavage of two arginines in prothrombin.
  D. Blocking transamidase-catalyzed cross-linking of fibrin.
  E. Stimulation of antithrombin binding to factor IIa.

2 Clopidogrel is an irreversible inhibitor of which enzyme system?
  A. Adenylyl cyclase
  B. Purinergic receptor 2Y12
  C. PDE3A
  D. COX-1
  E. Antithrombinase III

3 The major advantage of heparin (UFH) over LMWH is _____.
A. more selectivity
B. better pharmacokinetics
C. more predictable results
D. greater affinity of UFH~AT complex for thrombin and factor X
E. less metabolism

4 Eptifibatide is a cyclic heptapeptide which has high specificity for what receptor?
A. PDE3A
B. COX-1 receptor
C. GP IIb/IIIa receptors ($\alpha_{IIb}\beta_3$)
D. P2Y$_{12}$ receptor
E. Thrombin-catalytic site/exosite 1

5 Which of the structures shown below represents the immediate active form of prasugrel?

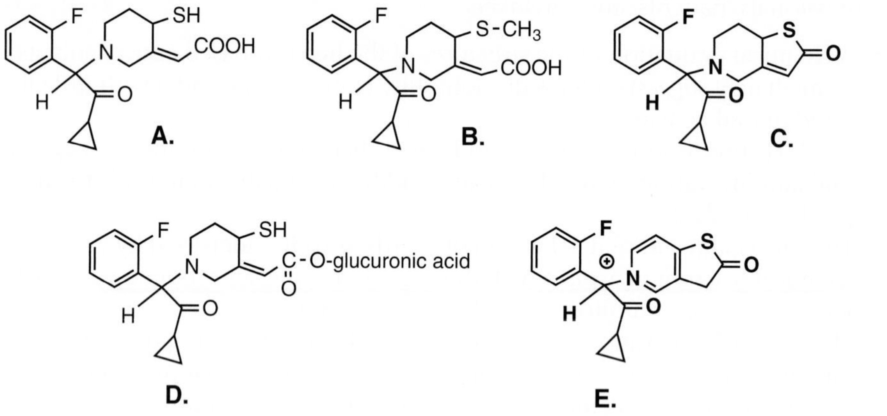

A.   B.   C.

D.   E.

Insulin and Drugs Used to Treat Diabetes

17

Learning Objectives

Upon completion of this chapter the student should be able to:

1. Describe the mechanism of action of antidiabetic drugs:
   a. Insulin and Insulin analogs
   b. Oral hypoglycemic agents
   c. Insulin secretagogues (sulfonylureas, meglitinides)
   d. Biguanides
   e. Insulin sensitizers (PPAR agonists, glitizones)
   f. α-Glucosidase inhibitors
   g. Incretins (GLP-1 agonists) and DPP-IV Inhibitors
   h. Amylin agonists
   i. SGLT2 inhibitors (gliflozins)

2. Discuss the structure activity relationships (SAR) of the antidiabetic drugs.

3. Identify any clinically relevant physicochemical properties (solubility, in vivo and in vitro chemical stability, absorption and distribution).

4. Discuss significant metabolism of each drug class.

5. Explain clinical uses and clinically relevant drug–drug, drug–food interactions.

Introduction

Diabetes (types 1 and 2) is a complex chronic disease characterized by uncontrolled glucose homeostasis (hyperglycemia) and morbidities associated with microvascular (blindness, end-stage renal disease, painful neuropathies) and macrovascular (cardiovascular disease, atherosclerosis, myocardial infarct, stroke, and limb amputation) complications. Therapy is directed at controlling hyperglycemia. Maintenance of euglycemia (normal blood sugar levels) is measured as the patient's glycosylated hemoglobin (HbA_{1c} below 7%). An HbA_{1c} <7% helps to moderate the microvascular pathologies; however, reduction of the risk for macrovascular pathologies also requires management of cardiovascular risk factors (e.g., smoking, dyslipidemias, hypertension, and antiplatelet therapy).

Treatment of type 1 diabetes requires insulin replacement therapy. Treatment of the type 2 diabetic utilizes several classes of oral hypoglycemic agents but may also require the use of insulin. They include:

- Insulin secretagogues (sulfonylureas, meglitinides)
- Biguanides (metformin)
- Insulin sensitizers (TZD, meglitizones)
- α-Glucosidase inhibitors
- Glucagon-like peptide 1 analogs (GLP-1)
- Dipeptidyl peptidase-IV (DPP-IV) inhibitors
- Amylin antagonists
- Sodium/glucose cotransporter 2 (SGLT2) inhibitors

Insulin and Insulin Analogs
Insulin Structure

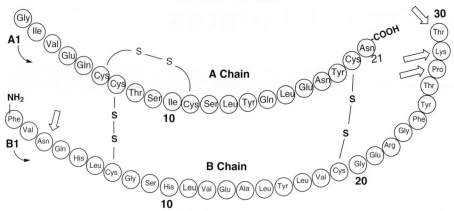

MOA

- Insulin facilitates the transport of glucose into adipose and muscle cells by stimulating the complicated cascade of events that result in the release of the glucose transporter 4 (GLUT4) from storage vesicles and its translocation to the cell membrane.
- The N- and C-termini of both the A and B chains are involved in binding to the insulin receptor.

Amino Acid Interactions with the Insulin Receptor		
Chain	N-Terminus	C-Terminus
A	Gly^1, Glu^4, Gln^5	Tyr^{19}, Asn^{21}
B	Val^{12}	Tyr^{16}, Gly^{23}, Phe^{24}, Phe^{25}, Tyr^{26}

SAR

- Insulin is a 51 amino acid protein consisting of an A chain (21 amino acids) and a B chain (30 amino acids) linked together by two disulfide bonds with an additional disulfide bond in the A chain.
- Insulin is formed in the β-cells of the pancreas from preproinsulin (110-amino acid chain) which loses a 24 amino acid unit to form proinsulin which is then converted to insulin by a series of proteolytic modifications.
- Regular insulin is considered a short-acting insulin with a 30 to 60 minutes onset.
- Intermediate-acting insulins are prepared by combining regular insulin with a stoichiometric amount of the positively charged polypeptide protamine to form neutral protamine Hagedorn or NPH insulin, or by preparing crystalline zinc insulin in an acetate buffer (Lente, Ultralente). In both cases, the limited solubility of these insulins in subcutaneous fluids leads to the intermediate onset and duration of action.
- All of the rapid-acting insulins are created by making a substitution in the amino acid sequence at the beginning or the end of the B chain (Table 17.1). This creates insulin analogs that are rapidly absorbed from the subcutaneous space, peak fast, and have a short duration of action.
- The long-acting insulins are created by preparing derivatives that form depot reservoirs upon subcutaneous administration.
  - Ultralente is a four-zinc acetate crystalline product with a slower dissolution rate than the two-zinc acetate (Lente) derivative.
  - Glargine is the result of the replacement of A^{21} Asn by Gly and the addition of two Args to the C-terminus of the B chain.
  - Detemir is prepared by the removal of B^{30} Thr and by adding a C14 fatty acid to the B^{29} Lys.

Table 17.1 Insulin Analogs (All Prepared by Recombinant DNA Technology)

Generic Name	Trade Name	Change in A Chain	Change in B Chain
Lispro	Humalog	None	$Pro^{28} \rightarrow Lys$ $Lys^{29} \rightarrow Pro^{29}$
Aspart	NovoLog	None	$Pro^{28} \rightarrow Asp$
Glulisine	Apidra	None	$Arg^{3} \rightarrow Lys$ $Lys^{29} \rightarrow Glu$
Glargine	Lantus	$Asn^{21} \rightarrow Gly$	Add: Arg^{31} and Arg^{32}
Detemir	Levemir	None	Remove: Thr^{30} Add: C14 fatty acid to Lys^{29}

Physicochemical and Pharmacokinetic Properties

- Fifty percent of the insulin secreted from the pancreas is degraded in the liver by a thiol metalloproteinase.
- Insulin is filtered through the glomerulus and reabsorbed by the renal tubules.
- Insulin is degraded by the renal tubules as well as at the surface of insulin-sensitive tissues.
- Insulin concentrations above 0.2 mM/L form hexamers even in the absence of zinc ions.
- Insulin takes a few hours to reach its peak activity, and then continues to be active for an additional several hours (Table 17.2).

Clinical Applications

- Pharmacologic treatment for type 1 diabetes requires intensive insulin therapy. The large number of short- and long-acting insulin analogs allows for the use of multiple doses of basal (intermediate or long-acting insulin) as well as prandial (rapid-acting) insulin. Prandial insulin dose is matched to carbohydrate intake, premeal plasma glucose and anticipated activity.
- Different types of insulin can be premixed in the same syringe allowing for a rapid-acting and long-acting insulin to be administered at the same time.
- Major adverse effects include hypoglycemia and mild to severe allergic reactions.

CHEMICAL NOTES

Hexamers tend to form viscous fibrils which do not resuspend, are inactive and can clog delivery pumps. Preparation in phosphate buffers can overcome the problem. Insulin tends to be absorbed onto tubing when in very dilute solution (<0.03 mmol/L) this can be overcome by addition of albumin. Chemical instability involves deamidation of the A terminal Asn^{21} with zinc insulin solutions at pH 2–3.

DRUG–DRUG INTERACTIONS

Drugs That May Require *Lowering* the Insulin Dose: Oral hypoglycemics; pramlintide; ACE inhibitors; disopyramide; fibrates; fluoxetine; MAO inhibitors; salicylates; sulfonamide antibiotics.

Table 17.2 Classification and Pharmacokinetics of Insulin Preparations

Type	Brand	Onset (min)	Peak (h)	Duration (h)
Rapid-Acting				
Lispro	Humalog	5–15	0.5–1.5	3–4
Aspart	NovoLog	5–15	0.5–1.5	3–4
Glulisine	Apidra	5–15	0.5–1.5	3–4
Short-Acting				
Human insulin (regular)	Humulin R	30–60	2–3	4–6
Intermediate-Acting				
Lente	Humulin L	2–4	6–12	18–26
NPH	Humulin N	2–4	4–12	18–26
Long-Acting				
Ultralente	Humulin U	4–8	10–30	>36
Glargine	Lantus	1–4	5–24	20–24
Detemir	Levemir	1–4	5–24	20–24
	Humalog Mix 75/25	5–15	Dual	10–16
	Humalog Mix 50/50	5–15	Dual	10–16
	NovoLog Mix 50/50	5–15	Dual	10–16
	NovoLog Mix 70/30	5–15	Dual	10–16
	70/30 (OTC)	30–60	Dual	10–16

DRUG–DRUG INTERACTIONS

Drugs That May Require Increasing the Insulin Dose: Corticosteroids and oral birth control pills; niacin; danazol; diuretics; β-adrenergic agonists; isoniazid; phenothiazines; thyroid hormones; atypical antipsychotics (olanzapine and clozapine).

Sulfonylurea Oral Hypoglycemic Agents

Sulfonylurea Pharmacophore

MOA

- The sulfonylureas stimulate the release of insulin from the pancreatic β-cells by binding to the sulfonylurea receptor type 1 (SUR1) site on the ATP-sensitive K^+ channel which opens the voltage-sensitive Ca^{2+} channels leading to an increase in intracellular Ca^{2+} causing the exocytotic release of insulin.
- The ATP-sensitive K^+ channel consists of four pore-forming K^+ subunits (Kir6.2) surrounded by four regulatory sulfonylurea receptor subunits (SUR1) (Fig. 17.1).

SAR

- Sulfonylureas were developed from the observed hypoglycemic effect of sulfonamides used to treat typhoid fever in 1942.
- The first-generation sulfonylureas (Fig. 17.2) have small lipophilic substituents at R_1 of the pharmacophore and alkyl or cyclic lipophilic substituents at R_2.
- First-generation sulfonylureas require high doses and have long plasma half-lives with short duration of action which increases their potential for adverse effects.
- Second-generation sulfonylureas have increased potency, rapid onset, shorter plasma half-lives and longer duration of action due to the strong binding affinity to the ATP-sensitive K^+ channel associated with the larger p-(β-arylcarboxyamidoethyl) group on R_1.

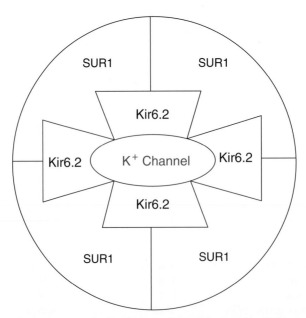

Figure 17.1 Idealized view of the ATP-sensitive K^+ Channel showing four pore-forming inwardly rectifying K^+ subunits (Kir6.2) surrounded by four regulatory sulfonylurea receptor subunits (SUR1).

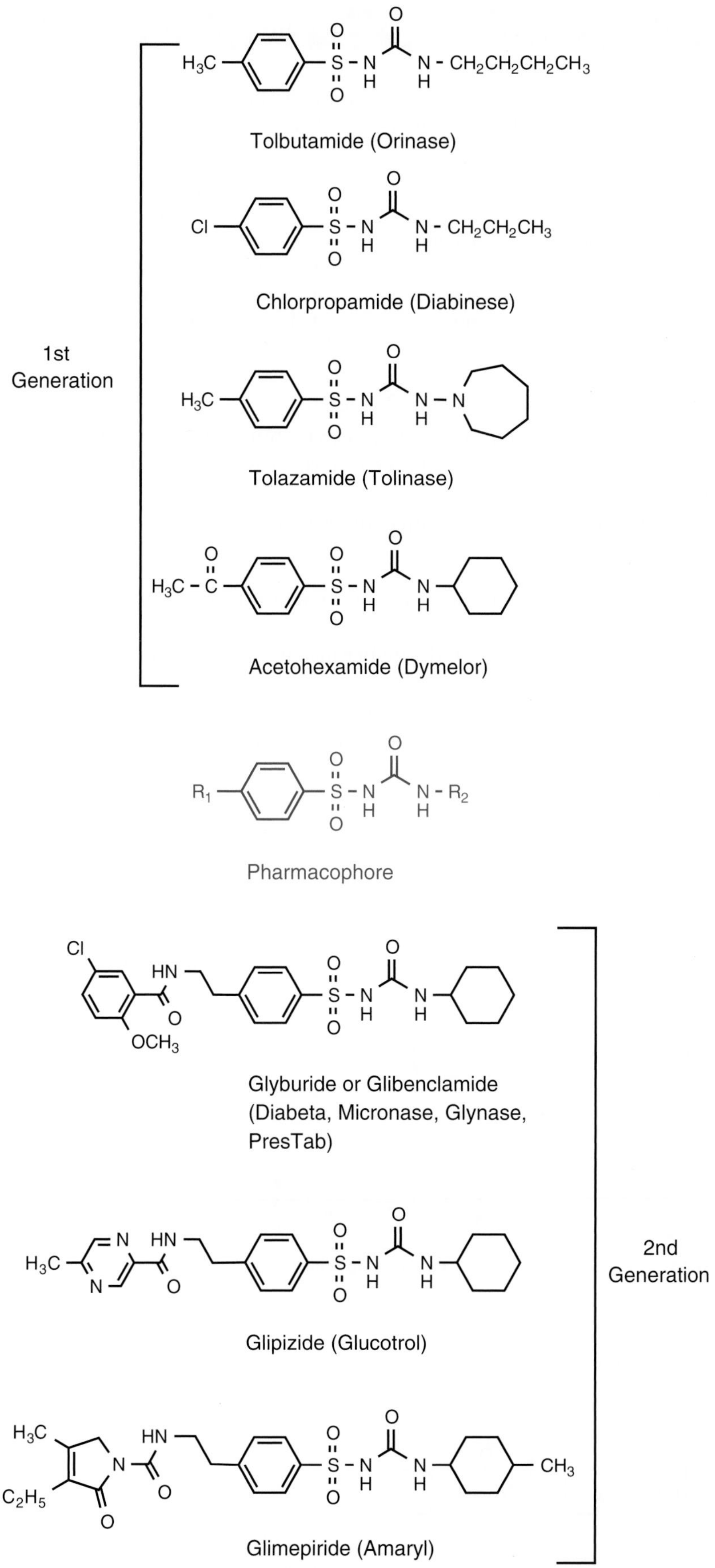

Figure 17.2 First- and second-generation sulfonylurea oral hypoglycemics.

Table 17.3 Pharmacokinetic Properties of the Sulfonylureas

Drug (Sulfonylureas)	Equivalent Dose (mg)	Serum Protein Binding (%)	Half-Life (h)	Duration (h)	Renal Excretion (%)
Tolbutamide	1,000	95–97	4.5–6.5	6–12	100
Chlorpropamide	250	88–96	36	up to 60	80–90
Tolazamide	250	94	7	12–14	85
Acetohexamide	500	65–88	6–8	12–18	60
Glyburide	5	99	1.5–3.0	up to 24	50
Glipizide	5	92–97	4	up to 24	68
Glimepiride	2	99	2–3	up to 24	40

Physicochemical and Pharmacokinetic Properties

- Sulfonylureas are highly protein bound to plasma albumin (Table 17.3).
- Dose, half-life, duration of action, and excretion decreases from first to second generation.
- Food delays absorption but does not affect bioavailability.

Metabolism

- Metabolism (Fig. 17.3) occurs in the liver and metabolites are excreted through the kidneys.
- Chlorpropamide has a 10× longer half-life because its ω and ω-1 propyl group hydroxylation is slow.
- Tolbutamide and tolazamide undergo sequential benzylic oxidation to an alcohol (active) and to an acid metabolite (inactive). Tolazamide also undergoes hydroxylation of its aliphatic ring to an active metabolite which increases its duration of action.

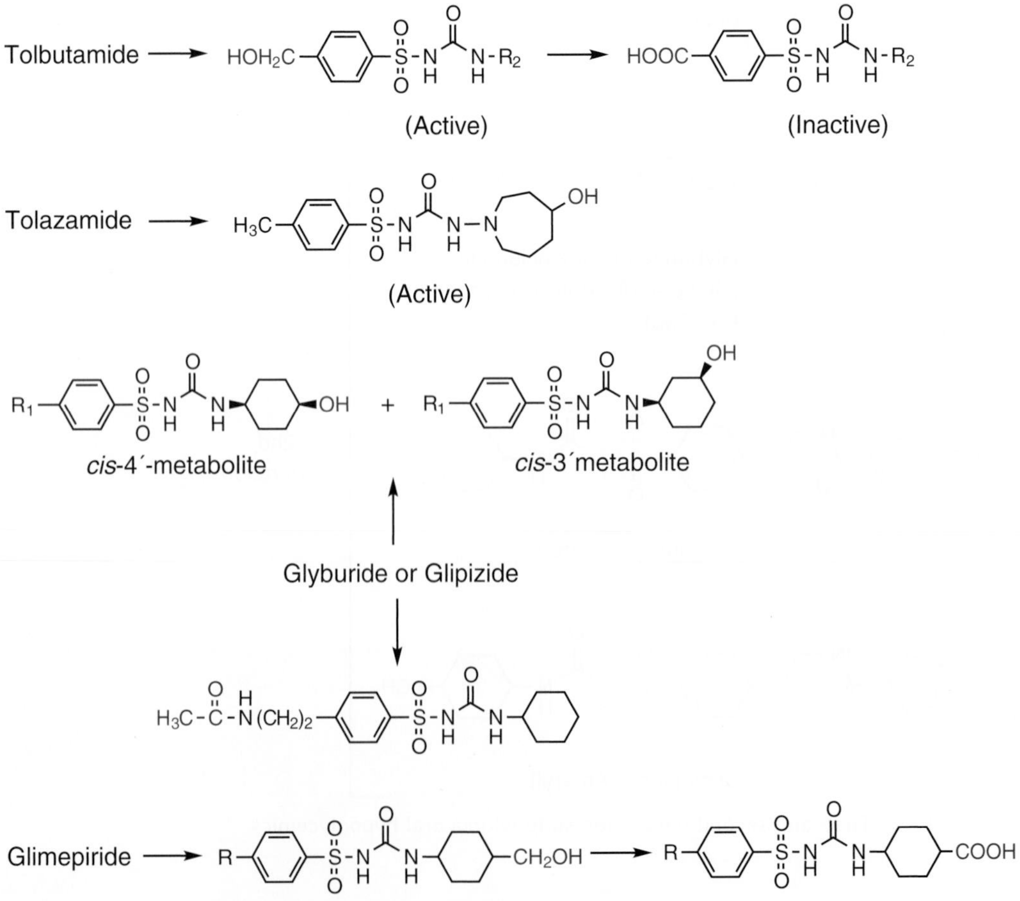

Figure 17.3 Metabolism of tolbutamide, tolazamide, glyburide, glipizide, and glimepiride.

- Acetohexamide is metabolized to an active compound via reduction of the keto group on the phenyl ring and is deactivated by hydroxylation of the cyclohexyl ring.
- Glyburide and glipizide are metabolized via hydroxylation of the cyclohexyl ring and amide hydrolysis of the R_1 group followed by acetylation.
- Glimepiride is metabolized by CYP2C9 sequentially at the cyclohexyl ring methyl group to alcohol (active) and to acid (inactive) metabolites.

Clinical Applications

- Sulfonylureas (second generation) are currently used as a second oral hypoglycemic agent when initial treatment with metformin and life style changes fail to achieve glycemic control (HbA_{1c} <7%).
- Sulfonylureas have reduced efficacy over time.

ADVERSE EFFECTS

⚠️ **Sulfonylureas:** Hypoglycemia (especially in liver disease and renal insufficiency); weight gain; rash; hemolytic anemia (glipizide).

DRUG–DRUG INTERACTIONS

Monitor with CYP2C9 Inhibitors: Sulfonamides; azole antifungals; amiodarone; cimetidine; diclofenac; fluvastatin, lovastatin; isoniazid; sertraline; zafirlukast.

Meglitinides

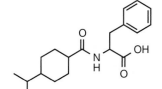

Meglitinide

MOA

- Stimulate the release of insulin from the pancreas in much the same way as the sulfonylureas by blocking the ATP-sensitive K^+ channel.
- Repaglinide binds to the SUR1 receptor on the β-cells but also to SUR1, SUR2A, and SUR2B found on cardiac and smooth muscle cells respectively, conferring extrapancreatic effects.
- Nateglinide binds selectively to the β-cell SUR1.

SAR

Glyburide/glibenclamide

- Meglitinide is the prototype structure and is the benzoic acid derivative of the non-sulfonylurea moiety of glibenclamide (Fig. 17.4).
- Repaglinide is an analog of meglitinide containing a number of additional substituents (*m*-ethoxy, isobutyl and piperidine ring).
- Nateglinide is a phenylalanine analog of meglitinide where the phenyl carboxyl group is transposed to the α-carbon of the ethyl side chain creating an amino acid functionality.

Note

The extrapancreatic effects of nonselective SUR binding include reduction of Ca^{2+} influx in cardiac cells and decreasing blood flow in cardiac vasculature both being detrimental to ischemic preconditioning and to the heart during an ischemic attack.

STEREOCHEMICAL NOTE

Nateglinide, glibenclamide, and glimepiride display a U-shaped conformation formed by the hydrophobic interaction versus the hydrocarbon moieties on either end placing the peptide bond at the bottom of the U which is believed to play an important role in binding to the SUR1.

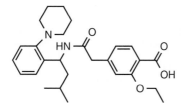

Nateglinide (Starlix)   Repaglinide (Prandin)

Figure 17.4 The meglitinide hypoglycemic agents.

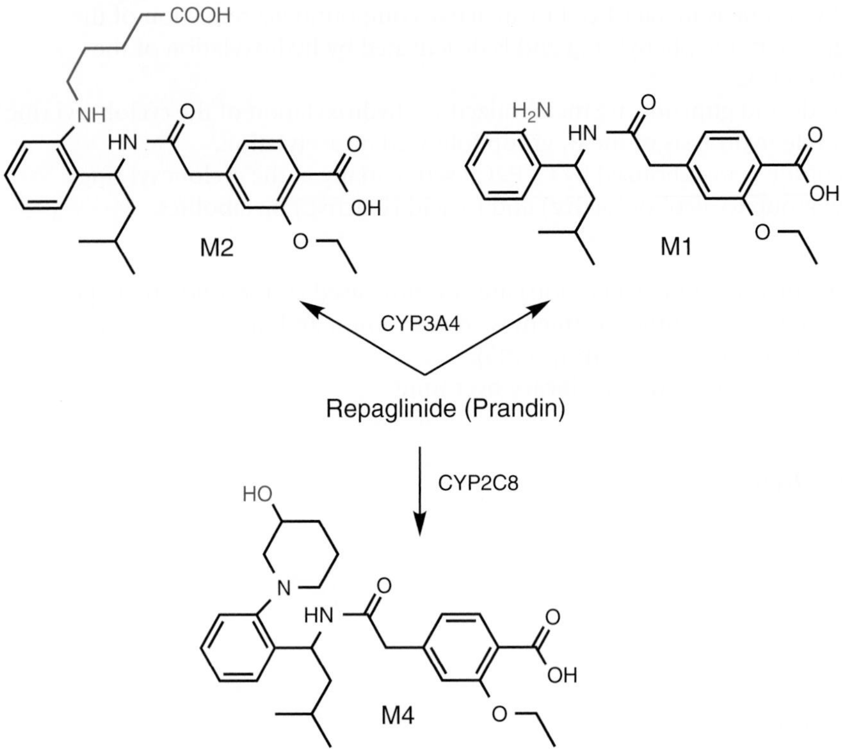

Figure 17.5 Metabolism of repaglinide.

Physicochemical and Pharmacokinetic Properties

- Repaglinide has a rapid onset and a short duration of action.
  - Does not produce prolonged hyperinsulinemia as can be seen with the sulfonylureas.
  - Bioavailability is 60% and 98% protein bound.
  - Fivefold more potent than glyburide (glibenclamide) via IV and 10-fold more potent on oral administration.
  - Approximately 90% of a single oral dose is eliminated in the feces and 8% in urine after extensive metabolism.
- Nateglinide is rapidly absorbed with onset beginning in 20 minutes.
  - Bioavailability is 73% and 98% protein bound.
  - Excreted in urine (83%) and feces (10%) after extensive metabolism.

Metabolism

Repaglinide

- Extensively metabolized via CYP2C8 and 3A4 hydroxylation of the piperidine ring.
- Eliminated as O-glucuronides (Fig. 17.5).

Nateglinide

- Extensive oxidative metabolism via CYP2C9 (70%) and 3A4 (30%).
- Hydroxylation occurs at the isopropyl group on the cyclohexyl ring.

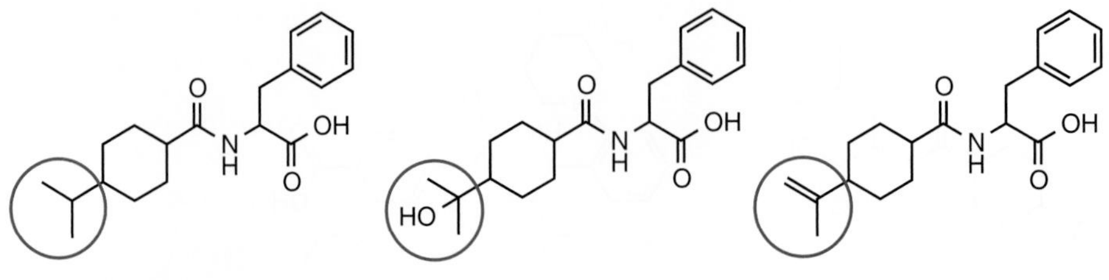

Nateglinide   Hydroxy metabolite   Isoprene metabolite

- The isoprene minor metabolite has antidiabetic activity comparable to that of nateglinide.
- The acyl glucuronide is also formed on the carboxyl group.

Clinical Applications

- Meglitinides are currently used as an alternative to the sulfonylurea oral hypoglycemic agents when initial treatment with metformin and life style changes fail to achieve glycemic control (HbA$_{1c}$ <7%).
- Can be used in patients with renal dysfunction, even in those with severe impairment.
- Monitor repaglinide when given with glucuronidation inhibitors.
- Take up to 30 minutes before meals.
- When taken with a high-fat meal, AUC and Cmax decrease.

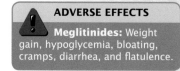

ADVERSE EFFECTS

Meglitinides: Weight gain, hypoglycemia, bloating, cramps, diarrhea, and flatulence.

Some Glucuronidation Inhibitors

Benzodiazepines, diclofenac, naproxen, and some dietary supplements (quercetin and ginseng).

Biguanides

Metformin (Glucophage) Buformin Phenformin

MOA

- Decrease hepatic gluconeogenesis and increase glycogenolysis.
- Enhance insulin sensitivity in liver and skeletal muscle.
- Have antihypertriglyceridemic activity.
- Block breakdown of fatty acids through activation of AMP-dependent protein kinases.

SAR

- Developed from galegine (isoamylene guanidine) first isolated from *Galega officinalis*.

Galegine

- The biguanides are formed by linking two guanidine groups together with different side chains.
- Metformin is the only biguanide available for use in the United States. Phenformin and buformin are derivatives of metformin but have been discontinued due to their potential to cause toxic lactic acidosis as a result of decreased gluconeogenesis as well as increased incidence of cardiac mortality.

Physicochemical and Pharmacokinetic Properties

- Quickly absorbed from the small intestines.
- Bioavailability is 50% to 60%; peak plasma levels in 2 hours.
- Not protein bound and widely distributed.
- Not metabolized and excreted unchanged in the urine (t$_{1/2}$ = 2 to 5 hours).

Clinical Applications

- Metformin (Glucophage, Glucophage XR) is the first-line drug for the treatment of type 2 diabetes along with life style modifications.
- Can also be used to treat insulin resistance in women with polycystic ovary syndrome.
- Do not use in patients with renal dysfunction (serum creatinine >1.4–5).

ADVERSE EFFECTS

Metformin: GI discomfort and diarrhea; lactic acidosis; metallic taste; impaired B$_{12}$ absorption.

DRUG–DRUG INTERACTIONS OF NOTE

Competitive renal excretion by cimetidine.

Insulin Sensitizers (Peroxisome Proliferator–Activated Receptor Agonists [Glitizones])

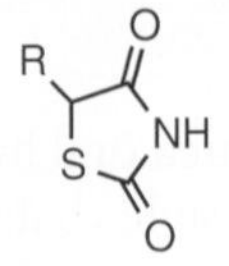

"Glitizone" pharmacophore
(a thiazolidinedione)

MOA

- Peroxisome proliferator–activated receptor (PPAR) family consists of three members: PPARα, PPARδ, and PPARγ.
- PPARs are ligand-regulated transcription factors that control gene expression.
- Bind as heterodimers with a retinoid X receptor.
- Upon binding, the agonists increase the rate of transcription of a number of gene products.
- PPARγ agonists act by increasing the sensitivity of muscle cells to insulin by increasing glucose transporter expression.
- Slow down gluconeogenesis and at the same time lower systemic fatty acids.

SAR

- The pharmacophore is thiazolidinedione.
- For agonist activity, R must be a *para* substituted phenyl ring attached to the central nucleus via a methylene bridge.
- Original drugs were troglitizone, rosiglitizone (Avandia), and pioglitizone (Actos).
- Troglitizone was withdrawn due to severe hepatotoxicity and rosiglitizone is only available from certified pharmacies because of its increased risk of cardio-vascular effects.
- Pioglitizone is the only glitizone in common use today.

Physicochemical and Pharmacokinetic Properties

- Highly protein bound to albumin (>97%).
- Absolute bioavailability is 83% not affected by food.
- Dosage adjustment not necessary in patients with renal failure.
- Maximum hypoglycemic effect not until 3 to 4 months.

Metabolism

- Metabolized by CYP2C9 and metabolites M1, M2, and M3 contribute to agonist activity (Fig. 17.6).

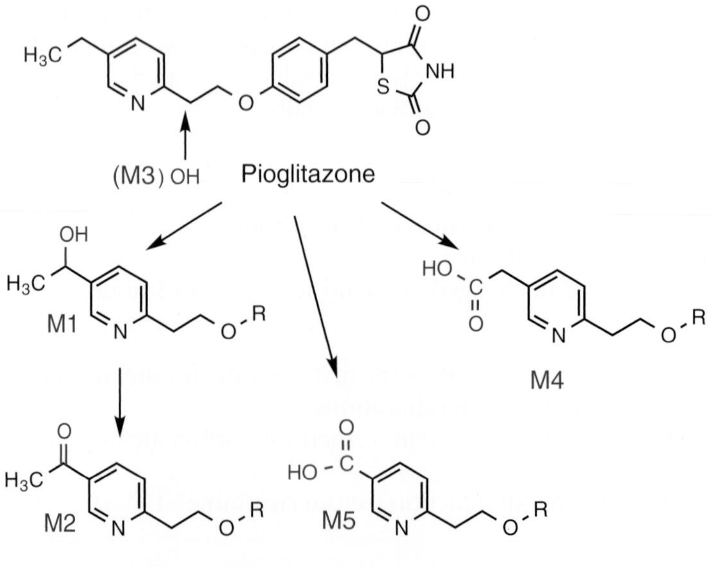

Figure 17.6 Metabolism of pioglitazone.

Clinical Applications

- Major role for pioglitizone is as an alternative for patients who do not respond or cannot tolerate sulfonylureas in combination with metformin to attain HbA$_{1C}$ <7%.
- Very low incidence of drug–drug interactions.

⚠ ADVERSE EFFECTS

Pioglitazone: Weight gain; peripheral edema; heart failure; increase risk of bladder cancer; Rosiglitazone has high risk of MI.

α-Glucosidase Inhibitors

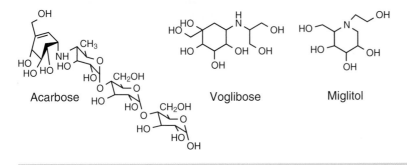

Acarbose   Voglibose   Miglitol

MOA

- α-Glucosidase consists of maltase, sucrase, isomaltase, and glucoamylase, and is a membrane-bound enzyme present in the brush border of the small intestines.
- α-Glucosidase catalyzes the breakdown of disaccharides to monosaccharides which are then actively absorbed.
- The disaccharides are formed by the action of salivary and pancreatic α-amylase on ingested complex polysaccharides.
- Inhibiting α-glucosidase delays the absorption of carbohydrate in the gut by moving undigested disaccharides into the distal sections of the small intestines and the colon thus reducing postprandial hyperglycemia.

SAR

- α-Glucosidase inhibitors mimic the natural disaccharide substrates by containing polyhydroxy groups.
- Acarbose (Precose) is an oligosaccharide isolated from *Actinomyces utahensis*.
- Voglibose (Voglib) and miglitol (Glycet) are respectively, cyclohexane and piperidine, polyhydroxy-containing molecules.

Physicochemical and Pharmacokinetic Properties

- Acarbose (Precose) has extremely low oral bioavailability with <2% absorbed.
- Miglitol (Glycet) is absorbed orally but not metabolized and is excreted unchanged in the urine.
- Absorption of miglitol is saturable at high doses with 25 mg being completely absorbed while a 100-mg dose is only 50% to 70% absorbed.
- No evidence exists to show that systemic absorption of miglitol adds to its therapeutic effect.
- Voglibose (Voglib) is poorly absorbed and no metabolites have been identified.

Metabolism

- Acarbose is metabolized solely in the GI tract by intestinal bacteria forming sulfate, methyl and glucuronide conjugates. (Fig. 17.7).
- About 34% of the intestinal metabolites are absorbed and excreted through the kidneys.
- An active metabolite is formed by the cleavage of one glucose molecule.

Clinical Applications

- α-Glucosidase inhibitors are used for treatment and management of type 2 diabetes in combination therapy as second- or third-line agents.
- Acarbose is the drug of choice in this category lowering HbA$_{1c}$ by 0.5% to 1%.
- Miglitol (Glycet) and voglibose (Voglib) have the same indication as acarbose.

⚠ ADVERSE EFFECTS

 α-Glucosidase inhibitors: Bloating, gas, nausea, and diarrhea.

☆ DRUG–DRUG INTERACTIONS

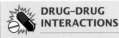 Acarbose: Digoxin absorption inhibited; Miglitol: Glucosamine possible hyperglycemia; Acarbose and Miglitol: Somatropin may antagonize the hypoglycemic effect; Voglibose: None known.

Figure 17.7 Metabolism of Acarbose.

GLP-1 Agonists

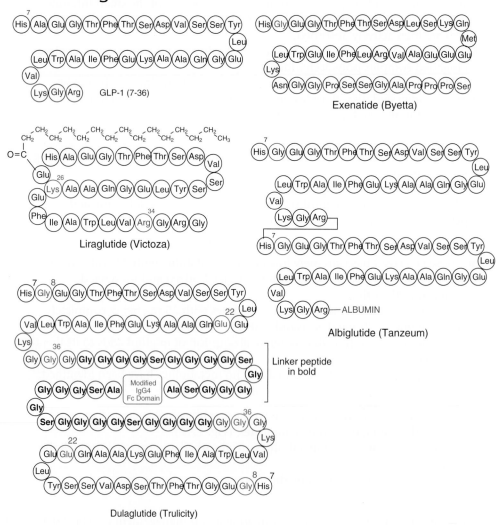

Dulaglutide (Trulicity)

- GLP-1 is a 30 or 31 amino acid peptide produced in the L-cells of the GI tract by posttranslational processing of proglucagon by the action of prohormone convertase enzymes.

MOA

- GLP-1 is released from the L-cells of the GI tract in response to a meal in two molecular forms, GLP-1-(7–36) and GLP-1-(7–37). Each form is equipotent but the GLP-1-(7–36) is predominant.
- GLP-1 promotes insulin secretion by the pancreatic β-cells.
- GLP-1 is classed as an incretin (along with GIP) since it augments the amount of insulin released by an oral dose of glucose.
- Release of GLP-1 and GIP from the L-cells of the GI tract has the same mechanism as the release of insulin by the glucose-stimulated insulin secretion from pancreatic β-cells (i.e., metabolism of glucose in the L-cells leads to closure of the ATP-linked K^+ channels resulting in the influx of Ca^{2+} causing the release of insulin).

SAR

- GLP-1 agonist analogs were developed to resist metabolism by dipeptidyl peptidase IV (DPP-IV) which removes a His-Ala dipeptide from the N-terminus.
- Changing the penultimate Ala to Gly yields exenatide (Byetta) which is a 39 amino acid analog of GLP-1. Exenatide was first isolated from the saliva of the Gila monster lizard (*Heloderma suspectum*).
- Exenatide has an in vivo half-life of 3 hours as compared to GLP-1 with a half-life of 1 to 2 minutes.
- Liraglutide (Victoza) has Lys^{26} substituted with a α-glutamoyl-(N-α-hexadecanoyl) chain as well as replacement of Lys^{34} with an Arg^{34}.
- Liraglutide has a half-life of 11 hours and is suitable for once a day dosing.
- Albiglutide is generated through the fusion of two copies of modified GLP-1 to human albumin.
- Albiglutide's GLP-1 (97% amino acid sequence homology to human GLP-1 fragment 7–36) sequence has been modified to confer resistance to dipeptidylpeptidase IV (DPP-IV) mediated proteolysis.
- The human albumin moiety of albiglutide, together with the DPP-IV resistance, greatly extends its half-life to 5 days allowing once weekly dosing.
- Dulaglutide is a fusion protein that consists of 2 identical chains containing an N-terminal GLP-1 analog sequence covalently linked to a modified IgG4 Fc domain by a small peptide linker and is produced using mammalian cell culture.
- The GLP-1 analog portion of dulaglutide is 90% homologous with human GLP-1-(7–36).
- The IgG4-Fc moiety is modified to resist interaction with high-affinity Fc receptors, cytotoxicty, and immunogenicity.
- Dulaglutide's modifications increase its duration of action and allow for once weekly dosing.

Physicochemical and Pharmacokinetic Properties

- GLP-1 agonist analogs have an increased half-life since they are all resistant to DPP-IV metabolism as well as having decreased renal elimination.
- Renal elimination is decreased because of their enhanced binding to albumin.
- Typical responses to the GLP-1 analogs include:
  - Enhanced release of insulin.
  - Suppressed release of glucagon.
  - Depression of gastric emptying.
  - Reduced appetite.
- Exenatide is self-regulating in that it lowers blood sugar when levels are elevated but does not lower blood sugar when glucose levels return to normal.
- Exenatide reaches median peak plasma concentrations in 2.1 hours following SC administration.
- Liraglutide has the stronger albumin binding which is achieved by the presence of the α-glutamoyl-(N-α-hexadecanoyl) chain.
- Maximum plasma concentrations of liraglutide are achieved at 8 to 12 hours after SC dose.

- The average steady state concentration of liraglutide over 24 hours is approximately 128 ng/mL.
- Absolute bioavailability of liraglutide following subcutaneous administration is approximately 55%.
- Albiglutide reaches maximum plasma concentration within 3 to 5 days after SC administration.
- The average steady state concentration of albiglutide is reached in 4 to 5 weeks and its elimination $t_{1/2}$ is 5 days.
- Dulaglutide is administered SC and reaches maximum plasma concentration within 24 to 72 hours.
- The absolute bioavailability of dulaglutide following SC administration is 65% of a single 0.76 mg dose and 47% following a single 1.5 mg dose.
- The elimination $t_{1/2}$ of dulaglutide for both doses is approximately 5 days.

Clinical Applications

- The GLP-1 agonists are indicated as additional therapy for patients who have not achieved glycemic control on metformin or a sulfonylurea or both.
- Liraglutide, albiglutide, and dulaglutide all have a risk of causing thyroid C-cell cancer and the products have black box warnings with the recommendation that it be used only in patients for whom the benefits outweigh the risk.
- Albiglutide and dulaglutide also have a black box warning in regard to their contraindication in patients with a personal or family history of medullary thyroid cancer (MTC) or patients with multiple endocrine neoplasia syndrome type 2 (MEN 2).

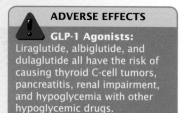

ADVERSE EFFECTS

GLP-1 Agonists:
Liraglutide, albiglutide, and dulaglutide all have the risk of causing thyroid C-cell tumors, pancreatitis, renal impairment, and hypoglycemia with other hypoglycemic drugs.

Dipeptidyl Peptidase IV Inhibitors (DPP-IV)

MOA

- Developed as small molecule, oral alternatives to the GLP-1 agonist analogs (Fig. 17.8).
- They prolong the activity of GLP-1 and GIP by reversibly inhibiting their metabolic breakdown by DPP-IV.
- DPP-IV is a proline-specific serine protease (aminopeptidase).
- The exact mechanism of DPP-IV inhibition varies, however the cyano group in saxagliptin, vildagliptin, and alogliptin appears to bind reversibly to the active site serine in DPP-IV (Fig. 17.9).

SAR

- There are currently 5 DPP-IV inhibitors used clinically: saxagliptin, vildagliptin, sitagliptin, linagliptin, and alogliptin.
- Developed to insert a basic amino group in the position equivalent to the penultimate Ala in GLP-1 (see Gly in Exenatide).
- There are three pharmacophores: α-aminoacylpyrrolidine, xanthine, and 2,4-pyrimidine dione.
- Three inhibitors (saxagliptin, vildagliptin, and alogliptin) contain a cyano group which most likely forms a reversible covalent imidate adduct with the enzyme active site Ser[630] deactivating DPP-IV (Fig. 17.9).
- Saxagliptin and vildagliptin contain the α-aminoacylpyrrolidine pharmacophore.
- Sitagliptin has a piperazine ring fused to a pyrazole in place of the pyrrolidine but maintains the α-aminoacyl moiety and the amide bond.
- Linagliptin contains a xanthine ring as the central pharmacophore.
  - The xanthine has a C8 aminopiperidine and an N7 butynyl substituent for binding to the DPP-IV enzyme catalytic site.
  - The C8 aminopiperidine's primary amine occupies the recognition site for the amino terminus of GLP-1 and hydrogen bonds with Glu[205], Glu[206], and Tyr[547] being held in place by π-stacking aromatic interactions with the phenol of the Tyr[547].

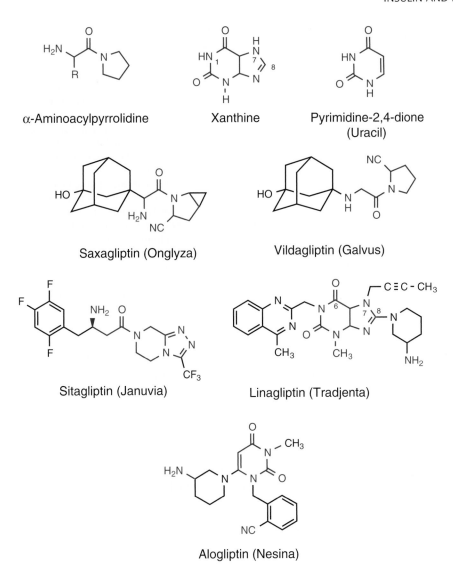

Figure 17.8 α-Aminoacylpyrrolidine, xanthine and uracil pharmacophores and currently available DPP-IV inhibitors.

- The N7 butynyl substituent occupies a hydrophobic pocket.
- The C6 xanthine carbonyl hydrogen bonds with the backbone NH of Tyr[631].
- The quinazoline ring interacts by π-stacking with the aromatic ring of Trp[629].
- Alogliptin contains the 2,4-pyrimidinedione pharmacophore with the essential basic amino group on the piperidine ring and a cyano group which most likely binds to the active site serine as in the saxagliptin example (Fig. 17.9).

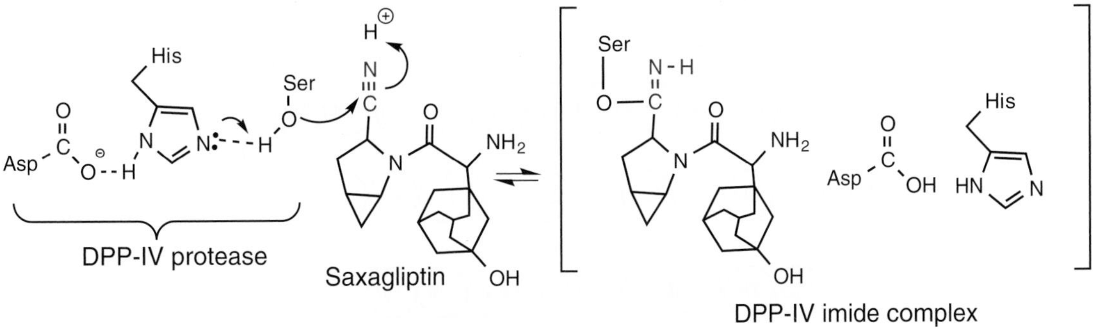

Figure 17.9 Proposed mechanism of reversible inhibition of DPP-IV by saxagliptin.

Table 17.4 Some Pharmacokinetic Parameters of DPP-IV Inhibitors			
Inhibitor	Bioavailability	Protein Binding	Metabolism
Sitagliptin (Januvia)	87%	38%	No significant metabolism. 79% excreted unchanged in urine; remainder in feces
Linagliptin (Tradjenta)	30%	70–80%	No significant metabolism. 90% excreted unchanged; 80% in feces; remainder in urine
Saxagliptin (Onglyza)	67%	<10%	50% of absorbed dose metabolized by CYP 3A4 yielding an active metabolite
Vildagliptin (Galvus)	90%	<10%	No significant metabolism. Elimination half-life of ~90 min
Alogliptin (Nesina)	63%	NS	Minimal metabolism (10–20%) to inactive metabolites. 63% excreted unchanged via the kidneys.

Physicochemical and Pharmacokinetic Properties

- DPP-IV inhibitors enhance GLP-1's ability to produce insulin in response to a meal (Table 17.4).
- Decrease glucagon.
- Slow the rate of nutrient absorption.
- Slow gastric emptying.
- Reduce food intake.
- Saxagliptin (Onglyza) is metabolized by CYP3A4 to 5-hydroxy saxagliptin which is half as potent. Therefore it should be monitored when coadministered with 3A4 inhibitors.

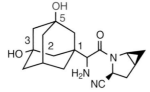

5-Hydroxy Saxagliptin

ADVERSE EFFECTS

DPP-IV Inhibitors:
Headache and nasopharyngitis. Linagliptin may be implicated in acute pancreatic and facial edema.

Clinical Applications

- The DPP-IV inhibitors can be used either alone or in combination with metformin or a thiazolidinedione for control of type 2 diabetes mellitus.
- They do not increase weight and rarely produce hypoglycemia.

Amylin Agonists

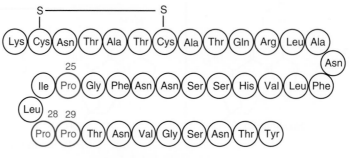

Pramlintide (Symlin)

MOA

- Amylin is a 37 amino acid hormone released along with insulin from the pancreatic β-cells.
- Amylin helps maintain glucose plasma levels by suppression of glucagon secretion, delaying gastric emptying, and modulating the appetite center of the brain.
- Amylin is unsuitable as a drug because it aggregates and is insoluble in solution.

SAR

- Pramlintide (Symlin) is an amylin agonist analog which has enhanced water solubility and reduced aggregation liability by replacing Ala^{25}, Ser^{28}, and Ser^{29} with prolines.

Physicochemical and Pharmacokinetic Properties

- SC administration shows peak effects in 15 minutes with a duration of action of 150 minutes.
- Can be given before meals.
- Elimination is via the kidneys which is also the site of metabolism.
- Elimination half-life of 30 to 50 minutes.
- Should not be mixed with insulin since the difference in pH of the insulin and pramlintide products is 4.0 and 7.8 respectively.

Clinical Applications

- Pramlintide is used together with insulin for those patients who are unable to achieve their postprandial glucose levels on insulin alone.
- Since it delays gastric emptying it may delay absorption of coadministered drugs.
- Use with caution in renal compromised patients since it is renal metabolized.

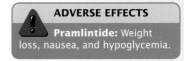

ADVERSE EFFECTS

Pramlintide: Weight loss, nausea, and hypoglycemia.

Sodium Glucose Cotransporter Inhibitors (SGLT2 Inhibitors)

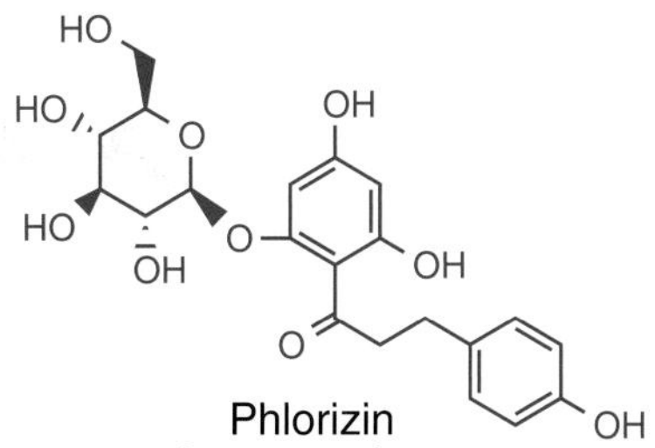

Phlorizin
pharmacophore

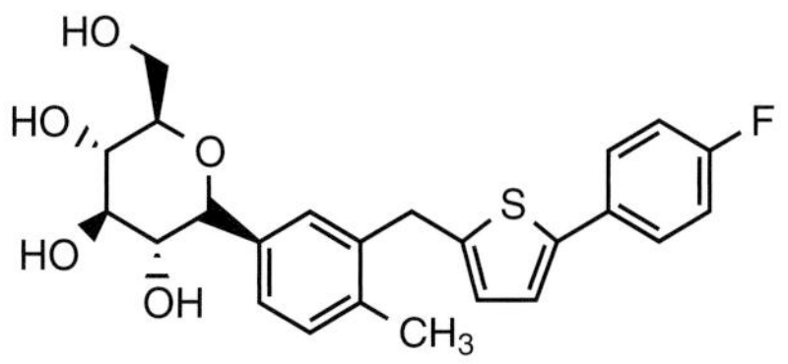

Canagliflozin (Invokana)

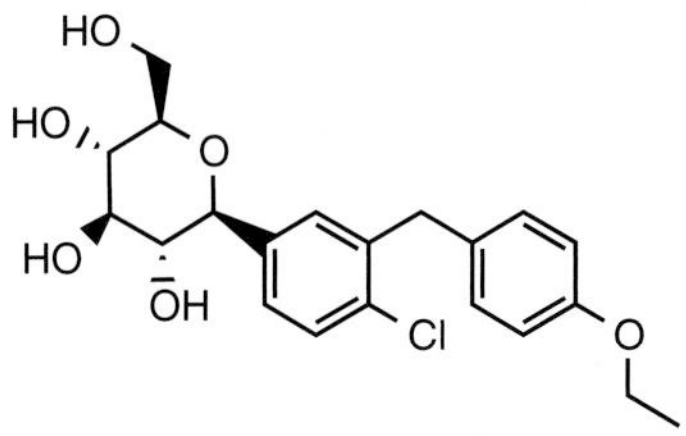

Dapagliflozin (Farxiga)

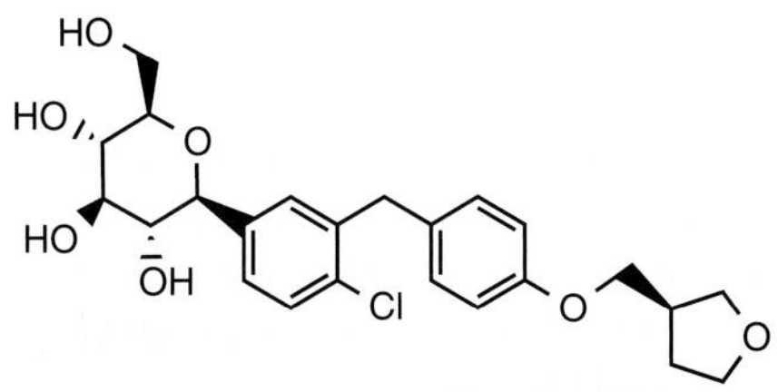

Empagliflozin (Jardiance)

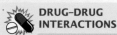

ADVERSE EFFECTS

Canagliflozin, Dapagliflozin, and Empagliflozin: Vaginal yeast infections. Urinary tract infection and increased urination. Increased LDLs. Hypoglycemia in patients on insulin or oral insulin secretagogues. Hypotension (dapagliflozin), bladder cancer (dapagliflozin), hyperkalemia with potassium-sparing diuretics (canagliflozin).

DRUG–DRUG INTERACTIONS

Nonselective inducers of UGT enzymes (phenobarbital, phenytoin, rifampin, ritonavir) may decrease levels of both drugs. Canagliflozin increases Digoxin AUC whereas Dapagliflozin does not.

MOA

- The SGLT2 inhibitors work by inhibiting the sodium-glucose transport protein found in the kidney and is responsible for reabsorbing the majority of filtered glucose from the renal proximal tubules.
- Inhibiting SGLT2 lowers the renal threshold for glucose (RT_G), and thereby increases urinary glucose excretion.
- SGLT2 is a transmembrane protein consisting of 672 amino acids.
- The SGLT2 channel is formed by 14 transmembrane (TM) α-helices.
- TMs 4–7 and TMs 10–13 form a hydrophilic cavity which is involved in substrate binding and translocation.
- Phlorizin and glucose bind to Thr[156] and Lys[157] in TM IV.

SAR

- The pharmacophore (phlorizin) is a β-D-glucoside first isolated from the bark of the apple tree.
- It consists of a glucose moiety linked to two phenyl rings (the aglycone) through an alkyl spacer.
- Glycosides are metabolically unstable being hydrolyzed easily by GI β-glucosidases thereby being poor drug candidates.
- There are currently 3 SGLT2 inhibitors (canagliflozin, dapagliflozin, and empagliflozin) approved for the treatment of diabetes.
- They are designed to be stable to hydrolysis by GI β-glucosidases.
- Canagliflozin replaces the O-glycoside with a C-glycoside thereby conferring metabolic stability as well as significant selectivity for SGLT2 over SGLT1.
- Canagliflozin replaces the alkyl spacer with a thienyl moiety to link the two phenyl rings thus retaining SGLT2 inhibiting activity.
- Dapagliflozin and Empagliflozin are also C-glycosides with a methylene alkyl spacer between the two phenyl rings.

Physicochemical and Pharmacokinetic Properties

Canagliflozin

- Ninety-nine percent protein bound.
- Sixty-five percent oral bioavailability with a t_{max} = 1 to 2 hours.
- Half-life ($t_{1/2}$) is between 10.6 hours and 13.1 hours depending upon the dose.
- Elimination: Feces – as canagliflozin (41.5%), a hydroxylated metabolite (7.0%), and an O-glucuronide (3.2%).
- Urine- as O-glucuronide (30.5%) metabolite with <1% as unchanged drug.

Dapagliflozin

- Ninety-one percent protein bound.
- Eighty percent oral bioavailability.
- Mean plasma half-life is 12.9 hours.
- Elimination: 75% of oral dose in feces (only 15% as parent drug) and 21% in urine (only 2% as parent drug).

Empagliflozin

- Eighty-six percent protein bound.
- ~60% oral bioavailability.
- Mean plasma half-life is 13.1 hours.
- Elimination: Primarily excreted as glucuronides via the kidney with 15% excreted unchanged.

Metabolism

- Canagliflozin is metabolized by both UGT1A9 and UGT2B4 to inactive glucuronides with minor oxidative metabolism via CYP3A4 (Fig. 17.10). Dapagliflozin is primarily metabolized by UGT1A9. Empagliflozin's most abundant metabolites are three glucuronides (2-O-, 3-O-, and 6-O-glucuronide) metabolized by UGT2B7, UGT1A3, UGT1A8, and UGT1A9.

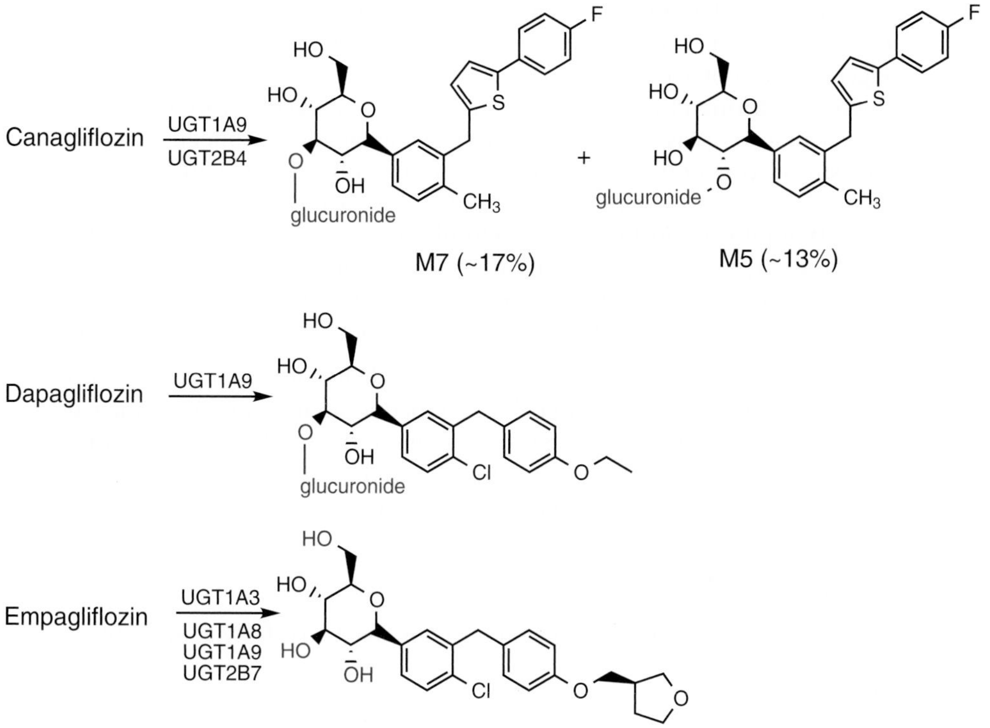

Figure 17.10 Metabolism of canagliflozin and Dapagliflozin by uridine glucuronyltransferase (UGT).

Clinical Applications

- Canagliflozin, dapagliflozin, and empagliflozin are indicated as adjuncts to diet and exercise to improve glycemic control in adults with type 2 diabetes mellitus.
- Not recommended in patients with type 1 diabetes.
- Should not be used in patients with severe renal impairment (GFR <30 mL/min) although empagliflozin has been shown to be safe in patients with mild renal impairment.

Chemical Significance

Diabetes mellitus is a global health problem. It is estimated by the WHO that 347 million worldwide have diabetes and it is projected to be the seventh leading cause of death by 2030. In the United States 25.8 million children and adults – 8.3% of the population – have diabetes and it is estimated to rise to 30.3 million by 2030. Diabetes increases the risk of heart disease and stroke; neuropathy (nerve damage); retinopathy and blindness, and is among the leading causes of kidney failure. Diabetes is a disease with no cure and therefore requires chronic drug treatment from insulin supplementation to a number of hypoglycemic agents with a variety of chemical structures and mechanisms of action. The pharmacist is in the forefront of patient and health care provider information and counseling when it comes to questions about the specific drugs prescribed for the treatment of diabetes.

As an integral part of a pharmacist's education, medicinal chemistry gives the pharmacist insight into how chemical structure relates to a drug's mechanism of action as well as its adsorption, distribution, metabolism, excretion, and adverse effects all of which relates the choice of an appropriate medication. Some examples:

- If Insulin is formulated with zinc acetate then it will be a long-acting insulin.
- Sulfonylurea structures can be recognized as having the ability to release insulin from pancreatic β-cells by binding to the ATP-sensitive K+ channel leading to an intracellular increase in Ca^{2+} causing exocytotic insulin release.

- If a sulfonylurea has a large p-(β-arylcarboxyamidoethyl) group at the R_1 position then it will have increased potency, rapid onset, shorter plasma half-life and longer duration of action.
- A biguanide formed by linking 2 guanidino groups together with different side chains will lower plasma glucose by decreasing hepatic gluconeogenesis and increasing glycogenolysis. Also will enhance insulin sensitivity in liver and skeletal muscle, but have the potential to cause toxic lactic acidosis.
- A DPP-IV inhibitor must have an electrophilic group (usually cyano) in the 2 position of the pyrrolidine ring in order to form an irreversible covalent imidate adduct with the enzyme active site Ser^{630} deactivating it.
- SGLT2 inhibitors have a C-glycoside which prolongs their duration of action.

The role of the pharmacist requires a vast knowledge of facts in combination with skills in order to deliver competent therapeutic advice and care to the patient. Understanding the role of the chemical structure in the mechanism of action of drug molecules gives the pharmacist a most useful handle upon which to grab when recalling facts to apply to their role as drug experts.

Review Questions

1 Which of the following insulin analogs contains a C14 fatty acid chain?
   A. Lispro
   B. Detemir
   C. Aspart
   D. Glulisine
   E. NPH

2 Which of the following oral hypoglycemic drugs stimulate the release of insulin from the pancreas by blocking the ATP-sensitive K+ channel?
   A. Metformin
   B. Canagliflozin
   C. Nateglinide
   D. Pioglitizone
   E. Acarbose

3 Which of the following hypoglycemic drugs is a PPARγ agonist?
   A. Voglibose
   B. Miglitol
   C. Pioglitizone
   D. Exenatide
   E. Canagliflozin

4 Which of the following DPP-IV inhibitors contains a central xanthine pharmacophore?
   A. Sitagliptin
   B. Linagliptin
   C. Saxagliptin
   D. Vildagliptin
   E. Alogliptin

5 Which of the following drugs is based on the phlorizin pharmacophore?
   A. Glibenclamide
   B. Canagliflozin
   C. Liraglutide
   D. Repaglinide
   E. Metformin

Steroids: Adrenocorticoids, Androgens, Estrogens, and Related Nonsteroidal Compounds

18

Learning Objectives

Upon completion of this chapter the student should be able to:

1. Discuss steroid nomenclature including the numbering system and differentiate between α and β, and *cis* and *trans* stereochemistry.

2. Identify specific sites in the biosynthesis of steroid hormones that are critical to understanding the mechanism of action of the therapeutic agents.

3. Outline the metabolism of the end products of steroid hormone biosynthesis.

4. Describe the action of the steroidal hormones at their respective receptors as they relate to the mechanisms of action of therapeutic hormones.

5. Identify the structural features (SAR) important to the steroidal drugs; specifically, the corticosteroids, androgenic–anabolic steroids, and estrogenic and progestin steroids.

6. Recognize metabolic reactions involving steroidal hormones and label the resulting metabolites as active or inactive.

7. Discuss the physicochemical and pharmacokinetic properties of the steroidal hormones.

8. Describe the unique dosage forms of the steroidal hormones and identify the chemical reasons for these dosage forms.

9. Discuss the ancillary classes of drugs associated with men's and women's health including mechanism of action, physicochemical and pharmacokinetic perimeter, and metabolism of:
   a. erectile dysfunction (ED) drugs
   b. α_1-Adrenergic antagonists
   c. 5α-Reductase inhibitors
   d. luteinizing hormone–releasing hormone (LHRH) therapy
   e. antiandrogens (nonsteroidal and steroidal)
   f. selective estrogen receptor modulators (SERMs)
   g. aromatase Inhibitors

Steroids – General Characteristics

Introduction

The steroids of biologic and medicinal significance generally fall into one of three groups: adrenocorticoid (corticosteroid), androgen, or estrogen hormones, which include the progestins. The corticosteroids include the glucocorticoids, which regulate carbohydrate, lipid, and protein metabolism, and the mineralocorticoids, which influence salt balance and water retention.

The corticosteroids and androgens have much in common including:

- structures, chemistry, and nomenclature.
- stereochemical features.
- chemical substituents which render water soluble, orally activity, or modify drug absorption.

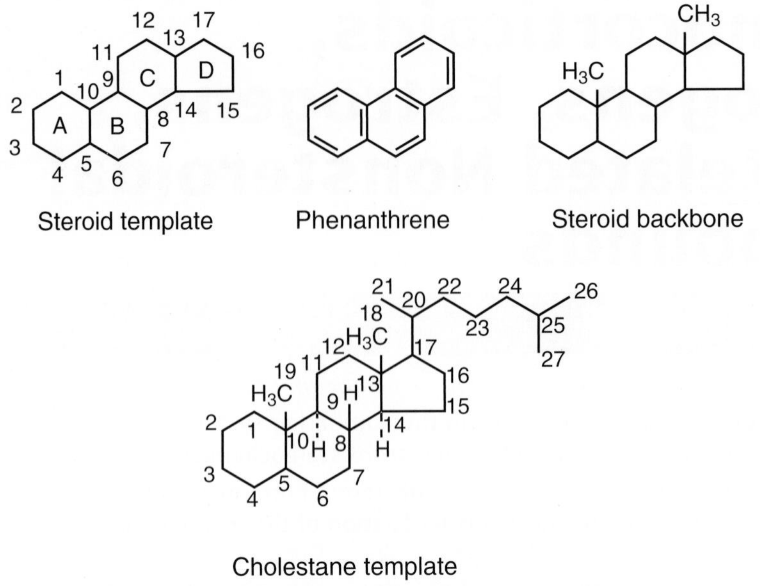

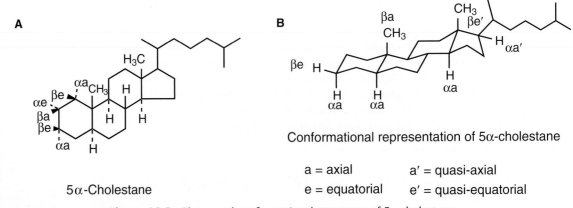

Figure 18.1 Basic steroid structure and numbering system.

- a biochemical origin in acetyl–coenzyme A and cholesterol.
- steroid hormone catabolism which occurs primarily in the liver through similar metabolic pathways.

Despite similarities, each class of steroids demonstrates unique and distinctively different biologic activities.

The first part of this chapter focuses on the similarities among the steroids and reviews steroid nomenclature, stereochemistry, biosynthesis, and general mechanism of action. The second portion of the chapter will focus on the specific classes of steroids (e.g., adrenocorticoids, androgens, and estrogens).

Steroid Nomenclature and Structure

- Steroids consist of four fused rings (A, B, C, and D) (Fig. 18.1).
- Steroids are hydrocarbons with a cyclopentane (D) ring attached to a saturated phenanthrene (i.e., cyclopentanoperhydrophenanthrene).
- The steroid numbering system is illustrated by cholestane, a 27-carbon steroid (Fig. 18.1).
- Note that the angular methyl groups are numbered 18 (attached to C13) and 19 (attached to C10).
- The steroid structure has two planes: the top or β surface which points upward and the bottom or α surface pointing downward (Fig. 18.2A) (5α-cholestane).

Figure 18.2 Planar and conformational structures of 5α-cholestane.

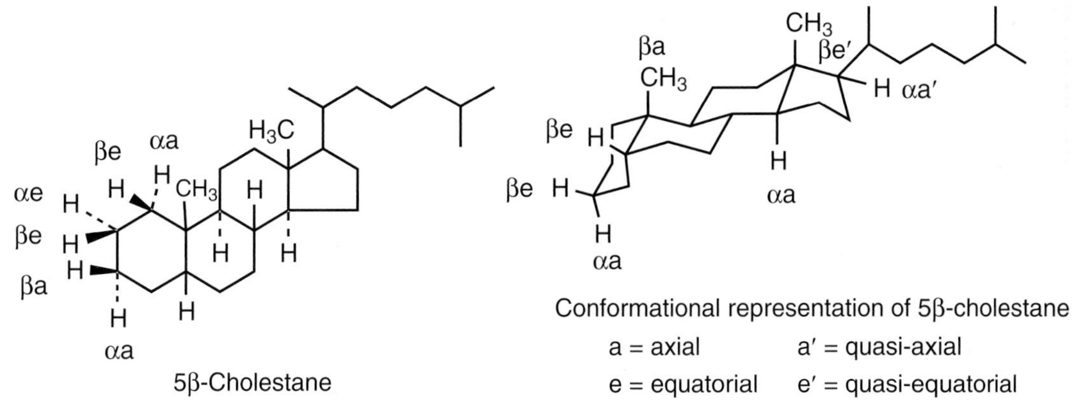

Figure 18.3 Planar and conformational structures for 5β-cholestane.

- Hydrogen atoms or functional groups on the β surface are denoted by solid lines; whereas, those on the α surface are designated by dotted lines.
- The 5α notation in Figure 18.2A is used to denote the configuration of the hydrogen atom at C5, which is opposite from the C19 angular β-methyl group, making the A/B ring juncture *trans*.
- Similarly, the configuration of the 8β and 9α hydrogens and the 14α hydrogen and C18 angular β-methyl group denote *trans* fusion for rings B/C and C/D.
- The side chains at position 17 are always β in the endogenous steroids unless indicated by dotted lines.
- Hydrogen atoms, methyl groups, or other substituents can be found on the α face in either an equatorial arrangement (α-equatorial [αe] or an axial arrangement α-axial [αa]) or on the β face in either an equatorial arrangement (β-equatorial [βe] or an axial arrangement β-axial [βa]) (Fig. 18.2B).
- Steroids are considered rigid ring structures.
- Substituents on adjacent carbons are either *trans* and are drawn as a broken line and solid line (i.e., C5 hydrogen and C19 methyl) or *cis* (i.e., C18 methyl and C17 side chain) and are drawn with heavy lines (Fig. 18.2A).
- The term *anti* defines the orientation of rings that are connected to each other and have a *trans* type relationship (i.e., the bond equatorial to ring B, at C9, is anti to the bond equatorial to ring B, at C10; thus, 5α-cholestane has a *trans-anti-trans-anti-trans* backbone).
- 5β-cholestane (Fig. 18.3) has a *cis-anti-trans-anti-trans* backbone in which the A/B rings are fused *cis*.
- The conventional drawing of the steroid nucleus does not show the hydrogen atoms at 8β, 9α, or 14β positions, and the C18 and C19 methyl groups are shown as solid lines. If the carbon at C5 is saturated, the hydrogen is added and drawn as either 5α or 5β.

Steroid Biosynthesis

The biosynthetic pathway leading to the corticosteroid, androgen, and estrogen hormones begins with cholesterol esters and proceeds through pregnenolone as shown in Figure 18.4. Key points of these biosynthetic pathways are listed below, but the reader is referred to pages 913 to 914, 1350 to 1352, and 1389 to 1390 in *Foye's Principles of Medicinal Chemistry, Seventh Edition*, for a complete detailed discussion. The site of the terminal step in the biosynthesis of the various steroids is shown in Table 18.1.

Glucocorticoid/Mineralocorticoid Biosynthesis

- Corticotropin-releasing factor (CRF) released by the hypothalamus stimulates the anterior pituitary to produce adrenocorticotropic hormone (ACTH).
- ACTH traveling via the bloodstream binds to and stimulates the adrenal glands (adrenal cortex) to release cholesterol.

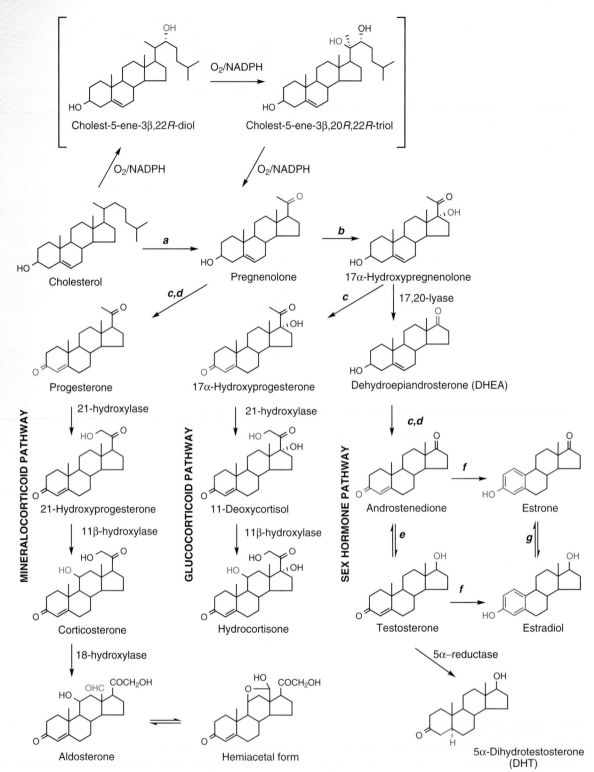

Figure 18.4 Biosynthesis of the adrenocorticoids and sex hormones from cholesterol. The enzymes involved in these biosynthetic pathways are (**a**) adrenodoxin reductase, (**b**) 17α-hydroxylase, (**c**) 5-ene-3β-hydroxysteroid dehydrogenase, (**d**) 3-oxosteroid-4,5-isomerase, (**e**) 17β-hydroxysteroid dehydrogenase type 5, (**f**) aromatase, and (**g**) estradiol dehydrogenase. Oxidative reactions involve CYP450 isoforms (CYP11A, 11B2, 17, and 21).

- This process involves the ACTH/G-protein/adenylyl cyclase–coupled receptors to produce cAMP. The cAMP activates cholesterol esterase, which releases free cholesterol from the adrenal cortex gland.
- Circulating levels of glucocorticoids have a negative inhibitory feedback loop on the hypothalamus-pituitary-adrenal (HPA) axis to reduce their synthesis.

Table 18.1 Sites of Steroid Synthesis

Steroid	Major Endocrine	Stimulating Hormones
Corticosteroids/ glucocorticoids	Adrenal cortex	Hypothalamus (corticotropin-releasing factor [CRF])
		Anterior pituitary (adrenocorticotropic hormone [ACTH])
Mineralocorticoids	Adrenal cortex	Angiotensin II
Estrogens	Ovary/placenta/adipose tissue	Anterior pituitary (gonadotropins – follicle-stimulating hormone [FSH], luteinizing hormone [LH], etc.)
Androgens	Testes	Anterior pituitary (gonadotropins – FSH and LH)

- The oxidative conversion of cholesterol to pregnenolone occurs in the mitochondria of the adrenal cortex.
- Pregnenolone is the precursor to the mineralocorticoids, glucocorticoids, androgen, and estrogen hormones (Fig. 18.4).
- Oxidation and isomerization converts pregnenolone to progesterone.
- Oxidation at the C21, C11, and on the C18 methyl group gives rise to corticosterone and aldosterone, with these latter two steps occurring under the influence of the renin–angiotensin system.
- 17α-Hydroxypregnenolone formed from pregnenolone is the precursor to both the glucocorticoids and the androgen–estrogen hormones (Fig. 18.4).
- Oxidation, isomerism, and hydroxylation of 17α-hydroxypregnenolone give rise to hydrocortisone (Fig. 18.4).

Androgen Biosynthesis

- Regulation of androgen biosynthesis begins with hypothalamic-pituitary-testicular (HPT) axis. LHRH secreted by the hypothalamus stimulates the pituitary gland to release follicle-stimulating hormone (FSH) and luteinizing hormone (LH).
- LH binds to its receptor on the surface of the Leydig cells to initiate testosterone biosynthesis through the release of cholesterol and formation of 17α-hydroxypregnenolone (Table 18.1) (Fig. 18.5).
- FSH binds to membrane receptors on Sertoli cells to regulate spermatogenesis and testis development.

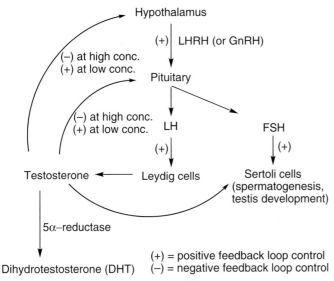

Figure 18.5 HPT axis positive feedback loop control (+); negative feedback loop control (−).

- Increased testosterone has a negative inhibitory action on the hypothalamus and pituitary to inhibit release of gonadotropin-releasing hormone (GnRH) and FSH/LH, respectively.
- Dehydroepiandrosterone (DHEA) (a C19 steroid) is produced in the endoplasmic reticulum membrane through the action of **17,20-lyase** followed by dehydrogenase/isomerase-catalyzed c and d formation of androstenedione (Fig. 18.4). (Bold lettering identifies the key step associated with ancillary drug sites of action.)
- Testosterone and androstenedione are interconvertible (NOTE: testosterone is also synthesized in the ovaries and placenta of women).
- 5α-Dihydrotestosterone (DHT), with high binding affinity to androgen receptors (ARs), is formed by **5α-reductase**, type 1 and type 2 (type 1 expressed in sebaceous glands, type 2 expressed in various tissues including the prostate, hair follicles, and liver).
- Loss of the C19 angular methyl group and aromatization, via the **aromatase** enzyme, of the A ring leads to formation of 17β-estradiol or estrone, which are interconvertible (Fig. 18.4, step g).
- Common estrogen synthesizing sites include ovaries, follicle of the corpus luteum, adipose tissue, and placenta in females.
- FSH initiates the process of estrogen synthesis via its action in the ovaries (Table 18.1).

End Product Metabolism: Hydrocortisone

- The corticosteroid end product hydrocortisone is rapidly but reversibly metabolized to inactive cortisone by 11β–hydroxysteroid dehydrogenase type 1 (11β-HSD1).
- The 11β-HSD1, the "liver" isozyme, limits the oral use of hydrocortisone (Fig. 18.6) and regulates hepatic gluconeogenesis in the liver and fat production in adipose tissues.
- 11β–hydroxysteroid dehydrogenase type 2 (11β-HSD2), the "kidney" isozyme, is unidirectional, oxidizing hydrocortisone to cortisone, preventing hydrocortisone from binding to mineralocorticoid receptors present in the kidney tissues.
- Liver metabolism of hydrocortisone gives rise to inactive o-glucuronides and o-sulfate conjugates of urocortisol, 5β-dihydrocortisol, and urocortisone, terminating glucocorticoid activity (Fig. 18.6).
- NOTE: End product metabolism of androgen and estrogen hormones will be covered later in this chapter.

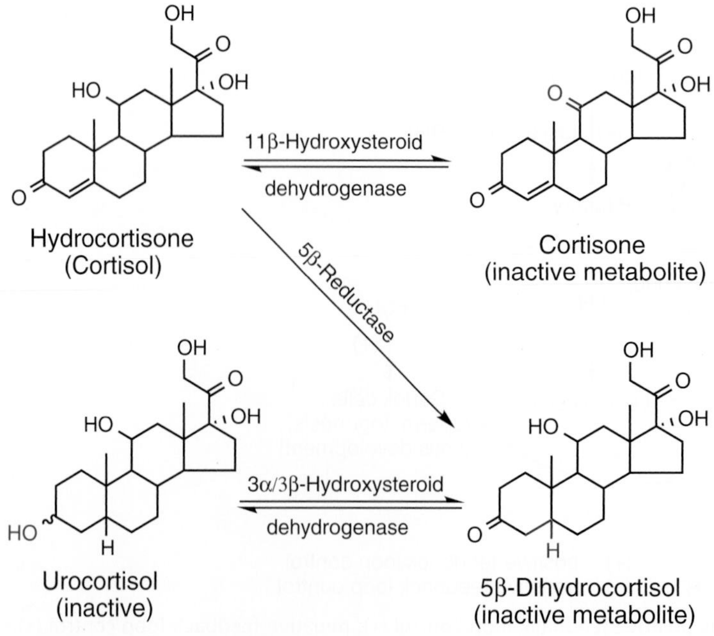

Figure 18.6 Major routes of metabolism for hydrocortisone.

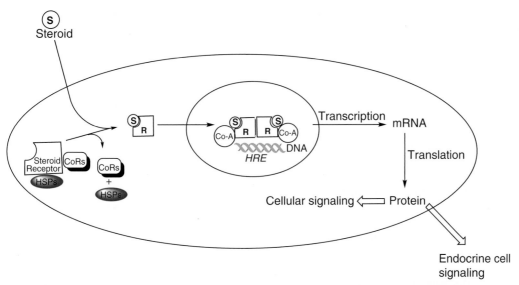

Figure 18.7 General steroid mechanism of action: The steroid receptor (R) is maintained in an inactive complex by heat shock proteins (HSPs) (HSPs 70/90) and corepressor proteins (CoRs). During ligand binding with a steroid (S), the HSPs and CoRs are lost. The steroid–receptor complex enters the nucleus and homodimerization, phosphorylation, and complexation occurs with coactivators (Co-A) and the hormone response element (HRE) of DNA. Gene transcription leads to mRNA synthesis and translation to form proteins.

General Mechanism of Steroid Hormone Action

The steroids, corticosteroids, and androgen–estrogen hormones share a common mode of action. Present at extremely low concentrations (e.g., 0.1 to 1 nmol/L), these steroids bind with high affinity to intracellular receptors regulating gene expression and the biosynthesis via the formation of steroid–receptor complexes and subsequent transactivation gene expression, as outlined in Figure 18.7. Key points in this process include:

- The lipophilic steroid hormones are carried in the bloodstream, reversibly bond to serum carrier proteins, and are released for diffusion through the cell membrane.
- The steroid–receptor complex is found in the cytosol of target cells (Fig. 18.7).
- The steroid initiates a conformational change in the receptor with dissociation of the heat shock proteins and corepressors.
- The steroid–receptor complex translocates to the nucleus where it undergoes dimerization and binding to cellular DNA and nuclear protein factors.
- DNA transcription and mRNA production is followed by an increase in protein synthesis in the rough endoplasmic reticulum.
- The resulting proteins include enzymes, receptors, and secreted factors that subsequently result in the steroid hormonal responses regulating cell function, growth, and differentiation.

Corticosteroids

It should be noted that the steroids covered in this section of the chapter may be classified as either adrenocorticoids or corticosteroids. The drugs listed in Table 18.2 have both glucocorticoid and mineralocorticoid activity, but are used primarily for their glucocorticoid activity. Only fludrocortisone acetate is used therapeutically for its mineralocorticoid activity.

Introduction

Addison disease, Cushing disease, and Conn syndrome are pathologic conditions related to the adrenal cortex and the hormones produced by this gland.

Table 18.2 Corticosteroids, Trade Names, and Their Routes of Administration

Adrenocorticoid	Trade Name	PO	IV	IM	Inhaled/Intranasal	Topical
Alclometasone dipropionate	Aclovate					•
Amcinonide						•
Beclomethasone dipropionate	Qnasl, Quar				•	
Beclomethasone dipropionate monohydrate	Beconase AQ				•	
Betamethasone sodium phosphate/betamethasone acetate	Celestone Soluspan			•		
Betamethasone dipropionate	Diprolene					•
Betamethasone valerate	BetaVal, Dermabet, Luxiq, Valnac					•
Budesonide	Rhinocort				•	•
Ciclesonide	Alvesco, Omnaris				•	
Clobetasol propionate	Clobex, Cormax, Olux, Embeline					•
Clocortolone pivalate	Cloderm					•
Cortisone acetate		•		•		
Desonide	Desonate, Desowen, Verdeso					•
Dexamethasone	Generic, Ozurdex[a]	•				•
Dexamethasone acetate				•		
Dexamethasone sodium phosphate	Maxidex[b]		•	•		•[b]
Desoximetasone	Topicort					•
Diflorasone diacetate						•
Fluocinonide	Lindex, Vanos					•
Fludrocortisone acetate		•				
Flunisolide	Aerospan				•	
Fluocinolone acetonide	Capex, Retisert, Synalar, Dermotic					•[c]
Fluorometholone	FML					•[b]
Fluorometholone acetate	Flarex					•[b]
Flurandrenolide	Cordran, Cordran SP					•
Fluticasone propionate	Cutivate, Flovent Diskus, Flonase				•	•
Halcinonide	Halog					•
Halobetasol propionate	Ultravate					•
Hydrocortisone		•				•[c]
Hydrocortisone acetate						•[c,d]
Hydrocortisone butyrate						•
Hydrocortisone cypionate	Cortef	•				
Hydrocortisone sodium succinate			•	•		
Hydrocortisone valerate						•
Methylprednisolone	Medrol	•				
Methylprednisolone acetate	Depo-Medrol			•		•
Methylprednisolone sodium succinate	Solu-Medrol, A-Methapred		•	•		
Mometasone furoate	Elocon, Asmanex-HFA				•	•
Mometasone furoate monohydrate	Nasonex					
Prednicarbate	Dermatop					•
Prednisolone		•				
Prednisolone acetate	Flo-Pred	•		•		
Prednisolone acetate	Pred-Mild, Pred-Forte, Omnipred					•[b]
Prednisolone sodium phosphate	Pediapred, Orapred, Orapred ODT	•		•		•[b]
Prednisone		•				
Triamcinolone acetonide	Kenalog, Triderm, Nasacort			•	•	•
Triamcinolone hexacetonide	Aristospan			•		

PO, orally; IV, intravenous; IM, intramuscular.
[a]Intravitreal implant.
[b]Ophthalmic formulations.
[c]Also in otic, ophthalmic, and rectal formulations (Cortenema).
[d]Also in intrarectal foam (Cortifoam).

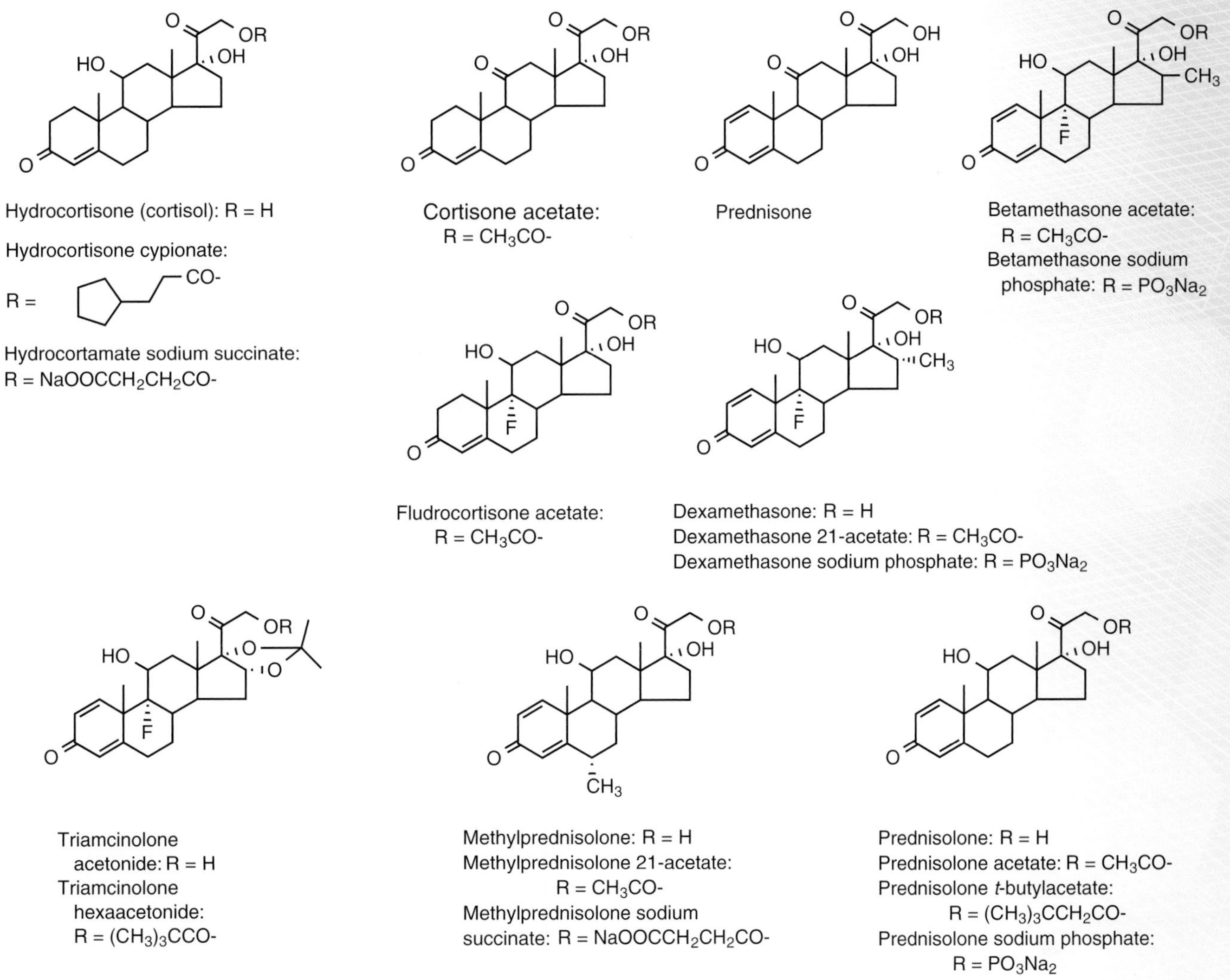

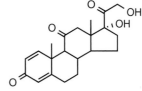

Hydrocortisone (cortisol): R = H

Hydrocortisone cypionate:

R =

Hydrocortamate sodium succinate:
R = NaOOCCH₂CH₂CO-

Cortisone acetate:
R = CH₃CO-

Prednisone

Betamethasone acetate:
R = CH₃CO-
Betamethasone sodium
phosphate: R = PO₃Na₂

Fludrocortisone acetate:
R = CH₃CO-

Dexamethasone: R = H
Dexamethasone 21-acetate: R = CH₃CO-
Dexamethasone sodium phosphate: R = PO₃Na₂

Triamcinolone
   acetonide: R = H
Triamcinolone
   hexaacetonide:
   R = (CH₃)₃CCO-

Methylprednisolone: R = H
Methylprednisolone 21-acetate:
   R = CH₃CO-
Methylprednisolone sodium
succinate: R = NaOOCCH₂CH₂CO-

Prednisolone: R = H
Prednisolone acetate: R = CH₃CO-
Prednisolone t-butylacetate:
   R = (CH₃)₃CCH₂CO-
Prednisolone sodium phosphate:
   R = PO₃Na₂

Figure 18.8 Systemic corticosteroids.

These diseases are associated with hypoadrenalism, hyperadrenalism, and defects in the biochemical synthesis of adrenocorticoids, respectively. These early observations and discovery had led to considerable interest in the functioning of this gland, which ultimately opened the door to the development of adrenocorticoid drugs.

Systemic Corticosteroids

The systemic corticosteroids are shown in Figure 18.8.

MOA

- The corticosteroids act as anti-inflammatory agents through binding to the glucocorticoid receptor (GR), which in turn complexes with DNA to regulate gene expression.
- It is proposed that the corticosteroids inhibit or switch off inflammatory genes by recruiting the enzymatic protein histone deacetylases (HDACs) to reverse the effects of histone acetyltransferase (HAT) (Fig. 18.9).
- Inflammatory diseases (e.g., asthma, rheumatoid arthritis, COPD, inflammatory bowel disease) are characterized as conditions in which inflammatory proteins are produced through the action of HAT on chromatin and are associated with production of proinflammatory transcription factors (e.g., nuclear factor κB [NFκB] and activator protein-1).

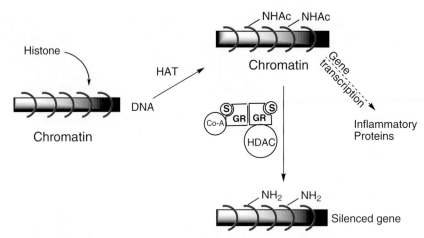

Figure 18.9 Proposed mechanism whereby glucocorticoid receptor~corticosteroid (GR~corticosteroid) complexes with HDAC and coactivators (CoA) are capable of repressing gene-activated inflammatory protein synthesis through deacetylation of histone-wrapped DNA (chromatin) (NOTE: HAT, histoneacetyltransferase; S, corticosteroid).

SAR (the result of animal and not human studies)

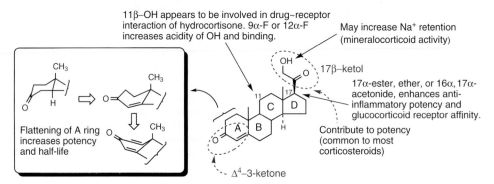

- Steroids~receptors binding may involve the β-surfaces of rings C and D and the **17**β-ketol side chain.
- Bulky substituents on the β-surface abolishes glucocorticoid activity, but similar substitution on α-surface does not.
- 2,6, or 16α-CH$_3$ groups as well as 6α-F group increase glucocorticoid activity.
- Increasing hydrophilicity through addition of α-OH groups decreases glucocorticoid activity.

Physicochemical and Pharmacokinetic Properties
- The corticosteroids can be administered by injection (IV, IM, intravitreal), oral (tablets or solutions), topical, intra-articular, or nasal inhalation (Table 18.2).
- The route of administration depends on the disease being treated and the physicochemical, pharmacologic, and pharmacokinetic properties of the drug (Table 18.3).
- The oral corticosteroids are well absorbed, undergo little first-pass metabolism in the liver, and demonstrate oral bioavailability of 70% to 80%, except for triamcinolone (Table 18.3).
- Glucocorticoids are primarily bound to corticosteroid-binding globulin (transcortin).
- Glucocorticoids cross the placenta and can be distributed into breast milk.

Metabolism
- C21 esters, whether hydrophilic or lipophilic, are hydrolyzed to the active drug by plasma esterases. The nature of the ester affects the onset and duration of action.

Table 18.3 Pharmacokinetics of Commonly Used Oral Corticosteroids

| | Potency Relative to Hydrocortisone | | | | | |
| | Glucocorticoid | Mineralocorticoid | | | Protein | Duration of |
Corticosteroid	Activity[a]	Activity[b]	Bioavailability (%)	Half-Life (h)	Binding (%)	Action (d)
Hydrocortisone	1	2+	96	1.7	90	1–1.5
Betamethasone	20–30	0[b]			>90	2.8–3
Cortisone	0.8	2+	21–94	0.5	95	1–1.5
Dexamethasone	20–30	0[b]	78	3.0	90–95	2.8–3
Methylprednisolone	5	0[b]	90	2.3	90–95	1–1.5
Prednisone	3.5	1+	80	3.6	~90	1–1.5
Prednisolone	4	1+	82	2.8	90–95	1–1.5
Triamcinolone	5	0[b]	23	2.6	~90	1–1.5

h, hours.
[a]Anti-inflammatory, immunosuppressant, metabolic effects.
[b]Sodium and water retention, potassium depletion effects.

- Regardless of the route of administration, the corticosteroids are metabolized in the liver and excreted into the urine primarily as glucuronides.
- Hepatic oxidative metabolism (primarily, CYP3A4-catalyzed reactions) rapidly converts many of the systemic and topical corticosteroids to inactive metabolites protecting patients from the HPA-suppression of endogenous steroid production (Fig. 18.10).

Topical Corticosteroids

- Topically applied glucocorticoids are capable of being absorbed systemically, although to a much smaller extent than oral administration.
- The extent of absorption of topical corticosteroid is determined by concentration in the formulation, the vehicle in which it is applied (cream or ointment), the condition of the skin, and the use of occlusive dressings.
- The intrinsic activity of the steroid strongly influences the systemic effects observed. The commonly used topical corticosteroids are listed in Table 18.2 and Figures 18.8 and 18.11.
- Once absorbed, topical corticosteroids are metabolized through pathways similar to the systemically administered corticosteroids. Metabolism occurs primarily in the liver and the metabolites are excreted into the urine or in the bile.
- While present at very low plasma levels the risk for potential adverse effects from systemic exposure of topical corticosteroids is still significant (see adverse effects).

High-Potency Topical Corticosteroids

Concern exists with the systemic adverse effects associated with the chronic use of topically absorbed potent corticosteroids. Key structural features which increase lipophilicity and potency include: 7α-, 9α-, or C21 chloro groups (alclometasone, beclomethasone dipropionate, and halcinonide, clobetasol propionate, halobetasol propionate, respectively), the propionate and acetate groups, the acetonide group such as that found in triamcinolone, desonide, and amcinonide, and the fluoro groups present in diflorasone diacetate. Topical use could produce adverse effects through suppression of the HPA axis.

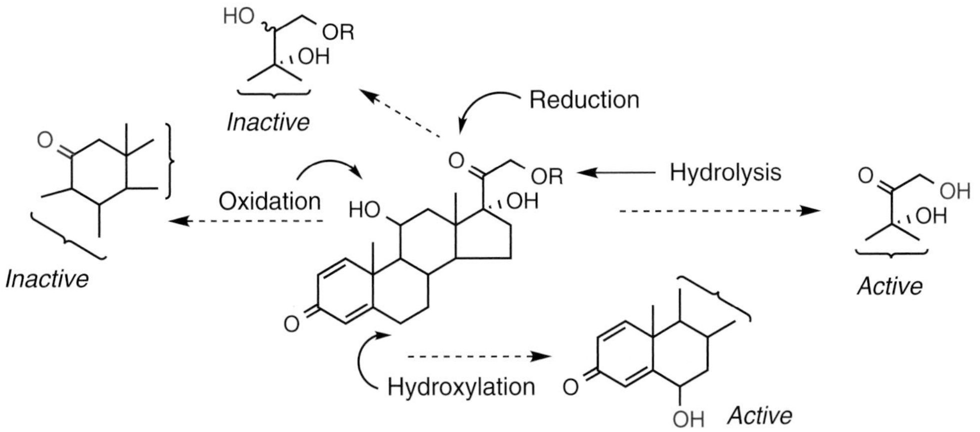

Figure 18.10 General corticosteroid metabolism: C21 esters undergo hydrolysis to the active 21-hydroxy derivative, C11 oxidation to the ketone, C20 reductions give inactive products, and 6-hydroxylation results in an active drug.

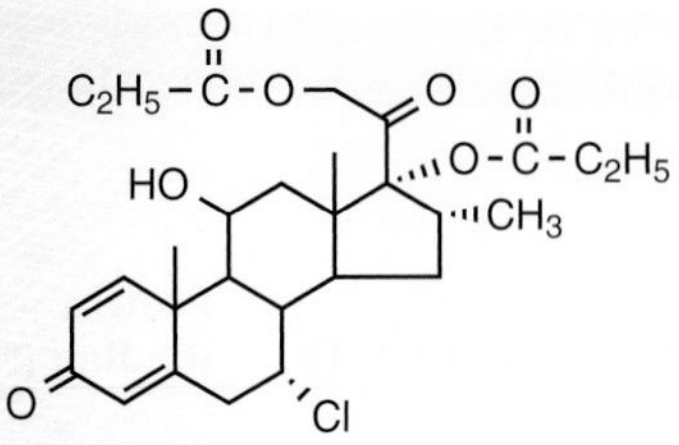

Figure 18.11 Topical corticosteroids.

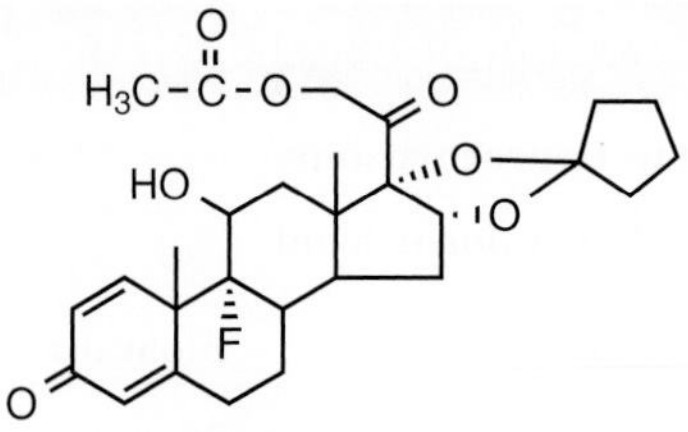

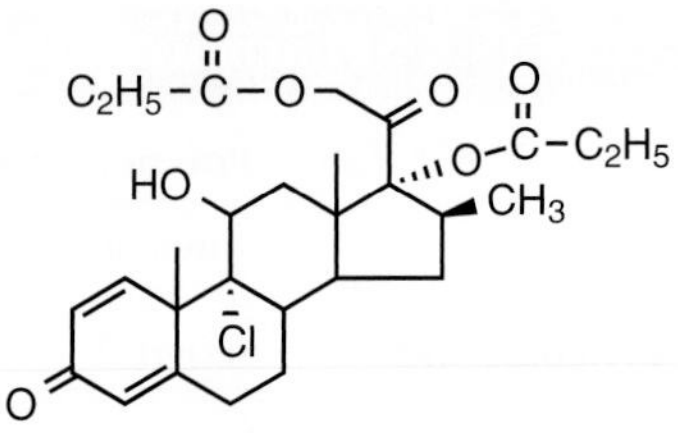

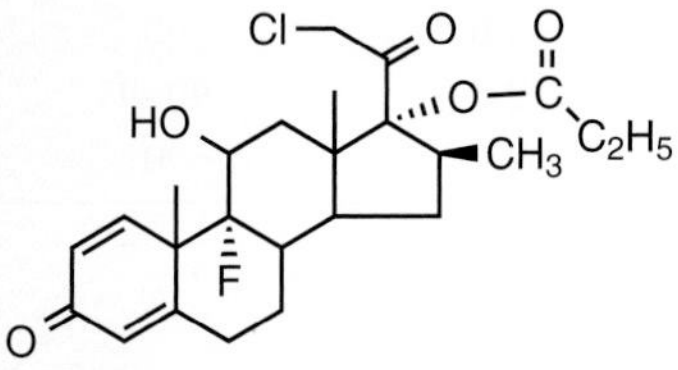

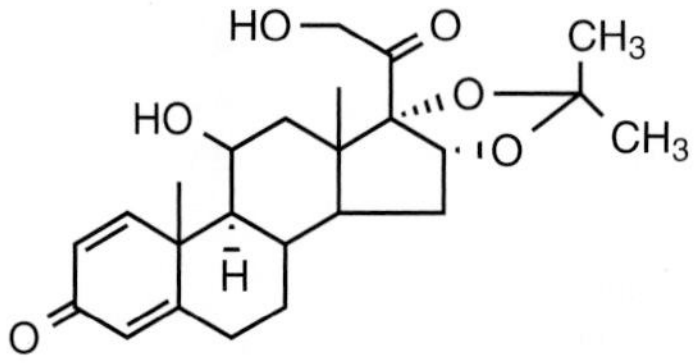

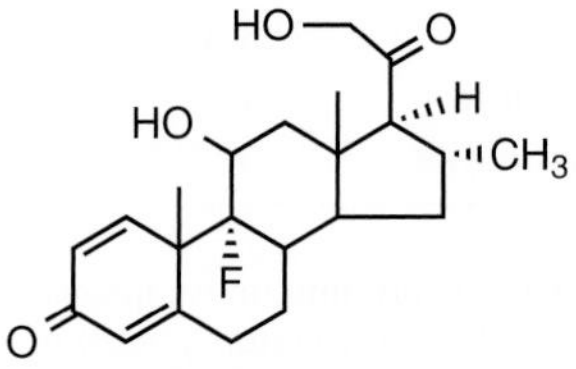

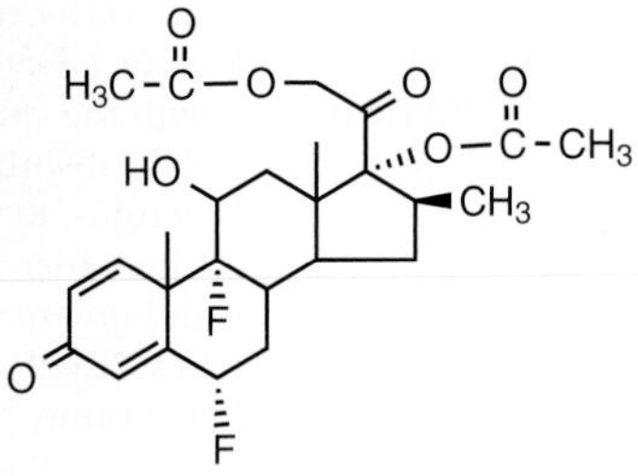

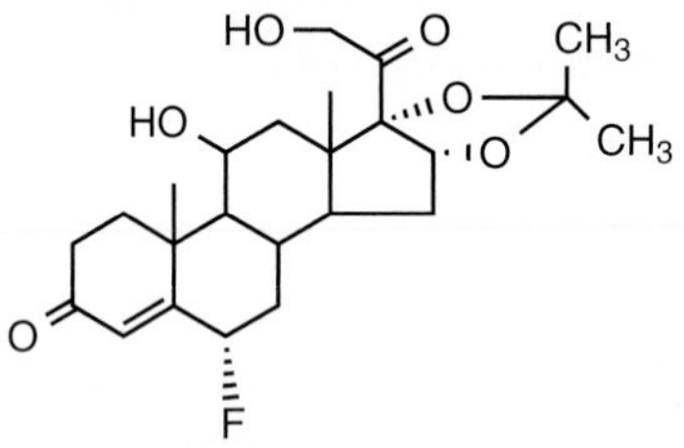

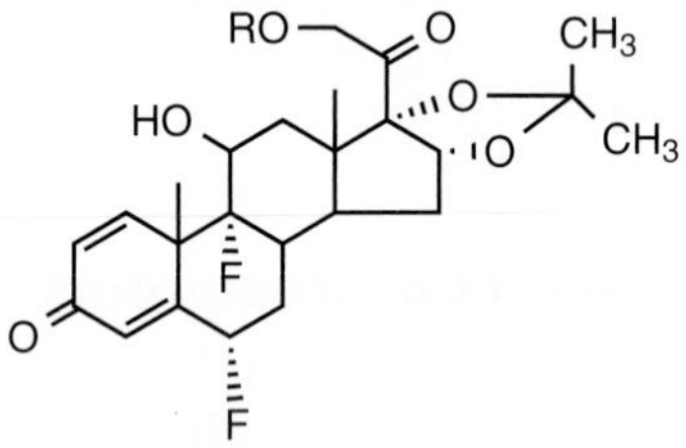

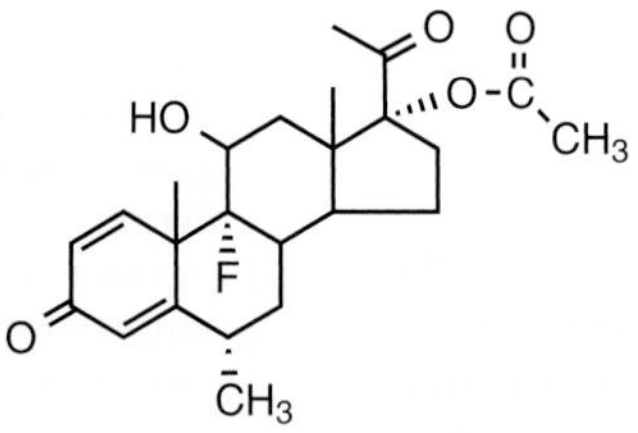

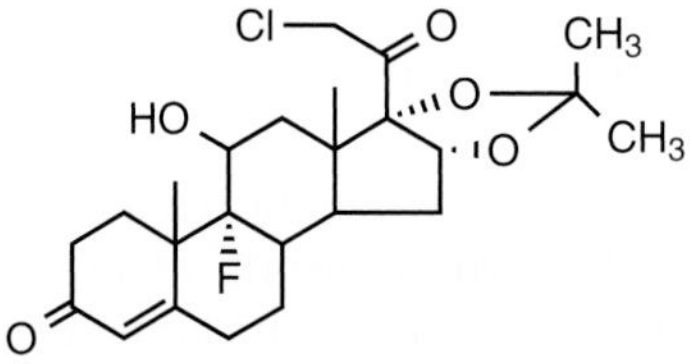

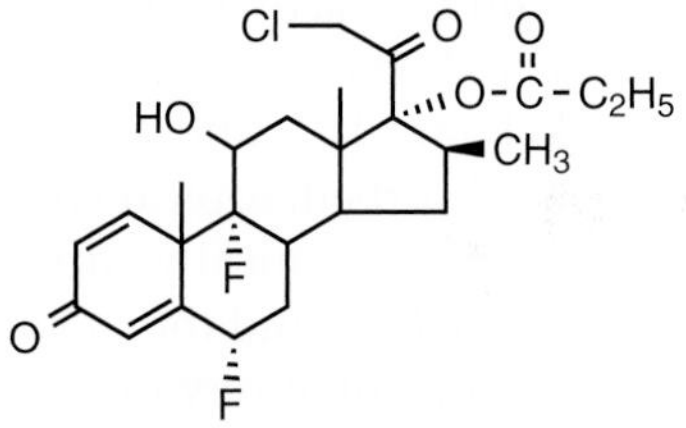

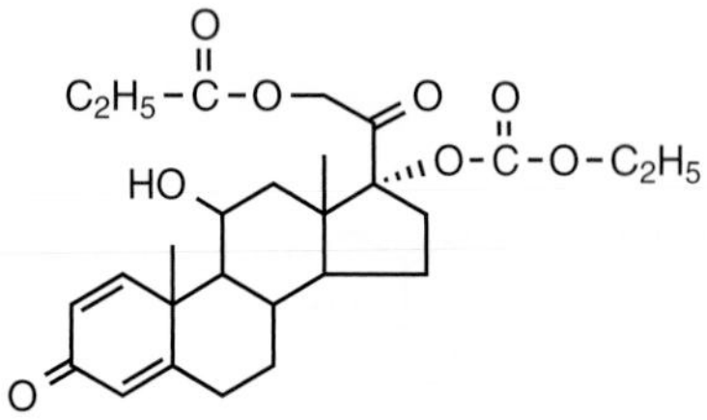

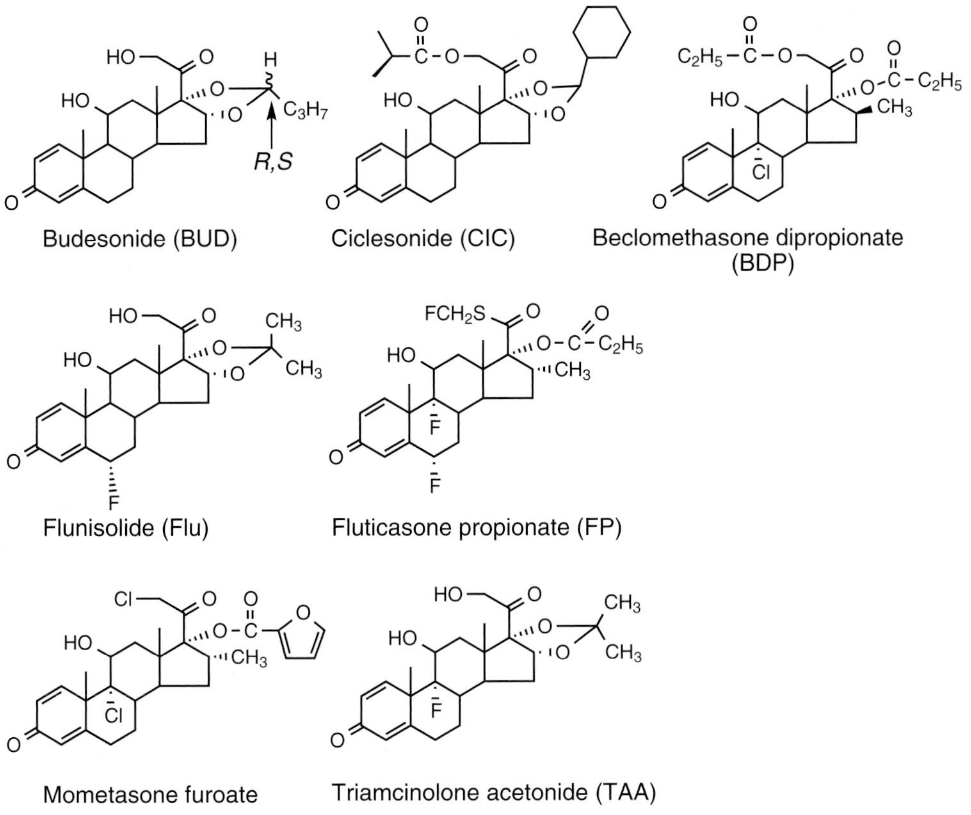

Budesonide (BUD)   Ciclesonide (CIC)   Beclomethasone dipropionate (BDP)

Flunisolide (Flu)   Fluticasone propionate (FP)

Mometasone furoate   Triamcinolone acetonide (TAA)

Figure 18.12 Inhaled and intranasal corticosteroids.

Inhaled and Intranasal Corticosteroids

Site and MOA

- While the MOA does not differ from the oral corticosteroids the site of action is quite different.
- Inhaled and intranasal corticosteroids are expected to exhibit their anti-inflammatory action in the lungs (Table 18.2) (Fig. 18.12). Systemic absorption will elicit adverse effects.
- The potency or affinity of corticosteroids is: CIC>mometasone furoate>FP)>BUD ~ BDP>TAA ~ Flu (see Fig. 18.12 for abbreviations).

SAR

- Substitutions and their effects on glucocorticoid activity are summarized in Figure 18.13.

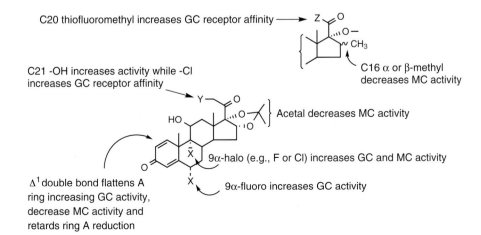

C20 thiofluoromethyl increases GC receptor affinity

C16 α or β-methyl decreases MC activity

C21 -OH increases activity while -Cl increases GC receptor affinity

Acetal decreases MC activity

9α-halo (e.g., F or Cl) increases GC and MC activity

Δ^1 double bond flattens A ring increasing GC activity, decrease MC activity and retards ring A reduction

9α-fluoro increases GC activity

Figure 18.13 SAR effects associated with steroid substitution on glucocorticoid (GC) and mineralocorticoid (MC) activity or receptor affinity.

Table 18.4   Pharmacokinetic Properties of Inhaled and Intranasal Corticosteroids

Parameters	Beclomethasone Dipropionate[a] (BDP)	Budesonide (BUD)	Ciclesonide[a] (CIC)	Flunisolide (Flu)	Triamcinolone Acetonide (TAA)	Fluticasone Propionate[a] (FP)	Mometasone Furoate
Bioavailability (%):							
Pulmonary	~20[b]	~39	~60 desCIC	40	25[c]	~30	<1
Nasal	~20[b]	<20	—	50	25[c]	13–16	ND
Protein binding (%)	87	85–90	99	Moderate	68[c]	91	90
Half-life (h)	~0.17 inhaled				1–7 nasal	~14 inhaled	ND
Onset of action (d)	3–7	2–3		3–7	4–7	2–3	7 h

h, hours; ND, Not detected
[a]Prodrug.
[b]Estimated for inhaled beclomethasone aerosol.
[c]Data from oral inhalation administration.

Physicochemical and Pharmacokinetic Properties

- A comparison of the properties of inhaled and intranasal corticosteroids is difficult due to significant variables. The desired properties include:
  - high lipophilicity (i.e., log $P \geq 4$, BDP, CIC, mometasone furoate, Flu).
  - low bioavailability, high systemic clearance leading to short half-life, high protein binding (Table 18.4).
  - rapid systemic clearance with inactive metabolites (see below).
- 10% to 30% of an inhaled dose of the corticosteroid is deposited in the respiratory tract, with the remainder of the dose deposited in the mouth and throat where it is swallowed. The swallowed drug is eliminated in the feces usually without significant absorption.
- The device for administering the drug (metered dose inhaler or dry powder inhaler [DPI], and spacer use), and patient skill produce variable activity.

Metabolism

- BDP and CIC are prodrugs which are metabolized via hydrolysis of the C21 ester to the active drug (Fig. 18.14).

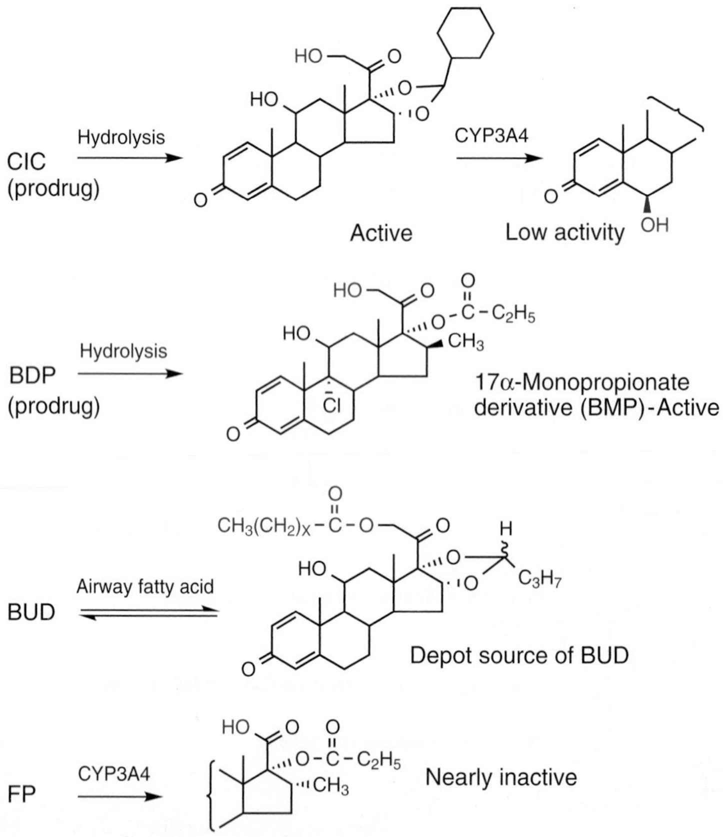

Figure 18.14 Metabolic pathways associated with inhaled and intranasal corticosteroids. Generally, all of these drugs undergo CYP3A4 oxidation at β–C6 reducing activity (shown for CIC), while BUD is reversibly esterified in airways with a fatty acid to give a drug depot. FP undergoes C20 oxidative hydrolysis.

- Further metabolism for most of these inhaled and intranasal corticosteroids involves 6β-hydroxylation (e.g., Flu, BUD, mometasone furoate, triamcinolone acetonide) to a product with greatly reduced activity.

Adrenocorticoid Antagonists

Several adrenocorticoid antagonists have been developed including eplerenone and spironolactone (mineralocorticoid antagonist) and mifepristone (glucocorticoid antagonist). These drugs are discussed in Chapter 13 or later in this chapter. These drugs are competitive inhibitors of the endogenous steroid receptor interfering with mRNA and protein synthesis.

Clinical Applications

Two major uses of glucocorticoids are in the treatment of rheumatoid diseases and allergic manifestations. They are used in the treatment of severe asthma, in sepsis (associated with gram-negative bacteremia), and acute respiratory distress syndrome. They are effective in the treatment of rheumatoid arthritis, acute rheumatic fever, bursitis, spontaneous hypoglycemia in children, gout, rheumatoid carditis, sprue, and allergy and contact dermatitis. The treatment of chronic rheumatic diseases and allergic conditions with glucocorticoids is symptomatic and continuous with symptoms returning after withdrawal of the drug. In addition, these drugs are moderately effective in the treatment of ulcerative colitis, dermatomyositis, periarteritis nodosa, idiopathic pulmonary fibrosis, idiopathic thrombocytopenic purpura, regional ileitis, acquired hemolytic anemia, nephrosis, cirrhotic ascites, neurodermatitis, and temporal arteritis. The newer analogs such as diflorasone diacetate, desoximetasone, flurandrenolide, and fluocinonide are effective topically in the treatment of psoriasis. Glucocorticoids can be combined with antibiotics to treat pneumonia (i.e., AIDS patients with *Pneumocystis carinii* infection), peritonitis, typhoid fever, and meningococcemia.

Fludrocortisone acetate (a drug with high mineralocorticoid activity) along with hydrocortisone is intended for use as partial replacement therapy for adrenocortical insufficiency in Addison's disease (hypoadrenalism).

> **ADVERSE EFFECTS**
>
> **⚠ Corticosteroids:** Serious adverse effects are commonly associated with long-term therapy with the oral corticosteroids. These effects are related to suppression of the HPA axis and involve glucocorticoid-induced adrenocortical insufficiency, glucocorticoid-induced osteoporosis, and generalized protein depletion. Facial mooning, flushing, sweating, acne, thinning of scalp hair, abdominal distention, and weight gain may be seen. Osteoporosis may be quite severe leading to bone loss and increased fracture. With acute adrenal insufficiency, the drugs must not be removed abruptly. Topical and inhaled dosage forms should be chosen whenever possible.

Physiological Effects of Corticosteroids

Corticosteroids influence all tissues of the body and produce numerous effects in cells. These steroids regulate carbohydrate, lipid, and protein biosynthesis and metabolism (glucocorticoid effects), as well as influence water and electrolyte balance (mineralocorticoid effects). The primary physiological function of the glucocorticoids is to maintain blood glucose levels through mobilization and promotion of amino acid metabolism and gluconeogenesis. Additional effects include preventing or minimizing inflammatory reactions and suppressing immune responses through a decreased expression of: platelet-activating factor, interleukin (IL)-1, tumor necrosis factor, and inducible nitric oxide synthase. The primary physiologic function of mineralocorticoids is to maintain electrolyte balances in the body by enhancing Na^+ reabsorption and increasing K^+ and H^+ secretion in the kidney.

Androgens and Men's Health

Introduction

Androgens (i.e., male sex hormones), their pharmacological target (i.e., the AR), and the tissues that rely on the androgens represent the basis of this section of the chapter. The three significant diseases and the drugs that target these diseases will be addressed, with emphasis being placed upon the drugs.

- Aging-related androgen insufficiency (male andropause and hypogonadism) involving the inability of the testes to produce sufficient testosterone to maintain sexual function, muscle mass and strength (sarcopenia), bone mass (osteoporosis), and fertility (spermatogenesis). Testosterone replacement therapy (TRT) is commonly called for.

- Prostate problems, commonly seen in men 50 years and older, include infection of the prostate (prostatitis), a noncancerous enlargement of the prostate (benign prostatic hyperplasia [BPH]), and prostate cancer.
- Testicular cancer is the most common form of cancer among males (aged 15 to 44 years). Testicular cancer is androgen-dependent, with survival rates increasing significantly if treatment begins before the cancer metastasizes.

Androgens

- Testosterone and its metabolite 5α-DHT are the primary endogenous androgens and play crucial physiological roles in establishing and maintaining male characteristics.

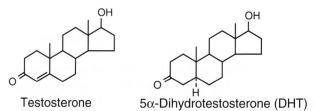

Testosterone   5α-Dihydrotestosterone (DHT)

- Androgens are essential for differentiation and growth of male reproductive organs, initiation and regulation of spermatogenesis, and control of male sexual behavior.
- The androgens (testosterone and DHT) are important for the development of male attributes in other tissues such as muscle, bone, hair, larynx, skin, lipid tissue, and kidney, and their level is regulated via negative feedback loop on the HPT axis (Fig. 18.5).

Androgen Biosynthesis

- Androgens are primarily synthesized in the testes (Table 18.1), but adipose tissue is also a source of androgens in men and women.
- The conversion of pregnenolone via 17α-hydroxypregnenolone, DHEA, and androstenedione is shown in Figure 18.4.
- DHT is formed from testosterone (approximately 6% to 8%) via 5α-reductase isoforms, type 1 and type 2, with a higher affinity for ARs in comparison to testosterone.

Drugs Used to Treat Aging-Related Androgen Insufficiency

MOA

- The androgens (testosterone and DHT) produce their effects through binding to the ARs, which function as an intracellular transcriptional factor (Fig. 18.7).
- The AR targets tissue location:
  - Testosterone AR tissue: muscle, bone, brain, and bone marrow.
  - DHT major tissue: genitalia, prostate, skin, and hair follicles.
- The AR is a soluble protein of 919 amino acids consisting of three major functional domains: N-terminal domain serving a modulatory function, the DNA-binding domain (DBD), and the ligand-binding domain (LBD) (Fig. 18.15).

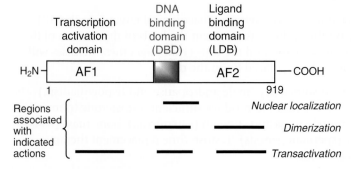

Figure 18.15 Structural features of the AR.

Androgenic Steroids:

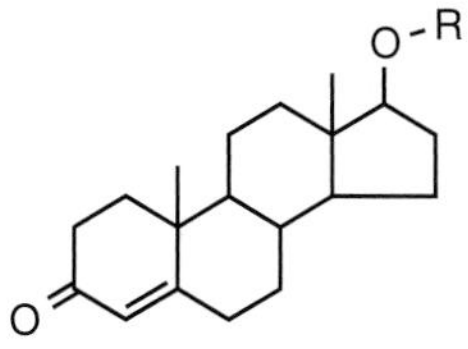

Testosterone 17β-enanthanate
R = CH₃(CH₂)₅CO

Testosterone 17β-undecanoate
R = CH₃(CH₂)₉CO

Testosterone 17β-cyclopentylpropionate
R = ⬠—CH₂CH₂CO
(cypionate)

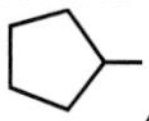

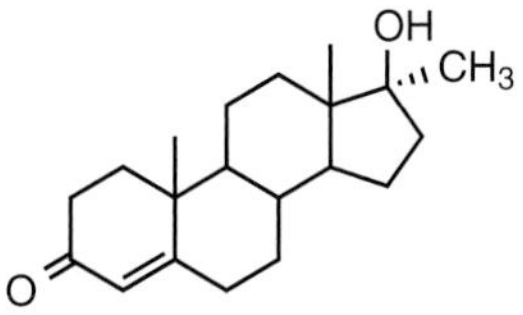

17α-Methyltestosterone 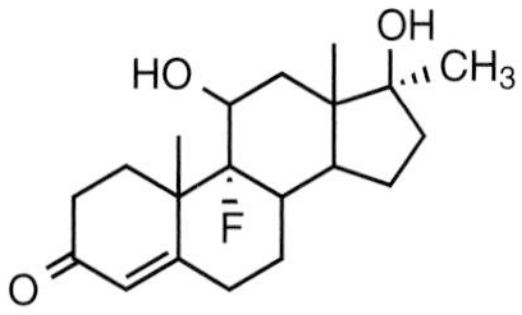   Fluoxymesterone

Anabolic Steroids:

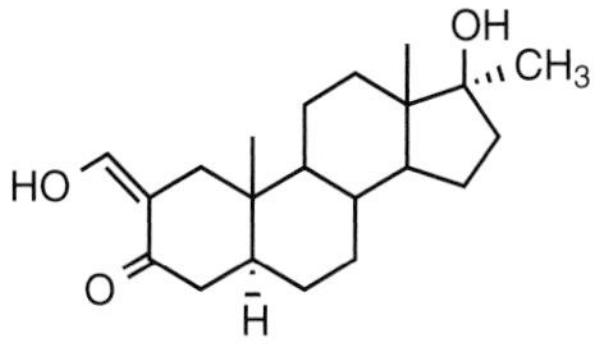

Oxymetholone
(Anadrol)

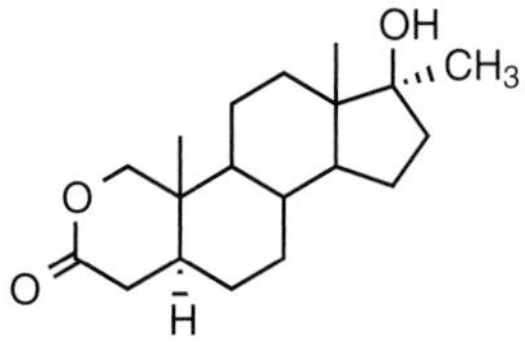

Oxandrolone
(Oxandrin)

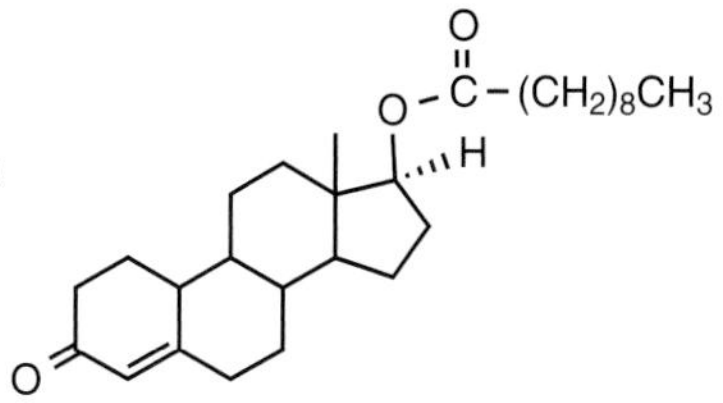

Nandrolone decanoate

Figure 18.16 Androgenic and anabolic steroids.

- Two transactivation functions have been identified: N-terminal activation function (AF1) is ligand-independent and the C-terminal activation function (AF2), which functions in a ligand-dependent manner.
- Upon cytoplasmic agonist binding: AR undergoes conformational changes; heat shock proteins dissociate; dimerization, phosphorylation, and translocation to the nucleus occurs (Fig. 18.7).
- The translocated receptor binds to androgen-response elements (AREs, HRE in Fig. 18.7) in DNA resulting in ligand-induced AR conformational changes.
- A nongenomic pathway of AR also has been reported in various tissues.
- For additional details see pages 1353 to 1354 in *Foye's Principles of Medicinal Chemistry, Seventh Edition*.

SAR

A complete dissociation of anabolic and androgenic effects is not possible, leaving drugs with a gradation of anabolic and androgenic activity. The androgenic steroids are best represented by different testosterone-derived dosage forms plus a limited number of anabolic steroids shown in Figure 18.16.

- Structural modifications that improve androgenic and anabolic activity are shown in Figure 18.17.

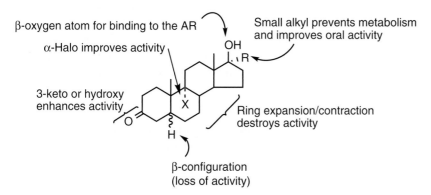

β-oxygen atom for binding to the AR

Small alkyl prevents metabolism and improves oral activity

α-Halo improves activity

3-keto or hydroxy enhances activity

Ring expansion/contraction destroys activity

β-configuration (loss of activity)

Figure 18.17 Structural changes that improve androgenic and/or anabolic activity.

Table 18.5 Testosterone Products and Properties

Drug	Trade Name	Onset of Peak Response	Duration of Action	Time to Peak Conc.	Bioavailability (%)	Elimination Half-Life
Methyltestosterone	Android Testred	—	24 h	2 h	70	3 h
Fluoxymesterone		—	24 h	—	80	9 h
Testosterone undecanoate	Aveed			7 d		
Testosterone cypionate	Depo-Testosterone	6–24 h	2–4 wks	24 h	—	8 d
Testosterone enanthate	Delatestryl	6–24 h	2–4 wks	24 h	—	8 d
Testosterone pellets	Testopel	1–2 mo	3–6 mo	1 mo	—	—
Transdermal patches	Androderm	3–6 mo	24 h	4–12 h	—	10–100 min
Transdermal gels	AndroGel Fortesta Testim	3–6 mo	5 d	4 (2–8) h	~10	10–100 min
Transdermal solution	Axiron	~2 wks		2 h		
Buccal mucosal	Striant	—	24 h	5 (0.5–12) h	—	6 h

h, hours

⚠ **ADVERSE EFFECTS**

Testosterone Replacement Therapy (TRT): Increasing testosterone levels with TRT may pose problems by stimulating the growth of the prostate. Long-term TRT could exacerbate BPH or fuel the growth of undiagnosed prostate cancer and could cause breast enlargement in men (gynecomastia). Men with a palpable prostate nodule, a prostate-specific antigen level greater than 4 ng/mL, or severe lower urinary tract symptoms associated with BPH are usually advised to avoid TRT. Oral preparations (i.e., fluoxymesterone and 17α-methyltestosterone) have very limited use because of potential liver toxicity commonly associated with the 17α-alkyl group. Additional adverse effects include: stomach upset, headache, acne, increased hair, anxiety, change in sex drive, sleeplessness, increased urination, depression, and increased frequency and duration of erections.

- Esters at 17 β–OH must be hydrolyzed to give the active compound.
- Ring A–hydroxylated analogs or oxygen insertion, oxymetholone, and oxandrolone respectively, increase in anabolic over androgenic activity (Fig. 18.16).
- The 19-norsteroids (nandrolone) exhibit a more favorable anabolic to androgenic activity ratio.

Physicochemical and Pharmacokinetic Properties

- Orally administered testosterone is ineffective in the treatment of male androgen insufficiency syndromes due to extensive presystemic first-pass metabolism (~90% – see metabolism section below) and high binding (97% to 98%) to sex hormone-binding globulin (SHBG). Thus the need for other routes of administration.
- A variety of dosage forms are available for TRT with their pharmacokinetic properties shown in Table 18.5. The dosage forms include:
  - **Injection (IM):** Bypass the problems of first-pass metabolism with depot formulations. Testosterone esters undergo differing rates of in vivo ester hydrolysis to release free testosterone over an extended period of time.
  - **Implants:** Subcutaneous testosterone pellets reduce dose flexibility while giving a longer duration of action.
  - **Reservoir-Type Transdermal Systems:** Transdermal TRT systems deliver testosterone bypassing the rapid first-pass metabolism associated with oral products. These products are applied to the abdomen, back, thighs, or upper arms resulting in continuous skin absorption over 24 hours to achieve normal testosterone levels.
  - **Gel:** Delivers testosterone from a hydroalcoholic gel for 24 hours from the skin on the lower abdomen, upper arm, or shoulder. As the gel dries, approximately 10% of the testosterone is absorbed through the skin.
  - **Solution:** Delivered to the underarm which following evaporation of ethanol and isopropyl alcohol leaves the testosterone residue for absorption. Steady state levels of DHT/testosterone result after absorption along with estradiol.
  - **Buccal Mucosal:** In a gel-like substance that adheres to the gumline delivering physiological amounts of testosterone for systemic circulation, especially in hypogonadal males.

Metabolism

- Metabolism occurs in target tissues or the liver, with extensive first-pass hepatic metabolism to inactive metabolites as shown in Figure 18.18.
- Primary metabolites, androsterone, and etiocholanolone are found in the urine (~90%).

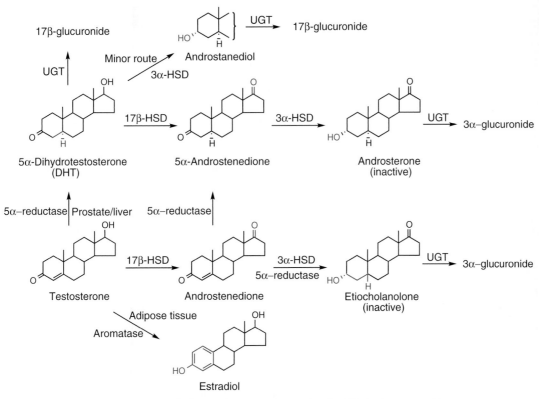

Figure 18.18 Testosterone metabolism. The enzymes involved: *HSD*, hydroxysteroid dehydrogenase; *UGT*, UDP - glucuronosyltransferase.

- In the prostate gland, skin, and liver, testosterone is reduced to DHT by 5α-reductase (types 1 and 2).

Clinical Applications

Testosterone and structurally related steroidal androgens have been used to treat male hypogonadism (testosterone plasma levels <280 to 300 ng/dL) and aging-related androgen insufficiency. Low endogenous testosterone plasma concentrations are associated with sarcopenia and frailty arising from decreased fat-free mass, lessened muscle strength, and reduced bone mineral density (osteoporosis) as well as decreased sexual libido and erectile dysfunction (ED). TRT is available for treatment of these conditions.

The anabolic steroids are approved for use in treatment of anemias caused by deficient red blood cell production (hypoplastic anemia) especially due to myelotoxic drug administration. In addition, these agents are indicated as adjunctive therapy to promote weight gain following extensive surgery, chronic infections, or severe trauma.

Phosphodiesterase (PDE) Inhibitors – Drugs Used for Erectile Dysfunction (ED)

The inability to develop or maintain an erection of the penis sufficient for satisfactory sexual performance is referred to as ED. While hormonal insufficiencies (hypogonadism) may be the cause of ED it is more likely that ED is associated with psychogenic or organic causes (i.e., depression, psychological stress, diabetes, hypertension, spinal cord injuries, and drug adverse effects). Currently, the first-line treatment consists of first- and second-generation oral PDE5 inhibitors.

Physiological Process of Penile Erection

- The physiological mechanism of penile erection is mediated via an NO/cyclic guanosine monophosphate (cGMP) pathway.
- Parasympathetic neurons and vascular endothelial cells release NO, which activates soluble guanylyl cyclase increasing cGMP.

⚠ ADVERSE EFFECTS

Associated with TRT Dosage Forms: The major problem with testosterone ester injections is a saw-toothed pattern of testosterone plasma concentrations, with initial supraphysiological levels followed by subphysiological levels before the next injection. IM injection also causes local irritation with erratic absorption. Pellet implants limit dosage flexibility, while skin reactions have been reported for reservoir-type transdermal systems especially if site of application is not rotated. A potential side effect of the gel is possible transfer of the medication through skin-to-skin contact until the gel is completely dry or covered. This is of particular concern with children and is included as a "black box" warning for these products. The buccal preparation may lead to gum irritation or pain, bitter taste, or headache.

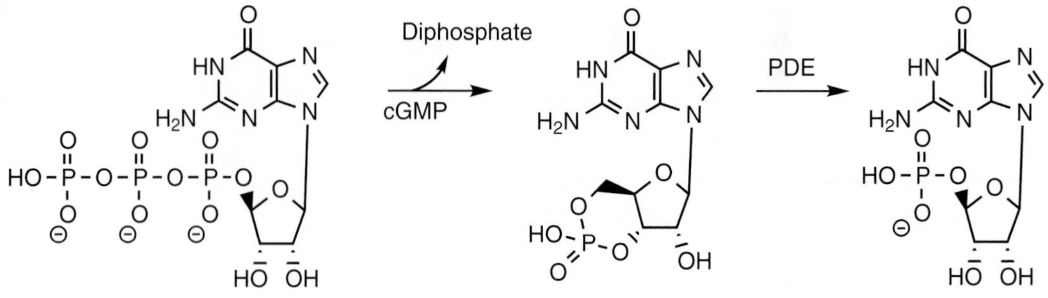

Figure 18.19 Biosynthesis of cGMP and its hydrolytic metabolism.

ADVERSE EFFECTS

Anabolic Steroids:
Performance-enhancing anabolic substances are a point of major interest for athletes, government, and news media. Strong antidoping legislation has resulted in removal of several of the previously available anabolic steroids (i.e., stanozolol, chlorotestosterone, testolactone, and norethandrolone). Adverse effects include shrinking of testicles, gynecomastia, low sperm count, increased hair growth, high blood pressure, heart attack, stroke, high cholesterol, increased aggression, and live disease.

- cGMP of the corpus cavernosum of the penis leads to relaxation of the vascular smooth muscle, increasing blood flow to the penis and penile erection.
- Metabolic hydrolysis of cGMP is catalyzed by PDE enzyme inhibition of which results in a maintained erection (Fig. 18.19).

MOA

- The PDEs are a family of regulatory enzymes found in various tissue of the body.
- ED drugs inhibit PDE5 enzymes found predominately in corpus cavernosum penis.
- PDE5 selectivity is important for reducing adverse effects. The selectivity ratio, PDE5/PDE6, reported for avanafil is 120, vardenafil is 16, and sildenafil is 7.4.
- There are no obvious chemical similarities among the PDE inhibitors, although it has been suggested that the 3-chloro-4-methoxyphenylmethylamino moiety may be important for PDE selectivity (Fig. 18.20).

Physicochemical and Pharmacokinetic Properties

- The PDE5 inhibitors have limited oral bioavailability because of extensive presystemic metabolism in the intestine and hepatic first-pass metabolism via CYP3A4 isoform family (see metabolism below).
- The drugs are quite lipophilic and exhibit rapid absorption after oral administration (Table 18.6).
- High-fat meal had no significant effect on the rate and extent of absorption of tadalafil or avanafil, but decreased the rate of absorption for sildenafil and vardenafil.
- All the drugs are highly bound to plasma protein (94–96%).

Nonsteroidal Androgens

Selective androgen receptor modulators (SARMs) are actively being pursued. The desire is to have a drug with agonist action at muscle and bone ARs resulting in increased muscle and bone mass and antagonist or partial agonist action at prostate ARs, thus sparing prostate effects. Preclinical and early clinical investigations suggest a number of drug classes with potential for SARMs. No drug is approved at this time.

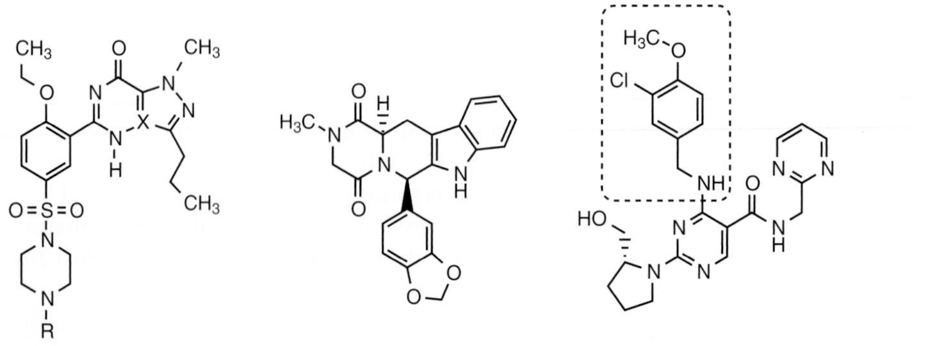

Sildenafil (Viagra)
(X = C, R = CH₃)
Vardenafil (Levutra)
(X = N, R = C₂H₅)

Tadalafil
(Cialis)

Avanafil
(Stendra)

Figure 18.20 Phosphodiesterase 5 inhibitors.

Table 18.6 Some Properties and Pharmacokinetics of the Phosphodiesterase 5 Inhibitors				
Drugs	**Sildenafil**	**Vardenafil**	**Tadalafil**	**Avanafil**
c log P^a	2.3 ± 0.7	3.0 ± 0.7	1.4 ± 0.8	—
Oral bioavailability (%)	38–40	15 (8–25)	~36	38–41
Onset of action (h)	<0.5	<1	0.5–1	~0.25
Duration of action (h)	<4	<1	<36	
Time to peak concentration (h)	0.5–2	0.5–3	0.5–6	
Elimination half-life (h)	3–5	4–5	18	~5
Active metabolites	N-desmethyl	N-desethyl	None	None
Excretion (%) feces/metabolites	~80	>90	~60	~63
Urine/metabolites	~13	<10	~35	~21

h, hours

[a]Chemical Abstracts, American Chemical Society, calculated using Advanced Chemistry Development (ACD/Labs) Software V8.14 for Solaris (1994–2006 ACD/Labs).

Metabolism

- The PDE5 inhibitors all undergo extensive first-pass hepatic metabolism via CYP3A4 and CYP2C9/19 (minor) isoforms to active and inactive metabolites (Fig. 18.21).
- Hepatic CYP3A and CYP2C activity is reduced in the elderly warranting dose reductions for sildenafil and vardenafil in these patients.

Clinical Applications

- The first line of treatment of ED is PDE5 inhibition because these drugs can be given orally.
- If the PDE5 inhibitors are not effective, then the cause may be low libido, and men should have their testosterone blood levels checked (in some instances, TRT may help to resolve ED).
- Other alternative drugs currently available for the treatment of ED include prostaglandin E_1, which is given by injection at the base of the penis or by suppository into the tip of the penis, as well as the α_1-adrenergic blocker and the nonselective PDE inhibitor papaverine.
- Vacuum devices and penile implants are also available.

DRUG–DRUG INTERACTIONS

All inducers and inhibitors of CYP3A activity have the potential to interfere with the elimination of the ED drugs. This interaction potential has been verified clinically for several inducers of CYP3A4 (e.g., rifampicin and tadalafil). The strong inhibitors of CYP3A4 (ritonavir, indinavir, saquinavir, erythromycin, and ketoconazole) increase the plasma levels for sildenafil, vardenafil, tadalafil, and avanafil. Grapefruit juice, a selective inhibitor of CYP3A4 intestinal metabolism, also increases the plasma concentrations of sildenafil and vardenafil but not of tadalafil and avanafil. Ritonavir, as an inhibitor of CYP3A4 and CYP2C9, increases the plasma levels for vardenafil by 50 to 300 times, most likely as a consequence of the simultaneous inhibition of both cytochromes.

ADVERSE EFFECTS

PDE5 Inhibitors: A number of adverse effects have been reported with these drugs, including nausea, cutaneous flushing, headache, and retinal effects, including a bluish haze and increased light sensitivity, and indigestion. It has been suggested that some of these side effects result from the inhibition of other PDEs, including the isoform PDE6. PDE6 is expressed only in the retina and its inhibition has been associated with visual sharpness and color tinge in vision. In addition, consult a physician before combining an α-adrenergic blocker for high blood pressure and an ED inhibitor. Other vasodilators associated with regulating the intracellular levels of cGMP, such as nitroglycerin, should not be used in combination with the PDE5 inhibitors.

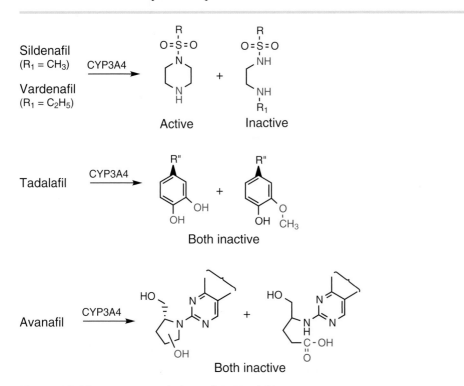

Figure 18.21 Primary metabolites of PDE5 inhibitors.

Drugs Used to Treat Prostate-Related Diseases

Diseases of the prostate represent a threat to men's health with the incidence of BPH escalating rapidly with age. Drugs that inhibit the metabolism of testosterone to DHT (i.e., 5α-reductase inhibitors) or block urethral constriction (α₁-adrenergic receptor antagonists) are used as front-line treatment for urinary obstruction associated with BPH.

Prostate cancer is the most common noncutaneous cancer and remains the second leading cause of death from cancer in American men. AR antagonists (i.e., antiandrogens) and LHRH (also called gonadotropin-releasing hormone) analogs are routinely used followed by anticancer chemotherapy.

α₁-Adrenergic Antagonists

MOA

- Three α₁-adrenoceptor subtypes exist (α_{1A}, α_{1B}, and α_{1D}) (Fig. 18.22).
- α_{1A}-adrenoceptor subtype is the predominate receptor expressed in the prostatic and urethral smooth muscle.
- Uroselective antagonists to these receptors provide relief from the symptoms of BPH.
- Drugs with high affinity for α_{1A}-adrenoceptor, as expressed by K_i (nmol/L), are shown in Table 18.7, although predictable pharmacological profiles based on receptor affinity is not clear at this time.

Physicochemical and Pharmacokinetic Properties

- The α-adrenergic antagonists have good bioavailability as shown in Table 18.7.
- These drugs are highly bound to plasma proteins (82% to 98%).
- Their long half-life permits once-a-day dosing.

Metabolism

- Inactivation of the quinazoline α₁-adrenergic antagonists generally consists of O-dealkylation and N-dealkylation (Fig. 18.23).
- Silodosin undergoes glucuronidation to an active metabolite, which has four times the serum level of silodosin with an extended half-life.
- A second major metabolite of silodosin results from alcohol and aldehyde dehydrogenase oxidation (Fig. 18.23).

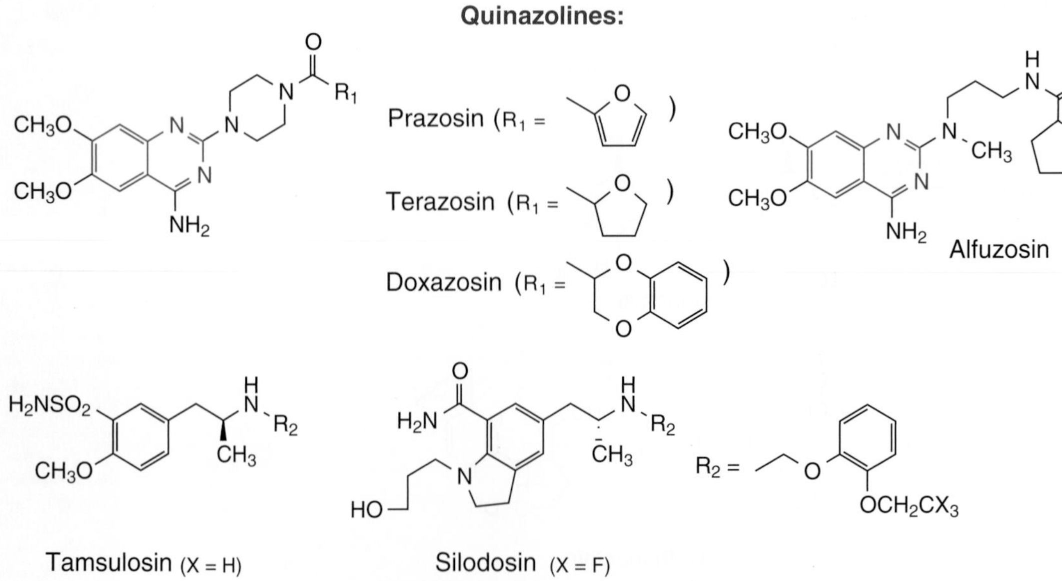

Figure 18.22 α₁-Adrenergic antagonists for treatment of BPH.

Table 18.7 Properties and Pharmacokinetics of the α-Adrenergic Antagonists

Drug[a]	Alfuzosin	Doxazosin	Silodosin	Tamsulosin	Terazosin
Trade name	Uroxatral	Cardura	Rapaflo	Flomax	Generic
Oral bioavailability (%)	65	65 (62–69)	32	>90 fasted	90
Improve urine flow (wks)	<2	1–2	<1	1	2
Duration of action (h)	>48	18–36	24	>24	>18
Time to peak concentration (h)	1–2	2–5	2–3	4–5	1
Elimination half-life (h)	3–10	18–22	13	9–15	9–12
Excretion (%)...feces	69	63–65	55	21	55–65
...urine	24–30	~20	33	76	30
K_i (nmol/L) for α_{1A}	2.4	2.7	1.0	0.5	2.5

h, hours

[a]Prazosin is commonly used for treatment of hypertension and not usually used for BPH.

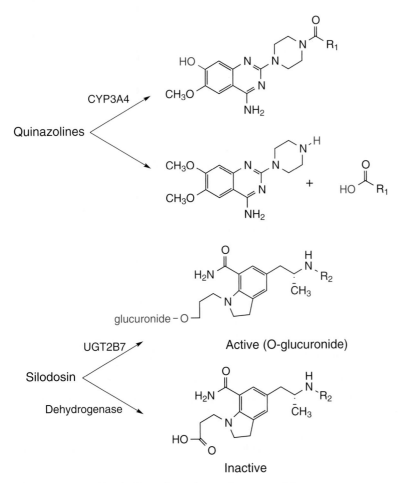

Figure 18.23 Metabolism of α_1-adrenergic antagonists.

5α-Reductase Inhibitors

Finasteride
(Proscar)

Dutasteride
(Avodart)

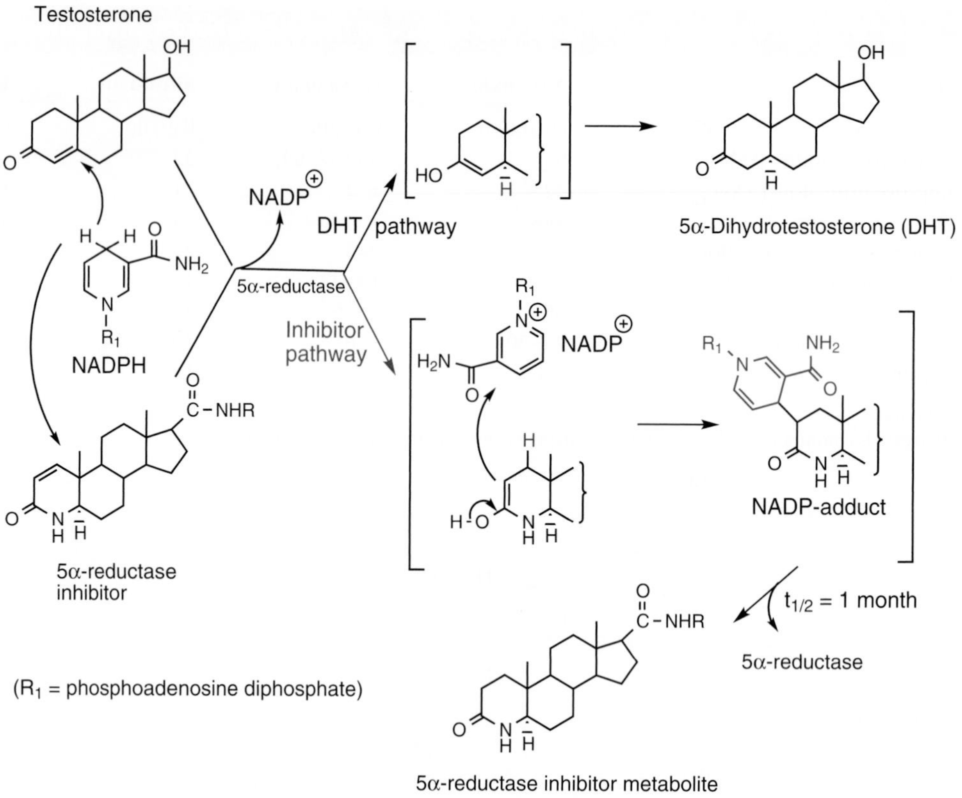

Figure 18.24 5α-Reductase, with its NADPH coenzyme component, reduces testosterone to DHT (DHT pathway), as well as finasteride or dutasteride (inhibitor pathway). The reduced 5α-reductase inhibitor is proposed to alkylate NADP⁺ giving the NADP-adduct that only slowly regenerates the 5α-reductase and the reduced drug.

MOA

- Inhibitors of 5α-reductase are effective in the treatment of BPH.
- Inhibitors suppress the metabolism of testosterone to DHT via a mechanism-based irreversible inhibition of type 1 and type 2 5α-reductase isoenzymes (Fig. 18.24).
- Finasteride is 30 times more selective for type 2 5α-reductase than dutasteride.
- The reduction of finasteride to dihydrofinasteride proceeds through an enzyme-bound, nicotinamide adenine dinucleotide phosphate (NADP)-dihydrofinasteride adduct, thus explaining the exceptional potency and specificity of finasteride and dutasteride.

Physicochemical and Pharmacokinetic Properties

- 5α-Reductase inhibitors are highly bound to plasma proteins (90% to 99%).
- The drugs exhibit good bioavailability and the metabolites are excreted primarily in the urine (Table 18.8).
- Finasteride exhibits a high degree of selectivity for type 1 reductase as indicated by its IC_{50}.

Table 18.8   Pharmacokinetic Properties of 5α-Reductase Inhibitors

Drugs	Finasteride	Dutasteride
Oral bioavailability (%)	65 (34–108)	60 (40–94)
Onset of action (h)	<24	—
Duration of action (wks)	—	>5
Time to peak concentration (h)	—	2–3
Elimination half-life (h)	5–6	5 (wks)
Active metabolite	None	6-*p*-OH
Excretion (%) feces/urine	57/40	40/5
IC_{50} (nmol/L)	313 type 1	3.9 type 1
	11 type 2	1.8 type 2

h, hours

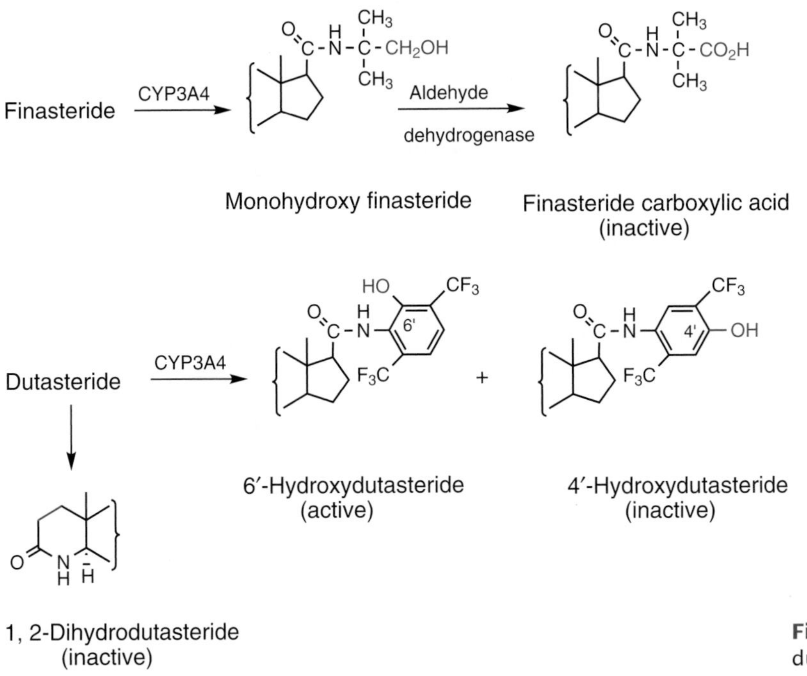

Finasteride $\xrightarrow{\text{CYP3A4}}$ Monohydroxy finasteride $\xrightarrow[\text{dehydrogenase}]{\text{Aldehyde}}$ Finasteride carboxylic acid (inactive)

Dutasteride $\xrightarrow{\text{CYP3A4}}$ 6'-Hydroxydutasteride (active) + 4'-Hydroxydutasteride (inactive)

1, 2-Dihydrodutasteride (inactive)

Figure 18.25 Metabolites of finasteride and dutasteride.

Metabolism

- 5α-Reductase inhibitors are extensively metabolized with CYP3A4 oxidation at the C17 amide group representing a major route of metabolism (Fig. 18.25).
- 6'-Hydroxydutasteride exhibits activity comparable to the parent drug.

Phytotherapy

- A number of plant extracts are popularly used to alleviate BPH. Formal evidence that they are effective is often scanty (see *Foye's Principles of Medicinal Chemistry, Seventh Edition* p. 1369, for a complete detailed discussion).
- Extracts of the saw palmetto berry (*Serenoa repens*) are widely used for the treatment of BPH. The American dwarf saw palmetto plant (*S. repens*) is rich in fatty acids and plant sterols (primary active constituents).
- The mechanism of action for the saw palmetto is not established, but the inhibition of 5α-reductase and a decrease in DHT production is suggested. At best, saw palmetto may induce mild to moderate improvements in urinary symptoms and flow measures.
- Many other Western herbs have been investigated for the treatment of BPH, and in most cases, the main active components are sterols, such as β-sitosterol.

DRUG–DRUG INTERACTIONS

Because finasteride and dutasteride are metabolized primarily by CYP3A4, the CYP3A4 inhibitors, such as ritonavir, ketoconazole, verapamil, diltiazem, cimetidine, and ciprofloxacin, may increase the drugs' blood levels and, possibly, cause drug–drug interactions. Clinical drug interaction studies have shown no pharmacokinetic or pharmacodynamic interactions between dutasteride and tamsulosin or terazosin, warfarin, digoxin, and cholestyramine.

Clinical Applications

- DHT plays a critical role in determining prostate size, and multiple lines of evidence suggest the importance of DHT in the development of BPH.
- In addition, an increased estradiol/DHT ratio was observed in the aging human prostate, which can be relevant to the development of BPH.
- Therefore, the use of α_1-adrenergic antagonists and 5α-reductase inhibitors is indicated.
- Tamsulosin, alfuzosin, and silodosin are the first-line drugs for the treatment of BPH because they have fewer cardiovascular effects than terazosin and doxazosin.
- Finasteride and dutasteride are used singly or in combination with an α_1-adrenergic antagonist (dutasteride plus tamsulosin [Jalyn]) to treat BPH.

Prostate Cancer

Prostate cancer is the second leading cause of cancer death and the most commonly diagnosed cancer in American men. The incidence of prostate cancer

⚠ ADVERSE EFFECTS

α₁-Adrenergic Antagonists: The most common adverse effects for α₁-adrenergic antagonists are related to abnormal ejaculation and orthostatic hypotension, with vasodilation, dizziness, headache, and tachycardia also occurring in some patients during early treatment. These cardiovascular side effects are attributed to a nonselective blockade of α₁-adrenoceptors present in vascular smooth muscle in addition to the required blockade of α₁-adrenoceptors in the prostate. The most common adverse effect reported for the 5α-reductase inhibitors included sexual dysfunction such as impotence and decreased libido. The 5α-reductase inhibitors decrease PSA concentrations by ~50% and could mask prostate cancer.

increases with increase in age. Traditional treatments for localized prostate cancer include watchful waiting, surgery (radical prostatectomy), and external beam radiation, while surgical and pharmacological approaches to induce androgen deprivation are used for advanced prostate cancer.

Deprivation therapy consists of LHRH agonists and antagonists, antiandrogens, and chemotherapy. Because prostate cancer often causes no symptoms, digital rectal examination and the serum prostate-specific antigen (PSA) tests are most often used to screen for prostate cancer. The cancer cells make excessive amounts of PSA protein under the control of testosterone and DHT.

Advanced prostate cancer in its earliest stages is thought to be dependent on androgens (i.e., testosterone) for growth. As such, androgen deprivation therapy, either by surgical castration or pharmacological suppression of LH, is used as first line treatment for advanced prostate cancer.

Luteinizing Hormone–Releasing Hormone (LHRH) Therapy

Both LHRH agonists and LHRH antagonists are used in treatment of prostate cancer.

MOA

- The LHRH agonists, leuprolide acetate, goserelin acetate, and triptorelin acetate or pamoate (Fig. 18.26), when administered in a continuous, nonpulsatile manner, suppress serum testosterone levels to a similar extent as surgical castration (i.e., less than 50 ng/dL).

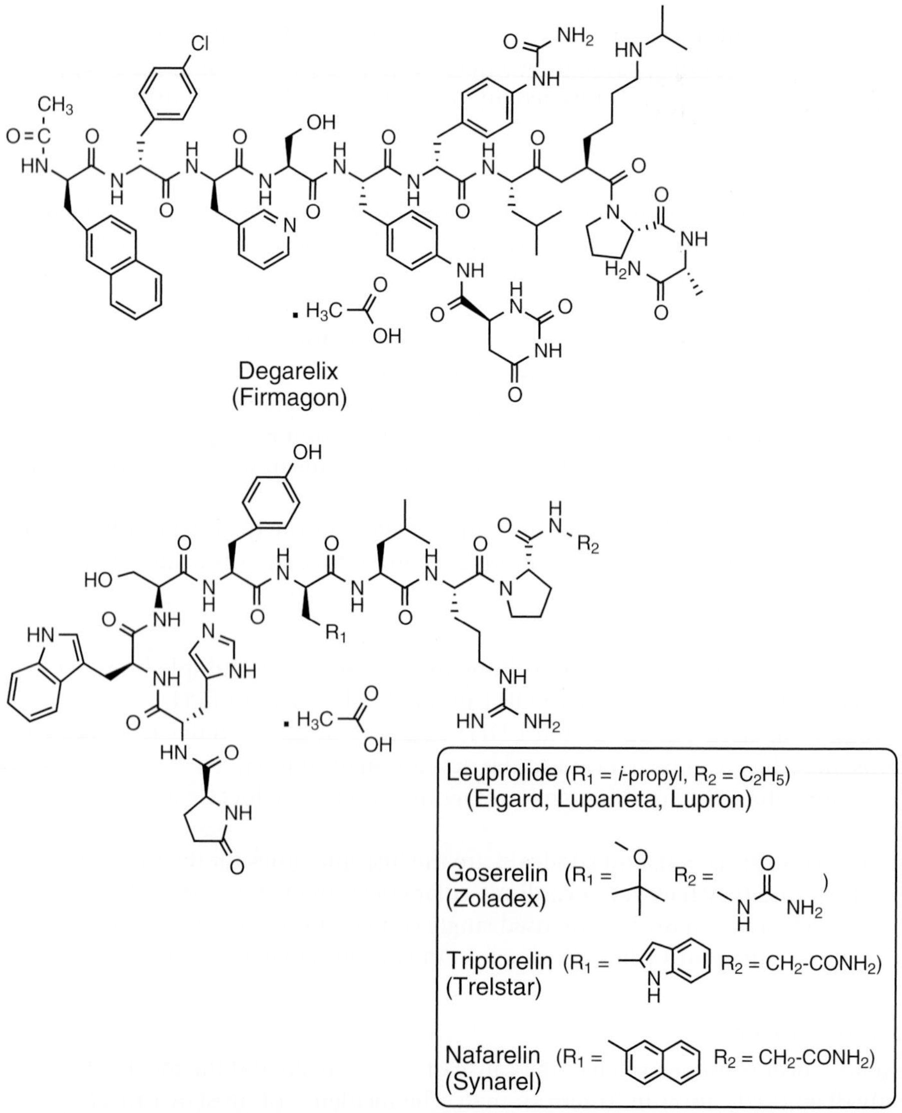

Degarelix
(Firmagon)

Leuprolide (R₁ = *i*-propyl, R₂ = C₂H₅)
  (Elgard, Lupaneta, Lupron)

Goserelin (R₁ = ... R₂ = ...)
(Zoladex)

Triptorelin (R₁ = ... R₂ = CH₂-CONH₂)
(Trelstar)

Nafarelin (R₁ = ... R₂ = CH₂-CONH₂)
(Synarel)

Figure 18.26 LHRH therapeutic agents.

- The agonists bind with greater affinity to the LHRH receptor than LHRH causing desensitization of the LHRH receptor complex in the anterior pituitary gland and a down-regulation of testosterone production.
- After a surge of LH, pituitary biosynthesis of LH stops and testicular production of testosterone ceases.
- The LHRH antagonist, degarelix acetate (Fig. 18.26), is a competitive inhibitor of endogenous LHRH for receptor binding on the pituitary gonadotropic cells, and directly inhibits LH production.
- Degarelix acetate does not cause a surge in LH or testicular androgen production.

Physicochemical and Pharmacokinetic Properties

- A variety of depot formulations for intramuscular or subcutaneous administration of leuprolide, goserelin, and triptorelin are available.
- Durations of action ranges from 1 to 4 months.
- Degarelix acetate is available as a lyophilized powder for injection.

Clinical Applications

- Androgen deprivation therapy significantly reduces serum testosterone levels depriving prostate cancer cells of their primary signal for growth-alleviating symptoms associated with advanced prostate cancer.
- LHRH agonists and antagonists are considered for palliative care and not a cure for prostate cancer.

> **⚠ ADVERSE EFFECTS**
>
> **LHRH Therapy:** Hot flashes, impotence, loss of libido, and changes in serum lipids are common. Long-term use of LHRH therapy is associated with osteoporosis, decreased cognitive abilities, fatigue, and vascular stiffness. Symptoms may worsen over the first few weeks of treatment. Tumor flare reactions may occur transiently, but can be prevented by antiandrogens or by short-term estrogens at low dose for several weeks.

Antiandrogens

Several classes of drugs are generally considered antiandrogens including the nonsteroidal agents (Fig. 18.27) and the androgen biosynthesis inhibitor abiraterone.

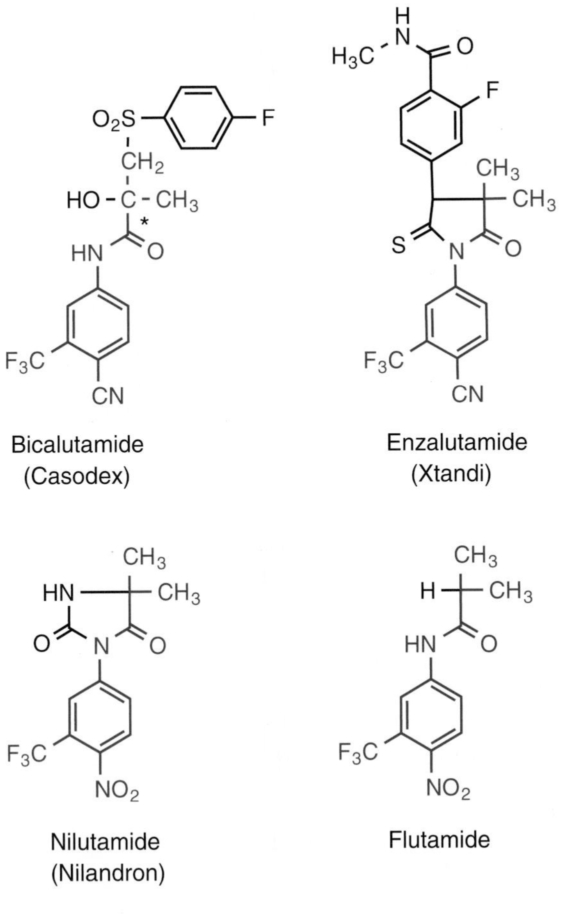

Bicalutamide
(Casodex)

Enzalutamide
(Xtandi)

Nilutamide
(Nilandron)

Flutamide

Figure 18.27 Nonsteroidal antiandrogens.

Table 18.9 Some Properties and Pharmacokinetics of the Antiandrogens

Drugs	Bicalutamide	Enzalutamide	Flutamide	Nilutamide
Oral bioavailability (%)	80–90	~82	>90	—
Onset of action (wks)	8–12		2–4	1–2
Duration of action	8 d		>3 mo	1–3 mo
Time to peak conc. (h)	31	~1	2–3	1–4
Elimination half-life (h)	~6 d	5.8 d	8	40–60
Excretion (%) feces/urine	43/34	14/71	<10/~28	<10/62

h, hours

NONSTEROIDAL ANTIANDROGENS

MOA

- The antiandrogen-specific target is the C-terminal of the LBD of the 919 amino acid AR protein (Fig. 18.15).
- Binding to the AR prevents AR dimerization, which in turn blocks translocation into the nucleus.
- This results in reduced coactivator peptide recruitment and DNA binding leading to inhibition of AR gene transcription.
- Enzalutamide and flutamide are pure antagonists, although the flutamide metabolite exhibits agonist activity.
- Enzalutamide may also cause cleavage of poly(ADP-ribose) polymerase leading to apoptotic effects.

Physicochemical and Pharmacokinetic Properties

- Bicalutamide is administered as a racemate although its antiandrogenic activity resides exclusively in the *R*-enantiomer.
- The nonsteroid antiandrogens are highly bound to plasma protein (80% to 96%).
- Additional pharmacokinetic data are reported in Table 18.9.

Metabolism

- Nonsteroidal antiandrogens are subject to amide hydrolysis and CYP oxidation with the potential for drug–drug interactions through inhibition or enhancement of their respective CYP isoform.
- Metabolic reactions are shown in Figure 18.28.

STEROIDAL ANTIANDROGEN (ABIRATERONE ACETATE)

MOA

- Abiraterone is the active drug formed from the prodrug abiraterone acetate.
- Abiraterone is an irreversible inhibitor of 17α-hydroxylase and 17,20-lyase blocking the sequential formation of 17α-hydroxypregnenolone and DHEA (Fig. 18.4) as well as the conversion of progesterone to 17-hydroxyprogesterone and on to androstenedione.
- Inhibition of CYP17 reduces the level of testosterone to castration levels.
- The site of action involves adrenal, prostatic, and testicular tumor tissue.
- The 16,17 double bond in abiraterone is essential for activity.

Physicochemical and Pharmacokinetic Properties

- Abiraterone acetate should not be administered with food.
- The drug is rapidly absorbed and converted to abiraterone.
- Abriaterone is highly bound to plasma proteins (>99%).
- Abriaterone acetate (55%) and abriaterone (22%) are eliminated via feces.

Metabolism

- Inactive metabolites found in the plasma consist of abiraterone sulfate formed via sulfotransferase and N-oxide formed by CYP3A4.

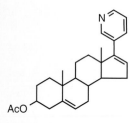

Abiraterone acetate
(Zytiga)

Bicalutamide $\xrightarrow{\text{CYP3A4}}$ R–NH ... Inactive + R–NH$_2$

Enzalutamide $\xrightarrow{\text{CYP2C8}}$ N-Desmethylexzalutamide (Active) + Inactive

Flutamide $\xrightarrow{\text{CYP1A2}}$ Active + R'–NH$_2$

Nilutamide $\xrightarrow[\text{CYP2C}]{\text{FMO}}$ Inactive (Potential toxin) + Active

Figure 18.28 Metabolism of the nonsteroidal antiandrogens.

Clinical Applications

- Both the nonsteroidal and steroidal antiandrogens are approved for and commonly used for treating the latter stages of metastatic castration-resistant prostate cancer (CRPC).
- The nonsteroidal drugs are only indicated for combined use with an LHRH analog.
- Abiraterone is administered with prednisone for metastatic CRPC, which is reported to significantly prolong overall survival.
- A major advantage of abiraterone may be associated with its ability to block androgen production from nongonadal sources.

Immunotherapy and Chemotherapy

SIPULEUCEL-T (PROVENGE)

- Sipuleucel-T is a therapeutic personalized cancer vaccine.
- Prepared from the patient's white blood cells (antigen-presenting cells [APCs]).
- Incubated with fusion protein consisting of:
  - antigen prostatic acid phosphatase (PAP), an enzyme commonly found in prostate cancer.
  - an immune signaling factor–granulocyte-macrophage colony stimulating factor (GM-CSF).
- This activated blood product is reintroduced into the patient to produce an immune response against prostate cancer cells containing PAP.
- The patient is commonly treated with an antihistamine/acetaminophen mixture 30 minutes prior to treatment to reduce an acute infusion reaction seen in 70% of the patients.

CABAZITAXEL

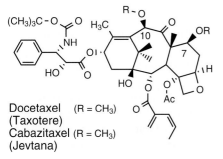

Docetaxel (R = CH$_3$)
(Taxotere)
Cabazitaxel (R = CH$_3$)
(Jevtana)

ADVERSE EFFECTS

Nonsteroidal Antiandrogens: A variety of adverse effects are reported with these drugs. These include pain (general), back pain, asthenia, hot flashes, constipation, nausea. In the case of nilutamide, rare, but potential fatal, severe lung problems have been reported. Flutamide has been reported to cause liver failure possibly associated with the agonist action of the metabolite hydroxyflutamide. This may lead to antiandrogen withdrawal syndrome. Newer members of this class of drugs are recommended. It should be noted that these drugs are also used with an LHRH analog, which complicates the adverse effects profile.

ADVERSE EFFECTS

Steroidal Antiandrogen: The common adverse effects reported for abiraterone acetate include fatigue, joint discomfort, edema, hot flash, diarrhea, vomiting, cough, hypokalemia, hypertension, and fluid retention. The drug is free of agonist actions often reported for the nonsteroidal antiandrogens.

MOA

- The taxanes, in general, are active by virtue of their ability to bind to β-tubulin subunit of microtubules promoting a stable tubulin conformation.
- Disassembling of the tubulin into mitotic spindles is inhibited resulting in inhibition of cell mitosis (antimitotic action).
- Tumor resistance to docetaxel is associated with high binding of docetaxel to P-glycoprotein (P-gp), an ATP-dependent drug efflux mechanism (resulting from an overexpression of the multidrug resistance gene that encodes for P-gp).
- The C7-C10 dimethoxy groups inhibit binding of cabazitaxel to P-gp, thus preventing ATP-dependent drug efflux.

Physicochemical and Pharmacokinetic Properties

- The methoxy of cabazitaxel improves lipid solubility and penetration of the CNS.
- Cabazitaxel in polysorbate 80 is administered as an intravenous infusion diluted in 13% ethanol.
- The drug is highly bound to plasma proteins (~82%).
- Excretion occurs primarily in the feces (76%) with numerous metabolites.
- Metabolism is catalyzed primarily by CYP3A4/5 (80% to 90%) as well as CYP2C8.
- Mono- and didemethylation, leading to docetaxel, and t-butyl hydroxylation are common metabolic pathways.

Clinical Applications

- Cabazitaxel in combination with prednisone is used for the treatment of hormone-refractory metastatic CRPC previously treated with a docetaxel-containing regimen.
- The patient is commonly treated with an antihistamine, a corticosteroid, an H_2 antagonist, and possibly an antiemetic 30 minutes prior to treatment.

> **⚠ ADVERSE EFFECTS**
>
> **Cabazitaxel:** Common adverse effects consist of neutropenia, diarrhea, nausea, vomiting, neurotoxicity, and fatigue. Renal failure has also been reported leading to discontinuation of therapy.

Estrogens, Progestins, and Women's Health

Introduction

The two principal classes of female sex hormones are estrogens and progestins. These agents will serve as the bases for discussion in this section of the steroid chapter. The naturally occurring estrogens are C_{18} steroids and contain an aromatic A ring with a phenolic group at the C3 position. The most potent endogenous estrogen is estradiol. The naturally occurring progestins are C_{21} steroids and contain a 3-keto-4-ene structure in the A ring and a ketone at the C20 position. The most potent endogenous progestin is progesterone.

Estrogen–Progestin Hormones: Female Reproductive Cycle

- Similar to the male reproductive cycle (Fig. 18.5), the female reproductive cycle is controlled by an integrated system involving the hypothalamus, anterior pituitary gland, ovary, and reproductive tract (Fig. 18.29).
- LHRH exerts its action on the pituitary gland in a pulsatile fashion, which in turn releases FSH and LH.
- FSH and LH regulate production of estrogen and progesterone.
- FSH promotes the initial development and growth of ovarian follicles, which secrete estradiol.
- As estradiol levels increase, the production of FSH decreases via a negative feedback loop inhibition stimulating LH release.
- The level of LH rises to a sharp peak (LH surge) at the midpoint of the menstrual cycle causing the dominant follicle to rupture and release its egg (ovulation). The luteal phase follows ovulation and ends at menses. During this phase, the endometrium shows secretory activity and cell proliferation declines.

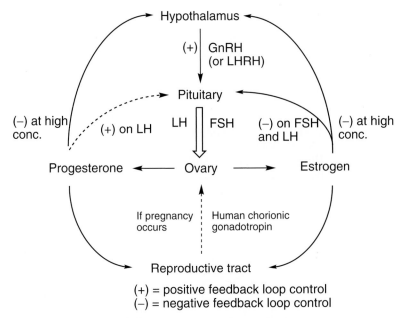

Figure 18.29 Hypothalamic, ovarian, and reproductive tract interrelationships.

- The drugs effecting female hormonal control generally act at the level of the hypothalamus or pituitary gland. For a discussion of the role of LHRH, LH, or FSH as it relates to pregnancy or menses see Chapter 41 in *Foye's Principles of Medicinal Chemistry, Seventh Edition*.

Estrogen Biosynthesis

- Estrogens are primarily synthesized in the ovary, placenta, and adipose tissue (Table 18.1).
- The biosynthesis of estradiol and estrone involve aromatization of the A ring from the precursors testosterone and androstenedione, respectively (Fig. 18.4).
- The enzyme aromatase is involved in the loss of the C19 methyl group and aromatization. The interconversion of estradiol and estrone is significant.

Estrogens

Endogenous Estrogens

Present in women are the following three estrogens (Fig. 18.30):

- Estradiol, the most potent, represents 10% to 20% of the circulating estrogen. It plays a key role in development of secondary sex characteristics during puberty (i.e., hair growth, skin softening, breast growth, and accumulation of fat in the thighs, hips, and buttocks), stimulation of the mammary glands during pregnancy, as well as having thermoregulatory capacity.
- Estrone is 10-fold less potent and accounts for 60% to 80% of the circulating estrogen.
- Estriol, a very weak estrogen representing 10% to 20% of the estrogens. Estriol is not used therapeutically.

Synthetic and Natural Steroidal Estrogens

- Several esters of estradiol (i.e., valerate and cypionate) as well as sulfate conjugates of estrone, equilin, and equilenin are commercially available (Fig. 18.30).
- Ethinyl estradiol (EE) is an important estradiol analog used primarily in birth control products. The addition of the 17-α-ethinyl group is added to reduce the oxidative metabolism of estradiol to estrone.

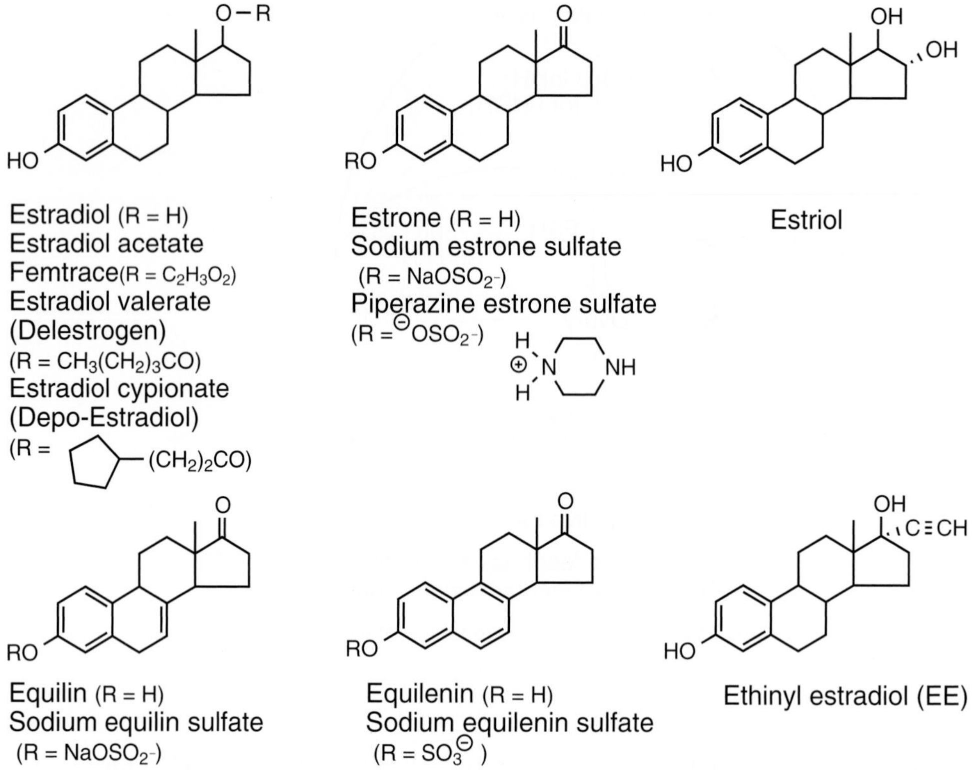

Estradiol (R = H)
Estradiol acetate
Femtrace(R = $C_2H_3O_2$)
Estradiol valerate
(Delestrogen)
(R = $CH_3(CH_2)_3CO$)
Estradiol cypionate
(Depo-Estradiol)
(R = —(CH₂)₂CO)

Estrone (R = H)
Sodium estrone sulfate
(R = $NaOSO_2^-$)
Piperazine estrone sulfate
(R = $^\ominus OSO_2^-$)

Estriol

Equilin (R = H)
Sodium equilin sulfate
(R = $NaOSO_2^-$)

Equilenin (R = H)
Sodium equilenin sulfate
(R = $SO_3^\ominus$)

Ethinyl estradiol (EE)

Figure 18.30 Estrogenic steroids.

- Conjugated estrogens are made up of mixtures of sodium salts of sulfate esters of estradiol derived from equine urine or prepared synthetically from estrone and equilin. These include:
  - Premarin (sodium estrone sulfate [52.5% to 61.5%] and sodium equilin sulfate [22.5% to 30.5%]).
  - Menest (estrone sodium sulfate [75% to 85%] and sodium equilin sulfate [6% to 15%]) (Fig. 18.28).
  - Synthetic conjugated estrogens A (Cenestin) and estrogens B (Enjuvia) with 9 of the 10 known synthetic conjugated estrogenic substances.
  - An orally effective sulfate conjugate estropipate (piperazine estrone sulfate; Ortho-Est).

MOA of Estrogens (Estradiol)

- Estradiol's biological target, the estrogen receptor (ER) consists of two subtypes (ERα and ERβ).
- ERα is composed of 595 amino acids and is found in the female reproductive tract and mammary glands.
- ERβ found primarily in vascular endothelial cells, bone, and male prostate tissues is composed of 485 amino acids and is regulated hormonally via estradiol.
- Estradiol has similar affinities for both ERα and ERβ. Binding of estradiol to either ERα or ERβ leads to receptor phosphorylation and a conformational change to produce either homo- or heterodimers (ERα/ERα, ERβ/ERβ, or ERα/ERβ).
- Dimeric ER~estradiol complex migrates to the cell nucleus binding to specific estrogen-response elements (ERE) within DNA, leading to DNA transcription (Fig. 18.7) (substituting ER for steroid receptor and ERE for HRE, thus depicting the ER~complex).
- Structural features of the ER are similar to those of the AR shown in Figure 18.15 with AF1, DBD, and AF2.
- In addition to the nuclear ER, the target cell has a second ER within the cell membrane, where a nongenomic mechanism prevails.
- The cell membrane receptors are G-protein coupled and are linked to a cascade of intracellular signals.

SAR

Structural features important for estrogenic activity.

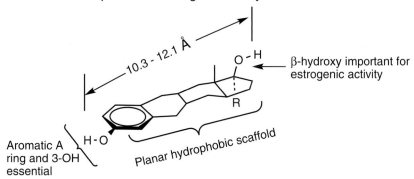

10.3 - 12.1 Å

β-hydroxy important for estrogenic activity

Aromatic A ring and 3-OH essential

Planar hydrophobic scaffold

- Functionality at the C_1 position greatly reduces activity and hydroxyl groups at positions 6, 7, and 11 reduces activity.

- The presence of unsaturation in ring B boosts the estrogenic potency.

- Certain modifications at the 17α and 16 positions can lead to enhanced activity.

Physicochemical and Pharmacokinetic Properties

- While extremely potent with high affinity for the ER, estradiol when administered orally is promptly conjugated in the intestine to glucuronide and sulfate esters, and oxidatively metabolized by the liver (see below) resulting in its low oral bioavailability and therapeutic effectiveness.
- Prodrug esters of estradiol (i.e., 17β-acetate, 17β-valerate, and 17β-cyclopentylpropionate) are administered orally, in a vaginal ring, or intramuscularly and slowly hydrolyzed releasing free estradiol in some cases over a prolonged period of time with a duration of action of 2 to 4 weeks (Fig. 18.30).
- Alkylation of the C17 position with a chemically inert alkyne group gives rise to an orally active, rapidly absorbed product ethinylestradiol (EE) (Fig. 18.30), with an improved oral bioavailability and a longer elimination half-life (6 to 20 hours vs. 20 minutes for estradiol) (Table 18.10).
- EE undergoes extensive first-pass metabolism but with enterohepatic recycling.

Metabolism

- Estradiol's activity is significantly reduced by protein binding and biochemical oxidation to estrone (Fig. 18.4) as well as conversion to estriol (Fig. 18.31).
- Water-soluble glucuronide and sulfate conjugates occur mainly in the liver. Intestinal secretion, bacterial hydrolysis, and enterohepatic recycling follow.
- Aromatic hydroxylation and O-methylation are metabolic reactions affecting most estrogens (Fig. 18.31).

Clinical Applications

- Menopause is described as the loss of ovarian follicle function, followed by the cessation of menses. This is preceded by perimenopause, characterized

Table 18.10   Some Pharmacokinetic Properties of Steroidal Estrogens

Estrogen	Bioavailability	Bound to SHBG%	Bound to Albumin%	Metabolic Clearance (L/day/m²)
17β-Estradiol	Poor	37	63	580
Estrone		16	80	1,050
Estrone sulfate		0	99	80
Equilin sulfate				175
Ethinyl estradiol	40	0	98.5	

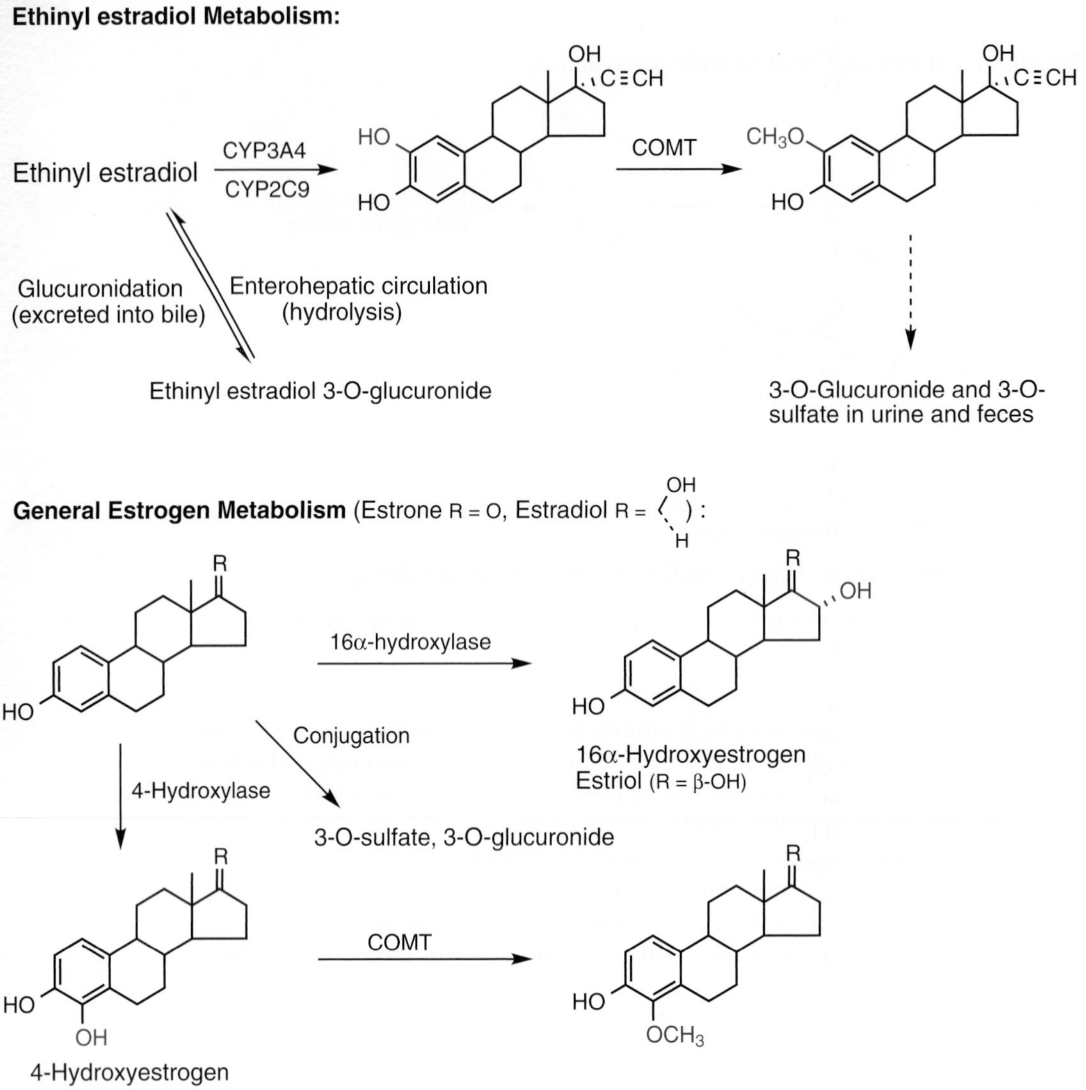

Ethinyl estradiol Metabolism:

General Estrogen Metabolism (Estrone R = O, Estradiol R = OH / H):

Figure 18.31 Metabolic pathways for estrogenic steroid metabolism.

⚠ **ADVERSE EFFECTS**

Estrogen Replacement Therapy (ERT): Breast tenderness and swelling, weight gain, and vaginal bleeding are the more common adverse effects of ERT. The controversial box warnings with estrogen-alone therapy include an increased risk of endometrial cancer in women with a uterus (adding progestin to the therapy reduces the risk of endometrial hyperplasia). An increased risk of stroke, deep vein thrombosis, and dementia with conjugated estrogen-alone therapy. The risk of breast cancer has also been reported to increase in ERT patients. It is suggested that women be prescribed the lowest effect dose over the shortest time period of estrogen.

by a gradual decline in the secretion of estrogen and progesterone resulting in physiological changes: bone loss, urogenital atrophy, and perhaps, incontinence, vasomotor symptoms, enhanced risk of cardiovascular disease, sexual dysfunction including decreased libido, as well as reduced skin elasticity.

- Medical management of menopause (e.g., burning, itching, and dryness of the vagina) and the vasomotor symptoms (e.g., hot flashes, night sweats, mood and sleep disturbances) revolves largely around symptomatic treatment. Hormone replacement therapy (HRT) primarily with estrogens (estrogen replacement therapy [ERT]) in menopausal patients usually gives symptomatic relief and is considered safe when used short term.
- Estrogens remain the gold standard in the relief of menopausal symptoms. Alternative dosage forms (e.g., transdermal sprays, creams, gels, patches, intravaginal tablets, rings, injections, and oral tablets) that deliver the minimum dose of estrogen to minimize adverse effects and maximize therapeutic benefit through normalizing the estrone to estradiol ratio.
- The conjugated estrogens such as estrone, equilin, and equilenin sulfate ester salts are used in HRT.
- Estradiol and the estradiol esters may be used to treat vasomotor symptoms and the symptoms of vulvar and vaginal atrophy in menopausal women.

- Micronized progesterone in combination with oral conjugated estrogens provides valuable therapeutic benefit in the treatment of menopausal symptoms without the risk of endometrial hyperplasia.

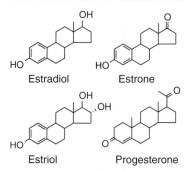

Estradiol   Estrone

Estriol   Progesterone

- For a comprehensive discussion of HRT and the drug therapy the reader is referred to pages 1417 to 1422 and Table 41.16 in *Foye's Principles of Medicinal Chemistry, Seventh Edition.*

CHEMICAL NOTE

Diethylstilbestrol (DES): A nonsteroid estrogen, with the *trans* configuration, exhibits a 10-fold increase in estrogenic potency over the *cis* isomer due to its structural resemblance to estradiol. Studies suggested that women exposed to DES prenatally have a higher incidence of breast cancer after age 40. It was hypothesized that the catechol metabolites of DES (i.e., O-quinones) could depurinate DNA-generating mutations which could initiation cancer.

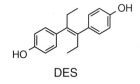

DES

Progestins

The most potent of the progestins is the endogenous hormone progesterone. Its biosynthetic pathway is shown in Figure 18.4, while its role in the female reproductive system was previously outlined in Figure 18.29. Progesterone biosynthesis in the female is initiated in the ovary when LH, secreted from the anterior pituitary, binds to the target cell surface LH receptor.

Progesterone is secreted specifically by the corpus luteum (~30 mg/day) in reproductive-age women and by the placenta during pregnancy. Compared to the ovaries, the adrenal gland produces only a small fraction of progesterone (1 mg/day).

Progesterone has limited medicinal utility due in part to its low bioavailability and rapid metabolism in the liver regardless of the route of administration, and has a half-life of 5 to 10 minutes. The clinically more important synthetic progestins commonly used in the combination oral contraceptives (OCs) are discussed in this context and are shown in Figure 18.32.

MOA of Progestins in the Presence of Ethinyl Estradiol

- The exact role of both the progestin and EE is not known with certainty, although the physiological outcome is well characterized.
- These drugs act synergistically with the progestin creating a negative feedback on the hypothalamus, decreasing the pulsated frequency of LHRH release, thus decreasing secretion of FSH and LH by the anterior pituitary and inhibiting follicular development.
- The progestin specifically prevents midcycle LH surge and blocks ovulation.
- Progestins also increase the viscosity of the cervical mucus, thus inhibiting sperm penetration through the cervix.
- EE through a negative feedback on the anterior pituitary suppresses secretion of FSH and inhibits follicular development.

SAR

- Progestin activity is restricted to molecules with a steroid nuclei of either the androstane (or 19-norandrostane) or pregnane (or 19-norpregnane) class.

Androstanes (R_1 = H; R_2 = CH$_3$)
19-norandrostanes (R_1 = R_2 = H)
Pregnanes (R_1 = C$_2$H$_5$; R_2 = CH$_3$)
19-norandropregnanes (R_1 = C$_2$H$_5$; R_2 = H)

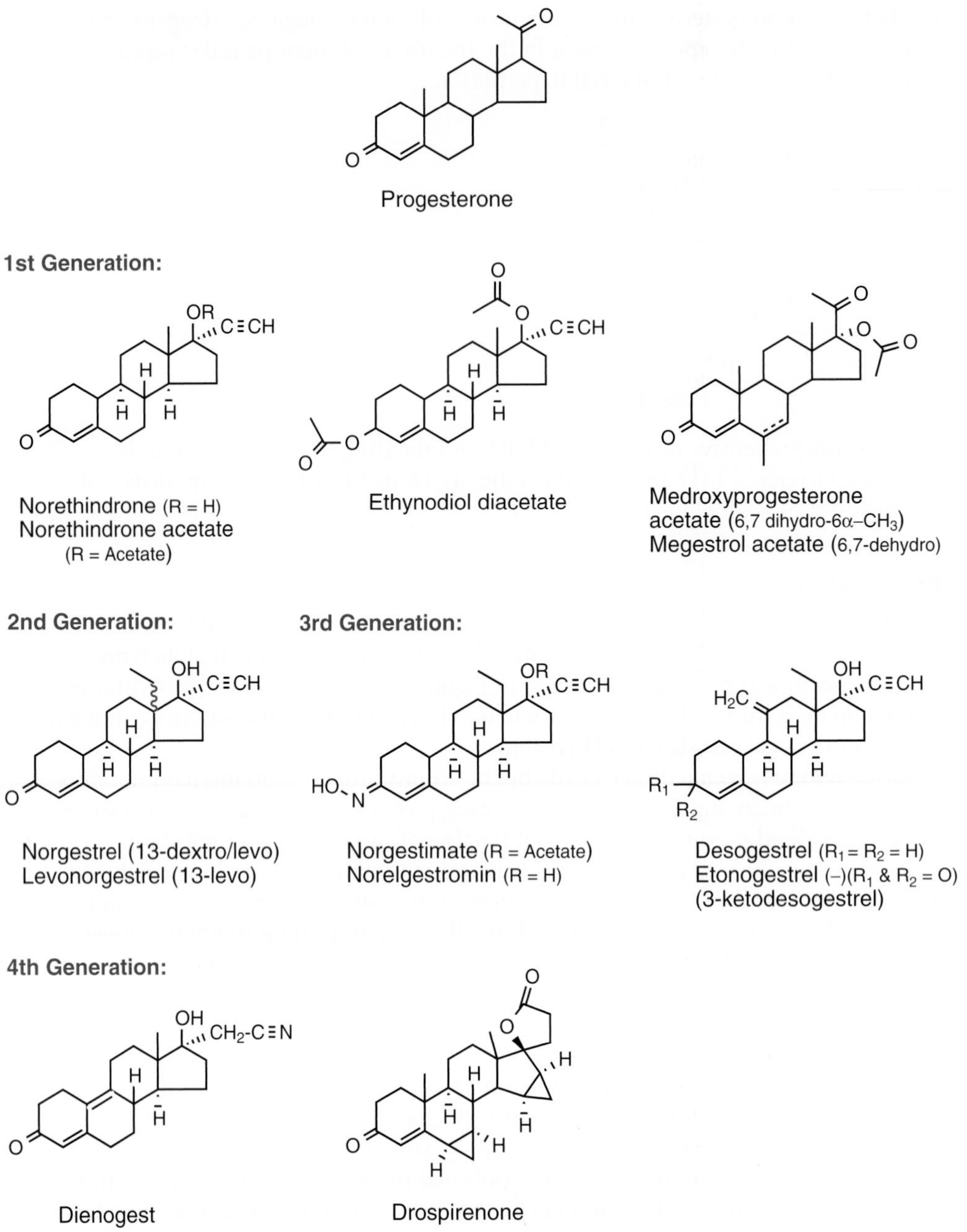

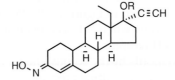

Progesterone

1st Generation:

Norethindrone (R = H)
Norethindrone acetate
(R = Acetate)

Ethynodiol diacetate

Medroxyprogesterone
acetate (6,7 dihydro-6α–CH₃)
Megestrol acetate (6,7-dehydro)

2nd Generation:

Norgestrel (13-dextro/levo)
Levonorgestrel (13-levo)

3rd Generation:

Norgestimate (R = Acetate)
Norelgestromin (R = H)

Desogestrel (R₁ = R₂ = H)
Etonogestrel (–)(R₁ & R₂ = O)
(3-ketodesogestrel)

4th Generation:

Dienogest

Drospirenone

Figure 18.32 Progestins.

- The androstanes normally have a 17α-substituent (i.e., ethynyl, methyl, ethyl) to provide oral bioavailability.
- Removal of the C19 CH₃ group increases progestin activity by 5- to 10-fold.
- The C18 ethyl group also increases activity (norgestrel).
- Acetylation of the 17β-OH of norethindrone increases drug duration of action.
- Removal of the 3-keto function of norethindrone allows retention of potent progestin activity without androgenic effects.
- Activity of the pregnane class is enhanced by unsaturation at C6 and C7, by addition of a halogen or methyl group at C6, and by addition of a methyl group at C11.
- These substitutions prevent metabolic reduction of the two carbonyl groups and metabolic oxidation at C6.
- A C21 fluoro group prevents metabolic hydroxylation and enhances oral effectiveness.
- As noted in Table 18.11, the progestins exhibit activity in addition to their anti-estrogen activity due to binding to other steroid receptors (Table 18.12).

Table 18.11   Hormonal Activities of Representative Progestins

Progestogen	Antiestrogenic	Androgenic	Antiandrogenic	Glucocorticoid	Antimineralocorticoid
Progesterone	++	–	+	++	++
Medroxyprogesterone acetate (1)	++	+	–	++	–
Norethisterone (1)	++	++	–	–	–
Levonorgestrel (2)	++	++	–	–	–
Etonogestrel (3)	++	++	–	+	–
Norgestimate (3)	++	++	–	?	?
Drospirenone (4)	++	–	++	–	++
Dienogest (4)	++	–	++	–	–

Number in parentheses indicates the generation of progestogen. ++, effective; +, weekly effective; –, not effective; ?, unknown.

- The fourth generation progestins are unique in their antiandrogenic activity.
- Glucocorticoid receptor (GR) activity can be decreased by small substituents on the B-ring as well as substituents at the C17 position.

Physicochemical and Pharmacokinetic Properties

- The progestins are generally well absorbed with good bioavailability and are highly bonded to plasma proteins.
- The pharmacokinetic data are presented in Table 18.13.

Metabolism

- The progestins are extensively metabolized and share several common metabolic pathways as shown in Figure 18.33.
- Generally, the acetate derivatives are rapidly hydrolyzed to an active drug.
- The norgestimate and norelgestromin are prodrugs which give rise to the active agents via removal of the oxime and the acetate ester.

Clinical Applications

- The progestins are used most commonly in combination OCs with EE.
- The progestins are classified by generation (first to fourth) based upon receptor binding selectivity for the progesterone receptor (Tables 18.11 and 18.12) and reduced androgenic agonist activity.
- The OCs are classified by whether they are monophasic (a single dose of estrogen/progestin) or multiphasic (varying doses) in their hormonal composition.
- OCs with unique dosage forms containing a progestin include: transdermal patch (EE plus norelgestromin), the vaginal ring (EE plus etonogestrel), subdermal implants (etonogestrel), and the intrauterine device (IUD) (progestin only: levonorgestrel or norethindrone).
- Levonorgestrel (plan B) is also an emergency contraceptive if administered within 72 hours of intercourse and again 12 hours later.

Table 18.12   Steroid Receptor Binding Affinity

Progestogen	Progesterone	Androgen	Glucocorticoid	Mineralocorticoid
Progesterone	50	0	10	100
Medroxyprogesterone acetate (1)	115	5	29	160
Norethisterone (1)	75	15	0	0
Levonorgestrel (2)	150	45	1	75
Etonogestrel (3)	150	20	14	0
Norgestimate (3)	15	0	1	0
Drospirenone (4)	35	65	6	230
Dienogest (4)	5	10	1	0

Number in parentheses indicates the generation of progestogen.

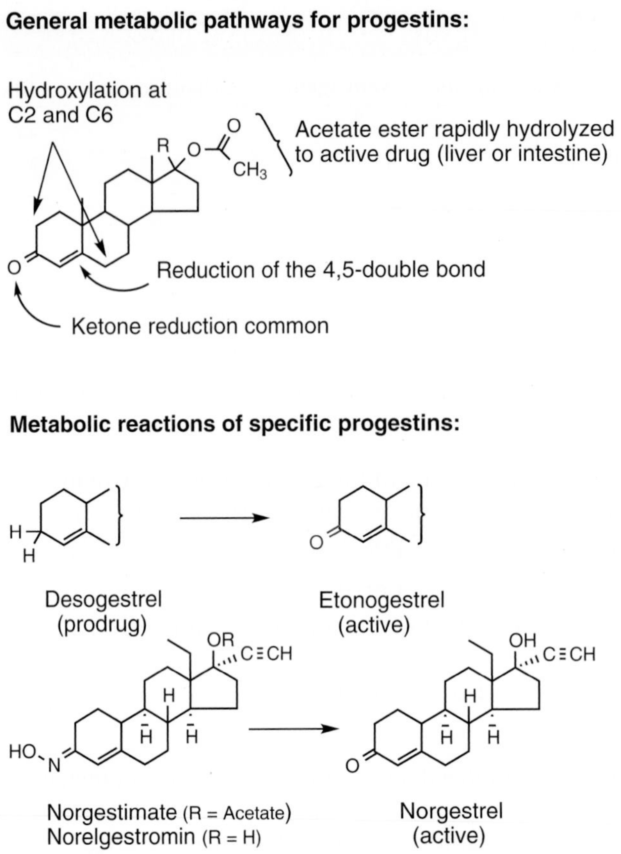

General metabolic pathways for progestins:

Metabolic reactions of specific progestins:

Figure 18.33 Metabolic reactions of progestins.

DRUG–DRUG INTERACTIONS

Drug efficacy may be reduced in the presence of one or more drugs that induce liver enzymes (e.g., carbamazepine, phenytoin, topiramate, barbiturates) or that can alter the enterohepatic recycling of the hormones. Some antibiotics may reduce the effectiveness of OCs (e.g., β-lactams, macrolides, tetracyclines, and sulfas) and several protease inhibitors (e.g., amprenavir, nelfinavir, ritonavir). The herbal product St. John's Wort may lessen the effectiveness of OCs.

- For a more comprehensive discussion of the various OC combinations and their dosage forms the reader is referred to Chapter 41 in *Foye's Principles of Medicinal Chemistry, Seventh Edition*, as well as the FDA website.
- Medroxyprogesterone and megestrol acetate are not used in OCs, but rather, medroxyprogesterone is used to treat vasomotor symptoms during menopause and secondary amenorrhea and abnormal uterine bleeding, while megestrol is used to treat anorexia, cachexia, and unexplained weight loss in AIDS patients.

Table 18.13 Pharmacokinetic Properties for Some Progestins

Drug	Oral Bioavailability	Protein Binding	Elimination Half-Life (h)	Time to Peak Conc. (h)	Elimination Renal/Fecal (%)
Progesterone	<10%	>90%	<5 min	2–4	50–60/10
Oral micronized			34.8	6.8	
Vaginal gel					
Norethindrone acetate	65	>80	8 (5–14)	0.5–4	50/20–40
Medroxyprogesterone acetate	High	>90%	30	2–4	15–22/45–80
Megestrol acetate	ND	>90	38 (13–104)	2–3	66/20
Norgestrel	60%	>90%	20	24	45/32
Levonorgestrel	60	>90	16 (8–30)	24	45/32
Norgestimate (desacetyl)	60	>50–60	37	1–2	47/37
Desogestrel (as etonogestrel)	76	>90	12–58	1–2	43/50
Dienogest	~90	~90	9–10	3–8	~70/11–12
Drospirenone	76	98.5			

Protein binding may involve SHBG and/or albumin.
Renal metabolites are primarily conjugates.
h, hours. *ND*, no data available.

> ⚠️ **ADVERSE EFFECTS**
>
> **OC Therapy:** Adverse effects of OC therapy include: increased risk of stroke, acute myocardial infarction, and venous thromboembolism (VTE) (especially during the first three months of use). The incidence of these cardiovascular conditions increases in women aged >35 years. The risk of VTE in a third generation OC (EE + desogestrel or norgestimate) is about twice that found in patients who use a second generation OC (EE + levonorgestrel) and is five times greater than in nonusers of OCs. A metabolic adverse effect of the estrogen component is an increase in hepatic production of proteins, including those that enhance venous and arterial thromboembolism, while the progestin component has an adverse effect on the lipid profile (i.e., elevation of serum triglyceride and a decrease in HDL levels). Generally, the estrogen component of a combination OC balances out this negative impact on a patient's lipid profile. Neoplastic effects associated with OCs include an increased risk of cervical cancer but a decreased risk of colorectal, endometrial, and ovarian cancers and these can persist long after the discontinuation of therapy. Intracyclic bleeding (spotting) is more prevalent with low-dose EE. The drospirenone component of several OCs is reported to have a significantly higher risk of dangerous blood clots as well as the potential of causing hyperkalemia due to its diuretic action through its affinity for mineralocorticoid receptors.

Progestin Antagonists and Progesterone Receptor-Modulators

Built on the same steroid nucleus as the progestins, the drugs mifepristone and ulipristal acetate offer a different therapeutic utility (Fig. 18.34).

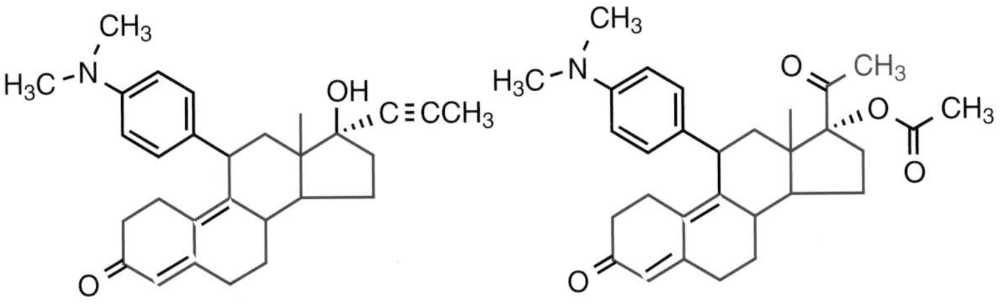

Mifepristone
(Korlym, Mifeprex)
(19-desmethylandrostane)

Ulipristal acetate
(Ella)
(19-desmethylpregnane)

Figure 18.34 Progestin antagonists/progesterone receptor modulators.

MOA

- Mifepristone competes with progesterone for the progesterone and glucocorticoid receptors (antagonist) producing what is characterized as a progesterone receptor-modulator effect.
- In a similar fashion ulipristal acetate is a progesterone receptor-modulator, but also possesses a partial agonist activity at the progesterone receptors.

Physicochemical and Pharmacokinetic Properties

- Both drugs are readily absorbed following oral administration (bioavailability of mifepristone ~70%).
- Both drugs are highly bound to plasma proteins (94% to 98%) and exhibit a $t_{1/2}$ of 40 to 90 hours with excretion occurring in the feces.

Metabolism

- Both mifepristone and ulipristal acetate are substrates for CYP3A4 resulting N- and N, N-demethylation with both metabolites possess biologic activity.
- Drug–drug interaction can be expected between these drugs and CYP3A4 inducers.
- Mifepristone appears also to be an inhibitor of CYP2B6.

Clinical Applications

- Mifepristone when used in combination with a prostaglandin for medical termination of intrauterine pregnancies of up to 49 days of gestation.
- Ulipristal acetate is used for emergency contraception if used within 120 hours of unprotected intercourse with 60% efficiency (similar to levonorgestrel).

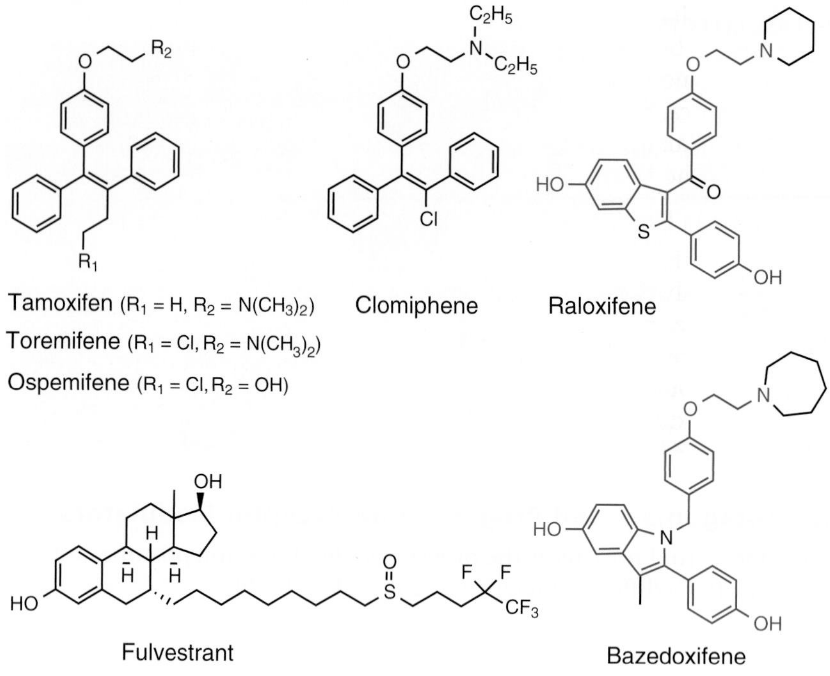

Figure 18.35 SERMs and antiestrogens.

- Additional uses include treatment of uterine fibroids with ulipristal and Cushing's syndrome for mifepristone.

Treatment of Breast Cancer

Selective Estrogen Receptor Modulators (SERMs)

By definition, SERMs are drugs, which are active at ERs, but depending on the tissue the drugs may exhibit either agonist or antagonist activity (Fig. 18.35). Upon binding to the ERs (e.g., ERα and ERβ) the SERMs will produce various therapeutic activities (Table 18.14). Three of the SERMs (i.e., raloxifene, bazedoxifene, and ospemifene) are discussed in Chapter 21 for their role in calcium homeostasis. The binding and treatment of breast cancer is discussed below.

MOA

- Tamoxifen, or more accurately its active metabolite 4-hydroxytamoxifen, has high affinity for the ER in ER-positive breast cancer cells where it complexes with corepressor proteins to block DNA-modulated gene expression.
- The hydroxytamoxifen-corepressor~DNA complex suppresses proproliferative protein production necessary for breast cancer cell growth, and by binding causes conformational changes in drug~ER complex.
- Tamoxifen, toremifene, and fulvestrant act as antagonists at the estrogen-positive ERs and are most effective in postmenopausal women.

Table 18.14 SERMs and Their Agonist–Antagonist Sites of Action

Drug	Trade Names	Action and Location	
		Agonist	Antagonist
Tamoxifen	Soltamox	Bone/uterus	Breast
Toremifene	Fareston	Pituitary	Breast
Ospemifene	Osphena	Bone	Breast/uterus
Clomiphene	Clomid		All tissues
Raloxifene	Evista	Bone	Breast/uterus
Bazedoxifene	Duavee	Bone	Endometrium
Fulvestrant	Faslodex	None	All tissues

- Tamoxifen causes cells to remain in the G_0-G_1 phase and is cytostatic.
- A similar action can be postulated for toremifene.
- Fulvestrant with a binding affinity of 89% versus estradiol impairs ER dimerization, nuclear translocation, and accelerates ER~protein degradation.
- Clomiphene acts as an antagonist at ERs in the hypothalamus by blocking inhibitor feedback mechanisms.

Physicochemical and Pharmacokinetic Properties

- The triarylethylene SERMs are all readily absorbed via the oral route.
- These drugs are highly bound to plasma protein (>98%) and with the exception of ospemifene have extended $t_{1/2}$ of several days.
- All of these agents are extensively metabolized, with the major route of excretion being via the fecal route.
- Clomiphene is a mixture of E and Z isomers and while commonly drawn in the E configuration the Z isomer represents the more active isomer.

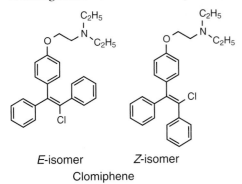

E-isomer   *Z*-isomer

Clomiphene

- Fulvestrant is administered via injection (IM) and is also extensively metabolized and excreted in the feces.

Metabolism

- The triarylethylene SERMs exhibit similar types of CYP450-catalyzed metabolism as shown in Figure 18.36.
- The metabolism of fulvestrant involves O-sulfate and O-glucuronide conjugation at the C3 and C17 positions as well as ketone formation at C17.

Clinical Applications

- Tamoxifen, toremifene, and fulvestrant are used to treat metastatic breast cancer in patients with tumors that are ER positive, but not in ER-negative tumors. The drugs are most effective in postmenopausal women.
- The structurally similar triarylethylene clomiphene is approved for use in initiating ovulation through FSH/LH release by the pituitary.
- Ospemifene, the fourth of the triarylethylenes, is approved primarily for treatment of dyspareunia following menopause through its estrogen agonistic activity on the endometrium.
- Raloxifene and bazedoxifene, as discussed in Chapter 20, are used to prevent postmenopausal osteoporosis.

Aromatase Inhibitors

Estrogens play an important role in the development and progression of hormone receptor–positive breast cancer. It therefore follows that estrogen deprivation would have a significant therapeutic benefit in treatment of such cancers. The aromatase inhibitors (Fig. 18.37) treat breast cancer by reduction of estrogen levels.

MOA

- The indirect precursor to estradiol involves aromatization of androstenedione, or aromatization of the direct precursor testosterone involves reactions catalyzed by the enzyme aromatase (Fig. 18.4).

> **⚠ ADVERSE EFFECTS**
>
> **SERMs:** The adverse effects of SERMs, in general, are mild, and commonly consist of hot flashes and GI disturbance. With tamoxifen fluid retention, hypercalcemia, vaginal discharge or dryness, nausea, and increased bone and tumor pain have been reported. Ospemifene use can result in an increased risk of endometrial hyperplasia and endometrial cancer, which can be reduced by adding a progestin to the therapy (see Chapter 20 for additional discussion of adverse effects).

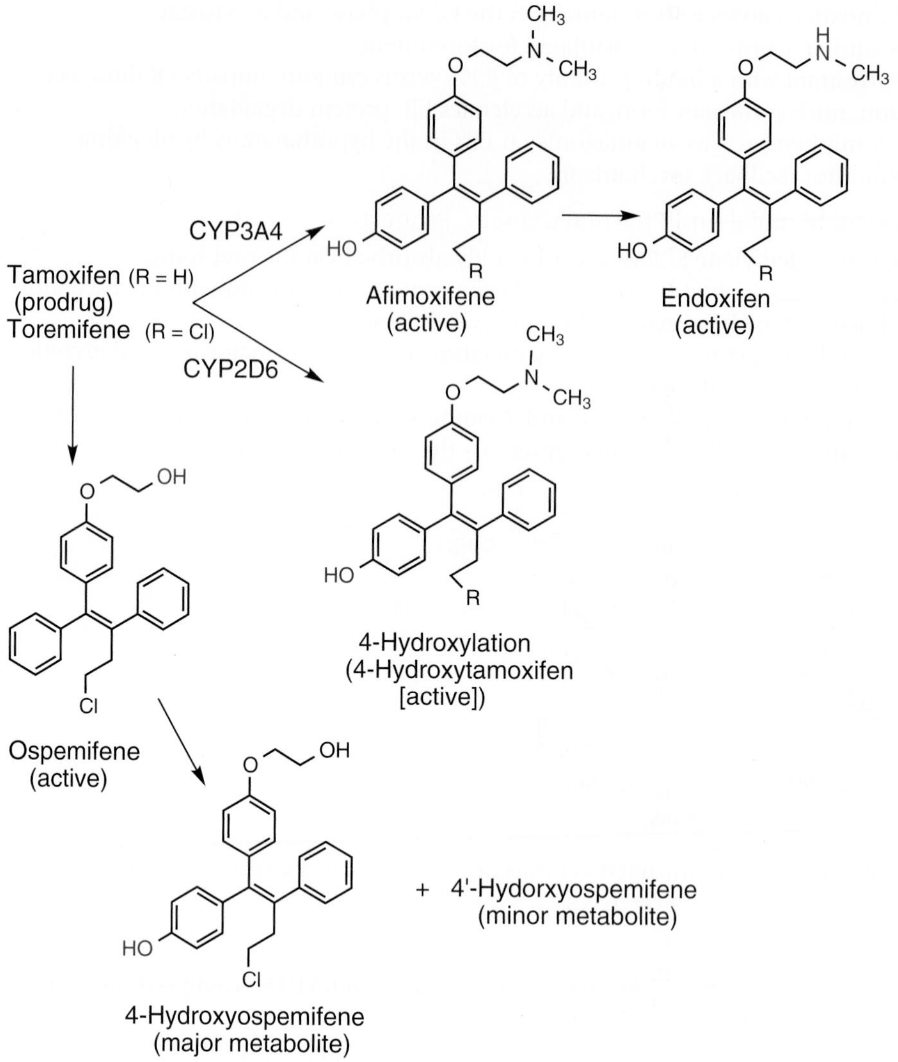

Figure 18.36 Metabolism of triaryl SERMs.

- The nonsteroidal inhibitors of aromatase reversibly bind to the active site of the aromatase involving methionine[374], tryptophan[224], phenylalanine[221], coordination with the heme iron of a CYP450, and possibly the N4 nitrogen of the triazole.
- Exemestane, a steroidal aromatase inhibitor, is an irreversible suicide inhibitor, which binds to the active site of aromatase and involves aspartic acid[309], and serine[478] (Fig. 18.38).
- Following binding of exemestane to aromatase, degradation of the enzyme occurs.

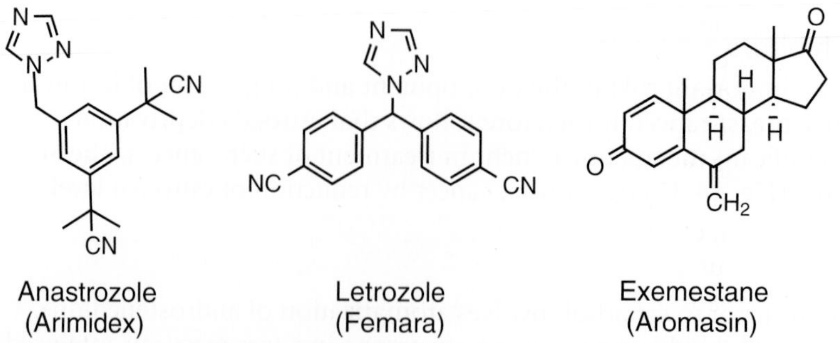

Anastrozole
(Arimidex)

Letrozole
(Femara)

Exemestane
(Aromasin)

Figure 18.37 Aromatase inhibitors.

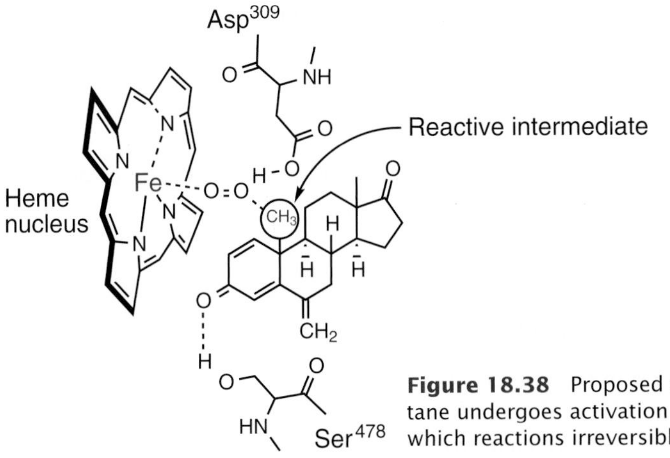

Figure 18.38 Proposed site and mechanism of action of exemestane in which exemestane undergoes activation via a heme catalyzed oxidation at C19 forming an intermediate which reactions irreversible with the Asp[309] of the aromatase enzyme.

Physicochemical and Pharmacokinetic Properties

- The aromatase inhibitors are readily absorbed via the oral route.
- These drugs show moderate binding to plasma proteins (40% to 60%).
- The primary route of excretion of anastrozole is in the feces, while letrozole is excreted via urine.
- Metabolic inactivation is associated with N-dealkylation to triazole.

Clinical Applications

- The aromatase inhibitors are considered first-line treatment of postmenopausal women with hormone receptor–positive metastatic breast cancer.
- All three drugs are reported to be valuable for treatment of advanced breast cancer.

> **ADVERSE EFFECTS**
>
> ⚠ **Aromatase Inhibitors:** Patients on letrozole may experience a decrease in bone mineral density suggesting that bone density should be monitored. In addition, serum cholesterol should be monitored since hypercholesterolemia has been reported in a high percentage of patients. The adverse effects reported for anastrozole are minor and consist of hot flashes, asthenia, arthritis, pain, and skin reactions.

Luteinizing Hormone–Releasing Hormone (LHRH) Therapy

As previously discussed on p. 338 the LHRH agonists are prescribed for treatment of prostate cancer, but are also effective in the treatment of endometriosis through a common mechanism of action. Through down-regulation of the hypothalamic–pituitary axis, the decreased secretion of FSH and LH causes suppression of ovarian function. The peptides leuprolide, goserelin, triptorelin, and nafarelin are commonly prescribed for treatment of endometriosis (Fig. 18.26). Often a progestin, such as norethindrone, and/or a bisphosphonate is added to the therapy to offset the vasomotor symptoms, urogenital atrophy, or bone mineral adverse effects.

Chemical Significance

The term steroid represents a type of organic compound generally composed of four cycloalkane rings joined together. The prime example of a biologically active steroid is cholesterol, a dietary lipid, which in fact serves as the starting material for the glucocorticoid, mineralocorticoid, androgen, and estrogen hormones. This chapter deals with the endogenous adrenocortical hormones and, more specifically, with the therapeutic aspects of these hormones, their clinically important derivatives, and agonists and antagonists of these agents. The chemistry basically defines the type of biological activity that specific steroids have; the instability of the endogenous adrenocortical hormones accounts for the many chemical derivatives that are therapeutically used, and the need to accentuate or block the actions of the endogenous hormones accounts for the synthetic hormonal agonists and antagonists utilized as drugs.

A practitioner drawing on their chemical knowledge can easily identify a steroid nucleus, recognize specific pharmacophores which define which class of adrenocortical, androgen, estrogen, or progestin hormone one is dealing with, and appreciate how the addition of unique chemical groups can improve specificity of action, reduce metabolic inactivation or activation, and change a drug from an agonist to an antagonist. Metabolic inactivation can often alert the practitioner of potential drug–drug interactions. The broad physiological effects of the adrenocorticoids, androgens, estrogens, or progestins, and their drug counterparts can also help explain the potential adverse effects experienced by the patient.

Review Questions

1 The presence of an aromatic A ring in a steroid suggests what type of action?
   A. Glucocorticoid action
   B. Mineralocorticoid action
   C. Androgenic action
   D. Estrogenic action
   E. Cholesterol biosynthesis inhibitor

2 The role of a progestin in the combination OC is:
   A. Direct inhibition of FSH and LH release.
   B. Direct inhibition of LHRH release from the hypothalamus.
   C. Antagonistic action on estradiol binding to the ER.
   D. Complexation with dimeric progesterone reception to prevent translocation to the cell nucleus.
   E. Block ovulation by preventing ovarian follicles from releasing the egg.

3 Treatment of your patient with avanafil results in what direct action?

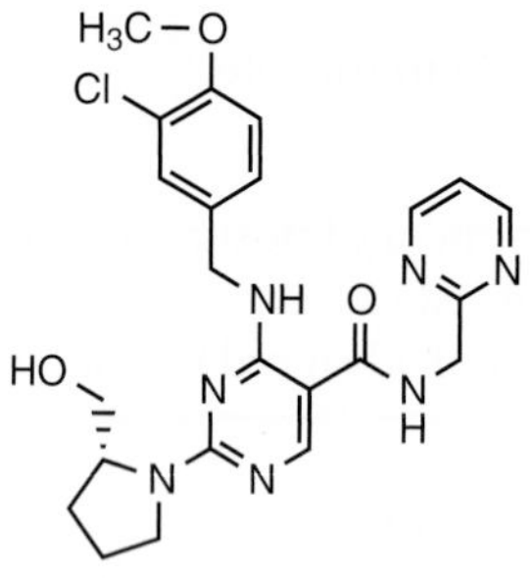

Avanafil

   A. Increased release of NO from vascular endothelial cells.
   B. Increase in cGMP.
   C. Increased blood flow to the corpus cavernosum.
   D. Increased synthesis of cGMP.
   E. Selective inhibition of PDE5 preventing cGMP hydrolysis.

4 What is the role of the circled group in the drug shown below?

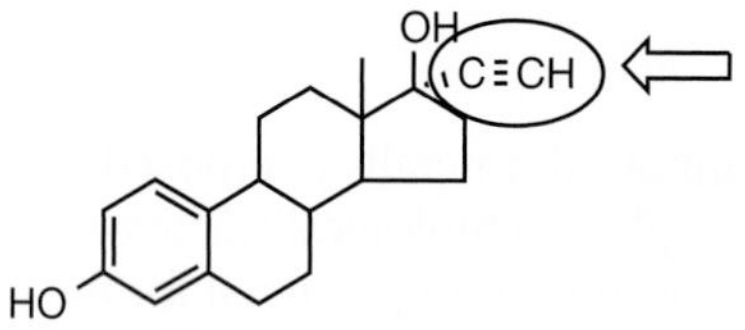

   A. Prolong activity by inhibition of drug metabolism.
   B. Improve selectivity by preventing binding to ERβ while facilitating binding to ERα.
   C. Add lipophilicity to improve penetration of breast tissue.
   D. Increase receptor binding by increasing acidity of the OH group.
   E. Change the drug from an agonist to an antagonist.

5 Dutasteride treatment of BPH is if effective by virtue of the drugs ability to:
   A. Inhibit urethral constriction via blocking adrenergic nerves.
   B. Block 17β-hydroxysteroid dehydrogenase, thus decreasing testosterone synthesis.
   C. Irreversibly inhibit 5α-reductase, thus decreasing DHT synthesis.
   D. Reversibly inhibit aromatase to increased estrone synthesis.
   E. Irreversibly inhibit 17,20-lyase decreasing overall synthesis of androgens.

Thyroid Function and Thyroid Drugs

Learning Objectives

Upon completion of this chapter the student should be able to:

1 Understand the production and physiological role of the thyroid hormones.

2 Discuss the biochemistry of the thyroid hormones.

3 Discuss the drugs used to treat thyroid gland disease.
   a. Thyroid replacement agents
   b. Antithyroid drugs

4 Identify any clinically relevant physicochemical properties (solubility, in vivo and in vitro chemical stability, absorption and distribution).

5 Discuss clinically significant metabolism of each drug class.

6 Explain clinical uses and clinically relevant drug–drug, drug–food interactions.

Introduction

The thyroid gland is the source of the hormones thyroxine (T_4) and triiodothyronine (T_3). These hormones are necessary for normal fetal development as well as adult metabolism. The thyroid gland is the only mammalian organ that can incorporate iodine into organic molecules. The thyroid hormones are derived from the amino acid tyrosine. They are biosynthesized in the follicular cells of the thyroid gland. The release of thyroid hormones is under control of the hypothalamus and pituitary glands (Fig. 19.1). Simply, the hypothalamus releases TRH which stimulates the pituitary to secrete TSH which in turn stimulates the thyroid to secrete T_4 and T_3 in a 4:1 ratio, respectively. T_4 is converted to the active T_3 peripherally by the action of deiodinase enzymes. Plasma levels of T_4 form a feedback loop inhibiting mainly TSH release.

Thyroid disorders are most commonly the result of glandular overactivity (hyperthyroidism) or underactivity (hypothyroidism) with the latter being most common. In either condition, life-long pharmacotherapy is needed. Thyroid dysfunction is diagnosed based upon plasma levels of TSH, T_4 and T_3. Hypothyroidism is usually treated by hormone replacement whereas hyperthyroidism may require treatment with antithyroid agents usually thioamides.

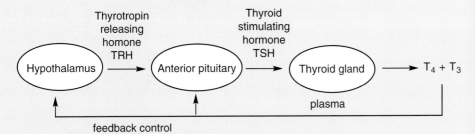

Figure 19.1 Physiological control of thyroid hormone release.

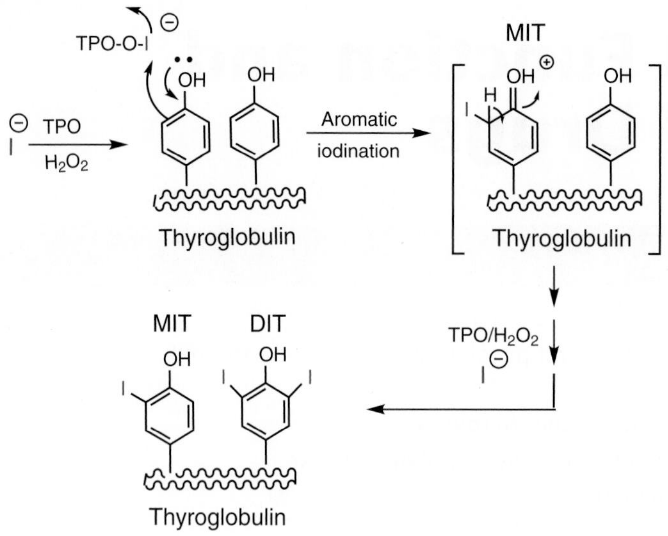

Figure 19.2 Iodination of tyrosine residues on thyroglobulin by hypoiodate formed via action of thyroperoxidase and hydrogen peroxide.

Biosynthesis of Thyroid Hormones

- Ingested iodine is absorbed via the small intestines and transported to the thyroid gland.
- The iodine is concentrated, oxidized and incorporated into the tyrosine residues of thyroglobulin (Tg) in the follicles of the thyroid gland.
- The follicle is a spherical structure consisting of a luminal cavity surrounded by a one-cell thick layer of follicular (acinar) cells.
- Tg is a glycoprotein that is formed in the follicular cells and transported to the lumen of the follicle where its tyrosine residues are iodinated.
- Iodination occurs under the influence of thyroperoxidase (TPO) and H_2O_2 to form mono- and diiodinated tyrosine residues (MIT and DIT) via reaction with hypoiodate anion (Fig. 19.2).
- Coupling of MIT and DIT is also catalyzed by TPO and H_2O_2.
- Iodinated Tg undergoes proteolysis within the follicular cells to release T_3 and T_4 and to a lesser extent rT_3 (reverse T_3) which are secreted into the plasma (Fig. 19.3).

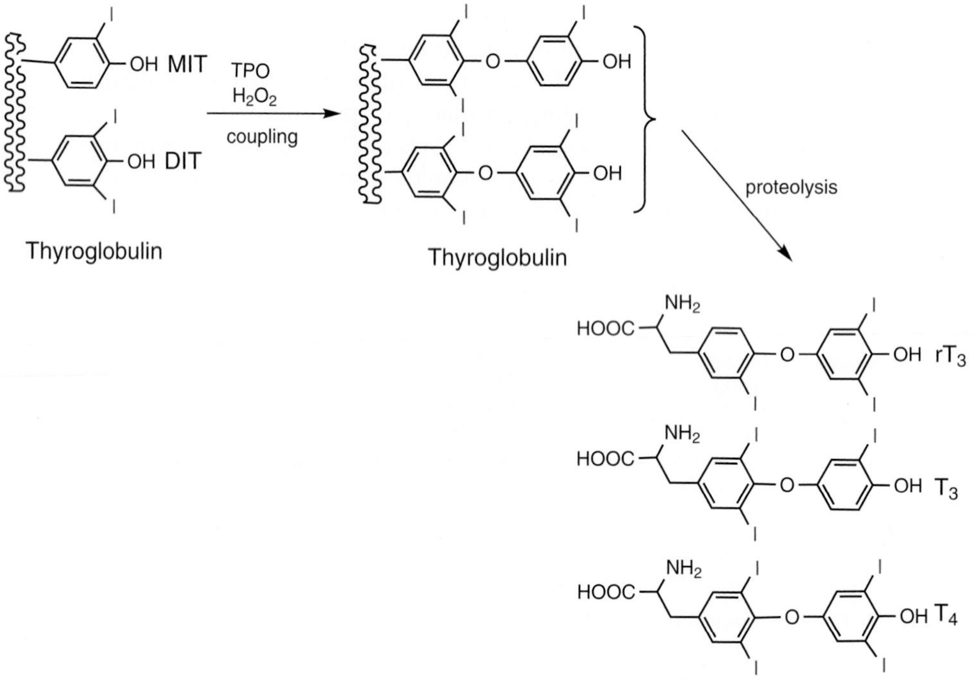

Figure 19.3 Coupling of the iodinated tyrosine residues (MIT and DIT) on thyroglobulin and the proteolytic release of thyroxine (T_4) and triiodothyronine (T_3) and reverse T_3 (rT_3).

Transport of the Thyroid Hormones

- >99% of circulating hormones are found bound to plasma proteins and are readily available for release and entry into cells.
- The plasma proteins include, thyroxine-binding globulin (TBG, ~80%), transthyretin (TTR, ~15%), and human albumin (~5%).
- A very minor amount of thyroid hormones also bind to plasma lipoproteins.

Thyroid Replacement Agents

MOA

- Thyroid hormones (T_3 and T_4) enter cells via membrane transporters. Once inside the cell T_4 is converted to T_3 and T_3 crosses into the nucleus and binds to its receptor. The T_3-receptor complex interacts with specific sequences of DNA which has the effect of either stimulating or inhibiting transcription of specific genes. This process is very similar to how steroids bind and effect gene expression.
- All cells are targets for the thyroid hormones where they have a profound effect on many significant physiological processes such as development, growth, and metabolism. Metabolic effects include increasing the basal metabolic rate which increases body heat as a result of increased oxygen consumption as well as by enhancing carbohydrate and lipid metabolism.

SAR

- The fundamental structure of the thyroid hormones consists of two phenyl rings coupled through an X group.
- The phenyl rings must be appropriately substituted for hormone action.
- Although many derivatives have been prepared it has been found that none were more active than the natural hormones T_4 and T_3.
- Therefore the following substituents are required for maximal activity.
  - X must be an oxygen forming a phenoxyphenyl pharmacophore.
  - R_1 is an alanine moiety.
  - R_5 and R_3 are iodine.
  - $R_{3'}$ must be iodine.
  - $R_{4'}$ is a hydroxyl group.
  - This defines the structure of triiodothyronine (T_3).
  - The L-isomers are more active than the D-isomers.
  - If $R_{5'}$ is an iodine, then this is thyroxine (T_4) which is considered a prohormone for T_3.

Physicochemical and Pharmacokinetic Properties

- Conformational properties of the thyroid hormones.

Structures of the thyroid hormones

Thyroid hormone pharmacophore

Conformational orientation of aromatic rings of 3,5-diiodothyronine

- Molecular modeling studies revealed that the 3,5-diiodothyronines would favor a perpendicular orientation of the planes of the aromatic rings in order to minimize the interaction between the bulky 3,5-iodines and the 2',6'-hydrogens.
- In this orientation the 3'- and 5'-positions of the ring are not conformationally equivalent, and the 3'-iodine of T_3 could be oriented either distal (away from) or proximal (closer) to side chain–bearing ring. The converse can be said of the 5'-iodine.

- There are four thyroid hormone products available for clinical use: Desiccated Thyroid USP, Levothyroxine (T_4), Liothyronine (T_3), and Liotrix.

Desiccated Thyroid USP

- Desiccated thyroid USP is derived from either hog or beef thyroid glands and standardized to contain 0.17% to 0.25% iodine.
- The desiccated preparations release T_4, T_3, DIT, and MIT via proteolytic activity of GI tract enzymes.
- The product has unacceptable variability in potency and ratios of T_4 and T_3.
- Contraindicated in patients hypersensitive to beef or pork.

Levothyroxine (T_4)

- Drug of choice for the treatment of hypothyroidism.
- Physicochemical and Pharmacokinetic properties.
  - pKa = 6.7.
  - Oral bioavailability is 50% to 80%; food decreases bioavailability.
  - Protein binding is 99%.
  - Duration of action is several weeks.
  - Elimination half-life is 3 to 4 days; 50% feces, 50% urine.

Liothyronine (T_3)

- Synthetic form of T_3 and is used when a rapid onset and cessation of action is required, as for example in patients with heart disease.
- Physicochemical and pharmacokinetic properties.
  - pKa = 8.4.
  - Oral bioavailability is 95%.
  - Weakly bound to protein.
  - Duration of action is several days.
  - Elimination half-life is 1 day; 100% via the urine.

Liotrix

- A mixture of T_4 and T_3 in a ratio of 4:1.
- No advantage over levothyroxine and costs more.

Metabolism

- T_4 is considered a pro-hormone since it must be converted to the active T_3 form (Fig. 19.4).
- The outer ring of T_4 is deiodinated by the enzyme 5′ deiodinase (5′-D) to yield T_3 (~33%).
- The inner ring of T_4 is deiodinated by 5′-D to yield rT_3 (~40%) which has no known biological function.
- In addition, both T_3 and T_4 undergo sulfonation and glucuronide conjugation of the outer ring phenolic hydroxyl group.
- The ether bond can be cleaved as well and decarboxylation and oxidative deamination occurs at the alanine side chain.

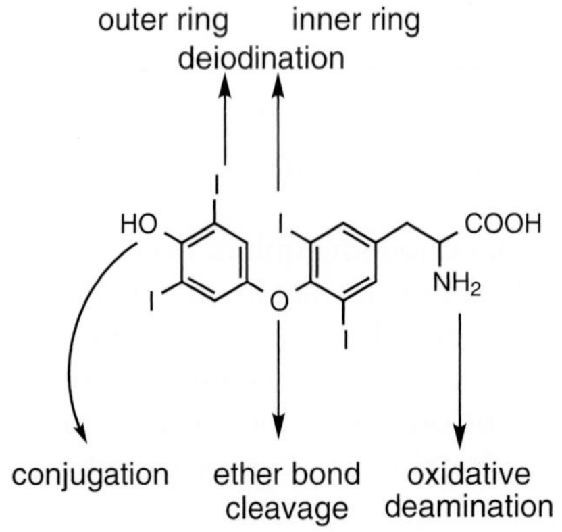

Figure 19.4 Metabolic pathways for thyroxine.

Clinical Applications

- Hypothyroidism is diagnosed by symptoms and by thyroid function tests: TSH levels (elevated), T_4 and T_3 plasma levels.
- Hypothyroidism is more common in women than men.
- Levothyroxine is the drug of choice for the treatment of hypothyroidism.
- Treatment of hypothyroidism generally requires life-long pharmacotherapy.
- There are no contraindications for thyroid replacement drugs, however if excess supplement is taken symptoms similar to hyperthyroidism may appear (see following section).
- There are some drug–drug interactions that may require dosing adjustments.
- Food may decrease adsorption so supplements should be taken ½ hour before or 2 hours after eating.

> ### Clinical Presentation of Hypothyroidism
>
> Difficulty concentrating; loss of interest in pleasurable activities; feeling of "worthlessness" and fatigue. Hypertension. Hair, nail, and skin changes. Unexplained weight gain; cold intolerance and muscle aches.

Antithyroid Drugs

Thiouracil (R = H)
Methylthiouracil (R = CH_3)
Propylthiouracil (PTU, R = n-C_3H_7)

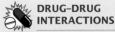

Methimazole (MMI, R = H)
Carbimazole (R = C_2H_5OCO)

> ### DRUG–DRUG INTERACTIONS
>
> **Thyroid Hormones:** Thyroxine binds to aluminum antacids, sucralfate, cholestyramine, colestipol, ferrous sulfate, and kayexalate. Thyroxine metabolism increased by phenobarbital, phenytoin, rifampin, and carbamazepine. Amiodarone and β-blockers inhibit the conversion of T_4 to T_3.

MOA

- Thioamides are the major drugs used to treat hyperthyroidism.
- They are potent inhibitors of TPO thereby preventing the iodination of tyrosine residues on Tg and the coupling reactions that form the iodothyronines.
- Some of them are also capable of inhibiting the peripheral deiodination of T_4 and T_3 by the enzyme 5'-DI.

SAR

- Chemically the antithyroid drugs are cyclic derivatives of thiourea and are generally called thioureylenes.
- They can exist as tautomers however the role of tautomerization is not clear in the activity of these derivatives.

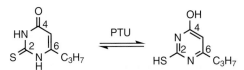

keto/enol tautomers of PTU

> ### CHEMICAL NOTE
>
> Iodide decreases the vascularity and lowers BMR and inhibits the release of thyroid hormones from the thyroid gland. It is no longer used to treat hyperthyroidism since the advent of the antithyroid drugs. It is administered as Lugol's solution (Strong Iodine Solution USP) for 2 weeks prior to thyroid surgery to ensure decreased vascularity and firming of the gland.

- The most useful thioureylenes are five- and six-membered heterocyclic derivatives, that is, thiouracils and thioimidazoles.
- Both heterocycles inhibit TPO within the thyroid gland however only the thiouracils can also inhibit peripheral deiodination T_4 and T_3.
- SAR of thiouracils as inhibitors of peripheral deiodination revealed:
  - The C2 thioketo/enol and an unsubstituted N1 are essential for activity.
  - Activity is enhanced by a C4 enol and alkyl groups at C5 and C6.
- The thioimidazoles are more inhibitory of TPO and are longer acting than the thiouracils.
- The fact that the thioimidazoles are not inhibitory of peripheral deiodination may be due to the presence of the methyl group at N1.
- Only *n*-propyl thiouracil (PTU) and 1-methyl-2-mercaptoimidazole (MMI) are useful clinically.

Physicochemical and Pharmacokinetic Properties

- PTU can exist as keto/enol tautomers where the six-membered ring becomes aromatic.

MMI   3-Methyl-2-thiohydantoin

- PTU is readily absorbed with a 50% to 80% bioavailability. The variability may be due to a large first pass effect.
- PTU plasma half-life is 1.5 hours yet has little influence on the duration of anti-thyroid action or dosing interval since it is accumulated in the thyroid gland.
- PTU is 80% to 90% plasma bound.
- Excretion is primarily via the kidney.
- MMI is rapidly absorbed orally with 93% bioavailability.
- MMI plasma half-life is 5 to 6 hours yet like PTU has little influence on the duration of antithyroid action or dosing interval since it is accumulated in the thyroid gland.
- MMI does not bind to serum albumin and excretion is primarily via the kidney.

Metabolism

- PTU is excreted as the inactive N1-glucuronide.
- MMI is extensively metabolized in the liver, most likely to 3-methyl-2-thiohydantoin.

Thiourea

Thioketo   Thioenol (SH form)

Thiourea tautomers

Clinical Applications

- Hyperthyroidism is diagnosed by elevated T_4 and lowered TSH plasma levels.
- Goal of therapy for the treatment of hyperthyroidism is to reestablish the euthyroid metabolic state.
- Onset: 2 to 4 weeks and therapy with antithyroid agents is usually for 6 months to 2 years.
- Other treatments include surgery; radioactive iodine (I^{131}) and β-blockers to block the effect of catecholamine on the heart.

Tyrosine Kinase Inhibitors for the Treatment of Thyroid Cancer

MOA

- Tyrosine kinases (TK) are important enzymes for the modulation of growth factor signaling. Deregulated TKs play a significant role in many neoplastic diseases.
- Inhibiting TKs prevents their effects on cell division stimulation and cell longevity.

Clinical Presentation of Hyperthyroidism

Anxiety; irritability; insomnia; fatigue. Palpitations and variable blood pressure. Heat intolerance; diaphoresis; enlarged goiter. Weight loss with increased appetite. Exophthalmia; diplopia and osteoporosis on long-term disease.

ADVERSE EFFECTS

Antithyroid Drugs: Occur in 3% to 12% of treated patients and include nausea and gastrointestinal distress and more commonly maculopapular pruritic rash. MMI may alter taste. More serious adverse effects include agranulocytosis and hepatotoxicity especially associated with PTU.

DRUG–DRUG INTERACTIONS

Antithyroid Drugs: MMI may potentiate oral anticoagulants. Digoxin, theophylline, and β-blocker dosing may have to be lowered when hyperthyroid patients become euthyroid.

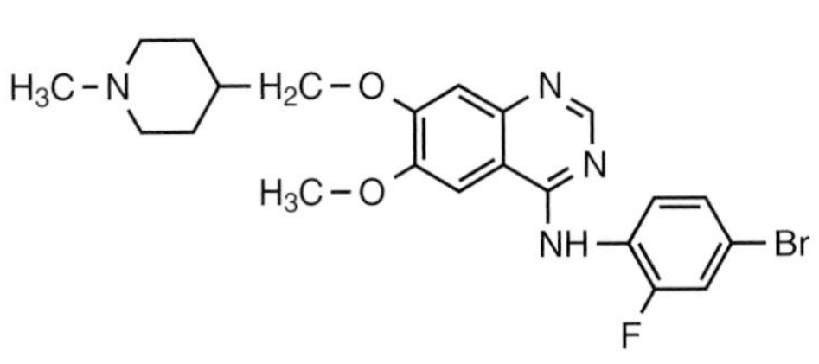

Vandetanib (Caprelsa)
(EGFR TKI)

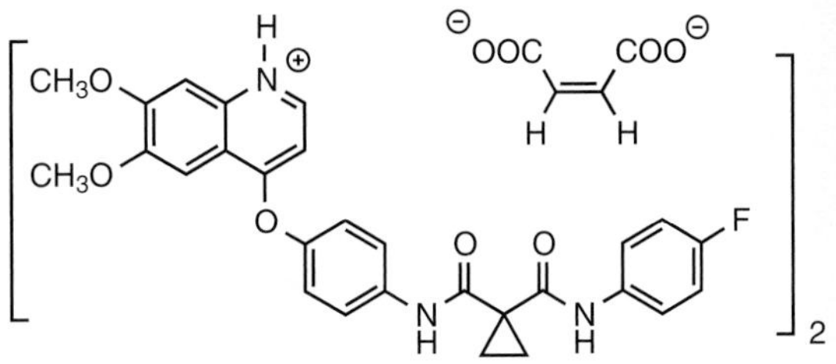

Cabozantinib malate (Exelixis)
(VEGFR TKI)

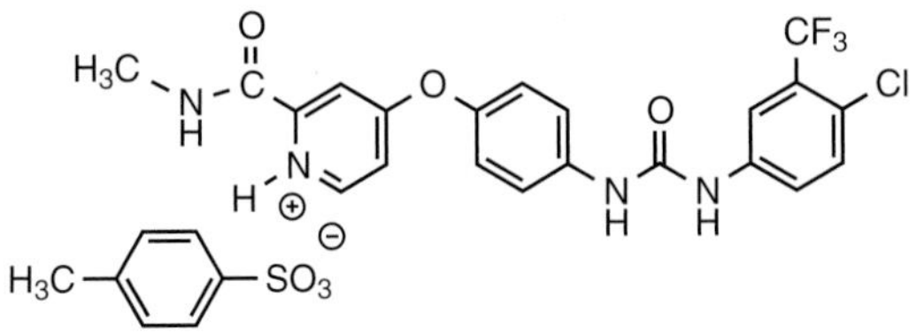

Sorafenib tosylate (Nexavar)
(Multikinase TKI)

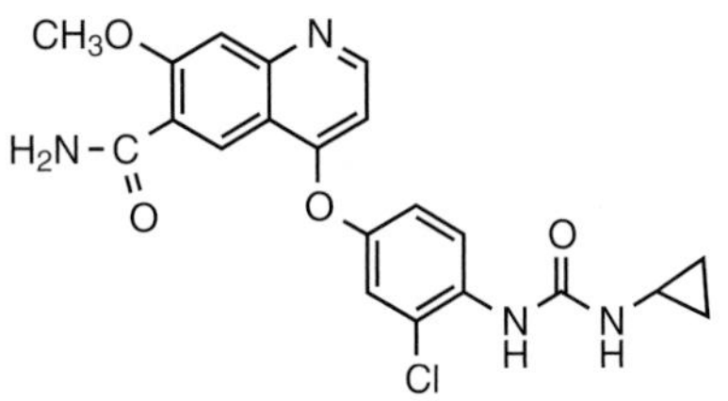

Lenvatinib (Lenvima)
(Multikinase TKI)

Figure 19.5 Tyrosine kinase inhibitors for thyroid cancer.

- TK inhibitors compete with the ATP-binding site of the catalytic domain of several oncogenic tyrosine kinases.
- TK targets for tyrosine kinase inhibitors (TKIs) for the treatment of thyroid cancer include epidermal growth factor receptor (EGFR/HER1), vascular endothelial growth factor receptor (VEGFR1, 2 and 3), or multikinase inhibitors that are nonselective and inhibit other TKs, for example, platelet-derived growth factor receptor (PDGFR) and human epidermal growth factor 2 (HER2).
- See Chapter 26 in this text as well as Chapter 37 in *Foye's Principles of Medicinal Chemistry, Seventh Edition,* for a more complete discussion of these types of inhibitors.

SAR

- Vandetanib is an EGFR TKI (Fig. 19.5).
  - EGFR TKIs have a 4-anilinoquinazoline substitution at C6 or C7 with an oxygen-containing moiety.
  - EGFR TKIs bind to the active conformation of the kinase (type 1 TKI).
- Cabozantinib is a VEGFR TKI.
  - VEGFR TKIs bind to the active, inactive, or both forms of VEGFR kinase.
  - VEGFR TKIs structures contain a potentially ionizable amine (the quinoline nitrogen in cabozantinib).
- Sorafenib is primarily a VEGF TKI but it also has affinity for PDGFR. And therefore classified as a multikinase inhibitor.
  - It contains an ionizable amine on the pyridine ring for binding to the ATP site on the kinase.
- Lenvatinib is classified as a multikinase inhibitor with action at both VEGFR2 and VEGFR3 kinases as well as PDGFR and fibroblast growth factor receptors (FBFs).
  - It contains a quinoline ring system substituted at C4 with an oxygen-containing moiety.
  - It is an analog of cabozantinib where the 4-phenoxy ring is substituted with a urea moiety and the 6-methoxy changed to an amide group.

Metabolism

- Cabozantinib and sorafenib undergo N-oxidation via CYP3A4.
- Vandetanib undergoes N-dealkylation by CYP3A4.

- Lenvatinib is metabolized via CYP3A4 to N-oxide and O-dealkylated products.

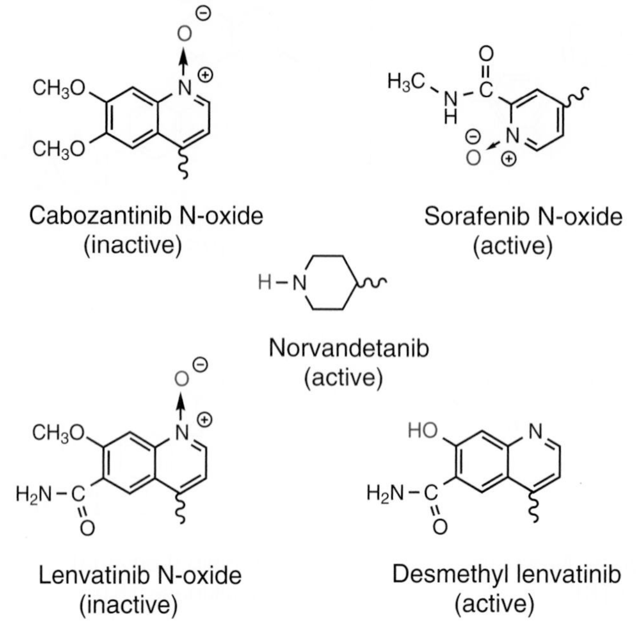

Cabozantinib N-oxide
(inactive)

Sorafenib N-oxide
(active)

Norvandetanib
(active)

Lenvatinib N-oxide
(inactive)

Desmethyl lenvatinib
(active)

Tyrosine kinase inhibitor major metabolites

Clinical Applications

Thyroid cancer is a disease where malignant cells grow in thyroid gland tissue. There are four main types of thyroid cancers: papillary, follicular, medullary, and anaplastic. Treatment can be by surgery (most common), radioactive iodine therapy, and targeted chemotherapy such as TKIs for radioactive resistant and medullary type thyroid cancers.

- Cabozantinib is indicated for treatment medullary thyroid cancer. There is a black box warning for its use indicating it can cause the development of perforations, fistulas, and hemorrhage.
- Sorafenib is indicated for thyroid cancer as well as other nonthyroid cancers. There is the possibility of cardiac ischemia, hemorrhage, hypertension, and dermatologic toxicity.
- Vandetanib is indicated for the treatment of medullary thyroid cancer and there is a black box warning for the development of QT interval prolongation leading to sudden death.
- Lenvatinib is approved for the treatment of patients with locally recurrent or metastatic, radioactive iodine-resistant thyroid cancer. Adverse effects include hypertension, fatigue, cardiac failure, QT interval prolongation, and kidney failure.

Chemical Significance

Understanding the chemistry of the thyroid hormones allowed medicinal chemists to overcome the variability in the T_4 and T_3 levels in crude desiccated thyroid gland preparations by the synthesis of both T_3 and T_4. In addition knowing that T_4 is converted to the active T_3 in vivo has enhanced the treatment of hypothyroidism. The discovery that the thioureylenes inhibited TPO gave a therapeutic alternative to surgery and iodide therapy for the treatment of hyperthyroidism.

Everyday the pharmacist contributes to the discussion of appropriate drug therapy and often relies on the chemical facts necessary to the appropriate choice of medication for a given therapeutic decision. Medicinal chemistry provides the basis for understanding the mechanism of action as well as the pharmacokinetic parameters needed to get the right drug for the right condition to improve patient outcomes.

Review Questions

1 The thyroid hormones are derived from which one of the following amino acids?
A. Threonine
B. Lysine
C. Tyrosine
D. Phenylalanine
E. Alanine

2 In the pharmacophore of the thyroid hormones all of the substituents are necessary for maximal activity *except:*

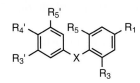

A. X must be oxygen
B. $R_{5'}$ and $R_{3'}$ must be iodine
C. $R_{4'}$ must be a hydroxyl group
D. R_5 and R_3 must be iodine
E. L-isomers more active than D-isomers

3 The thyroid gland secretes T_4 and T_3 in a ratio of:
A. 1:4
B. 1:1
C. 2:1
D. 4:1
E. 3:2

4 The thioimidazole antithyroid agents are shorter acting than the thioureas.
A. True
B. False

5 All of the following statements are true for both PTU and MMI *except:*
A. Potentiate oral anticoagulants
B. Inhibit TPO in the thyroid gland
C. β-Blocker dosing may have to be lowered when hyperthyroid patients become euthyroid.
D. Readily absorbed from the GI tract
E. Concentrated in the thyroid gland

20

Calcium Homeostasis

Learning Objectives

Upon completion of this chapter the student should be able to:

1. Define importance of maintaining calcium homeostasis.
2. Explain the roles of calcitonin, parathyroid hormone (PTH), and vitamin D in maintaining calcium homeostasis.
3. Explain the mechanism of action (MOA), structure activity relationships (SAR), and clinically significant physicochemical properties of the selective estrogen receptor modulators (SERMs).
4. Discuss the MOA, SAR, and clinically significant physicochemical properties of the bisphosphonates.
5. Explain the difference between human calcitonin and salmon calcitonin.
6. Discuss the structural relationship between teriparatide and PTH.
7. Explain the MOA and metabolism of cinacalcet and denosumab.

Introduction

Calcium homeostasis (serum Ca^{2+} concentration) is important for a number of physiological systems and processes, including regulation of bone growth, blood coagulation, neuromuscular excitability, plasma membrane structure and function, muscle contraction, glycogen and ATP metabolism, neurotransmitter/hormone secretion, and enzyme catalysis. Calcitonin (CT), parathyroid hormone (PTH), and vitamin D are the primary hormones involved in homeostatic regulation of calcium. Calcium in the body is found primarily (99%) as teeth and bones and exists as a calcium hydroxyphosphate crystalline matrix called hydroxyapatite $[Ca_{10}(PO4)_6(OH)_2]$. Calcium is also found in the plasma either bound to protein or complexed with (HPO_4^{2-}) counterions.

CT, PTH, and vitamin D, work in concert to maintain calcium homeostasis. Calcitonin is a 32 amino acid peptide secreted by the C cells of the thyroid gland in response to a hypercalcemic state. CT stimulates urinary secretion of Ca^{2+}, prevents Ca^{2+} resorption (breakdown) from bone by inhibiting osteoclast activity, and inhibits intestinal absorption of Ca^{2+}. PTH is an 84 amino acid peptide formed from a 115 amino acid preprohormone produced in the parathyroid gland. PTH is secreted in response to the hypocalcemic state and opposes the action of CT, that is, it inhibits renal secretion of Ca^{2+}, stimulates bone resorption (breakdown), and enhances intestinal absorption of Ca^{2+}. PTH stimulates the biosynthesis, activation and release of vitamin D (Fig. 20.1). Cholecalciferol (D_3) is derived from 7-dehydrocholesterol in the skin by the action of sunlight. D_3 is then converted in the liver to 25-hydroxycholecalciferol which is the major circulatory form of vitamin D. When circulatory levels of Ca^{2+} are low and PTH levels high, 25-hydroxycholecalciferol is converted into 1,25-dihydroxychole-calciferol in the kidney which is the active form of vitamin D, that is, it works in concert with PTH to enhance active intestinal absorption and stimulate bone resorption to produce Ca^{2+}.

The disease states resulting in or from calcium homeostatic imbalance include osteoporosis, hypocalcemia, hypercalcemia, hypoparathyroidism, and hyperparathyroidism. The drugs used to treat these diseases discussed in this

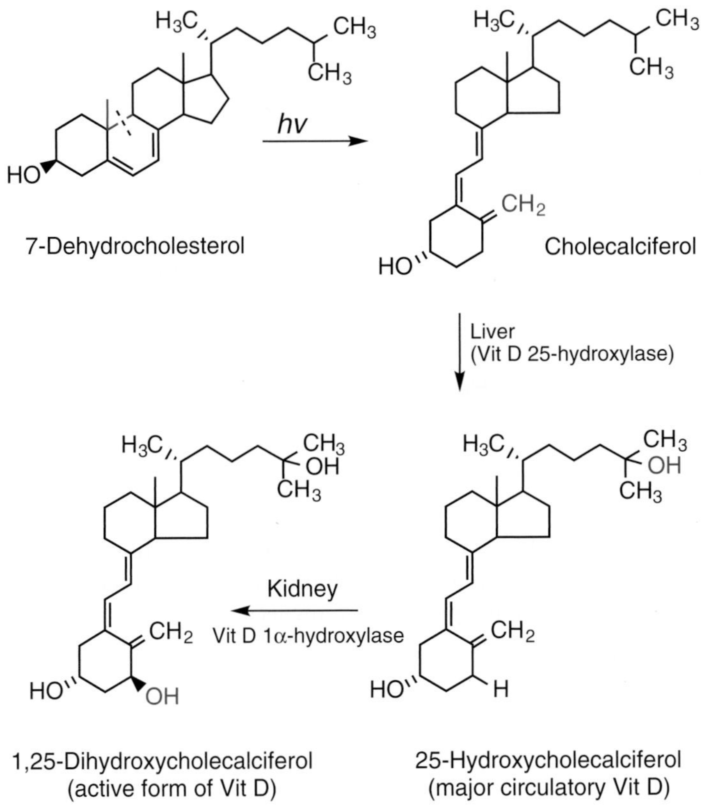

Figure 20.1 Bioactivation of vitamin D.

7-Dehydrocholesterol

Cholecalciferol

Liver
(Vit D 25-hydroxylase)

Kidney
Vit D 1α-hydroxylase

1,25-Dihydroxycholecalciferol
(active form of Vit D)

25-Hydroxycholecalciferol
(major circulatory Vit D)

chapter are the selective estrogen receptor modulators (SERMs), bisphosphonates, calcitonin, cinacalcet, teriparatide, and denosumab.

Selective Estrogen Receptor Modulators

MOA

- SERMs are both estrogen agonists and antagonists.
- They are agonists on receptors in osteoblasts and osteoclasts but antagonists at breast and uterine estrogen receptors.
- This dual action makes them useful for the treatment of osteoporosis as well as metastatic breast cancer in postmenopausal women with estrogen-positive tumors.
- Binding to the estrogen receptor activates the receptor to either increase or decrease gene expression with resulting effects on protein biosynthesis.

> Osteoclasts and osteoblasts along with osteocytes are responsible for the bone remodeling process. Osteocytes release cytokines and growth factors. Osteoclasts originate in the hematopoietic system and carry out bone resorption (breakdown). Osteoblasts are formed in the bone marrow and stimulate bone formation.

SAR

- The SERMs are triarylethylenes based on the structure of tamoxifen (Figs. 20.2 and 20.3).
- A phenol or phenoxy ring system is necessary to mimic the A-ring C3 phenol of estrogen in order to bind with the estrogen receptor activation factor-2 (AF-2) region (Fig. 20.2).
- The orientation of the 3-aryl rings in a propeller-type arrangement is necessary for tight receptor binding and biologic activity.

Physicochemical and Pharmacokinetic Properties

- Raloxifene and bazedoxifene exhibit poor bioavailability (2% and 6% respectively) perhaps due to presystemic O-glucuronidation of their phenolic hydroxyl groups.
- Ospemifene and toremifene are well absorbed and distributed orally however their absolute bioavailability has not been determined.

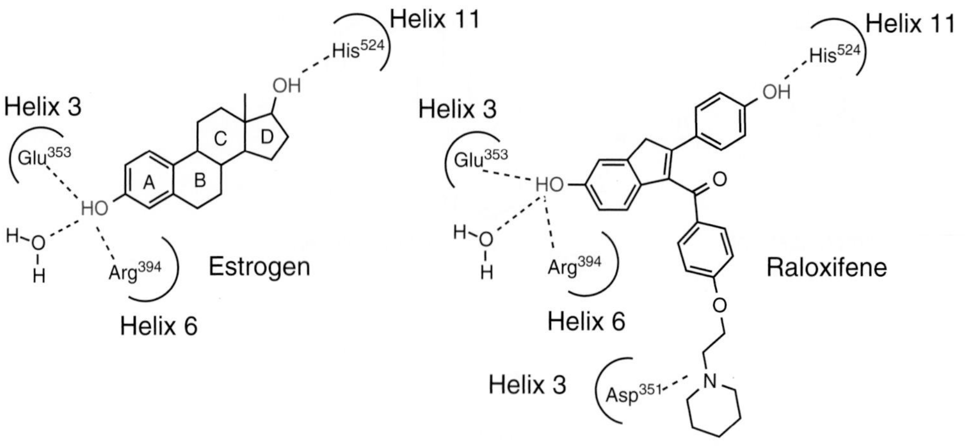

Figure 20.2 Comparison of the binding of estradiol and raloxifene in the binding domain (activation factor-2 [AF-2] region; Helices 3, 6, and 11) of the estrogen receptor.

Metabolism

- Raloxifene and bazedoxifene metabolism primarily occurs in the intestines and consists of O-glucuronide conjugation (Fig. 20.4). Little or no CYP450 metabolism occurs and their metabolites are excreted primarily in the feces.
- Ospemifene is metabolized by CYP3A4, 2C9, and 2C19. The primary metabolite is 4-hydroxy ospemifene.
- Toremifene is metabolized to ospemifene as well as being converted to N-desmethyl toremifene by CYP3A4.

Clinical Applications

- Raloxifene is indicated for the treatment of osteoporosis as well as risk reduction of invasive breast cancer.
- Bazedoxifene is indicated for osteoporosis treatment as well as decreasing vasomotor symptoms of menopause.
- Ospemifene is indicated for the treatment of vulvar and vaginal atrophy associated with menopause.

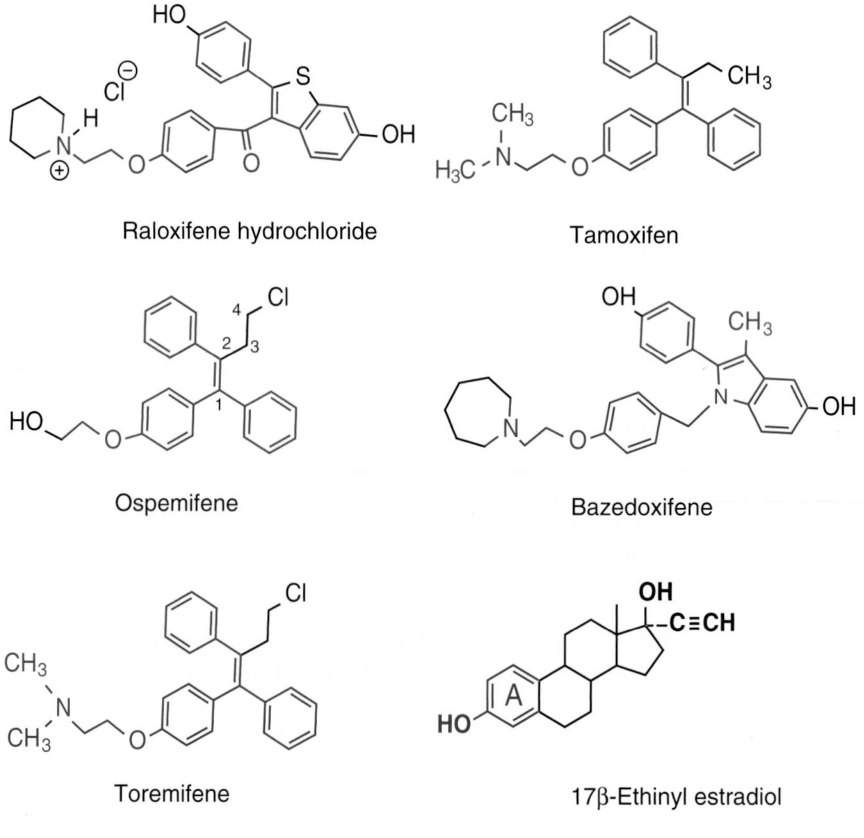

Figure 20.3 Structures of the currently approved selective estrogen receptor modulators (SERMs) highlighting their structural similarities with ring A of 17-ethinyl estradiol.

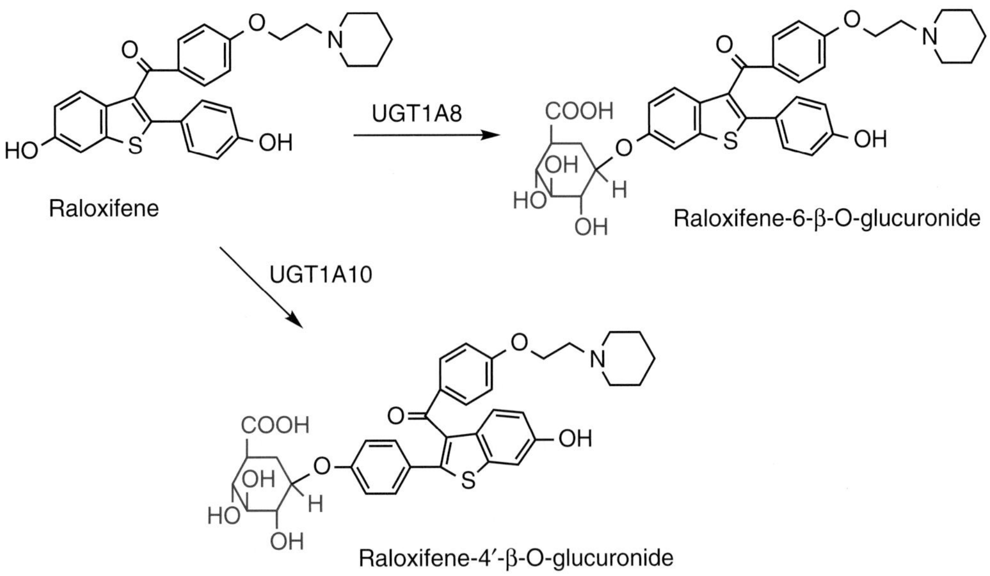

Raloxifene

UGT1A8 →

Raloxifene-6-β-O-glucuronide

UGT1A10 ↓

Raloxifene-4'-β-O-glucuronide

Figure 20.4 Metabolism of raloxifene.

- Toremifene is indicated for treatment of estrogen receptor–positive breast cancer in postmenopausal women.

BLACK BOX WARNING			
SERMs:			
SERM	**VTE/DVT/Stoke**	**Risk of Endometrial Cancer**	**QT Prolongation (Torsades de pointes)**
Raloxifene	X	—	—
Ospemifene	X	X	—
Bazedoxifene	X	X	—
Toremifene	—	—	X

Bisphosphonates

MOA

- Bisphosphonates are nonhydrolyzable analogs of pyrophosphate that bind to the hydroxyapatite portion of bone and effectively inhibit osteoclast activity.
- They also decrease the number of sites along the bone surface where bone resorption occurs by reducing osteoclast proliferation and life span.
- Proliferation and life span reduction is a result of inhibition of the mevalonate pathway and ATP-dependent enzymes within the osteoclasts.
- The result is that they limit bone turn over and allow osteoblasts to form well-mineralized bone.

SAR

- Bisphosphonates are nonhydrolyzable pyrophosphate analogs where the oxygen in P-O-P is replaced with a carbon atom (Fig. 20.5).
- A hydroxyl substituent at R_1 maximizes affinity for hydroxyapatite and improves the antiresorptive action.
- Substituents at R_2 influence potency.
  - Aminoalkyl substituted bisphosphonates are more potent than alkyl substituted.
  - Aminoalkyl chain length affects potency: 3>2.
  - Amino substitution also improves potency.
  - Combining the N into a heterocyclic ring increases potency.

DRUG–DRUG INTERACTIONS

SERMs
- Raloxifene is decreased when coadministered with ampicillin, cholestyramine, and warfarin. Increased with highly protein bound drugs, for example, phenytoin
- Ospemifene is increased with azole antifungals, rifampin, and highly protein bound drugs, for example, phenytoin
- Toremifene is increased with thiazide diuretics. Drugs that prolong the QT interval, for example, amiodarone, warfarin, CYP3A4 inhibitors, for example, azole antifungals, and CYP2C9 substrates, for example, phenytoin. Decreased with CYP3A4 inducers, for example, phenobarbital.

ADVERSE EFFECTS

Bisphosphonates
- Increased risk of atypical femur fracture
- Increased incidence of chemical esophagitis leading to increased risk of esophageal cancer
- Severe and possible incapacitating bone, joint, and musculoskeletal pain

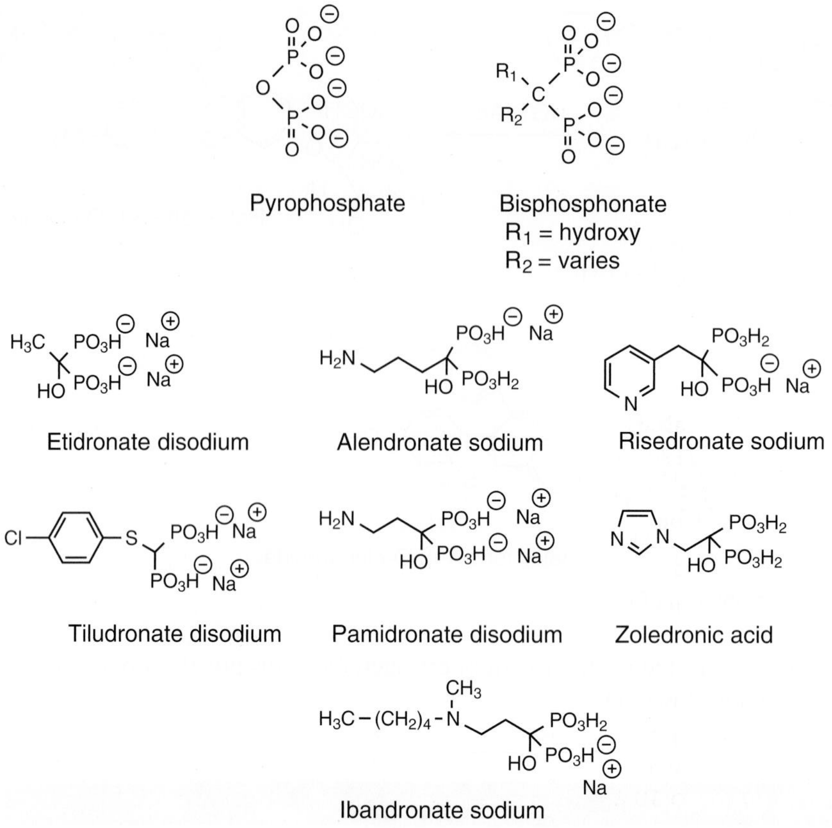

Figure 20.5 Bisphosphonate SAR related to pyrophosphate and clinically used bisphosphonates.

Physicochemical and Pharmacokinetic Properties

- Oral absorption is poor (1% to 5%).
- Up to 50% of an absorbed dose is taken up into bone within 4 to 6 hours, especially in the areas of bone that are remodeling.
- Bisphosphonates have a short half-life due to rapid rate of clearance. However once incorporated into bone they have a half-life of 1 to 10 years.
- The bisphosphonates are not metabolized to any great extent since they are either excreted unchanged or incorporated into bone.

Clinical Application

- Table 20.1 shows the clinically useful bisphosphonates, their generation and clinical application.

Table 20.1 Clinical Application of Bisphosphonates

Bisphosphonate	Generation	Clinical Indication
Etidronate	1st	Paget disease of the bone but not for osteoporosis
Alendronate	2nd	Prevention and treatment of osteoporosis Paget disease of the bone
Tiludronate	2nd	Paget disease of the bone but not for osteoporosis
Pamidronate	2nd	Prevention and treatment of osteoporosis Paget disease of the bone Hypercalcemia of malignancy
Residronate	3rd	Prevention and treatment of osteoporosis Paget disease of the bone
Ibandronate	3rd	Prevention and treatment of osteoporosis
Zoledronic acid	3rd	Prevention and treatment of osteoporosis Paget disease of the bone Hypercalcemia of malignancy

Calcitonin

- Calcitonin (CT) is a 32 amino acid peptide secreted by the C cells of the thyroid gland in response to a hypercalcemic state.

⚠ **ADVERSE EFFECTS**

Calcitonin
- Nausea, vomiting, anorexia, flushing, and possible allergic reaction
- Adverse effects more pronounced when administered IV
- Patients prone to allergies should carry an epi-pen

MOA

- Calcitonin in the presence of Ca^{2+} has three major effects.
  - Decreased bone resorption as a result of osteoclast brush borders disappearance and osteoclasts movement away from bone surfaces undergoing remodeling.
  - Decreased intestinal Ca^{2+} absorption.
  - Increased Ca^{2+} excretion.

SAR

- Commercially available calcitonin is a synthetic CT-salmon which contains the same 32 amino acid sequence as CT-human but differs structurally at 16 of the 32 amino acids (Fig. 20.6).
- The disulfide bond between Cys-1 and Cys-7 is critical for activity.

Physicochemical and Pharmacokinetic Properties

- CT-salmon is 50 times more potent than CT-human with longer duration of action.
- Duration of action is 8 to 24 hours (IM; SC) and 0.5 to 12 hours (IV).
- Available in parenteral, nasal spray and rectal suppository dosage forms.
- Parenteral (IM; SC) peak plasma levels attained in 15 to 25 minutes.
- Bioavailability of the nasal preparation is variable with peak plasma levels reached within 30 to 40 minutes.

Metabolism

- CT-salmon is metabolized in the kidney with an elimination half-life of 43 minutes.
- Presumably CT-salmon is metabolized by brush-border alanyl aminopeptidase and in the region of the cytosolic enzyme phosphoglucomutase in a manner similar to CT-human.

Clinical Applications

- Calcitonin is indicated for the treatment of postmenopausal osteoporosis, hypercalcemia of malignancy, and Paget disease of the bone.
- Calcitonin therapy requires concomitant oral administration of calcium salts and vitamin D.
- Calcitonin posses a potent analgesic action (~30 to 50× morphine) presumably due to endogenous opioid release (endorphins, enkephalins, and dynorphins).

Figure 20.6 Primary structures of salmon and human calcitonin (CT). Similarities are highlighted in *red*.

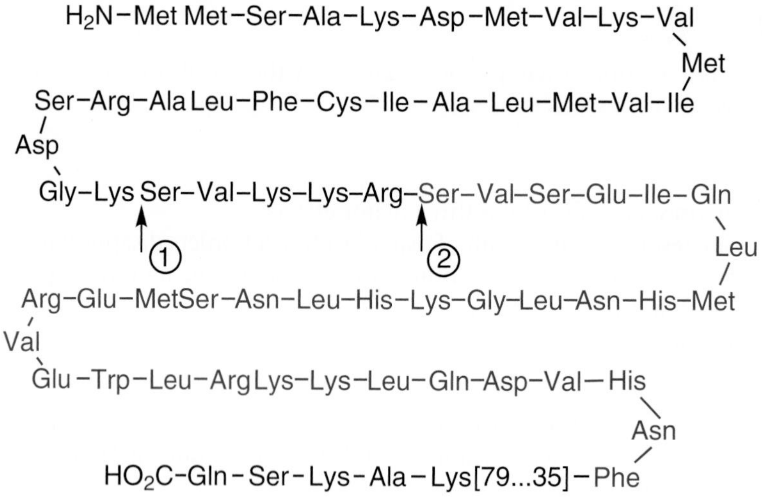

H₂N–Met Met–Ser–Ala–Lys–Asp –Met–Val–Lys–Val
                                                        \
                                                         Met
Ser–Arg–Ala Leu–Phe–Cys–Ile – Ala–Leu–Met–Val–Ile
      /
Asp
      \
       Gly–Lys Ser–Val–Lys–Lys–Arg–Ser–Val–Ser–Glu–Ile–Gln
              ↑①   ↑②   \
                                                           Leu
Arg–Glu–MetSer–Asn–Leu–His–Lys–Gly–Leu–Asn–His–Met   /
Val
    \
     Glu –Trp–Leu–ArgLys–Lys–Leu–Gln–Asp–Val— His
                                                 \
                                                  Asn
                                                 /
HO₂C-Gln – Ser–Lys–Ala – Lys[79...35] – Phe

Figure 20.7 Preproparathyroid hormone is the 115 amino acid protein indicated above. Cleavage at site 1 gives rise to proparathyroid hormone (89 amino acids) while cleavage at site 2 gives parathyroid hormone (PTH, 84 amino acids). The protein shown in *red* is teriparatide (34 amino acids).

> **⚠ ADVERSE EFFECTS**
>
> **Teriparatide**
> - Dizziness and leg cramps
>
> Contraindications:
> - patients predisposed to hypercalcemia
> - patients on digoxin because it increases calcium
> - patients at risk for osteosarcoma (FDA Black Box)

Teriparatide

- Teriparatide is recombinant human parathyroid 1–34, which is the biologically active portion of the endogenously produced preprohormone (Fig. 20.7).
- Its MOA is to increase the number of osteoblasts thereby influencing bone formation.
- It is rapidly absorbed via SC injection with a bioavailability of 95%.
- It is eliminated via hepatic and extrahepatic routes with a half-life of 1 hour.
- No metabolic studies have been done however the entire PTH preprohormone has been shown to undergo enzyme-mediated transformations in the liver.

Clinical Application

- Teriparatide is approved for the treatment of postmenopausal osteoporosis in patients who have a high risk of fracture, for the treatment of glucocorticoid-induced osteoporosis, and to increase bone mass in men with primary or hypogonadal osteoporosis.

Calcium Salts

- Appropriate dietary intake of Ca^{2+} during childhood, adolescence, and adulthood increases bone mineral density (BMD) and reduces the risk of osteoporosis later in life.
- The recommended amount of daily consumption of Ca^{2+} for teenagers is 1,300 mg, for men and premenopausal women is 1,000 mg, and for postmenopausal women is 1,200 mg.
- Table 20.2 gives the amount of Ca^{2+} in various commercial products.
- Calcium absorption from the GI tract is 25% to 40% which is increased in acidic environment of the stomach. Therefore taking drugs that lower stomach acid should be avoided such as H_2 antagonists or proton pump inhibitors.
- The citrate, lactate, and gluconate salts are more water soluble and depend less on acidity for absorption.

Cinacalcet

- Hypercalcemia can result from an over secretion of PTH most commonly resulting from secondary hyperparathyroidism in patients with chronic kidney disease (CKD) on dialysis.

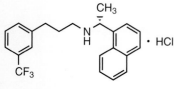

Cinacalcet hydrochloride
(Sensipar)

Table 20.2 Percent of Elemental Calcium Content in Various Salts

Salt	Calcium (%)	Elemental Calcium mg/Tablet
Calcium carbonate	40	
Tums (500 mg chewable)		200 mg
Titralac (1 g/5 mL suspension)		400 mg/5 mL
Alka-Mints (850 mg chewable)		340 mg
Os-Cal 500 (1,250 mg tablets)		500 mg
Viactiv (1,250 mg chewable)		500 mg
Tricalcium phosphate	39	
Calcium chloride	27	
Tribasic calcium phosphate	23	
Posture (1,565.2 mg tablets)		600 mg
Calcium citrate	21	
Citracal (950 mg tablets)		200 mg
Citracal Liquitab (2,376 mg effervescent tablets)		500 mg
Calcium lactate	13	
Generics (325 mg tablets)		42 mg
Generics (650 mg tablets)		84 mg
Calcium gluconate	9	
Neo-Calglucon (1.8 g/5 mL syrup)		115 mg/5 mL

MOA

- Cinacalcet is a drug that acts as a calcimimetic.
- The calcium-sensing receptors (CaSRs) on the surface of the parathyroid gland regulate PTH secretion.
- Cinacalcet lowers PTH levels by increasing the sensitivity of the CaSR resulting in the inhibition of PTH secretion leading to a decrease in serum calcium levels.

Metabolism

- Primarily CYP3A4, CYP2D6, and CYP1A2.
  - Oxidative N-dealkylation to hydrocinnamic acid and hydroxycinnamic acid.
  - Oxidation of the naphthalene ring on the parent drug to form dihydrodiols, which are further conjugated with glucuronic acid (Fig. 20.8).

Clinical Application

- Treat hyperparathyroidism (elevated PTH levels), a consequence of parathyroid tumors and chronic renal failure.

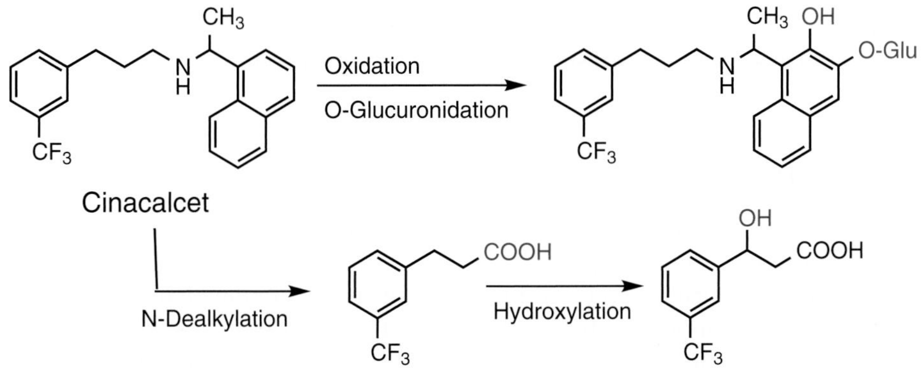

Figure 20.8 Oxidative metabolism of cinacalcet.

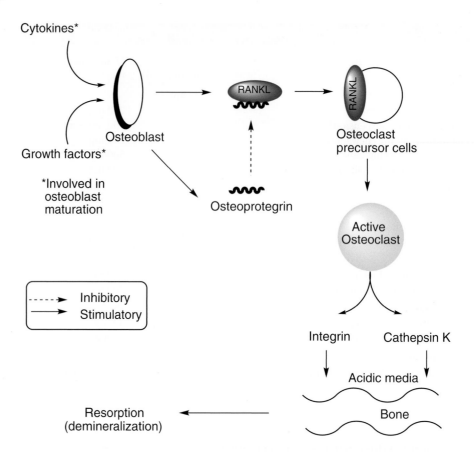

Figure 20.9 Bone resorption involving receptor activator of nuclear factor-kB ligand (RANKL).

Denosumab

MOA

- Denosumab is an antibody to the receptor activator of nuclear factor-kB ligand (RANKL).
- It functions as a RANKL inhibitor in much the same way as the natural ligand osteoprotegerin (Fig 20.9).
- Inhibition of RANKL modulates the production and activation of osteoclasts thereby maintaining a balance between osteoclasts and osteoblasts allowing good bone formation.

Clinical Applications

- Denosumab is approved for the treatment of postmenopausal women with osteoporosis at high risk for fracture.
- Denosumab has been shown to be more effective in improving BMD (4% to 7%) than weekly administration of alendronate (5%).
- It can reduce the risk of vertebral fraction (68%), hip fracture (40%), and nonvertebral facture (20%).
- Denosumab is administered subcutaneously once every 6 months.

⚠ ADVERSE EFFECTS

Denosumab: Back pain (35%), serious infections, hypocalcemia, and osteonecrosis of the jaw.

Chemical Significance

Understanding medicinal chemistry principles gives the pharmacist insight into the differences between the variety of drugs in a specific pharmacological class. Small changes to structure can result in significant changes in the pharmacokinetic properties of a drug molecule including decreased side effects, increased duration of action, enhanced absorption and distribution as well as explain the mechanism of action. For example, changing the oxygen in pyrophosphate (P-O-P) to a carbon atom (P-C-P) results in the nonhydrolyzable bisphosphonates that can bind to hydroxyapatite of bone and inhibit osteoclasts thereby limiting bone turnover and allowing osteoblasts to form well-mineralized bone. Therefore, understanding the chemistry of drugs used to treat and maintain calcium homeostasis contributes to the effective therapeutic decisions for safe and effective drug use.

Review Questions

1 Calcitonin is a 32 amino acid peptide secreted by
A. The parathyroid gland
B. The renal cells of the kidney
C. The thyroid gland
D. The hypothalamus
E. The pancreas

2 Which of the following structures of the selective estrogen receptor modulators (SERMs) would have the most effective binding to the AF-2 region of the estrogen receptor?

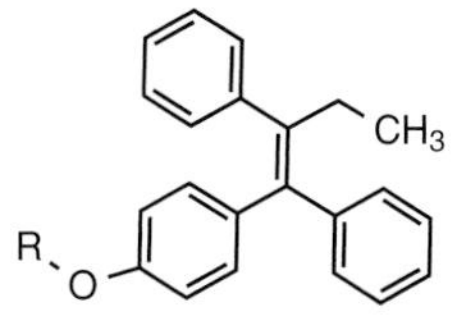

A 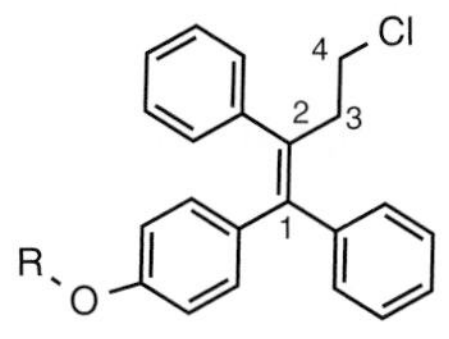   **B**

3 Bazedoxifene is primarily metabolized by which of the following metabolic pathways?
A. Aromatic hydroxylation in the liver
B. O-Glucuronide conjugation in the intestines
C. N-dealkylation in the liver
D. O-Sulfate conjugation in the liver
E. O-methylation in the liver

4 Based on the SAR of the bisphosphonates which of the following compounds would be expected to be the most potent?

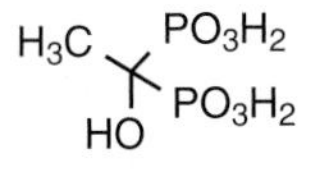

A 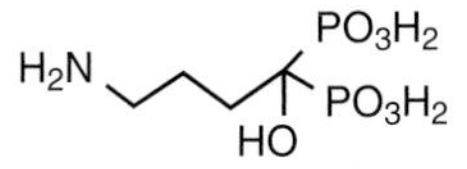   **B** 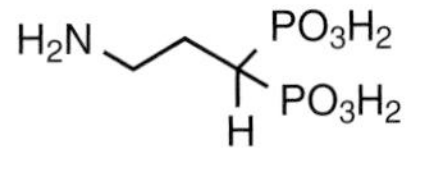   **C**

5 Which of the following statements about teriparatide is incorrect?
A. Its MOA involves increasing the number of osteoblasts
B. Rapidly absorbed orally with 95% bioavailability
C. Has $t_{1/2}$ of 1 hour
D. Eliminated by both hepatic and extrahepatic routes
E. Is a recombinant human parathyroid analog consisting of amino acids 1 to 34 of parathyroid hormone.

21 Nonsteroidal Anti-Inflammatory Drugs

Learning Objectives

Upon completion of this chapter the student should be able to:

1. Discuss the role of arachidonic acid, cyclooxygenase (COX) and prostaglandins in mediating inflammation.

2. Describe the anti-inflammatory and/or analgetic mechanism of action of:
   a. Salicylic acid derivatives
   b. Arylanthranilic acids (fenamates)
   c. Arylalkanoic acids
   d. Oxicams
   e. COX-2 selective inhibitors
   f. Acetaminophen
   g. Disease-modifying antirheumatic drugs (DMARDs)
   h. Biologics

3. Describe the mechanism of action of drugs used to treat gout.

4. Describe the mechanism of action of drugs used to treat multiple sclerosis.

5. Discuss applicable structure activity relationships (SAR) (including key aspects of receptor binding) of the above classes of anti-inflammatory, analgesic, or antigout agents.

6. List physicochemical and pharmacokinetic properties that impact in vitro stability and/or therapeutic utility of anti-inflammatory analgetic and antigout agents.

7. Diagram metabolic pathways for all biotransformed drugs, identifying enzymes and noting the activity, if any, of major metabolites.

8. Apply all of above to predict and explain therapeutic utility.

Introduction

Nonsteroidal anti-inflammatory drugs (NSAIDs) are competitive and reversible inhibitors of two closely related cyclooxygenase enzymes, COX-1 and COX-2. These are the rate-limiting enzymes in the biosynthesis of inflammatory PGE and PGF prostaglandins and the vasoactive prostanoids thromboxane A_2 (TXA_2) and prostacyclin (PGI_2). Inhibition of COX enzymes results in an attenuation of the inflammatory response, but can also lead to unwanted gastrointestinal and generalized bleeding and renal insufficiency.

In addition to NSAIDs, disease-modifying antirheumatic drugs (DMARDs, including immunosuppressants), and biologics, are available to mitigate the underlying pathology of serious and disabling inflammatory diseases such as rheumatoid arthritis, ulcerative colitis and multiple sclerosis. The inflammatory disease of gout has its own therapeutic arsenal of drugs that, via a variety of mechanisms, decrease pathologically high serum of uric acid levels.

Arachidonic Acid Metabolism: Prostaglandins and Leukotrienes

- Arachidonic acid, a 20-carbon fatty acid, is released from membrane phospholipids by phospholipase A_2.
- Released arachidonic acid is oxidized by two distinct enzymes, cyclooxygenase (COX) and lipoxygenase.

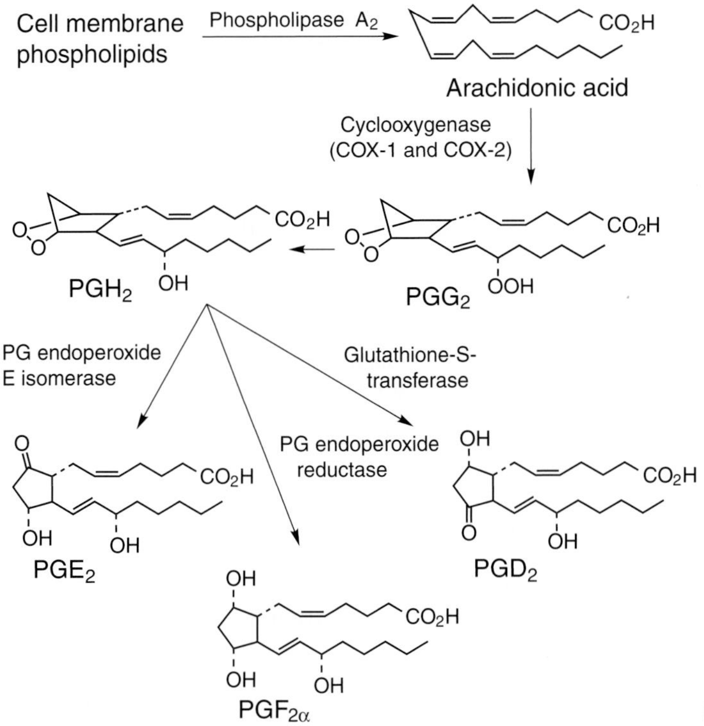

Figure 21.1 Prostaglandin biosynthesis.

- COX oxidation generates PGG_2 which is quickly reduced to PGH_2. Both are highly inflammatory, but short lived, prostaglandins.
- PGH_2 is converted into several prostaglandins and prostanoids (Figs. 21.1 and 21.2).
    - PGE_1 and PGE_2 (inflammatory, smooth muscle stimulant, gastric cytoprotective).
    - $PGF_{2\alpha}$ (inflammatory, smooth muscle stimulant).
    - TXA_2 (vasoconstrictor, platelet aggregation stimulant).
    - PGI_2 (vasodilator, platelet aggregation inhibitor).

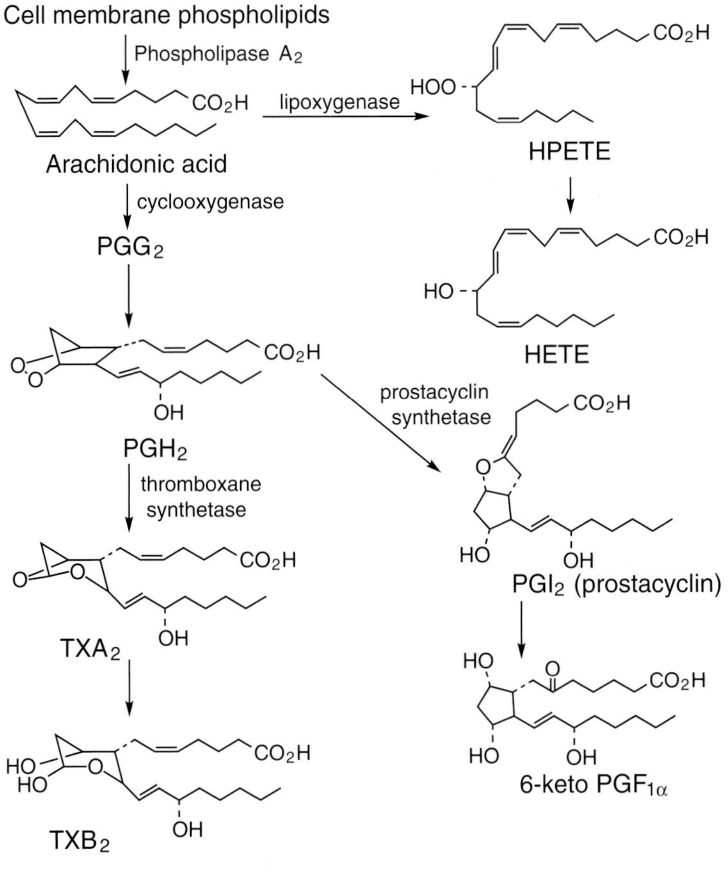

Figure 21.2 Biosynthesis of thromboxanes, prostacyclin, and leukotrienes.

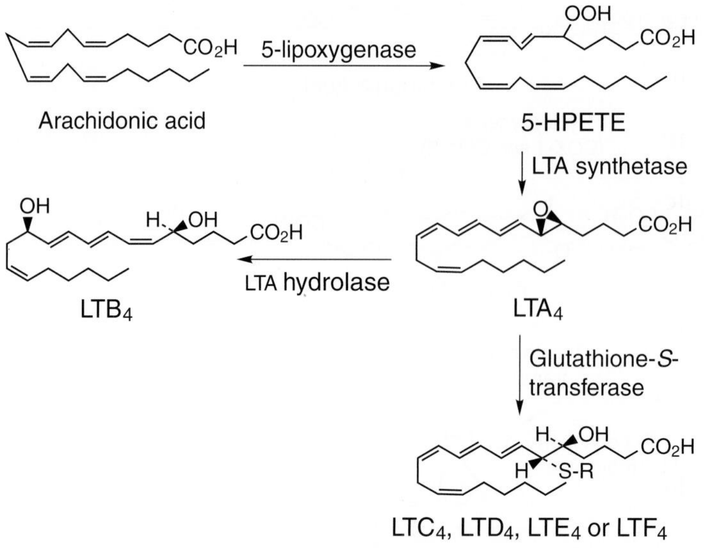

Figure 21.3 Biosynthesis of leukotrienes.

NSAID-Induced Asthma

NSAIDs induce asthmatic reactions in some patients. COX inhibition by NSAIDs causes arachidonic acid levels to rise and it has nowhere else to go except down the lipoxygenase pathway. Lipoxygenase-generated LTC_4 and LTD_4 are found in the Slow Reacting Substance of Anaphylaxis (SRS-A). If levels of these leukotrienes are sufficiently high, it results in bronchoconstriction and the symptoms of acute asthma.

- Lipoxygenase oxidation yields hydroperoxide HPETE (Fig. 21.3). HPETE is metabolized to highly inflammatory cysteinyl leukotrienes.
- Leukotrienes induce bronchoconstriction.
  - Leukotriene receptor antagonists (e.g., zafirlukast, montelukast) and biosynthesis inhibitors (zileuton) are used in the treatment of chronic asthma.

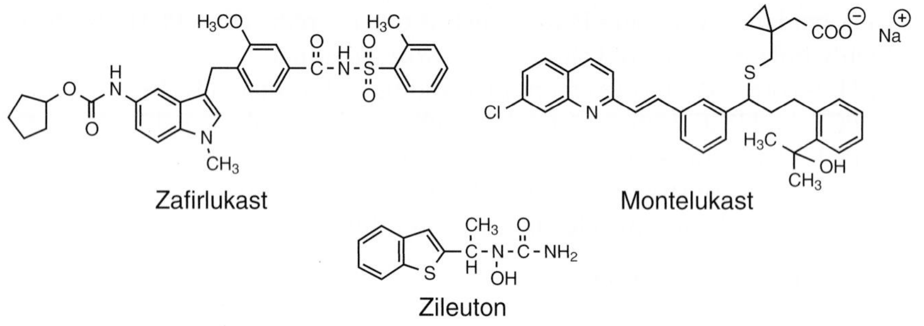

Zafirlukast

Montelukast

Zileuton
Leukotriene modifiers

CHEMICAL NOTE

Prostaglandins as Drugs

$PGF_{2\alpha}$ (dinoprost): Stimulate uterine smooth muscle.

PGE_2 (dinoprostone) and a stable 15-CH_3 derivative of $PGF_{2\alpha}$ (carboprost tromethamine): Abortifacients.

PGE_1 (alprostadil): Treat patent ductus arteriosus in newborns awaiting surgery, Raynaud phenomenon and erectile dysfunction.

15-CH_3-PGE_1 methyl ester (misoprostol): Prevent and/or heal NSAID-induced gastric ulcers.

$PGF_1\alpha$ phenyl analog (latanoprost): Lower intraocular pressure in open angle glaucoma.

PGI_2 analog (iloprost sodium): Lower blood pressure in pulmonary hypertension, promote vasodilation in Raynaud phenomenon or ischemia.

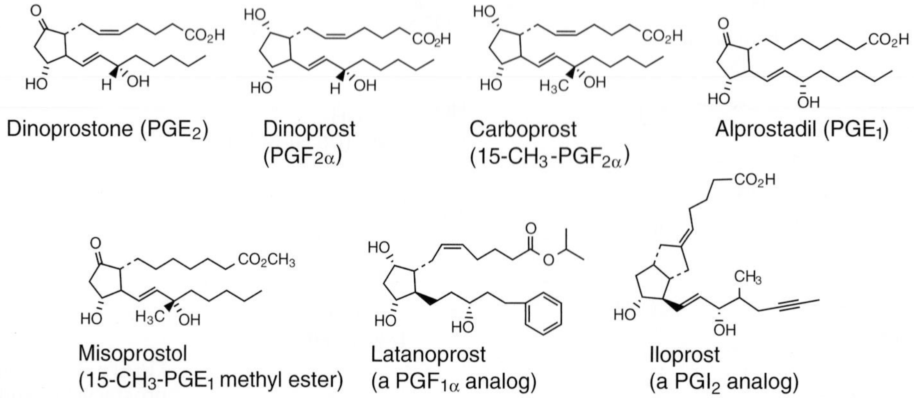

Dinoprostone (PGE_2)

Dinoprost ($PGF2\alpha$)

Carboprost (15-CH_3-$PGF_{2\alpha}$)

Alprostadil (PGE_1)

Misoprostol (15-CH_3-PGE_1 methyl ester)

Latanoprost (a $PGF_{1\alpha}$ analog)

Iloprost (a PGI_2 analog)

Cyclooxygenase (COX): The NSAID Receptor (Fig. 21.4)

- While structurally similar, COX-1 and COX-2 have different physiologic functions.
- COX-1 is continuously expressed (constitutive) and is homeostatic in gut, uterus, kidney, and platelets.
- COX-1 is not the intended target for anti-inflammatory NSAIDs. Inhibition of this isoform leads to adverse effects commonly associated with NSAIDs (e.g., gastrointestinal distress and hemorrhage).
- Intentional inhibition of COX-1 in platelets decreases risk of myocardial infarction and stroke. Many patients take low-dose aspirin daily for this cardiovascular benefit.
- COX-2 is inducible and expressed during inflammatory episodes and in some neoplastic diseases (e.g., colorectal cancer).
- COX-2 is the intended target for anti-inflammatory NSAIDs. Most NSAIDs are nonselective and inhibit COX-1 and COX-2 with equal ease.

COX Topography

- Five residues are critical to binding nonselective NSAIDs to the active site of COX-1 and COX-2.
  - Arg^{120} anchors anionic NSAIDs through ion–ion bonding.
  - Tyr^{348}, Val^{349}, Tyr^{385}, and Trp^{387} interact with NSAID aromatic rings through hydrophobic and van der Waals forces.

> **⚠ ADVERSE EFFECTS**
>
> **COX-1 Inhibitors (NSAIDs)**
> - GI distress, ulceration, and bleeding (inhibition of gastric COX-1)
> - Renal insufficiency (inhibition of renal COX-1 and COX-2)
> - Tinnitus (aspirin, indomethacin)
> - Hypersensitivity reactions (aspirin)
> - Reye syndrome (aspirin in children and young adults)

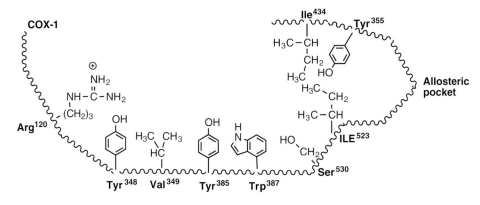

Active site cavity for COX-1

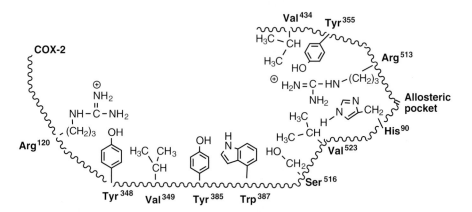

Active site cavity for COX-2

Figure 21.4 Representations of COX-1 and COX-2 active and allosteric sites. (Reprinted with permission from the *American Journal of Pharmaceutical Education*. Roche VF. A receptor-grounded approach to teaching nonsteroidal antiinflammatory drug chemistry and structure-activity relationships. *Am J Pharm Educ.* 2009;73(8):143.)

- An allosteric binding site is adjacent to the COX active site. The COX-2 allosteric site is larger and more flexible than the COX-1 allosteric site.
  - COX-2 has Val residues at positions 434 and 523, while COX-1 has bulkier Ile residues.
  - COX-1 Ile residues restrict access to the allosteric site. The Val residues of COX-2 allow allosteric site access by COX-2 selective NSAIDs.
  - The conformation of Tyr[355] is also less sterically restrictive on COX-2 compared to COX-1.
- COX-2 selective inhibitors bind predominantly within the COX-2 allosteric site. They are denied access to the COX-1 allosteric site.

Nonselective NSAIDs

SAR

- Nonselective NSAIDs must be anionic at pH 7.4 to bind to cationic Arg[120].
- Two aromatic rings mimic the double bonds of arachidonic acid. The rings should be conjugated but allowed to assume a noncoplanar conformation.
- Structural features that force noncoplanarity, block inactivating metabolism, and/or increase lipophilicity increase anti-inflammatory potency.
- Substitution of aromatic rings with metabolically stable functional groups increases duration of action.

Physicochemical Properties

- NSAIDs commonly have pKa values between 3.5 and 4.5.
- All NSAIDs are orally active and significantly bound to serum proteins.
- NSAIDs are physically irritating to gastric mucosa in free acid form.
- Selected physicochemical properties of nonselective NSAIDs are provided in Table 21.1.

Table 21.1 Selective Properties of Nonselective NSAIDs

Drugs	Onset (Duration) of Action	Peak Plasma Levels (h)	Protein Binding (%)	Biotransformation	Elimination Half-Life (h)	pKa
Aspirin	ND	2	90	Plasma hydrolysis and hepatic	<30 min	3.5
Diclofenac	30 min (~8 h)	1.5–2.5	99	Hepatic; first-pass metabolism: 3A4	1–2	4.0
Diflunisal	1 h (8–12 h)	2–3	99	Hepatic	8–12	3.3
Etodolac	30 min (4–6 h)	1–2	99	Hepatic: 2C9	6–7	4.7
Fenoprofen calcium	NR	2	99	Hepatic: 2C9	3	4.5
Flurbiprofen	NR	1.5	99	Hepatic: 2C9	6 (2–12)	4.2
Ibuprofen	30 min (4–6 h)	2	99	Hepatic; first-pass metabolism: 2C9, 2C19	~2	4.4
Indomethacin	2–4 h (2–3 d)	2–3	97	Hepatic: 2C9	5 (3–11)	4.5
Ketoprofen	NR	0.5–2	99	Hepatic: 2C9, 3A4	~2	5.9
Meclofenamate sodium	1 h (4–6 h)	4.0	99	Hepatic: 2C9	2–3	NR
Mefenamic acid	NR	2–4	79	Hepatic: 2C9	2	4.2
Meloxicam	NR	4–5	99	Hepatic: 2C9	15–20	1.1, 4.2
Nabumetone*	NR	2.5 (1–8), 6-MNA	99	Hepatic: 3A4, 1A2	6-MNA, 23	Neutral
Naproxen	NR	2–4	99	Hepatic: 2C9	13	4.2
Oxaprozin	NR	3–5	99	Hepatic: 2C9	25	4.3
Piroxicam	2–4 h (24 h)	2	99	Hepatic: 2C9	50	1.8, 5.1
Sulindac*	NR	2–4	93	Hepatic: sulfide metabolite active	50	4.5
Tolmetin	NR	<1	99	Hepatic	5	3.5

h, hours
*Prodrug.

Metabolism

- Phase I metabolism of NSAIDs is commonly catalyzed by CYP2C9. Coadministered drugs that induce, compete for, or inhibit CYP2C9 have the potential to impact anti-inflammatory action.
- Most NSAIDs are excreted as glucuronic acid conjugates of the carboxylic acid commonly found on the parent drug, or of a carboxylic acid or phenol generated through Phase I metabolism.

Salicylic Acid Derivatives

- Available salicylic acid derivatives are shown in Figure 21.5. The most widely used drugs in this class are aspirin and diflunisal.

Aspirin

MOA

- Aspirin is acetylsalicylic acid. Its unique mechanism of COX inhibition involves irreversible acetylation of a Ser residue (Fig. 21.6). It is the only NSAID to act irreversibly.
- The acetoxy ester is essential to aspirin's unique irreversible mechanism. The acetyl moiety is transferred to the oxygen of Ser[530] of COX-1 and, to a much lesser extent, Ser[516] of COX-2.
- Anionic aspirin abstracts proton from COX Ser, which enhances its nucleophilic character.
- Serine attacks the electrophilic carbonyl carbon of aspirin's acetoxy group, acetylating the enzyme and liberating salicylic acid.
  - The acetylated enzyme is irreversibly inhibited.
  - The liberated salicylate decreases COX-2 expression.
- Low-dose (81 mg) aspirin targets COX-1 in platelets. Irreversible acetylation results in a decrease in levels of pro-aggregatory and vasoconstrictive TXA_2.
  - In higher doses, COX-2 in the vessel wall is inhibited, decreasing production of PGI_2. This counteracts the cardiovascular benefits of platelet COX-1 inhibition.

SAR

- An anionic carboxylate group adjacent to the acetoxy ester is essential. Altering the relative position of these two groups inactivates the structure.

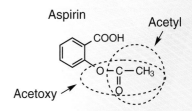

Aspirin

Aspirin acetoxy and acetyl moieties

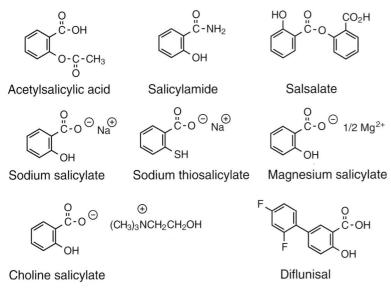

Acetylsalicylic acid   Salicylamide   Salsalate

Sodium salicylate   Sodium thiosalicylate   Magnesium salicylate

Choline salicylate   Diflunisal

Figure 21.5 Salicylic acid derivatives.

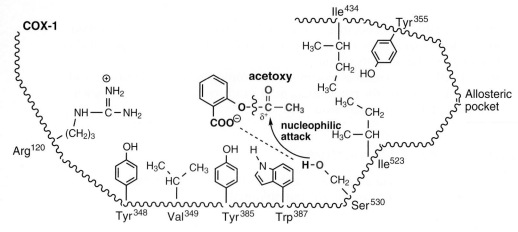

COX-1: Initial interactions with acetylsalicylic acid (aspirin)

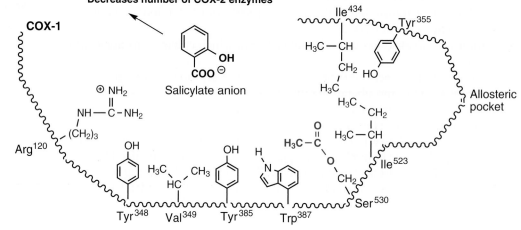

COX-1: Irreversibly acetylated by aspirin

Figure 21.6 Irreversible COX-1 acetylation by aspirin.

- The aromatic ring is essential. It provides a sterically unhindered scaffold for the carboxylate anion and acetoxy ester to react in a concerted fashion with COX Ser.

Physicochemical Properties

- Aspirin is strongly acidic with a pKa of 3.5 (COOH). It is passively absorbed from the gastrointestinal tract.
- Increasing gastric pH in the area around the tablet through buffering promotes dissolution and gastric absorption. This has the potential to decrease gastrointestinal distress.
- Increasing total gastric pH by coadministration of H_2 antagonists or proton pump inhibitors decreases gastric absorption by increasing ionization.
- The rate of aspirin absorption from the gut is dosage form-dependent. Absorption is faster in nonenteric coated formulations of small particle size.

Metabolism (Fig. 21.7)

- Aspirin can hydrolyze in plasma before it reaches the COX target. The hydrolyzed drug is inactive.
- Glucuronide conjugation of carboxylic acid and phenol groups is possible (one per molecule) and represents 15% of urinary metabolites.

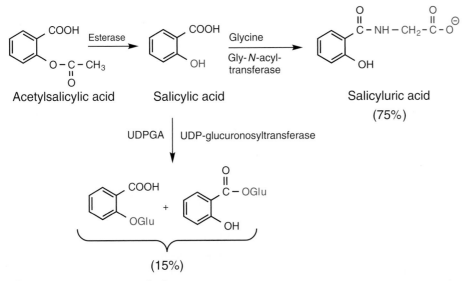

Figure 21.7 Aspirin metabolism.

- The predominant Phase II reaction of salicylic acid is glycine conjugation. The glycine conjugate of aspirin is called salicyluric acid.

Clinical Applications

- Aspirin is administered every 4 to 6 hours for relief of mild–moderate pain due to inflammation. It also has an antipyretic action.
- A variety of oral dosage forms, including enteric coated tablets, are available.
- Low-dose aspirin (baby aspirin) inhibits platelet aggregation and decreases the risk of negative cardiovascular events.
- Aspirin can induce a potentially fatal syndrome in children and teenagers who have, or have recently recovered from, flu or chickenpox. Aspirin should be strictly avoided in these patients.
- Gastrointestinal distress from aspirin therapy can be attenuated by enteric coating, coadministration of PGE_1 replacement therapy (e.g., misoprostol), or buffering.

Diflunisal

SAR

- Diflunisal anchors reversibly to COX through ion–ion binding with Arg^{120}.
- The *o*-F interacts sterically with the 6-H of the nonhalogenated ring to promote the desired noncoplanar orientation of aromatic rings.
- The *p*-F blocks inactivating aromatic hydroxylation. The duration of action is approximately two to three times that of aspirin, allowing twice daily dosing.

Metabolism

- Diflunisal is excreted in urine as the glucuronide conjugate of the carboxylic acid or phenol.

N-Arylanthranilic Acids (Fenamates)

SAR (Fig. 21.8)

- Anthranilic acid is the nitrogen isostere of salicylic acid. The two fenamate NSAIDs have all structural features needed for high affinity binding to the COX active site.

DRUG–DRUG INTERACTIONS

Aspirin
- Warfarin and other anticoagulants (hemorrhage)
- Probenecid (antagonism of uricosuric action)
- Aldosterone (antagonism of diuretic action)
- Corticosteroids (attenuated salicylate blood levels)
- Alcohol (increased risk of gastrointestinal bleeding)
- Ketorolac (increased risk of serious toxicity)
- Other drugs that bind strongly to serum proteins (e.g., oral hypoglycemic-increased risk of hypoglycemia; methotrexate-increased risk of methotrexate-induced lung, skin, and hemorrhagic toxicity)

Coadministration of Aspirin and Other NSAIDs

Patients should be cautioned against taking OTC or prescription nonselective NSAIDs with low-dose aspirin, as they may counteract aspirin's desired antiplatelet effects by sterically blocking access to COX-1 Ser^{530}. If both NSAIDs are warranted, aspirin should be taken 2 hours prior to, or 8 hours after, an isolated dose of the other NSAID.

Aspirin Therapy in Colon Cancer

Aspirin continues to earn its reputation as "the wonder drug." The positive impact of aspirin therapy on survival in patients diagnosed with COX-2 expressing colorectal cancer (stage I, II, or III) has been documented in the literature. NSAID therapy (commonly aspirin) has also been shown to reduce the risk of colon cancer diagnosis.

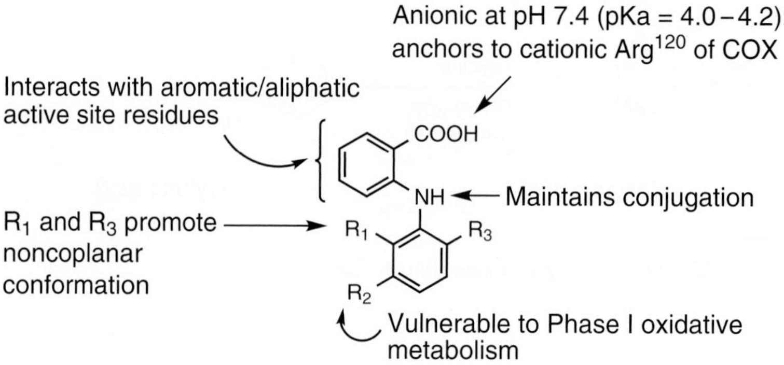

Figure 21.8 Fenamate SAR.

- The 2,3,6-trisubstituted meclofenamic acid is 25 times more potent than the 2,6-disubstituted mefenamic acid. This can be attributed to:
  - Higher lipophilicity provided by two chlorine atoms.
  - Lower pKa, resulting in stronger ion–ion anchoring to Arg^{120}.
  - Promotion of the required noncoplanar conformation between rings.

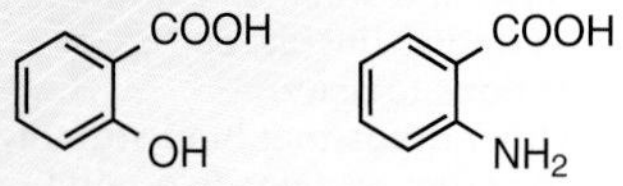

Salicylic acid and anthranilic acid

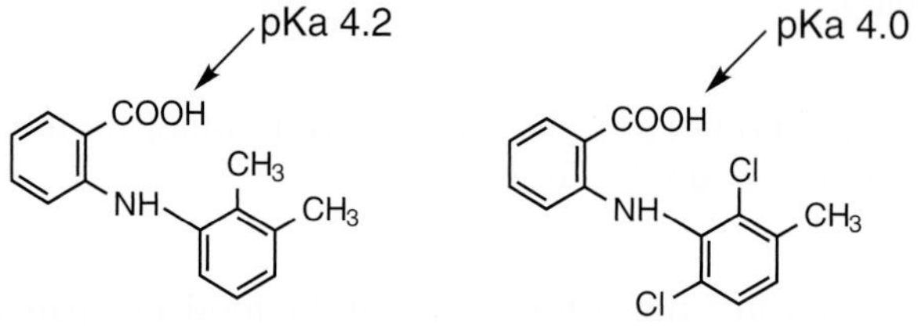

Mefenamic acid　　Meclofenamic acid

Fenamates

Fenamate Metabolism (Fig. 21.9)

- Both fenamates undergo CYP2C9-catalyzed benzylic hydroxylation followed by cytosolic oxidation to the carboxylic acid. The least sterically hindered of mefenamic acid's two CH_3 groups is preferentially oxidized.
- Phase II glucuronic acid conjugation can occur on the original or Phase I-generated COOH.

Clinical Applications

- Fenamates are available by prescription for the treatment of mild/moderate pain due to inflammation. Primary dysmenorrhea is a common indication.
- The duration of mefenamic acid therapy should not exceed 1 week.
- The more potent meclofenamate sodium is used in the treatment of osteo- and rheumatoid arthritis. Longer courses of therapy may be needed to achieve therapeutic outcomes.

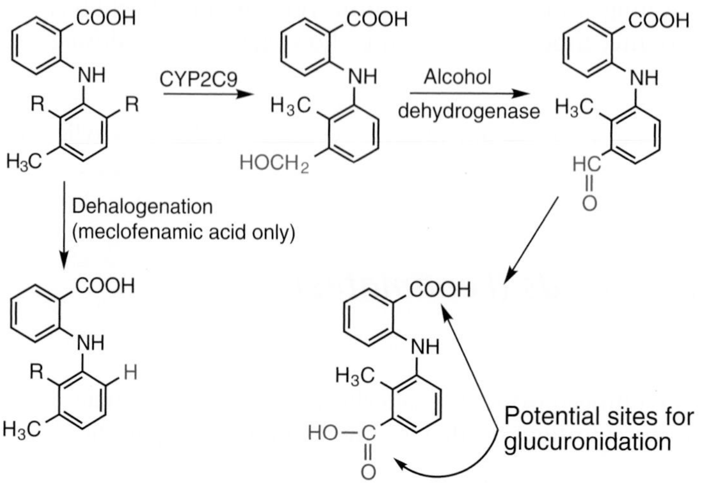

Figure 21.9 Fenamate metabolism.

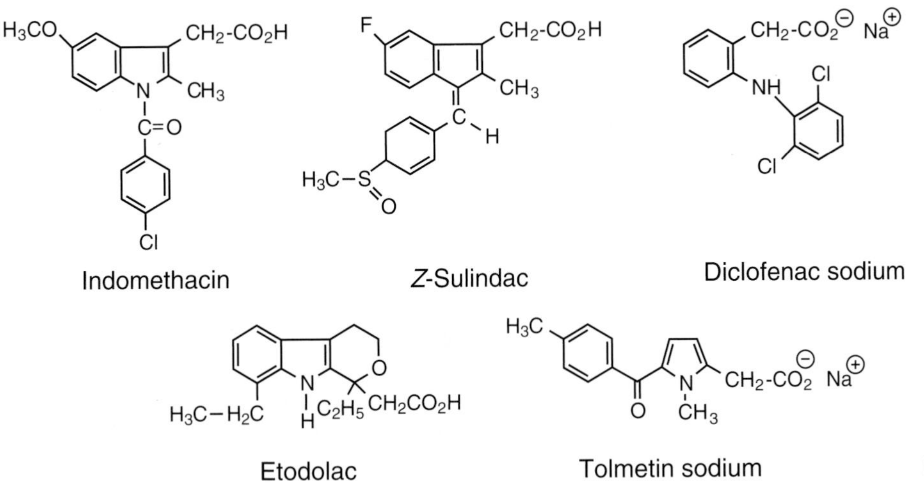

Indomethacin

Z-Sulindac

Diclofenac sodium

Etodolac

Tolmetin sodium

Figure 21.10 Arylacetic acid NSAIDs.

Arylalkanoic Acids

- The majority of marketed NSAIDs are arylalkanoic acids. These NSAIDs are classified by the nature of the aryl nucleus bearing the alkanoic acid moiety.

SAR

- The essential carboxylic acid must be positioned one carbon away from an aromatic nucleus.
- The carbon connecting the aromatic ring to the acidic group is referred to as the α-carbon. It can be unsubstituted (arylacetic acids; Fig. 21.10) or substituted with a single CH_3 group (arylpropionic acids; Fig. 21.11).
  - Phenylacetic acids where the aryl group is phenyl are referred to as "fenacs" (e.g., diclofenac). Phenylpropionic acids are collectively called "profens" (e.g., ibuprofen).
- Disubstitution or monosubstitution of the α-carbon with groups larger than methyl abolishes activity, presumably due to steric hindrance to ion–ion anchoring.
- Arylpropionic acid NSAIDs are chiral. The S isomer is active but products are commonly marketed as racemic mixtures.
- A second aromatic ring conjugated with the ring bearing the acetic or propionic acid moiety provides optimal activity. Substituents that force a noncoplanar conformation, augment lipophilicity, and/or inhibit inactivating CYP-mediated metabolism enhance activity.

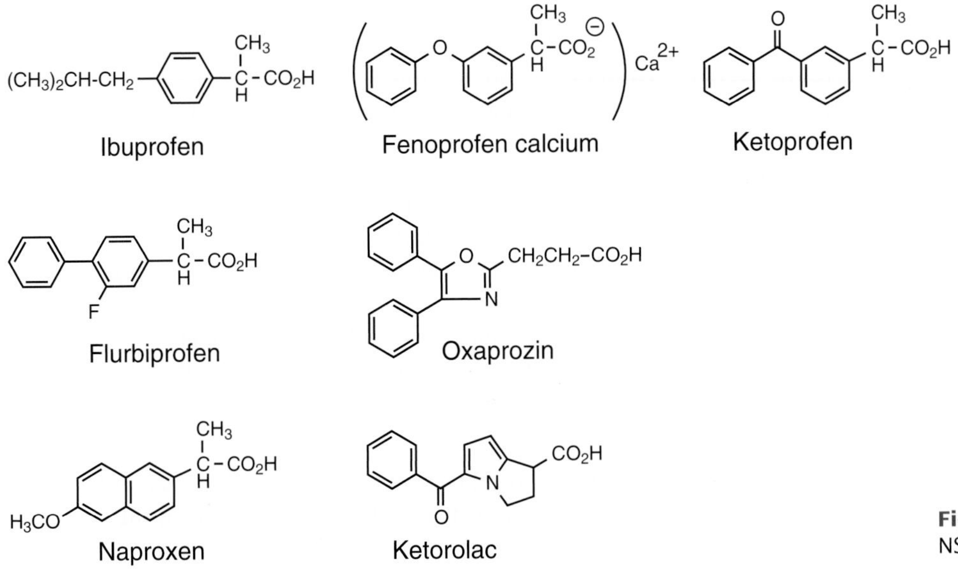

Ibuprofen

Fenoprofen calcium

Ketoprofen

Flurbiprofen

Oxaprozin

Naproxen

Ketorolac

Figure 21.11 Arylpropionic acid NSAIDs.

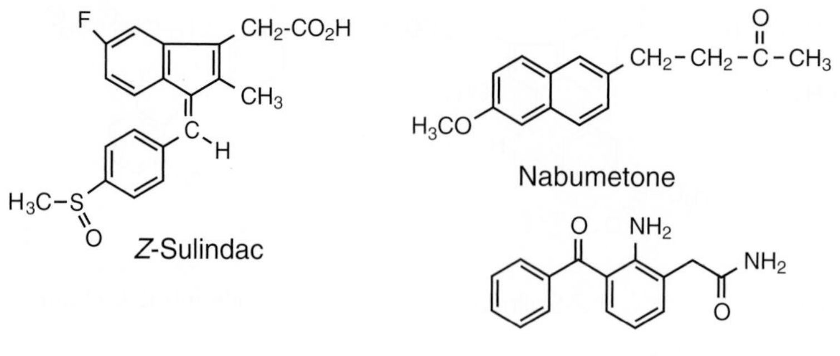

Figure 21.12 Prodrug NSAIDs.

- Arylalkanoic acids with metabolically stable *para* substituents on phenyl rings have longer durations of action than unsubstituted structures, and can be dosed less frequently.
- Prodrug NSAIDs are either nonacidic (nabumetone, nepafenac) or contain a sulfinyl moiety that must be reduced to the sulfide (sulindac) (Fig. 21.12).

Selected Arylalkanoic Acid NSAIDs

Indomethacin

- Indomethacin is an indoleacetic acid. It shows a preference for COX-1 enzymes (Table 12.2) and, as a result, is highly ulcerogenic.
- The active conformation of indomethacin places the benzoyl moiety in a *cis*-like orientation with respect to the benzo portion of the indole ring.
- A steric interaction between the 2-CH_3 and the benzoyl moiety promotes the desired noncoplanar conformation.
- The 2-CH_3 also promotes a time-dependent pseudoirreversible inhibition of COX by inserting into a hydrophobic pocket adjacent to the active site.
- The major inactivating metabolic reaction is CYP2C9-catalyzed O-dealkylation (Fig. 21.13).

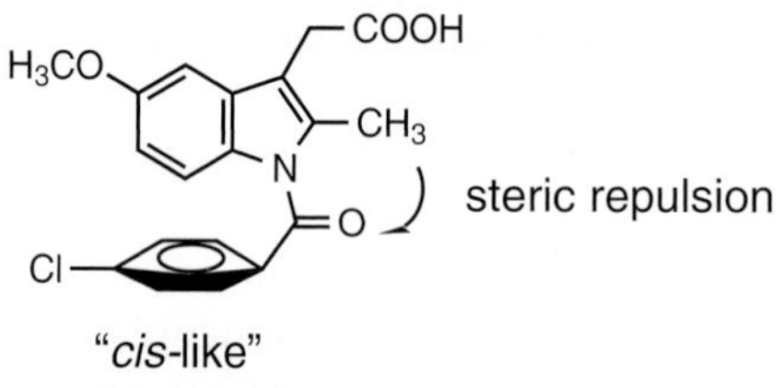

Indomethacin conformation

Sulindac

- Sulindac is an indene acetic acid. The active *Z* isomer places the phenyl substituent *cis* to the benzo portion of the indene ring.
- Steric interaction between hydrogen atoms at the indene C7 and the *ortho* position of the phenyl ring promotes the desired noncoplanar conformation.
- Sulindac is prodrug that is activated by a reversible reduction of the sulfoxide (also called a sulfinyl) to the methylsulfide (Fig. 21.14). This increases duration of action and allows dosing once or twice daily.
- Unlike indomethacin, sulindac's 2-CH_3 does not promote time-dependent pseudoirreversible binding to COX.
- Sulindac is acidic and can directly irritate gastrointestinal mucosa. As a prodrug, it does not inhibit gastric COX-1, so it shows a lower level of gastrointestinal distress than other nonselective NSAIDs.

Etodolac

- Etodolac has a dihydropyran ring fused to the indole ring and no second aromatic substituent. The carboxylic acid is two carbons removed from the aromatic indole ring.

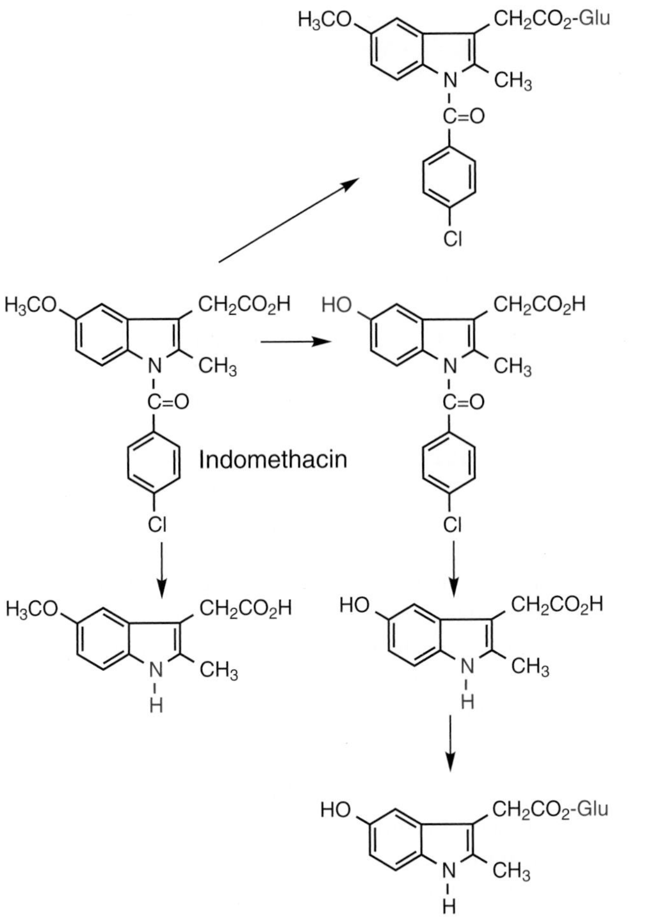

Figure 21.13 Indomethacin metabolism.

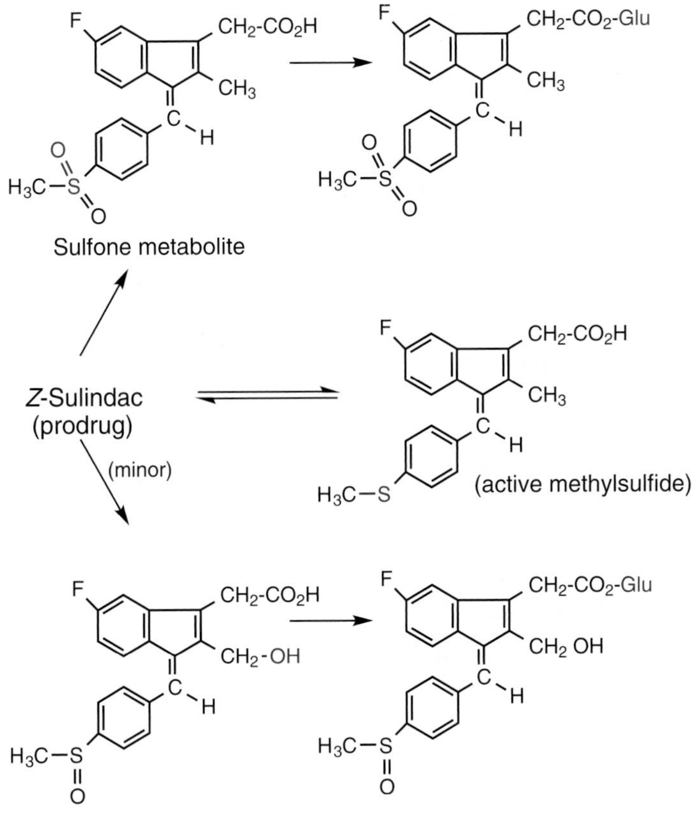

Figure 21.14 Sulindac metabolism.

Table 21.2   A Comparison of IC_{50} (μmole) Binding Constants for Selective Versus Nonselective Cycloxygenase (COX) Inhibitors

Drug	COX-1	COX-2	COX-1/COX-2 Ratio
Etoricoxib	116	1.1	106
Rofecoxib	18.8	0.53	35
Valdecoxib	26.1	0.87	30
Celecoxib	6.7	0.87	7.6
Nimesulide	4.1	0.56	7.3
Diclofenac	0.15	0.05	3.0
Etodolac	9.0	3.7	2.4
Meloxicam	1.4	0.70	2.0
Indomethacin	0.19	0.44	0.4
Ibuprofen	4.8	24.3	0.2
6-MNA	28.9	154	0.2
Piroxicam	0.76	9.0	0.08

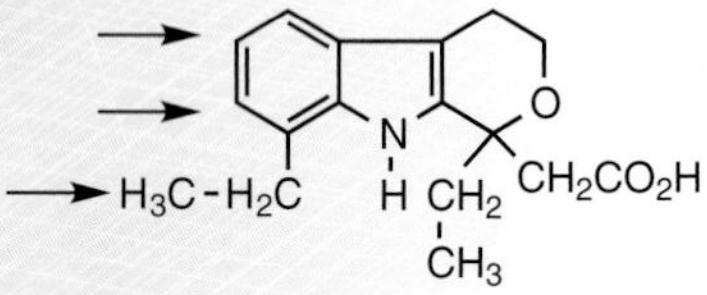

Sites of etodolac hydroxylation

- Despite these deviations from arylacetic acid SAR, etodolac retains significant anti-inflammatory activity (50 times that of aspirin).
- Etodolac is the most COX-2 preferential of the nonselective NSAIDs (Table 21.2), leading to relatively mild gastrointestinal side effects.
- Etodolac is inactivated by aromatic and aliphatic hydroxylation.

Ibuprofen

- In lieu of a second aromatic ring, ibuprofen has a *p*-isobutyl group with a size and lipophilicity approximating a second aromatic ring. It is less potent than nonselective NSAIDs with two aromatic rings.
- Ibuprofen undergoes chiral interconversion in vivo which epimerizes the inactive *R* isomer to the active *S*. Giving the racemic mixture produces as much anti-inflammatory activity as administering the pure *S*+ isomer.
  - Fenoprofen also exhibits this property.
- Duration is short because of rapid inactivating CYP2C9 and 2C19-mediated ω and ω-1 hydroxylation (Fig. 21.15).
- Ibuprofen is one of two arylalkanoic acids available OTC.
  - Naproxen is the other.

Flurbiprofen

- Flurbiprofen's biphenyl moiety allows direct conjugation between aromatic rings.

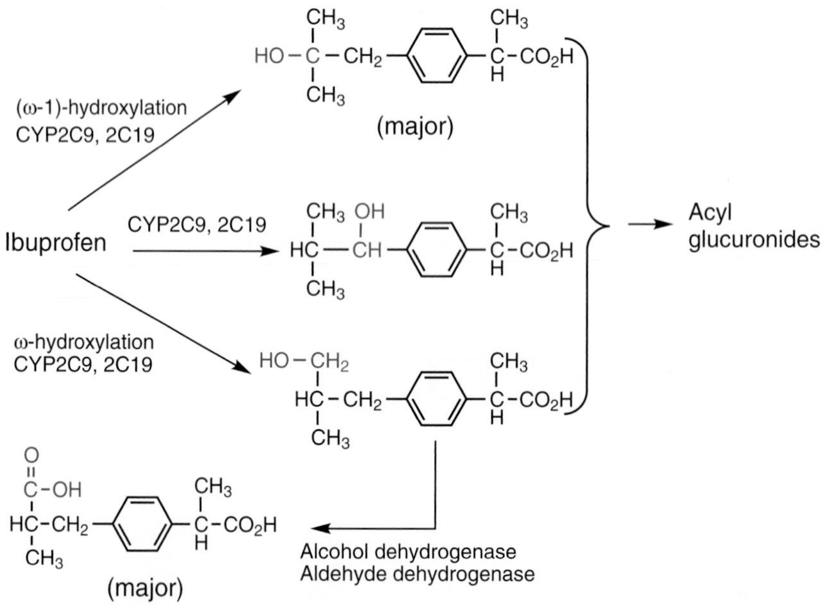

Figure 21.15 Ibuprofen metabolism.

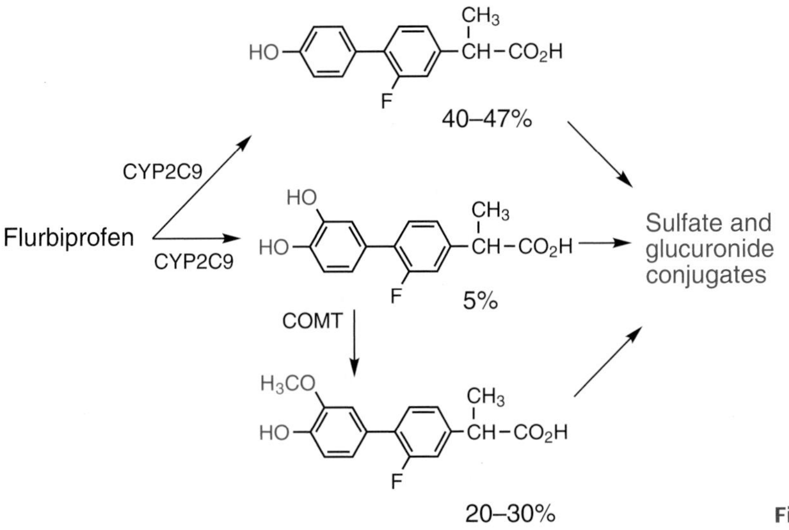

Figure 21.16 Flurbiprofen metabolism.

- Steric hindrance between the *ortho* F and H atoms on the two rings promotes a noncoplanar orientation.
- A CYP2C9-generated catechol metabolite can be further metabolized by catechol-O-methyltransferase (COMT). Conjugation with glucuronic acid or PAPS precedes excretion (Fig. 21.16).

Nepafenac

- Nepafenac is an arylacetic acid prodrug marketed as an ophthalmic suspension. It is used to treat ocular inflammation and pain postsurgery.
- In unionized parent drug form, nepafenac is well absorbed into ocular tissues. It subsequently hydrolyzes to generate the essential carboxylic acid, amfenac.
- Onset of anti-inflammatory action after instillation is approximately 15 minutes and the duration exceed 8 hours.
- Amfenac is vulnerable to CYP2E1-catalyzed hydroxylation at C5.

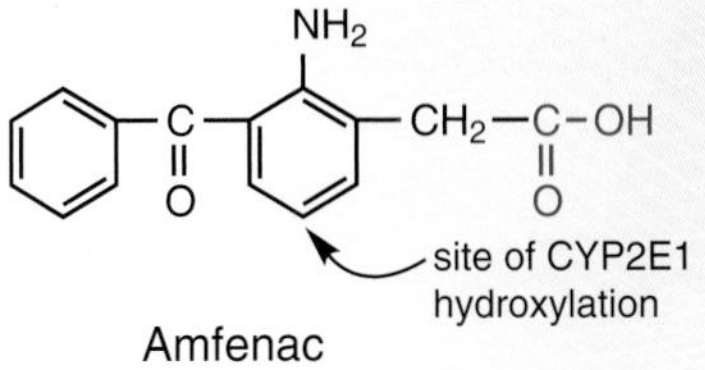

Amfenac

Tolmetin and Ketorolac

- Tolmetin is a COX-1 preferential pyrroleacetic acid NSAID. Gastrointestinal distress is common.
- Tolmetin's duration of anti-inflammatory action is predictably short due to rapid inactivating benzylic hydroxylation followed by cytosolic oxidation to the carboxylic acid (Fig. 21.17).

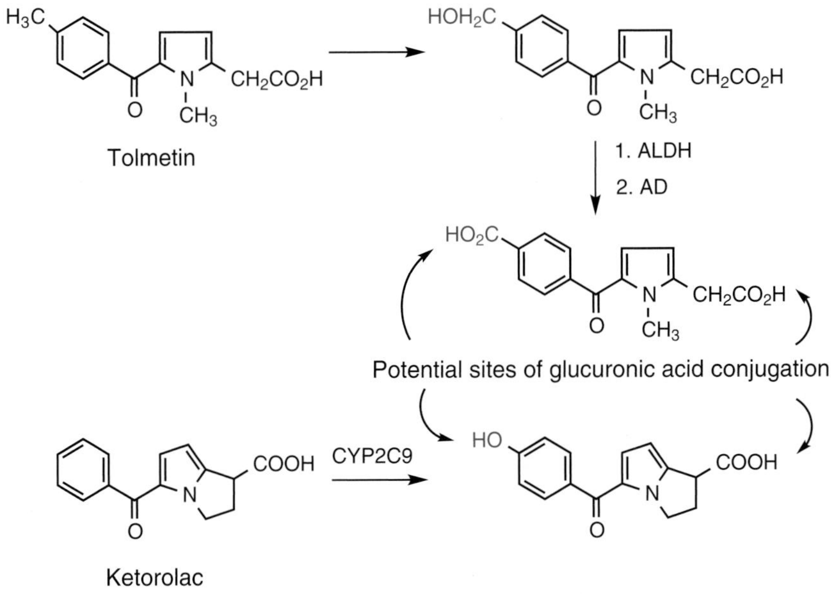

Figure 21.17 Tolmetin and ketorolac metabolism. ALDH, alcohol dehydrogenase; AD, aldehyde dehydrogenase.

Ketorolac Toxicity

- Ketorolac's high risk of GI ulceration and hemorrhage has been estimated at five times that of all other NSAIDs, and some patients have died from anaphylactic shock and cardiac arrest after IM or IV administration of the drug. Oral ketorolac should only be dispensed to patients who have responded safely to an initial IM or IV dose at their physician's office or in the hospital. Aspirin increases the risk of serious toxicity from ketorolac, so coadministration of these two NSAIDs is contraindicated.

- Moving tolmetin's aromatic CH_3 group to the α-carbon and joining that CH_3 to the N-CH_3 of the pyrrole ring provides ketorolac, a pyrrolepropionic acid with an analgesic action comparable to opioids.
- Ketorolac is indicated only for treatment of acute or breakthrough pain, and has a toxicity profile that limits therapy to no more than 5 days.

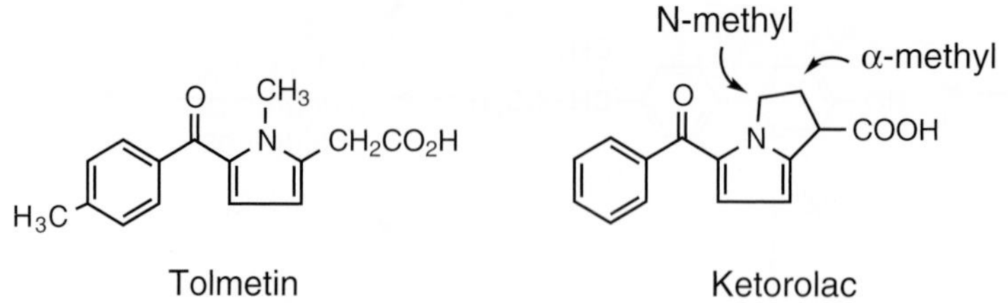

Tolmetin   Ketorolac

Naproxen and Nabumetone

- Naproxen and nabumetone are of predictably lower potency than many NSAIDs because of the single aromatic ring (naphthylene).
- Naproxen is an arylpropionic acid, and its potency is enhanced to some extent by the α-CH_3 group. The $S+$ isomer of naproxen is the most active.
- Naproxen's duration is prolonged because inactivating O-dealkylation is sluggish (Fig. 21.18).
- Nabumetone is nonacidic in parent form. Biotransformation to the essential acidic functional group is catalyzed by β-oxidase. 6-Methoxynaphthylene-2-acetic acid (6-MNA) is the active metabolite.
  - Many inactive and/or less active nabumetone metabolites are generated. The lack of an α-methyl and a second aromatic ring also contribute to nabumetone's relatively low anti-inflammatory potency.
- Because it is a nonacidic prodrug, nabumetone's risk of gastrointestinal distress is among the lowest of all nonselective NSAIDs.

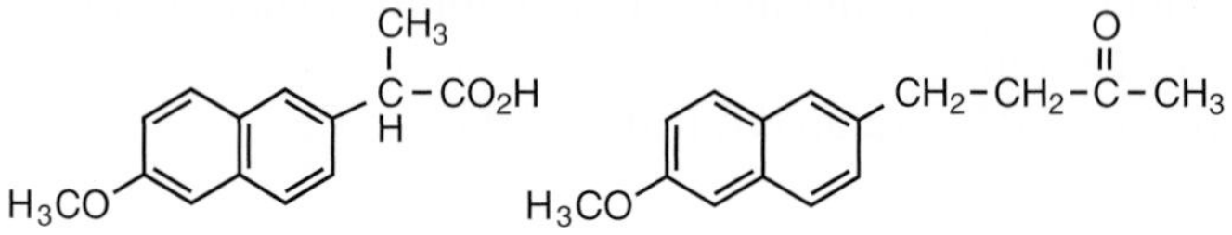

Naproxen   Nabumetone

Diclofenac

- With an anti-inflammatory action up to 1,000 times that of related structures, diclofenac is one of the most potent of all nonselective NSAIDs.

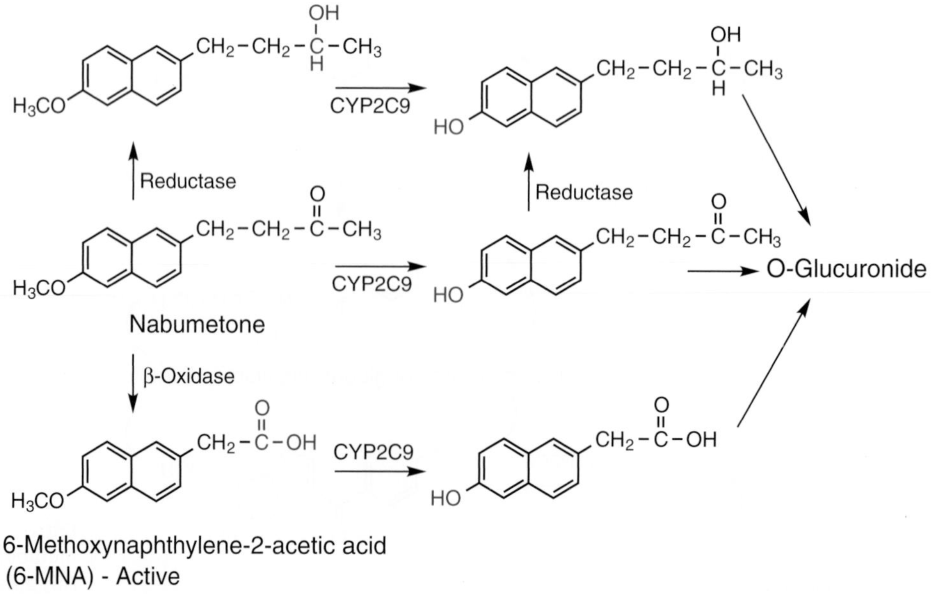

Figure 21.18 Nabumetone metabolism.

6-Methoxynaphthylene-2-acetic acid
(6-MNA) - Active

Figure 21.19 Diclofenac metabolism and CYS arylation.

- Diclofenac is structurally related to meclofenamate. In both NSAIDs, the two Cl atoms augment lipophilicity and ensure a noncoplanar orientation between aromatic rings.
- Diclofenac is associated with a hepatotoxicity risk, particularly when used in high doses. A highly electrophilic diclofenac quinoneimine metabolite capable of arylating hepatocyte Cys residues can cause significant tissue destruction over time or in patients with pre-existing hepatic damage (Fig. 21.19).
  - High lipophilicity assures rapid passive diffusion into hepatocytes on first pass.
  - When administered in high doses, diclofenac is also actively transported into the liver by the organic anion transporting protein OATP1B3.

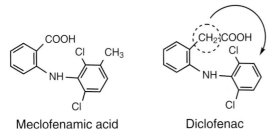

Meclofenamic acid   Diclofenac

CHEMICAL NOTE

Diclofenac's hepatotoxicity can be significant, but not every patient on the drug will be impacted. With conscientious attention and counseling from their pharmacist, most patients can take diclofenac safely. However, a closely related structural analog, lumiracoxib, was withdrawn from the market shortly after it was released due to unacceptably high hepatotoxicity. The carbon atom of lumiracoxib's quinoneimine metabolite adjacent to the fluorine substituent would be much more electrophilic (and, thus, more susceptible to attack by hepatocyte cysteine residues) than either of diclofenac's meta carbons because fluorine is much more electronegative than chlorine. Hepatocyte arylation and cell death would be predictably more extensive with lumiracoxib than diclofenac.

Diclofenac   Lumiracoxib   Lumiracoxib quinoneimine metabolite

- Diclofenac can be administered topically, as well as orally. The risk of hepatotoxicity from topical formulations is lower since first pass extraction into the liver is avoided.
- Transdermal formulations (diclofenac sodium or epolamine) are applied directly to arthritic and/or inflamed joints.
- Ophthalmic solutions of diclofenac sodium are available to treat postoperative ocular inflammation or ocular pain.

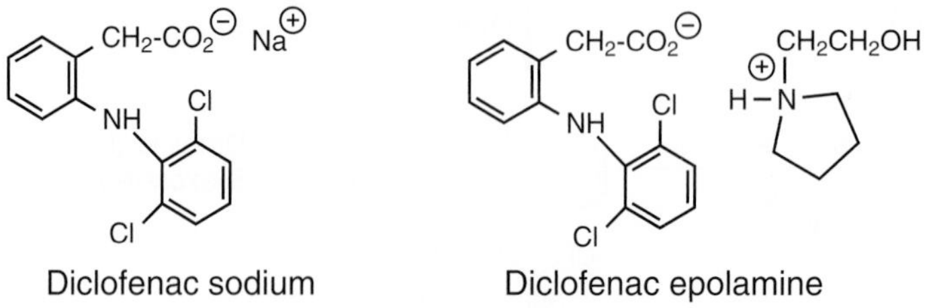

Diclofenac sodium

Diclofenac epolamine

Diclofenac salts

Oxaprozin

- Oxaprozin is the only oxazole-containing NSAID, and it breaks a few SAR rules.
  - The COOH is separated from the aromatic oxazole ring by two carbons.
  - The drug has a very prolonged duration (59 hour half-life) despite the presence of two unsubstituted phenyl rings.
- An adverse reaction unique to oxaprozin is photosensitivity. Rash can manifest with or without sun exposure, but is more common in the former environment.

Oxicams

- Oxicams are the only class of nonselective NSAIDs that do not have COOH as the requisite acidic functional group. The essential anion is generated through loss of proton from a highly acidic enol (pKa 4.2 to 4.6).
- The active anion is stabilized through resonance throughout the entire oxicam structure (Fig. 21.20).

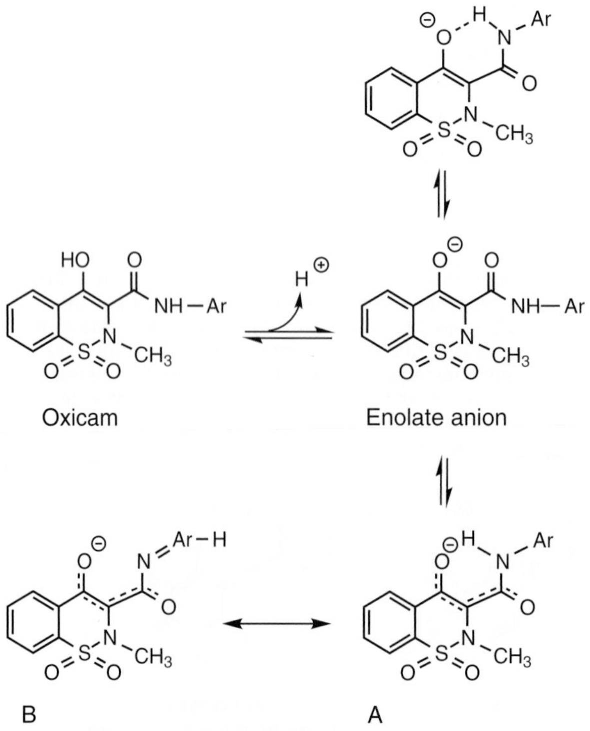

Oxicam

Enolate anion

B   A

Figure 21.20   Oxicam enolate anion stabilization.

Figure 21.21 Oxicam metabolism. ALDH, alcohol dehydrogenase; AD, aldehyde dehydrogenase.

SAR

- The oxicams meet all structural requirements for binding to the COX active site.
- Piroxicam shows a preference for COX-1 while meloxicam binds more effectively to COX-2 (Table 21.2). Meloxicam's thiazole sulfur may play a role in its selectivity profile.
  - COX-1 selective piroxicam is more ulcerogenic than COX-2 preferring meloxicam.

Physicochemical Properties

- Both oxicams undergo enterohepatic cycling which extends duration of action. Once daily dosing is the rule.

Metabolism (Fig. 21.21)

- Both oxicams are metabolized by CYP2C9 via aromatic hydroxylation (piroxicam) or hydroxylation at the aromatic CH_3 moiety (meloxicam). Glucuronic acid conjugation precedes excretion.

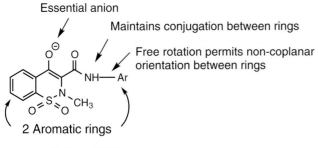

COX-2 Selective Inhibitors

- COX-2 expressed in inflamed tissue is the intended target of NSAIDs when used as anti-inflammatory agents.
- COX-2 inhibition in the vessel wall can lead to a decrease in PGI_2-mediated vasorelaxation. Unless there is a compensatory inhibition of COX-1 in platelets,

negative cardiovascular consequences, including myocardial infarction and stroke, can occur.
  • Two highly COX-2 selective inhibitors, rofecoxib and valdecoxib, were pulled from the US market due to an unacceptably high risk of cardiovascular morbidity.

Celecoxib   Rofecoxib   Valdecoxib

COX-2 selective inhibitors

  • COX-2 inhibition in the kidney reduces glomerular filtration and can lead to renal failure.
  • Celecoxib is the only COX-2 selective inhibitor currently available in the United States.

SAR

  • Celecoxib is a diarylheteroaromatic COX-2 selective inhibitor. These inhibitors bind predominantly in the allosteric pocket of the COX-2 enzyme.
  • Essential functional groups include an anion that can anchor to cationic Arg[513] and two conjugated aromatic rings (Fig. 21.22).

Celecoxib SAR

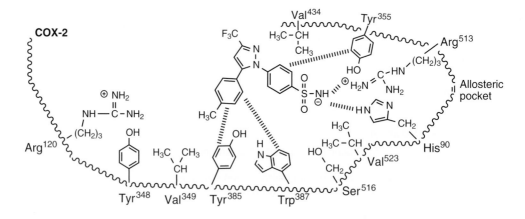

Figure 21.22 Celecoxib binding to COX-2.

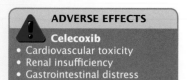

ADVERSE EFFECTS

Celecoxib
• Cardiovascular toxicity
• Renal insufficiency
• Gastrointestinal distress

Metabolism (Fig. 21.23)

  • Like most nonselective NSAIDs, celecoxib is metabolized by CYP2C9 and excreted in part as a glucuronide conjugate.

Clinical Applications

  • Celecoxib's primary use is as an anti-inflammatory agent in osteo- and rheumatoid arthritis. It is less likely to induce gastrointestinal distress than nonselective NSAIDs used in these disease states.
  • COX-2 is expressed by some neoplasms. An off-label use for celecoxib is in the prevention of lung and colorectal cancers.

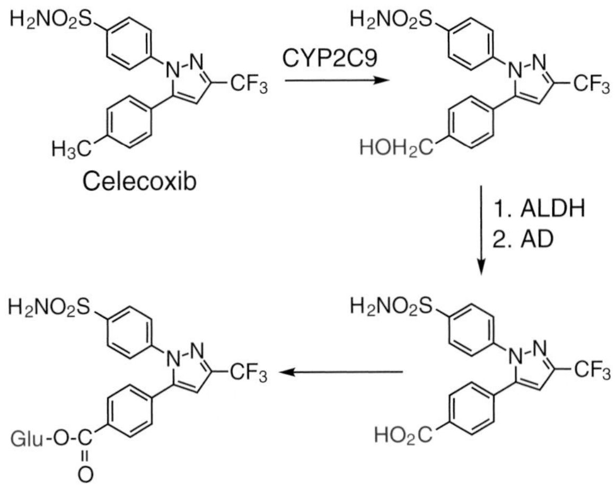

Figure 21.23 Celecoxib metabolism.

- Celecoxib inhibits CYP2D6. Appropriate caution should be used when coadministering CYP2D6 substrates with celecoxib.

Acetaminophen

- Acetaminophen is an analgetic/antipyretic agent. It is not an NSAID and has no clinically useful anti-inflammatory action.

MOA

- Despite widespread use over many years, the exact mechanism of acetaminophen's antipyretic and analgesic action is unknown. It is believed to work through both central and peripheral mechanisms.
- It has been proposed that acetaminophen augments endogenous cannabinoid receptor ligands (endocannabinoids) through inhibition of the enzyme fatty acid amide hydrolase (FAAH).

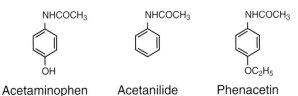

Acetaminophen   Acetanilide   Phenacetin

SAR

- Acetaminophen's acetanilide component is important to its analgesic/antipyretic profile.
- The *p*-OH decreases hemato- and nephrotoxicity. Both acetanilide (unsubstituted) and phenacetin (*p*-ethoxy) are too toxic for therapeutic use.

Metabolism (Fig. 21.24)

- Between 4% and 10% of the parent drug undergoes CYP2E1 and/or CYP1A2 mediated oxidation to N-acetyl-*p*-benzoquinoneimine (NAPQI), a highly electrophilic and hepatotoxic metabolite.
- NAPQI can be detoxified by reaction with endogenous reduced glutathione (GSH). If GSH stores are depleted, the sulfhydryl reagent N-acetylcysteine can protect hepatocytes from destruction.
  - N-acetylcysteine forms the same excretable acetaminophen conjugate as would have been formed with endogenous GSH.

Acetaminophen Hepatotoxicity

Due to the high risk of serious hepatotoxicity, the FDA has recently recommended that products containing acetaminophen restrict the amount of drug in each dose to 325 mg. This allows some margin of safety if patients unknowingly take more than one medication containing this analgesic. Adults should not take more than 4 g of acetaminophen in any 24-hour period.

Acetaminophen and Alcohol

Acetaminophen hepatotoxicity is significantly increased in the presence of alcohol. Ethanol induces CYP2E1, prolongs its half-life from 7 to 37 hours, and successfully competes with acetaminophen for this isoform if the two are taken together. However, if acetaminophen is taken up to 24 to 37 hours after alcohol consumption stops (e.g., to treat the symptoms of hangover), the risk of liver damage via the CYP2E1-generated NAPQI metabolite is increased. It has been reported that adults taking acetaminophen 8 hours after exposure to a significant amount of alcohol generated 22% more of the hepatotoxic NAPQI metabolite than adults who were not alcohol-exposed.

N-acetylcysteine and reduced glutathione

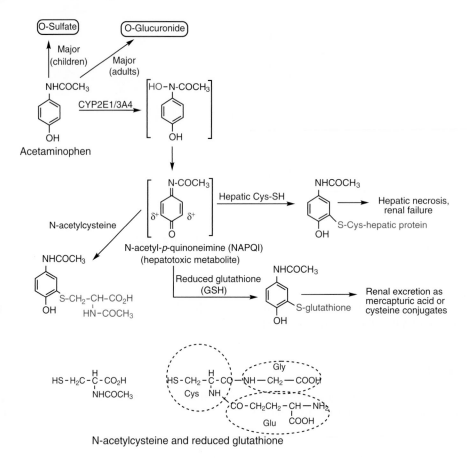

Figure 21.24 Acetaminophen metabolism.

Disease-Modifying Antirheumatic Drugs (Fig. 21.25)

- NSAIDs attenuate the symptoms of inflammatory disease processes but DMARDs address the "root cause" of the pathology. In doing so, they slow disease progression and morbidity.
- The mechanisms of action of DMARDs are complex, multifaceted, and not completely elucidated.

Gold Compounds (Gold Sodium Thiomalate, Auranofin)

MOA

- The primary mechanism of gold DMARDs is inhibition of sulfhydryl-containing lysosomal enzymes involved in mediating the inflammatory response. These enzymes include glucuronidase, acid phosphatase, collagenase, and acid hydrolase.

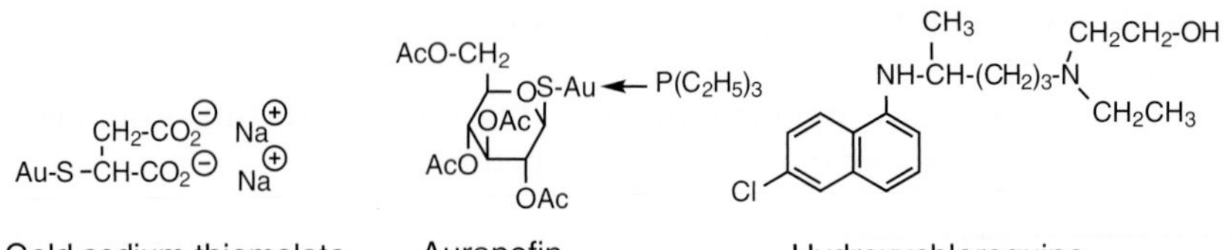

Gold sodium thiomalate   Auranofin   Hydroxychloroquine

Immunosuppressants:

Leflunomide   Methotrexate   Sulfasalazine

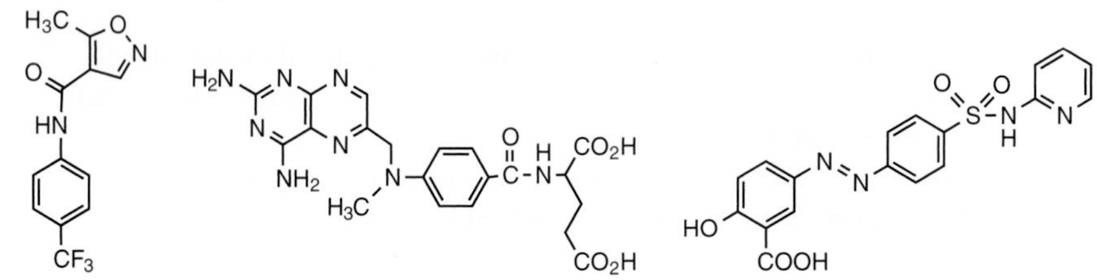

Figure 21.25 Disease-modifying antirheumatic drugs (DMARDs).

- All therapeutically useful agents have the gold atom bound to sulfur. This facilitates the reversible interaction of the gold with the enzymatic SH.
- Additional potential mechanisms include inhibition of mucopolysaccharide biosynthesis (gold sodium thiomalate) and suppression of COX-2 gene expression (auranofin).

SAR

- The active form of gold is the monovalent aurous ion (Au^+).
- The remainder of the gold compound structure is designed to promote water solubility for widespread distribution in vivo and preparation of injectable dosage forms.

Physicochemical Properties

- In solution, Au^+ is rapidly oxidized to auric ion (Au^{3+}) or converted to metallic gold. Aqueous preparations not otherwise stabilized decompose at room temperature.
- One approach to stabilizing gold in these therapeutic compounds is through complexation with phosphine ligands, such as triethylphosphine (auranofin). Gold compounds thus stabilized can be administered orally as well as by injection.
- Route of administration impacts drug distribution in vivo. Oral administration allows gold accumulation in red blood cells (40%) while injection leads to accumulation in the reticuloendothelial system, adrenal cortex, and renal cortex.
- Gold is sequestered in inflamed joints for 3 weeks or more and very slowly eliminated via the urine over the course of months.
- The gold in the injectable preparation (gold sodium thiomalate, administered weekly) is bound extensively to serum proteins (95% to 99%) while that in the oral auranofin preparation (administered daily) is only 60% bound. Onset of action is slow.

Clinical Applications

- The percentage of gold in the oral auranofin preparation is approximately 60% of that found in the injectable gold sodium thiomalate preparation (29% vs. 50%). This equates to a lowered toxicity risk for auranofin, but also to a potential decrease in efficacy.
- Patients must receive gold sodium malate injections while lying down. They must remain recumbent for 10 minutes postinjection to minimize dizziness and to be monitored for serious and immediate adverse drug reactions including anaphylactic shock, breathing or swallowing difficulties, and bradycardia.
- A metallic taste is an indication of toxicity and should be reported to the pharmacist or physician.
- Toxicity symptoms can be eased with corticosteroid therapy. In the case of serious toxicity, gold elimination can be facilitated by the use of metal chelating agents such as dimercaprol and penicillamine.
- Urinalysis must be performed before each injection of gold sodium thiomalate to rule out proteinuria or hematuria. Therapy should be discontinued if either condition is found.

H₂C — CH — CH₂OH
| |
HS   SH

CH₃
|
H₃C — C — CH — COOH
| |
HS   NH₂

Dimercaprol and penicillamine

⚠	**ADVERSE EFFECTS**

Gold
- Serious or life-threatening toxicities (rare)
  - Anaphylactic shock
  - Blood dyscrasias
  - Interstitial pneumonitis and fibrosis
  - Corneal inflammation and/or ulceration
- Additional toxicities
  - Pruritus, followed by dermatitis (sometimes severe), exacerbated by sun exposure
  - Alopecia
  - Dizziness
  - Gastrointestinal distress
  - Nephritis (may become serious if not treated early)
  - Hepatitis, jaundice
  - Stomatitis

Hydroxychloroquine

MOA

- Hydroxychloroquine is an aminoquinoline that inhibits the lysosomal enzymes cartilage chondromucoprotease and cartilage cathepsin B.
- Like gold compounds, it may take months for the realization of full therapeutic benefits.

ADVERSE EFFECTS

Hydroxychloroquine
- Immediate or delayed retin-opathy (including years after drug discontinuation)
- Renal toxicity
- CNS toxicity
- Hematologic toxicity
- Dermatologic toxicity

DRUG–DRUG INTERACTIONS

Hydroxychloroquine
- CYP2D6 vulnerable beta blocker antihypertensives (avoid or monitor closely)
- Cimetidine (reduce hydroxy-chloroquine dose)
- Magnesium-containing ant-acids (reduced hydroxychlo-roquine absorption; separate administration by 2 to 4 hours)

ADVERSE EFFECTS

Leflunomide
- Diarrhea
- Elevated liver enzymes
- Alopecia

Teriflunomide

Metabolism

- Hydroxychloroquine undergoes N-dealkylation to the primary amine, which is subsequently deaminated and further oxidized to the carboxylic acid. CYP2D6 has been implicated in the biotransformation.
- Hydroxychloroquine is also a known inhibitor of CYP2D6.

Hydroxychloroquine metabolite

Immunosuppressants

- Immunosuppressant DMARDs thwart the pathologic immune responses believed to underpin rheumatoid arthritis and other autoinflammatory processes.
- Immunosuppressant use is generally reserved for patients who have not responded to a less toxic DMARD, corticosteroid, and/or NSAID therapy.

Leflunomide

- Leflunomide is an orally active immunosuppressant prodrug that undergoes opening of the isoxazole ring to yield the active α-cyanoenol, teriflunomide.
- Teriflunomide is believed to lower immunoreactive antibody formation through the inhibition of de novo UMP biosynthesis. Specifically, the rate-limiting enzyme dihydroorotate dehydrogenase is inhibited, which halts B-lymphocyte proliferation.
  - Teriflunomide also blocks nuclear factor kappa B (NF-κB) and inhibits tyro-sine kinase enzymes. The drug is used in the treatment of relapsing multiple sclerosis (MS).
- Leflunomide's long duration is attributed to enterohepatic cycling.
- Both leflunomide and teriflunomide can be severely hepatotoxic and carry a teratogenicity risk.

Methotrexate

- Methotrexate is an orally active antifolate that blocks the production of pyrimi-dine nucleotides through the direct inhibition of 7,8-dihydrofolate reductase. As with leflunomide, B-lymphocytes proliferation is halted due to a lack of uridine nucleotides.
- The biosynthesis of tissue-damaging polyamines, such as spermine, is also inhibited by this immunosuppressant.
- Methotrexate is actively transported into cells, and the monoglutamate "tail" is polyglutamated by the enzyme folylpolyglutamate synthase. The polyanionic metabolites are active and trapped inside the cell.
- The 7-hydroxymethotrexate metabolite retains antirheumatic activity.
- Elimination of methotrexate into urine is also a carrier-mediated process, which explains its relatively short 3- to 10-hour elimination half-life.

← monoglutamate tail subject to polyglutamation

7-Hydroxymethotrexate

Sulfasalazine

- Sulfasalazine is an orally active prodrug immunosuppressant.

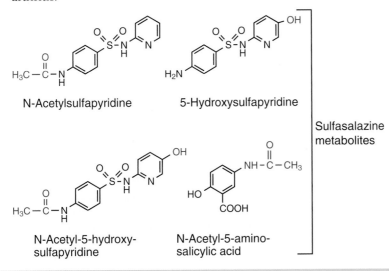

Figure 21.26 Sulfasalazine metabolism.

Metabolism (Fig. 21.26)

- Sulfasalazine undergoes bacteria-catalyzed hydrolysis in the bowel to release 5-aminosalicylic acid (a prostaglandin synthesis inhibitor) and sulfapyridine (a purine synthesis inhibitor and stimulator of adenosine release).
- Both hydrolysis products are acetylated prior to excretion. The rate of sulfapyridine acetylation (but not 5-aminosalicylic acid acetylation) is dependent on patient N-acetyltransferase (NAT) phenotype.

Clinical Applications

- Sulfasalazine works locally in the bowel to suppress inflammation in patients with ulcerative colitis.
- Sulfasalazine metabolites are absorbed from the colon in amounts ranging from 25% (5-aminosalicylic acid) to 80% (sulfapyridine).
- Although oral bioavailability of the intact prodrug is only about 15% to 20%, the delayed release formulation has found use in the treatment of rheumatoid arthritis.

<table>
<tr><td colspan="2" style="text-align:center">⚠️ ADVERSE EFFECTS</td></tr>
<tr><td colspan="2">Methotrexate and Sulfasalazine
• Renal and hepatotoxicity
• Dermatologic reactions
• Crystalluria
• Bone marrow suppression
• Risk of opportunistic infection (methotrexate)
• Pulmonary toxicity (methotrexate)
• Sulfa allergy (sulfasalazine)
• Photosensitivity (sulfasalazine)</td></tr>
</table>

N-Acetylsulfapyridine

5-Hydroxysulfapyridine

N-Acetyl-5-hydroxy-sulfapyridine

N-Acetyl-5-amino-salicylic acid

Sulfasalazine metabolites

Apremilast

- Apremilast is an orally active phosphodiesterase 4 (PDE4) inhibitor approved for use in psoriatic arthritis.
- Inhibition of PDE4 blocks the hydrolysis of cyclic adenosine monophosphate (cAMP).
- Increased levels of cAMP activate protein kinase A (PKA) in monocytes and macrophage. This results in inhibition of selected inflammatory mediators, including interleukins (i.e., IL-12 and IL-23) and NF-κB.
- CYP3A4/5 mediated O-demethylation produces a phenolic metabolite that is glucuronidated prior to excretion.

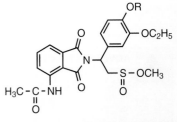

Apremilast (R = CH$_3$) and O-desmethyl metabolite (R = H)

Table 21.3	Tumor Necrosis Factor (TNFα) Blockers for Rheumatoid Arthritis	
TNFα Blocker	**Trade Name**	**Mechanism**
Etanercept	Enbrel	Soluble TNFα "decoy" receptor (recombinant human)
Infliximab	Remicade	Anti-TNFα antibody (chimeric)
Adalimumab	Humira	Anti-TNFα antibody (recombinant human)
Certolizumab pegol	Cimzia	Anti-TNFα antibody (humanized and linked to polyethylene glycol (PEG)[a]
Golimumab	Simponi	Anti-TNFα antibody (human)

[a]To reduce immunogenicity

Biologics

Tumor Necrosis Factor Blockers (Table 21.3)

- The release of cytokines, including tumor necrosis factor alpha (TNFα), from activated T lymphocytes in inflamed joints is a major causative factor of joint destruction in rheumatoid arthritis.
- Blocking the endogenous receptor-mediated action of TNFα through the provision of soluble "decoy receptors" or by targeting its destruction by specific anti-TNFα antibodies can ease and/or reverse the disabling symptoms and halt disease progression.
- As protein drugs, all TNFα blockers must be administered by injection. Patients receiving these drugs are commonly prescribed oral methotrexate, as each medication appears to promote the activity of the other.

T-Lymphocyte Activation Blockers (Abatacept)

- Abatacept is a chimeric protein that mimics an endogenous regulatory protein known as CTLA-4. CTLA-4 is expressed on the surface of T-lymphocytes during periods of activation.
- Abatacept binds costimulatory B7 and keeps it from binding to T-lymphocyte CD28.
- This stabilizes the T-cell and attenuates cytokine release, T-cell proliferation, and the debilitating symptoms of rheumatoid arthritis.

B-Lymphocyte Blockers (Rituximab)

- Rituximab is a chimeric monoclonal antibody that targets the destruction of developed peripheral B-lymphocytes. Its use is often reserved for patients who have not improved on TNFα blocker therapy.
- Serious adverse reactions to rituximab include potentially fatal infusion reactions characterized by cardiac and respiratory complications. Infusion reactions occur most commonly with the first dose. Mucocutaneous reactions and serious infections (hepatitis B reactivation and multifocal leukoencephalopathy) can also occur.

Interleukin Receptor Blockers (Anakinra and Tocilizumab)

- These therapeutic proteins block the destructive action of interleukin cytokines in arthritic tissues.
- Anakinra is a nonglycosylated analog of the endogenous IL-1Rα protein. It competitively prevents IL-1 binding to its receptor, IL-1R1. Levels of this antagonizing protein are known to be low in the synovium of rheumatoid arthritis patients.
  - IL-1R1 blockade decreases synovial NO, PGE_2, and collagenase. This, in turn, slows cartilage destruction and relieves symptoms.
- Tocilizumab blocks IL-6 receptors and breaks the immunological response cycle mediated by this cytokine.

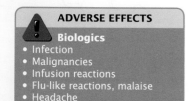

ADVERSE EFFECTS

Tumor Necrosis Factor (TNFα) Blockers
- Infection (primarily tuberculosis)
- Urinary tract infections
- Malignancy (T-cell lymphoma, leukemia, skin cancers)
- Injection/infusion site reaction
- Headache
- Rash

ADVERSE EFFECTS

Biologics
- Infection
- Malignancies
- Infusion reactions
- Flu-like reactions, malaise
- Headache

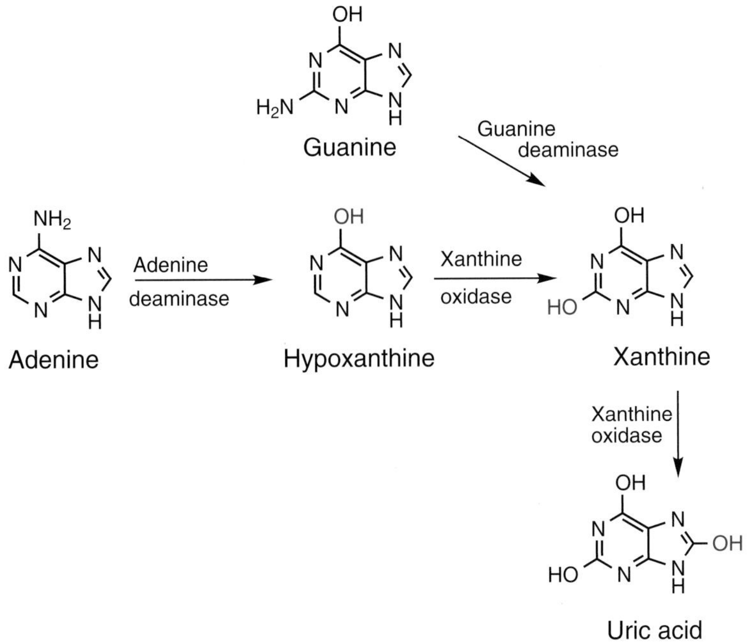

Figure 21.27 Uric acid biosynthesis.

Drugs for the Treatment of Gout

Etiology of Gout

- Gout refers to joint inflammation secondary to the accumulation of monosodium urate monohydrate crystals in synovia, synovial fluid, and surrounding tissue.
- Uric acid is produced by oxidation of purine bases guanine and adenine (Fig. 21.27). The rate-limiting enzyme, xanthine oxidase, is a target for some antigout drugs.
- Uric acid and monosodium urate levels can be elevated due to excessive biosynthesis and/or compromised urinary excretion via the urate anion transporting protein (URAT1). When synovial fluid becomes supersaturated, needle-like crystals of monosodium urate monohydrate precipitate in the joints.
- Gout affects more men than women, and African American men are twice as likely as Caucasian men to be afflicted. Incidence is also positively related to weight, elevated blood pressure, a purine-rich diet, and alcohol consumption.
- Acute gout can flare during times of stress, fatigue, and/or feasting. Drugs that increase uric acid levels (e.g., penicillin, insulin) can also precipitate attacks.
- Chronic gout is diagnosed when the disease has progressed to the joint destruction and deformity stage (tophaceous gout).

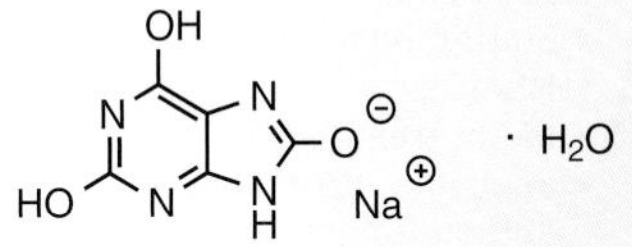

Specific Drugs (Fig. 21.28)

Colchicine

- Colchicine is a natural product that decreases inflammation by increasing synovial tissue pH. This increases uric acid solubility and facilitates crystal dissolution.
- Colchicine can be used to abort or prevent acute attacks of gout.
- CYP3A4 generates only minor amounts of 2- and 3-desmethylcolchicine, but the parent drug can accumulate (with serious and potentially fatal toxicity risks) if strong CYP3A4 inhibitors are coadministered.
- The primary metabolite is N-deacetylcolchicine, resulting from hydrolysis in liver.

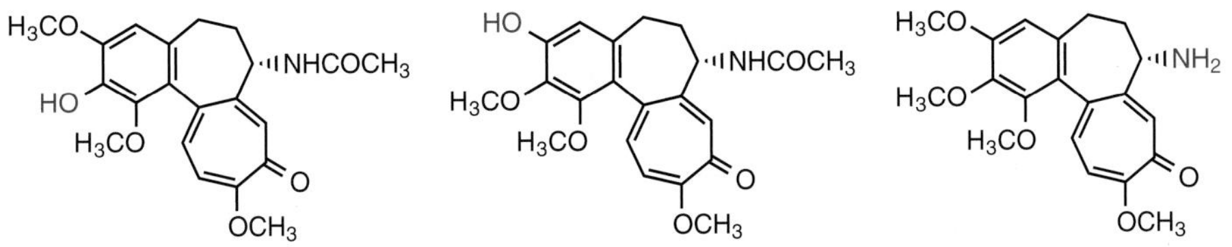

Monosodium urate monohydrate

ADVERSE EFFECTS

Colchicine
- Severe diarrhea with dehydration and pain
- Thrombocytopenia, aplastic anemia (chronic therapy)

DRUG–DRUG INTERACTIONS

Colchicine
- Uricosurics (enhanced excretion of the active alloxanthine metabolite)
- ACE Inhibitors, thiazide diuretics, ampicillin, amoxicillin (increased risk of hypersensitivity and rash)
- Oral anticoagulants (augmented anticoagulant action; may not include warfarin)
- 6-Mercaptopurine (increased antineoplastic efficacy allowing for dose reduction and possible decreased risk of mercaptopurine adverse effects)

2-Desmethylcolchicine   3-Desmethylcolchicine   N-Deacetylcolchicine

Colchicine metabolites

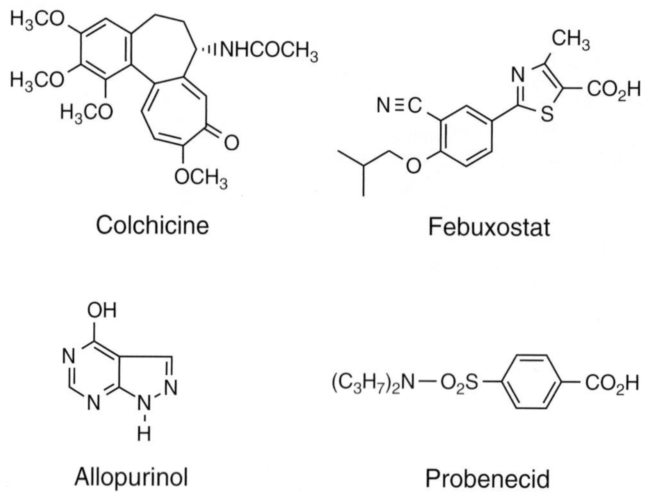

Colchicine

Febuxostat

Allopurinol

Probenecid

Figure 21.28 Drugs for the treatment of gout.

DRUG–DRUG INTERACTIONS

Probenecid
- Allopurinol (synergistic uric acid lowering action)
- Aspirin (antagonism of the uric acid lowering action of each drug)
- Probenecid inhibits the renal clearance of:
  - Sulfinpyrazone
  - NSAIDs
  - Methotrexate
  - Penicillin, cephalosporin, and sulfonamide antibiotics
  - Sulfonylurea hypoglycemic agents

Probenecid

- Probenecid is classified as a uricosuric. It enhances the rate of uric acid excretion by the kidney.
- Probenecid's carboxylic acid ensures affinity for the URAT1 transporter protein that returns excreted uric acid to the bloodstream. By blocking URAT1, proben-ecid causes more uric acid to be retained in urine.
- The *n*-propyl sulfonamide substituents provide optimal URAT1 blocking activity.
- Probenecid's ability to nonselectively block active secretion and reabsorption of organic acids via organic anion transporters (OATs) leads to the potential for drug–drug interactions (e.g., loop diuretics, penicillins, NSAIDs).
- Phase I and II metabolic reactions are predictable (Fig. 21.29). Metabolites retain some uricosuric activity due to the presence of a carboxylic acid.

Allopurinol

- Allopurinol is an inhibitor of uric acid biosynthesis.
- Structurally, it is the pyrazole-containing analog of hypoxanthine, one of two endogenous purine substrates for xanthine oxidase. It binds strongly to this rate-limiting enzyme and inhibits the normal biosynthesis of urate.

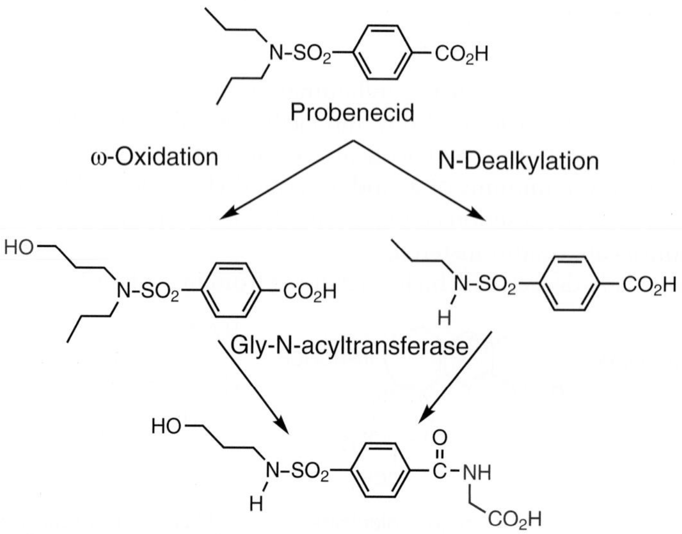

Probenecid

ω-Oxidation

N-Dealkylation

Gly-N-acyltransferase

Figure 21.29 Probenecid metabolism.

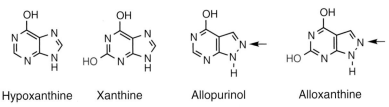

Figure 21.30 Febuxostat metabolism.

- Alloxanthine, the product of allopurinol oxidation by xanthine oxidase, is the pyrazole-containing analog of xanthine. It is a long-acting inhibitor of xanthine oxidase.
  - Alloxanthine accumulation over time explains the augmented decreases in uric acid biosynthesis observed with chronic allopurinol use.

Hypoxanthine	Xanthine	Allopurinol	Alloxanthine

> **⚠ ADVERSE EFFECTS**
> **Allopurinol**
> - Hypersensitivity manifested by maculopapular rash (discontinue if observed)
> - Gastrointestinal distress
> - Alkaline phosphatase elevation
> - Acute gout flares (manage with colchicine)

Febuxostat

- Febuxostat is a nonpurine xanthine oxidase inhibitor that noncompetitively inhibits access to the enzyme's catalytic site by the endogenous hypoxanthine and xanthine substrates. Uric acid biosynthesis is halted.
- Febuxostat is extensively metabolized by several CYP and glucuronidating isoforms (Fig. 21.30). The hydroxylated metabolites may retain some therapeutic activity.
- Coadministration of drugs that depend upon xanthine oxidase metabolism (e.g., mercaptopurine, azathioprine, theophylline) may result in unwanted toxicity. The concomitant use of the more toxic mercaptopurine and azathioprine is contraindicated.

Pegloticase

- Pegloticase is a recombinant pegylated uricase enzyme that metabolizes uric acid to a more water soluble metabolite, allantoin. Allantoin undergoes further hydrolysis to yield glyoxylic acid and urea (Fig. 21.31).

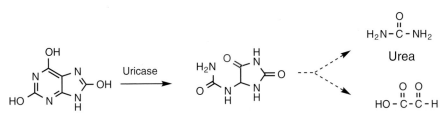

Uric acid   Allantoin   Urea   Glyoxylic acid   **Figure 21.31** Uric acid metabolism.

- As a protein-based drug, pegloticase is not orally active. It is administered by IV infusion. Premedication with antihistamines and corticosteroids decreases the risk of infusion-related anaphylaxis.
  - Patients with glucose-6-phosphate dehydrogenase deficiency are at particularly high risk for infusion reactions and anaphylaxis, and should not receive pegloticase.
- Acute gout flares stimulated by pegloticase can be treated proactively with NSAIDs and/or colchicine. Prophylaxis should begin at least 1 week prior to pegloticase administration.

Drugs for the Treatment of Multiple Sclerosis (Fig. 21.32)

- In addition to teriflunomide (previously discussed), two other immunomodulators are available to treat relapsing MS, fingolimod, and dimethyl fumarate.
- Fingolimod is an analog of sphingosine and requires phosphorylation for activity.
  - Once phosphorylated, fingolimod acts as an agonist at sphingosine-1-phosphate receptors (S1PR), specifically $S1P_1$.
  - SIP_1 activation sequesters lymphocytes, which prohibits their movement into the CNS and decreases the risk of central inflammatory lesions that lead to MS symptoms.
  - Fingolimod is metabolized via oxidation and amidation to inactive metabolites.

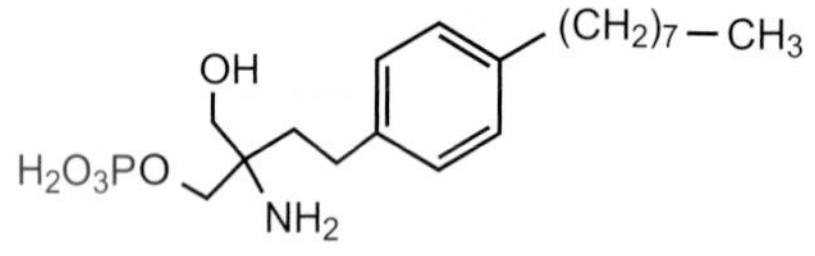

Fingolimod phosphate

- Dimethyl fumarate's mechanism is yet to be fully elucidated.
  - Known activities of relevance include inhibition of cytokine-induced nuclear translocation of NF-κB, augmentation of nuclear factor erythroid 2-related factor 2 (Nrf2) and its nuclear translocation, and increased levels of reduced glutathione (GSH), which reduces nerve fiber demyelination.

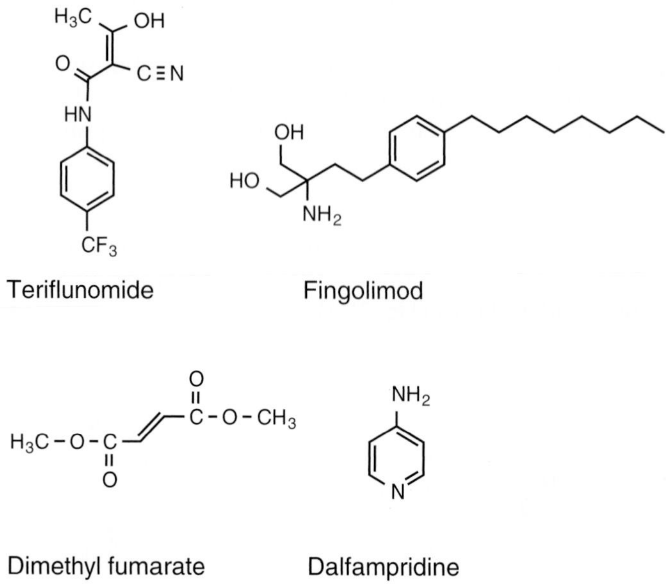

Teriflunomide   Fingolimod

Dimethyl fumarate   Dalfampridine

Figure 21.32 Drugs used in the treatment of multiple sclerosis.

- Dimethyl fumarate (DMF) is rapidly hydrolyzed to monomethyl fumarate (MMF), an active metabolite.
- Dalfampridine, is an aromatic amine used as an adjunct treatment in MS. It improves ambulation through blockade of CNS voltage-gated potassium channels, which increase functional nerve conductance in demyelinated axons.

Chemical Significance

Inflammatory processes contribute to the pathology of many common diseases, some obvious (e.g., acute injury and rheumatoid arthritis) and others less so (e.g., macular degeneration, stroke, Alzheimer's Disease). Anti-inflammatory drugs work to attenuate the symptoms of pathologic inflammation or address inflammation's root causes.

NSAID use in the United States is ubiquitous, with over 111 million prescriptions for these pain-relieving agents filled annually and many more patients self-medicating with OTC NSAIDs. Even OTC NSAIDs carry serious toxicity risks in selected patients, a fact not always recognized by the general public.

There are many ways a pharmacist's unique knowledge of drug chemistry can help keep patients on anti-inflammatory agents, antigout drugs, and acetaminophen safe while addressing their therapeutic needs. For example:

- Is your patient taking OTC ibuprofen for an acute injury while also on chronic low-dose aspirin "heart health" therapy? Make sure you counsel to separate the doses by several hours and ensure they understand that the dosing delay depends on which drug is taken first.
- Is your patient prone to hepatotoxicity due to ongoing liver disease or concomitant hepatotoxic drug therapy? Make sure they are not prescribed the potentially hepatotoxic NSAID diclofenac, and counsel on the hepatotoxic potential of OTC acetaminophen (Tylenol). Counsel all patients on the hepatotoxic risk of exceeding the recommended daily dose of acetaminophen.
- Is your patient prone to GI ulceration? Help the physician select one of the more COX-2 preferential (meloxicam, etodolac) or selective (celecoxib) NSAIDs to minimize the risk of bleeding. Make sure the patient avoids COX-1 preferential NSAIDs like indomethacin and piroxicam.
- Do you have a patient predisposed to or newly diagnosed with colon cancer? Advise the patient of the potential benefits of aspirin therapy and consider consultation with the physician on whether COX-2 selective inhibitor therapy (celecoxib) should be instituted.
- Does a patient stabilized on colchicine for gout prophylaxis present with a prescription for erythromycin ethylsuccinate? Consult with the physician on alternative antibiotic therapy to minimize the risk of colchicine accumulation and toxicity due to inhibited CYP3A4 metabolism.
- Is a patient on probenecid picking up a bottle of full strength aspirin from your pharmacy? Make sure that they are not the ones taking the salicylate and inform them of the nature of the contraindication between these two medications. Also make sure they are hydrating well to flush the excreted uric acid from their system.
- Does a patient on auranofin DMARD therapy complain of a bad taste when they take the medication? Alert the patient that the bad taste, if metallic in nature, may be a sign of serious toxicity and confirm appropriate follow-up with the physician.
- Is an elderly rheumatoid arthritis patient on infliximab infusion and methotrexate therapy in your pharmacy seeking OTC medications for fever and a persistent productive cough? Alert the physician to the potential of bacterial pneumonia so that, if warranted, the antiarthritis therapy can be suspended.

Many patients will be self-medicating or taking prescription medications for their anti-inflammatory benefits. As the most accessible health care professional, pharmacists are in the best position to ensure they do so in ways that optimize their complete health and well-being.

Review Questions

1 Which NSAID has the proper chemistry to irreversibly inhibit COX enzymes?

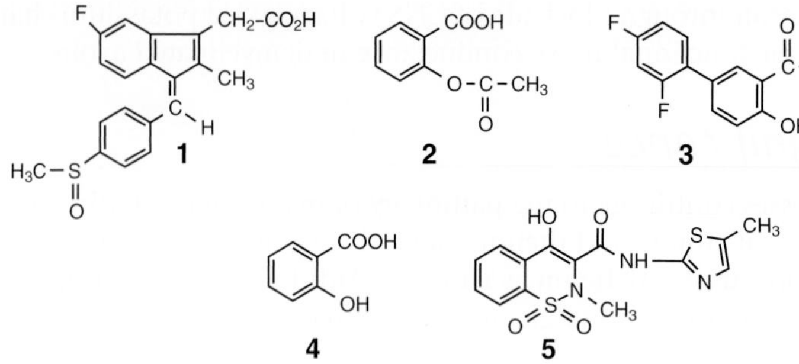

A. NSAID **1**
B. NSAID **2**
C. NSAID **3**
D. NSAID **4**
E. NSAID **5**

2 Which NSAID would be least likely to induce gastrointestinal ulceration and bleeding with chronic administration?

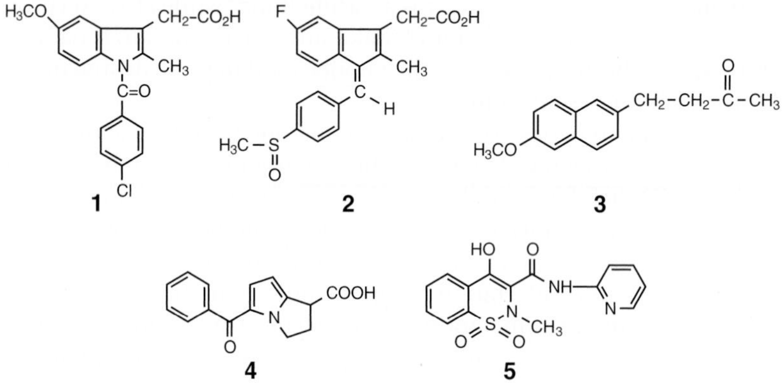

A. NSAID **1**
B. NSAID **2**
C. NSAID **3**
D. NSAID **4**
E. NSAID **5**

3 Which analgetic would form the most electrophilic (and, therefore, the most hepatodestructive) quinoneimine metabolite?

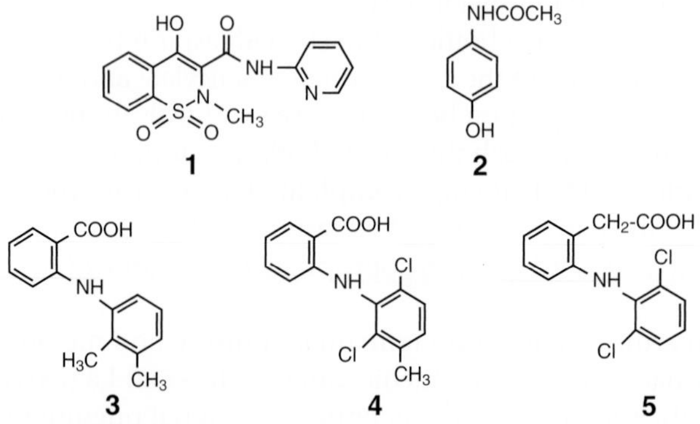

A. NSAID **1**
B. NSAID **2**
C. NSAID **3**
D. NSAID **4**
E. NSAID **5**

4 Which of the gold salts that are, or were formerly, marketed for the treatment of rheumatoid arthritis is stable enough for oral administration?

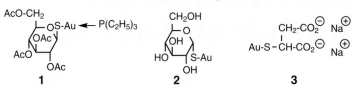

A. Gold salt **1**
B. Gold salt **2**
C. Gold salt **3**

5 Which antigout drug has the appropriate chemistry to successfully compete with the endogenous xanthine oxidase substrates xanthine and hypoxanthine for access to the active site of this rate-limiting uric acid biosynthesis enzyme?

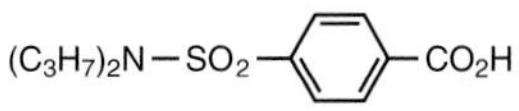

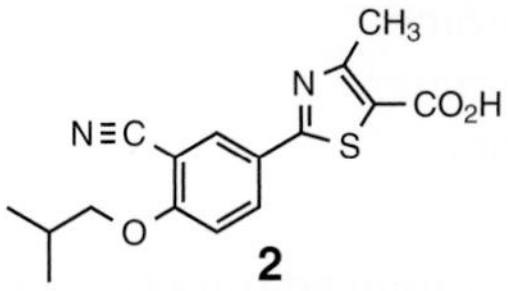

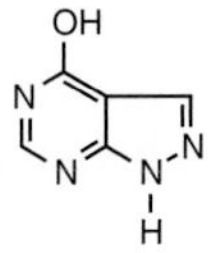

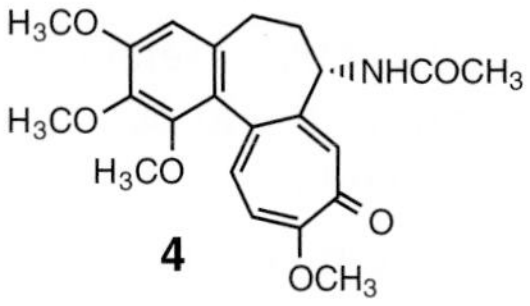

A. Antigout drug **1**
B. Antigout drug **2**
C. Antigout drug **3**
D. Antigout drug **4**

22

Antihistamines and Related Antiallergic and Antiulcer Agents

Learning Objectives

Upon completion of this chapter the student should be able to:

1. Discuss the importance of histamine tautomerization to receptor binding and activation.

2. Describe mechanism of action of:
   a. Inhibitors of histamine release
   b. H_1-selective antihistamines
   c. H_2-selective antihistamines
   d. Proton pump inhibitors
   e. Prokinetic agents

3. Distinguish sedative from nonsedative H_1-selective antihistamines by structure and receptor-binding properties.

4. Discuss applicable structure activity relationships (SAR) (including key aspects of receptor binding) of the above classes of antihistamine, antiallergic, and antiulcer agents.

5. List physicochemical and pharmacokinetic properties that impact in vitro stability and/or therapeutic utility of antihistamine, antiallergic, and antiulcer agents.

6. Diagram metabolic pathways for all biotransformed drugs, identifying enzymes and noting the activity, if any, of major metabolites.

7. Apply all of above to predict and explain therapeutic utility.

Introduction

Histamine is an endogenous autacoid, a term that means "self-remedy." Once produced, histamine is stored in mast cells and basophils. Beyond the low level of basal histamine secretion, release is primarily the result of the interaction of an antigenic stimulus with IgE antibodies on the cell surface. The result of the antigen–antibody reaction is stimulation of phospholipase C, which increases intracellular levels of inositol triphosphate and calcium, and elevates levels of diacylglycerol in the membrane. The increased concentration of these second messengers leads to exocytosis. In mast cells, this process is also known as degranulation.

Histamine is found in high concentration in the lungs, gastrointestinal tract, and skin, and it activates four main types of histamine receptors, H_1 through H_4. The histamine receptors currently of greatest interest in antihistamine-based treatment of allergic reactions and gastrointestinal distress and ulceration are H_1 and H_2, respectively.

In addition to blocking histamine action at H_1 receptors, inhibition of histamine release can be therapeutically beneficial in thwarting allergic inflammatory reactions. An alternative approach to mitigating gastrointestinal distress due to excessive gastric acidity is blocking the secretion of gastric acid from the parietal cells through the inhibition of H^+, K^+-ATPase, also known as the proton pump.

Chemistry of Histamine

- In unionized form, histamine has one neutral and two basic nitrogen atoms.
- The predominant form at pH 7.4 is the monocation, where the side-chain primary amine is protonated and both imidazole nitrogen atoms are unionized.

- Histamine is achiral but can exist in multiple conformational states in solution. The *trans* conformation is believed to be preferred at H_1 and H_2 receptors.

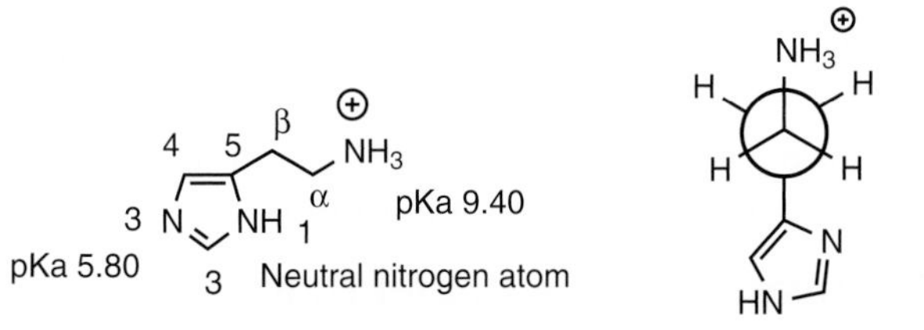

Histamine monocation

Histamine monocation in
trans conformation

Histamine monocation

- Histamine's imidazole ring is capable of tautomerization.
- Tautomerization is essential for histamine activity at H_1 receptors, with the N^τ-H tautomer important for initial receptor binding and the N^π-H tautomer important for receptor activation.
- Slight structural modifications can increase selectivity for one receptor subtype over the others. For example, 2-methylhistamine is selective for the H_1 receptor while 4-methylhistamine is H_2-selective.

N^τ–H tautomer

N^π-H tautomer

Tautomers of histamine

2-Methylhistamine
(H_1-selective agonist)

4-Methylhistamine
(H_2-selective agonist)

2-methylhistamine and 4-methylhistamine

- Histamine's half-life is very short. Once released from mast cell and basophil storage sites, it is rapidly metabolized through N-methylation and side-chain oxidation.
- Catalyzing enzymes include N-methyltransferase, monoamine oxidase, diamine oxidase, and aldehyde oxidase. (See Figure 32.3, Chapter 32 in *Foye's Principles of Medicinal Chemistry, Seventh Edition*, for a complete detailed discussion.)

Histamine Receptors

- H_1 receptors (a $G_{q/11}$ coupled).
  - The H_1 receptor is found in smooth muscle of gut, bronchi, uterus, and vasculature.
  - Stimulation results in relaxation and increased permeability of the vasculature and spontaneous contractions in the smooth muscle of the uterus, gut, and bronchi.
- H_2 receptors (G_s and G_q coupled).
- The H_2 receptor is found in the parietal cells of the stomach and in vascular smooth muscle, hepatocytes, and various blood cells.
- Stimulation of H_2 receptors in the stomach leads to an increase in gastric acid secretion.
- H_3 receptors (G_i and G_o coupled).
  - The H_3 receptor is found primarily in the central nervous system (CNS).
  - Stimulation by histamine results in a decrease in the release of other neurotransmitters.

- H_4 receptors (G_i and G_o coupled).
  - H_4 receptors are found on eosinophils, neutrophils, dendritic cells, T cells, and mast cells.
  - Stimulation promotes production and release of inflammatory cytokines and mediators of chemotaxis.
  - H_4 receptors mediate symptoms of autoimmune disorders and some allergic disorders (e.g., bronchial asthma, conjunctivitis, and rhinitis).

Histamine Binding and Activation of the H_1 and H_2 Receptors

H_1 Receptor (Fig. 22.1)

- Amino acid residues of the human H_1 receptor important to binding histamine and antiallergic antihistamines have been identified.
- The H_1 receptor contains the anionic Asp residue common to all G protein-coupled receptors. Asp^{107} is the site where cationic histamine (and antihistamines) anchor through ion–ion bonding.
- Histamine is recognized by the H_1 receptor in its N^τ-H tautomeric form. Hydrogen bonding with Asn^{198} and Lys^{191} is essential.
- Once recognized, a charge relay system involving a series of proton transfers between histamine and the H_1 receptor results in tautomerization to the N^π-H form and receptor stimulation.

H_2 Receptor (Fig. 22.2)

- The H_2 receptor contains the conserved anionic Asp residue (Asp^{98}) for anchoring histamine and antagonists.
- Tautomerization to the N^π-H form occurs in a manner analogous to that previously described, resulting in H_2 receptor activation.

Histamine receptor recognition phase:

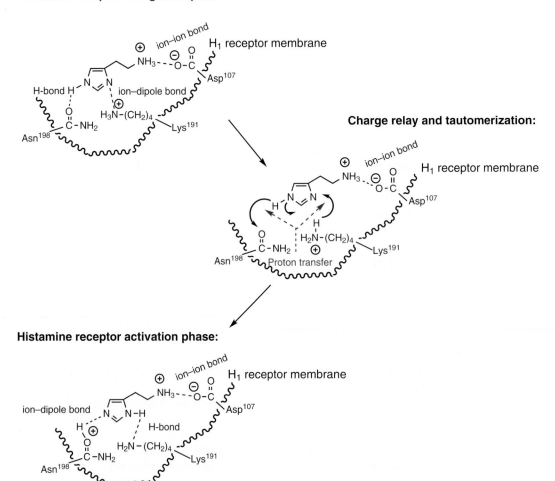

Charge relay and tautomerization:

Histamine receptor activation phase:

Figure 22.1 Histamine binding and activation of H_1 receptor.

Histamine receptor recognition phase:

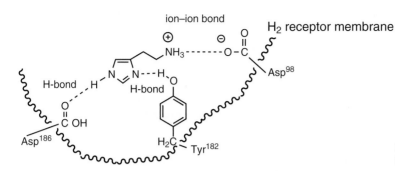

Figure 22.2 Histamine recognition at the H_2 receptor.

Inhibitors of Histamine Release

- Drugs that inhibit mast cell degranulation halt histamine-induced actions by depriving the receptors of endogenous agonist.
- They do not interact with histamine receptors and cannot reverse the actions of previously released histamine.

MOA

- The mechanism of cromolyn sodium and nedocromil sodium is phosphorylation of the protein moesin.
- Moesin phosphorylation promotes association with actin and other proteins found in secretory granules. This stabilizes mast cells to exocytosis.
- There is evidence to suggest that lodoxamide may act by inhibiting eosinophil infiltration, possibly by decreasing the number of T-helper 2 lymphocytes.

Specific Drugs (Fig. 22.3)

Cromolyn Sodium

- Dosage forms are targeted to specific indications.
  - Nebulization solution: prophylactic treatment of asthma- or exercise-induced bronchospasm.
  - Intranasal and ophthalmic solution: seasonal rhinitis and allergic conjunctivitis, respectively.
  - Oral formulation: mastocytosis (a rare disorder involving the overexpression of mast cells).
- Systemic absorption from sites of administration is poor.
- Cromolyn sodium is excreted unchanged in urine and feces.

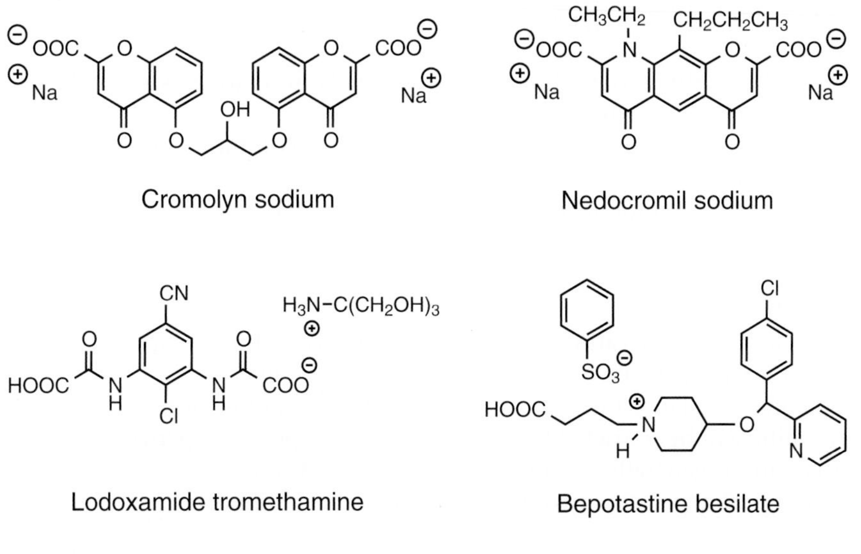

Cromolyn sodium

Nedocromil sodium

Lodoxamide tromethamine

Bepotastine besilate

Figure 22.3 Mast cell degranulation inhibitors.

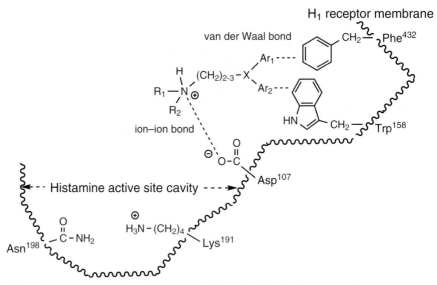

Figure 22.4 Representation of antihistamine binding to the H$_1$ receptor.

⚠ **ADVERSE EFFECTS**

Mast Cell Degranulation Inhibitors
- Administration-related stinging, coughing, gastrointestinal distress
- Hypersensitivity reactions
- Headache

Nedocromil Sodium, Lodoxamide Tromethamine, and Bepotastine Besilate
- These drugs are used exclusively by the ophthalmic route.
- Systemic absorption from the eye is minimal.
- The drugs are excreted essentially unchanged in urine.

Inhibitors of Released Histamine at H₁ Receptors (H₁-Selective Antihistamines)

- Drugs that block the action of released histamine at H$_1$ receptors are commonly referred to as antihistamines.
- Antihistamines treat allergic responses to pollen and other environmental allergens (seasonal allergies), foods, animal dander, and plant/animal toxins (bee stings, poison ivy).

Antihistamine Binding to Muscarinic Receptors

There is 40% homology between the H$_1$ receptor and muscarinic M$_1$ and M$_2$ receptors. Binding of antihistamines to muscarinic receptors is responsible, in part, for the adverse effects commonly associated with antihistamines, particularly those classified as first-generation (centrally active) agents. These adverse effects include sedation, dry mouth, and blurred vision.

Antihistamine Binding at H₁ Receptors (Fig. 22.4)

- Antihistamines block the binding of histamine at the H$_1$ receptor by one of two mechanisms.
  - Most are inverse agonists that bind to an inactive form of the receptor. This shifts the normal conformational equilibrium away from the active (G-protein coupled) to the inactive (G-protein uncoupled).
  - Others are classic antagonists that bind with both active and inactive receptor conformations, inhibiting the ability of histamine to reach the H$_1$ receptor active site.
- Because antihistamines do not bind with the same H$_1$ receptor residues as histamine, their structures are distinctly different.
  - The only binding residue all antihistamines have in common with histamine is Asp[107].
- Two hydrophobic receptor areas are important to antihistamine binding. One contains Phe[432] and another includes Trp[158].

⚠ **ADVERSE EFFECTS**

Antihistamines (particularly first-generation agents)
- Sedation
- Dry mouth
- Blurred vision
- Tachycardia
- Urinary and fecal retention

SAR (Fig. 22.5)
- A straight chain or cyclic cationic amine is essential to anchor to Asp[107].
- Tertiary amines provide the highest antihistaminic activity.
  - In potentially sedative first generation antihistamines, bulk around the tertiary amine will be restricted (e.g., N,N-dimethyl, piperidine, pyrrolidine).
- The spacer unit is generally unsubstituted and is 2 to 3 carbons in length.
  - Branching enhances activity only in phenothiazine antihistamines.

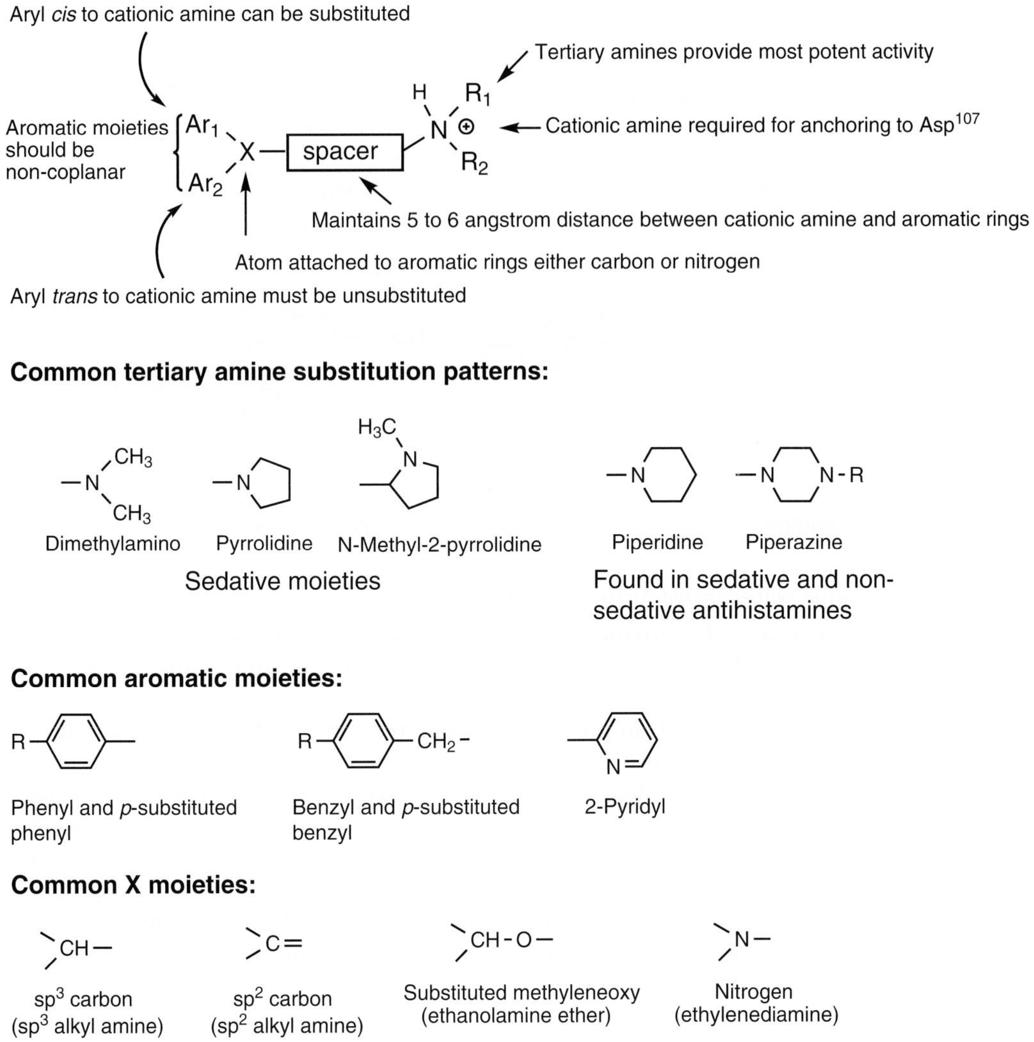

Aryl *cis* to cationic amine can be substituted

Tertiary amines provide most potent activity

Aromatic moieties should be non-coplanar

Cationic amine required for anchoring to Asp[107]

Maintains 5 to 6 angstrom distance between cationic amine and aromatic rings

Atom attached to aromatic rings either carbon or nitrogen

Aryl *trans* to cationic amine must be unsubstituted

Common tertiary amine substitution patterns:

Dimethylamino   Pyrrolidine   N-Methyl-2-pyrrolidine

Sedative moieties

Piperidine   Piperazine

Found in sedative and non-sedative antihistamines

Common aromatic moieties:

Phenyl and *p*-substituted phenyl

Benzyl and *p*-substituted benzyl

2-Pyridyl

Common X moieties:

sp³ carbon
(sp³ alkyl amine)

sp² carbon
(sp² alkyl amine)

Substituted methyleneoxy
(ethanolamine ether)

Nitrogen
(ethylenediamine)

Figure 22.5 H_1-selective antihistamine SAR.

- The X moiety determines the class of antihistamine.
  - Saturated or unsaturated carbon: alkyl amines.
  - Substituted methyleneoxy: ethanolamine ethers (also called aminoalkyl ethers).
  - Saturated nitrogen: ethylenediamines.
- When Ar_1 and Ar_2 are different, X becomes chiral if the aromatic moieties are bound to a carbon atom. In most cases the *S* isomer will be the active form.
- Two aromatic rings bind with high affinity to H_1 receptor residues.
  - The ring that is *trans* to the cationic amine binds to a sterically restricted receptor area that includes Phe[432]. This aromatic moiety must be unsubstituted.
  - The ring that is *cis* to the cationic amine binds via van der Waals forces to a receptor area that includes Tyr[158]. Substituted and/or benzyl aromatic moieties are accepted.
  - The bulkier ring assumes the required *cis* orientation through free rotation in sp³ alkyl amines, ethanolamine ethers, and ethylenediamines. The rings must be synthetically placed in rigid sp² alkyl amines.
  - Potency is enhanced if the two aromatic rings are noncoplanar.
- Lipophilic substituents on the *cis* ring enhance activity by promoting distribution. Duration increases if the substituent also blocks inactivating metabolism.

Nonsedative (Peripherally Selective) Antihistamine SAR (Fig. 22.6)

- These compounds are sometimes called second-generation antihistamines to distinguish them from the centrally active (sedative) first-generation agents.

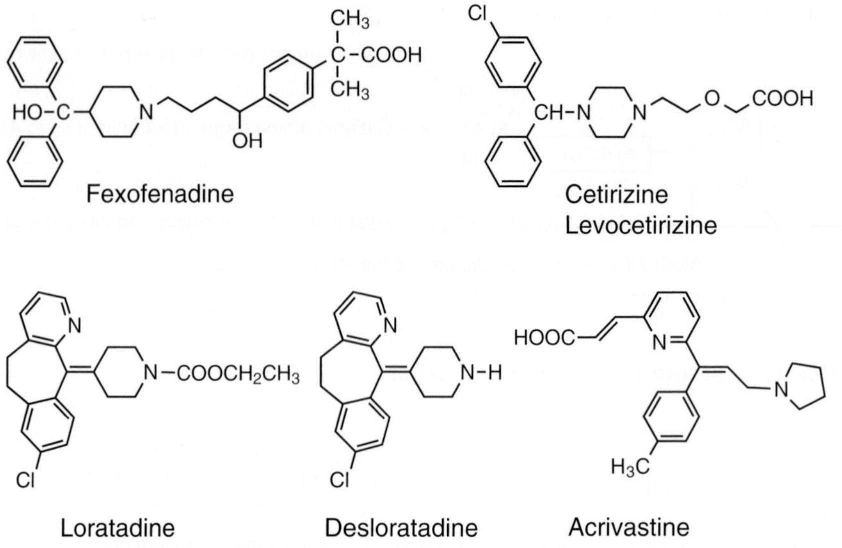

Figure 22.6 Nonsedative antihistamines.

Zwitterionic Nonsedative Antihistamines

- Fexofenadine and cetirizine contain a long, flexible substituent on the cationic amine that ends in an anionic carboxylate. The long, flexible chain allows an internal salt to form.
- The folded conformation predominating in the bloodstream minimizes affinity for CNS transporting proteins. Affinity for central P-glycoprotein (P-gp) efflux proteins is high, so molecules that passively enter the brain are readily ejected.
- At the peripheral H_1 receptor, the extended conformation allows two ionic bonds with charged receptor residues to form. The ionic bond that zwitterionic antihistamines form with Lys^{191} is strong and augments antihistaminic potency (Fig. 22.7).
- A third zwitterionic antihistamine (acrivastine) that follows a different structural motif is available in combination with pseudoephedrine.

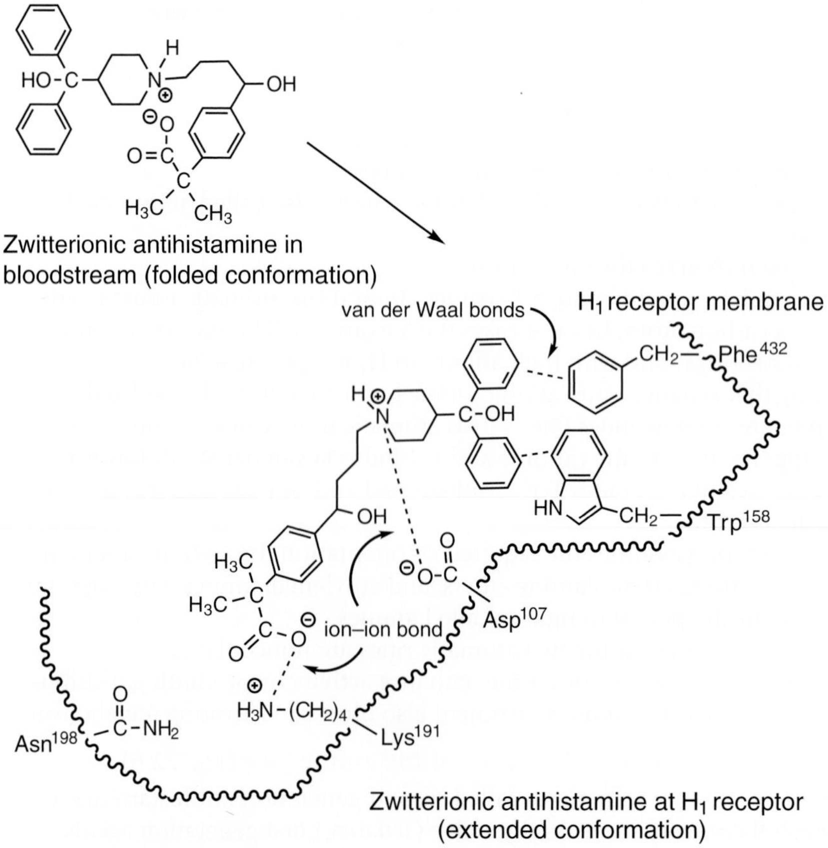

Figure 22.7 Zwitterionic antihistamine binding at the H_1 receptor.

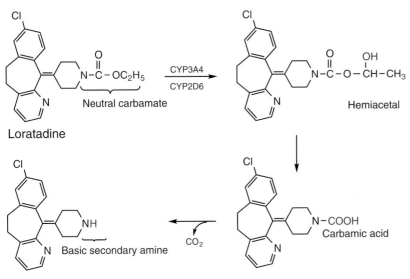

Figure 22.8 Loratadine metabolism.

CHEMICAL NOTE

A Tale of Two Antihistamines: Terfenadine and Astemizole: Two zwitterionic antihistamines terfenadine (Seldane) and astemizole (Hismanal) were pulled from the US market due to their ability to cause life-threatening arrhythmia, particularly when coadministered with CYP3A4 inhibitors (including grapefruit juice) or competing substrates (e.g., erythromycin, "conazole" antifungal agents). Both antihistamines are cardiotoxic in parent drug form but metabolized to nontoxic metabolites by this isoform. Note that one of terfenadine's nontoxic metabolites is fexofenadine, which is marketed as a peripherally selective antihistamine. When either parent drug is allowed to accumulate, prolongation of the QT interval and potentially fatal torsades de pointes can occur.

Loratadine and Desloratadine

- Loratadine has a neutral carbamate nitrogen atom that cannot protonate at physiologic pH. It is almost completely metabolized on first pass to the ionizable secondary amine (desloratadine, Fig. 22.8).
- Desloratadine is also a marketed nonsedative antihistamine.
- Loratadine and desloratadine are the only nontertiary amine-containing antihistamines.
- These two antihistamines do not assume a folded conformation in the bloodstream. Like zwitterionic antihistamines, they have a low affinity for CNS transporting proteins and a high affinity for central P-gp.

Physicochemical Properties

- With the exception of loratadine, antihistamines are cationic amines at pH 7.4. Amine pKa is commonly in the 8.5 to 9.5 range.
- Peripherally selective zwitterionic antihistamines also have a carboxylic acid functional group with a pKa ranging from 2 to 4.
- H_1 antihistamines are lipophilic and readily penetrate membranes.
- Many antihistamines administered for the treatment of allergic rhinitis are given orally. Absorption is efficient and onset of action swift.
- Drugs administered topically via nasal spray or ophthalmic solution are well absorbed from the nasal passages and conjunctiva, respectively.
- The duration of many sedative antihistamines falls in the 4- to 6-hour range.
- The duration of nonsedative antihistamines is much longer owing to slower H_1 receptor dissociation.
  - Elimination half-lives of nonsedative antihistamines range between 8.3 hours (cetirizine) to 27 hours (desloratadine).

Metabolism

- Common metabolic reactions that inactivate H_1 antihistamines include N-dealkylation of aliphatic amine side chains with subsequent deamination and oxidation to the carboxylic acid, and aromatic hydroxylation of phenyl rings.
- Phase I carboxylic acid or phenolic metabolites can undergo Phase II conjugation with PAPS (phenol) or glucuronic acid (phenol and carboxylic acid).
- Excretion of unchanged antihistamines and their metabolites is predominantly renal. Fecal excretion is more common in some nonsedative antihistamines.

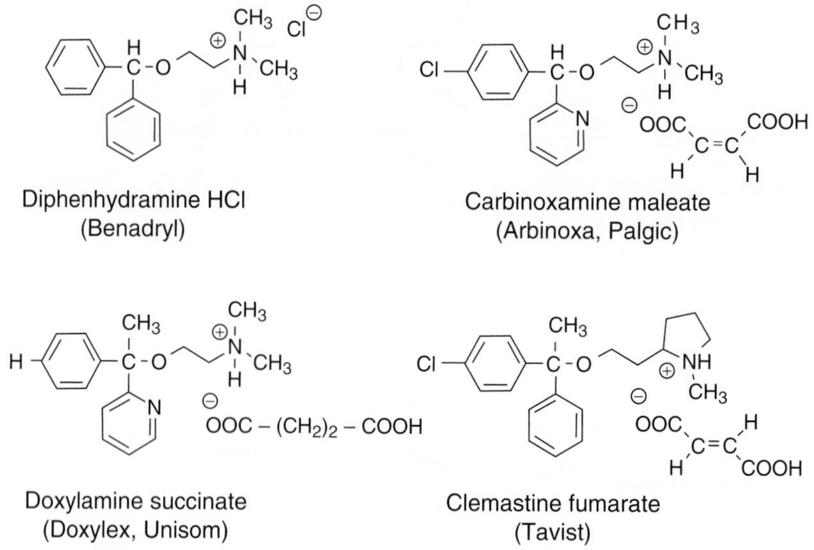
Diphenhydramine HCl
(Benadryl)
Carbinoxamine maleate
(Arbinoxa, Palgic)
Doxylamine succinate
(Doxylex, Unisom)
Clemastine fumarate
(Tavist)

Figure 22.9 Ethanolamine ether antihistamines.

Ethanolamine Ether Antihistamines

- Ethanolamine ethers are highly anticholinergic and sedative. Some are marketed as OTC sleep aids.
  - Cautious use in elderly, patients needing to remain alert, and those taking other sedative drugs (including alcohol) is warranted.
- An α- or *o*-CH_3 augments anticholinergic activity. A *para* substituent enhances antihistaminic action.

Specific Drugs (Fig. 22.9)

Diphenhydramine

- The anticholinergic activity of diphenhydramine allows use in Parkinsonism and as an antispasmotic. It is also used as an OTC sleep aid and an antimotion sickness antiemetic.
- The unsubstituted aromatic rings of this achiral antihistamine lead to a short duration. Diphenhydramine is commonly administered every 4 to 6 hours.
- Diphenhydramine metabolism serves as an example of the reactions common to this class of antihistamine (Fig. 22.10).
- A unique adverse effect is inhibition of CYP2D6-mediated inactivation of selected antihypertensive β-adrenergic blockers (e.g., metoprolol), leading to excessive drops in heart rate and blood pressure.

Carbinoxamine

- Carbinolamine is the *p*-chloro-2-pyridyl analog of diphenhydramine. The chlorine atom adds lipophilicity and blocks inactivating aromatic hydroxylation, resulting in higher potency and a longer duration of action.
- Carbinoxamine is chiral. The *S*(+) isomer has the majority of the antihistaminic activity.

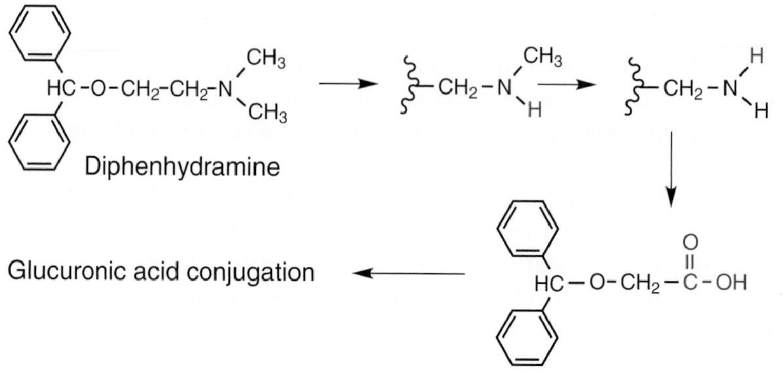
Diphenhydramine
Glucuronic acid conjugation

Figure 22.10 Diphenhydramine metabolism.

- Carbinoxamine is less sedative than diphenhydramine. It is used exclusively in the treatment of allergic rhinitis.

Doxylamine

- The α-CH_3 augments anticholinergic and sedative potential. Doxylamine is a commonly used OTC sleep aid.
- Its antiemetic activity makes it useful in the treatment of nausea and vomiting associated with early pregnancy.
- When used to treat allergic rhinitis, doxylamine is administered every 4 to 6 hours. As a sleep aid, it is administered 30 minutes before bedtime.

Clemastine

- Clemastine is administered every 12 hours for the treatment of allergic rhinitis, urticaria, and angioedema.

Sedative (Nonselective) Ethylenediamine Antihistamines

- Nonzwitterionic agents block H_1 and muscarinic receptors in the CNS and periphery. They reach the chemoreceptor trigger zone (CTZ) and are very effective antiemetic agents.
- Most currently available agents contain a piperazine ring as the ethylenediamine scaffold.
- Due to significant sedation, all precautions noted for the ethanolamine ether antihistamines apply.

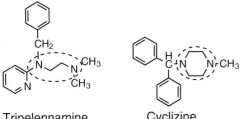

Tripelennamine   Cyclizine

Ethylenediamine antihistamines

Specific Drugs (Fig. 22.11)

Chlorcyclizine, Cyclizine, and Meclizine

- Chlorcyclizine is marketed for control of symptoms in upper respiratory allergies (e.g., hay fever).

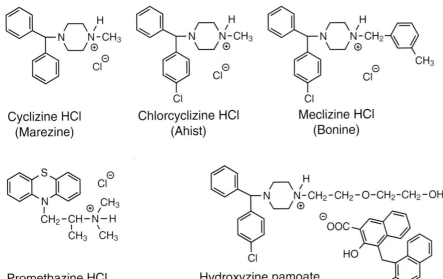

Cyclizine HCl
(Marezine)

Chlorcyclizine HCl
(Ahist)

Meclizine HCl
(Bonine)

Promethazine HCl
(Phenergen)

Hydroxyzine pamoate
(Vistaril)

Figure 22.11 Sedative (nonselective) ethylenediamine antihistamines.

- The *p*-chloro group blocks inactivating aromatic hydroxylation and promotes drug distribution across membranes. Chlorcyclizine can be administered every 6 to 8 hours.
- Cyclizine is marketed specifically for motion sickness and the control of dizziness associated with inner ear disturbances.
  - Cyclizine has a lower potency and a shorter duration than chlorcyclizine. It is administered at twice the dose, and more frequently (every 4 to 6 hours).
- Meclizine is used exclusively for motion sickness and vertigo. The *p*-Cl group serves the same potency and duration-promoting purpose as in chlorcyclizine.

Hydroxyzine

- Hydroxyzine's primary uses are control of histamine-induced pruritus, as an adjunct to anesthesia, and in the management of anxiety and stress-related agitation.
- It is marketed as the pamoate and hydrochloride salt forms, the latter being readily water soluble.
  - The hydrochloride salt can be given either orally or by injection. The more lipophilic pamoate salt is administered exclusively in capsule form.
- This highly sedative antihistamine (FDA approved in 1956) can be metabolized by cytosolic alcohol and aldehyde dehydrogenase enzymes into the nonsedative antihistamine cetirizine (FDA approved in 1995).

Promethazine

- Promethazine is a phenothiazine, which is a specialized form of ethylenediamine. The nitrogen "X" moiety of the pharmacophore is incorporated into the tricyclic phenothiazine ring system.
- Phenothiazine antihistamines are highly sedative. Promethazine is used for seasonal allergic rhinitis, pruritus, as an adjunct to anesthesia and an analgesic, and as a postsurgical, postanesthesia, motion sickness and pregnancy-related antiemetic.
- This antihistamine is marketed in oral tablets, syrups, and solutions, and is also available by injection or as rectal suppositories.

Sedative (Nonselective) Alkylamine Antihistamines

- Although potentially sedative, alkylamine antihistamines are less likely to induce drowsiness than other nonselective antihistamines. While they are less effective antiemetic agents, they show fewer anticholinergic adverse effects.
- Alkylamine antihistamines have structures that promote a long duration and high potency. They enjoy widespread use in seasonal allergy therapy.

Specific Drugs (Fig. 22.12)

Chlorpheniramine and Brompheniramine

- Chlorpheniramine is a popular OTC antihistamine with relatively low sedation. It is available in immediate and extended release tablets and other oral formulations and is administered every 4 to 6 hours.

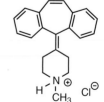

Chlorpheniramine maleate (R = Cl)
(Chlor-Trimeton)
Dexchlorpheniramine maleate
Brompheniramine maleate (R = Br)
(J-Tan PD)
Dexbrompheniramine maleate
(Ala-Hist IR)

Triprolidine HCl
(Histex)

Cyproheptadine HCl

Figure 22.12 Sedative (nonselective) alkylamine antihistamines.

BLACK BOX WARNING

Promethazine – While promethazine is approved for use (with caution) in children greater than 2 years of age, the lowest dose possible should always be used. In 2006 the FDA issued a warning for pediatric promethazine use, as fatal respiratory depression has been observed in patients under 2 years. Patients of any age with compromised respiratory function (e.g., asthma, COPD, sleep apnea) should not use promethazine, and it should not be administered to those on other respiratory depressant drugs. The drug also increases seizure risk in patients with seizure disorders.

- Brompheniramine's enhanced lipophilicity leads to a higher potency. Its duration is comparable to chlorpheniramine but it is dosed in half the strength. The $S(+)$ isomer of both drugs is active.
- A severe drop in blood pressure can occur if dexchlorpheniramine is administered to patients stabilized on monoamine oxidase inhibitor antidepressants (e.g., phenelzine, tranylcypromine, isocarboxazid).

Triprolidine and Cyproheptadine

- Triprolidine and cyproheptadine are sp^2 alkylamines.
- The marketed E isomer of triprolidine is 1,000 times as active as the Z isomer due to proper orientation at the H_1 receptor. It is available as an oral liquid or syrup and is administered every 6 hours.
- Cyproheptadine has the hydrocarbon spacer incorporated into a piperidine ring. The small CH_3 amine substituent allows for the prediction of sedation.
  - N-demethylation and reduction of one double bond would yield a structure reminiscent of the nonsedative antihistamine desloratadine.

Cyproheptadine   Desloratadine

- Off-label indications for cyproheptadine include appetite stimulation in chronic disease (e.g., cancer, cystic fibrosis), nightmare attenuation in posttraumatic stress disorder (PTSD), and prevention of migraine headaches in children and teens.
  - Cyproheptadine's efficacy in stimulating weight gain may be due, in part, to central antiserotonergic properties.

Nonsedative (Peripherally Selective) Antihistamines

- Nonsedative antihistamines penetrate poorly into the CNS. They are rapidly effluxed by the P-gp transporter if they passively cross the blood–brain barrier.
- Peripherally selective antihistamines are not totally devoid of sedation, and some patients will complain of drowsiness when taking them.
  - They are more likely to allow patients to remain alert during therapy than first-generation structures.
- Affinity for the H_1 receptor is very high. For zwitterionic agents, this enhanced affinity is due, in part, to a strong ionic bond between a carboxylate anion and the H_1 receptor Lys[191].
  - Slow receptor dissociation results in prolonged durations that often allow for once daily dosing.
- Nonsedative antihistamines are used in the treatment of seasonal allergies, hay fever, and generalized allergic rhinitis.

Specific Drugs (Fig. 22.13)

Cetirizine and Levocetirizine

- The nitrogen bearing the acidic side chain protonates to anchor to the H_1 receptor Asp[107]. The high affinity binding has been referred to as pseudoirreversible.
  - Levocetirizine's receptor dissociation half-life is 143 minutes.
- Five milligrams of $R(-)$-levocetirizine is approximately equipotent with 10 mg of racemic cetirizine. Both are given once daily.
- A wide variety of oral dosage forms of cetirizine are available, including chewable tablets. Levocetirizine is available in tablet and oral solution formulations.
- Despite being metabolized by CYP3A4-mediated O-dealkylation, drug–drug interactions are negligible since only 14% of a dose of levocetirizine is biotransformed. That percentage rises in racemic cetirizine but does not exceed 50%.
  - Metabolites are inactive.

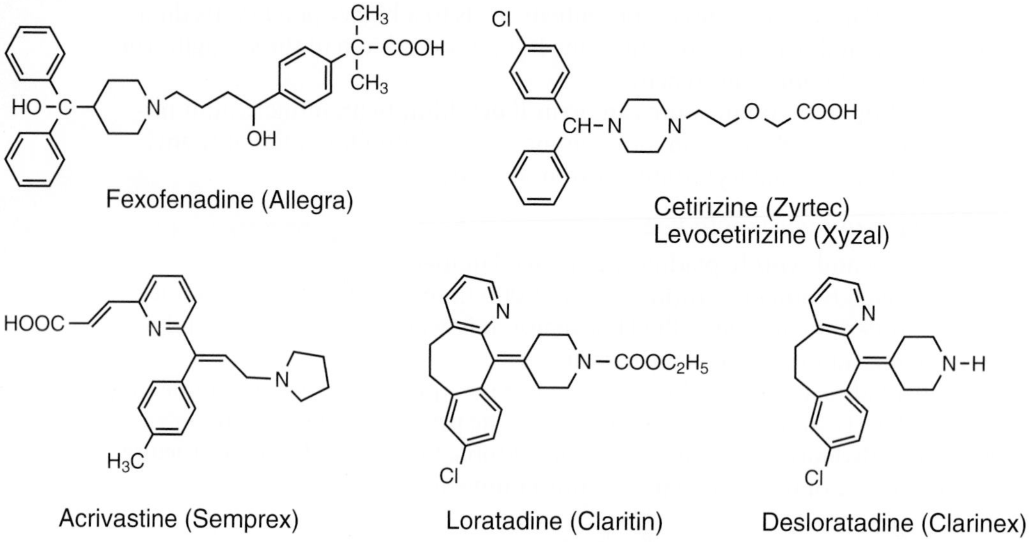

Figure 22.13 Nonsedative (peripherally selective) antihistamines.

- Cetirizine and levocetirizine can be classified as ethylenediamines but they are not effective antiemetics since they do not concentrate in the CNS.

Fexofenadine

- Fexofenadine's carbinol carbon is chiral. The racemic mixture is marketed as a single agent or in combination with the decongestant pseudoephedrine. Once or twice-daily administration is the norm.
- The length of fexofenadine's acidic side chain is longer than cetirizine's which results in a less optimal H_1 receptor fit. Daily doses are 18 to 36 times higher than racemic cetirizine.
- Fexofenadine is extensively protein bound and metabolism resistant. Approximately 5% of a dose is metabolized, and over 3% of that metabolism involves esterification of the terminal COOH group by gut flora. Fecal elimination predominates.
- The drug is transported out of the gut by organic anion transporting protein OATP1A2. Fruit juices inhibit these transporters and, if consumed within a few hours of fexofenadine, can drop serum concentrations up to 70%.
  - Fexofenadine should always be taken with water.
- Aluminum- or magnesium-containing antacids also decrease fexofenadine gastrointestinal absorption.

Acrivastine

- Acrivastine is a carboxylic acid-containing analog of the sedative alkylamine triprolidine.
- The positioning of the anionic center is different from cetirizine and fexofenadine, which both had the carboxylate containing side chain as a substituent on the cationic nitrogen. Nonsedative action is retained.
- Acrivastine is marketed only in combination with the decongestant pseudoephedrine. It is administered every 4 to 6 hours.

Loratadine and Desloratadine

- Loratadine and desloratadine are long acting and do not concentrate in the CNS.
- Desloratadine H_1 receptor affinity is higher (and its dissociation rate slower) than loratadine, possibly because of the inability of loratadine to take on a positive charge.
- Loratadine can be taken once daily or in two divided doses. Desloratadine, the more potent and persistent of the two agents, is administered once daily in half the strength of loratadine.

- Desloratadine is metabolized by CYP2C8 to an active 3-hydroxy metabolite that is subsequently O-glucuronidated prior to excretion. Desloratadine metabolism is subject to pharmacogenetic variability.
- Elimination of both agents is equally distributed between urine and feces.
- Potent inhibitors of CYP3A4 (e.g., erythromycin, ketoconazole), or of both loratadine metabolizing isoforms (e.g., cimetidine) can result in significantly elevated serum concentrations (40% to 307% increases).
- The impact of these CYP inhibitors on desloratadine serum concentrations is considerably less (6% to 73% increases).

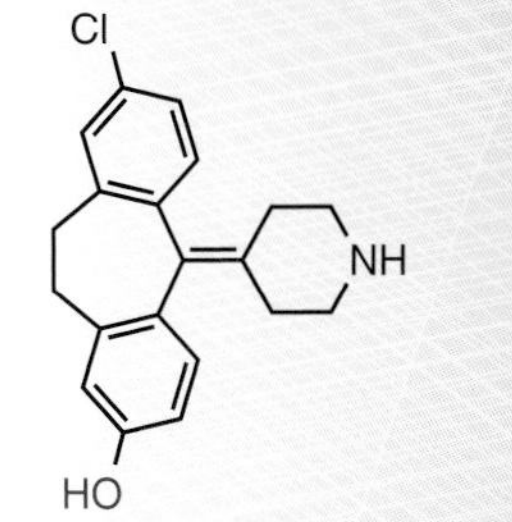

3-Hydroxydesloratadine

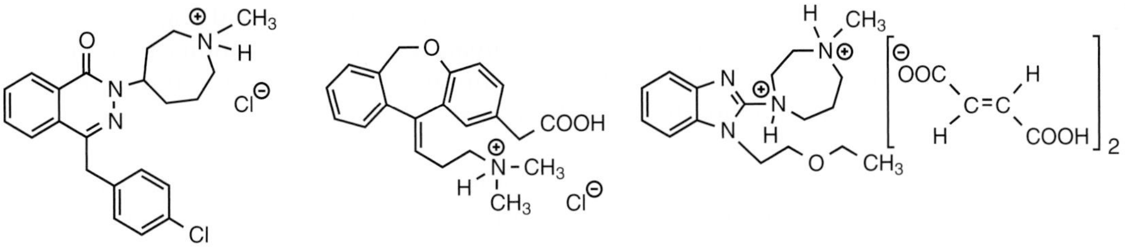

Azelastine HCl (Astelin) Olopatadine HCl (Patanol) Emedastine difumarate (Emadine)

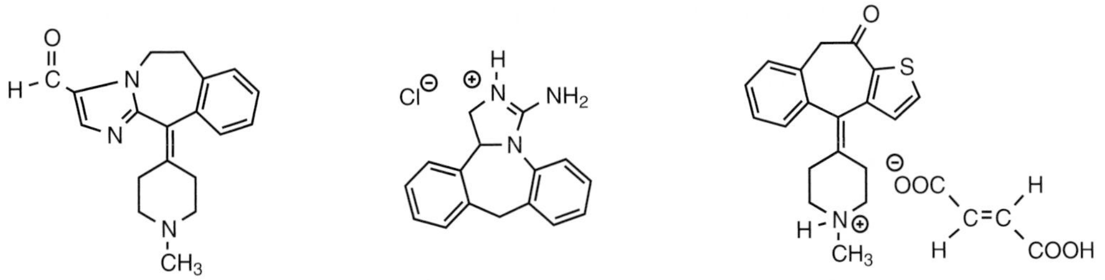

Alcaftadine (Lastacaft) Epinastine HCl (Elestat) Ketotifen fumarate (Alaway)

Figure 22.14 Topical antihistamines.

Topical Antihistamines (Fig. 22.14, Table 22.1)

- Topical antihistamines are instilled directly into the eye or nasal passages to reduce symptoms of allergic conjunctivitis and associated ocular itching or seasonal nasal allergies, respectively.
- Many agents have high H_1 receptor affinities, slow rates of receptor dissociation, and prolonged durations.
  - Alcaftadine is a prodrug. Oxidation of the aldehyde substituent to the carboxylic acid is required for activity.
- These drugs may also act by inhibiting mast cell degranulation. The conjunctiva is rich in these histamine-containing cells.
- Absorption from the conjunctiva and nasal mucosa is effective. Less than 5% of a dose penetrates into the cornea and much of the drug reaches the gut as a result of swallowing fluid from tear ducts and nasal drainage.
- Ocular irritation from these agents is low, although some patients may note brief stinging or burning upon instillation.

Table 22.1	**Dosing Parameters for Topical Antihistamines**			
Topical Antihistamine	**Ophthalmic Solution Strengths (%)**	**Ophthalmic Dosing (each eye)**	**Nasal Solution Strengths**	**Nasal Dosing (each nostril)**
Azelastine	0.05	1 drop twice daily	137 µg/spray 15%	1–2 sprays twice daily
Olopatadine	0.1, 0.2	1 drop once (0.2% or twice (0.1%) daily	0.6%	2 sprays twice daily
Emedastine	0.05	1 drop up to 4 times daily	NA	NA
Alcaftadine	0.25	1 drop once daily	NA	NA
Epinastine	0.05	1 drop twice daily	NA	NA
Ketotifen	0.025	1 drop twice daily	NA	NA

- When delivered topically, sedation that is prevalent with orally administered first-generation antihistamines is minimized.
- The only contraindication to the use of topical antihistamines is hypersensitivity to the drug or any product used in its formulation.
  - Some patients complain of a bad taste or a mild headache after use.

Antiulcer Agents

- Gastric acid (pH 1.4 to 2.1 in the fasting state) can be highly erosive to mucosal surfaces. When secreted in excess (e.g., in *Helicobacter pylori* infection) or when the normal mechanisms to protect gastric mucosa are inhibited (e.g., through the use of oral nonsteroidal anti-inflammatory agents), ulceration can result.
- Histamine, gastrin, and acetylcholine stimulate gastric acid secretion through receptor-mediated mechanisms. Each agent potentiates the action of the others.
- Histamine stimulates parietal H_2 receptors in the gastric mucosa. This activates a sulfhydryl (SH) containing enzyme called H^+, K^+-ATPase, or the proton pump.
  - It is the proton pump that is ultimately responsible for actively secreting hydrochloric acid into the stomach lumen.
- Two therapeutic approaches to protecting the gastric mucosa from erosion secondary to excessive acid secretion are blockade of released histamine at H_2 receptors (H_2-selective antihistamines) and inhibition of the proton pump.

H_2-Selective Antihistamines (Fig. 22.15)

H_2 Receptor Binding (Fig. 22.16)

- The molecules commonly called H_2 antagonists are inverse agonists that bind to and promote the inactive conformation of the H_2 receptor. This decreases the basal level of histamine agonist action.

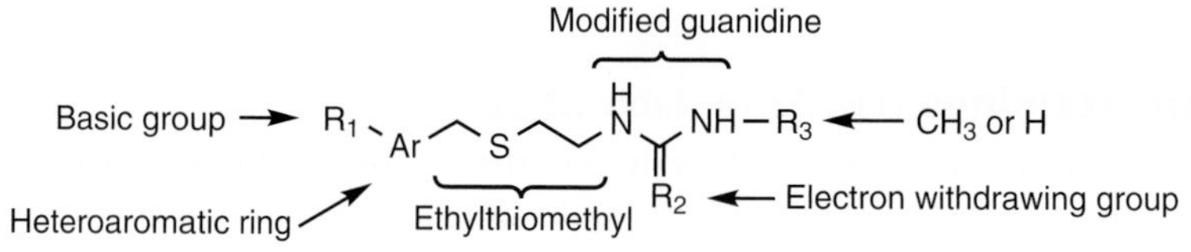

H_2-selective antihistamine pharmacophore

- The methythioethyl moiety found on all H_2-selective antihistamines binds hydrophobically to aromatic and aliphatic residues, including Val[99] and Phe[254].
- Asp[98] forms an essential hydrogen bond with a polar (e.g., cimetidine N^π-H) or protonated nitrogen atom.
- Asp[186] forms a hydrogen bond with one NH group of the modified guanidine moiety in all structures except cimetidine.
- Asn[159] also forms stabilizing hydrogen bonds with the NO_2 component of the nitromethylene (R_2) moiety of ranitidine and nizatidine. Cimetidine's R_2 group, N-cyanoimino, binds with Thr[103].

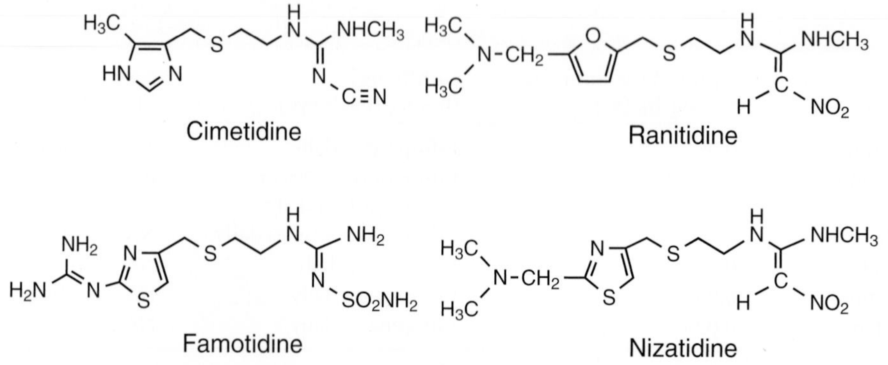

Figure 22.15 H_2-selective antihistamines.

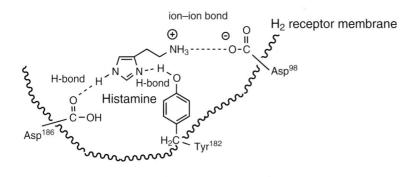

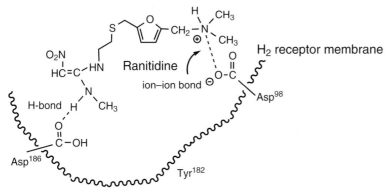

Figure 22.16 Histamine and antihistamine binding at H_2 receptors.

SAR

- A heteroaromatic ring that either incorporates an ionizable basic center or has one as a substituent is required.
  - Cimetidine's imidazole N^π is weakly basic.
  - Ranitidine and nizatidine have a basic dimethylaminomethyl substituent on a furan and thiazole ring, respectively.
  - Famotidine contains a basic diaminomethyleneamino (guanidino) thiazole substituent.
- A modified guanidine moiety substituted with a powerful electron withdrawing group to destroy basicity is found in the same relative molecular area as histamine's primary amine.
  - The electron withdrawing groups include cyano (cimetidine), nitro (ranitidine and nizatidine), and sulfamoyl (famotidine).
- The heteroaromatic ring must be separated from the substituted guanidine moiety by the equivalent of a 4 carbon chain (spacer unit).
  - The ethylthiomethyl spacer unit found in all H_2-selective antihistamines is isosteric with a butyl group. The position of the sulfur atom in this spacer is important.

Physicochemical Properties

- While H_1-selective antihistamines are lipophilic, H_2-selective antihistamines are polar. Experimental Log P values range from −0.64 (famotidine) to 0.40 (cimetidine). By comparison, the experimental Log P values for the H_1-selective antihistamines diphenhydramine and chlorpheniramine are 3.27 and 3.38, respectively.
- The calculated pKa values of cimetidine, and nizatidine are 6.8. Ranitidine and famotidine pKa values are 8.1 and 8.4, respectively.
- The oral bioavailability of H_2-selective antihistamines is moderate. At 70%, nizatidine shows the highest oral bioavailability and famotidine has the lowest at 40 to 45%.
- Half-lives are in the 1.5- to 4-hour range.

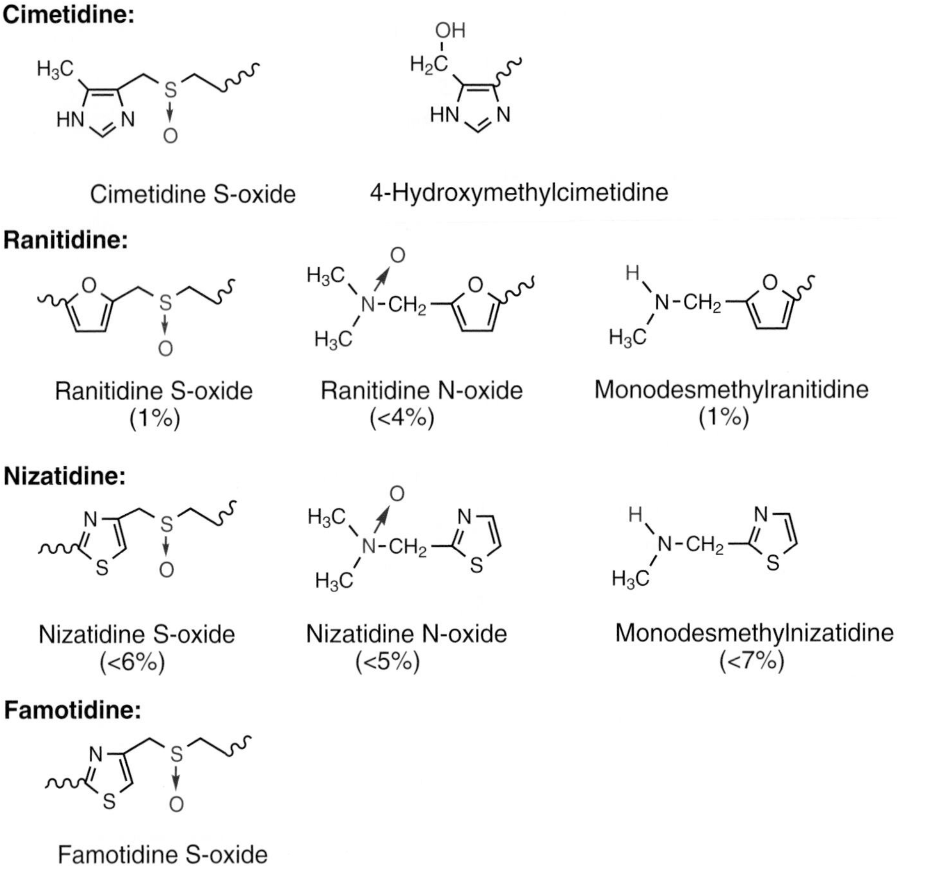

Cimetidine:

Cimetidine S-oxide

4-Hydroxymethylcimetidine

Ranitidine:

Ranitidine S-oxide
(1%)

Ranitidine N-oxide
(<4%)

Monodesmethylranitidine
(1%)

Nizatidine:

Nizatidine S-oxide
(<6%)

Nizatidine N-oxide
(<5%)

Monodesmethylnizatidine
(<7%)

Famotidine:

Famotidine S-oxide

Figure 22.17 Metabolites of H$_2$-selective antihistamines.

DRUG–DRUG INTERACTIONS

Chemical Basis for Drug–Drug Interactions for H$_2$-Selective Antihistamines

- Cimetidine's imidazole N$^\pi$ competes with a CYP His residue for cytochrome heme ferrous ion (Fe^{2+}). This complexation results in reversible, but significant inhibition of many CYP isoforms, including 3A4, and sets the stage for potentially serious drug–drug interactions.
- Ranitidine (furan-containing) is a weak CYP inhibitor while famotidine and nizatidine (thiazole ring) do not inhibit these metabolizing enzymes.
- All H$_2$-selective antihistamines increase gastric pH and interfere with coadministered drugs that depend upon strong gastric acidity for absorption (e.g., iron salts, cephalosporins, ketoconazole).

Metabolism (Fig. 22.17)

- Between 30% and 60% of the dose of each H$_2$-selective antihistamine is excreted unchanged in urine. Cimetidine, ranitidine, and famotidine undergo first pass biotransformation to a greater extent than nizatidine.
- Ranitidine and nizatidine undergo identical biotransformation reactions, although the fraction of the dose that is vulnerable to each differs.
- Sulfoxidation is common to all H$_2$-selective antagonists. Methyl hydroxylation is unique to cimetidine.

Specific Drugs

Cimetidine

- OTC cimetidine's 2-hour half-life requires frequent administration for continuous control of gastric hyperacidity. It is often administered with meals and at bedtime.
- Cimetidine increases prolactin secretion that can result in reversible gynecomastia in patients treated longer than 1 month.
- Elderly patients metabolize and excrete the drug at a slower rate. Reversible sedation and mental confusion have been noted.

Ranitidine

- Ranitidine is available OTC and by prescription in a wide variety of oral dosage forms and strengths. A solution for injection is also marketed.
- The elimination half-life of ranitidine is 2.5 to 3 hours, but it can be prolonged in elderly or others with compromised renal function. The drug is commonly given twice daily or at bedtime.
- Drug–drug interactions with ranitidine are rare, but it can increase serum levels of the short-acting benzodiazepine triazolam. Facilitated absorption in the basified gastric media may be involved.
- Adverse effects are rare, but ranitidine should be immediately discontinued if signs of hepatotoxicity are noted.

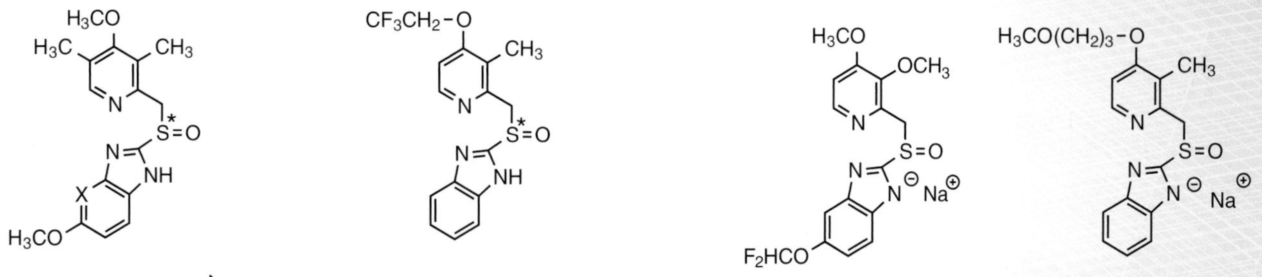

Omeprazole (Prilosec)
Esomeprazole (Nexium) } X = CH
Tenatoprazole - investigational X = N

Lansoprazole (Prevacid)
Dexlansoprazole (Dexilant)

Pantoprazole sodium
(Protonix)

Rabeprazole sodium
(Axiphex)

Figure 22.18 Proton pump inhibitors.

Famotidine

- Famotidine is available OTC and by prescription in several oral dosage forms and one intravenous injectable formulation.
- The 4.5-hour elimination half-life allows the drug to be given once at bedtime or twice daily. Patients taking the drug for hypersecretory disease may require an every 6-hour dosing regimen.
- Famotidine may additionally prolong the QT interval if coadministered with other drugs known to have this effect.

Nizatidine

- Nizatidine is available OTC and by prescription as tablets, capsules, and in oral solution.
- The 1- to 2-hour half-life of nizatidine, along with rapid clearance, protects patients from drug accumulation. It is commonly administered twice daily or at bedtime.
- Coadministration of antacids should be avoided due to a 10% decrease in the intestinal absorption of nizatidine.

Proton Pump Inhibitors (PPIs, Fig. 22.18)

- H^+, K^+-ATPase (the proton pump) is located in the canaliculus, the acid secreting network of the parietal cell.
- PPIs stop the basal secretion of gastric acid, as well as secretion stimulated by histamine, gastrin, and acetylcholine.

Proton Pump Chemistry

- The proton pump catalyzes the exchange of cytoplasmic protons for extracytoplasmic potassium ions.
- Cys^{813} of the catalytic alpha subunit is the residue most critical to the inhibiting action of the PPIs.
  - Cys^{813} can interact with all activated PPIs, regardless of their chemical reactivity.
- PPIs that are more slowly activated have time to reach and react with Cys^{822}, which is buried much deeper within the ATPase protein.
- The covalent disulfide bond formed between PPIs and the proton pump is considered irreversible.
  - Regaining full acid-secreting capability requires the de novo synthesis of new ATPase enzymes.
- Reduced glutathione (GSH) is capable of regenerating the inhibited pump if it can reach the inactivating disulfide bond. Only Cys^{813} is accessible to GSH.
  - GSH stores are limited so regeneration of the proton pump is incomplete.

PPI Chemistry

- The PPI pharmacophore is 2-pyridylmethylsulfinylbenzimidazole. The sulfinyl moiety is chiral, resulting in *R* and *S* isomers.
  - Isomers are equipotent but have different metabolic profiles, impacting bioavailability and plasma clearance.

DRUG–DRUG INTERACTIONS

Cimetidine
- Amiodarone (conduction disturbances, pulmonary toxicity)
- Antacids (decreased absorption of cimetidine due to chelation)
- Benzodiazepines (CNS depression)
- Beta adrenergic blocking agents (hypotension, bradycardia, cardiac arrest)
- Calcium channel blockers (hypotension, bradycardia, cardiac arrest)
- Carbamazepine (blood dyscrasias, Stevens-Johnson syndrome)
- Carmustine (myelosuppression)
- Hydantoins (CNS depression and dysfunction)
- Procainamide (seizure, arrhythmia, blood dyscrasias)
- Serotonin-selective reuptake inhibitors (serotonin syndrome)
- Sulfonylureas (hypoglycemia)
- Warfarin (hemorrhage)

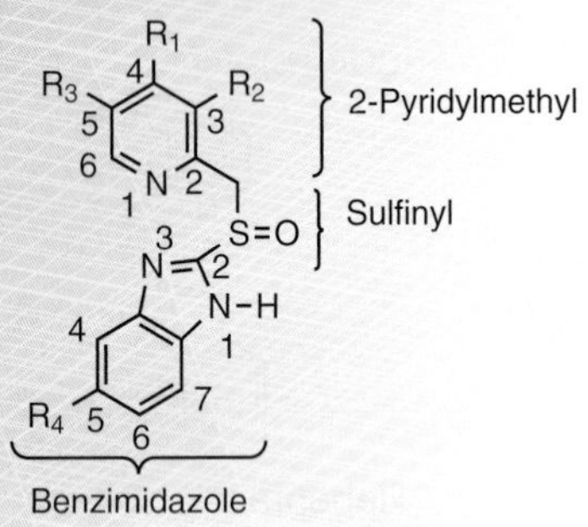

Proton pump inhibitor
pharmacophore

- The sulfinyl moiety is activated through two protonation reactions and a spontaneous rearrangement to generate reactive sulfenic acid and sulfenamide structures. These intermediates form the inactivating disulfide bond with the proton pump Cys residues (Fig. 22.19).
- The pKa of pyridine (3.8 to 4.5) and benzimidazole N_3 (0.11 to 0.79) ensures the unreactive unionized conjugate predominates in the bloodstream (pH 7.4).
- At the canaliculus pH of 1.5, a very small fraction of the conjugate with a cationic benzimidazole and an unionized pyridine moiety (BzH$^+$-Pyr) is generated.
- Any cationic conjugate is trapped within the canaliculus.
- Intramolecular nucleophilic attack by the unionized pyridine at the electrophilic benzimidazole C2 generates the unstable spiro intermediate.
- The spiro intermediate decomposes to generate sulfenic acid and sulfenamide structures that react with proton pump Cys residues.

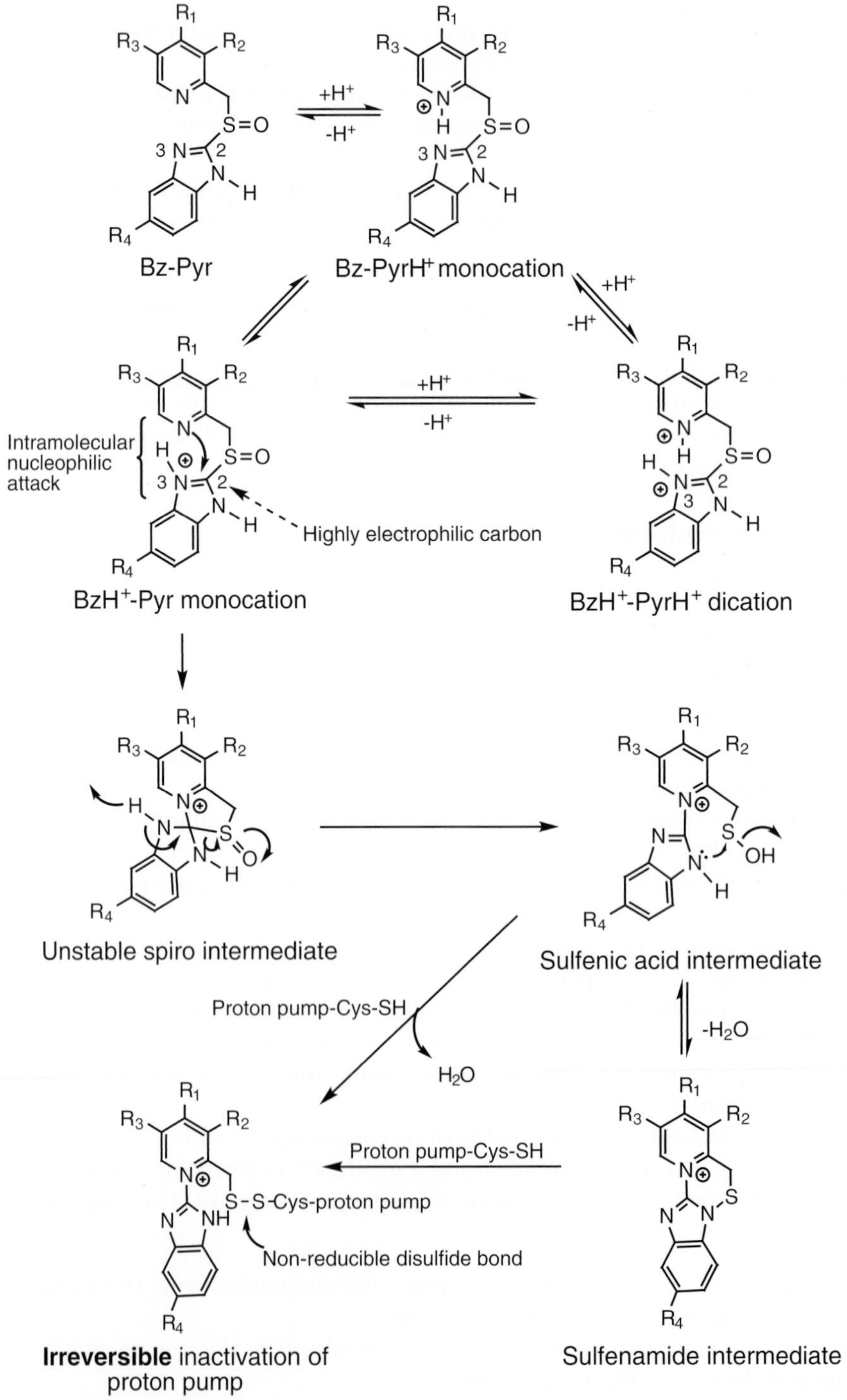

Figure 22.19 Proton pump inhibitor activation. (Reprinted with permission from the American Journal of Pharmaceutical Education. Roche VF. The chemically elegant proton pump inhibitors. *Am J Pharm Educ.* 2006;(5):101.)

SAR

- Electron donating substituents on the pyridine ring increase pyridine nitrogen nucleophilicity and enhance activity.
  - E.g., R_1 = OR; R_2 and/or R_3 = CH_3.
- Electron donating substituents at benzimidazole C5 increase C2 electrophilicity and enhance activity.
  - E.g., R_4 = OCH_3.
- Electron withdrawing substituents at any substitution site decreases activity
  - E.g., R_2 = OCH_3; R_4 = $OCHF_2$.

Physicochemical Properties

- In unionized form, PPIs are very weak bases. They only protonate in strongly acidic media.
- To avoid premature activation in stomach lumen, PPIs must be administered in enteric coated tablets or encapsulated enteric coated granules.
- Raising gastric pH by adding sodium bicarbonate to the formulation eliminates the need for enteric coating and allows faster absorption from the intestine.
- Taking PPIs with food can delay the rate, but not the extent, of absorption.
- Because PPIs act irreversibly, their plasma elimination half-life does not correlate with the duration of therapeutic action.
  - Drugs with a plasma half-life of 2 hours provide antisecretory action that lasts beyond 24 hours.
  - Gastric acid secretory recovery after drug discontinuation takes days.
- PPIs are extensively protein bound (95% to 98%).

Metabolism (Figs 22.20 to 22.23)

- PPIs are extensively metabolized. Excretion is predominantly renal. Fecal excretion is significant for lansoprazole and (to a lesser extent) omeprazole.
- Polymorphic CYP2C19 metabolizes all PPIs. Rabeprazole shows the lowest dependence on CYP2C19 metabolism.
  - CYP2C19-generated metabolites are inactive.
  - Extensive CYP2C19 metabolizers may appear as "nonresponders" to some PPI therapy.
- *S*-Omeprazole (esomeprazole) is biotransformed by CYP2C19 at about one-third the rate of the *R* isomer. Racemic omeprazole and esomeprazole also inhibit CYP2C19.
- *R*-Lansoprazole (dexlansoprazole) provides more consistent activity in poor and extensive CYP2C19 metabolizers compared to the *S* isomer or the racemic mixture.

DRUG–DRUG INTERACTIONS

Proton Pump Inhibitors
- Azole antifungals (decreased absorption)
- Cephalosporin antibiotics (decreased absorption)
- Clarithromycin (increased absorption, decreased PPI metabolism, enhanced healing of *H. pylori*-induced ulcers)
- Clopidogrel (inhibition of clopidogrel to active metabolite; increased risk of stroke and/or myocardial infarction; omeprazole and esomeprazole)
- Diazepam (decreased clearance by 25% to 50%; omeprazole and esomeprazole)
- Digoxin (increased absorption)
- Iron salts (decreased absorption)
- Protease inhibitors (decreased absorption)
- Salicylates (increased enteric-coated tablet dissolution leading to an increase in gastric side effects)
- Theophylline (increased absorption from sustained release formulations)
- Vitamin B12 (decreased absorption)
- Sulcralfate (delayed lansoprazole/dexlansoprazole absorption)

ADVERSE EFFECTS

Proton Pump Inhibitors
- Increased risk of bone fracture (hip, wrist, and spine) with use greater than 1 year.
- *Clostridium difficile*–associated diarrhea in hospitalized patients
- Headache
- Generalized gastrointestinal distress

CHEMICAL NOTE

Clopidogrel–PPI Interaction: Clopidogrel, an antiplatelet prodrug requiring CYP2C19 activation, is commonly administered to patients who have suffered a stroke or who are at risk of an ischemic event. Patients on clopidogrel therapy are at higher risk for gastrointestinal ulceration and may be prescribed a PPI. Omeprazole and esomeprazole are strong CYP2C19 inhibitors and, by inhibiting clopidogrel activation, put patients at risk for a fatal cardiovascular event. If a PPI is indicated, rabeprazole (which does not inhibit and is not significantly metabolized by CYP2C19) or lansoprazole is the best choice. Omeprazole and esomeprazole should be strictly avoided.

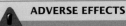

Omeprazole/Esomeprazole competition or inhibition

Clopidogrel (antiaggregatory prodrug) — 2C19 → 2-Oxoclopidogrel — 2C19 → Active metabolite

Clopidogrel activation

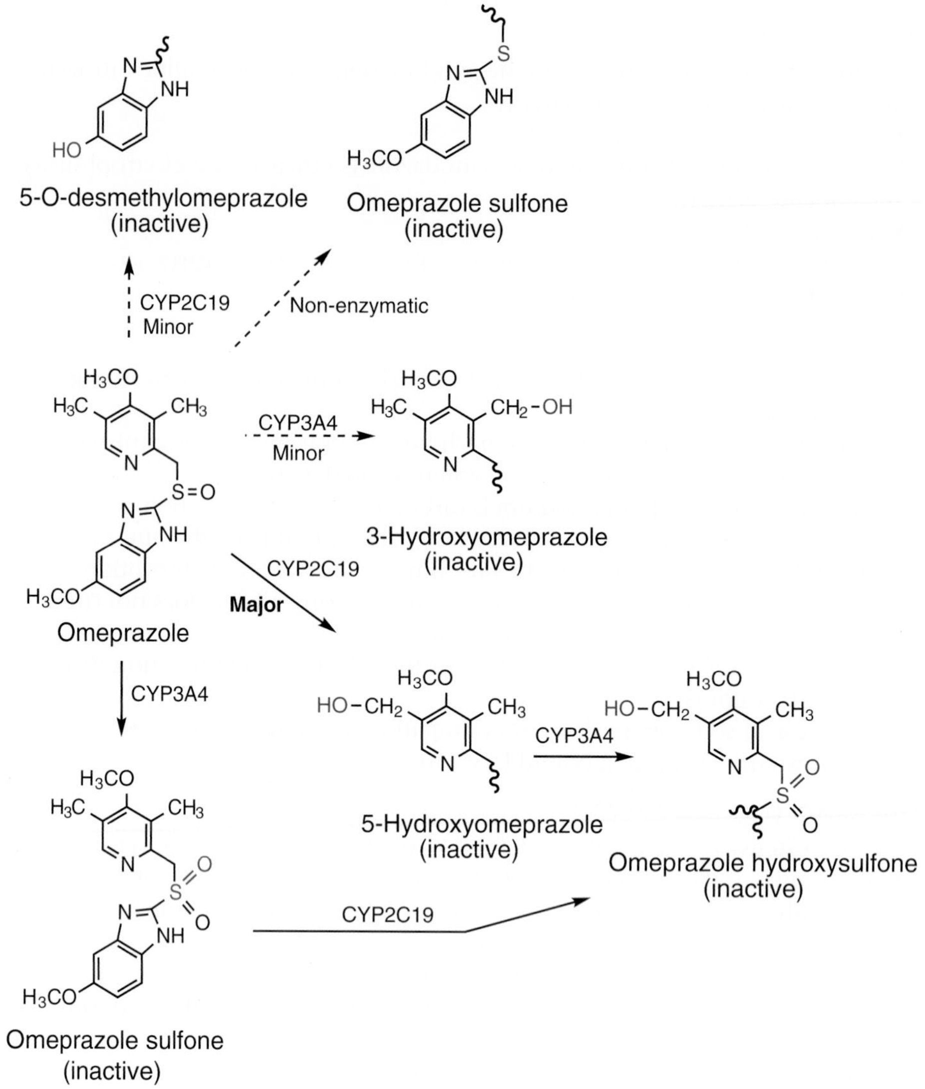

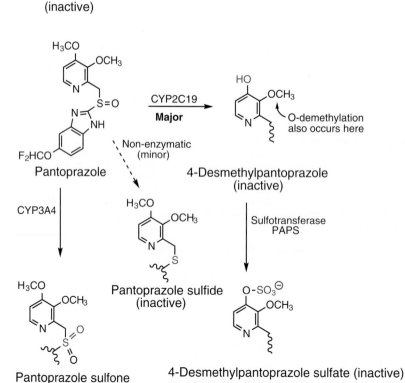

Figure 22.20 Omeprazole metabolism. (Reprinted with permission from the American Journal of Pharmaceutical Education. Roche VF. The chemically elegant proton pump inhibitors. *Am J Pharm Educ.* 2006;70(5):101.)

Figure 22.21 Pantoprazole metabolism. (Reprinted with permission from the American Journal of Pharmaceutical Education. Roche VF. The chemically elegant proton pump inhibitors. *Am J Pharm Educ.* 2006;70(5):101.)

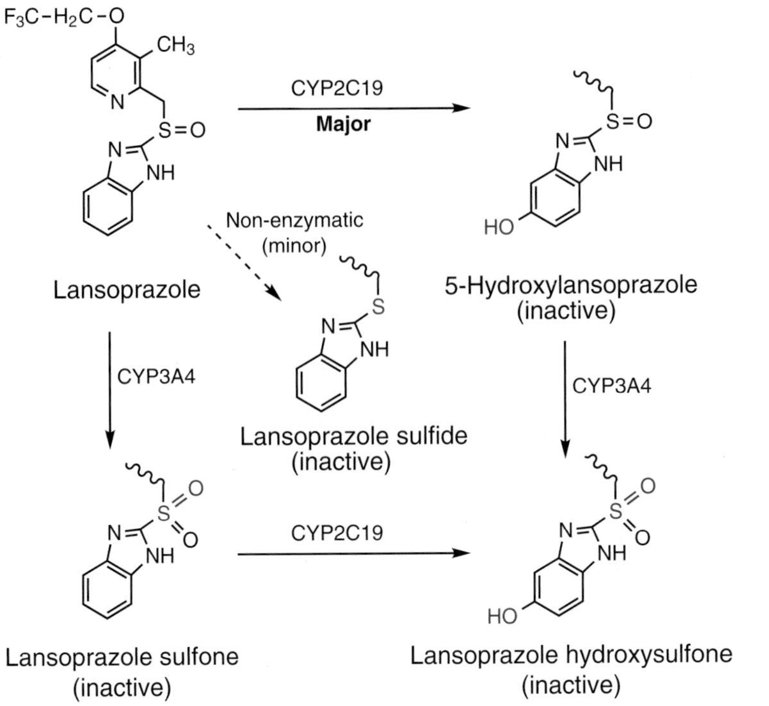

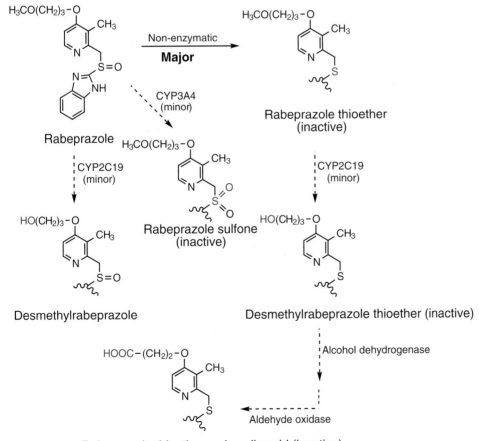

Figure 22.22 Lansoprazole metabolism. (Reprinted with permission from the American Journal of Pharmaceutical Education. Roche VF. The chemically elegant proton pump inhibitors. *Am J Pharm Educ.* 2006;70(5):101.)

Figure 22.23 Rabeprazole metabolism. (Reprinted with permission from the American Journal of Pharmaceutical Education. Roche VF. The chemically elegant proton pump inhibitors. *Am J Pharm Educ.* 2006;70(5):101.)

Clinical Application

- PPIs are potent therapeutic agents in the treatment of erosive esophagitis (GERD), gastric or duodenal ulcers, and in the eradication of ulcer-inducing *H. pylori*.

Specific Drugs

- PPIs are available as enteric coated tablets, capsules, or IV solutions.
- Only omeprazole and lansoprazole are available OTC.

Omeprazole and Esomeprazole

- A nonenteric coated combination product of omeprazole and sodium bicarbonate is available. The increase in gastric pH prevents premature decomposition.
- Reactivity is high and Cys[813] is the targeted proton pump residue. The 20-hour duration is shorter than would be anticipated without GSH rescue.

Lansoprazole and Dexlansoprazole

- The negative inductive effect of the 4-trifluoroethoxy substituent decreases pyridine nucleophilicity compared to all other PPIs except pantoprazole.
- Reactivity is still high and Cys[813] is the inactivated residue, leading to a 20-hour duration.

Pantoprazole Sodium

- The benzimidazole 5-difluoromethoxy substituent decreases pantoprazole reactivity. A significant fraction interacts with Cys[822], which is resistant to GSH rescue.
- The 47-hour duration is significantly longer than observed with other PPIs.

Rabeprazole Sodium

- Rabeprazole's pyridine nitrogen has the highest pKa (4.53) of all marketed PPIs, resulting in extensive trapping in the parietal cell (10 times that of omeprazole).
- The unionized pyridine's high nucleophilicity results in a 10-fold faster formation of the sulfenic acid and sulfenamide intermediates. Onset of action is approximately 1 hour.
- Rabeprazole exhibits the highest level of gastric acid secretion control of all PPIs within the first 24 hours of therapy.
  - In combination with amoxicillin and clarithromycin to treat *H. pylori*-induced ulcers, rabeprazole produces positive results in 7 days, compared to 10 or 14 days for omeprazole and lansoprazole, respectively.

Tenatoprazole (Investigational)

- Tenatoprazole is the *S*-imidazopyridine derivative of esomeprazole.
- Imidazopyridines are activated at a slower rate than benzimidazole-containing PPIs. This permits reaction with Cys[822] of the ATPase. Pump inactivation is GSH-insensitive, leading to a long duration.
- Tenatoprazole's 7- to 9-hour plasma half-life is longer and its potency is higher than esomeprazole. It is claimed to offer superior control of nighttime hyperacidity.
- *S*-tenatoprazole is metabolized predominantly by CYP3A4. This is in contrast to the *R* enantiomer, which shows a greater dependence on CYP2C19.

Prokinetic Agents

- Prokinetic agents (GI stimulants) facilitate gastric emptying through the promotion of peristalsis.
- 5-HT$_4$ receptor stimulation in the gut promotes acetylcholine release, which induces gastrointestinal smooth muscle contraction.
- Metoclopramide and dexpanthenol are marketed in the United States. Prokinetic agents used internationally and/or in clinical trials are shown in Chapter 32, Table 32.12 in *Foye's Principles of Medicinal Chemistry, Seventh Edition*.

Metoclopramide

- Metoclopramide is given orally or via injection in diabetic gastroparesis, gastroesophageal reflux (GERD), or to control nausea and vomiting secondary to cancer chemotherapy.
- In addition to 5-HT$_4$ agonism, metoclopramide antagonizes dopamine D$_2$ and 5-HT$_3$ receptors. The D$_2$ antagonist action puts patients at risk for the development of extrapyramidal adverse effects, including irreversible tardive dyskinesia.
  - Treatment should be restricted to no longer than 3 months.
  - Discontinue at the first signs of involuntary and uncontrollable muscle movement (e.g., rhythmic tongue protrusion, facial grimacing).

Dexpanthenol

- Dexpanthenol is a precursor of coenzyme A (CoASH), which is used in the biosynthesis of acetylcholine.
- Dexpanthenol is administered IM for the treatment of paralytic or post-operative paralytic ileus and intestinal atony.

Metoclopramide   Dexpanthenol

Insoluble Gastrointestinal Protectants

- Sucralfate is a complex of sucrose sulfate and the antacid aluminum hydroxide. The sulfate anions complex with exposed proteins of gastric ulcers and protect tissue from further damage by pepsin and bile acids.
- Sulcralfate acts locally and can inhibit absorption of coadministered drugs (e.g., H$_2$-selective antagonists, quinolone antibiotics).
- Like sucralfate, bismuth subsalicylate (e.g., Pepto-Bismol) coats ulcerated tissue to provide a physical barrier that shields the injury from irritants found in the stomach.

R = SO$_3$[Al$_2$(OH)$_5$]

Sucralfate   Bismuth subsalicylate

Chemical Significance

The Center for Disease Control and Prevention estimated that 17.6 million adults were diagnosed in 2012 with allergic rhinitis, representing 7.5% of the US population at that time. The percentage of children who presented with respiratory, food, and skin-related allergies during this period was 10.6, 5.6, and 12, respectively. The Asthma and Allergy Foundation of America claim that 50 million US citizens (approximately 20%) suffer from one or more types of allergy. They list allergy as the fifth leading chronic disease among adults and the third among children less than 18 years of age. The American Academy of Allergy, Asthma and Immunology reports that, world-wide, the incidence of allergic diseases is on the rise. This includes potentially fatal allergic responses to drugs.

Likewise, peptic ulcer disease in the United States is increasing, due primarily to environmental factors such as stress, smoking, *H. pylori* infection, and the use of nonsteroidal anti-inflammatory drugs (NSAIDs). It has been estimated that 50% of US citizens older than 60 years of age show evidence of *H. pylori*. In the United States, the cost of peptic ulcer disease in terms of hospitalization, nondrug related outpatient care, and lost work is in the range $5.65 billion per year.

Fortunately, effective drug therapy is available (often OTC) to help patients manage symptoms associated with these distressing disorders and, in some cases, effect healing. However, patients and prescribers will require the wise counsel of pharmacists to help them select the medication(s) that best meet therapeutic needs and ensure their safe and appropriately use. For example:

- Does a patient requiring an antihistamine for seasonal allergies drive a truck cross-country for a living? Make sure she selects a nonsedative agent, such as fexofenadine (Allegra), cetirizine (Zyrtec), or desloratadine (Clarinex). Dispersible or chewable tablets could be recommended for convenience on the road.

- Is a patient who will soon be cruising the Caribbean on his honeymoon worried about motion sickness? Suggest a nonselective ethylenediamine (e.g., meclizine), as they are generally of lower sedation than antiemetic ethanolamine ether antihistamines. Guide the patient away from second generation (nonsedative) antihistamines since, as peripherally selective agents, they do not have an antiemetic effect.

- Is a patient with problematic seasonal allergies and exercise-induced asthma planning a hiking vacation in the Rockies during the spring when the colorful mountain flowers are at their peak? Proactively inhibiting histamine release with cromolyn sodium might be an appropriate approach to helping her enjoy her time away.

- Does a patient on a number of other CYP-metabolized medications bring a box of OTC Tagamet (cimetidine) to the pharmacy counter for purchase? Direct him away from this potent CYP inhibitor and toward an OTC H_2-selective antagonist that does not contain an imidazole ring (e.g., famotidine).

- Is a patient who has safely taken cimetidine for periodic gastrointestinal upset purchasing a box of Mylanta antacid tablets? Counsel this patient to separate the doses by several hours so as to avoid cimetidine–aluminum and magnesium complexation that would decrease the therapeutic effectiveness of the H_2-selective antihistamine.

- Does a patient with chronic hypersecretory disease ask for the longest-acting PPI she can get? Consider recommending that the patient talk with her physician about pantoprazole (Protonix), as it provides the highest degree of non-GSH reversible action by forming a disulfide bond with inaccessible Cys^{822} rather than the more easily reached Cys^{813}.

- Has a GERD patient known to be an extensive CYP2C19 metabolizer complained that he is getting no relief from his OTC omeprazole? Consider contacting the patient's physician to recommend the pure *S* isomer, esomeprazole, which is not as dependent on this isoform for inactivation. Alternatively, rabeprazole, which is not extensively metabolized by any CYP isoform, could be recommended.

- Has a patient taking prescription esomeprazole for severe erosive esophagitis suffered a stroke and now requires clopidogrel therapy? Contact the physician to stop the CYP2C19-inhibiting esomeprazole and evaluate the patient's need for ongoing PPI therapy. If it is warranted, select a PPI that is not an inhibitor of this isoform (e.g., rabeprazole).

Clearly the pharmacist's expertise, including an in-depth understanding of drug chemistry, will reap significant therapeutic and patient safety benefits in the clinical setting. Conscientious and proactive counseling on medications patients may mistakenly view as uncomplicated will help drive home the importance of engaging the pharmacist as a pivotal member of the care team.

Review Questions

1 Which of the H_1-selective antihistamine drawn below carries the highest risk of sedation?

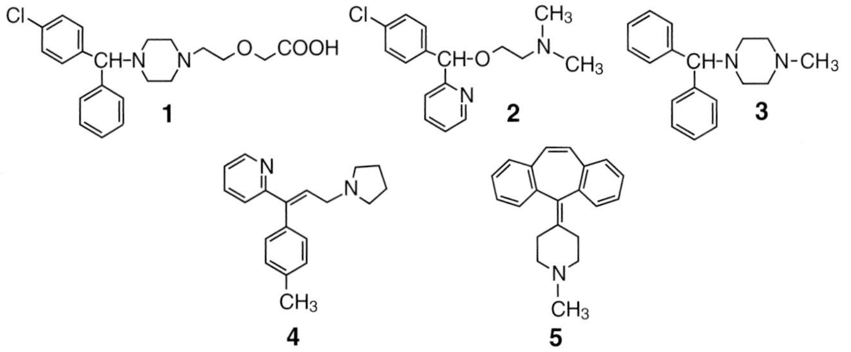

A. Antihistamine **1**
B. Antihistamine **2**
C. Antihistamine **3**
D. Antihistamine **4**
E. Antihistamine **5**

2 Which of the H_1-selective antihistamine would provide the most effective antiemetic therapy in a patient who needs to remain alert while taking it?

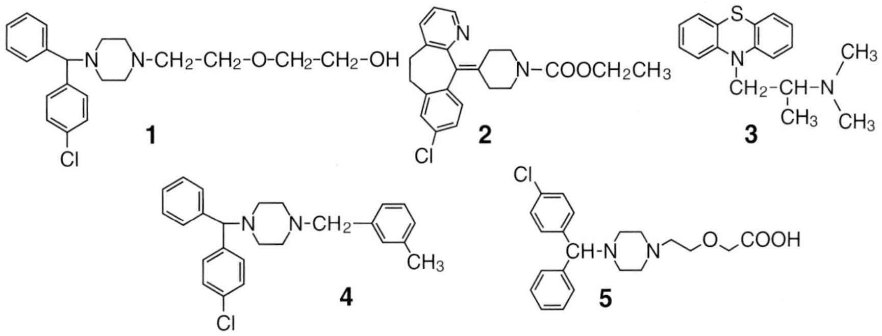

A. Antihistamine **1**
B. Antihistamine **2**
C. Antihistamine **3**
D. Antihistamine **4**
E. Antihistamine **5**

3 Which H_1-selective antihistamine has the highest affinity for H_1 receptors?

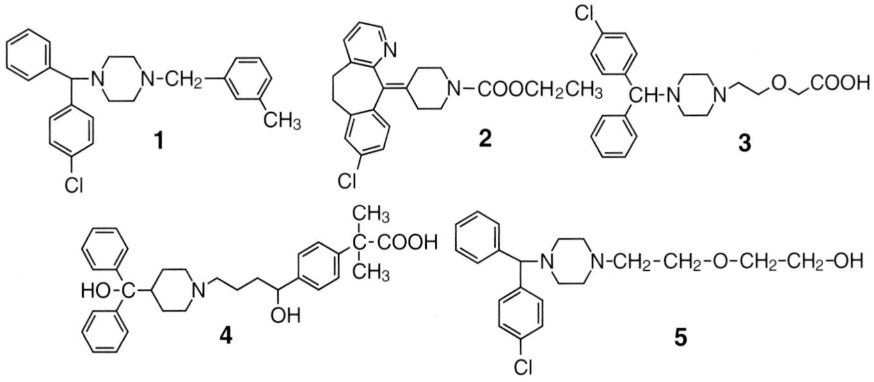

A. Antihistamine **1**
B. Antihistamine **2**
C. Antihistamine **3**
D. Antihistamine **4**
E. Antihistamine **5**

4 Which H_2-selective antihistamine is most likely to precipitate a potentially serious interaction when coadministered with CYP-vulnerable drugs?

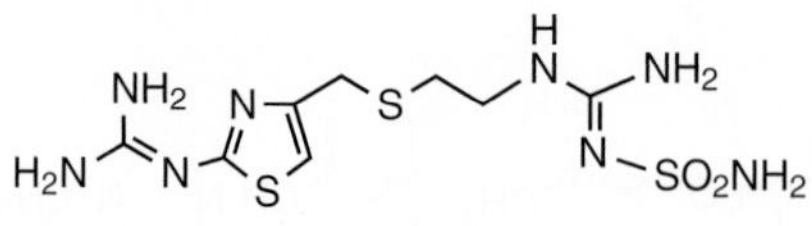

1

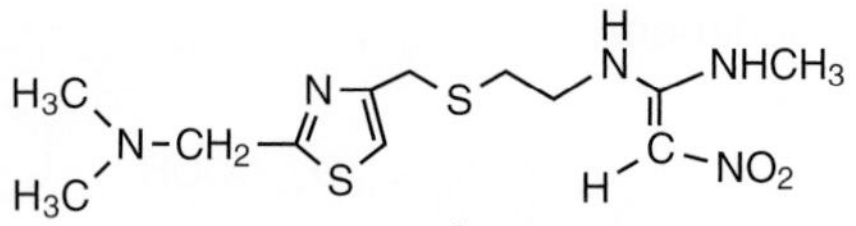

2

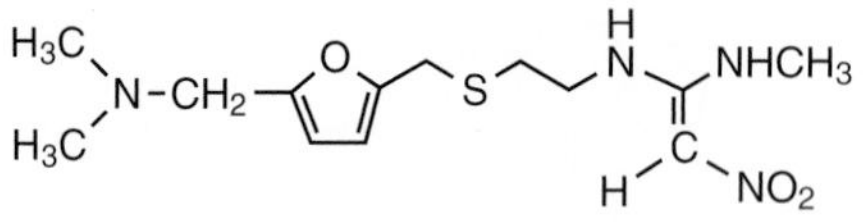

3

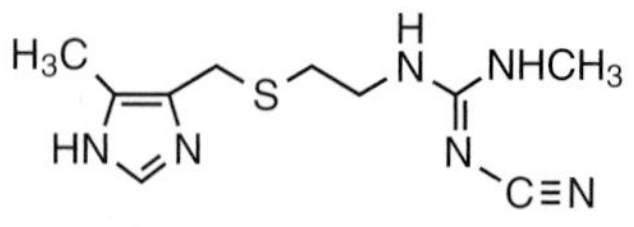

4

A. Antihistamine **1**
B. Antihistamine **2**
C. Antihistamine **3**
D. Antihistamine **4**

5 Which proton pump inhibitor would be most likely to form disulfide bonds with the GSH-inaccessible Cys^{822} of H^+, K^+-ATPase?

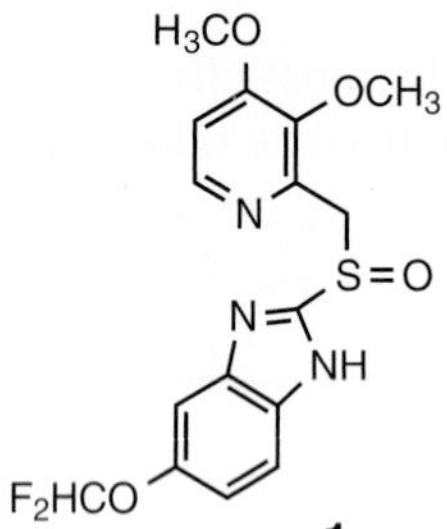

1

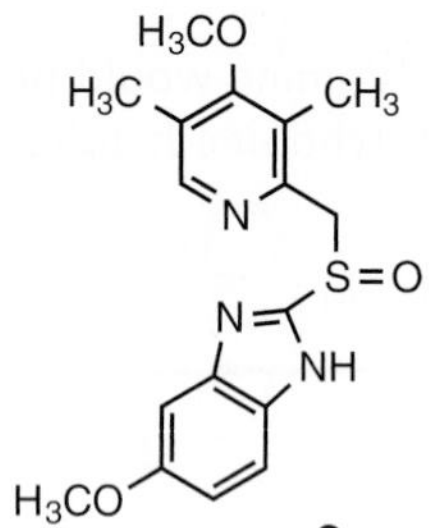

2

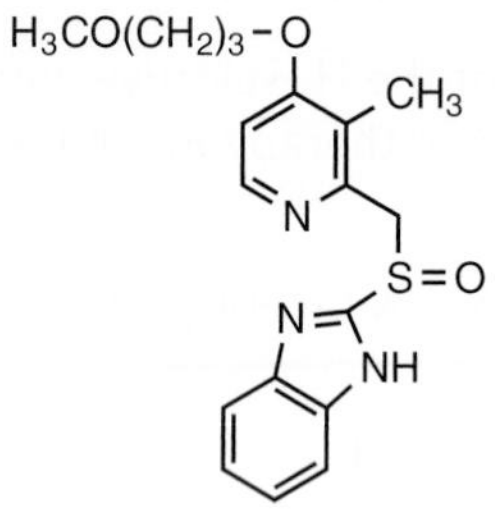

3

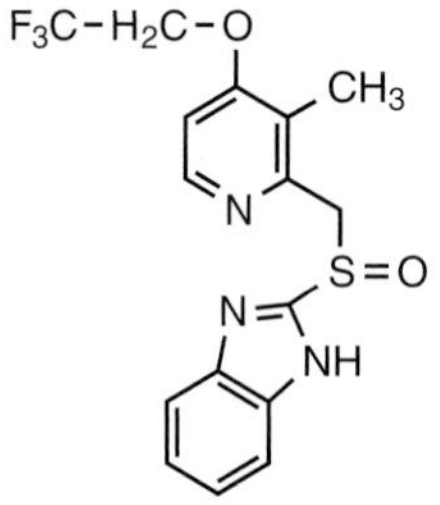

4 (racemic)

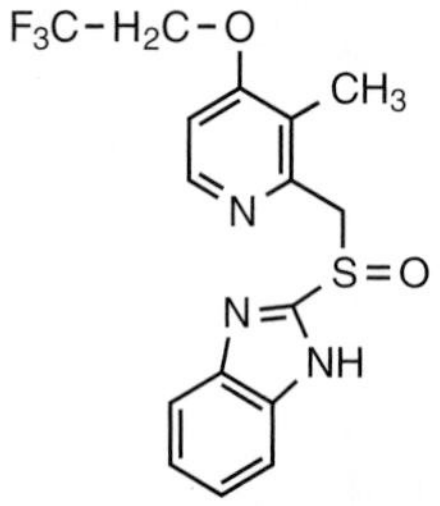

5 (*R* isomer)

A. PPI **1**
B. PPI **2**
C. PPI **3**
D. PPI **4**
E. PPI **5**

Antimicrobial Agents

23

Part I
TREATMENT OF GENERAL BACTERIAL INFECTIONS

Learning Objectives

Upon completion of this chapter the student should be able to:

1 Describe the sites and mechanisms of action of the antibacterial and antibiotic agents:
   a. Sulfonamides and trimethoprim
   b. Fluoroquinolones
   c. Nitrofuran
   d. Miscellaneous antibacterial drugs (metronidazole, methenamine, phosphomycin)
   e. β-Lactams (penicillins, cephalosporins, and other β-lactams)
   f. Aminoglycosides
   g. Macrolides
   h. Lincosamide (clindamycin)
   i. Tetracyclines
   j. Amphenicols (chloramphenicol)
   k. Peptide antibiotics (vancomycin, streptogramins, bacitracin, daptomycin)
   l. Linezolids.

2 Discuss the structure activity relationship (SAR) of antimicrobial agents.

3 Discuss the clinically relevant physicochemical and pharmacokinetic properties of the antimicrobial agents including in vitro and in vivo chemical instability.

4 Discuss clinically significant metabolism, both human and microbial, of the various classes of antimicrobial agents.

5 Discuss the mechanisms of bacterial resistance to the various antimicrobials and, where known, draw representative structures of the metabolic products responsible for resistance.

6 Identify chemically derived clinically significant adverse effects.

7 Broadly compare the clinical applications of the various antimicrobial drug classes.

Definitions

Chemotherapy is a branch of pharmacodynamics, which is differentiated from general drug therapy in that the goal of chemotherapy is to destroy pathogenic cells in the presence of normal human cells. Specific examples of the application of chemotherapy include antibacterial, antifungal, antiparasitic, antiviral, and anticancer drugs. There is a unique and significant difference in the underlying diseases for which chemotherapy is prescribed. Bacteria, fungus, and the various human parasites infections involve self-contained cellular organisms, while viruses are strands of genetic information (e.g., RNA or DNA), which rely on human cellular machinery for multiplication and infection. Cancer cells are actually human undifferentiated cells, which grow and divide in an unregulated manner. The basis for drug design in chemotherapy is to identify uniqueness in the pathogenic cell, which differentiates it from the host cell such that the pathogenic cell can be selectively destroyed. While this has been quite successful over the years with bacterial, fungal, and human parasitic infections, the same cannot be said for viral or cancer diseases due to the close relationship between the pathogenic cell and the normal human cell.

A common approach used in chemotherapy involves designing drugs that work against metabolic processes unique to the pathogenic cell. Such drugs are referred to as antimetabolites, and as such may produce reversible or irreversible blocking actions on the unique metabolic process of the pathogen. The result of these actions is that the drugs are either capable of slowing the growth or killing the pathogenic cell. In the former case, these agents are referred to as "static" agents and inhibit the growth of the pathogenic cell, while in the latter case they are said to act as "cidal" agents destroying the cell. Where known, the antimetabolite and its metabolite will be highlighted. For a more detailed discussion of the historical background and development of antibacterial chemotherapeutic agents the reader is referred to Chapter 33 in *Foye's Principles of Medicinal Chemistry, Seventh Edition.*

Synthetic Antibacterial Agents
Sulfonamides

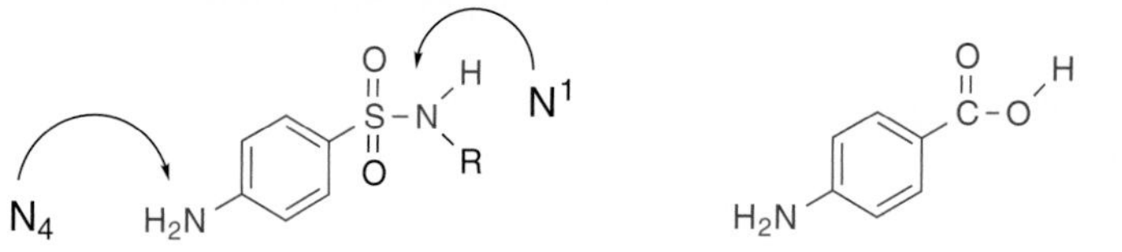

Sulfonamides
(R = H, sulfanilamide)

p-Aminobenzoic acid (PABA)
(bacterial metabolite)

The first marketed antibacterial drugs were the sulfanilamides.

MOA

- The sulfonamides are reversible competitive inhibitors of bacterial biosynthesis of folic acid.
- Sulfonamides inhibit the incorporation of *p*-aminobenzoic acid (PABA) into the precursor of folic acid (Fig. 23.1). Some evidence suggests incorporation of the sulfonamide into bacterial biosynthesis of folic acid in place of PABA.

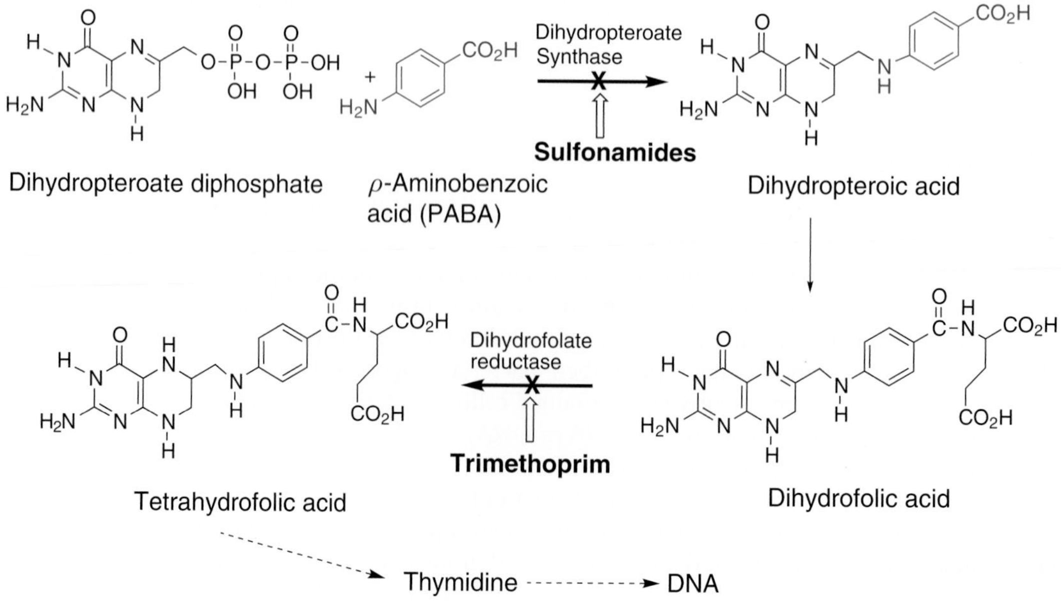

Figure 23.1 Microbial biosynthetic pathway leading to tetrahydrofolic acid and sites of action of antibacterial drugs.

- Selective inhibition results from the fact that human cells require preformed folic acid (a member of the vitamin B complex).
- The sulfonamides are generally considered to be bacteriostatic in their action.

SAR

- The sulfonamide nitrogen must have an attached hydrogen with a pKa similar to that of PABA (~6.5).
- Substitution on N1 with an acetyl group (e.g., sulfisoxazole acetyl) must be hydrolyzed in the gut to release the active drug.
- The aromatic amine must be unsubstituted and located in the 4 position of the aromatic ring.
- No additional substitution on the aromatic ring is permitted.

Physicochemical and Pharmacokinetic Properties

- Water-soluble alkaline salts can be made at the sulfonamide nitrogen.
- The systemic sulfonamides are rapidly absorbed and rapidly excreted into the urine.
- Metabolism occurs via acetylation of the N4 nitrogen.

> **CHEMICAL NOTE**
>
> N4 acetylation reduces drug solubility, which may result in precipitation in the urine leading to crystalluria. Increasing the pH of urine with a systemic alkalizer along with increased water intake will decrease the risk of this potential adverse effect.

Clinical Application

The sulfonamides were the first effective antibacterial drugs discovered in the 1930s. The drugs are effective against both gram-positive and gram-negative organisms, but bacterial resistance and newer more effective drugs have replaced the majority of the previously available sulfonamides. Today, many of the sulfonamides have been discontinued, but some are still available and are primarily limited to treatment of susceptible gram-negative organisms. Products containing sulfonamides are shown in Table 23.1. Sulfisoxazole, in the form of the prodrug

Table 23.1   Clinically Relevant Sulfonamides

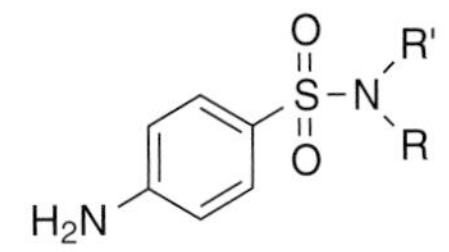

Drug: Generic Name	Product	R	R'	pKa
Sulfisoxazole acetyl (prodrug)	In combination with erythromycin ethylsuccinate	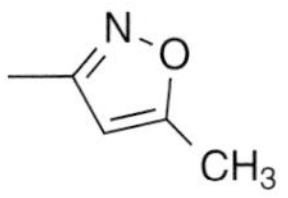 (isoxazole with CH₃)	$-\overset{O}{\underset{\|\|}{C}}-CH_3$	5.6 after hydrolysis
Sulfamethoxazole	In combination with trimethoprim	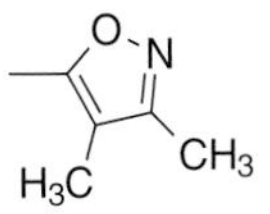 (isoxazole with H₃C, CH₃)	—H	5.0
Sulfadiazine	Oral dosage form	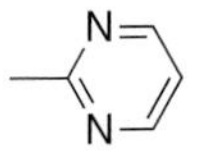 (pyrimidine)	—H	6.52
Silver sulfadiazine	Topical dosage form	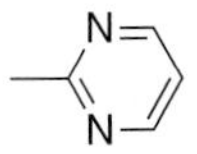 (pyrimidine)	⊖ Ag ⊕	
Sulfacetamide sodium	Ophthalmic dosage form	$-\overset{O}{\underset{\|\|}{C}}-CH_3$	⊖ Na ⊕	5.4 free acid
Sulfasalazine	Gastrointestinal oral dosage form	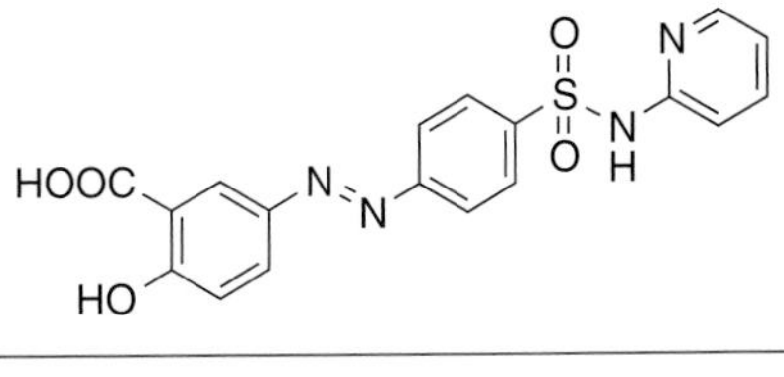 (full structure shown below)		

Antibacterial Spectrum of Sulfonamides

S. pyogenes, S. pneumonia, H. influenza and *Haemophilus ducreyi, Nocardia, Actinomyces, Calymmatobacterium granulomatis, Chlamydia trachomatis,* and in some instances, *E. coli.*

 ADVERSE EFFECTS

Sulfonamides: Hypersensitivity to sulfonamide nucleus (skin and mucous membrane reactions). GI disturbance: nausea, vomiting, and anorexia.

N1-acetylsulfisoxazole, is used in combination with erythromycin ethylsuccinate (EES) and indicated for the treatment of otitis media. Sulfamethoxazole in combination with trimethoprim (see below) is used to treat uncomplicated urinary tract infections, while sulfadiazine when combined with the antiprotozoal agent pyrimethamine is used to treat *Toxoplasma gondii* infections. Silver sulfadiazine is used topically to treat burns, with both the sulfa drug and the silver ion having antibacterial activity. Sodium sulfacetamide is a water-soluble preparation used to treat ophthalmic infections, while sulfasalazine is effective in the treatment of ulcerative colitis. It is only poorly absorbed from the GI tract where it is hydrolyzed by intestinal bacteria giving rise to sulfapyridine and 5-aminosalicylate, both of which may exhibit activity.

Trimethoprim

Trimethoprim (Proloprim, Trimpex)
Trimethoprim + sulfisoxazole (Co-Trimoxazole)
Trimethoprim + sulfamethoxazole (Bactrim, Septra)

Dihydrofolic acid

Antimicrobial Spectrum for Trimethoprim

E. coli, Staphylococcus saprophyticus, P. mirabilis, Salmonella typhi, Klebsiella pneumonia, and *Pneumocystis jiroveci.*

MOA

- A reversible competitive inhibitor of dihydrofolate reductase (Fig. 23.1) and, when used in combination with a sulfonamide, a synergistic response results because of sequential blocking of tetrahydrofolic acid biosynthesis.
- Bacterial and human dihydrofolate reductase have different structural requirements.

Clinical Application

Clinically, trimethoprim can be used singly or in combination to treat various microbial infections associated with urinary tract infections (UTIs) (see combination with sulfamethoxazole in Table 23.1). In addition, the combination drug regimen is useful for treating acute exacerbations of chronic bronchitis and in high doses to treat *Pneumocystis jiroveci,* a common complicating infection seen in AIDS patients.

4-Quinolones

4-Quinolone
(pharmacophore shown in red)

The fluoroquinolones have been found to be effective in treatment of various bacterial infections depending on the nature of the substitution on the 4-quinolone pharmacophore.

MOA

- Irreversible inhibitors of DNA gyrase and topoisomerase IV, key enzymes involved in DNA-dependent RNA polymerase (DDRP).
- Binding of the drug to DNA gyrase involves the carboxyl and the ketone.

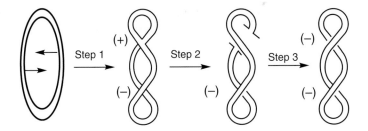

A. View from the top: Step 1. Stabilize positive node. Step 2. Break both strands of the back segment. Step 3. Pass unbroken segment through the break and reseal on the front side.

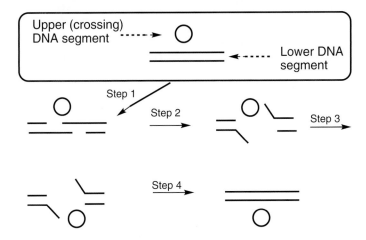

B. View from the side: Step 1. Staggered cuts in each strand. Step 2. Gate opens. Step 3. Transverse segment passed through the break. Step 4. Reseal cut segment.

Figure 23.2 Schematic depicting supercoiling of circular DNA catalyzed by DNA gyrase.

- Bacterial DNA gyrase alters the conformation of the DNA by catalyzing double-strand cuts, passing the uncut portion of the molecule through the break, and resealing the DNA (Fig. 23.2) (supercoiling of DNA).
- Without the process described above, the genetic information is not accessible and bacterial growth is inhibited.

SAR

- The quinolone pharmacophore is essential for activity through binding to the DNA gyrase (Table 23.2).
- R_1 is important for potency and commonly consists of an ethyl or cyclopropyl.
- Fluoro at C6 improves penetration of the bacterial cell wall through improved hydrophobicity.
- Heterocyclic substitution at C7 affects the spectrum of activity against gram-negative bacteria.
- R_8 affects spectrum of activity as does R_1/R_8 linked forming a third ring in the molecule (finafloxacin).

Physicochemical and Pharmacokinetic Properties

- 4-Quinolones are incompatible with heavy metals (e.g., Ca^{2+}, Mg^{2+}, Zn^{2+}, Fe^{2+}, Al^{3+}) due to an insoluble chelate resulting from bonding between the metal and the C3 carboxyl and C4 ketone.
- 4-Quinolones may cause skin phototoxicity upon exposure to the sun (UV A radiation).

Metal chelation

Table 23.2 Second-, Third-, and Fourth-Generation Quinolones

Drug: Generic Name	Trade Name	R_1	 ⟨N (CH₂)x⟩	R_8
Norfloxacin	Noroxin	C_2H_5	 (piperazine)	H
Ciprofloxacin	Cipro	(cyclopropyl)	 (piperazine)	H
Gatifloxacin	Tequin	(cyclopropyl)	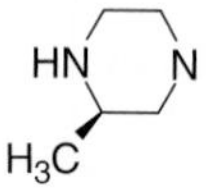 (methylpiperazine)	CH_3-O
Moxifloxacin	Avelox	(cyclopropyl)	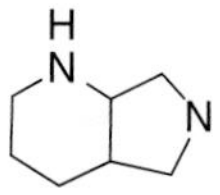 (octahydropyrrolopyridine)	CH_3-O
Gemifloxacin	Factive	(cyclopropyl)	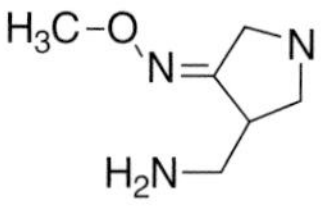 (H_3C-O-N= ... H_2N)	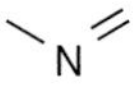 (N=)
Besifloxacin	Besivance	(cyclopropyl)	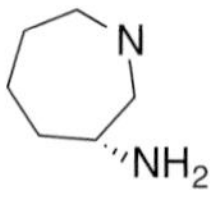 (azepane-NH₂)	Cl
Finafloxacin[a]	Xtoro	(cyclopropyl)	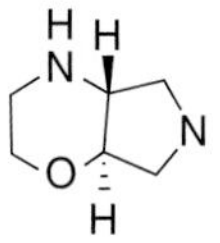 (oxa-pyrrolopyridine)	N≡C

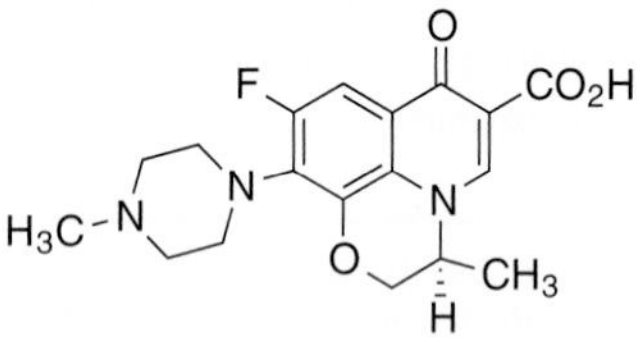

Ofloxacin (Racemic)(Floxin)
Levofloxacin (1-S)(Levaquin)

[a]Recently approved for treatment of otitis externa (swimmer's ear) caused by *P. aeruginosa* and *S. aureus*.

⚠ **ADVERSE EFFECTS**

4-Quinolones: GI disturbance: nausea, vomiting, and abdominal discomfort. CNS effects: headache and dizziness, but may also include hallucinations, insomnia, and visual disturbances due to binding of lipophilic drugs to GABA receptors. Several analogs caused QT prolongation leading to their removal from the market.

Contraindication

With heavy metal antacids or laxatives, where precipitated chelates prevent absorption.

- 4-Quinolones have good bioavailability orally, are moderately bound to plasma protein, and have a half-life of 3 to 12 hours (see Chapter 33 in *Foye's Principles of Medicinal Chemistry, Seventh Edition,* for a complete detailed discussion).
- Distribution of the drug to the CNS, due to hydrophobicity, results in binding to central $GABA_A$ receptors.

Metabolism

- Glucuronidation at the C3 carboxyl group.
- Hydroxylation in the C7 heterocycles.

Clinical Application

The fluoroquinolones represent a potent class of bactericidal agents with utility in a variety of infectious conditions. The most common indications include UTIs caused by sensitive organisms; prostatitis; some sexually transmitted infections; respiratory infections; and bone, joint, and soft tissue infections.

Nitroheteroaromatic Compounds

Nitrofurantoin

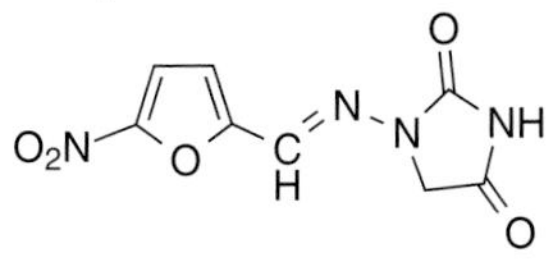

Nitrofurantoin
(Furadantin, Macrodantin)

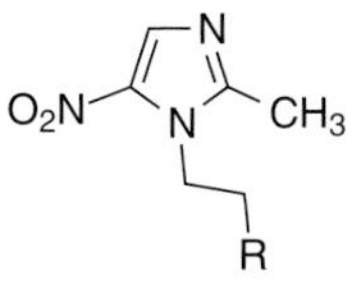

Metronidazole, R = OH (Flagyl)
Tinidazole R = $SO_2C_2H_5$ (Fasigyn)

**Antibacterial Spectrum
4-Quinolones**

E. coli, species of Salmonella, Shigella, *Enterobacter,* Campylobacter, and *Neisseria. S. pneumonia, H. influenza, Moraxella catarrhalis, S. aureus, M. pneumoniae, Chlamydia pneumonia,* and *Legionella.*

- Nitrofurantoin is the only nitrofuran which remains available and is used for treatment of uncomplicated UTIs.

Metronidazole

- Metronidazole and tinidazole are used to treat some bacterial infections (e.g., GI tract peptic ulcer, pseudomembranous colitis) and protozoal infections (e.g., giardiasis, trichomoniasis).

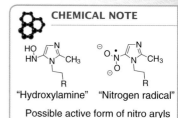

CHEMICAL NOTE

"Hydroxylamine"   "Nitrogen radical"

Possible active form of nitro aryls

MOA

- Most likely, the nitroheteroaromatic compounds are prodrugs in which the nitro group is reduced to the active hydroxylamine or nitrogen radical which interferes with DNA and or RNA.

Methenamine and Phosphomycin

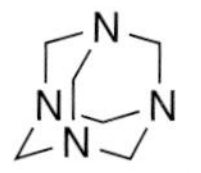

Methenamine
(Prosed, Urimax, Urised, Uroqid-Acid)

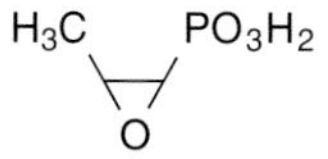

Phosphomycin
(Monurol)

- Both methenamine and phosphomycin have limited values and are used in uncomplicated UTIs.

MOA

- Methenamine is a prodrug, which in acidic urine generates ammonia and formaldehyde. The latter forms a Schiff's base with bacterial protein resulting in antibacterial action.
- Phosphomycin, through alkylation of a key sulfhydryl group in a bacterial transferase essential in cell wall glycoprotein synthesis, inhibits bacterial growth.

Antibiotic Antibacterial Agents

β-Lactams

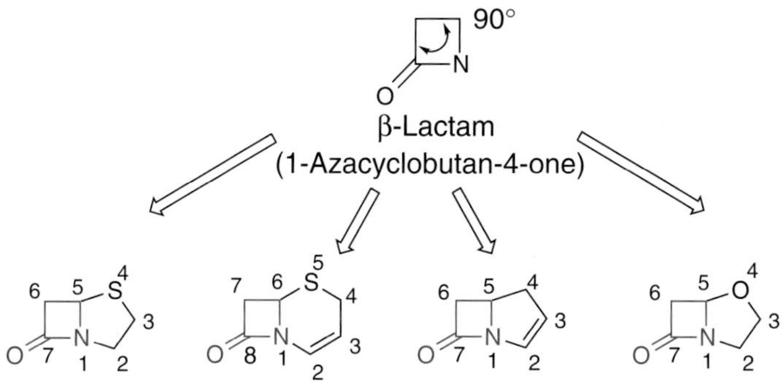

90°

β-Lactam
(1-Azacyclobutan-4-one)

| Penam | Cephem | Carbapenem | Clavam |

The discovery of the β-lactams in the late 1930s represents a milestone in drug therapy, which led to the development of the penicillins (penams), cephalosporins

CHEMICAL NOTE

The intact β-lactam ring of the penicillins, cephalosporins, and other similar derivatives is the key pharmacophore essential to antimicrobial action. The highly strained nature of the β-lactam ring results from its small size.

(cephems), carbapenems, and clavulanic acids (clavams, adjuncts to the β-lactams). The β-lactam (azetidinone ring) with additional ring substituents influences the spectrum of activity of the β-lactams. The β-lactam ring is also the site of action associated with the development of bacterial resistance.

MOA

A common mechanism of action exists for the penicillins, cephalosporins, and carbapenems, and involves the irreversible acetylation of bacterial cell wall transamidase (also referred to as transpeptidase), which carries out cell wall cross-linking.

- Cell wall cross-linking is essential for the integrity of both gram-positive and gram-negative bacterial cell walls and without this cross-linking the bacteria cell is destroyed.
- The effectiveness of the -lactams as bactericidal agents results from the -lactam ring structurally mimicking the D-alanine-D-alanine portion of the peptidoglycan cell wall with which the transamidase (a penicillin-binding protein [PBP]) normally reacts (Fig. 23.3).
- Reaction Pathway A leads to normal cross-linking in the cell wall.
- Reaction **Pathway B** leads to irreversible destruction of acetylated transamidase.
- The β-lactams are suicide substrates of the PBPs. The PBPs are present in the plasma membrane of gram-positive and gram-negative bacteria.

Penicillin β-Lactams

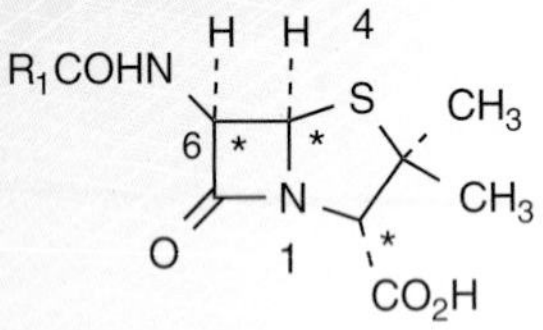

Physicochemical and Pharmacokinetic Properties

- The penicillins consist of the bicyclic penam ring system with stereochemical centers at C2, C5, and C6 positions.
- Salt formation at the C2 carboxyl with organic bases (procaine or benzathine) results in insoluble drugs which serve as drug depots (IM injection) with long durations of action.

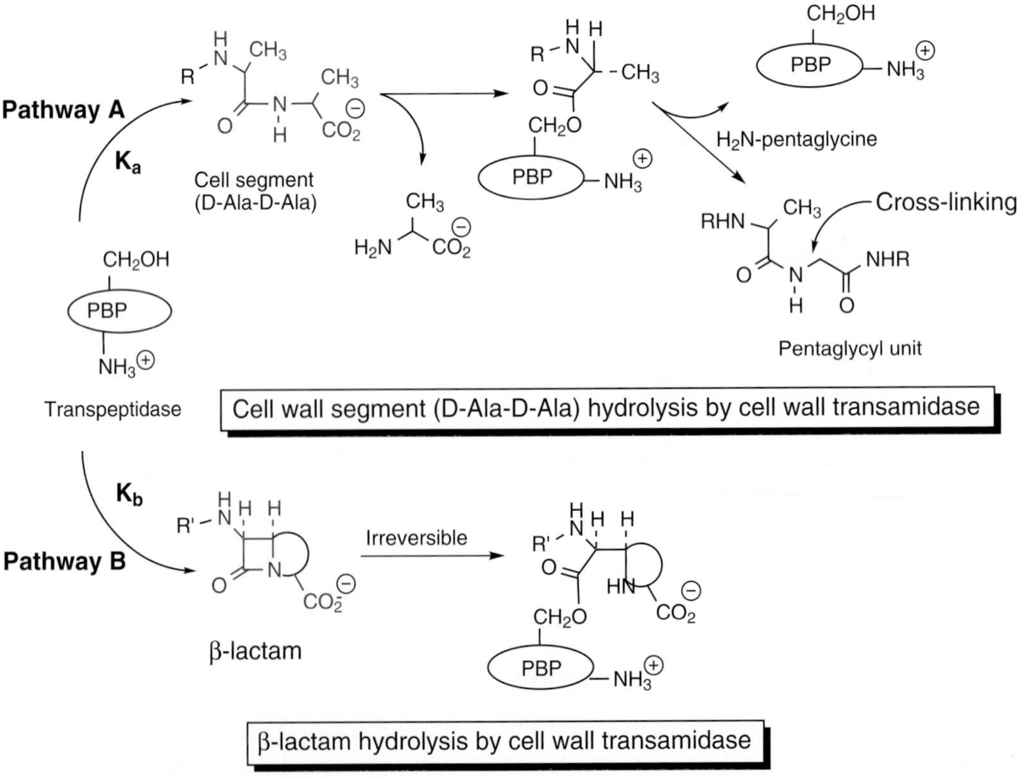

Penicillins

Figure 23.3 Cell wall cross-linking and MOA of β-lactams.

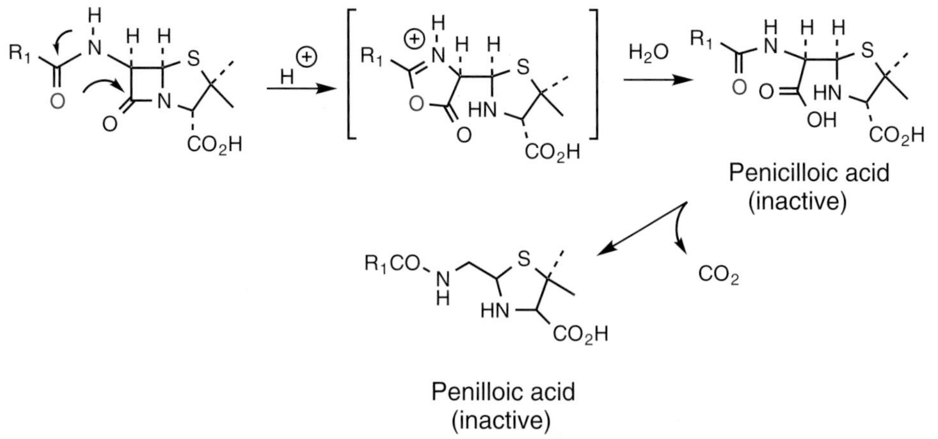

Figure 23.4 Penicillin chemical instability to acid media.

- Salt formation with inorganic bases (KOH) enhances water solubility and dissolution.
- Amide substituents at the C6 position influence in vitro chemical stability and thus, the route of administration (oral or IM), metabolic stability of the β-lactam, and spectrum of activity (i.e., gram-positive and gram-negative bacteria).
- Acid instability of penicillins (i.e., stomach acid) results from a nucleophilic attack at the C6 amide carbonyl (Fig. 23.4). This reaction has clinical significance and affects the route of administration.
- Penicillins exhibit instability in basic media giving rise to penicilloic acid, but the clinical significance of this reaction is minimal.

Metabolism

- The major mechanism of drug resistance to the β-lactams is associated with hydrolysis of the β-lactam ring catalyzed by the bacterial enzyme β-lactamase (penicillinase and cephalosporinase).
- β-lactamase enzymes are produced in the cell wall by both gram-positive and gram-negative organisms. In gram-positive bacteria, the enzyme is secreted into the region surrounding the cell, while in gram-negative bacteria, the enzyme is found within the periplasmic space within the cell wall.
- The β-lactamase–catalyzed hydrolysis product forms from opening the β-lactam, giving rise to inactive penicilloic acid and penilloic acid (Fig. 23.4).

SAR

- Stability to acid media with improved oral absorption results from the addition of electron withdrawing groups in the R_1 substituent at C6, which reduces the electron density around the carbonyl oxygen (e.g., -O- in Pen-V; $-NH_3^+$ and aromatic ring in ampicillin and amoxicillin, but not $-COO^-$ in ticarcillin) (Table 23.3).
- Addition of bulky substituents at R_1 hinders access of penicillinase (a β-lactamase) to the β-lactam carbonyl to reduce β-lactam hydrolysis, thus improving activity of the drug to β-lactamase producing bacteria (Fig. 23.5).
- Addition of an α-ionizable group to R_1 (carboxylate or ammonium) improves activity toward both gram-positive and gram-negative organisms.

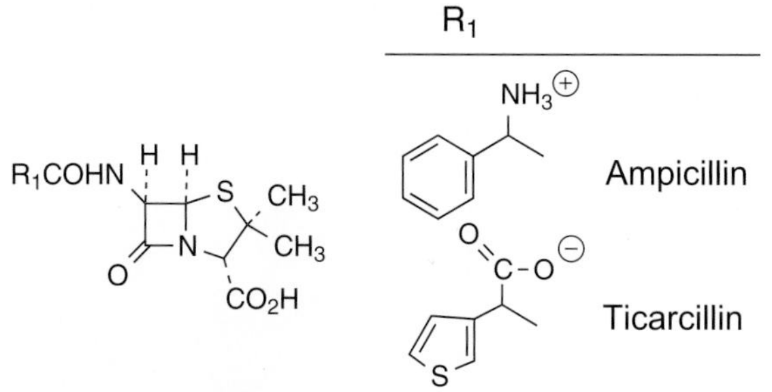

- An ureido group at the α position of R_1, as in mezlocillin and piperacillin, increases the spectrum of activity toward both gram-positive and gram-negative organisms.

Table 23.3   Improved Acid Stability and Absorption of Substituted Penicillins

Drug	R_1	% Absorption Intact Drug
Benzylpenicillin	phenyl—CH_2—	15–30
Pen V	phenyl—O←CH_2—	60–73
Ampicillin	phenyl—CH(NH_3^+)—	30–50
Amoxicillin	HO—phenyl—CH(NH_3^+)—	75–90

Space filling model of Benzylpenicillin

Bulky R_1 produces steric hindrance

β-lactamase   Penicillins

Space filling model of Dicloxacillin

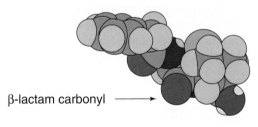

Benzylpenicillin (no hindrance)

β-lactam carbonyl →

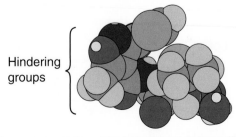

Dicloxacillin (steric hindrance)

Hindering groups

Figure 23.5 Benzylpenicillin prone to β-lactamase–catalyzed hydrolysis and dicloxacillin resistant to β-lactamase–catalyzed hydrolysis.

- The ureido side chain does not protect the β-lactam from penicillinase nor are any of the nitrogen atoms in the ureido basic, and therefore, they are not ionizable at physiological pH.

Mezlocillin
(not available in the United States)

Piperacillin

Clinical Applications

- The benzyl penicillins (penicillin G and penicillin V) are used primarily against susceptible gram-positive cocci (Table 23.4). Infections of the upper and lower respiratory tract and genitourinary tract are commonly treated with these penicillins although resistance is a common problem.

Table 23.4 Commercially Significant Penicillins and Related Molecules

Generic Name	Trade Name	R_1
Benzyl Penicillins (Fermentation-Derived)		
Benzylpenicillin (Penicillin G)	Generic	C_6H_5—CH_2
Phenoxymethylpenicillin (Penicillin V)	Generic	C_6H_5—CH_2
Ampicillins (Semi-synthetic)		
Ampicillin (X = H)	Generic	
Amoxicillin (X = OH)	Amoxil, Lorotid	
Methicillins (Semi-synthetic, Penicillinase Resistant)		
Nafcillin	Nallpen	
Oxacillin (x = y = h)	Bactocill	
Dicloxacillin (x = y = ci)	Generic	
Carboxypenicillins (Semi-synthetic, Broad Spectrum, Parenteral Penicillin)		
Ticarcillin (in combination with clavulanic acid)	Timentin	
Uridopenicillin (Semi-synthetic, Broad Spectrum, Parenteral Penicillin)		
Piperacillin (in combination with tazobactam)	Generic Zosyn	

ADVERSE EFFECTS

Penicillins: Hypersensitivity reactions are the most common adverse effect associated with the penicillins (0.7% to 4%). Indications include maculopapular rash, fever, bronchospasm, dermatitis, Stevens–Johnson syndrome, and the most severe reaction being anaphylaxis. Hypersensitivity reactions are associated with hapten formation resulting from a protein/β-lactam breakdown product.

A penicillin

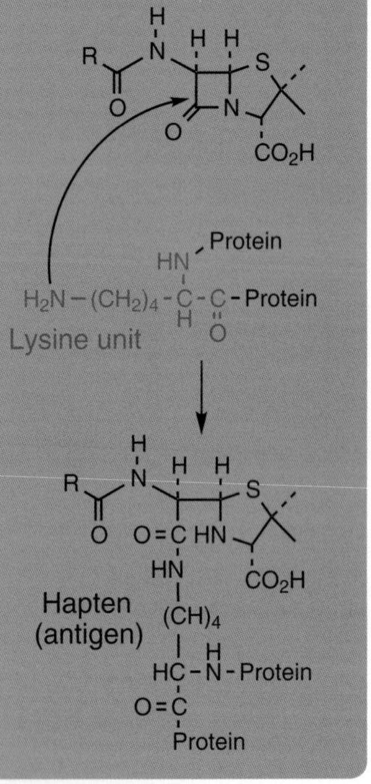

- The ampicillin class of penicillins (i.e., ampicillin and amoxicillin) show increased activity against gram-negative bacteria because of the α-amino substituent. These drugs find benefit in treating *Haemophilus influenza* and *Neisseria gonorrhoeae* infections in addition to common gram-positive infection associated with pneumococcal, streptococcal, and meningococcal infections.
- Drugs of the methicillin class (i.e., dicloxacillin, nafcillin, oxacillin) are valuable to treat penicillinase-producing bacteria, although resistant strains of bacteria are now common (i.e., methicillin-resistant *Staphylococcus aureus* [MRSA]). The methicillin class of drugs is generally less potent than benzyl penicillins against gram-positive bacteria.
- Ticarcillin, a member of the carboxypenicillins, has improved gram-negative activity and can be used in combinations for treatment of *Pseudomonas aeruginosa* infections.
- The ureidopenicillin piperacillin is a broad spectrum antibiotic with activity against both gram-positive and gram-negative bacteria.

β-Lactamase Inhibitors – Clavam, Penams, and Diazabicyclocarbamoyl Sulfate

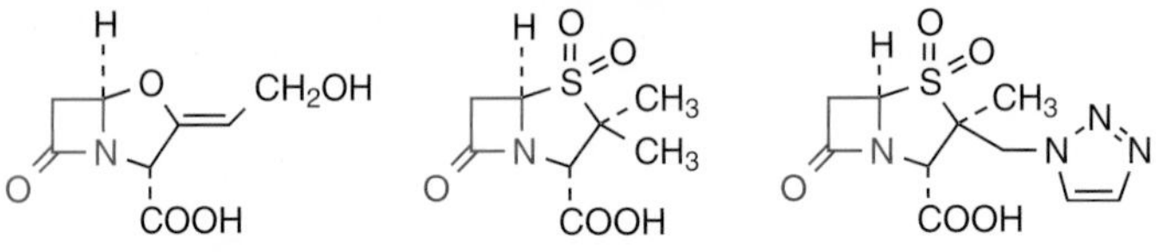

Clavulanic acid   Sulbactam   Tazobactam

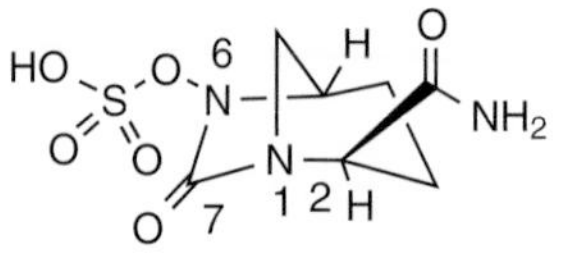

Avibactam

- The clavam (clavulanic acid), the penams (sulbactam and tazobactam), and diazabicyclocarbamoyl sulfate (avibactam) share the property of little or no antimicrobial activity, but all four are strong inhibitors of β-lactamase.
- The pharmacophore of clavulanic acid, sulbactam, and tazobactam is that of the β-lactam which reacts irreversible with an OH nucleophile in the β-lactamase, while avibactam reacts reversibly with presumably the same serine-OH of β-lactamase (Fig. 23.6).

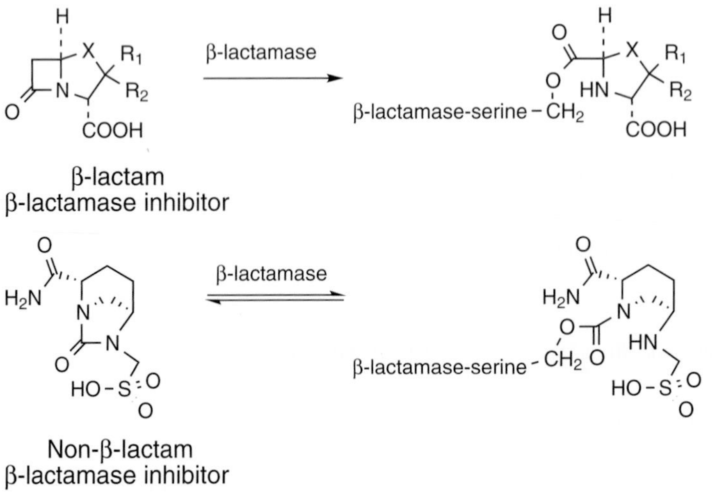

Figure 23.6 MOA of β-lactamase inhibitors.

- The structure of non–β-lactam avibactam overlaps nicely with the pharmaco-phore of the cephalosporin antibiotics.

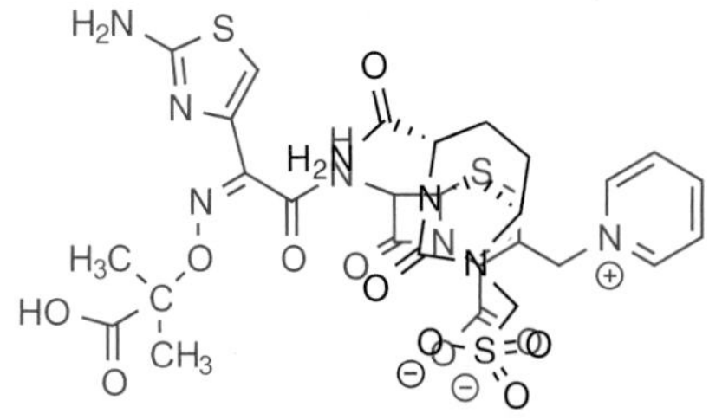

Overlapping similarities between ceftazidime and avibactam. NOTE: C2 carboxylate and N6 sulfate, N7 amide nitrogen and N2 amide nitrogen, and C8 carbonyl of β-lactam and C7 carbonyl of urea in ceftazidime and avibactam, respectively.

- Clavulanic acid, sulbactam, and tazobactam are used in combination with the penicillins to expand the spectrum of action of the penicillin to β-lactamase–producing organisms (e.g., Augmentin [amoxicillin/clavulanic acid], Timentin [ticarcillin/clavulanic acid], Unasyn [ampicillin/sulbactam], and Zosyn [piper-acillin/tazobactam]). The same combination approach is used with cephalo-sporins plus β-lactamase inhibitors (e.g., Zerbaxa [ceftolozane/tazobactam] and Avycaz [ceftazidime/avibactam]) (Table 23.5).
- These combinations increase the activity of the respective β-lactam against gram-positive bacilli which produce various β-lactamase enzymes.

Cephalosporins

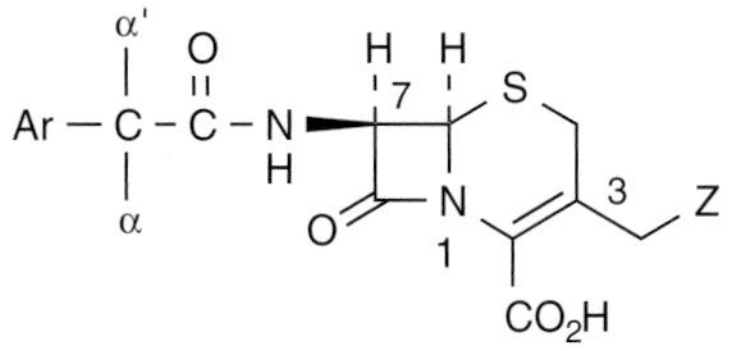

Cephalosporins

Cephalosporins differ from the penicillin by the expansion of the five-member tetrahydrothiazole ring into a six-member dihydrothiazine ring (cephem) and the addition of a substituent at the 3 position. Both of these changes affect the biologic reactivity (see SAR).

MOA

- The mechanism of action for cephalosporins does not differ significantly from that of the penicillins (Fig. 23.3).
- Because of the reduced ring strain in the cephem ring system, the cephalosporins are less reactive toward transpeptidase (i.e., slower K_b) than the penicillins.
- The Z group at C3, which acts chemically as a leaving group, improves the rate of the transpeptidase acylation (K_b, Pathway B) (Figs. 23.3 and 23.7).

CHEMICAL NOTE

The methylthiotetrazole (MTT) heterocycle has been associated with prothrombin deficiency and bleeding prob-lems. In addition, this same heterocycle has been linked to alcohol intolerance.

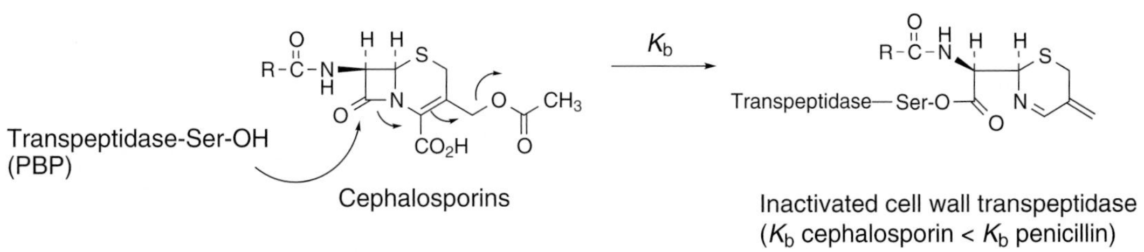

Transpeptidase-Ser-OH (PBP)

Cephalosporins

Inactivated cell wall transpeptidase (K_b cephalosporin < K_b penicillin)

Figure 23.7 Acyla-tion of bacterial cell wall transpeptidase with cephalosporins.

Table 23.5 Cephalosporins

Generic Name	Trade Name	R	X	Y
First Generation				
Cefazolin	Ancef Kefzol Zolicef			H
Cephalexin	Keflex Biocef Keftab		—CH₃	H
Cefadroxil	Cefadroxil		—CH₃	H
Second Generation				
Cefuroxime	Ceftin Kefurox Zinacef		—CH₂OCONH₂	H
Cefoxitin	Mefoxin		—CH₂OCONH₂	OCH₃
Cefotetan	Cefotan			OCH₃
Cefaclor	Ceclor		—Cl	H
Cefprozil	Cefzil			H

Generic Name	Trade Name	R	X
Third Generation			
Cefotaxime	Claforan		—CH₂OAc
Ceftizoxime	Cefizox	Same	—H
Ceftriaxone	Rocephin	Same	

Table 23.5 Cephalosporins (*Continued*)

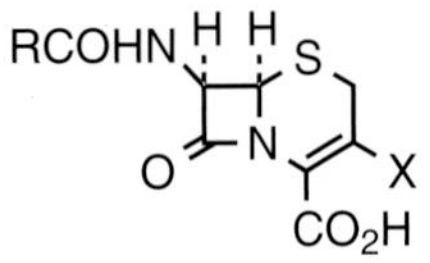

Generic Name	Trade Name	R	X
Third Generation (Continued)			
Ceftazidime	Fortaz, Tazicef Avycaz (combination of ceftazidime plus avibactam)[a]	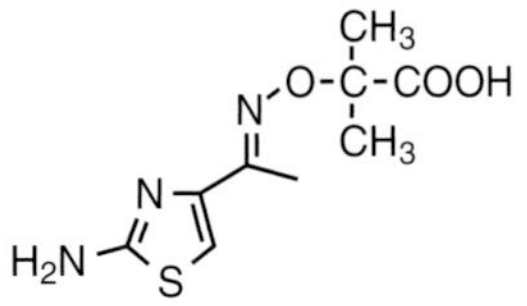	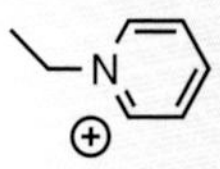
Cefixime	Suprax	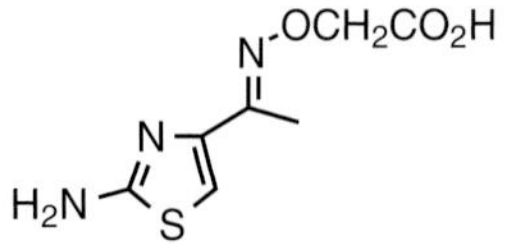	—CH=CH₂
Ceftibuten	Cedax	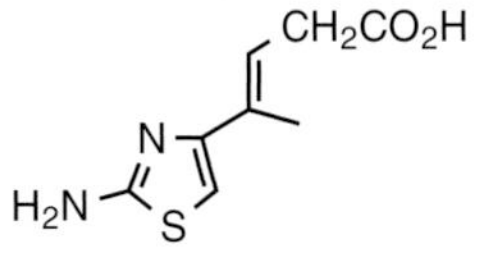	—H
Cefpodoxime proxetil	Vantin		—CH₂OCH₃
Cefdinir	Omicef	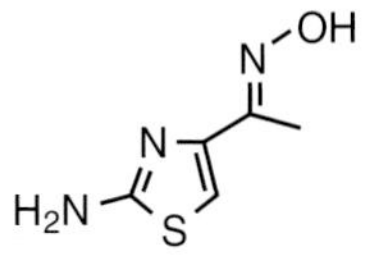	—CH=CH₂
Cefditoren pivoxil	Spectracef 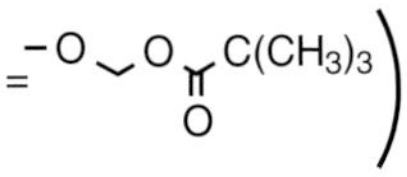	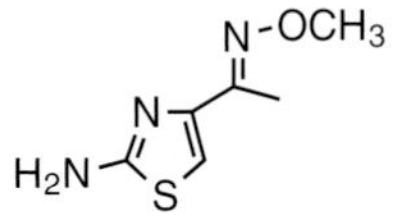	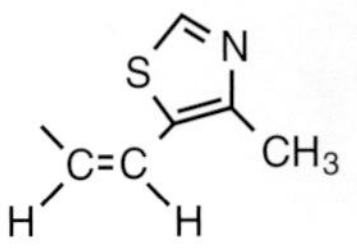
Ceftolozane	Zerbaxa (combination of Ceftolozane plus tazobactam)[a]	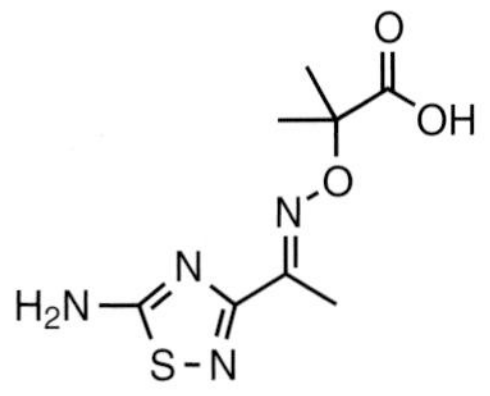	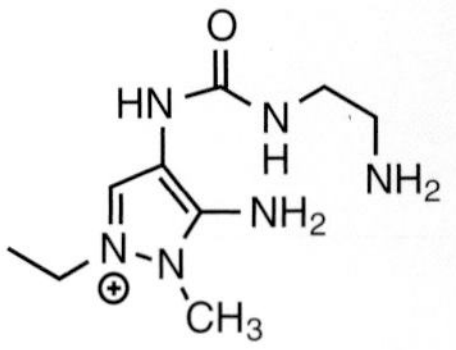
Fourth Generation			
Cefepime	Maxipime	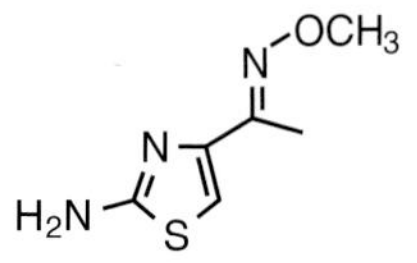	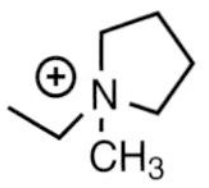
Fifth Generation			
Ceftaroline fosamil	Teflaro	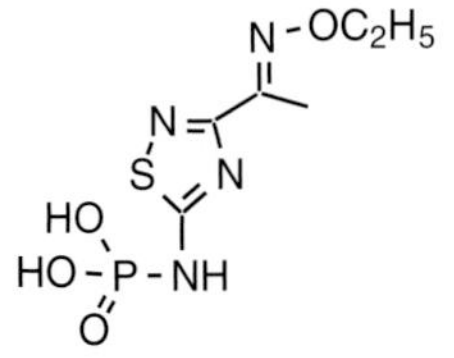	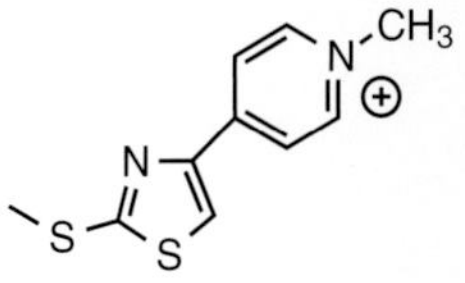

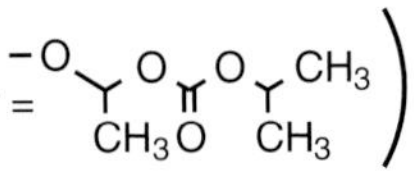

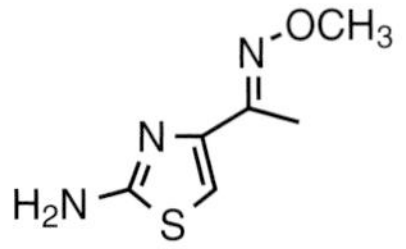

[a]See earlier discussion of β-lactamase inhibitor plus a β-lactam.

- The physiochemical properties toward acid and base and the antibacterial mechanism of action of the cephalosporins do not differ from those of the penicillins.

SAR

- Various leaving groups at C3 (Z group in cephalosporin) in addition to the acetoxy are effective in promoting antibacterial action.

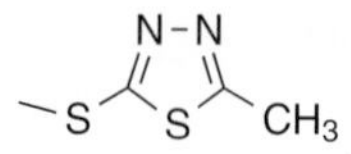

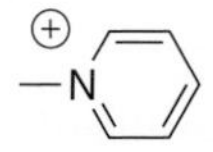

N-Methyl-5-thiotetrazole (MTT)　2-Methyl-5-thiothiadiazine　Pyridinium

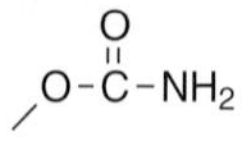

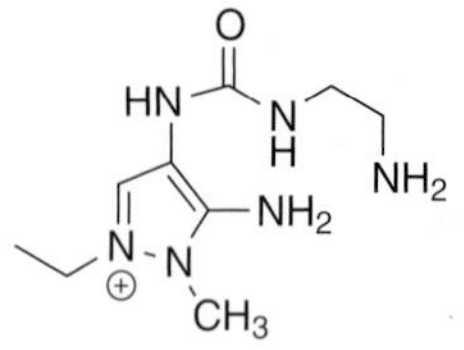

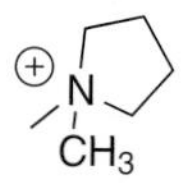

Carbamate　Methylpyrazolium　Methylpyrrolidinium

- The addition of an α-amino (α = NH$_2$, α′ = H), similar to what was seen in the penicillin series, gives rise to orally active cephalosporins (see first- and second-generation cephalosporins) (Table 23.5).
- When α and α′ are combined and converted into an oxime increased stability toward cephalosporinase (β-lactamase) is exhibited, provided the Z isomer is used. This is attributed to steric hindrance around the β-lactam carbonyl.

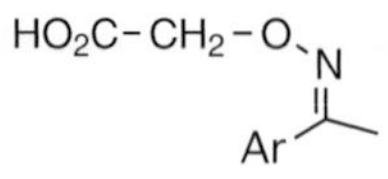

E-isomer 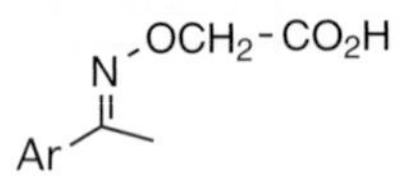　　*Z*-isomer

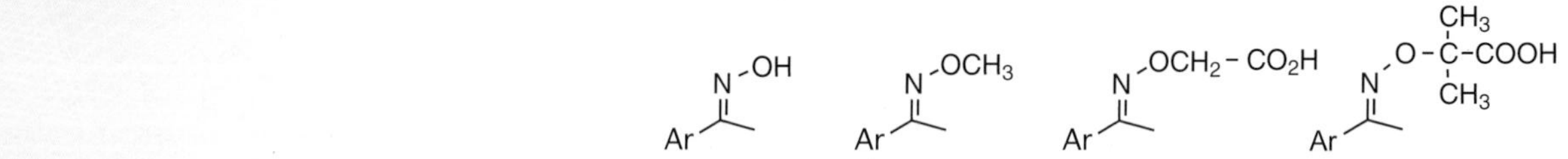

Z-Oxime containing cephalosporins: ceftriaxone, ceftizoxime, ceftazidime, ceftibuten, cefdinir, cefepime, cefuroxime, cefotaxime, cefixime, cefpodoxime, cefditoren, ceftaroline, ceftolozane

- A C7 alkoxy (present in cefoxitin and cefotetan) confers β-lactamase resistance.
- A variety of aryl (Ar) groups are found in the cephalosporins, the most common of which is 3- substituted 5-aminothiazole. The third-, fourth- and fifth-generation cephalosporins contain this aryl group.
- Prodrugs resulting from substitution of the C2 carboxylic acid give rise to orally active cephalosporins. The prodrug is hydrolyzed enzymatically in the wall of the intestine to generate the active cephalosporin as shown in Figure 23.8.
- A fourth prodrug cephalosporin is ceftaroline fosamil in which a phosphoamide derivative is used to confer in vitro stability and water solubility necessary for parenteral administration of the drug.

Clinical Applications

- First-generation cephalosporins show activity against gram-positive bacteria such as *S. aureus* and several *Streptococci,* several gram-negative cocci such as *Neisseria gonorrhorae* and *Neisseria meningitidis,* as well as several enterobacteriaceae (i.e., *Escherichia coli, Klebsiella pneumoniae, Proteus mirabilis*). Drugs in this class include cefadroxil, cefazolin, and cephalexin (Table 23.5).
- Second-generation cephalosporins show activity against aerobic and anaerobic bacilli such as *E. coli, K. pneumoniae, Proteus, H. influenza, Enterobacter, Bacteroides fragilis,*

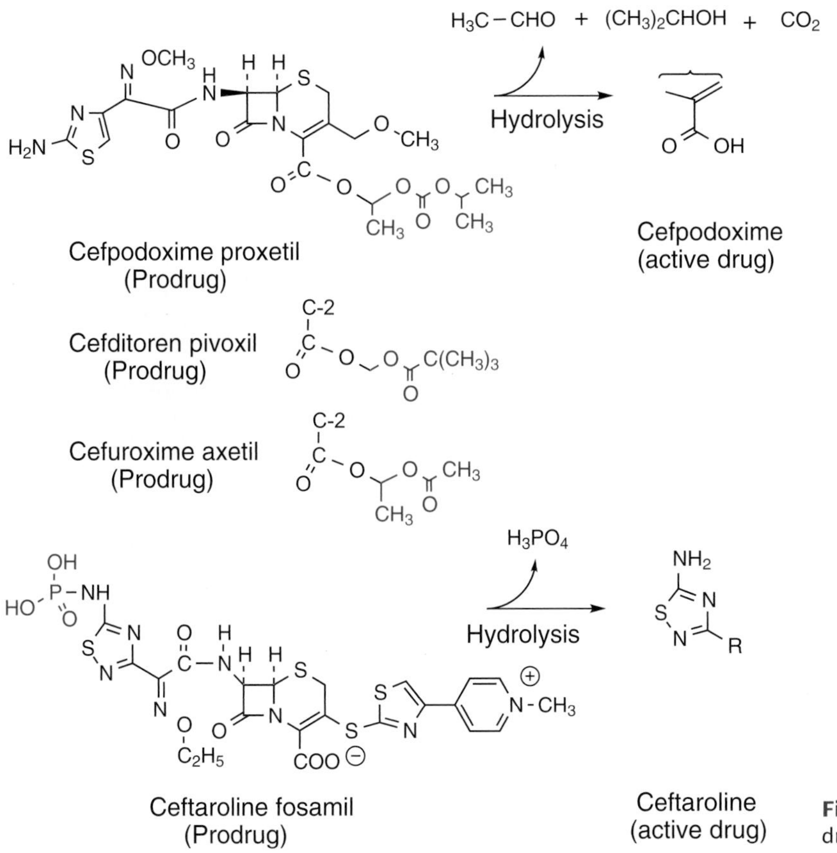

Figure 23.8 Activation of cephalosporin prodrugs via hydrolysis to the active drug.

and the gram-negative bacteria *N. gonorrhorae.* The drugs in this class are shown in Table 23.5.

- Third-generation cephalosporins show activity against gram-negative *N. gonorrhorae, H. influenzae* and various enterobacteriaceae including *E. coli, K. pneumoniae, Proteus, Enterobacter,* and have variable effects on *P. aeruginosa.* See Table 23.5 for drugs in this class.
- Fourth-generation cephalosporin cefepime exhibits activity against β-lactamase–producing *Staphylococcus, Streptococci,* a wide variety of gram-negative aerobes, and some anaerobes.
- Fifth-generation cephalosporin (i.e., ceftaroline fosamil) has reported activity against β-lactamase–producing MRSA and *Streptococcus pneumoniae.*

> **⚠ ADVERSE EFFECTS**
>
> **Cephalosporins:** Hypersensitivity reactions similar to those reported with the penicillins are also common with the cephalosporins. Cross-sensitivity to the penicillins is possible, but is probably far less than immunological test might indicate. Skin tests are not dependable.

Carbapenems

The carbapenems are synthetic derivatives of the natural product thienamycin (Fig. 23.9) ($R_1 = H$). The mechanism of action of the carbapenems is the same as that of other β-lactams.

SAR

- The side chain basic amine is commonly incorporated into a five-member heterocyclic ring (Fig. 23.9).
- When R_2 is a hydrogen (imipenem), the drug is a substrate for renal dehydropeptidase–1 (DHP-1), leading to hydrolysis of the β-lactam (see metabolism below).
- When R_2 is a methyl, with *R* configuration, the drug is stable to DHP-1 hydrolytic inactivation.
- Replacement of the ring sulfur (seen in the penicillins) with an sp^2 carbon increases the reactivity of the β-lactam to both the transpeptidase as well as the β-lactamase.
- The *trans* stereochemistry at positions C5–C6 along with the hydroxyethyl at C6 results in stability to β-lactamase.

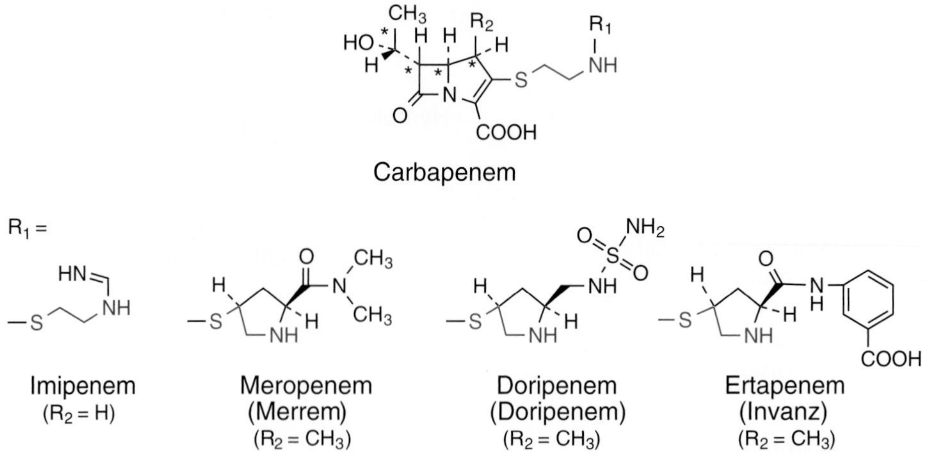

Carbapenem

R₁ =

Imipenem	Meropenem	Doripenem	Ertapenem
(R₂ = H)	(Merrem)	(Doripenem)	(Invanz)
	(R₂ = CH₃)	(R₂ = CH₃)	(R₂ = CH₃)

Figure 23.9 Carbapenem-derived antibiotics.

- The carbapenems usually have four or more asymmetric centers in the molecule, with the stereochemistry playing an important role in the biologic activity.

Physicochemical and Pharmacokinetic Properties

- None of the carbapenems are active via the oral route of administration.
- All of the carbapenems are soluble at alkaline pH allowing for IV solutions.

Metabolism

- The only clinically significant human metabolism of the carbapenems is the β-lactam hydrolysis catalyzed by DHP-1, and while this is not a significant problem for the C4 methylated carbapenems, it is a problem for imipenem.
- β-Lactam hydrolysis can be blocked by the addition of cilastatin to the formulation of imipenem.

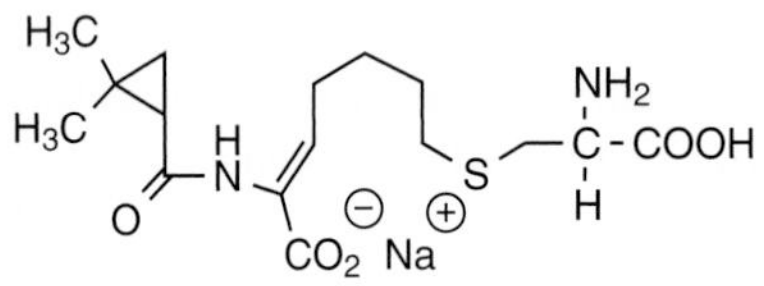

Cilastatin sodium

Clinical Applications

- The carbapenems are broad-spectrum antibacterial agents with activity against gram-negative bacilli, anaerobes, and *S. aureus.*
- Common infections for which carbapenems are prescribed include UTIs; lower respiratory infections; intra-abdominal and gynecological infections; and skin, soft tissue, bone, and joint infections.

Monobactam

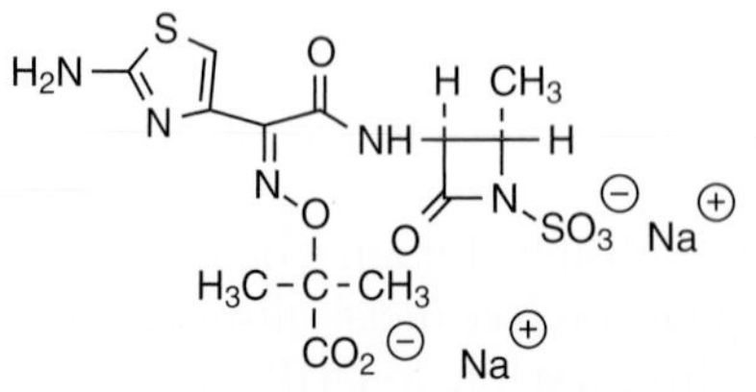

Aztreonam disodium (Azactam)

- A single monobactam, aztreonam, is presently marketed in the United States.
- Binding to PBP through the β-lactam pharmacophore results in a mechanism of action similar to the other β-lactams.
- The N1 sulfonic acid replaces the carboxylic acid common to the other β-lactams.

- The *S*-2-methyl and the *Z*-oxime are reported to confer stability toward β-lactamase.
- Aztreonam is not orally active and is used parenterally.
- The spectrum of activity is exclusively gram-negative aerobes and the drug is commonly used in severe hospital-acquired infections.

⚠ **ADVERSE EFFECTS**

Aztreonam: Normally, few side effects have been reported. Gastrointestinal effects include nausea, vomiting, and diarrhea. Cross-sensitivity to penicillins or cephalosporins is not expected with the possible exception of ceftazidime.

Aminoglycosides

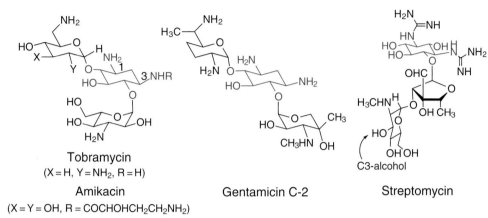

Tobramycin
($X = H$, $Y = NH_2$, $R = H$)

Amikacin
($X = Y = OH$, $R = COCHOHCH_2CH_2NH_2$)

Gentamicin C-2

Streptomycin

- The aminoglycosides are characterized as glycoside substituted 1,3-diaminoinositols.
- The aminoglycosides are natural products or semisynthetic derivatives of the natural product.

MOA

- Aminoglycosides inhibit protein synthesis by virtue of their ability to bind to the 16S unit within the 30S ribosomal subunit of bacteria (Fig. 23.10).
- Irreversible binding ultimately leads to formation of nonsense proteins killing the bacteria.

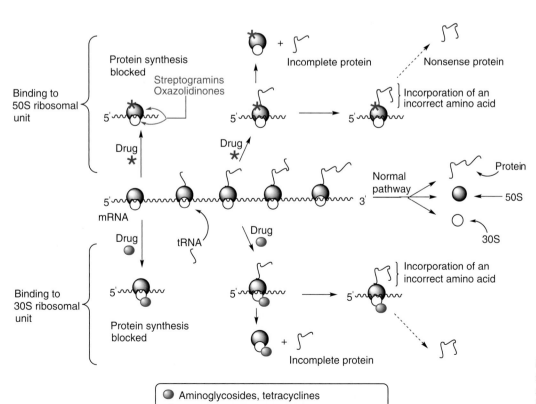

Figure 23.10 General scheme of mechanism and sites of action of protein-blocking drugs on ribosomal units.

- Aminoglycosides penetrate gram-negative bacterial cells via a process of passive and active transport, but do not enter gram-positive bacteria and, therefore, are not active against these bacteria.

SAR

- The addition of an α-hydroxy-γ-aminobutyl group to the amino group in the 1,3-diaminoinositol blocks bacterial metabolic inactivation (i.e., amikacin) by blocking binding to the *R* factor–mediated enzyme.

Physicochemical and Pharmacokinetic Properties

- The aminoglycosides are quite hydrophilic and are thus water soluble.
- Because of the hydrophilic nature of these drugs they are not absorbed via the oral route.
- Orally administered aminoglycosides exhibit antibacterial effects locally in the gastrointestinal tract.
- The drugs have a high pKa and form acid salts at a basic nitrogen.
- Primary route of elimination is through the kidney via glomerular filtration (patients with renal impairment exhibit potential for nephrotoxicity – accumulation of aminoglycoside).

Metabolism

- The aminoglycosides are generally excreted unchanged.
- Bacterial resistance is associated with acetylation at a primary amine in ring I or II or adenylation at a secondary alcohol in tobramycin or gentamicin.

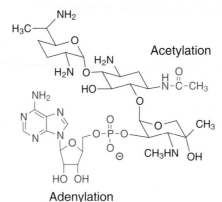

- Phosphorylation at the C3 alcohol in streptomycin leads to resistance.

Clinical Applications

- The aminoglycosides are commonly reserved for serious gram-negative infections.
- Streptomycin has limited utility today and is primarily used to treat mycobacterial infections (i.e., tuberculosis [TB]).

Topical/Oral Aminoglycosides

Neomycin B (R = NH₂)
Paromomycin (R = OH)

- Both neomycin and paromomycin are used orally to suppress intestinal flora or treat amebiasis, or applied to the skin to prevent or treat topical infections.

ADVERSE EFFECTS

Aminoglycosides:
Ototoxicity – potential for hearing loss and loss of balance. Nephrotoxicity – accumulation of aminoglycosides in proximal tubular cells.

CHEMICAL INCOMPATIBILITY

A β-lactam and an aminoglycoside should not be physically mixed since the nucleophilic amino group of the aminoglycoside will catalyze the hydrolysis of the β-lactam ring.

- In either dosage form, the drugs are not intended to be absorbed. Potential toxicity can result from absorption through significant intestinal ulceration or if used on large areas of denuded skin.

Macrolide Antibiotics

Erythromycin Class

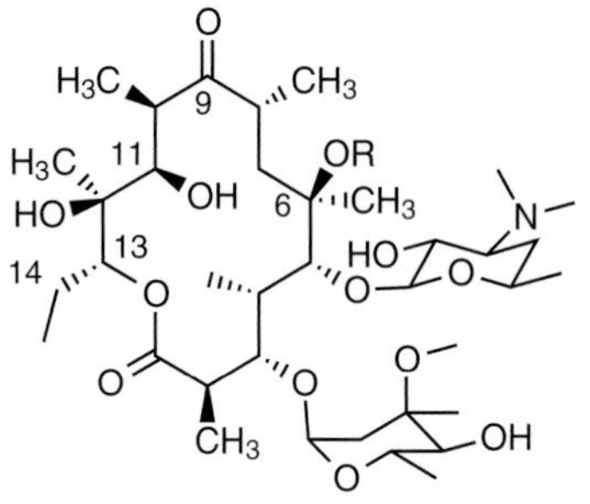

Erythromycin A (R = H)
Clarithromycin (R = CH₃)
(Biaxin)

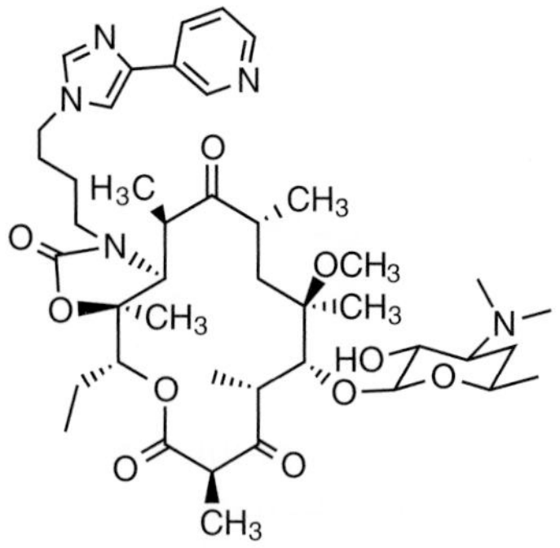

Telithromycin
(Ketek)

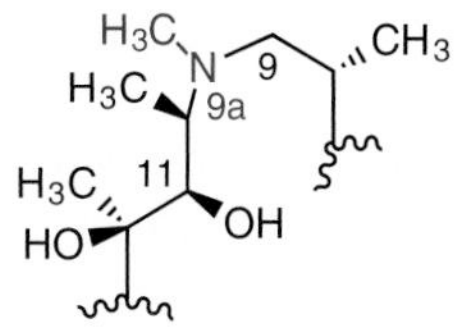

Azithromycin
(Zmax, Zithromax)

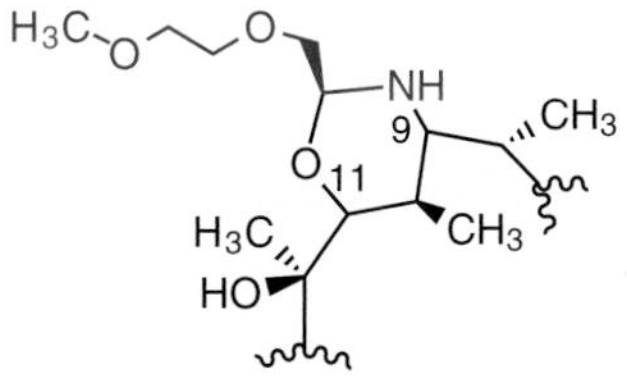

Dirithromycin
(discontinued)

- The erythromycin class of macrolides is based upon the 14-unit lactone or synthetic expansions of this macrolide at C9–C10.

MOA

- The macrolides are bacteriostatic drugs.
- The macrolides bind to the 23S of ribosomal RNA (rRNA) within the 50S ribosomal subunit of bacteria involved in prokaryotic protein synthesis. They interfere with peptidyl transferase, preventing protein elongation (Fig. 23.10).
- Primary binding sites in rRNA appear to be at Domain II adenine 752 and Domain V adenines at 2058 and 2059.

SAR

- Methylation of the C6 alcohol improves acid stability without reducing biologic activity.
- Reductive amination of the C9 ketone with incorporation of the C9 amino and C11 alcohol into a ring acetal gives dirithromycin, a prodrug which is metabolized in vivo to the active amino alcohol erythromycin.
- Removal of the C3 sugar cladinose from erythromycin does not reduce activity (see telithromycin). The cladinose appears to be associated with the release of motilin, which increases gastric motility.
- Addition of the side chain group at C9–C11 greatly increases binding of telithromycin to Domain II of bacterial rRNA, reducing potential for bacterial resistance.

Physicochemical and Pharmacokinetic Properties

- Erythromycin is unstable upon oral administration. The instability results from an acid-catalyzed intramolecular cyclization involving the C6 OH and the C9 ketone leading to anhydroerythromycin (Fig. 23.11). The anhydroerythromycin is also associated with GI irritation.

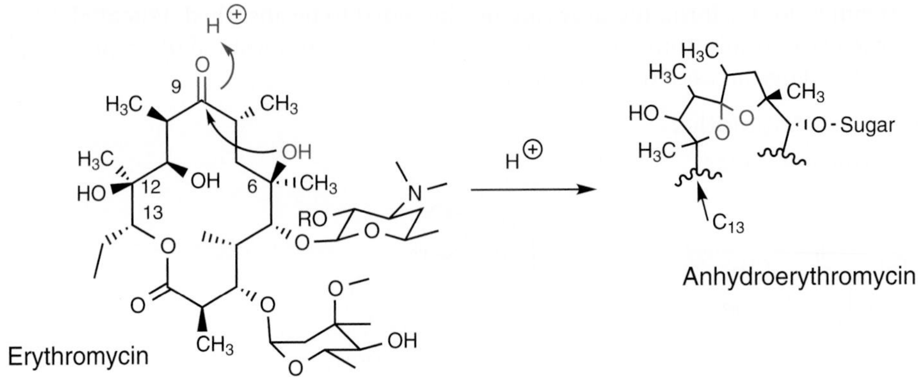

Figure 23.11 Stomach acid catalyzed instability of erythromycin.

- Soluble (hydrochloride) and insoluble (stearate) salts of erythromycin can be prepared at the tertiary amine in the C5 sugar (desosamine). Many of the salt formulations have been discontinued.
- Ester formation with ethyl succinate as the R group (C6 OH) gives rise to a tasteless prodrug (EES), requiring hydrolysis to the free OH for activity.

Bacterial Resistance

- Bacterial mutation in rRNA occurs at Domain V adenine 2058 (i.e., adenine to guanine).
- Bacterial mutation involving methylation at adenine 2058 also increases resistance through reduced binding of erythromycin and clarithromycin to the 23 rRNA.

Clinical Applications

- The macrolides are quite potent and safe when used to treat gram-positive cocci infections of the upper and lower respiratory tract, otitis media (*S. pneumoniae, S. aureus, Staphylococcus*), various atypical organisms (*Mycoplasma pneumoniae, Legionella pneumophila, Chlamydia pneumonia*), and *Mycobacterium avium* complex infections.
- The macrolides are bacteriostatic in action.
- Drug–drug interactions may be associated with the inhibitory action of the macrolides on CYP3A4 for drugs commonly metabolized by this cytochrome isoform.

> ⚠️ **ADVERSE EFFECTS**
>
> **Macrolides:** GI disturbance – more common with erythromycin. Cardiovascular toxicity – QT prolongation.

Fidaxomicin (18-Member Macrolide)

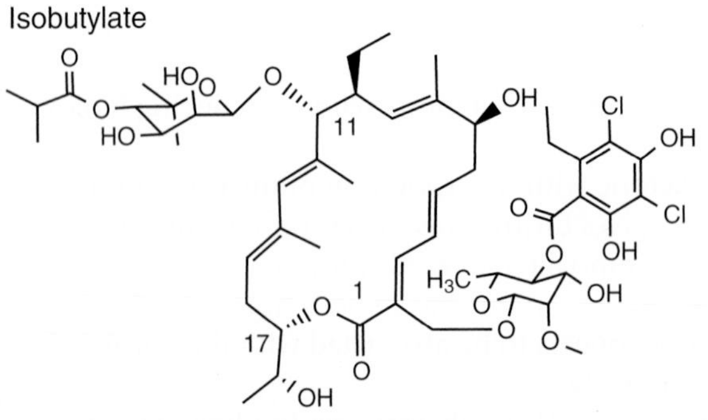

Isobutylate

Fidaxomicin (Dificid)

Fidaxomicin is a macrolide antibiotic (18-member ring) isolated from the fermentation broth of *Dactylosporangium aurantiacum*.

MOA

- Fidaxomicin inhibits DDRP, thus interfering with protein synthesis.

- The drug binds to the RNA polymerase in a noncompetitive manner leading to bactericidal action.

Physicochemical and Pharmacokinetic Properties

- Fidaxomicin is poorly absorbed following oral administration, which makes it useful to treat intestinal infections.

Metabolism

- The major metabolite of fidaxomicin results from hydrolysis of the isobutyl ester from the C11 sugar.
- The resulting product exhibits bactericidal activity although less than that of fidaxomicin.

Clinical Applications

- Fidaxomicin has a very narrow spectrum of activity with primary activity against *Clostridium difficile*–associated diarrhea (CDAD) (88% effectiveness).
- *C. difficile* releases toxins which can cause bloating and diarrhea, which can be life-threatening.
- Fidaxomicin exhibits minimum adverse effects.

Lincosamides

Two lincosamides are marketed: the natural-occurring antibiotic lincomycin and clindamycin, a semisynthetic derivative. The lincosamides are derivatives of an eight-carbon sugar with a thiomethyl-mixed acetal at the C1 and a substituted pyrrolidinecarboxamide at C6.

MOA

- The lincosamides block bacterial protein synthesis by virtue of the ability to bind to the 50S ribosomal subunit (Fig. 23.10).
- The site of binding of the lincosamides is similar to that of the macrolides and cross-resistance can be expected.

Physicochemical and Pharmacokinetic Properties

- Clindamycin is available as a water-soluble hydrochloride salt formed at the basic nitrogen in the pyrrolidinecarboxamide or the water-soluble phosphate ester formed at the C2 alcohol.
- Clindamycin palmitate is the water-insoluble and tasteless palmitate ester formed at the C2 alcohol.

Metabolism

- Metabolic inactivation is associated with N-demethylation and S-oxidation to the sulfoxide.

Clinical Applications

- Clindamycin is used most commonly to treat gram-positive cocci infections and anaerobes and is used when β-lactams cannot be used.
- The drug is used for the treatment of skin and soft tissue infections and respiratory tract infections associated with susceptible organisms.
- Clindamycin is used topically for the treatment of *Propionibacterium acne*.
- Lincomycin is only recommended for serious infections when less toxic agents are not available.

Lincosamides

Lincomycin (Lincocin) (R = OH, R' = H)   Clindamycin (Cleocin) (R = H, R' = Cl)

> **BLACK BOX WARNING**
>
> *Associated with the development of pseudomembranous colitis associated with toxin released from C. difficile. Severe diarrhea with potential for death is more common with lincomycin.*

Tetracyclines

Tetracycline   Tetracycline derivatives

Table 23.6 Commercially Available Tetracyclines

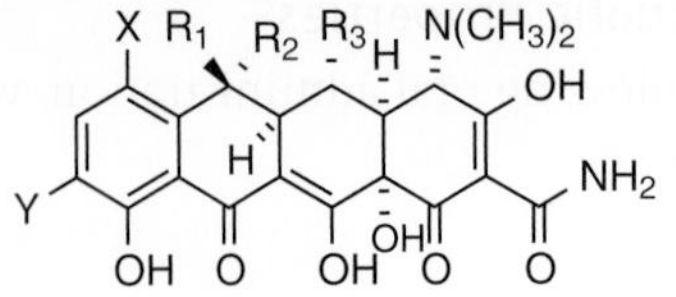

Generic Name	Trade Name	X	Y	R₁	R₂	R₃
Tetracycline	Generic	H	H	OH	CH₃	H
Demeclocycline	Declomycin	Cl	H	OH	H	H
Minocycline	Minocin	N(CH₃)₂	H	H	H	H
Doxycycline	Vibramycin	H	H	H	CH₃	OH
Tigecycline	Tygacil	N(CH₃)₂	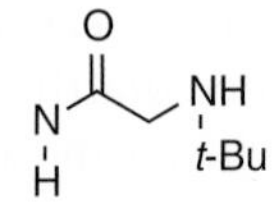 O (t-Bu glycylamino)	H	H	H

- The tetracycline family of drugs is derived from a natural-occurring antibiotic composed of a four linearly fused six-member ring system.
- A limited number of drugs from this family are available commercially today (Table 23.6).

MOA

- The tetracyclines are reversible inhibitors (bacteriostatic) of protein synthesis.
- The tetracyclines are thought to bind to rRNA at the 16S rRNA portion of the 30S subunit preventing subsequent binding of tRNA (Fig. 23.10).

SAR

- The C4 dimethylamino group with the α-stereochemistry is essential for activity, but this group is not involved in binding to the rRNA.
- Binding at the site of action requires the groups at C1 through C3 and C10 through C12 as well as the D ring (Fig. 23.12).
- The *t*-butyl glycine of tigecycline is important by virtue of its effectiveness against resistant organisms with gene-expressing efflux pumps.
- Allowable substitution is shown in Table 23.5, which indicates marketed tetracyclines.

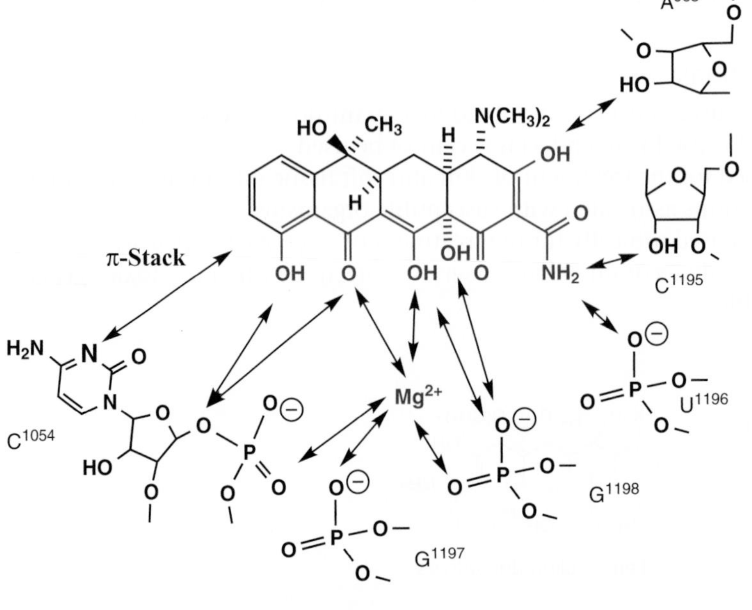

Figure 23.12 Schematic representation of the primary binding site for a tetracycline and the sugar phosphate groups of 16S rRNA which also involves a magnesium ion and the critical functional groups on the "southern" and "eastern" face of the tetracycline (Modified with permission from Brodersen DE, Clemons WM Jr, Carter AP, et al. The structural basis for the action of the antibiotics tetracycline, pactamycin, and hygromycin B on the 30S ribosomal subunit. *Cell*. 2000;103:1143–1154. Copyright 2000 Elsevier.)

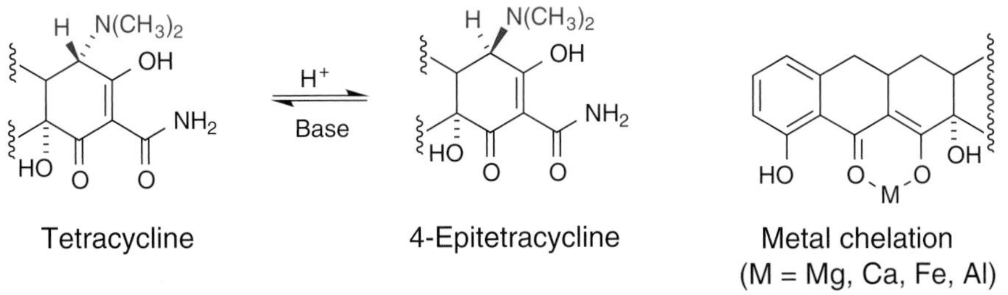

Figure 23.13 Chemical instability of tetracyclines to acid and heavy metals.

Physicochemical and Pharmacokinetic Properties

- The tetracyclines are amphoteric in nature with basicity associated with the C4 α-tertiary amine and acidity associated with the enol–keto functionality at C1 to C3 and enol–keto–phenol at C10 through C12 (pKa = 2.8 to 3.4, 7.2 to 7.8, and 9.1 to 9.7, respectively).
- Water-soluble tetracyclines salts (e.g., hydrochlorides), if allowed to stand in solution, are unstable due to epimerization occurring at C4 (Fig. 23.13).
- The tetracyclines form insoluble chelates with divalent and trivalent heavy metals (M) at the oxygen substituents at C11 and C12 (e.g., Ca^{2+} in dairy products; Ca^{2+}, Mg^{2+}, Al^{3+} in antacids; and Mg^{2+} in saline laxatives).
- Tetracyclines are contraindicated during pregnancy or in children due to binding to tooth enamel calcium leading to permanent tooth discoloration (Fig. 23.13).
- The tetracyclines have asymmetry at C4, C4a, C5, C5a, C6, and C12a all of which are important for biologic activity.
- Because of extended conjugation, several of the tetracyclines may cause a phototoxic reaction in skin exposed to sunlight.

Clinical Applications

- The tetracyclines are considered to be broad-spectrum antibiotics, but their use has been limited by the development of resistant organisms.
- The tetracyclines are useful in the treatment of a wide variety of infections from community-acquired urinary tract infections to topical use to control of acne. The reader is referred to the Spectrum of Action.
- The tetracyclines may cause drug–drug interactions via induction of CYP450 metabolizing enzymes.

Amphenicols

Chloramphenicol sodium succinate

- Because of the potential for severe toxicity associated with the amphenicol class of drugs, only chloramphenicol sodium succinate, a prodrug, is available today.
- The active drug results from hydrolytic removal of the succinate-generating chloramphenicol.
- The site and mechanism of action of the amphenicols is similar to that of clindamycin and the macrolides (reversible binding to 50S ribosomal subunit) (Fig. 23.10).
- Limited clinical use includes treatment of typhoid fever, *Haemophilus,* and rickettsial infections when other drugs are not available.

Spectrum of Action

The tetracyclines are useful in the treatment of a variety of infectious organisms including *E. coli* (community-acquired urinary tract infections), *Brucella* (responsible for brucellosis), *Borreli* (cause of borreliosis and Lyme disease), *Chlamydia* (a sexually transmitted disease), *Rickettsia* (Rocky Mountain spotted fever), *Mycoplasma pneumonia, Bacillus* (cause of anthrax) and *Enterobacter* (cause of various opportunistic infections).

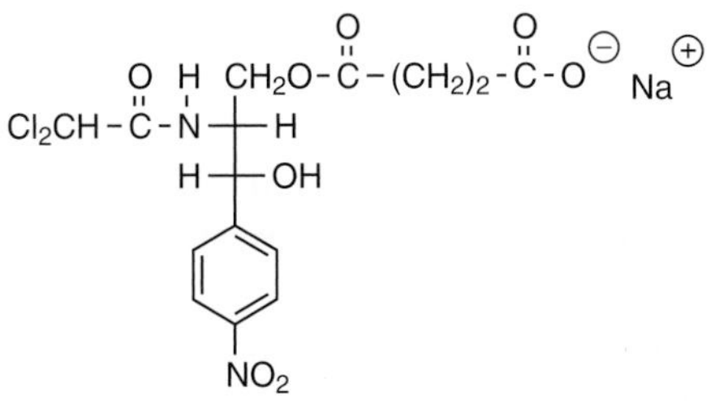

ADVERSE EFFECTS

Tetracyclines: GI irritation, which may include abdominal distress, nausea, vomiting, and diarrhea. Permanent brown discoloration of tooth enamel via calcium tetracycline precipitation (Fig. 23.13). Depressed bone growth in infants. Photosensitivity when patient is exposed to sunlight.

ADVERSE EFFECTS

Amphenicols: Fatal pancytopenia associated with bone marrow effects including anemia, leukopenia, thrombocytopenia, and aplastic anemia.

Table 23.7 Structures of the Glycopeptide Antibiotics

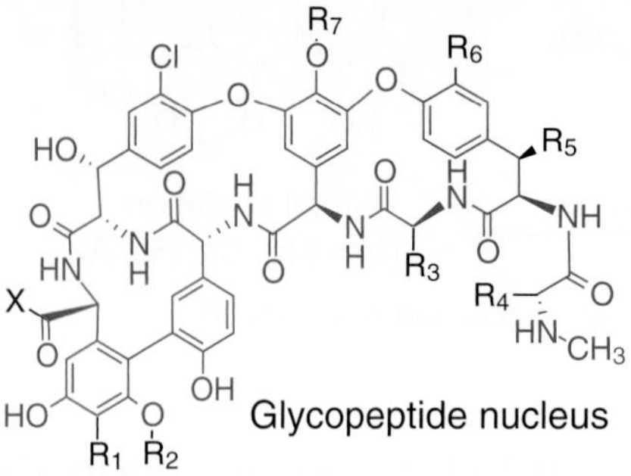

Glycopeptide nucleus

	X	R₁	R₂	R₃	R₄	R₅	R₆	R₇
Dalbavancin (Dalvance)	H 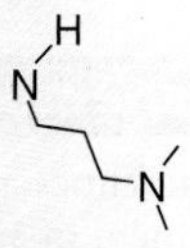	H	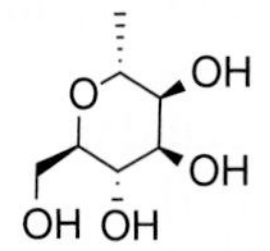 [sugar structure]	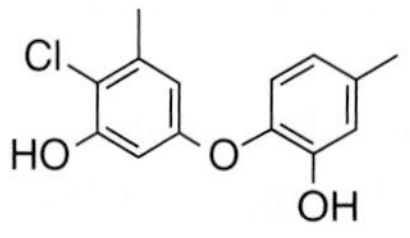 [structure]		H	H	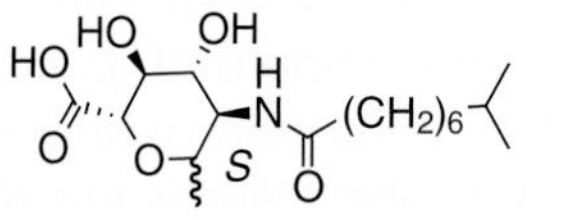 [lipid sugar structure]
Televancin (Vibativ)	OH	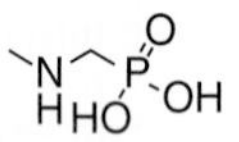 [phosphomethyl structure]	H	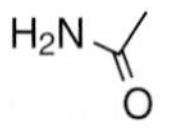 [structure]	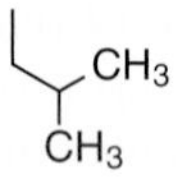 [isobutyl]	OH	Cl	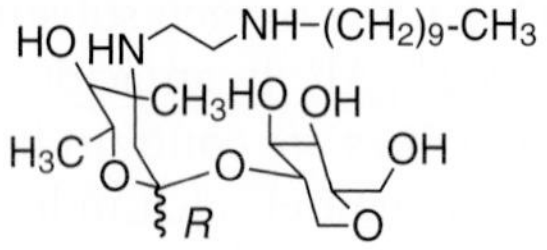 [sugar structure]
Vancomycin (Vancocin)	OH	H	H	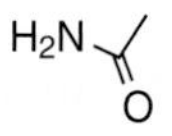 [structure]	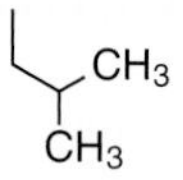 [isobutyl]	OH	Cl	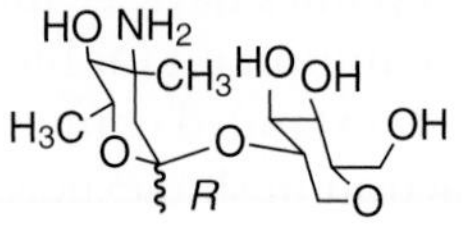 [sugar structure]
Oritavancin (Orbactiv)	OH	H	H	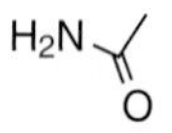 [structure]	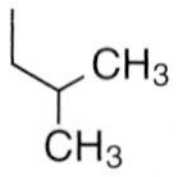 [isobutyl]	OH	Cl	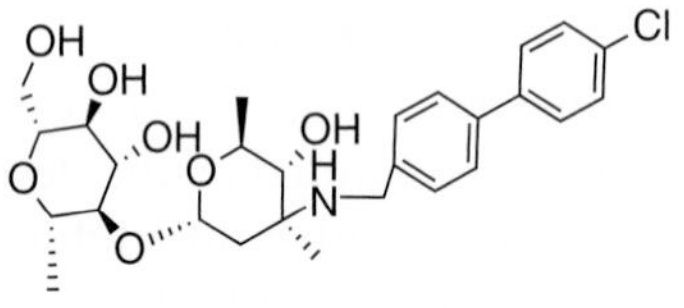 [biphenyl sugar structure]

Peptide Antibiotics

Glycopeptides

The glycopeptides are a group of antibiotics represented by the natural-occurring vancomycin and three semisynthetic agents (televancin, dalbavancin, and oritavancin), which exhibit a common basic heptapeptide core (Table 23.7).

MOA

- These glycopeptides block cell wall synthesis by binding to the D-alanine-D-alanine portion of the developing cell wall inhibiting both transglycosylation and cross-linking transpeptidation of the cell wall.
- Electrostatic binding is proposed to occur between electron-rich regions and key hydrogens as shown in Figure 23.14.
- In addition, the lipid portion found in the R₇ moiety of televancin, dalbavancin, and oritavancin improve surface membrane binding and increase membrane permeability. This lipophilic portion is missing in vancomycin resulting in less effectiveness against the gram-positive bacteria.
- Bactericidal activity is associated with osmotic insult through both increased membrane permeability and inhibition of cell wall cross-linking.
- It is reported that oritavancin has dual binding sites not only to both the normal binding sites (i.e., D-alanine-D-alanine), but also to the pentaglycyl bridging segment, thus providing activity in vancomycin-resistant *Staphylococcus aureus* (VRSA).
- Bacterial resistance occurs when the bacteria replace the D-alanine-D-alanine portion of the cell wall with D-alanine-D-lactate resulting in reduced binding affinity (~1000 fold) of the glycoproteins to the developing stable cell wall.

Figure 23.14 Binding of glycopeptide antibiotics to bacterial cell wall D-alanine-D-alanine moiety.

SAR

- The R_1 substituent in telavancin not only improves water solubility, but also decreases the drug's half-life.
- Studies suggest that the lipophilic substituent attached to the R_7 sugars serve as membrane-anchoring groups.
- These lipophilic groups in dalbavancin and telavancin may also prolong the half-life.
- The lipid side chains found in R_7 may also be key to destabilizing the bacterial cell wall though their detergent-like solubilization of the wall.

Physicochemical and Pharmacokinetic Properties

- Due to the complex glycopeptide structures, these drugs must be administered intravenously.
- Dalbavancin with its long half-life (~12 days) allows dosing on day 1 and day 8, while telavancin is dosed once daily and vancomycin requires dosing every 6 to 12 hours.
- Both dalbavancin and telavancin are highly bound to plasma proteins (>90%), while vancomycin is poorly bound (10% to 55%).

Clinical Applications

- The glycoproteins are approved for the treatment of acute bacterial skin and skin structure infections caused by susceptible organisms (gram positive) such as *S. aureus* (MRSA) including methicillin-susceptible and methicillin-resistant *Streptococcus pyogenes*.
- In addition, vancomycins find clinical utility in treatment of MRSA (respiratory tract infection) as well as *S. pneumoniae* (meningitis) when used systemically (IV) and *C. difficile* when used orally.

Daptomycin

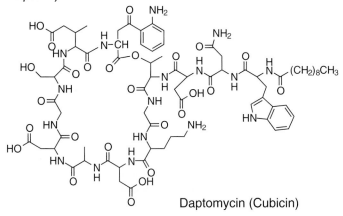

Daptomycin (Cubicin)

> **ADVERSE EFFECTS**
>
> **Vancomycin:** Effects of vancomycin associated with higher doses and a rapid infusion rate included anaphylactoid reactions (hypotension, wheezing, dyspnea, urticaria, and pruritus); drug rash (the so-called Red man syndrome) can also occur. These events are much less frequent with a slower infusion rate. In addition, vancomycin can cause nephrotoxicity and auditory nerve damage associated with elevated, prolonged concentrations of vancomycin. During clinical trials with dalbavancin and telavancin the following were reported: nausea, headaches, pyrexia, and diarrhea without indications of auditory toxicity or vestibular dysfunction. Telavancin was reported to exhibit taste disturbance in addition to the GI disturbances.

> **ADVERSE EFFECTS**
>
> **Daptomycin:** Nephrotoxicity, ototoxicity, extreme flushing ("Red man" syndrome). Adverse reactions appear to be related to excessively high blood concentrations.

- Daptomycin is a lipoprotein, which is highly bound to plasma protein (89% to 94%).
- Daptomycin binds to the cytoplasmic membrane (phospholipid bilayer) of bacteria disrupting electrical potential of the membrane.
- The drug is bactericidal.
- Effective against VRSA and MRSA.

Streptogramins

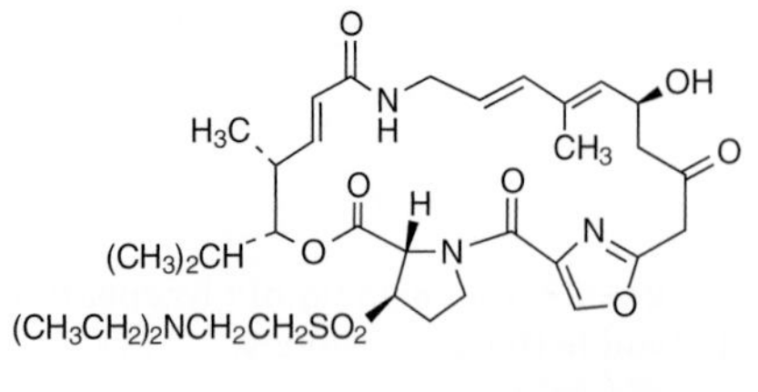

Dalfopristin
(Streptogramin A)

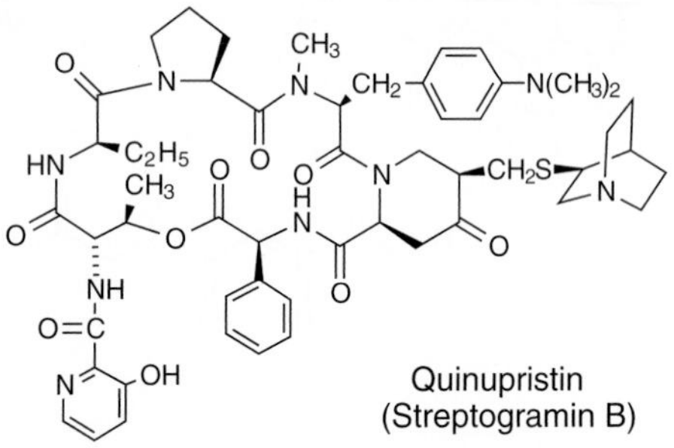

Quinupristin
(Streptogramin B)

- The combination of streptogramin A/streptogramin B (70/30) (Synercid), a macrolide and cyclic peptide, respectively, are synergistic and bactericidal when used in combination.
- Both drugs bind to the 70S rRNA with dalfopristin creating a high-affinity binding site for quinupristin to inhibit protein synthesis (Fig. 23.10).
- The combination is useful for treatment of vancomycin-resistant *Enterococcus faecium* and methicillin-sensitive *S. aureus* and *Streptococcus pyrogenes*.
- The combination (Synercid) is a potent inhibitor of CYP2A4 isozyme.

Polymyxins and Bacitracin A

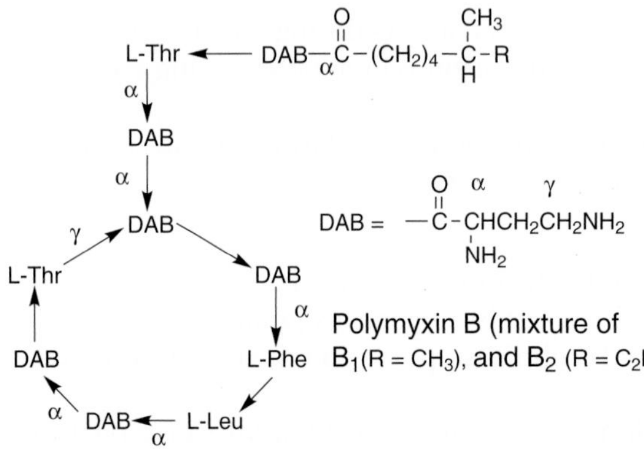

Polymyxin B (mixture of B$_1$(R = CH$_3$), and B$_2$ (R = C$_2$H$_5$)

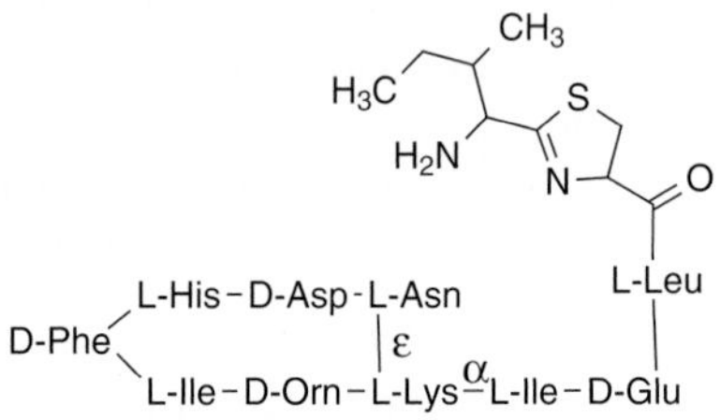

Bacitracin A

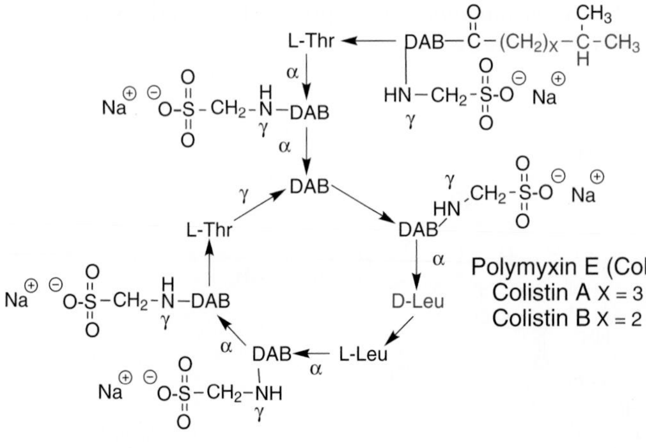

Polymyxin E (Colistimethate sodium)
Colistin A X = 3
Colistin B X = 2 } Coly-Mycin M

- Polymyxins B$_1$, B$_2$, plus bacitracin are cyclic peptides found in various OTC topical preparations with both antibiotics exhibiting activity against gram-negative bacteria and gram-positive bacteria.

- The site of action of both drugs is on the cytoplasmic membrane where they cause membrane disruption.
- Colistimethate sodium is intended for use via injection for treatment of gram-negative bacterial infections.

Physicochemical and Pharmacokinetic Properties of Peptide Antibiotics

- Systemic peptide antibiotics are not effective if administered orally and must be administered IV or IM.
- Peptide antibiotics administered orally are meant for treatment of local GI tract infections (i.e., polymyxin B and bacitracin).

Oxazolidinone Class of Antibacterial Agents

Generic (Trade) Names	R₁	R₂
Linezolid (Zyvox)	(morpholine N-methyl)	—NHAc
Tedizolid phosphate disodium (Sivextro)(Prodrug)	H₃C (tetrazole-tolyl)	phosphate disodium
Tedizolid	H₃C (tetrazole-tolyl)	—OH

The oxazolidinones represented by linezolid and tedizolid phosphate disodium/tedizolid are bacteriostatic (bactericidal against some bacteria) agents effective against primarily gram-positive bacteria.

MOA

- The oxazolidinones block protein synthesis at the initiation stage of the synthesis.
- The oxazolidinones bind to the 23S rRNA of the large 50S subparticle, thus preventing formation of the 70S complex needed for protein synthesis (Fig. 23.10).
- It is believed that these drugs distorts the binding site for the initiator-tRNA which overlaps both 30S and 50S ribosomal subunits preventing the translation of the mRNA.
- The larger size of tedizolid increases the binding of the drug resulting in increased potency over linezolid.
- Resistance to oxazolidinones is thought to occur through mutation in the 23S rRNA leading to distortion of the linezolid-binding site.

Physicochemical and Pharmacokinetic Properties

- The oxazolidinones are well absorbed orally (bioavailability 90% to 100%).
- Tedizolid phosphate disodium is a prodrug, which is rapidly and completely hydrolyzed to the active metabolite by phosphatases located in the intestinal brush border membrane.
- Linezolid is excreted primarily intact with minor inactive metabolites resulting from oxidation of the morpholine ring. Tedizolid is excreted primarily as the sulfate conjugate of the primary alcohol.

Clinical Applications

- The drugs are effective against MRSA, nosocomial pneumonia, community-acquired pneumonia, complicated and uncomplicated skin and skin structure infections, and vancomycin-resistant *E. faecium* infections.
- The oxazolidinones are not effective against gram-negative bacteria.
- Tedizolid is specifically approved for treatment of acute bacterial skin and skin structure infections.

ADVERSE EFFECTS

The oxazolidinones are reported to be well tolerated although severe cases of reversible blood dyscrasias have been noted for linezolid resulting in a package insert warning that complete blood counts should be monitored weekly, especially in patients with poorly draining infections and who are receiving prolonged therapy with the drug. Both drugs may inhibit monoamine oxidase, so patients should be cautious about eating tyramine-containing foods.

Topical Miscellaneous Antibiotics
Mupirocin and Retapamulin

Mupirocin
(Bactroban)

Retapamulin
(Altabax)

- Mupirocin and retapamulin are used topically to treat staphylococcal and streptococcal skin infections (both are used to treat impetigo caused by MRSA although resistance may develop over a period of time).
- Both drugs are bacteriostatic and block protein synthesis, but at different steps.
  - Mupirocin binds to bacterial isoleucyl tRNA synthase preventing isoleucine incorporation into protein.
  - Retapamulin binds to 50S ribosomal subunit blocking protein synthesis by blocking peptidyl transferase.
- Mupirocin is a mixture of acids of which pseudomonic acid A makes up ~90% of the mixture.

Chemical Significance

The impact of antimicrobial therapy is nearly impossible to appreciate in the 21st century since it has become so commonplace. In less than 75 years we have witnessed a time when bacterial infections were untreatable to a day when common infections are preventable or cured in a matter of a weeks through drug therapy. What is considered the modern anti-infectious era dawned with the discovery credited to Gerhard Domagk of the effectiveness of the sulfonamides (1936) in treating bacterial infections, which was followed by the introduction of the penicillins (1940s) through the work of Alexander Fleming and Howard Florey and Ernst Chain. In rapid progression the modern classes of antibiotics, semisynthetic antibiotics, and synthetic antibacterials have nearly conquered common infectious diseases only to be followed by today's challenges of pathogens resistant to multi-classes of antimicrobial therapy. The so-called antibiotic-resistant infections (ARIs) are said to cost US patients in excess of $20 billion a year with an estimate of over $35 billion annually in costs to society for hospitalization and lost wages.

The ability to apply medicinal chemistry concepts and knowledge to explain antibacterial drug therapy are numerous and include areas such as:

- Relating a chemical structure such as a sulfonamide to its site and mechanism of action of blocking a natural bacterial metabolite, aminobenzoic acid, into folic acid.
- Recognizing that electron-rich oxygens (β-keto carboxylic acid) can chelate a di- or trivalent metal thus accounting for drug–drug interaction which prevents absorption of the 4-quinolones.
- Identifying the presence of a β-lactam in the structure of a new antibiotic allows for a prediction of a mechanism of action, pharmacokinetic properties, potential adverse effects, and in some cases, its spectrum of activity of the new drug.
- Chemically identifying an antibiotic as an aminoglycoside which in turn indicates a mechanism of action as well as limitations on the route of administration.
- Recognizing an antibiotic as a peptide and as a result dictating the injectable nature of this product. Trade names or generic names offer little value in identification of pharmaceutical properties of a drug, while chemical structures often allow the pharmacist to predict the mechanism of action, routes of administration, chemical stabilities or instabilities, spectrum of activity, and potential adverse effects. It is imperative that the pharmacist take advantage of their unique training and understanding of chemical knowledge and apply this knowledge to future therapeutic advances.

Part II
TREATMENT OF MYCOBACTERIAL INFECTIONS

Many antimicrobial drugs are meant for treatment of very specific pathogenic organisms which include bacterial, fungus, protozoan, helminthes, and ectoparasites. This second section of antimicrobial drugs will describe drugs designated for the more specific organisms and, as such, these drugs normally have a very narrow spectrum of activity.

Learning Objectives

Upon completion of this chapter the student should be able to:

1 Describe the sites and mechanisms of action of:
   a. first-line antituberculin drugs (INH, rifamycins, pyrazinamide, and ethambutol [EMB]).
   b. second-line antituberculin drugs (aminoglycosides, ethionamide, *p*-aminosalicylic, cycloserine, and capreomycin).
   c. antileprosy drugs (sulfones, clofazimine).

2 Discuss the mechanism of resistance seen in first-line antituberculin drugs where known.

3 Discuss the SAR of first-line antituberculin drugs.

4 Discuss the physicochemical and pharmacokinetic properties of first-line antituberculin drugs and antileprosy drugs and, where known, identify the chemical group responsible for these properties.

5 Discuss the clinical use of drug combinations in treatment of TB and leprosy.

Introduction to Acid-Fast Bacilli (Mycobacterial Infections)

- Mycobacterial infections are unique in that the mycobacteria are a genus of acid-fast bacilli with an unusual cell envelope.
- In addition to the peptidoglycan layer and plasma membrane common to other bacteria, the mycobacterium outer layer of the cell envelope is rich in lipids (α-mycolic acid) and polysaccharides (poly α-D-arabinose, poly β-D-galactan, and poly α-D-mannan).
- Many of the antibacterials used to treat mycobacterial infections target the cell envelope of the pathogen because of features not found in human cell membranes.
- Infections associated with the mycobacteria include TB (pathogen: *Mycobacterium tuberculosis*), leprosy or Hansen disease (pathogen: *Mycobacterium leprae*), and disseminated *M. avium* complex (MAC) disease (pathogen: *M. avium* and *M. intracellulare*).
- A commonality exists between TB and leprosy in that the diseases are slow developing, may lie dormant for long periods of time, and are difficult to treat because of low antigenicity.
- Disseminated MAC disease is an opportunistic infections commonly found in immunocompromised individuals.

Antibacterial Agents – Tuberculosis

- Combination therapy for treatment of TB targets multiple sites in the mycobacterial cell envelope, Figure 23.15.

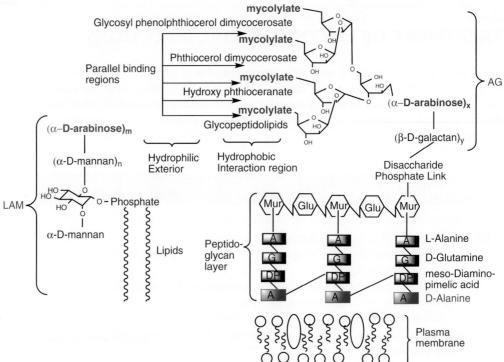

Figure 23.15 Diagrammatic representation of the cell wall/cell envelope of mycobacterium with drug sites of action highlighted in red.

First-Line Agents

Isoniazid (INH)

Isoniazid
(Isonicotinic acid hydrazide, INH)

Nicotinamide adenine dinucleotide (NAD⁺)

MOA

- INH inhibits mycobacterium cell wall synthesis affecting cell permeability.
- INH is a prodrug which is activated within the bacillus through a peroxidase-catalyzed oxidation to a potent radical acylating agents (Fig. 23.16).
- Mycolic acid, an essential component of the mycobacterium cell envelope, is biosynthesized in a step that requires an NADH reduction. NADH plus the enzyme inhA enoyl reductase reduces a double bond in one arm of mycolic acid.
- INH inactivates the reduction via acylation of NAD⁺ thus blocking cell wall synthesis (Fig. 23.16).
- INH is bactericidal.
- Resistance is associated with mutation in peroxidase enzyme.

SAR

- Movement of the acid hydrazide to another location or additional substitution on the pyridine ring decreases activity.
- Replacement of any of the hydrogens on the hydrazide does not improve activity.

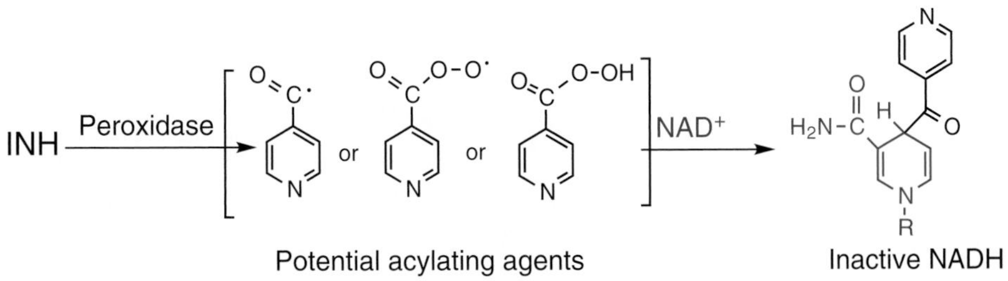

INH $\xrightarrow{\text{Peroxidase}}$ [Potential acylating agents] $\xrightarrow{\text{NAD}^+}$ Inactive NADH

Figure 23.16 Acylation of NAD⁺ by INH.

Physicochemical and Pharmacokinetic Properties

- INH is water soluble, readily absorbed (~100%), and is distributed to the CNS
- Poorly bound to plasma protein

Metabolism

- Inactivation of INH is catalyzed by N-acetyltransferase leading to the acetylated INH. Fast acetylators may require a dose increase to offset this inactivation (Fig. 23.17).
- Acetohydrazide has been associated with liver necrosis possibly catalyzed by hydroxyacetohydrazide.

INH $\xrightarrow{\text{Metabolism}}$ Inactive → Acetohydrazide → [Hydroxyhydrazide (potential toxin)]

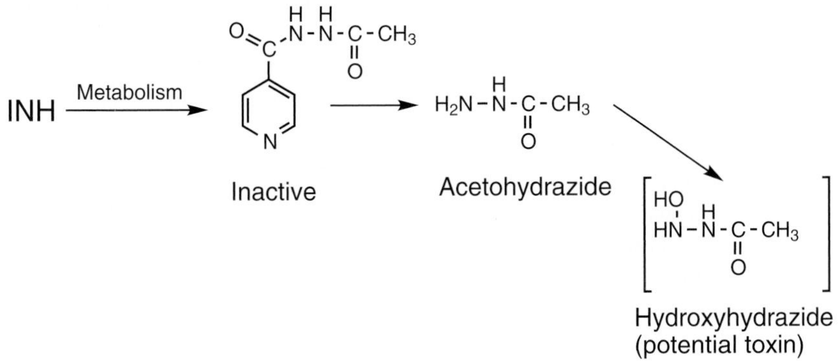

Figure 23.17 Metabolism of INH to inactive drug and potential liver toxin.

> **Fast/Slow Acetylators**
>
> Inuit's and Japanese show higher incidence of fast acetylation (half-life ~70 minutes). Scandinavians, Jews, and North African whites show higher incidence of slow acetylation (half-life ~2 to 5 hours).

Clinical Applications

- Monotherapy with INH is used in patients with dormant or latent mycobacterial infections.
- For active TB, INH is used in combination therapy – see below.

> **DRUG–DRUG INTERACTIONS**
>
> INH is an inhibitor of CYP2C19 and CYP3A and could increase toxicity of acetaminophen and diazepam. Aluminum ion reduces absorption of INH. Dose 2 hours apart to prevent this interaction.

Rifamycin Antibiotics

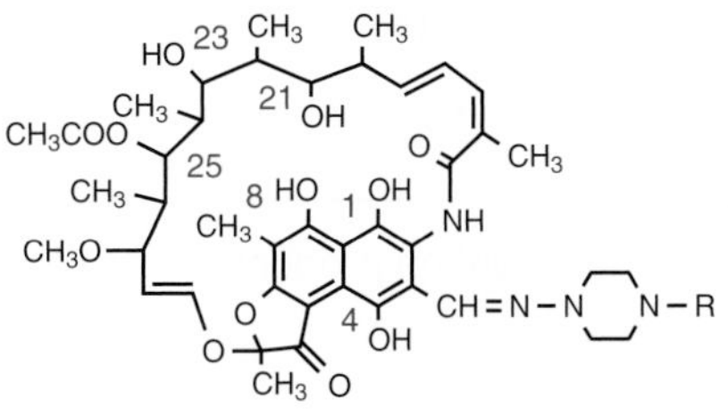

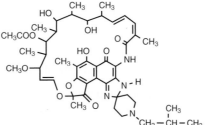

Rifampin (Rifadin, Rimactane) R = CH₃
Rifapentine (Priftin) R =

Rifabutin (Mycobutin)

> **ADVERSE EFFECTS**
>
> Rash, abnormal liver function tests, hepatitis, hepatotoxicity, and peripheral neuropathy. The neuropathy is related to the structural similarity of INH and pyridoxine/pyridoxal. Pyridoxine can be administered with INH.
>
> INH   Pyridoxal

MOA

- The rifamycins block bacterial DDRP resulting in inhibition of RNA synthesis.
- Binding of the rifamycins to DDRP occurs through (Fig. 23.18):
  - π–π bonding between naphthalene and aromatic amino acids of DDRP.
  - Chelation occurs between the C1 and C8 phenolic groups of the drug and zinc ion, a component of DDRP.
  - Hydrogen bonding occurs between C21 and C23 alcohols to the DDRP.

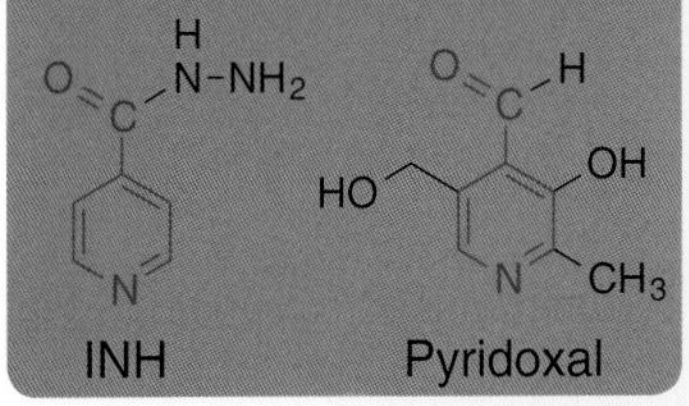

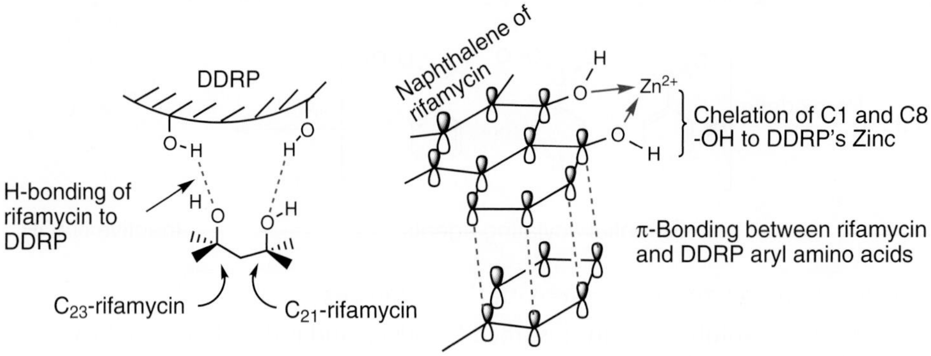

Figure 23.18 Binding of rifamycin to DNA-dependent RNA polymerase (DDRP).

CHEMICAL NOTE

Rifaximin: An analog of rifamycin, structural modification at the 3,4 position, is approved to treat travelers diarrhea and hepatic encephalopathy (Xifaxan). The drug is administered orally.

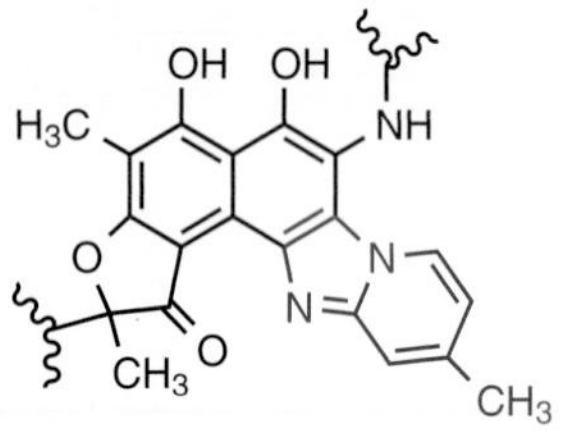

- The rifamycins are active against rapidly dividing intra- and extracellular bacilli leading to bactericidal activity.
- Mutations in DDRP often lead to drug resistance.

SAR

- Free –OH groups are required at C1, C8, C21, and C23 (i.e., acetylation of C21 or C23 destroys activity).
- Reduction of any of the double bonds in the macrocyclic ring progressively reduces biologic activity.
- Opening the macrocycle results in an inactive compound.
- Substitution at C3 and/or C4 gives varying degrees of antibacterial activity.

Physicochemical and Pharmacokinetic Properties

- Due to the extended conjugation the rifamycins cause a red or orange coloration of urine, stools, and sweat and tears (Fig. 23.19).
- Rifampin (RIF) and rifapentine are prone to air oxidation converting the hydroquinone portion of the drugs (1,4-dihydroxynapthaline) into a 1,4-quinone (Fig. 23.19).
- The rifamycins are zwitterionic in nature.
- The Schiff's base portion (at C3) of rifamycin is prone to hydrolysis in acid media.
- Food effects absorption (decrease with rifampin; increase with rifapentine).
- Highly bound to plasma protein (rifampin 89%, rifabutin 85%, rifapentine ~98%).
- Variable bioavailability (rifampin ~90%, rifabutin ~20%, rifapentine depends on meal).

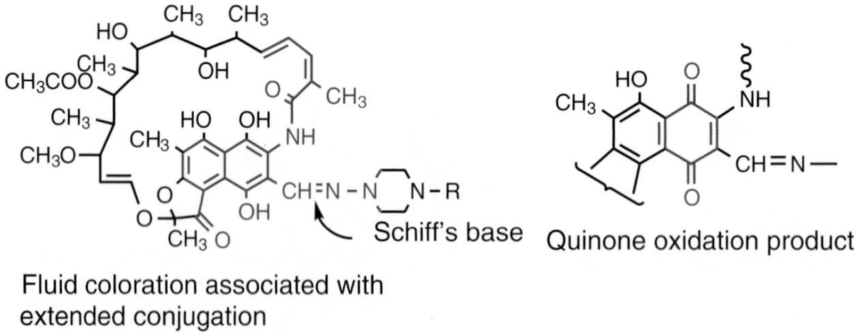

Figure 23.19 Physicochemical features of rifamycin indicated in red.

Fluid coloration associated with extended conjugation

Schiff's base   Quinone oxidation product

Metabolism

- The major route of human metabolism is C25 deactylation. The metabolite is still active.

Clinical Applications

- Active against most gram-positive and many gram-negative organisms.
- Commonly used in combination therapy for TB (see below).

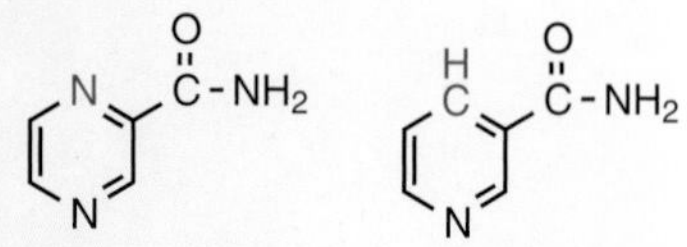

Pyrazinamide   Nicotinamide

Pyrazinamide

- Pyrazinamide is a bioisostere of nicotinamide.

MOA

- Pyrazinamide is thought to function as a prodrug. Susceptible organisms hydrolyze pyrazinamide through the action of pyrazinamidase to the pyrazinoic acid.
- Pyrazinoic acid lowers the pH in the *M. tuberculosis* environment inhibiting its growth and can also penetrate the cell membrane lowering the pH in the cytoplasm with a bactericidal effect.
- Resistant *M. tuberculosis* exhibits a point mutation leading to reduced hydrolysis of pyrazinamide to pyrazinoic acid.
- As a bioisostere of nicotinamide pyrazinamide/pyrazinoic acid may interfere with mycolic acid synthesis.
- Pyrazinamide is active at pH 5.5, but inactive at pH 7.

Physicochemical and Pharmacokinetic Properties

- Pyrazinamide is rapidly absorbed following oral administration with excellent bioavailability.
- The drug is also rapidly metabolized to the active drug, pyrazinoic acid.

Metabolism

- The common metabolites formed following oral administration of pyrazinamide are pyrazinoic acid and 5-hydroxypyrazinoic acid which is conjugated with glycine.

Clinical Applications

- Pyrazinamide is active against the semidormant intracellular form of *M. tuberculosis*.
- When used in combination with other first-line drugs pyrazinamide has significantly lowered the patient treatment time (see combination therapy below).

Ethambutol (EMB)

MOA

- EMB inhibits the enzyme arabinosyl transferase which catalyzes the polymerization of β-D-arabinofuranosyl-1-monophosphate to various polyarabinose components of the mycobacterial cell envelope (arabinogalactan [AG] and lipid arabinomannan [LAM])(Fig. 23.15).
- Resistance is associated with overexpression of arabinosyl transferase.
- Damage to the cell envelope confers antibacterial action (bacteriostatic at low dose).

Physicochemical and Pharmacokinetic Properties

- (+) EMB is the biologically active drug. The (−) isomer and the symmetrical isomer are not effective drugs.
- EMB is water soluble with most of drug excreted unchanged.
- Good bioavailability (~80%).
- Metabolic inactivation is associated with oxidation of the primary alcohol to an aldehyde and carboxylic acid via alcohol and aldehyde dehydrogenase, respectively.

Clinical Applications—Multidrug Therapy

- TB is difficult to treat because the mycobacterium exists in three pools: extracellular, intracellular, and necrotic pools within macrophages.
- The *M. tuberculosis* is a slow growing organism requiring long-term therapy.
- Drug resistance to first-line drugs is quite common leading to multiple drug-resistant TB (MDR-TB) or extensively drug-resistant TB (XTR-TB).
- Multidrug therapy including second-line drugs may be part of the treatment regimen.
- Drug resistance to both INH and RIF calls for the use of an aminoglycoside, a fluoroquinone, along with EMB, ethionamide, and pyrazinamide.

DRUG–DRUG INTERACTIONS

The rifamycins, especially rifampin, are strong cytochrome P-450 inducers. As a result, the potential exists for drug–drug interaction with those agents which are substrates for specific P-450 metabolism. Among the isoforms of P-450 induced by rifampin are 1A2, 2C9, 2C19, and 3A4. As a result, clinically significant drug–drug interactions may occur with a variety of drugs including oral contraceptives, barbiturates, several azole antifungals, cardiac agents (quinidine, digoxin, mexiletine, tocainide, metoprolol, verapamil), the anticoagulant coumarin, and the hypoglycemic sulfonylureas.

ADVERSE EFFECTS

Pyrazinamide: The most serious adverse effect associated with pyrazinamide is hepatotoxicity which may occur in up to 15% of the long-term users of the drug. The drug inhibits the excretion of uric acid which could lead to gout. Other adverse effects include GI disturbance leading to nausea, vomiting, and anorexia.

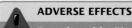

Ethambutol

- Resistance to all first-line drugs results in using a treatment combination of an aminoglycoside, a fluoroquinone, ethionamide, PAS, and cycloserine.
- "The cardinal rules" for drug therapy in the treatment of TB are:
  - Get drug susceptibility data on all drugs as soon as possible.
  - Drug therapy begins with at least three of the four first-line drugs (INH, RIF, and pyrazinamide) and preferably all four first-line agents.
  - Avoid using a single drug for TB therapy.
  - Always add at least two drugs to a failing regimen.
- The initial phase of drug therapy normally lasts for 8 weeks and is followed by a continuous phase of treatment lasting 4 to 7 months. During this phase INH and RIF are used.
- Patient compliance is extremely important and direct observation therapy (DOT) is recommended.
- INH may be used as a single agent when prophylaxis therapy is called for. This may last for 6 or 12 months.

Second-Line Drugs

- Resistance to INH and rifampin is the defining factor in MDR-TB which calls for the use of second-line drugs.
- Second-line drugs tend to be less well tolerated or cause a high incidence of adverse events.

Aminoglycosides

Streptomycin

Kanamycin A (R₁ = H)
Amikacin (R₁ = COCHOHCH₂CH₂NH₂)

MOA

- Inhibit protein synthesis by virtue of their ability to bind to the 16S unit within the 30S ribosomal subunit of bacteria (see aminoglycosides and Fig. 23.10).
- Irreversible binding ultimately leads to formation of nonsense proteins killing the bacteria.
- For additional information see discussion of these aminoglycosides in Part I of this chapter.

Clinical Applications

- In case of resistance to INH streptomycin may be added to the multidrug therapy.
- In MDR-TB one of the injectable aminoglycosides (amikacin or kanamycin) is commonly administered.
- High- or low-level resistance to the aminoglycosides may occur leading to XTR-TB.
- Resistance may involve genetic mutations leading to increased efflux of drug, mutations leading to reduced binding of drug to 16S, or increased activity of acetyltransferase leading to inactive aminoglycosides.

Ethionamide

MOA

- Ethionamide, a derivative of isonicotinamide, is a prodrug which requires oxidative activation via a flavin monooxygenase enzyme (EtaA) to an acylating agent (Fig. 23.20).

Ethionamide

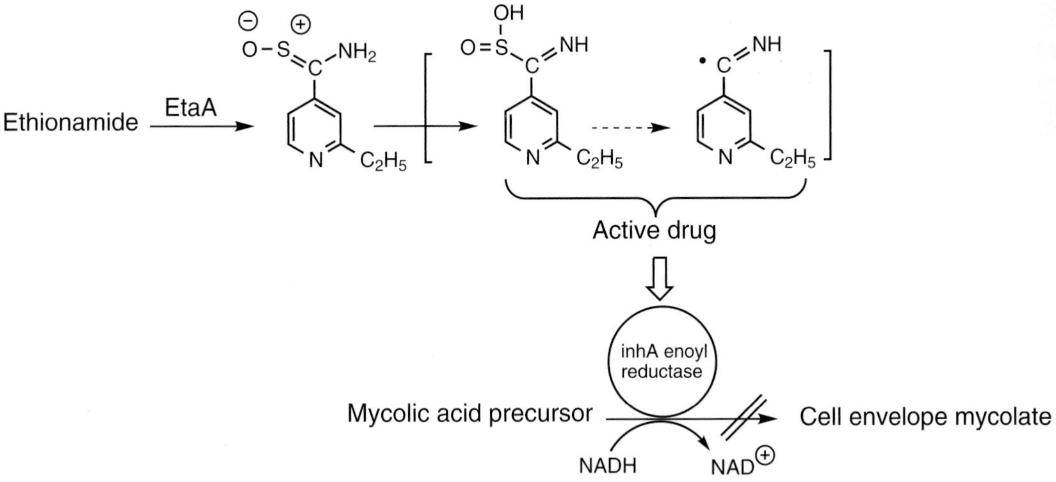

Figure 23.20 Activation and acylation of inhA enoyl reductase by ethionamide.

- Similar to INH, ethionamide acetylates within inhA enoyl reductase blocking mycolic acid synthesis.
- Site of action of ethionamide appears to be the same as that of INH.

Physicochemical and Pharmacokinetic Properties
- Orally active with nearly 100% bioavailable.
- Extensively metabolized (<1% excreted unchanged) with metabolites eliminated via the kidney.

Metabolism
- Inactive metabolites include 2-ethylisonicotinamide, N-methyl-2-ethyl-6-hydroxyisonicotinamide, and N-methyl-2-ethyl-6-hydroxythioisonicotinamide.

> **ADVERSE EFFECTS**
> GI irritation (nausea and vomiting), hepatitis, hypersensitivities, dizziness, depression and fatigue.

---

p-Aminosalicylic Acid (PAS)
- Once a very popular drug, but resistance, and adverse effects have reduced its value.

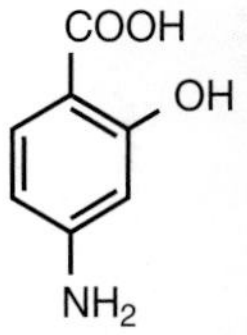

MOA
- PAS is an antimetabolite of PABA blocking folic acid synthesis with bacterio-static action (see mechanism of action of sulfonamides).
- Blocks acetylation of INH when used in combination with INH and thus reducing inactivation of INH.

Para-aminosalicylic acid

D-Cycloserine

D-Cycloserine
(Seromycin)

D-Alanine

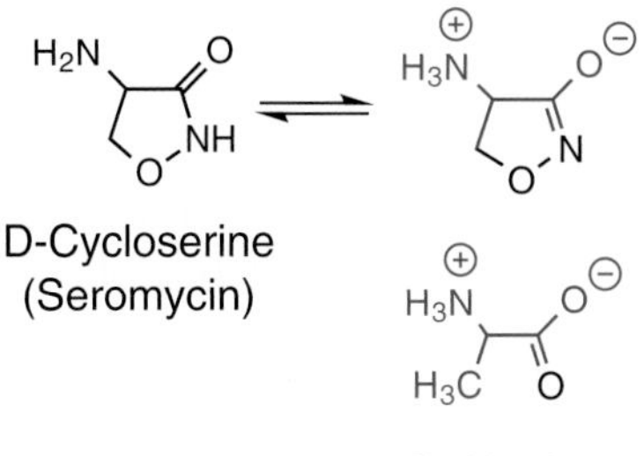

> **ADVERSE EFFECTS**
> High incidence of side effects including hypersensitivities leading to skin eruptions, hepatitis, and liver failure, and possible death.

MOA
- D-Cycloserine resembles D-alanine and is an antimetabolite of D-alanine.
- D-Cycloserine is a competitive inhibitor of two key enzymes unique to myco-bacteria, D-alanine racemase, and D-alanine ligase.
- D-Cycloserine blocks the racemase enzyme involved in the conversion of L-alanine to D alanine. The resulting D-alanine couples to itself via D-alanine ligase.

ADVERSE EFFECTS

Commonly involve CNS effects: headache, depression, psychosis, convulsions.

- D-Alanine-D-alanine is an intermediate in the cell wall biosynthesis. D-alanine is an important component in the mycobacterium cell wall (Fig. 23.15).

Capreomycin

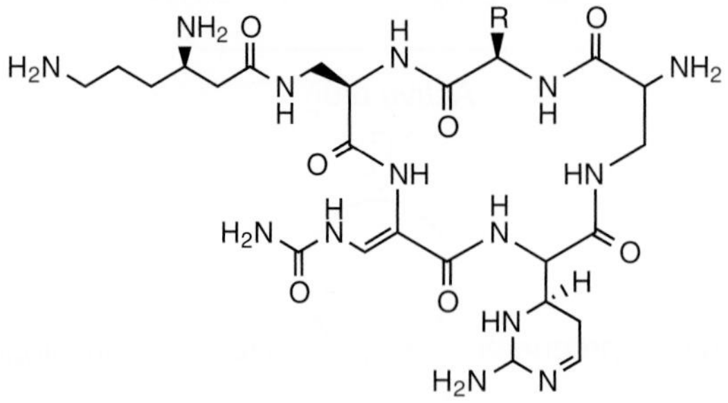

Capreomycin (Capastat)

- Capreomycin is a mixture of cyclic peptides with capreomycin Ia (R = OH) and capreomycin Ib (R = H) being the major components.
- The action of capreomycin may be similar to that of viomycin (a discontinued cyclic peptide), which blocks protein synthesis within bacteria by interfering with protein chain elongation.
- Capreomycin is administered IV or IM.

ADVERSE EFFECTS

Potential nephrotoxicity, ototoxicity, auditory damage, especially in patients with renal impairment. Hypersensitivity also possible.

Fluoroquinolones

- Fluoroquinolones have not received USFDA approval for treatment of TB, but several are being used in clinical studies.
- Several fluoroquinolones have proven to be of value as second-line drugs in treating mycobacterium infections especially in MDR-TB.
- The antibacterial mechanism of action of fluoroquinolones has previously been discussed (see synthetic antibacterials in Part I of this chapter).
- Moxifloxacin and levofloxacin are the most commonly used second-line anti-TB drugs.

Newly Approved Drug

Bedaquiline Fumarate

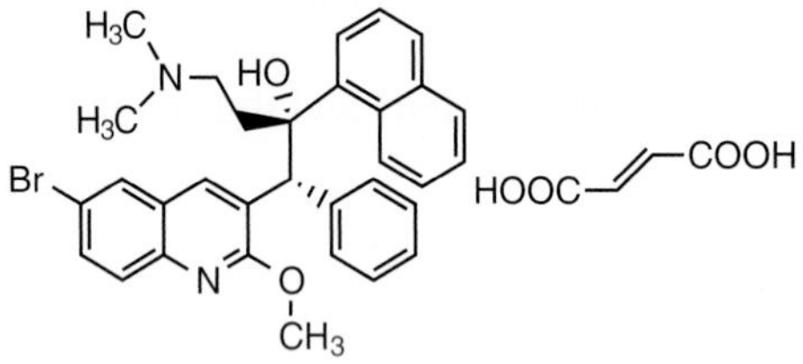

Bedaquiline (Sirturo)

- Approved for treatment of MDR-TB
- Should be administered in combination with three or four other drugs (i.e., ethionamide, kanamycin, pyrazinamide, ofloxacin, and cycloserine/terizidone – not available in the United States).

MOA

- Inhibits the cell membrane–bound ATP synthase C subunit of *M. tuberculosis* without blocking action of human mitochondrial ATP synthase.
- Prevents conformational changes in the synthase and blocks proton flow necessary for production of ATP and bacterial energy.

SAR

Critical structural features required for binding to the ATP synthase C subunit include tertiary amine, tertiary alcohol, halide, and ether oxygen (Fig. 23.21).

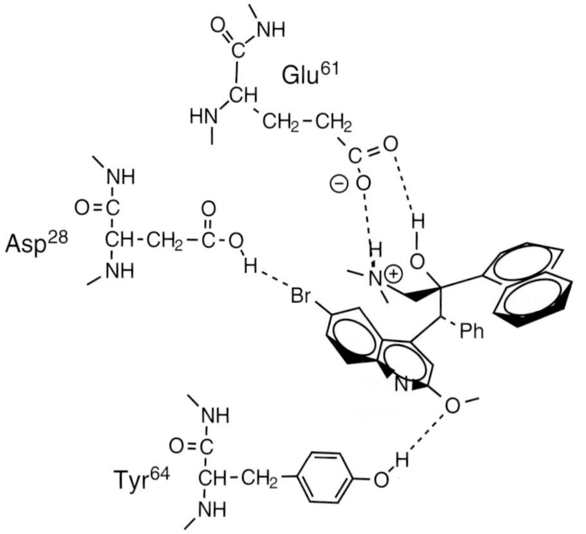

Figure 23.21 Binding of bedaquiline and amino acids in ATP synthase C subunit of *M. tuberculosis*.

Physicochemical and Pharmacokinetic Properties
- Readily absorbed orally and highly bound to plasma protein (>99%).

Metabolism
- CYP3A4 metabolized demethylation to desmethylbedaquiline with reduced activity.
- Bedaquiline should not be used with CYP3A4 inducers.
- Desmethylbedaquiline and bedaquiline have additive activity.

BLACK BOX WARNING

Has been attached to bedaquiline warning of possible additive QT interval prolongation. The patient should refrain from use of alcohol and the drug has the potential for hepatic-related adverse reactions.

MAC (Mycobacterium Avium-Intracellulare Complex) Antibacterials

- Infections by the *M. avium-intracellulare* complex, an opportunistic infection, is quite common in HIV patients.
- The protocol calls for prophylaxis therapy for MAC disease.
- Azithromycin or clarithromycin (macrolide antibiotics – see Part I of this chapter) is commonly prescribed.
- Treatment of MAC in HIV patients usually consists of clarithromycin plus EMB. A third drug can be added to this regimen and may include rifabutin, amikacin, or a fluoroquinone.

Leprosy (Hansen's Disease) Antibacterials

- Leprosy is a slow-developing infection caused by *M. leprae* and *Mycobacterium lepromatosis*.
- The disease has a very long incubation period and a low level of transmission which usually occurs via nasal droplets.
- There has been a significant drop in the incidence of leprosy over the last 50 years with the highest incidence seen in India, Brazil, and parts of Africa.

Dapsone (DDS)

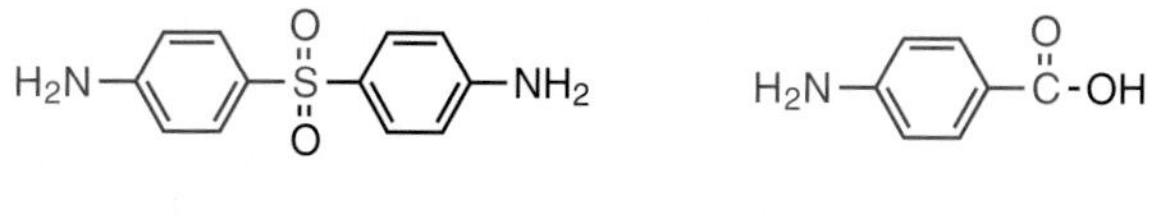

Diaminodiphenylsulfone (Dapsone)   *p*-Aminobenzoic acid

MOA
- Competitive inhibitor of PABA incorporation into folic acid (see sulfonamides Part I of this chapter).
- Dapsone is leprostatic in action.

SAR

- Replacement of one aromatic ring with a heterocyclic aromatic group gives a less active product.
- Substitution on the aromatic ring reduces activity.
- Addition of water-solubilizing groups to the *para*-amino groups give solubility, but requires reformation of the amino group for activity.

Physicochemical and Pharmacokinetic Properties

- Nearly water insoluble.
- pKa ~1.0, because of low pKa water-soluble salts are not practical.

Metabolism

- DDS is inactivated via acetylation or hydroxylation at one of the *para*-amino groups.
- The N-hydroxydiaminodiphenylsulfone may be responsible for the adverse effect of methemoglobinemia.

Clofazimine (Lamprene)

Clofazimine

- Clofazimine is an insoluble red dye, which is used in combinations for treatment of leprosy.
- The imino group at C2 and the *p*-chloroanilines found at C3 and C10 are essential for activity.

MOA

- Clofazimine is thought to act by generation of reactive oxygen species (i.e., superoxide anion) which have destructive affects on the pathogen.

Clinical Applications

- Chemotherapy for leprosy consists of combination therapy consisting of DDS, clofazimine, and rifampin.
- Therapy may take from 1 to 2 years or until the organism is not identified in skin smears.

Thalidomide (Thalomid)

Adjunct Therapy for Leprosy

Thalidomides

- A complication associated with the chemotherapy of leprosy is the condition of erythema nodosum leprosum (ENL) which consists of subcutaneous nodules which are painful, tender, and inflamed. ENL appears to be a hypersensitivity reaction associated with the chemotherapy of leprosy.
- Thalidomide has proven effective in treatment of ENL.
- Thalidomide's mechanism of action is thought to be associated with its ability to control cytokines which can produce inflammation.
- Thalidomide inhibits tumor necrosis factor-α.
- Thalidomide is a potent teratogen and its human use must be tightly controlled in women of childbearing age.

Chemical Significance

Mycobacteria are acid-fast bacilli that belong to the Mycobacteriaceae family. This family of organisms differs from other bacterial pathogens and are responsible for the human diseases of TB, leprosy, and MAC disease commonly associated with AIDS. This class of organisms has a unique cell wall, the diseases tend to be chronic in nature, and the drug therapy designed to treat the mycobacterial infections differs from the therapy used to treat other bacterial infections. It is estimated that approximately one-third of the world's population has been infected with *M. tuberculosis* although most of these infections are never diagnosed. Worldwide it was estimated that in 2012, there were 8.6 million new cases with 1.2 to 1.4 million deaths, of which 24% of those death occurred in individual coinfected with HIV. In the last

25 years, concern has been mounting about the development of multidrug-resistant TB (MDR-TB) and extensively drug-resistant TB (XTR-TB). Leprosy (Hansen's disease) caused by *M. leprae* and *M. lepromatosis* is similar to TB in that it is also a slow-developing chronic disease. In 2012, it was estimated that less than 200,000 new cases of the disease were diagnosed worldwide. Worldwide travel suggests an increased need for treatment options to these diseases and thus, the expectation for new therapies in the future.

Although the direct use of medicinal chemistry to clinical situations involving mycobacterial medications is somewhat limited due in part to the diversity of therapeutic agents there are two immediate applications of where an understanding of medicinal chemistry translates to a prediction of physicochemical properties of several drug classes. The rifamycin antibiotics and clofazimine are drugs, which produce coloration of body fluids of the patient. Extended conjugation, that is alternating single and double bonds and unshared pairs of electrons on nitrogen, is responsible for their coloration and is a hallmark associated with these drugs. Another distinctly chemical property of the rifamycins is their instability in the presence of air in which oxidation occurs at the *para*-hydroquinone nucleus embedded within the structure of these antibiotics. Identification of this chemical moiety allows for a prediction of this chemical instability.

Review Questions

1 What portion of the 4-quinolone is essential for binding of the drug to its site of action (pharmacophore)?

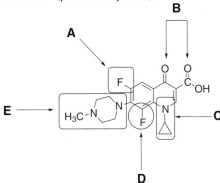

2 The mechanism of action of trimethoprim:
   A. Blocks the synthesis of dihydropteroic acid.
   B. Blocks the electron transport system of the bacteria.
   C. Blocks dihydrofolate reductase.
   D. Blocks t-RNA binding to m-RNA.
   E. Trimethoprim is incorporated into folic acid in place of *p*-aminobenzoic acid.

3 The basis for the antibacterial action of β-lactams is that these drugs become **bound** to what portion of the cell wall?
   A. Pentaglycine
   B. Mycolic acid
   C. D-alanine-glycine cross-linked
   D. D-alanine-D-alanine
   E. Transpeptidase (PBP)

4 The commonly accepted mechanism of action of the tetracyclines is:
   A. Damage to the cell wall by binding to D-alanine-D-alanine in the peptidoglycan portion of the cell wall.
   B. Inhibition of protein synthesis via binding to the 50S rRNA.
   C. Inhibition of protein synthesis via binding to the 30S rRNA.
   D. Inhibition of protein synthesis via binding to the 70S rRNA.
   E. Damage to the cell wall by blocking glycan polymerization in the peptidoglycan portion of the cell wall.

5 The cladinose in clarithromycin is thought to promote what side effect?
A. Bad taste
B. Ototoxicity
C. Heptatotoxicity
D. Stomach irritation
E. Hypersensitivity

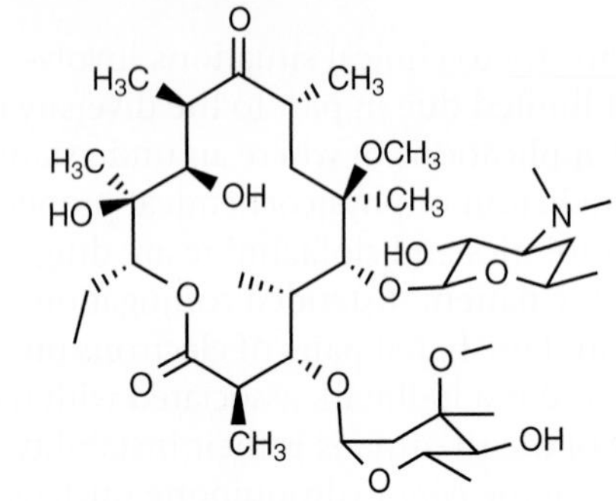

Clarithromycin

6 INH is a prodrug. Which of the following structures is considered the active intermediate of INH?

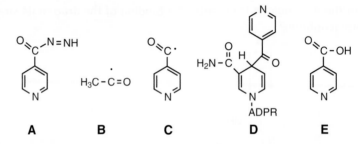

7 If you are dispensing rifampin, what counseling statement should be made to the patient concerning the use of this medication?
A. There may be a red discoloration of body fluids.
B. The drug may produce GI disturbances and it is best to be taken with a meal.
C. Take with large quantities of water to prevent possible damage to the kidney.
D. Report any changes in your hearing since this may be an early sign of a side effect.
E. Do not administer with metal containing antacids.

8 In the drug combination: INH, rifampin, pyrazinamide, and EMB, where is the site of action of EMB?
   A. Site A
   B. Site B
   C. Site C
   D. Site D
   E. Site E

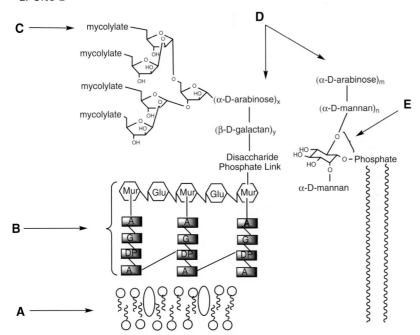

9 Simply by viewing the structure of clofazimine one would predict a _____.
   A. very water-soluble drug.
   B. drug with dye characteristics.
   C. drug readily hydrolyzed by nonspecific esterase.
   D. drug which is unstable to the acidic conditions of the GI tract.
   E. drug prone to air oxidation.

10 Diaminodiphenylsulfone is prescribed for the treatment of Hansen's Disease. The mechanism/site of action of this drug is ____.
   A. within the cell wall of the organism.
   B. in the respiratory enzyme system of the organism.
   C. in the pathway leading to folic acid
   D. in the peptidoglycan portion of the cell wall.
   E. in the mycolic acid synthesis pathway.

24

Antifungal Agents

Learning Objectives

Upon completion of this chapter the student should be able to:

1. Identify, by name, the fungal organisms and diseases responsible for topical and/or systemic fungal infections.

2. Describe the sites and mechanisms of action of the following classes of antifungal drugs:
   a. polyenes
   b. azoles
   c. allylamines
   d. echinocandins
   e. miscellaneous antifungal drugs including: flucytosine, griseofulvin, tolnaftate, and tavaborole.

3. Identify clinically significant physicochemical and pharmacokinetic properties of the various classes of antifungal agents.

4. Identify metabolic instabilities with chemical structures indicating the effects of metabolism on activity of the systemic antifungal agents.

5. Describe the basis for drug–drug interactions associated with the systemic azoles.

6. Discuss the clinical application of the antifungal drugs.

Introduction to Fungal Infections

- Fungal kingdom includes yeast, molds, rusts, mushrooms.
- Most fungi are not pathogenic to humans, but those with infectious properties are shown in Table 24.1.
- Immunocompromised patients are most susceptible to systemic fungal infections (e.g., transplant, HIV, cancer patients).

Table 24.1 Infectious Fungal Organisms

Disease State	Common Organisms	Topical/Systemic
Dermatomycosis: • Tinea capitis • Tinea pedis (athlete foot) • Tinea cruris (jock itch) • Tinea unguium • Ringworm	*Epidermophyton* *Microsporum* *Trichophyton*	Topical
Aspergillosis	*Aspergillus fumigatus, niger, flavus*	Systemic
Blastomycosis	*Blastomyces*	Systemic
Candidiasis	*Candida albicans*	Topical/systemic
Coccidioidomycosis, valley fever	*Coccidioides immitis*	Systemic
Crytococcosis	*Cryptococcus neoformans*	Systemic
Histoplasmosis	*Histoplasma capsulatum*	Systemic
Mucormycosis	*Rhizopus, Mucor, Absidia*	Topical/systemic
Pneumocystosis	*Pneumocystis jirovecii*	Systemic

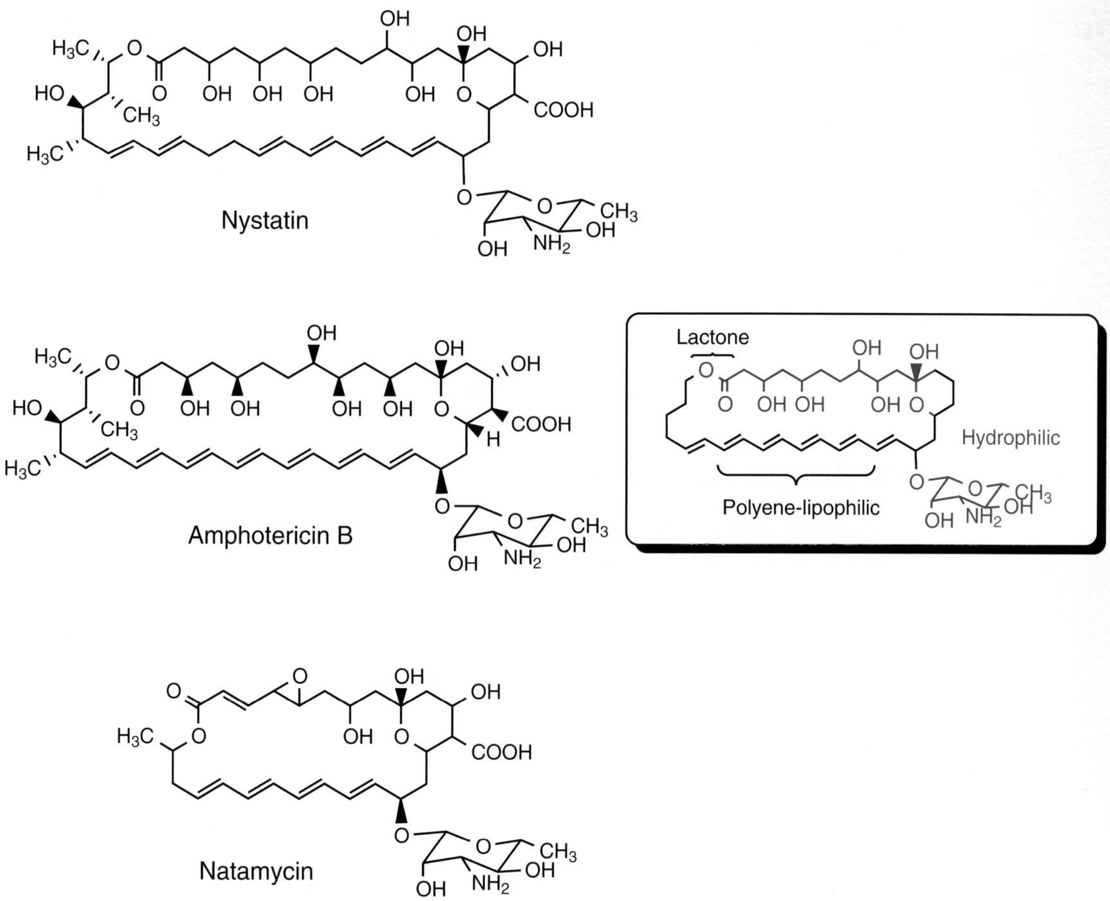

Figure 24.1 Commercially available polyenes. Inset indicates chemical features of this class of antifungal agents.

- Unique characteristics of fungal cells which offer sites for antifungal drugs include:
  - Cell wall composed of chitin and β–1,3-D-glucans (1,3-β-links of D-glucose polysaccharides).
  - Cell membrane composed of ergosterol and zymosterol rather than cholesterol found in human cells.

Antifungal Polyenes

- Macrocyclic lactones with four or more conjugated alkenes (polyene).
- Drugs possess lipophilic and hydrophilic regions (Fig. 24.1). As a result, these drugs are considered to display amphoteric properties.
- Polyenes tend to have severe systemic toxicity.

MOA

- The polyenes bind to ergosterol in the fungal cell membrane through the lipophilic portion of the polyene.
- They form a pore in the membrane, involving the hydrophilic portion of the drug, resulting in cell leakage and death.

Physicochemical and Pharmacokinetic Properties

- Not absorbed orally (creams and lotions used topically – nystatin).
- IV for systemic use (amphotericin B).
- Highly bound to plasma protein.
- Selective delivery through complexation (Table 24.2). Liposomal and lipid complexes decrease toxicity and may improve topical penetration at the site of infection.
- Poor penetration of the CNS.

> **Polyene Antifungal Spectrum**
>
> *Active against yeast: Candida albicans* (topical and systemic), *Cryptococcus neoformans;* mycoses: *Histoplasma capsulatum, Blastomyces dermatitidis, Coccidioides immitis;* mold: *Aspergillus fumigatus;* filamentous fungi: *Fusarium solani.*

⚠️ **ADVERSE EFFECTS**

Amphotericin B: Potential nephrotoxicity associated with amphotericin B nephrotoxicity: C-AMB>ABLC>L-AMB>ABCD; fever and chills, anemia, headache, nausea, vomiting, and weight loss.

Table 24.2   Polyene Formulations

Generic Name	Trade Name	Dosage Form
Nystatin	Mycostatin, Nilstat, Nystex	Cream, ointment, powder
Natamycin	Natacyn	Ophthalmic solution
Amphotericin B	Abelcet[a] (ABLC)	Injectables
Lipid complex	Amphotec[b] (ABCD)	
Liposomal	AmBisome[c] (L-AMB)	
Desoxycholate	Fungizone[d] (C-AMB)	

[a]*Complexation* with L-α-dimyristoylphosphatidylcholine (DMPC) and L-α-dimyristoylphosphatidylglycerol (DMPG).
[b]*Complexation* with sodium cholesteryl sulfate.
[c]*Intercalated* into a liposomal membrane of phosphatidylcholine, cholesterol, distearoylphosphatidylglycerol.
[d]*Discontinued.*

Antifungal Azoles

- Drugs based upon the azole pharmacophore in which the heterocycle is either a triazole or an imidazole.

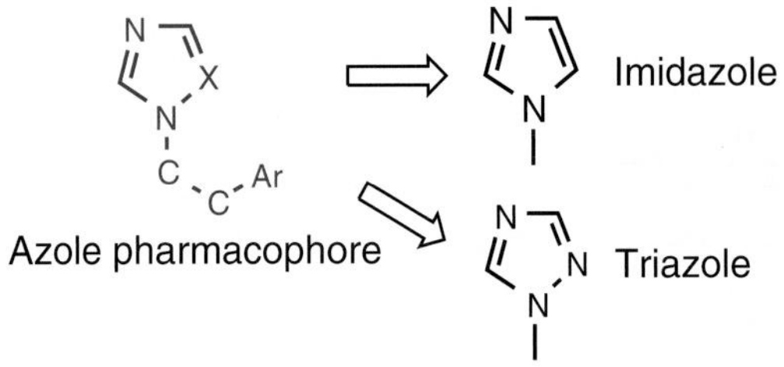

MOA

- Azoles inhibit the oxidative 14-demethylation of lanosterol blocking ergosterol synthesis (Fig. 24.2 Step 2).

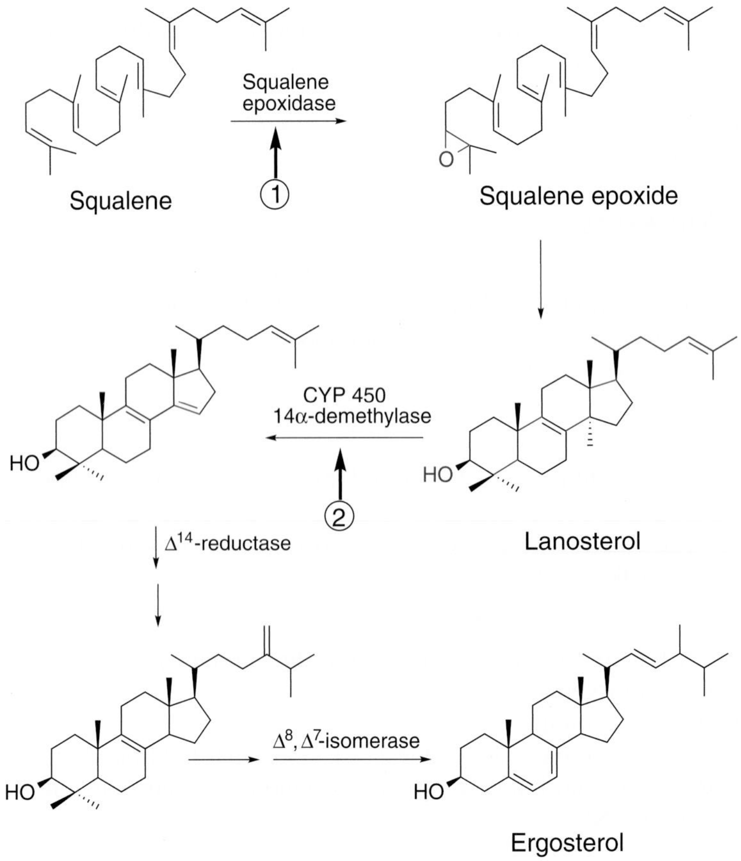

Figure 24.2 Key steps in the biosynthesis of ergosterol by fungi. Enzymatic steps known to be the site of action of currently employed antifungal agents are indicated by a *heavy black arrow* and a *number*.

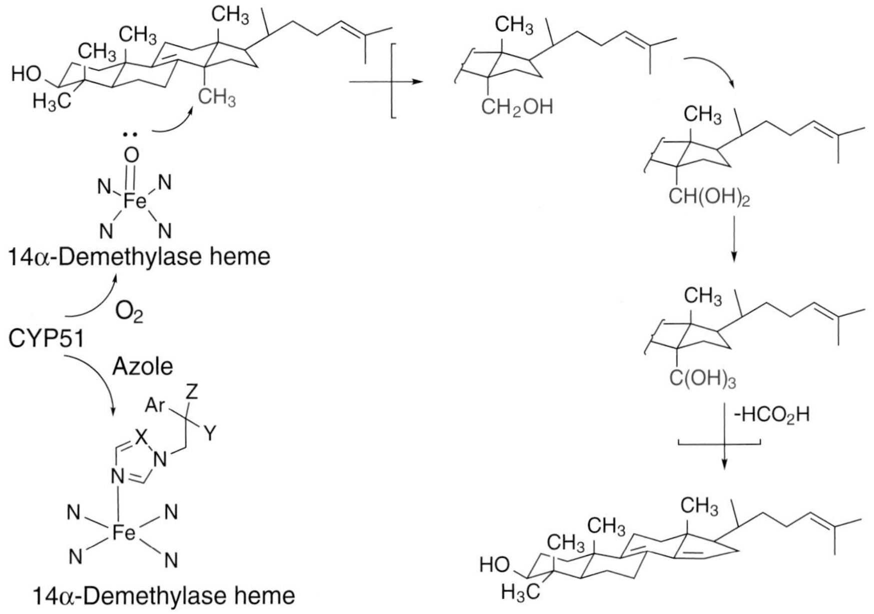

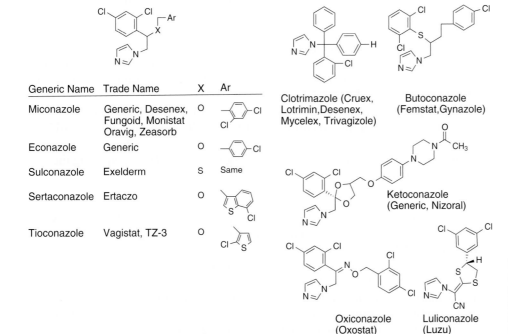

Figure 24.3 Demethylation of the 14α-methyl group from lanosterol via the CYP450 enzyme sterol 14α-demethylase, CYP51. Three successive heme-catalyzed insertions of activated oxygen into the three carbon–hydrogen bonds of the 14α-methyl group which raises the oxidation state of the methyl group to a carboxylic acid. The azoles bind to CYP51 through the N3 atom of the azole preventing oxygen transfer.

- Azoles bind to 14α-demethylase (CYP51) inhibiting the catalytic oxidation of the 14-methyl group (Fig. 24.3).
- The N3 atom is essential for binding to the 14α-demethylase.
- Increased levels of 14α-methyllanosterol, 4,14-dimethylzymosterol, and 24-metheylene dihydrolanosterol result from this inhibition weakening the fungal cell membrane leading to a leaky membrane.
- Median inhibitory concentration (IC_{50}) for fungal 14α-demethylase is much lower than for human 14α-demethylase.

Physicochemical and Pharmacokinetic Properties

- Topical imidazoles (Fig. 24.4) tend to be rapidly metabolized if used systemically, increasing toxicity, and as a result, are not used systemically.
- Topical products are available as creams, powders, gels, solutions, suppositories, lotions, ointments.
- Systemic products are available as tablets, capsules, and powders for oral suspension, injectable solutions.

Generic Name	Trade Name	X	Ar
Miconazole	Generic, Desenex, Fungoid, Monistat Oravig, Zeasorb	O	2,4-diClphenyl
Econazole	Generic	O	4-Clphenyl
Sulconazole	Exelderm	S	Same
Sertaconazole	Ertaczo	O	benzothiophene-Cl
Tioconazole	Vagistat, TZ-3	O	Cl-thiophene-methyl

Clotrimazole (Cruex, Lotrimin, Desenex, Mycelex, Trivagizole)

Butoconazole (Femstat, Gynazole)

Ketoconazole (Generic, Nizoral)

Oxiconazole (Oxostat)

Luliconazole (Luzu)

Figure 24.4 Imidazole antifungal agents. All imidazoles are used topically with the exception of ketoconazole which is available in both topical and systemic dosage forms.

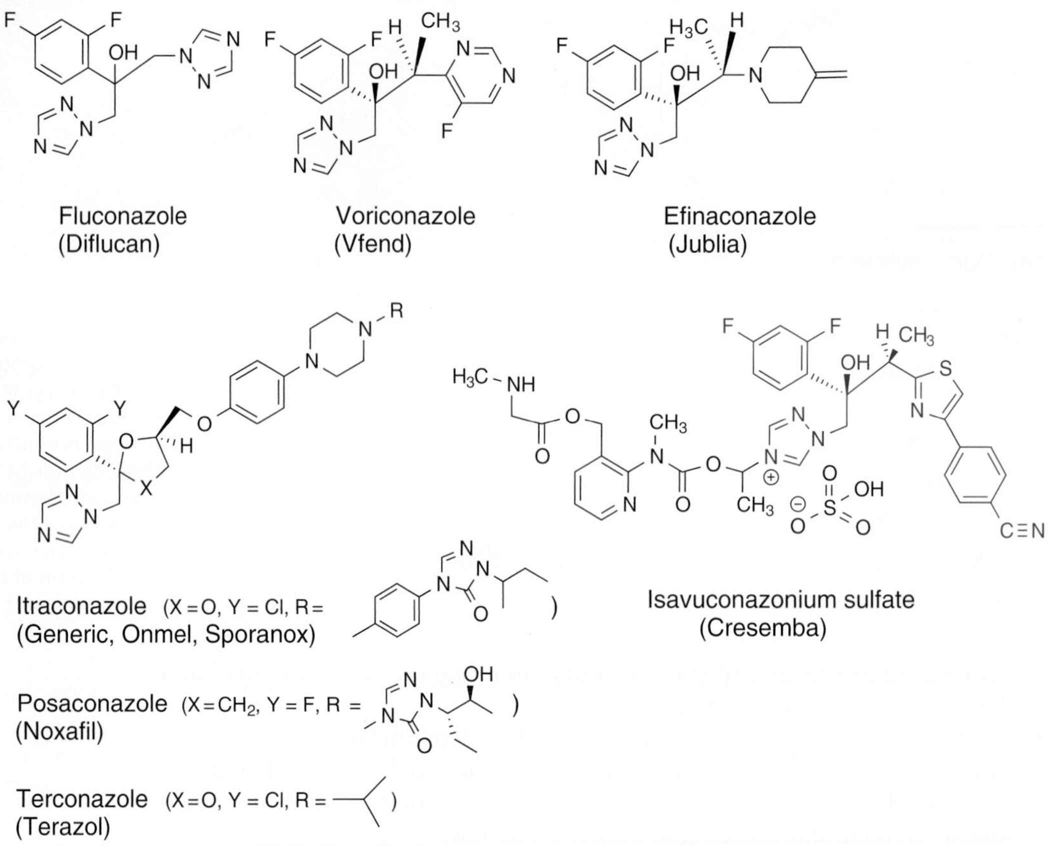

Fluconazole
(Diflucan)

Voriconazole
(Vfend)

Efinaconazole
(Jublia)

Itraconazole (X = O, Y = Cl, R =
(Generic, Onmel, Sporanox)

Posaconazole (X = CH₂, Y = F, R =
(Noxafil)

Terconazole (X = O, Y = Cl, R =
(Terazol)

Isavuconazonium sulfate
(Cresemba)

Figure 24.5 Triazole antifungal agents. While efinaconazole and terconazole are used topically, the remaining triazoles are used systemically.

DRUG–DRUG INTERACTIONS

As indicated in Table 24.3, systemic azoles are inhibitors of various CYP450 isoforms and several azoles are substrates for CYP450 isoforms. Ketoconazole's action is significantly reduced in the presence of phenytoin, carbamazepine, and rifampin. Ketoconazole greatly increases the activity of triazolam due to its inhibition of CYP3A4 through one or more mechanisms including enzymatic inhibition and gene transcription regulation at nuclear receptors. Both of these actions result in decreased drug metabolism by CYP3A4. Potential interactions between fluconazole and warfarin results in an increase in the area under the curve of warfarin prolonging prothrombin time. This effect is associated with CYP2C9 inhibition by fluconazole. The drug–drug interaction of itraconazole and the statins, lovastatin, and simvastatin, increases the potential toxicity of the statins. Ketoconazole and itraconazole also bind to and serve as substrates for P-gp. These actions suggest additional potential for drug–drug interactions. For additional discussion of drug–drug interactions the reader is referred to Chapter 4 in *Foye's Principles of Medicinal Chemistry, Seventh Edition.*

- Systemic azoles (e.g., ketoconazole – imidazole, and the triazoles: fluconazole, voriconazole, itraconazole, posaconazole, and isavuconazonium sulfate [Figs. 24.4 and 24.5]) are slowly metabolized.
- Isavuconazonium sulfate is unique among the azoles in that it is a prodrug, which undergoes hydrolysis to isavuconazole the active drug (see metabolism below).
- Isavuconazonium sulfate is water soluble and can be used as such for IV administration (also used orally).
- Other azole parenteral preparations require cyclodextrin for solubility. Cyclodextrin has been associated with nephrotoxicity.
- Ketoconazole and itraconazole require low stomach pH for oral absorption and, as a result, drug–drug interactions associated with administration of an antacid should be anticipated.
- Systemic products have good bioavailability (~90%) with high protein binding (~90% with the exception of voriconazole – 56% and fluconazole ~12%).
- Azoles are substrates and/or inhibitors of P-glycoprotein (P-gp) effecting their own excretion and excretion of P-gp substrates.

Metabolism

- The systemic azoles are inhibitors and substrates for various CYP450 isoenzymes (Table 24.3), although for posaconazole and isavuconazonium sulfate the effects of CYP450 isoforms are minimal.
- Ketoconazole, itraconazole, and voriconazole are extensively metabolized by CYP450 isoforms (Fig. 24.6).
- Posaconazole is metabolized to its glucuronide via uridine glucuronyltransferase (UTG), while isavuconazonium sulfate undergoes butylcholinesterase catalyzed hydrolysis to isavuconazole the active drug (Fig. 24.7).

Table 24.3 Azole as Inhibitors and Substrates for Cytochrome P450 Enzymes

Drug	CYP Inhibitor	Substrate	Drug–Drug Interaction
Ketoconazole	1A2, 2C19, 3A4	3A4	Yes
Fluconazole	2C19, 2C9, 3A4	Low metabolism	Yes
Itraconazole	3A4	3A4	Yes
Voriconazole	2C19, 2C9, 3A4	2C19, 2C9, 3A4	Yes
Posaconazole	3A4	Low metabolism	Lower potential
Isavuconazole	3A4, 2C8, 2C19	Low metabolism	Lower potential

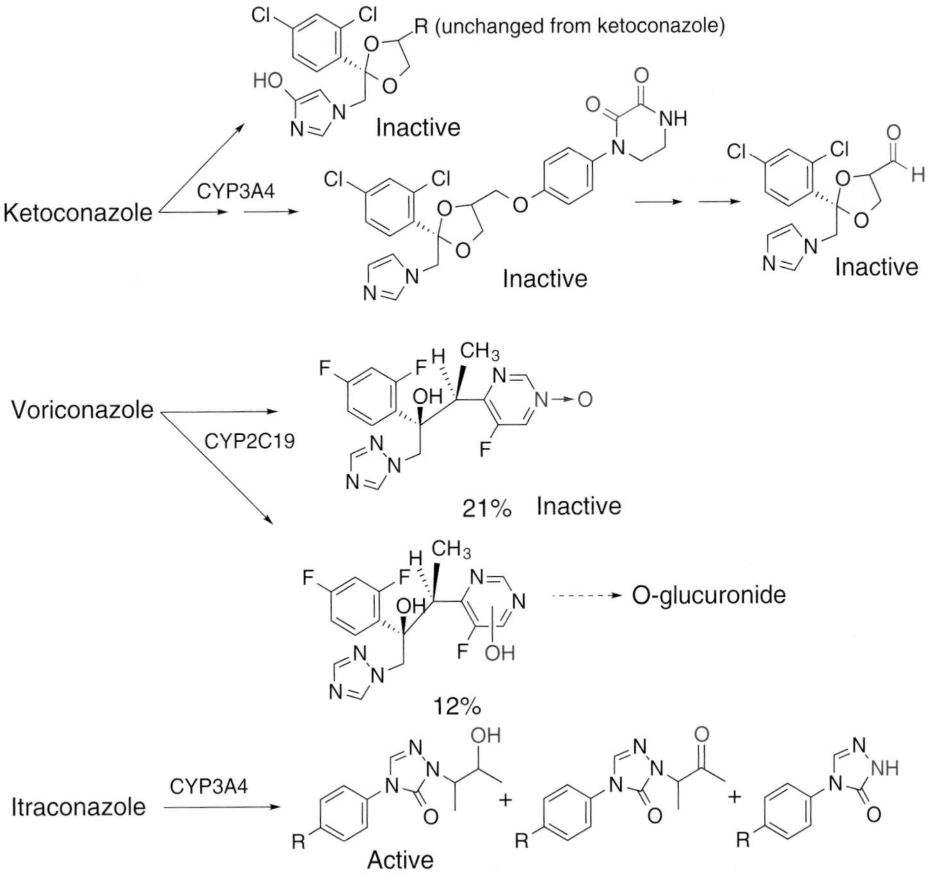

Figure 24.6 Major metabolic products formed from metabolism of systemic azoles.

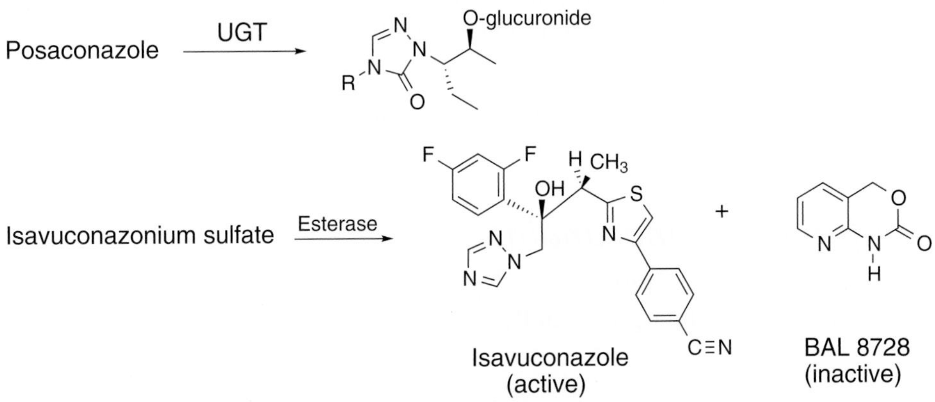

Figure 24.7 Major metabolic product of posaconazole and isavuconazonium sulfate. Little or no oxidative products are formed.

Clinical Applications

- The topical azoles are generally effective in the treatment of tinea corporis, pedis, cruris, versicolor, and cutaneous candidiasis.
- The newest azoles luliconazole and efinaconazole can effectively penetrate the nail plate to treat onychomycosis (also known as tinea unguium) caused by *Trichophyton rubrum* and *mentagrophytes*, *Epidermophyton*, and *Microsporum* nail infections. Published literature indicates that luliconazole, as a 10% solution, is effective in the treatment of onychomycosis of the nail, although presently not FDA approved for this use.
- Vaginal candidiasis is treatable with the imidazole antifungals available in vaginal creams, suppositories, and tablets.
- Oropharyngeal candidiasis is treatable with clotrimazole troches.
- Systemic azoles are effective against most of the human fungal infections including candida, cryptococcosis, coccidioidomycosis, blastomycosis, sporotrichosis, aspergillosis. Isavuconazole has been granted orphan drug approval for invasive candidiasis, aspergillosis, and mucormycosis.

> **⚠ ADVERSE EFFECTS**
>
> **Systemic Azoles:** The most common adverse effects involve the GI track and include nausea, vomiting, diarrhea. Rash has been reported with some azoles. Several azoles should be limited during pregnancy (Category C) or contraindicated (Category D – voriconazole). Isavuconazonium sulfate should not be used during pregnancy and serious hepatic reactions also have been reported for this drug.

> **Systemic Azoles Antifungal Spectrum**
>
> Yeast: *Candida albicans*, *Cryptococcus neoformans*; Mycoses: *Histoplasma capsulatum*, *Blastomyces dermatitidis*, *Coccidioides immitis*; Mold: *Aspergillus fumigatus* (not fluconazole); Filamentous fungi: *Fusarium solani*.

Antifungal Allylamines

Allylamine pharmacophore

Naftifine   Terbinafine   Butenafine

- The allylamines are based upon the allylamine pharmacophore.

MOA

- Inhibit cell membrane synthesis by blocking epoxidation of squalene (step 1, Fig. 24.2).
- Decreases levels of cell membrane ergosterol leading to cell wall leakage.
- Increases membrane squalene content which is abnormal.

Physicochemical and Pharmacokinetic Properties

- Naftifine and butenafine only used topically.
- Terbinafine used both topically and systemically.
- Terbinafine is very lipophilic resulting in distribution to nail beds.

Clinical Application

- Topical – treatment of dermatophytic infections.
- Systemic – treatment of onychomycoses.

Antifungal Echinocandins

- Cyclic peptides (Fig. 24.8).
- Possess large lipophilic groups at R_5.

MOA

- Block the synthesis of 1,3-β-D-glucan linkage in the fungal cell wall (important for cell wall rigidity).

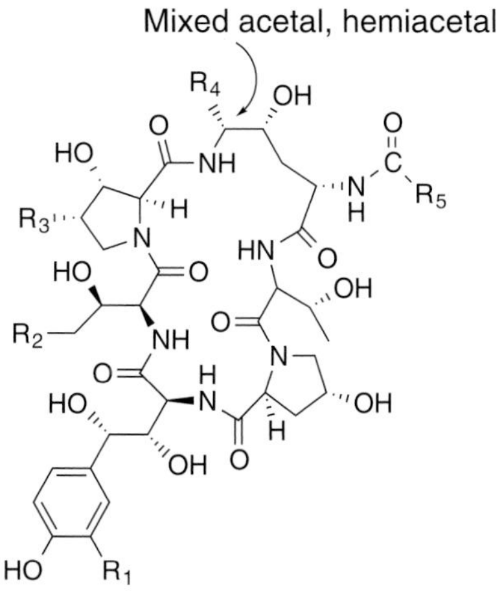

Mixed acetal, hemiacetal

Figure 24.8 Antifungal echinocandins.

Drug	R_1	R_2	R_3	R_4	R_5
Caspofungin (Cancidas)	H	CH_2NH_2	H	$NHCH_2CH_2NH_2$	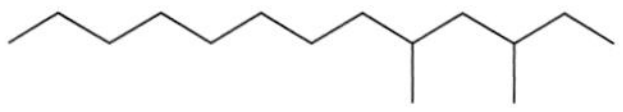
Anidulafungin (Eraxis)	H	H	CH_3	OH	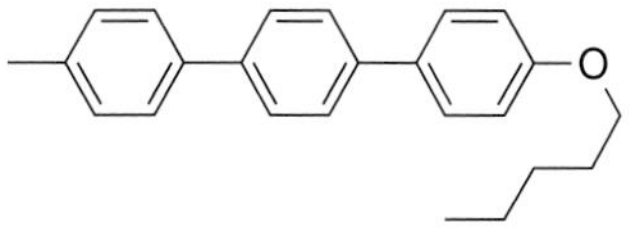
Micafungin (Mycamine)	OSO_3H	$CONH_2$	CH_3	OH	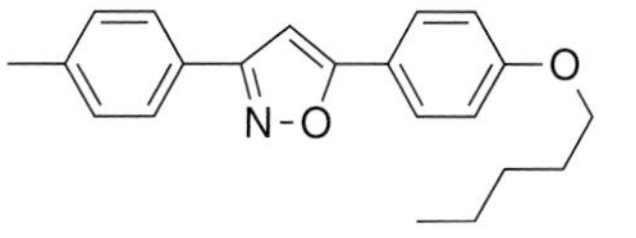

- Block β-1,3-D-glucan synthase (GS), a membrane-associated protein complex, which reduces cell wall integrity leading to cell lysis.

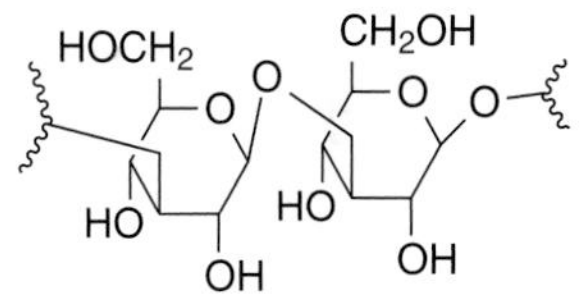

Physicochemical and Pharmacokinetic Properties
- Administered IV since they are not orally active.
- Highly bound to plasma protein (~96%).
- Echinocandins will not enter CNS.

Metabolism
- Caspofungin undergoes hydrolysis at the mixed acetal and at the aromatic amino acid located at the bottom of the cyclic peptide (Fig. 24.9).
- Anidulafungin stable to metabolism.
- Micafungin is O-sulfated or O-methylated on the aromatic phenol.

Clinical Applications
- The echinocandins are primarily beneficial for treatment of deeply invasive candidiasis.
- Potentially useful for treatment of aspergillosis.
- Not effective for *Cryptococcus neoformans* or *Trichosporon* spp.

CHEMICAL NOTE

1,3-β-D-glucan unit

Miscellaneous Antifungal Agents

Flucytosine

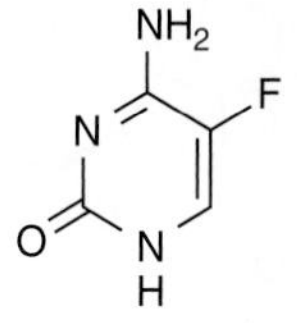

MOA
- Flucytosine is a prodrug which is transported into fungal cells via cytosine permease and activated through deamination to 5-flurouracil (5-FU) (Fig. 24.10).
- The active drug is 5-fluorodesoxyuridine monophosphate (5-FdUMP) formed from 5-FU.

Flucytosine (Ancobon)

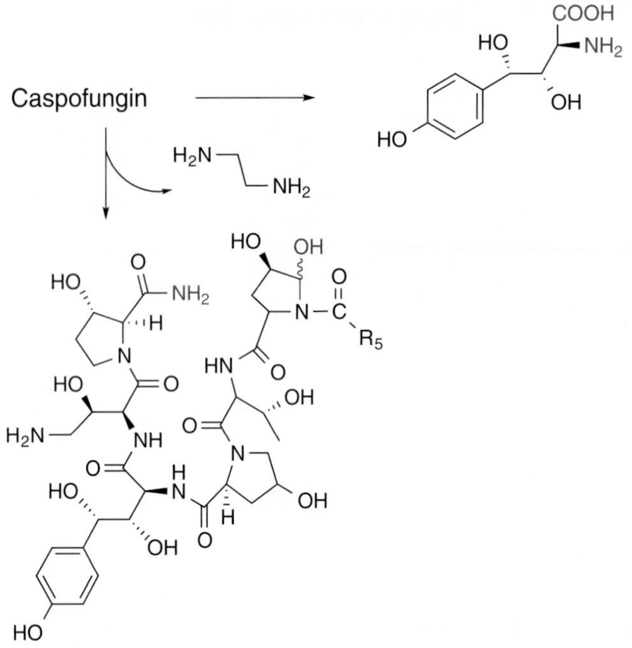

Figure 24.9 Metabolic products formed from caspofungin.

- 5-FdUMP blocks thymidylate synthase, thus blocking DNA synthesis.
- Deamination and activation do not occur in mammalian cells.

Physicochemical and Pharmacokinetic Properties

- Readily absorbed following oral administration.
- Distributed to the CSF.
- Resistance is associated with a decreased cytosine permease or uracil phosphoribosyltransferase activity.

Clinical Applications

- Treatment of cryptococcal meningitis especially in AIDS patients.
- Effective in treatment of some candida infections.
- Commonly used in combination with amphotericin B.

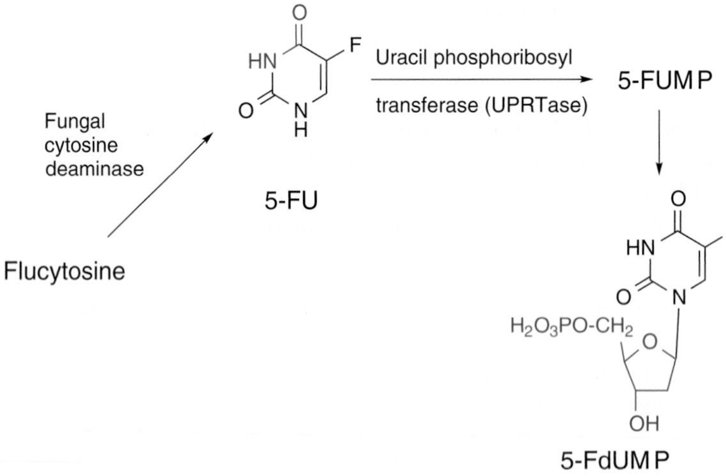

Figure 24.10 Metabolic activation of flucytosine by deamination, conjugated with ribosylphosphate to 5-fluorouracil monophosphate (5-FUMP) and on to 5-fluorodeoxyuridine monophosphate (5-FdUMP).

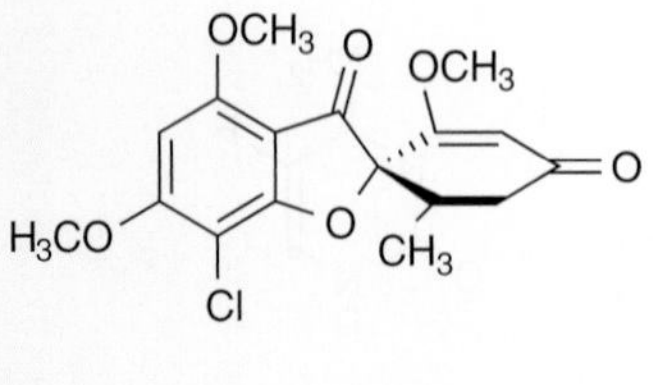

Griseofulvin

Griseofulvin

MOA

- Disrupts the mitotic spindle.
- Binds to microtubule protein inhibiting mitosis.

Physicochemical and Pharmacokinetic Properties

- Administered orally in the form of tablets or as a suspension.
- To improve dissolution the microcrystal and ultra-microcrystal forms are used.

Metabolism

- Inactivation through O-demethylation and O-glucuronidation.

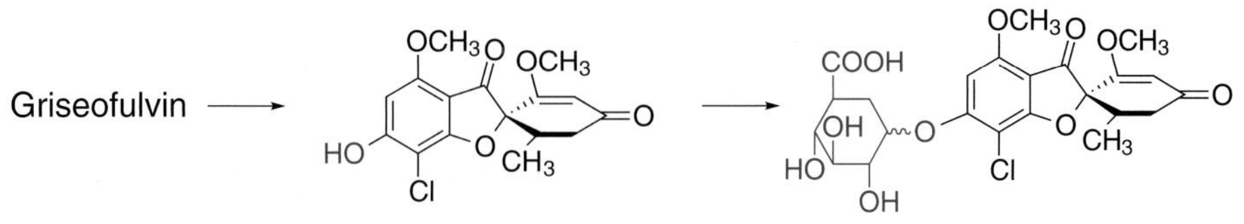

Griseofulvin ⟶

- CYP3A4 inducer (see *Foye's Principles of Medicinal Chemistry, Chapter 4, Seventh Edition*).

Clinical Application

- Fungistatic used to treat infections of skin, hair, and nails caused by *Microsporum*, *Epidermophyton*, or *Trichophyton*.

Ciclopirox

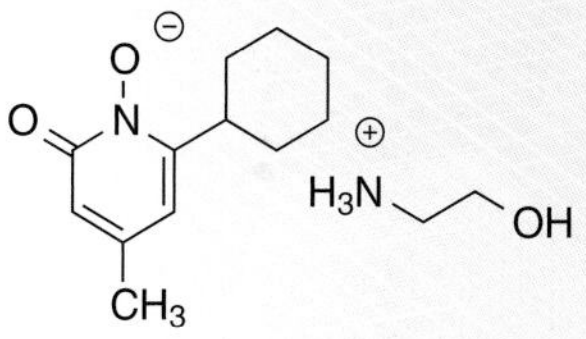

MOA

- The MOA remains uncertain, although its action appears to involve reactive oxygen species (ROS) and its ability to chelate Fe^{3+}.

Clinical Applications

- Used topically to treat *Candida*, *Epidermophyton*, or *Trichophyton* infections.
- A liquid formulation is effective in the treatment of onychomycosis nail infections.

Ciclopirox olamine

Tolnaftate

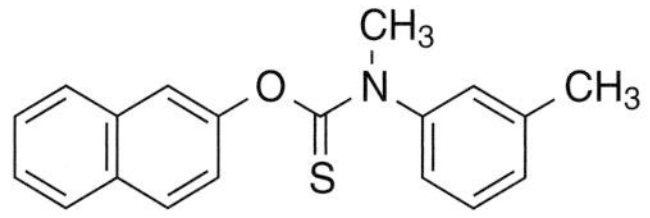

Tolnaftate (Absorbine Jr, Aftate, Tinactin)

- Although chemically unrelated to the allylamines, this drug inhibits cell membrane synthesis by blocking epoxidation of squalene (step 1, Fig. 24.2) (see allylamines above).
- Applied topically and used for many of the tinea infections.

Undecylenic Acid

$H_2C{=}CH(CH_2)_8COOH$

Undecylenic acid

- Topical fungistatic agent.
- Often available as a calcium or zinc salt which adds astringent value to the pharmaceutical.
- Especially valuable in the treatment of tinea pedis, tinea cruris, and diaper rash.

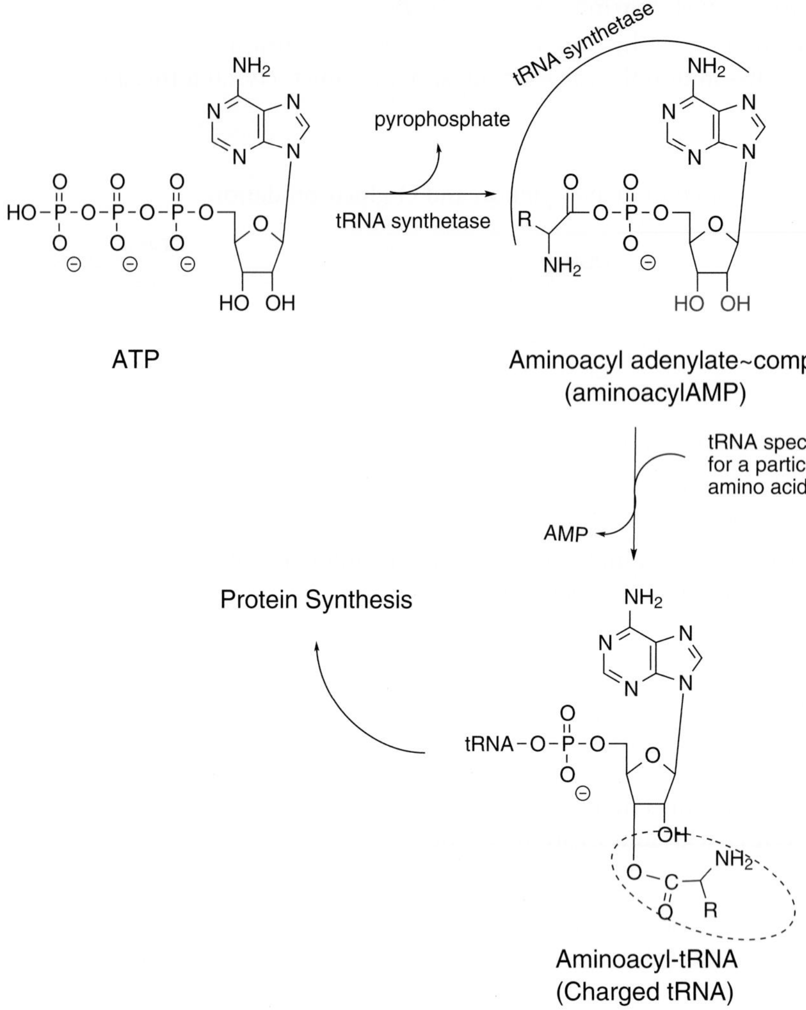

Figure 24.11 Biochemical process involving activation of a specific amino acid for transfer to the developing protein.

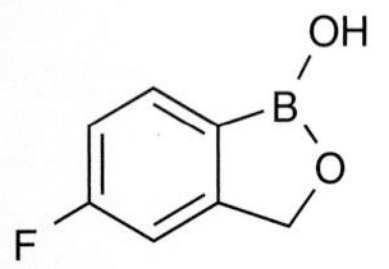

Tavaborole (Kerydin)

Tavaborole

MOA

- Protein synthesis involves the attachment of individual amino acids to the growing protein via tRNA synthetase. In this process the amino acid is initially attached to the 5′-phosphate of AMP in the aminoacyl adenylate complex (Fig. 24.11) followed by transfer to the 2′ position giving the "charged tRNA," and then on to the developing protein.
- Tavaborole is a noncompetitive inhibitor of cytoplasmic leucyl-tRNA through formation of a stable adduct between tRNA^{Leu} and tavaborole (Fig. 24.12).

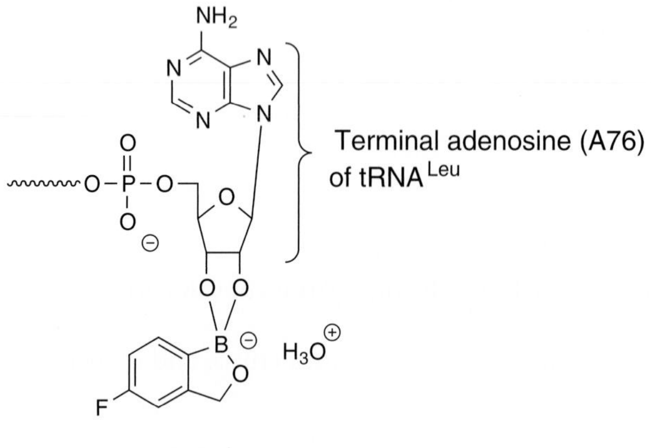

Figure 24.12 Binding between tavaborole and tRNA in the editing active site of leucyl-tRNA synthetase (LeuRS).

Physicochemical and Pharmacokinetic Properties

- Tavaborole is soluble in ethyl acetate–propylene glycol (5% solution) for topical application.
- The drug readily penetrates the full thickness of the human nail at antifungal concentrations.
- Tavaborole has a good hydrophilic/lipophilic balance with a clog P of 1.24. Less active derivatives are the 5-chloro (clog P 1.81) and 1-phenyl-5-fluoro (clog P 3.55) derivatives.

Clinical Applications

- Tavaborole has been approved for treatment of onychomycosis of the nail and nail bed involving *T. rubrum* or *T. mentagrophytes.*
- Negative fungal cultures were seen in >90% of patients within 2 weeks although the drug is used once daily for 48 weeks.

> **⚠ ADVERSE EFFECTS**
>
> **Tavaborole:** Adverse effects were mild and consisted of exfoliation at application site as well as erythema and dermatitis in a small number of patients.

Chemical Significance

In general, fungal infections and their treatment has been a low-priority therapeutic area since the most common human infections involve the skin and mucosal membranes. Systemic fungal infections are not common in healthy individuals because the body's immune system is capable of warding off the fungal pathogen. The same cannot be said for individuals with an immunocompromised system. For patients with AIDS, those who have undergone an organ transplant, or many cancer patients, the incidence of life-threatening fungal infections is quite significant. The need for systemic, safe antifungal drugs will continue to increase for the treatment of patients with compromised immune systems. In fact, there are really only three classes of antifungals available for systemic fungal infections: azoles, polyenes, and echinocandins (the allylamines are limited to dermatophytes).

The difficulties experienced by two of the classes of systemic antifungals are quite apparent from the chemical structures of these drugs. The polyenes and echinocandins are not available in orally active forms for systemic infections for two different reasons. The polyenes do not have a good hydrophilic/lipophilic balance and are not absorbed orally; yet do not penetrate the blood–brain barrier to be effective against fungal infections of the CNS. The difficult physicochemical properties recognized from a chemical observation results in pharmacokinetic challenges and unique dosage forms. The echinocandins can be recognized as cyclic peptides which lead to the prediction that such agents are not effective orally due to amide hydrolysis in the GI tract. In addition, such drugs will not enter the CSF limiting their effectiveness against specific types of fungal infections. New drugs that appear in either of these classes can be identified via their chemical structures and would be expected to have similar limitations.

The third class of systemic antifungals, the azoles, are plagued by a metabolic issue and that is their oxidation by CYP450 isoforms and their inhibitory action on various CYP450 isoforms, leading to a potential for drug–drug interactions. These problems are especially significant when treating fungal infections in cancer patients on anticancer drugs, organ transplant patients on anticoagulant medications, and AIDS patients on antiviral drugs. Knowing that a drug is an azole antifungal with a mechanism of action which requires blocking a specific CYP450-catalyzed oxidation should lead one to suspect inhibitory action of the azole drug on other drugs which are CYP450 substrates.

In summary, the relationship of physicochemical and pharmacokinetic properties of the various classes of antifungal agents to their chemical structure may be a useful predictor for understanding the value and potential limitations of newly developed antifungal agents, especially those that are based upon the presently used classes of agents.

Review Questions

1 The echinocandins are effective systemic antifungal agents by virtue of their ability to _____.
A. block cell membrane synthesis at the ergosterol level
B. damage the cell membrane leading to "leakage" of the membrane
C. block cell membrane synthesis at the β-glucan level
D. interfere with zymosterol synthesis

2 The topical antifungal allylamines (i.e. naftifine) are active by virtue of their ability to:
A. Block squalene epoxide synthesis leading to defective cell membranes.
B. Interfere with the conversion of folic acid to tetrahydrofolic acid.
C. Inhibit cell mitosis.
D. Bind to the fungal cell membrane leading to cell hemolysis.
E. Inhibit synthesis of membrane 1,3-β-glucans.

3 Which of the atoms or circled groups is <u>essential</u> for the azole antifungal to bind to CYP450?

4 The site of action of amphotericin B is in the:
A. Peptidoglycan portion of the cell membrane.
B. Cell membrane of the fungus binding to the ergosterol.
C. LAM portion of the cell wall.
D. Cytosol of the organism binding to the ribosome.
E. Demethylase enzyme.

5 A clinically significant drug–drug interaction that could reduce the effectiveness of Sporanox (itraconazole) for the treatment of blastomycosis is

_____.
A. the use of Zantac (ranitidine) to treat upset stomach.
B. the use of a multivitamin preparation consumed at the same time as Sporanox.
C. the consumption of alcoholic drinks with Sporanox.
D. the coadministration of an oral contraceptive.
E. in actual fact, a major advantage of Sporanox is that there is very little likelihood of drug–drug interactions with this azole.

Antiparasitic Drugs

25

Learning Objectives

Upon completion of this chapter the student should be able to:

1. Match the disease with a specific parasite and identify the stage of the pathogen for which therapy is effective.

2. Describe the mechanisms of action and site/stage of action, where appropriate, for drugs use.

3. Identify drugs to treat:
   a. amebiasis, giardiasis, and trichomoniasis (imidazoles, nitazoxanide [NTZ])
   b. pneumocystis (pentamidine, atovaquone)
   c. trypanosomiasis and leishmaniasis (suramin, eflornithine, nifurtimox, benznidazole, melarsoprol, stibogluconate, miltefosine).
   d. malaria (quinolones, pyrimethamine, artemisinins).
   e. helminthes (benzimidazoles, diethylcarbamazine [DEC], ivermectin [IVM], praziquantel [PZQ], pyrantel).
   f. ectoparasites (lindane, pyrethrums/pyrethroid, spinosad).

4. Identify clinically significant physicochemical and pharmacokinetic properties which limit or define the use of the various antiparasitic agents.

5. Identify metabolic instabilities with chemical structures indicating the effects of the metabolism on activity of the antiparasitic drugs.

6. Discuss the clinical application of the antiparasitic drugs.

Introduction to Antiparasitic Agents

Parasitic diseases are reported to infect over one billion people worldwide, with most of those infected living in developing countries. This fact accounts for the somewhat low interest in developing treatment modalities since health care resources are limited. Some of the parasitic infections affect in excess of 80% of the population. In recent years, this has begun to change because of global travel, improved communications, humanitarian efforts, and growing expectation of improving health outcomes in these developing countries. Unlike bacterial infections, parasitic infections often are quite complex involving vectors, which transmit the disease and multiple life stages of the parasite (only one of which may produce the human disease). Therefore, it is essential to be aware of the life cycle of these parasitic organisms as they relate to transmission and infectious stages as well as which stage(s) drug treatment is effective. The drugs used to treat parasitic infections are divided into those used to treat protozoal, helminth, and ectoparasitic infections.

Protozoal Diseases

Amebiasis

- Can lead to life-threatening dysentery (Table 25.1).
- Cyst form responsible for transmission in contaminated food and water. Intestinal form found in wall and extraintestinal form found in liver (trophozoite stage).

Table 25.1 Diseases Associated with Protozoal Infections and Their Characteristics

Disease	Organism	Life Stages	Infected Organ/Cells	Transmitter
Amebiasis	*Entamoeba histolytica*	Cyst/trophozoite	Intestine/liver	Contaminated food/water
Giardiasis	*Giardia lamblia*	Cyst/trophozoite	Intestine/liver	Contaminated water
Trichomoniasis	*Trichomonas vaginalis*	Trophozoite	Vagina/urethra/prostate	Sexual contact
Pneumocystis pneumonia (PCP)	*Pneumocystis jirovecii*	Yeast-like	Lung	Airborne
Trypanosomiasis:				
Sleeping sickness	*Trypanosoma brucei*	Trypomastigotes	CNS	Tsetse fly
Chagas disease	*Trypanosoma cruzi*	Trypomastigotes/amastigote	Heart	Reduviid bug
Leishmaniasis	*Leishmania spp.*	Promastigote/amastigote	Skin/systemic	Female sandflies
Malaria	*Plasmodium spp.*	Sporozoite/merozoite/trophozoite/gametes	Liver/red blood cells	*Anopheles* mosquito

Giardiasis

- Transmitted (cyst) via untreated drinking water – case of "campers diarrhea".
- Trophozoite found in small intestine leading to symptoms of diarrhea.
- Usually a self-limiting infection.

Trichomoniasis

- Sexually transmitted infection; usually asymptomatic in males.
- May be confused with bacterial (*Haemophilus vaginalis*) or fungal (*Candida albicans*) infections.

Pneumocystis Pneumonia (PCP)

- Organism has characteristics of both protozoa and fungus.
- Common in immunocompromised patients (AIDS/HIV, organ transplant, and cancer patients).
- Life threatening.

Trypanosomiasis/Leishmaniasis

- Not common in western countries.
- Depending on invasion site these diseases can be debilitating.

Malaria

- Four species of plasmodium responsible for human infections (e.g., *Plasmodium falciparum*, *Plasmodium vivax*, *Plasmodium malariae*, *Plasmodium ovale*).
- May be responsible for as many as two million deaths a year.
- *P. falciparum* most commonly associated with deadly outcome if untreated.
- Trophozoite form multiplies within the erythrocytes, which upon rupture leads to symptoms.
- Most drugs are effective within the erythrocytes (blood schizonticides) (for life cycle of plasmodium organisms, see Fig. 25.1).

Drug Therapy for Protozoal Infections

Treatment of Amebiasis, Giardiasis, and Trichomoniasis

Nitroimidazoles (Metronidazole and Tinidazole)

Common Symptoms of *P. falciparum* Malaria

Chills, fever, sweating, headache, fatigue, anorexia, nausea, vomiting, diarrhea (occurring on alternate days – tertian malaria).

Metronidazole (R = OH)
(Flagyl, Metryl, Satric)
Tinidazole (R = $SO_2CH_2CH_3$)
(Tindamax)

MOA

- The nitroimidazoles are thought to function as prodrugs.

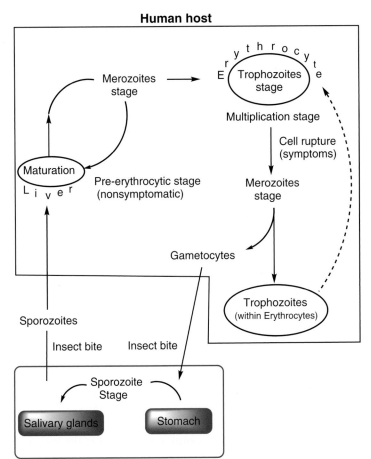

Figure 25.1 Life cycle of malarial protozoa.

- One electron reduction of the nitroimidazoles gives a radical species, which generates a superoxide radical anion and other reactive oxygen species (ROS) (Fig. 25.2).
- The ROS fragment key cellular components (i.e., DNA, RNA) of the protozoa leading to cellular death.

Physicochemical and Pharmacokinetic Properties

- Nearly 100% absorption follows oral administration of the nitroimidazoles.
- Distribution occurs to semen and urine in active concentrations.

Metabolism

- Nitroimidazoles undergo metabolism involving CYP450-catalyzed oxidation to active metabolites (Fig. 25.3).

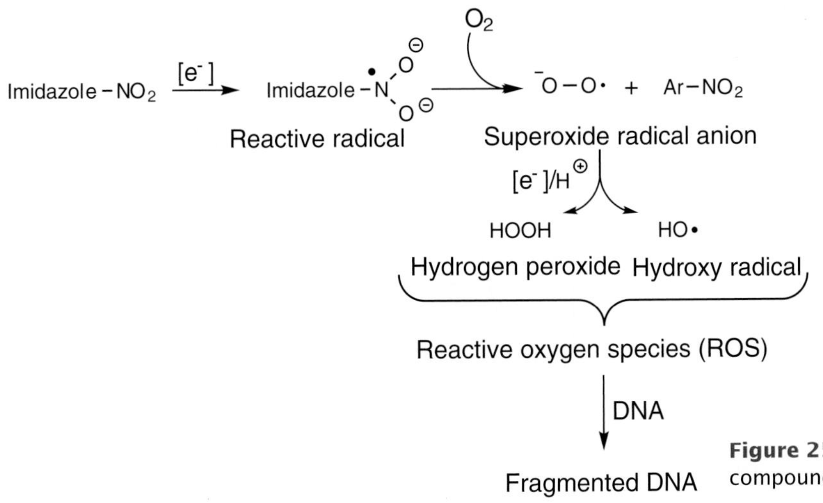

$$Imidazole-NO_2 \xrightarrow{[e^-]} Imidazole-\overset{\bullet}{N} \underset{O^\ominus}{\overset{O^\ominus}{}} \xrightarrow{O_2} {}^-O-O\bullet \; + \; Ar-NO_2$$

Reactive radical   Superoxide radical anion

$$[e^-]/H^{\oplus}$$

HOOH   HO•

Hydrogen peroxide   Hydroxy radical

Reactive oxygen species (ROS)

$\downarrow$ DNA

Fragmented DNA

Figure 25.2 Formation of ROS from nitroimidazole compounds.

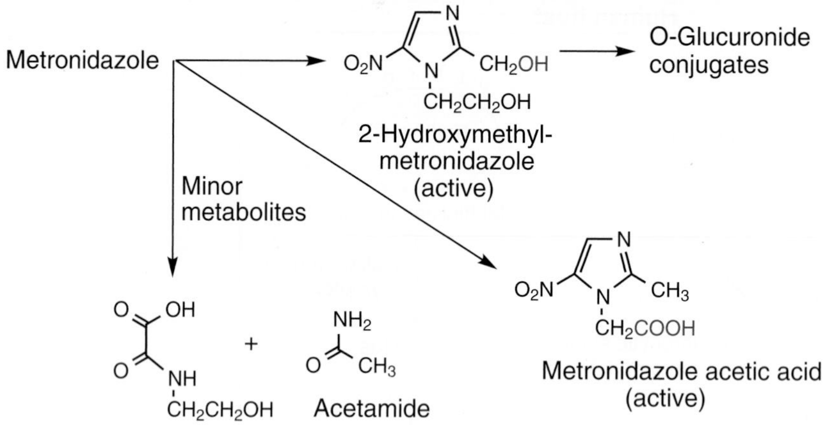

Figure 25.3 CYP450-catalyzed metabolism of metronidazole.

- A minor route of metabolism involves imidazole ring cleavage giving rise to acetamide, an animal carcinogen.

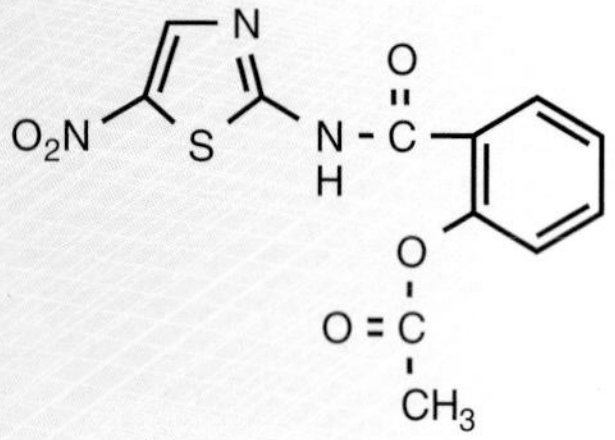

Nitazoxanide
(Alinia)

Nitazoxanide (NTZ) (a thiazole)

MOA

- NTZ is a prodrug, but unlike the nitroimidazoles does not appear to be active via a radical mechanism.
- NTZ is metabolized to tizoxanide (TIZ) and the hydroxylamine intermediate, both of which appear to be the active forms of the drug (Fig. 25.4).
- The drug blocks the action of pyruvate: ferredoxin oxidoreductase (PFOR) and formation of lactyl thiamine pyrophosphate (TPP) complex, which blocks the formation of lactic acid, a process essential to the protozoa (Fig. 25.4).

Physicochemical and Pharmacokinetic Properties

- TIZ is highly bound to plasma protein.
- TIZ is conjugated with glucuronic acid (O-glucuronide) and excreted in the urine, bile, and feces.

Clinical Applications

- The nitroimidazoles are effective in the treatment of amebiasis, giardiasis (metronidazole is not approved), and trichomoniasis.

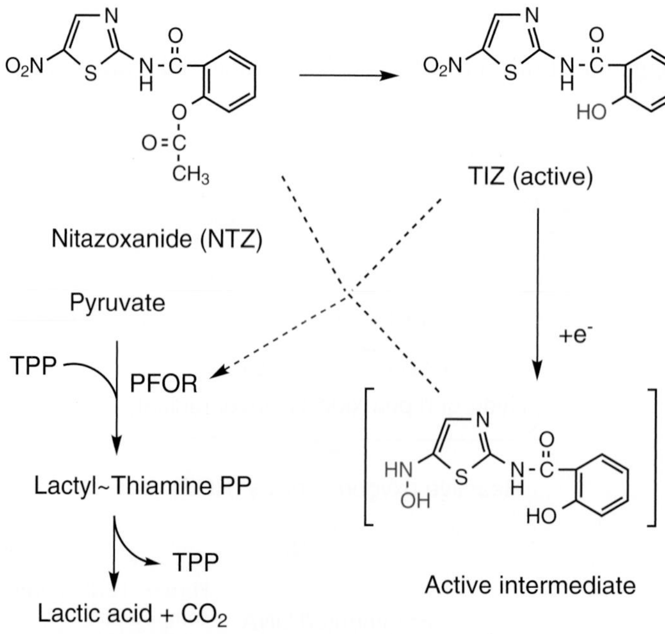

Figure 25.4 Metabolic activation of NTZ resulting in inhibitory action (indicated by dashed lines) on pyruvate: ferredoxin oxidoreductase (PFOR) enzyme. *TPP*, thiamine pyrophosphate.

- Metronidazole is also effective in the treatment of various anaerobic bacteria.
- NTZ is effective in the treatment of a number of parasites including protozoa (giardia and cryptosporidium) involved in GI infections and diarrhea as well as in the treatment of intestinal helminthes.

Treatment of Pneumocystis Pneumonia (PCP)
Pentamidine Isethionate

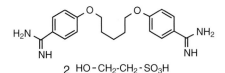

Pentamidine isethionate
(Pentam 300, Nebupent)

MOA

- Site of action is unknown and may differ dependent upon the organism.
- Thought to bind to DNA through the protonated amino groups interfering with DNA synthesis.
- Reported to exhibit inhibitory action on topoisomerase II.

Physicochemical and Pharmacokinetic Properties

- Not readily absorbed orally, but used parenterally (IV).
- Bound to plasma protein (~70%).
- Distributed to the lungs, but not the CNS.
- Available as an aerosol formulation.

Atovaquone

Atovaquone
(Mepron)

Coenzyme Q_{10}

MOA

- Inhibitor of the mitochondrial respiratory chain at cytochrome bc_1 complex.
- Possible structural similarity to coenzyme Q_{10} (ubiquinone) may account for activity.
- Inhibits nucleic acid/ATP synthase.

Physicochemical and Pharmacokinetic Properties

- Poorly absorbed from the GI tract.
- Improved bioavailability results from reduced particle size and if taken with a fatty meal.
- Highly bound to plasma protein (~99%).
- Recovered unchanged in bile fluid.
- The drug exhibits a bright yellow coloration.

Clinical Applications

- Both pentamidine and atovaquone are indicated for treatment of PCP when the patient is intolerant to sulfamethoxazole/trimethoprim (TMP-SMX).
- Useful for prevention of PCP in HIV patients.
- Pentamidine is also used (off-label) for treatment of trypanosomiasis and leishmaniasis.

⚠ ADVERSE EFFECTS

Nitroimidazoles:
Metallic taste, abdominal distress, disulfiram-like effect if taken with alcohol. Headache, nausea, and dry mouth have also been reported. Potential neurotoxicity is also possible. Due to the reactive intermediates these drugs are contraindicated during the first trimester of pregnancy.

Drug of Choice for Pneumocystis Pneumonia (PCP)

The combination of sulfamethoxazole–trimethoprim is most effective for treatment of PCP and extrapulmonary pneumocystosis caused by *Pneumocystis jirovecii*. It should be noted that previously it was thought that PCP was caused by *P. carinii* which proved to be incorrect. PCP is most common in immunocompromised patients.

Trypanosomiasis and Leishmaniasis (Tritryps)
Suramin Sodium

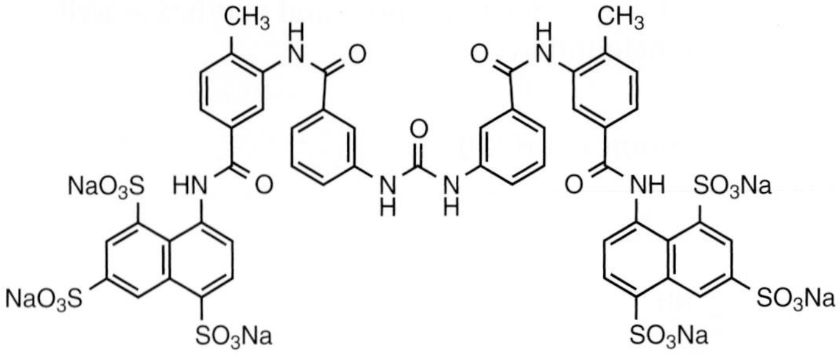

Suramin Sodium (Available from the CDC)

ADVERSE EFFECTS

Suramin Sodium:
Nausea, vomiting, and skin rash (i.e., hives) are commonly seen. A tingling or crawling skin sensation may be seen, and a nonserious clouding of urine may occur. Adrenal cortical damage (50% of patients) has been reported which may require corticosteroid replacement therapy.

MOA

- Thought to inhibit one or more of the following enzyme systems: dihydrofolate reductase (DHFR), thymidine kinase, glycolytic enzymes.

Physicochemical and Pharmacokinetic Properties

- Water soluble as sodium salt, but poorly absorbed.
- Used IV.
- Bound to plasma protein and will not enter the CNS, but is actively absorbed by parasites.

Eflornithine

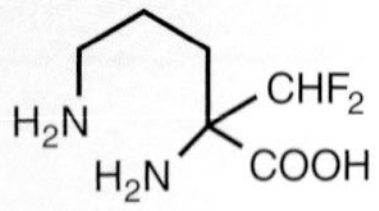

Eflornithine
(Ornidyl)

MOA

- Irreversible suicide inhibitor of ornithine decarboxylase (ODC) (Fig. 25.5).
- Blocks the synthesis of putrescine, which is the intermediate in the synthesis of spermidine and spermine.
- Polyamine (spermidine and spermine) responsible for maintaining membrane potential through various cation/ATPases.
- Mammalian ODC recovers rapidly whereas parasitic ODC does not.

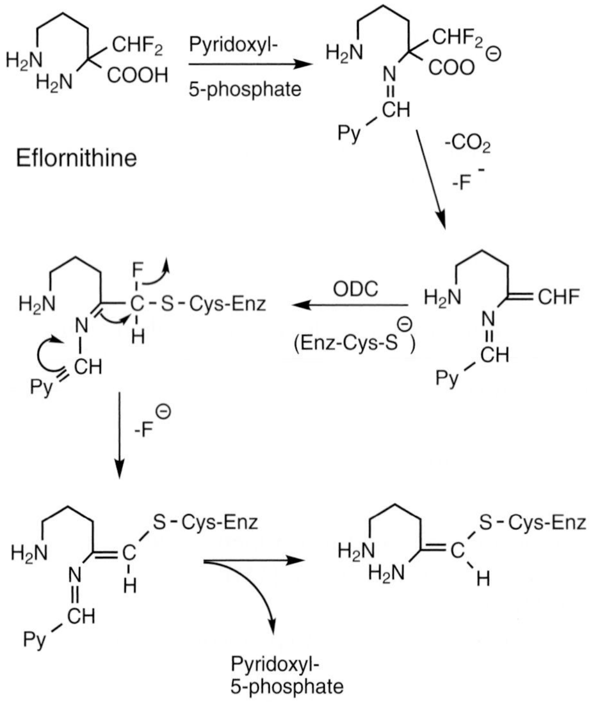

Figure 25.5 Inhibition of ODC (Enz-Cys-SH) by eflornithine. *ODC*, ornithine decarboxylase.

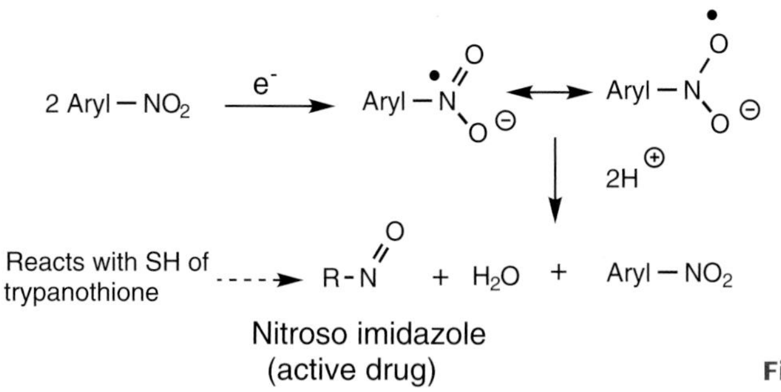

Figure 25.6 Proposed mechanism of action of benznidazole.

Physicochemical and Pharmacokinetic Properties

- Poorly absorbed orally and is used IV for treatment of parasites.
- Available as a cream for treatment of facial hirsutism in women (Vaniqa).

Nifurtimox

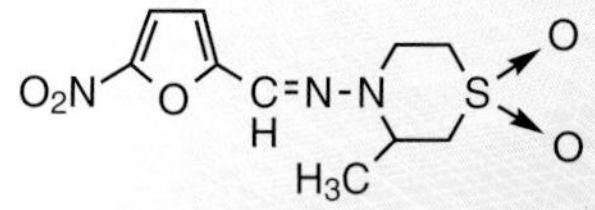

Nifurtimox
(Lampit)

MOA

- Similar to the nitroimidazoles (Fig. 25.2), nifurtimox is a prodrug, which undergoes an electron transfer leading to formation of ROS.
- The ROS damage trypanosome DNA and trypanothione reductase.

Benznidazole

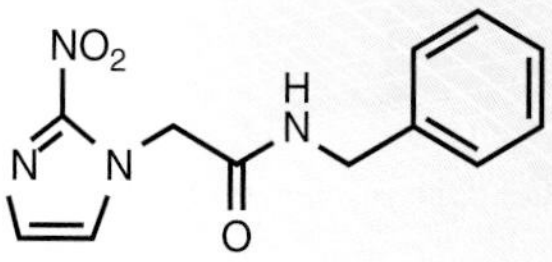

Benznidazole

MOA

- Unlike nifurtimox, benznidazole as a prodrug which through electron transfer gives rise to a nitroso intermediate which adds to trypanothione resulting in depletion of this essential parasitic metabolite (Fig. 25.6).

Melarsoprol

Melarsoprol
(Arsobal)

> **ADVERSE EFFECTS**
>
> **Benznidazole and Nifurtimox:** Both drugs are reported to cause skin rash, GI symptoms (e.g., nausea, vomiting, weight loss), and sleep disorders. In addition, benznidazole is reported to cause peripheral neuropathy, and nifurtimox may cause dizziness and vertigo.

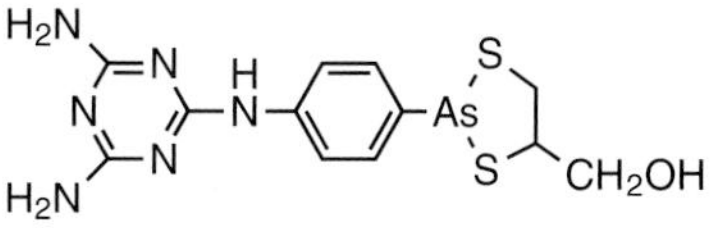

MOA

- Melarsoprol is active as a trivalent arsenic oxide (melarsen) (Fig. 25.7).
- Trivalent arsenic has a strong affinity for sulfhydryl groups present in many proteins and therefore has inhibitory action on a number of proteins.
- Proposed that melarsoprol through melarsen inhibits trypanothione reductase via reversible binding to trypanothione (Fig. 25.7).

Physicochemical and Pharmacokinetic Properties

- Administered IV.
- Enters the CSF.

> **ADVERSE EFFECTS**
>
> Significant toxic reactions have been reported including fever, encephalopathy, myocardial damage, peripheral neuropathy, and vomiting. Recommended that the drug be administered in a clinic setting.

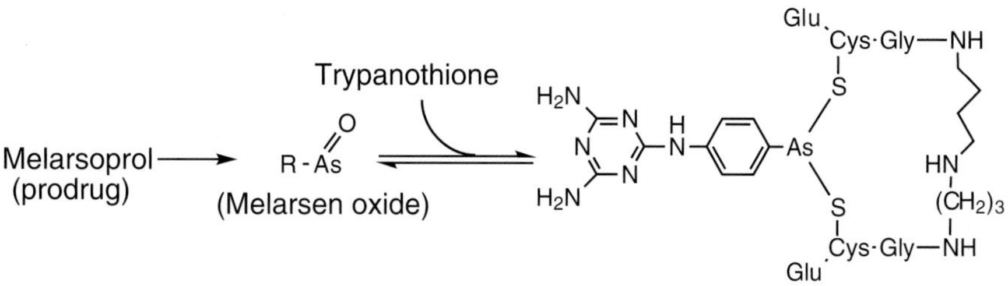

Figure 25.7 Melarsoprol reaction with trypanothione.

Sodium Stibogluconate

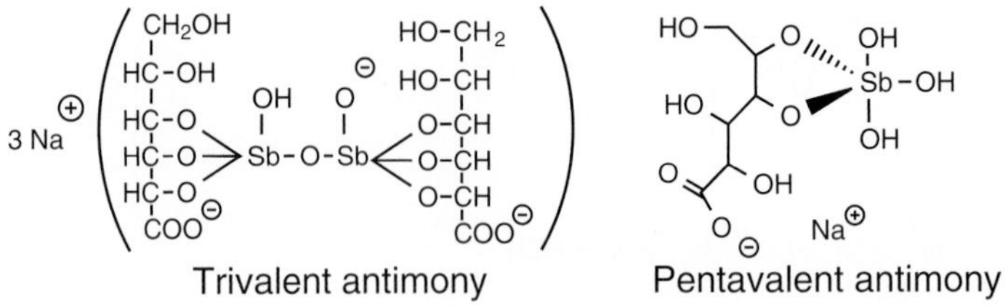

Trivalent antimony | Pentavalent antimony

Sodium stibogluconate
(Pentostam)

MOA

- It is unclear as to the action of sodium stibogluconate on *Leishmania* parasites.
- Both pentavalent [Sb^V] and trivalent [Sb^{III}] antimonials have been suggested as representing the active form of the drug (Fig. 25.8).
- Both Sb^{III} and Sb^V produce apoptosis through action on biopolymers (i.e., DNA).

Physicochemical and Pharmacokinetic Properties

- Administered IV or IM.
- Sb^V is less toxic than Sb^{III}.

ADVERSE EFFECTS

Sodium Stibogluconate: Nausea, vomiting, diarrhea, weakness and myalgia, skin rash, hepatotoxicity, and cardiotoxicity.

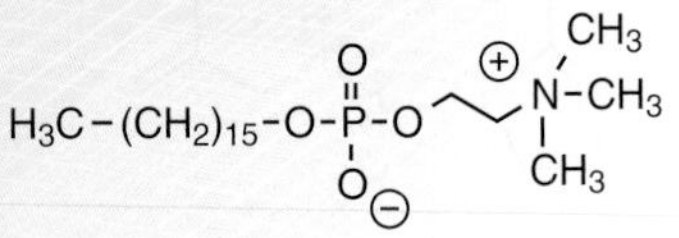

Miltefosine
(Impavido, Miltex–from CDC)

Miltefosine

- The drug's availability is limited.
- MOA remains unknown.
- Only orally active drug for treatment of the *Leishmania* parasites.

Clinical Applications of Tritryps Drugs

- Drugs used to treat trypanosomiasis/leishmaniasis have limited availability, but are available through the World Health Organization (WHO), which also supplies the Centers for Disease Control and Prevention (CDC) (Table 25.1).
- Suramin is only used for early stages of East African sleeping sickness caused by *Trypanosoma brucei rhodesiense* before the parasite enters the CNS.
- Eflornithine is effective against late stages of West African sleeping sickness caused by *T. b. gambiense* as is nifurtimox and melarsoprol.
- Melarsoprol is the only drug effective against late stages of East African sleeping sickness.
- Both nifurtimox and benznidazole are used to treat trypomastigote and amastigote forms of Chagas disease.
- Sodium stibogluconate and miltefosine are used primarily for the treatment of amastigote stage of visceral and cutaneous leishmaniasis.

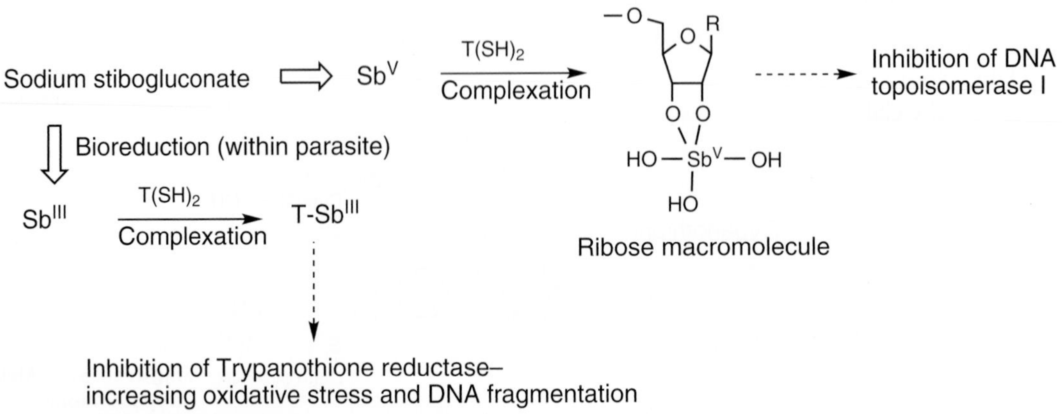

Figure 25.8 Proposed mechanisms of action of sodium stibogluconate. *T(SH)₂*, trypanothione.

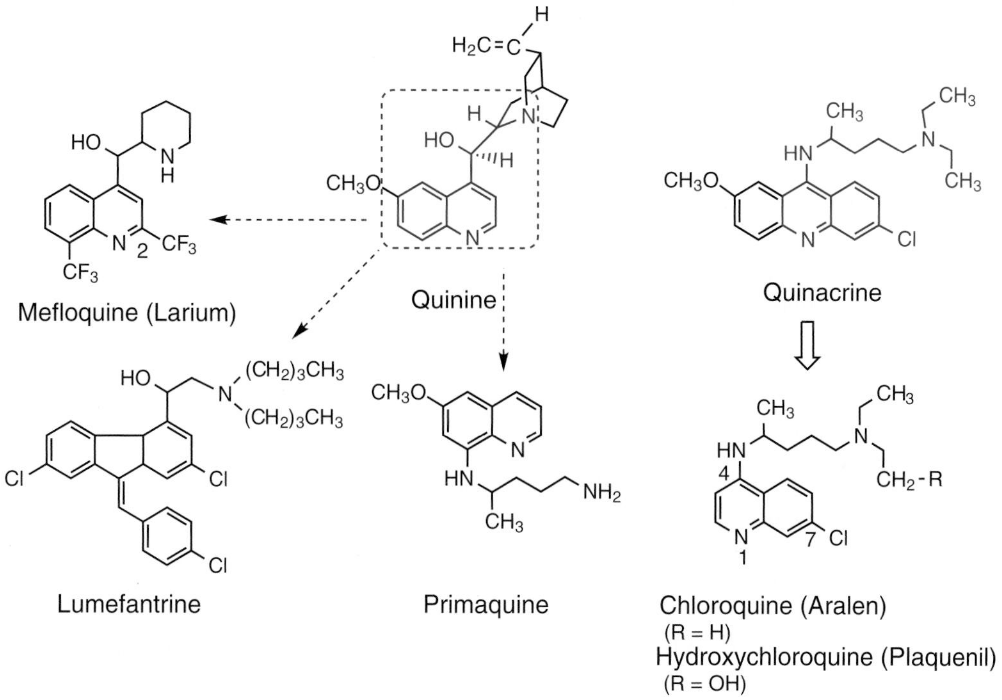

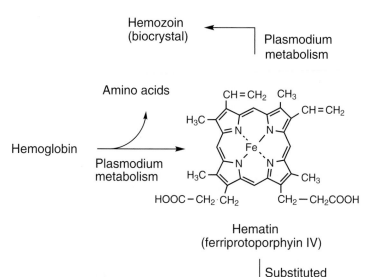

Figure 25.9 Structure similarity between the lead antimalarials (quinine and quinacrine) and the available antimalarials.

Treatment of Malaria

Quinine, isolated from the bark of the cinchona tree, and quinacrine, a synthetic derivative of quinine, were found to possess antimalarial activity and since have served as templates for the preparation of a large number of synthetic drugs. These structural relationships form the initial leads to drugs used today and are shown in Figure 25.9. These drugs are known as the quinolines.

Substituted Quinolines

MOA

- Ferriprotoporphyrin IX (FPIX) hypotheses: plasmodium utilizes hemoglobin as a source of amino acids producing hematin as a byproduct. Hematin is toxic to the plasmodium, but within the organism's food vacuole it is converted to the nontoxic inert biocrystal composed of dimeric hematin known as hemozoin (Fig. 25.10).

Hemozoin (biocrystal) ← Plasmodium metabolism

Amino acids

Hemoglobin → Plasmodium metabolism

Hematin (ferriprotoporphyin IV)

Hematin-quinoline-π-complex (toxic to plasmodium) ← Substituted quinolines

Figure 25.10 Proteolytic degradation of hemoglobin by the plasmodium organism to the toxic hematin and on to the nontoxic hemozoin. Site of action of substituted quinolines.

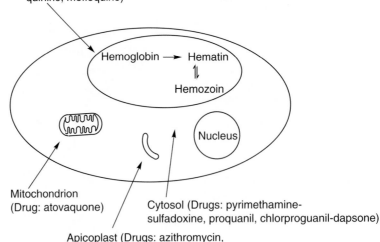

Figure 25.11 *Plasmodium* parasite cell as present within the erythrocyte and sites of drug action of various antimalarial drugs.

- Quinolines bind to hematin through π-bonding preventing plasmodium-catalyzed crystallization to the nontoxic hemozoin. The hematin~quinoline complex retains toxic to plasmodium.
- Weak base hypotheses: Quinolines as weak bases are concentrated in the acidic food vacuole (pH ~ 5.0) of the plasmodium, thus increasing their potential for toxicity to the plasmodium (Fig. 25.11).
- These mechanisms are based upon studies with chloroquine and it is assumed that other quinolones act similarly.

Mechanisms of Resistance

- Increased efflux of the quinolines from the food vacuole catalyzed via plasmodium transmembrane transporter protein. This is thought to occur through spontaneous mutation of a quinoline transporter gene (*pfcrt* gene).
- Increased CYP metabolism of quinolines.

Physicochemical and Pharmacokinetic Properties

- In general, the quinolines are well absorbed via the oral route and have similar pharmacokinetic properties as shown in Table 25.2.

SPECIFIC QUINOLINES
QUININE

SAR

- Stereochemical centers at C3, C4, C8, C9 are all important.
- Modification of C9 alcohol will reduce activity.
- Tertiary amine off of C9 is essential for activity.

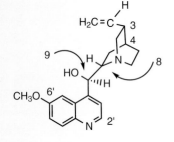

Quinine

Table 25.2	**Pharmacokinetic of Quinoline Antimalarials**			
Generic Name	**Protein Binding**	**Bioavailability**	**Half-Life (d)**	**Route of Excretion**
Quinine	69–92%	~85%	~0.5	Urine
Chloroquine	~55%	75%	3–9	Urine/feces
Hydroxychloroquine	30–40%	~74%	~40–50	Urine/feces
Mefloquine	>98%	~85%	20 median	—
Primaquine	NA	75%	~6 h	Urine
Lumefantrine	99%	Low (5–11%)	3–6	Feces

Metabolism

- Aromatic hydroxylation of quinine via CYP3A4 has been reported. Among the identified metabolites are the 2'-hydroxy- and 3'-hydroxyquinine.
- O-Demethylation leading to 6'-hydroxyquinine is also a metabolite of quinine.

CHLOROQUINE/HYDROXYCHLOROQUINE

SAR

- Chlorine at C7 improves activity (Fig. 25.9).
- C3 or C8 alkyl groups reduce activity.
- Stereochemistry appears to be unimportant – the drug is used as dl mixture.
- Hydroxylation of the N-ethyl group (e.g., hydroxychloroquine vs. chloroquine) does not significantly affect activity.

Metabolism

- N-dealkylation to desethyl and didesethyl via CYP3A4 and CYP2D6 gives active metabolites.

MEFLOQUINE

Metabolism

- The C2 CF_3 group was added to the quinoline nucleus with the intent of reducing metabolism via aromatic hydroxylation (Fig. 25.9).
- Inactivation does occur through oxidation of the secondary alcohol to a carboxylic acid group.

LUMEFANTRINE

Lumefantrine   Halofantrine

- Lumefantrine is an analog of halofantrine, a drug, which due to potential toxicity and pharmacokinetic properties, has little use today.
- Lumefantrine is quite lipophilic resulting in poor bioavailability but a long half-life (3 to 6 days) (Table 25.2), which is ideal when used with a short-acting artemisinin.
- Lumefantrine is used exclusively in combination with artemether (see artemisinins).

PRIMAQUINE

- A unique mechanism of action has been proposed for primaquine involving a radical intermediate (Fig. 25.9).
- The 8-aminoprimaquine radical may condense with plasmodium biomolecules to produce this toxic effect.
- The drug is a CYP1A2 inducer capable of drug–drug interaction with substrates for this cytochrome.

Metabolism

- Primaquine is metabolized through deamination and oxidation of the primary amine leads to the inactive carboxylic acid.

ADVERSE EFFECTS

Quinolines: Common adverse effects to most of the quinolines include GI disturbance: nausea, vomiting, along with rash, pruritus, urticaria, visual disturbance. A variety of neuropsychiatric effects have been reported including vertigo, tinnitus, nervousness, and irritability. The potential exists for hematologic reactions including various blood dyscrasias which are more likely in individuals with glucose-6-phosphate dehydrogenase deficiency. Some of these effects are more common with extended therapy. With halofantrine cardiovascular adverse effects have included prolonged QT interval, torsades de pointes, ventricular arrhythmias, and death.

Pyrimethamine

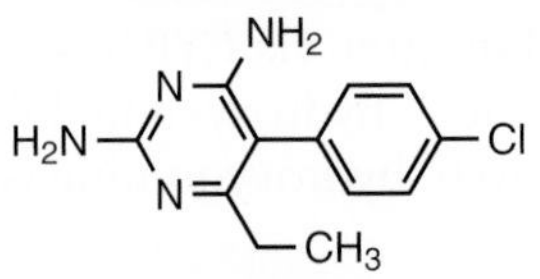

Pyrimethamine
(Daraprim)

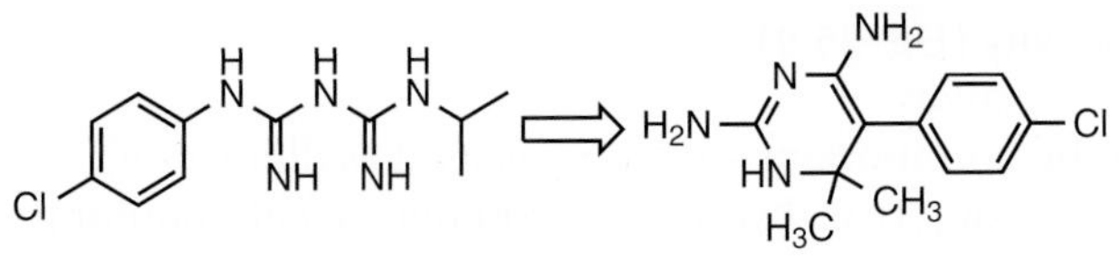

Proguanil ⟹ Cycloguanil

MOA

- Pyrimethamine was developed from the lead compound proguanil, which in turn is a prodrug giving rise to cycloguanil. Neither of the latter two is used today.
- Pyrimethamine is active by virtue of its ability to inhibit DHFR blocking folic acid synthesis (see Chapter 23, Fig. 23.1 – action of trimethoprim).
- Binds >1,000 times stronger to plasmodium DHFR than to host enzyme.
- Well absorbed via the oral route.
- Highly bound to plasma protein.
- Pyrimethamine plus sulfadoxine (sequential blocking of tetrahydrofolic acid), available in the past, has been discontinued due to drug resistance.

Artemisinins

Artemisinin is a natural product used in Chinese herbal medicine for thousands of years (Fig. 25.12). The lead compound artemisinin is isolated from *Artemisia annua* (quinghao, sweetworm wood). Its semisynthetic and synthetic derivatives

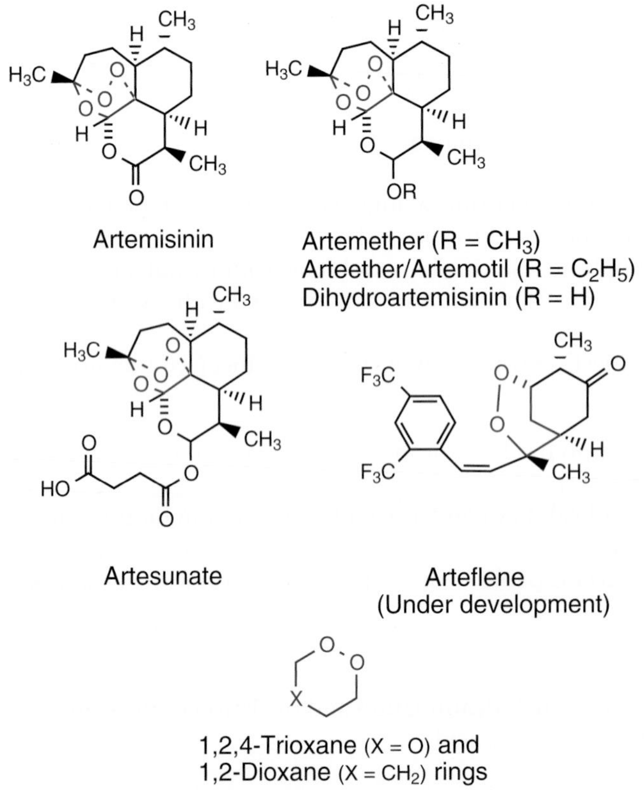

Artemisinin

Artemether (R = CH$_3$)
Arteether/Artemotil (R = C$_2$H$_5$)
Dihydroartemisinin (R = H)

Artesunate

Arteflene
(Under development)

1,2,4-Trioxane (X = O) and
1,2-Dioxane (X = CH$_2$) rings

Figure 25.12 Structures of artemisinin and artemisinin derivatives.

Table 25.3	**Artemisinins Used in Artemisinin-Based Combination Therapy (ACT)**	
Artemisinin	**2nd Component of ACT**	**Trade Name**
Artemether	Lumefantrine	Coartem
Artesunate	Amodiaquine	
Artesunate	Mefloquine	
Artesunate	Pyrimethamine/Sulfadoxine	
Dihydroartemisinin	Piperquine	

are members of the 1,2,4-trioxanes or 1,2-dioxane ring system in which the peroxide component is an essential feature of this drug class. The artemisinins are generally not available in the United States, with the exception of the combination of artemether–lumefantrine (Coartem).

MOA

- At present, the MOA of the artemisinins remains unknown although various mechanisms have been suggested and studied. Among these are the following:
  - The artemisinins act as prodrugs in which the endoperoxide reacts with Fe^{2+} from hemozoin to form Fe^{3+} and a free oxy radical.
  - The oxy radical gives rise to a carbon radical, which attacks the plasmodium.
  - The action occurs within the digestive vacuole involving hemozoin.
  - The endoperoxide targets the sarcoplasmic/endoplasmic reticulum Ca^{2+}-ATPase of the organism (PfATP6) altering calcium stores, thus affecting the plasmodium.
  - Possibly, a combination of both of the above mechanisms may be responsible for the lethal effects on plasmodium.

Physicochemical and Pharmacokinetic Properties

- Artemisinins with the exception of artesunate are hydrophobic and partition into plasmodium membranes.
- Available in oral (poor absorption), intramuscular, and rectal dosage forms.
- Artemisinins possess a very short half-life and should not be used singly, but rather used in combination with longer acting antimalarial drugs (artemisinin-based combination therapy [ACT]) (Table 25.3).

Metabolism

- Artemether, arteether, and artesunate are readily converted to dihydroartemisinin, an active metabolite via CYP2B6 and 3A4 (Fig. 25.13).
- An additional metabolism has been postulated and involves CYP450 catalyzed oxidations to deoxyartemisinins, with the loss of the endoperoxide.
- All of the metabolites can undergo O-glucuronidation and elimination in the urine.

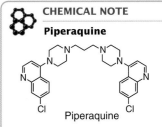

CHEMICAL NOTE

Piperaquine

Piperaquine

Piperaquine is an antimalarial drug used in China and recognized by WHO as valuable when used in combination with dihydroartemisinin.

Figure 25.13 Metabolism of artemisinins.

Table 25.4　Site and Spectrum of Action of Antimalarial Drugs

Drug/Drug Class	Site of Action			Plasmodium spp.			
	Liver	Erythrocyte	Gametocyte	*falciparum*	*ovale*	*vivax*	*malariae*
Artemisinins		✓	✓	✓	✓		✓
Chloroquine		✓		✓	✓	✓	✓
Hydroxychloroquine		✓		✓	✓	✓	✓
Mefloquine		✓		✓	✓	?	✓
Lumefantrine		✓		✓		✓	
Primaquine	✓		✓			✓	
Pyrimethamine		✓		✓	?	✓	?

Clinical Application of Antimalarials

- The objective in the treatment of uncomplicated malaria as defined by WHO is to cure the infection as rapidly as possible, while for severe malaria it is to prevent death.
- The goal is to eliminate the parasite from the blood and to do so WHO recommends ACT for uncomplicated *P. falciparum* malaria.
- The ACT drugs should be active against the erythrocytic form of the disease, while activity against the gamete form of the disease is also important to prevent the spread of the disease (Table 25.4).
- Antimicrobials commonly used in the treatment of malaria include tetracyclines, clindamycin, and atovaquone.

Suggested Reading

For more detailed information about drug therapy for malaria the reader is referred to Guidelines for the treatment of Malaria, Second Edition, WHO, 2010.

Helminth Infections

Helminthiasis is defined as a disease associated with parasitic worm infections, which in most cases involves worms in the intestinal track. The life cycle of most of the helminthes is quite complex and involves the human host, but also various organisms in the transmission of the diseases. It is estimated that as many as one-fourth of the world population may be infected with worms. Also unique to these parasitic organisms is the fact that in many instances the human host immune defenses do not recognize these pathogens. A list of common human helminthes is shown in Table 25.5.

Drug Therapy of Helminthiasis
Benzimidazoles

The benzimidazoles were discovered in the 1960s and although they are effective for the treatment of a variety of intestinal helminthes only two of these drugs are still available in the United States (Table 25.6).

MOA

- Several mechanisms have been proposed including inhibition of fumarate reductase in helminth mitochondria and binding to the organism's tubulin-inhibiting microtubule polymerization.

Physicochemical and Pharmacokinetic Properties

- The benzimidazoles (mebendazol and albendazole) exhibit poor water solubility and are poorly absorbed from the GI tract.
- The lack of absorption improves use against intestinal helminthes.
- Absorbed drug is bound to plasma protein and/or metabolized.

Metabolism

- Albendazole is converted to the S-oxide and retains activity.
- Mebendazole is hydrolyzed to the 2-aminobenzimidazole or reduced to the hydroxy derivative both of which are inactive.

Table 25.5 Human Infection Helminthes

Common Name	Phylum/Latin Name	Diseases	Location in Host
Round Worms	**Nematode**		
Hookworm	*Nector americanus*	Ancylostomiasis	Digestive tract
Hookworm	*Ancylostoma duodenale*	Ancylostomiasis	Digestive tract
Pinworm	*Enterobius vermicularis*	Enterobiasis	Digestive tract
Roundworm	*Ascaris lumbricoides*	Ascariasis	Small intestine
Whipworm	*Trichuris trichiura*	Trichuriasis	Intestinal wall
	Trichinella spiralis	Trichuriasis	Intestinal wall
Eyeworm	*Mansonella streptocerca*[a]	Filariasis	Subcutaneous layer
River blindness	*Onchocerca volvulus*	Onchoceriasis	Dermis
	Wuchereria bancrofti	Elephantiasis	Lymphatic-dwelling
	Brugia malayi	Elephantiasis	Lymphatic-dwelling
Timor filariasis	*Brugia timori*	Elephantiasis	Lymphatic-dwelling
Flat Worms	**Cestode**		
Beef tapeworm	*Taenia saginata*	Taeniasis	Digestive tract
Pork tapeworm	*Taenia solium*	Cysticercosis	Digestive tract
Dwarf tapeworm	*Hymenolepis nana*		Digestive tract
Fish tapeworm	*Diphyllobothrium latum*		Digestive tract
	Trematode		
	Schistosoma haemaetobium	Schistosomiasis	Urinary bladder
	Schistosoma mansoni	Schistosomiasis	Blood vessels of GI tract
	Schistosoma japonicum	Schistosomiasis	Blood vessels of GI tract

[a]*M. streptocerca, O. volvulus, W. bancrofti, B. malayi,* and *B. timori* are commonly referred to as lymphatic-dwelling filariae. An early stage of these parasites is the microfilariae stage.

Diethylcarbamazine (DEC)

MOA

- Unknown but three mechanisms have been suggested:
  - Drug promotes morphologic damage to the microfilaria including loss of the cellular sheath.

Diethylcarbamazine citrate
(Available from CDC)

Table 25.6 Benzimidazole Anthelmintics

Drugs	Trade Name	R₁	R₂	R₃
Thiabendazole	Discontinued in U.S.	(thiazole ring)	—H	H
Mebendazole		$-\overset{H}{N}-\overset{O}{\overset{\|}{C}}-OCH_3$	$-\overset{O}{\overset{\|}{C}}-$(phenyl)	H
Albendazole	Albenza	$-\overset{H}{N}-\overset{O}{\overset{\|}{C}}-OCH_3$	$-SCH_2CH_2CH_3$	H
Triclabendazole	Egaten[a] Fasinex[b]	$-S-CH_3$	$-O-$(2,3-dichlorophenyl)	Cl
Fenbendazole	Several brand names[b]	$-\overset{H}{N}-\overset{O}{\overset{\|}{C}}-OCH_3$	$-S-$(phenyl)	H
Flubendazole		$-\overset{H}{N}-\overset{O}{\overset{\|}{C}}-OCH_3$	$-\overset{O}{\overset{\|}{C}}-$(4-fluorophenyl)	H

[a]Egaten has recently been shown to be useful for treatment of fascioliasis by WHO.
[b]Used in veterinary practice for protection and treatment of parasite and worm infections.

- Inhibition of microtubular polymerization.
- Interference with arachidonic acid metabolism (blocking cyclooxygenase and leukotriene A synthase).

Physicochemical and Pharmacokinetic Properties

- Rapid onset of action following oral administration.
- Inactive in vitro, but no sign that DEC is a prodrug.
- Significant amount of the drug is recovered in the urine unchanged.

Metabolism

- N-oxide of the basic nitrogen
- N-deethylation

Ivermectin (IVM)

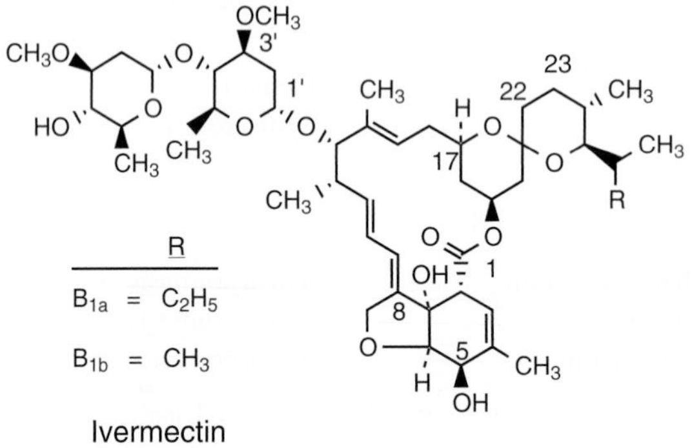

R

B$_{1a}$	=	C$_2$H$_5$
B$_{1b}$	=	CH$_3$

Ivermectin
(Stromectol)

MOA

- Decreases microfilaria motility via paralysis of the worm's muscle.
- This may be associated with the ability of the drug to induce Cl$^-$ influx and hyperpolarization or as a γ-aminobutyric acid (GABA) agonist.
- In addition, it has been suggested that IVM causes degeneration of microfilariae in utero reducing the number of microfilariae produced, reducing the chance for reinfection.

Physicochemical and Pharmacokinetic Properties

- The drug is readily absorbed following oral administration (NOTE: a topical dosage form has recently been approved for treatment of head lice – see below).
- The drug is highly bound to plasma protein (93%) resulting in a long half-life.

Metabolism

- C3' O-dealkylation.
- Loss of the terminal sugar.

Praziquantel (PZQ)

MOA

- Several mechanisms have been proposed:
  - Intestinal worms: drug may cause muscle contraction and paralysis via influx of Ca^{2+} leading to expulsion of the worm.
  - Intravascular worms: damage to the worm tegument exposing helminth to antigens leading to an antigen–antibody reaction and death of the worm.

Physicochemical and Pharmacokinetic Properties

- Most of the activity is found in the levo isomer, but it is administered as a mixture of isomers.
- The drug is rapidly absorbed, highly bound to plasma protein (~85%) and expensively metabolized to inactive metabolites.

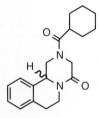

Praziquantel
(Biltricide)

Praziquantel (PZQ)

Urinary metabolites

Serum metabolite

Figure 25.14 Metabolism of PZQ.

Metabolism

- Hydroxylated by CYP 3A4 and 2B6, represents the major route of metabolism (Fig. 25.14).

Pyrantel Pamoate

Pyrantel pamoate
(Various OTC trade names)

MOA

- Pyrantel pamoate acts as a neuromuscular blocking agent at nicotinic receptors leading to worm paralysis.
- Also blocks cholinesterase of the worm.

Physicochemical and Pharmacokinetic Properties

- The pamoate salt is insoluble, not absorbed, and therefore, its use is limited to intestinal parasites.

Clinical Applications

- The spectrum of activity of the various anthelmintics is summarized in Table 25.7.
- A drug's activity is affected by the stage of the helminth (i.e., DEC is active against the microfilaria stage of the various filarie [Table 25.6], PZQ is not active against the liver stages of the cestode and trematode infections, but is active against other stages).
- Several of the anthelmintics are also valuable as veterinary products (e.g., pyrantel pamoate, triclabendazole, fenbendazole, flubendazole).

Ectoparasitic Infections

The ectoparasites are organisms that live on the surface of the skin of humans and other mammals. Three ectoparasites which fit this definition are the

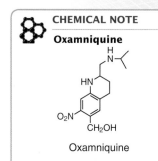

CHEMICAL NOTE

Oxamniquine

Oxamniquine

Oxamniquine was previously available with a spectrum of activity similar to that of praziquantel. The drug has been discontinued in the United States.

ADVERSE EFFECTS

Anthelmintics: Mebendazole and albendazole are not used during the first trimester of pregnancy because of possible teratogenic effects. DEC may cause anaphylactic reaction probably due to dead microfilariae (Mazzotti reaction). This same reaction may occur with IVM. Many of the anthelmintics cause GI distress.

Table 25.7 Therapeutic Application of Anthelmintics for Specific Helminth Infections

	M	A	DEC	IVM	PZQ	PP
Nematode Infections						
Necator americanus	✓	✓				
Ancylostoma doudenale	✓	✓				✓
Enterobius vermicularis	✓			✓		✓
Ascaris lumbricoides	✓	✓		✓		✓
Trichuris trichiura	✓	✓		✓		
Trichinella spiralis	✓	✓				
Wuchereria bancrofti			✓	✓		
Brugia malayi			✓	✓		
Brugia timori			✓			
Loa Loa		✓	✓	✓		
Onchocerca volvulus		✓	✓	✓		
Cestode Infections						
Taenia saginata	✓	✓			✓	
Taenia solium	✓	✓			✓	
Hymenolepis nana	✓				✓	
Diphyllobothrium latum					✓	
Trematode Infections						
Schistosoma hematobium					✓	
Schistosoma mansoni					✓	
Schistosoma japonicum					✓	

Benzimidazoles: M, mebendazole, A, albendazole; DEC, diethylcarbamazine; IVM, ivermectin; PZQ, praziquantel; PP, prantel pamoate.

Sarcoptes scabiei (itch mite), *Pediculus humanus,* and *Cimex lectularius.* The *S. scabiei* causes the condition of scabies which is characterized by intense itching, which results from the parasite burrowing into the epidermis. The human condition of lice is caused by any of the three *P. humanus* parasites: *P. humanus capitis* causes head lice, *P. humanus corporis* causes body lice, and *Phthirus pubis* causes crab lice. Lice are bloodsucking insects that leave skin punctures, which can lead to a hypersensitivity reaction leading to pruritus. *Cimex lectularius,* commonly known as bedbugs, also lives off of human blood causing skin irritation. There is no drug therapy for this condition.

Drug Therapy for Scabies and Pediculosis
Lindane (γ-hexachlorocyclohexane)

- Lindane blocks γ-aminobutyric acid (GABA) receptors leading to neuronal stimulation and paralysis of the parasite.
- Effective against the adult lice, its eggs, and its nits.
- Applied topically as a shampoo or crème.
- The lipophilicity of the drug can result in transdermal absorption and the possibility of systemic toxicity including neurotoxicity to the patient, especially children.

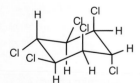

γ-Hexachlorocyclohexane
(Kwell)

Pyrethrums and Pyrethroids

The pyrethrums are natural-occurring products isolated from chrysanthemums with insecticidal activity and consisting of a mixture of material of which pyrethrin I and II represent major constituents (Fig. 25.15).

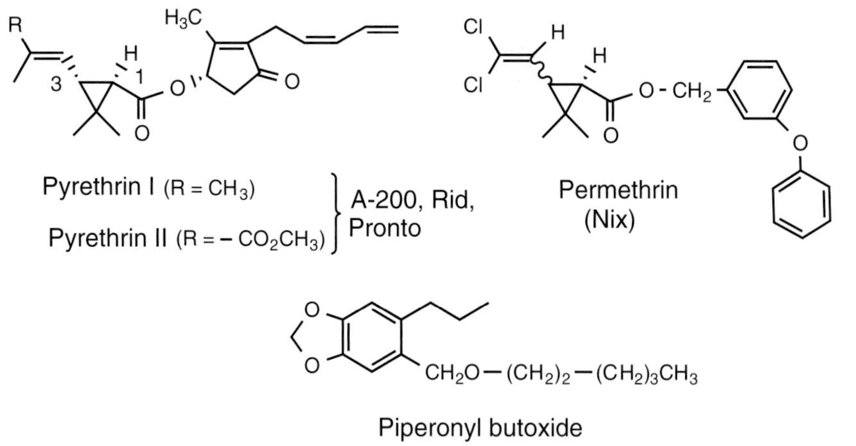

Figure 25.15 Structures of pyrethrum, pyrethroid, and piperonyl butoxide.

Pyrethrin I (R = CH₃)
Pyrethrin II (R = − CO₂CH₃) } A-200, Rid, Pronto

Permethrin (Nix)

Piperonyl butoxide

MOA

- Pyrethrins bind to proteins in the sodium channel slowing the rate of current inactivation.
- This binding leads to repetitive neuron firing, membrane depolarization, and prolonged opening of the sodium channel.

Physicochemical and Pharmacokinetic Properties

- Pyrethrins are quite unstable with a short duration of action as a result of hydrolysis, photolysis, and mixed function oxidation.
- The active isomers of the pyrethrins are the 1R, 3 cis and 1R, 3 trans isomers. The 1S isomers cis and trans are inactive.
- Piperonyl butoxide, a synergistic additive, does not have insecticidal activity, but blocks CYP450 enzymes in the organism to prolong stimulatory activity.
- Piperonyl butoxide lowers the amount of pyrethrins needed for cidal action (used in a 10:1 ratio).
- Permethrin, a synthetic derivative of the pyrethrins, is a potent insecticide, which is less expensive to produce.
- The pyrethrins and pyrethroids are lipophilic which increases the potential for penetration of the insects' lipid membranes.

Metabolism

- In humans the pyrethrins and permethrin undergo rapid ester hydrolysis, oxidation, or epoxidation to inactive and nontoxic chemicals (reduced potential for human toxicity).

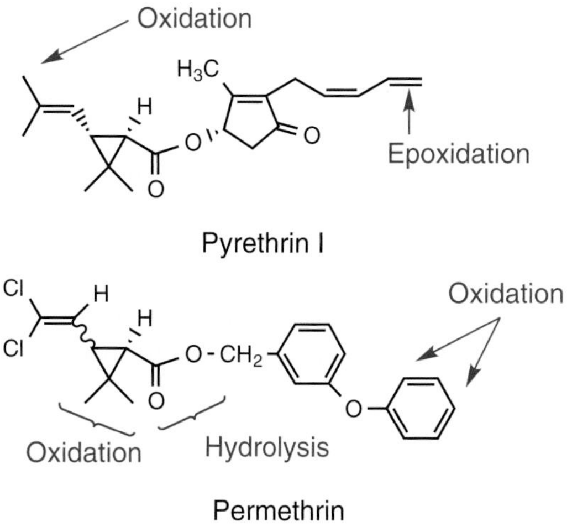

Oxidation

Epoxidation

Pyrethrin I

Oxidation

Oxidation   Hydrolysis

Permethrin

Spinosad

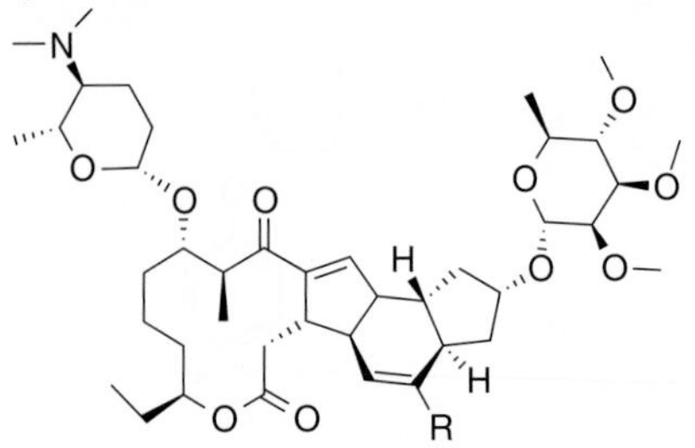

Spinosad (Natroba)

(Spinosyn A: R = H, Spinosyn D: R = CH₃)

Spinosad is a mixture of two natural-occurring macrocyclic lactones (spinosyn A and D – 5:1 ratio) isolated from actinomycete bacteria. The combination is reported to be effective against various insects including head lice.

MOA

- Spinosad causes hyperexcitation of nicotinic acetylcholine receptors (nAChRs).
- The action is on a unique nAChR (in *Drosophila* on the Dα6 subunit).
- Leads to paralysis of the insect and death.
- Also suggested that the spinosads may act on GABA receptors.

Physicochemical and Pharmacokinetic Properties

- Spinosad is available as a 0.9% topical suspension.
- Applied once to the scalp and removed.
- Minimal side effects – possible erythema.

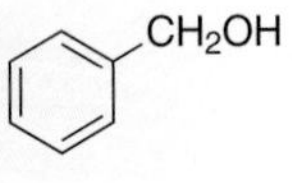

Benzyl alcohol
(Ulesfia)

Benzyl Alcohol

Benzyl alcohol is not insecticidal, but rather, asphyxiates lice by preventing the organism from closing their respiratory spiracles, thus causing the mineral oil vehicle to penetrate and obstruct the spiracles.

Ivermectin (Sklice)

- As previously indicated, IVM has been approved as a topical preparation (0.5%) for treatment of head lice.
- Stimulates release of GABA leading to chloride ion influx and hyperpolarization and paralysis of the parasite.

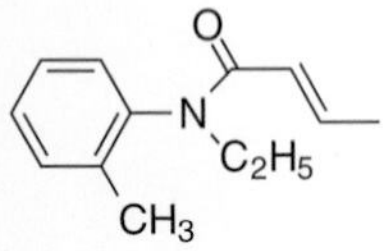

Crotamiton (Crotan, Eurax)

- Approved for use in treatment of scabies.
- Available as to topical cream (10%).

Crotamiton

Clinical Applications

- Drugs for the treatment of lice fall into two groups: those that kill both the lice and the nits (ovicid) and those that kill only the lice. The latter require multiple application of the drug.
- Drugs killing lice and nits: spinosad.
- Drugs killing only lice: benzyl alcohol, IVM, lindane, pyrethrins, permethrin.
- Lindane is considered a second line agent and is not recommended for pediatric use.
- Scabicides include permethrin, lindane (second line), crotamiton, IVM (not FDA approved).

Chemical Significance

Parasitic infections cover a wide range of conditions involving diverse organisms including protozoa, helminthes, and ectoparasites. It is estimated that nearly half the world's population is exposed to malaria and is therefore at risk of developing this single protozoan infection with as many as 655,000 deaths each year from malaria. The problem faced by those working to treat and cure this and other protozoan infections is that the vast majority of the infections and deaths occur in developing countries which cannot afford the cost of new medications for their treatment. Nearly 91% of all deaths from malaria occur in sub-Saharan Africa. While the numbers are much smaller, the incidence and deaths associated with amebiasis, trypanosomiasis, leishmaniasis, and many of the helminth infections are rare in the western world. The CDC estimates that 1,500 cases of malaria are diagnosed in the United States annually. World travel and human migration suggests that these rare infections may grow in years to come. We can expect significant changes in the prevention and treatment of malaria in the future since the Bill & Melinda Gates Foundation have specifically targeted malaria with research funds. The foundation has committed more than $2 billion in malaria grants and expansion of prevention and treatment tools as well as advocating for increased funding to treat this protozoal condition.

The utilization of chemistry for an understanding of the antiparasitic drugs can be demonstrated nicely by looking at the antimalarial quinine-derived drugs. It should not be surprising that treatment of malaria has suffered from the lack of new chemical entities onto the market. The treatment of malaria depends on two classes of drugs: the quinines and the artemisinins. Figure 25.9 demonstrates the chemical interrelationship between quinine and the substituted quinolines. Given the development of protozoan resistance, it is understandable that drug therapy with this extended class of agents leaves many patients without successful therapy. With the artemisinins the complexity of the chemistry has limited the availability of improved drugs. This may be especially significant in that medication costs pose a major hurdle in the treatment of patents in the developing countries. This same factor becomes important in the treatment of the more exotic diseases such as trypanosomiasis and leishmaniasis. Here, one sees the use of potentially toxic heavy metals such as organic arsenics and antimony derivatives for treatment. There is little economic incentive to drive pharmaceutical research to develop new and safer medications. This is also true for the development of new and safer anthelmintics. The available drugs often are natural products or structurally simple molecules, with limited innovative new drugs being developed.

Review Questions

1 Match the organism to the specific diseases listed below:

Enterobiasis _____	A. Protozoa
Malaria _____	B. Ectoparasite
Giardiasis _____	C. Helminth (round worm)
Lice _____	D. Helminth (flat worm)

2 The mechanism of action of eflornithine is:
  A. Inhibition of cell division as an irreversible inhibitor of ODC.
  B. Inhibits electron transfer in the mitochondrial respiratory chain.
  C. Reversible inhibitor of sulfhydryl proteins.
  D. Prevents DNA separation required for cell division.
  E. Inhibitor of microtubulin formation.

3 At what stage in the life cycle of the plasmodium are the quinine and quinolones most effective as antimalarials?
A. Merozoite stage
B. Trophozoite stage
C. Gametocyte stage
D. Sporozoite stage
E. Pre-erythrocytic stage

4 Shown below are the structures of several heterocyclic nitroaryl compounds. While their spectrum of activity may be quite different they all possess a similar mechanism of action. What is that similarity?

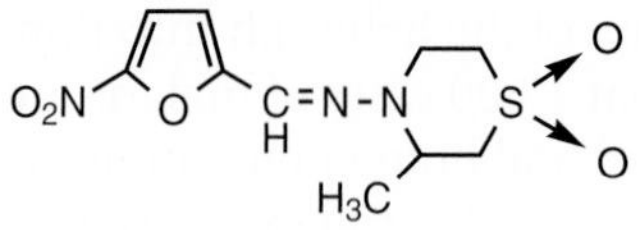

Nifurtimox

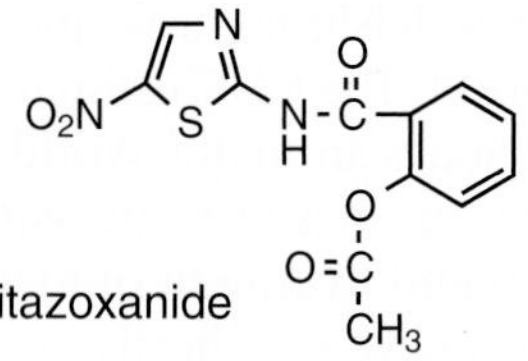

Nitazoxanide

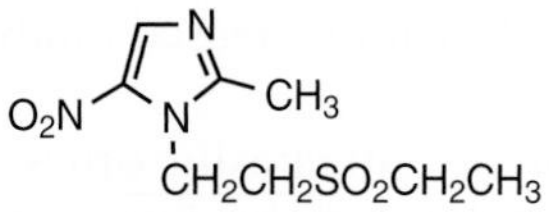

Tinidazole

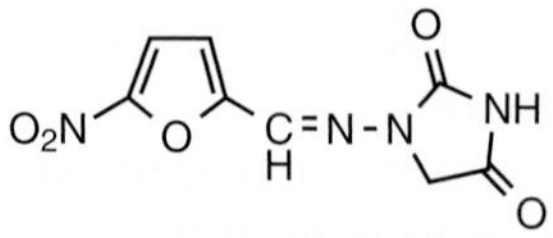

Nitrofurantoin

A. The heterocycles and nitro group are required binding units for attachment to the parasite's receptor.
B. All require metabolism to the aminoaryl group for activity.
C. The nitro group picks up and transports an electron which is involved in the mechanism of action of the drugs.
D. The nitro group is capable of binding to and fragmenting the DNA of the parasite.

5 What is the role of the pamoate in pyrantel pamoate?
A. Improves the stability of the drug.
B. Gives rise to an insoluble drug, which prevents the drug from being absorbed.
C. Insures a slow onset of action and long duration of action.
D. Improves duration of action of the intramuscular injectable dosage form of the drug.
E. Increases the solubility of the drug in nonaqueous media.

Cancer Chemotherapy

Learning Objectives

Upon completion of this chapter the student should be able to:

1. Describe mechanism of action of:
   a. DNA cross-linking agents
   b. Topoisomerase poisons
   c. Antibiotic antineoplastics
   d. Antimetabolites
   e. Mitosis inhibitors
   f. Tyrosine kinase inhibitors
   g. mTOR inhibitors
   h. Proteasome inhibitors
   i. Histone deacetylase inhibitors

2. Discuss structure activity relationships (SAR) (including key aspects of mechanism and/or receptor binding) of the above classes of antineoplastic agents.

3. List physicochemical and pharmacokinetic properties that impact in vitro stability and/or therapeutic utility of antineoplastic agents.

4. Diagram common metabolic pathways, identifying enzymes and noting the activity, if any, of major metabolites.

5. Apply all of the above to predict and explain therapeutic utility, toxicity, and mechanisms of resistance.

Introduction

Cancer is among the most feared diagnosis in human disease. However, great strides have been made to scientifically address the uncontrolled growth of cells that is the hallmark of the constellation of diseases known collectively as cancer.

Because of the multitude of complex biochemical mechanisms that initiate, promote, and propagate transformation of normal cells to malignant cells, an armamentarium of drugs old and new is at the clinician's disposal. The significant toxicity of these drugs, along with the high stakes of providing patients with optimum therapies from the point of diagnosis, demands a team-based approach to therapeutic decision making and the monitoring of holistic patient well-being. The pharmacist clearly needs to be at the epicenter of that team.

This chapter will focus on nonhormone and nonprotein drugs more commonly used in the treatment of various cancers. The reader is referred to Chapter 37 in *Foye's Principles of Medicinal Chemistry, Seventh Edition,* for a discussion of the causes, signs/symptoms, staging, and general therapeutic approaches to the treatment of cancer. Agents used less commonly in the treatment of selective neoplastic diseases, as well as an introduction to antineoplastic monoclonal antibodies, are also addressed in Chapter 37.

DNA Cross-Linking Agents (Fig. 26.1)

MOA

- DNA cross-linking agents are reactive electrophiles that interact with two nucleophilic residues on one or two DNA strands. Intrastrand and interstrand DNA cross-linking, respectively, results in DNA destruction and cell death.

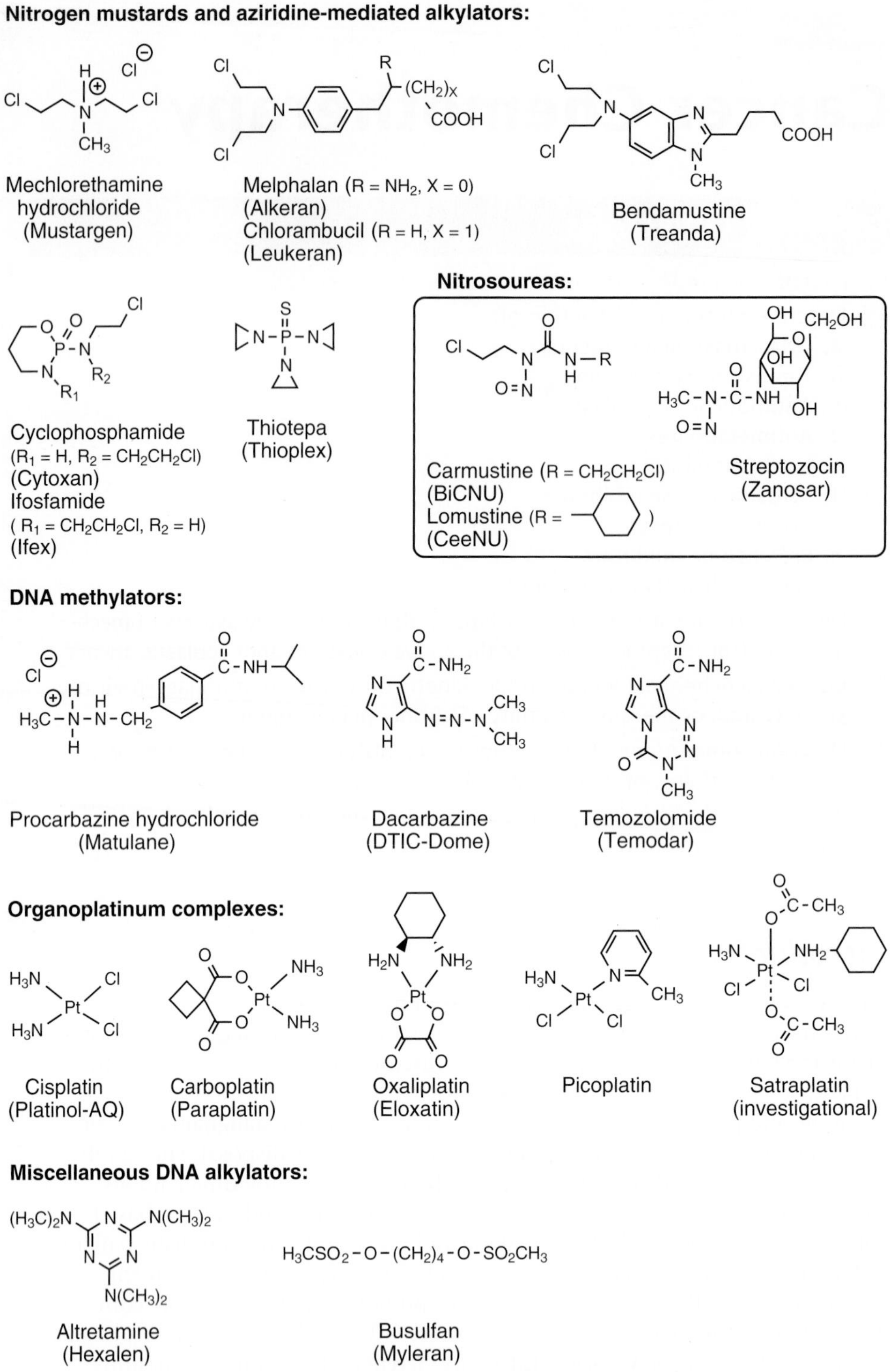

Nitrogen mustards and aziridine-mediated alkylators:

Mechlorethamine
hydrochloride
(Mustargen)

Melphalan (R = NH$_2$, X = 0)
(Alkeran)
Chlorambucil (R = H, X = 1)
(Leukeran)

Bendamustine
(Treanda)

Cyclophosphamide
(R$_1$ = H, R$_2$ = CH$_2$CH$_2$Cl)
(Cytoxan)
Ifosfamide
(R$_1$ = CH$_2$CH$_2$Cl, R$_2$ = H)
(Ifex)

Thiotepa
(Thioplex)

Nitrosoureas:

Carmustine (R = CH$_2$CH$_2$Cl)
(BiCNU)
Lomustine (R = ⬡)
(CeeNU)

Streptozocin
(Zanosar)

DNA methylators:

Procarbazine hydrochloride
(Matulane)

Dacarbazine
(DTIC-Dome)

Temozolomide
(Temodar)

Organoplatinum complexes:

Cisplatin
(Platinol-AQ)

Carboplatin
(Paraplatin)

Oxaliplatin
(Eloxatin)

Picoplatin

Satraplatin
(investigational)

Miscellaneous DNA alkylators:

$H_3CSO_2 - O - (CH_2)_4 - O - SO_2CH_3$

Figure 26.1 DNA cross-linking
agents.

Altretamine
(Hexalen)

Busulfan
(Myleran)

- Interstrand cross-linking is most common in nitrogen mustards. Intrastrand cross-linking predominates with organoplatinum complexes.
- The DNA nucleophile most vulnerable to attack is guanine N7.

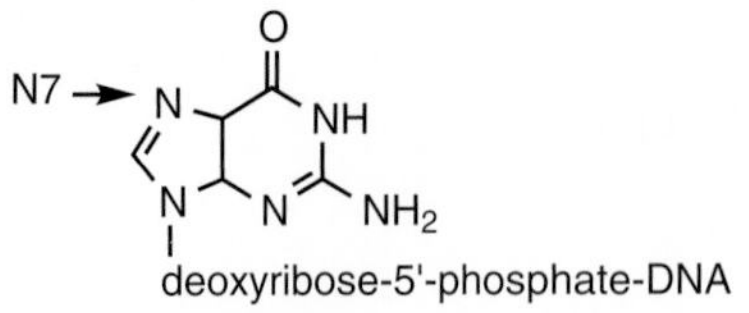

deoxyribose-5'-phosphate-DNA

Guanine N7

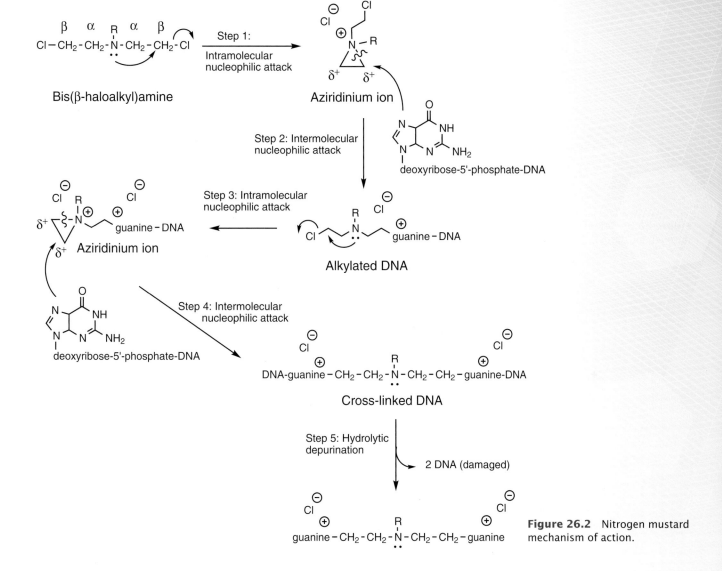

Figure 26.2 Nitrogen mustard mechanism of action.

Nitrogen Mustards

MOA

- Nitrogen mustards are bis-β-chloroalkylamines. Mustards spontaneously form the highly electrophilic quaternary aziridinium ion in aqueous media (Fig. 26.2).
- The aziridinium ion's strong δ^+ charge attracts the nucleophilic guanine N7. This breaks the highly strained ring and alkylates DNA.
- Since there are two β-chloroalkylamine "arms" to the structure, a second aziridinium ion will alkylate a guanine residue on another strand of DNA, resulting in interstrand cross-linking.

SAR (Fig. 26.3)

- True nitrogen mustards differ in structure only at the third substituent of the tertiary amine (R).
  - Aliphatic (R = CH_3) mustards are exceptionally reactive and toxic electrophiles. They are too reactive for oral administration and are given only by the IV route in cancers of the blood.
  - Aromatic (R = aryl) mustards are much weaker electrophiles. They are most commonly administered orally for the palliative care of patients with leukemia, lymphoma, or multiple myeloma.

ADVERSE EFFECTS

Nitrogen Mustards (the more reactive the mustard, the more severe the effect):
- Myelosuppression (lymphocytopenia, granulocytopenia)
- Nausea, vomiting
- Alopecia
- Tissue damage upon extravasation
- Induction of myelogenous leukemia with extended use

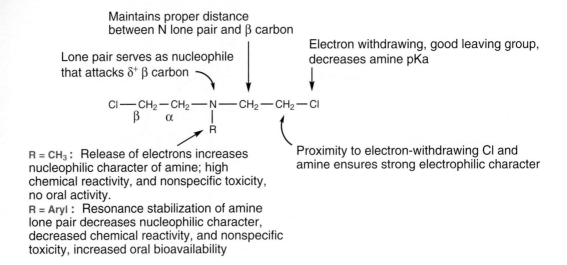

Maintains proper distance
between N lone pair and β carbon

Lone pair serves as nucleophile
that attacks δ⁺ β carbon

Electron withdrawing, good leaving group,
decreases amine pKa

$$Cl—CH_2—CH_2—N—CH_2—CH_2—Cl$$

R = CH₃: Release of electrons increases nucleophilic character of amine; high chemical reactivity, and nonspecific toxicity, no oral activity.
R = Aryl: Resonance stabilization of amine lone pair decreases nucleophilic character, decreased chemical reactivity, and nonspecific toxicity, increased oral bioavailability

Proximity to electron-withdrawing Cl and amine ensures strong electrophilic character

Figure 26.3 SAR of nitrogen mustards.

> ### CHEMICAL NOTE
>
> **Aqueous Decomposition of Nitrogen Mustards:** Nitrogen mustards are unstable in aqueous media. The β carbons can be attacked by the oxygen of water before or after cyclization to the quaternary aziridinium ion. In either case, the inactive dehalogenated diol is generated. Stability is enhanced in mildly acidic solutions.
>
> Active mustard
>
> Inactive dehalogenated diol

> ### CHEMICAL NOTE
>
> **Chemical Inactivation of Spilled or Extravasated Mustards:** Mustards, particularly aliphatic mustards, are strong vesicants and will blister and burn skin. If spilled or extravasated, the drugs can be inactivated with sodium thiosulfate. Skin exposed to nitrogen mustards should be treated intermittently with an ice compress for 6 to 12 hours to minimize serious tissue damage and the need for surgical resectioning.
>
> Mechlorethamine
>
> Inactive thiosulfate ester

Metabolism

- Aromatic mustards are predictably vulnerable to N-dealkylation (bendamustine), "benzylic" hydroxylation (bendamustine), and β-oxidation (chlorambucil).

- The β-oxidized metabolite of chlorambucil, phenylacetic acid mustard, is biologically active and responsible for some of its therapeutic efficacy.

Phenylacetic acid mustard

Nitrogen Mustard Prodrugs

- Cyclophosphamide and ifosfamide are bioactivated by a metabolic and spontaneous decomposition pathway initiated predominantly by CYP2B6 and CYP3A4, respectively (Figs. 26.4 and 26.5).
- Hydroxylation occurs on C4 of the oxazaphosphorine ring, and is more sluggish with ifosfamide due to steric hindrance from the adjacent chloroethyl group.
- The tertiary side chain amine of cyclophosphamide (a true mustard) generates the more reactive and cytotoxic aziridinium ion.
- Ifosfamide's secondary amine produces the less electrophilic and less cytotoxic aziridine species.
  - Thiotepa, another aziridine-based DNA cross-linker, is active intact. It is converted by oxidative desulfuration to an active triethylenephosphoramide metabolite.

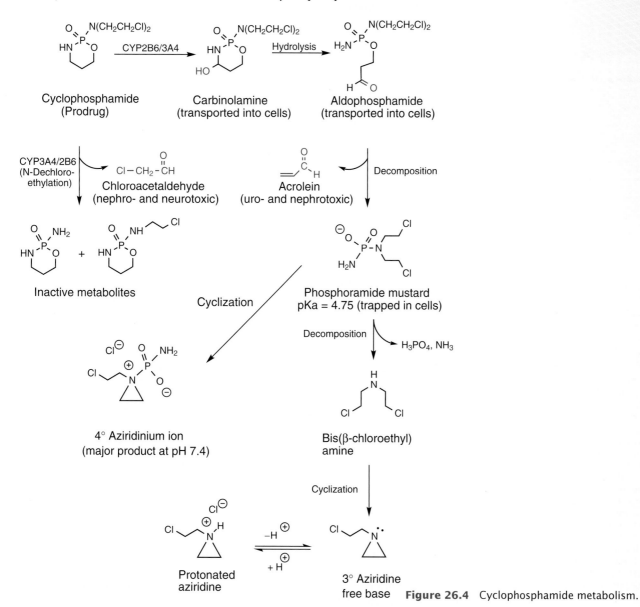

Figure 26.4 Cyclophosphamide metabolism.

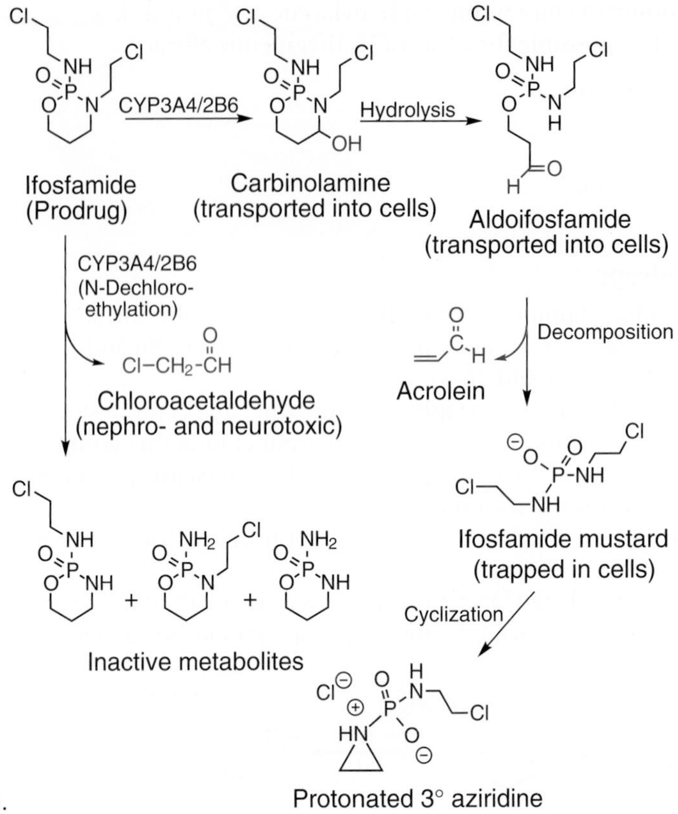

Figure 26.5 Ifosfamide metabolism.

- Inactivating metabolism involves N-dechloroethylation by the same two CYP isoforms. Nephro- and neurotoxic chloroacetaldehyde is formed in the process.

Thiotepa   Triethylenephosphoramide
(active metabolite)

Acrolein-Induced Urotoxicity

- Acrolein is an electrophilic byproduct of cyclophosphamide and ifosfamide activation. It destroys bladder cells by alkylating Cys residues.
- Mesna, a nucleophilic Cys decoy, concentrates in bladder. It reacts with and detoxifies acrolein (Fig. 26.6).
- The mesna–acrolein adduct is water soluble due to the highly ionized sulfate anion. It is readily excreted in urine.
- Mesna is marketed as a disulfide (dimesna) that requires metabolic reduction to the active sulfhydryl (thiol) moiety.

$^{\ominus}O_3S-CH_2-CH_2-S-S-CH_2-CH_2-SO_3^{\ominus}$ $\xrightarrow{\text{reduction}}$ $^{\ominus}O_3S-CH_2-CH_2-SH$

Dimesna   Mesna
(inactive disulfide)   (active sulfhydryl reagent)

Dimesna reduction to mesna

Chloroacetaldehyde-Induced Nephrotoxicity and Neurotoxicity

- Chloroacetaldehyde is an electrophilic byproduct of cyclophosphamide and ifosfamide dechloroethylation. Ifosfamide produces more chloroacetaldehyde per unit dose.
- Chloroacetaldehyde induces nephro- and neurotoxicity through alkylation of cell nucleophiles (e.g., Cys and Lys residues).
- Mesna does not concentrate in the nephron or nerve cells and cannot protect against chloroacetaldehyde-induced damage.

CHEMICAL NOTE

GSH Cytoprotection Against Acrolein and Chloroacetaldehyde Toxicity: Reduced glutathione (GSH) is an endogenous nucleophile that protects against damage by cytotoxic electrophiles. It reacts covalently with small amounts of chloroacetaldehyde such as that generated from cyclophosphamide, attenuating nephro- and neurotoxicity risk. GSH stores are limited and readily exhausted, and it is overwhelmed by the larger amounts of chloroacetaldehyde produced from the N-dechloroethylation of ifosfamide. Nephro- and neurotoxicity result.

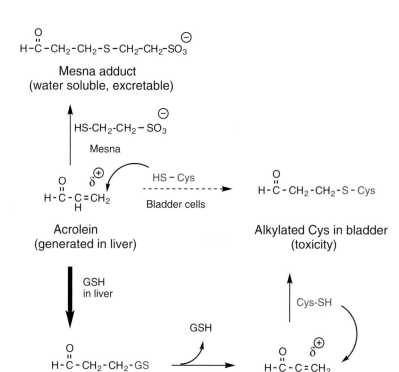

Figure 26.6 Bladder cell cysteine (Cys) alkylation by acrolein and rescue by mesna.

Nitrosoureas

MOA

- Two of the three nitrosoureas decompose in the cytoplasm to release an electrophilic 2-chloroethyl carbocation. Proton loss initiates the spontaneous and complex decomposition (Fig. 26.7).
- Each carbon of the chloroethyl carbocation is δ^+ and will be attacked by guanine N7 from two distinct DNA strands, resulting in cross-linked DNA.
- Another toxic electrophile capable of alkylating DNA, vinyl carbocation, forms via a different decomposition pathway.
  - Carmustine generates 2-chloroethylamine as a byproduct of this pathway, which also has DNA alkylating action.

SAR

- Carmustine and lomustine differ only in the group attached to the nonnitrosated urea nitrogen.
- Carmustine's second chloroethyl group confers a higher reactivity and cross-linking potency compared to lomustine's cyclohexyl moiety. Streptozocin is significantly less cytotoxic.

Physicochemical Properties (Table 26.1)

- Both carmustine and lomustine are lipophilic and readily cross the blood–brain barrier.

Table 26.1 Physicochemical Properties of Nitrosoureas

Nitrosourea	Melting Point (°C)	Log P	Water Solubility (mg/mL)	Dosage Form
Carmustine	30 (86°F)	1.53	4	IV solution in 10% ethanol; biodegradable wafer
Lomustine	90 (194°F)	2.83	<0.5	Capsule
Streptozocin	121 (250°F)	−2.80	50	IV solution in water

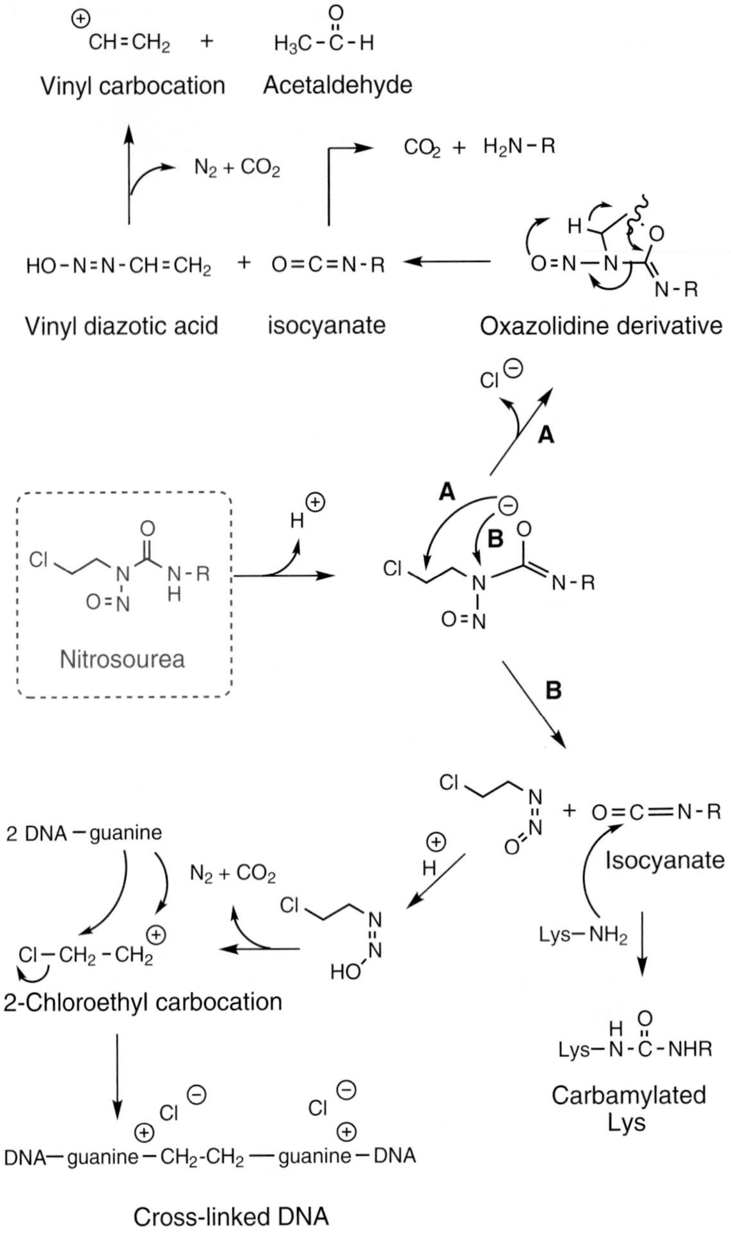

Vinyl carbocation   Acetaldehyde

$N_2 + CO_2$

$CO_2 + H_2N-R$

$HO-N=N-CH=CH_2$ + $O=C=N-R$

Vinyl diazotic acid   isocyanate   Oxazolidine derivative

$Cl^{\ominus}$

A

A

B

Cl

Nitrosourea

$H^{\oplus}$

B

Cl $O=C=N-R$

Isocyanate

2 DNA—guanine

$N_2 + CO_2$

$H^{\oplus}$

$Cl-CH_2-CH_2^{\oplus}$

2-Chloroethyl carbocation

$Lys-NH_2$

$Lys-N-C-NHR$

Carbamylated Lys

$DNA-guanine-CH_2\text{-}CH_2-guanine-DNA$

Cross-linked DNA

Figure 26.7 Nitrosourea decomposition to cytotoxic electrophiles.

- Carmustine is a low-melting solid that decomposes to an oil. Reconstituted IV solutions with an oily precipitate should be discarded.
- Carmustine is also formulated into a biodegradable wafer that is implanted directly into the brain after tumor resection.

Clinical Applications

- Chloroethyl-substituted nitrosoureas, carmustine and lomustine are used in brain cancers and Hodgkin's disease.
- Streptozocin, which lacks the chloroethyl moiety, is used in metastatic islet cell carcinoma.

Guanine O6-Methylators (Table 26.2)

MOA (Figs. 26.8 and 26.9)

- Procarbazine and the triazines dacarbazine and temozolomide methylate guanine at the O6 carbonyl. Methyl radical and methyl carbocation are the methylating species, respectively.

ADVERSE EFFECTS

Nitrosoureas
- Carmustine and Lomustine
  - Myelosuppression (thrombocytopenia, leukopenia), leading to infection
  - Dose-related acute and delayed pulmonary toxicity (primarily carmustine)
  - Seizures (carmustine wafer; may be related to nondrug component)
- Streptozocin
  - Dose-related renal toxicity

Table 26.2 Properties of Guanine O⁶-Methylators

Methylator	Toxic Species	CYP Isoform	Indication	Dosage Form	Adverse Toxicities (In Addition to Nausea and Vomiting)
Procarbazine	Methyl radical	1A, 1B	Hodgkin's disease	Capsule	MAO inhibition (leading to DDIs) Myelosuppression (leukopenia, anemia, thrombocytopenia)
Dacarbazine	Methyl carbocation	1A1, 1A2	Hodgkin's disease Metastatic melanoma	IV Sol	Hepatotoxicity Myelosuppression (leukopenia, thrombocytopenia)
Temozolomide	Methyl carbocation	None	Glioblastoma multiforme Anaplastic astrocytoma	Capsule	Fatigue Convulsions Thrombocytopenia

Sol, solution

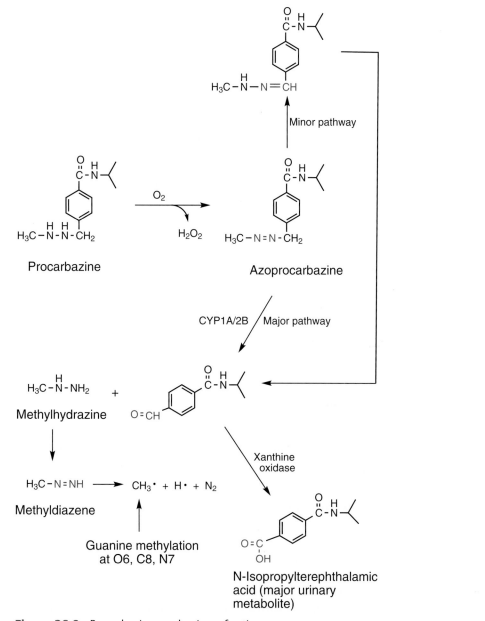

Figure 26.8 Procarbazine mechanism of action.

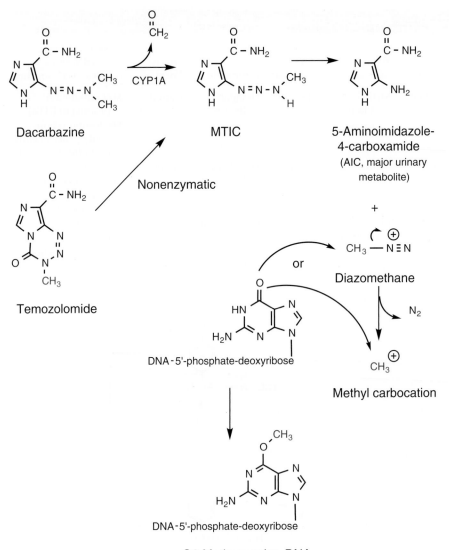

Figure 26.9 Triazine mechanism of action.

- Metabolism by isoforms of CYP1A (both triazines, procarbazine) and CYP1B (procarbazine) initiates molecular decomposition to the activated methyl groups. Temozolomide's conversion to methylcarbocation is CYP-independent.
- The unnatural 6-OCH₃ group of altered guanine pairs preferentially with thymine, rather than cytosine. This results in apoptosis via postreplication mismatch repair mechanisms.

Resistance Mechanisms

- O6-alkylguanine-DNA-alkyltransferase confers resistance. Patients who underexpress this repair enzyme respond best to these agents.

Organoplatinum Complexes

MOA (Fig. 26.10)

- Organoplatinum complexes are charge balanced. They access cells via the copper transporting protein CRT1 and/or by passive diffusion.
- Displacement of electron-accepting ligands (R) by cellular water provides monoaquo and diaquo analogs with a +1 or +2 charge, respectively, on the Pt(II).

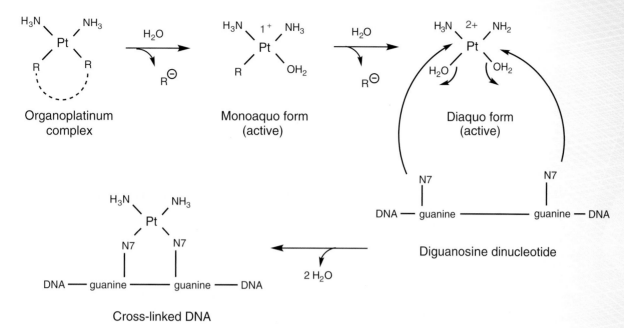

Figure 26.10 Organoplatinum mechanism of action.

- Cationic Pt attracts adjacent nucleophiles (e.g., guanine or adenine N7) on a single DNA strand. Only diaquo forms cross-link DNA.
- *Cis* orientation of the displaced ligands and water molecules on the active diaquo form is essential to DNA cross-linking. *Trans* isomers are therapeutically inactive.
- The most common intrastrand cross-linking occurs between diguanosine dinucleotides (60% to 65%), followed by guanine–adenine cross-linking (25% to 30%).
- Cross-linking results in altered base pairing that cannot be repaired by mismatch repair (MMR) proteins. After several failed attempts at repair, apoptosis results.

Resistance Mechanisms

- Downregulation of MMR, leading to decreased apoptosis (cisplatin and carboplatin).
- Downregulation of CTR1 (cisplatin and carboplatin).
- Conjugation with glutathione and metallothionein proteins (less problematic with picoplatin).
- Intracellular vesicle entrapment.
- Upregulation of nucleotide excision repair (NER) proteins, which successfully repair DNA damage.

SAR (Fig. 26.11)

- Two of the four organoplatinum ligands are commonly free or diaminocyclohexane (DACH)-incorporated amine groups.
  - DACH provides resistance to MMR proteins. This allows oxaliplatin to retain effectiveness in MMR deficient cancer cells.
  - DACH may also contribute to oxaliplatin's lack of dependence on CRT1 for intracellular transport.
  - Cisplatin and carboplatin have free amine moieties and rely on CRT1 for intracellular access.
- When R = Cl (cisplatin, picoplatin), displacement by water is facilitated in cancer cells, which are comparatively poor in Cl anion.
- When R = cyclobutyldicarboxylate (carboplatin), activating displacement by water is slowed compared to cisplatin. This decreases DNA cross-linking potency.

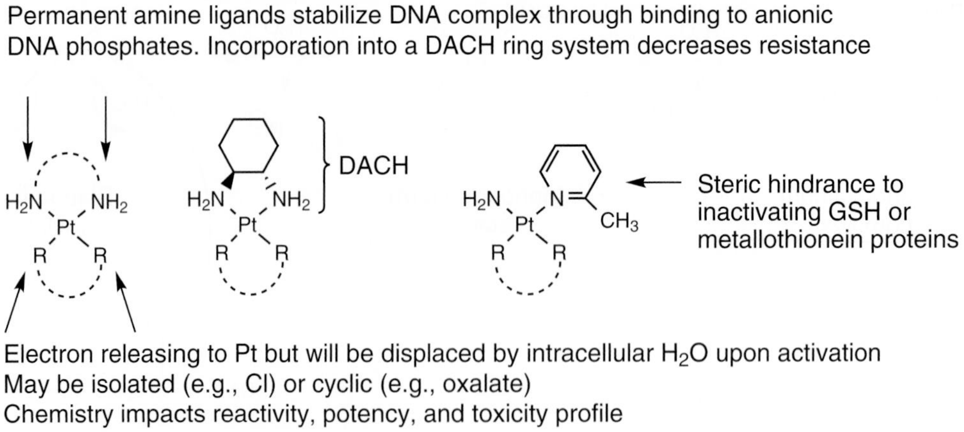

Permanent amine ligands stabilize DNA complex through binding to anionic
DNA phosphates. Incorporation into a DACH ring system decreases resistance

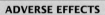

Electron releasing to Pt but will be displaced by intracellular H_2O upon activation
May be isolated (e.g., Cl) or cyclic (e.g., oxalate)
Chemistry impacts reactivity, potency, and toxicity profile

Figure 26.11 Organoplatinum SAR.

- When R = oxalate (oxaliplatin), the displaced oxalic acid dianion has the proper
  molecular geometry to chelate intracellular calcium, leading to use-limiting
  peripheral neuropathy.

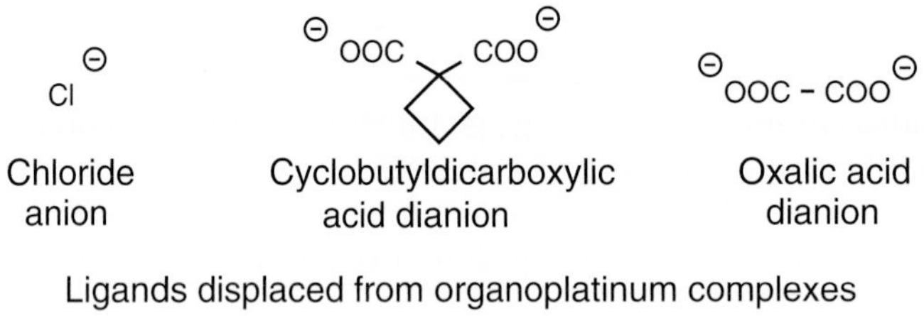

Chloride anion   Cyclobutyldicarboxylic acid dianion   Oxalic acid dianion

Ligands displaced from organoplatinum complexes

Cisplatin Toxicity

- Cisplatin is the most toxic organoplatinum complex. Administration of sodium
  thiosulfate inactivates and promotes the excretion of cisplatin (Fig. 26.12).
- Ototoxicity
  - Cisplatin concentrates in cochlea and can cause permanent hearing loss.
  - Prophylactic protection against cisplatin activation in the cochlea is achieved
    by amifostine pretreatment.
  - Amifostine, a thiophosphate prodrug, is activated by alkaline phosphatase.
    Healthy cells (e.g., cochlea) are preferentially protected from unwanted
    cisplatin-induced toxicity (Fig. 26.13).

Clinical Applications

- Organoplatinum complex indications include:
  - Cisplatin: metastatic testicular, ovarian, and advanced bladder cancer.
  - Carboplatin: Advanced ovarian cancer.
  - Oxaliplatin: Metastatic or advanced colorectal cancer.
  - Picoplatin: Small cell lung cancer (orphan drug).

ADVERSE EFFECTS

Cisplatin
- Vomiting (severe)
- Myelosuppression (leukopenia, thrombocytopenia, anemia)
- Anaphylaxis (all organoplatinum complexes)
- Nephrotoxicity (hydrate with chloride-containing solutions to promote diuresis)
- Tissue necrosis upon extravasation (treat with IV or subQ sodium thiosulfate)

CHEMICAL NOTE

Organoplatinum–Aluminum Incompatibility: Aluminum reacts with organoplatinum complexes (most notably cisplatin and carboplatin) to yield a discolored precipitate and loss of potency. Avoid administration of all organoplatinum complexes through aluminum-containing needles or IV sets.

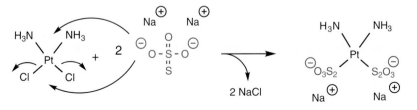

Cisplatin   Sodium thiosulfate   Water-soluble adduct

Figure 26.12 Cisplatin inactivation by sodium thiosulfate.

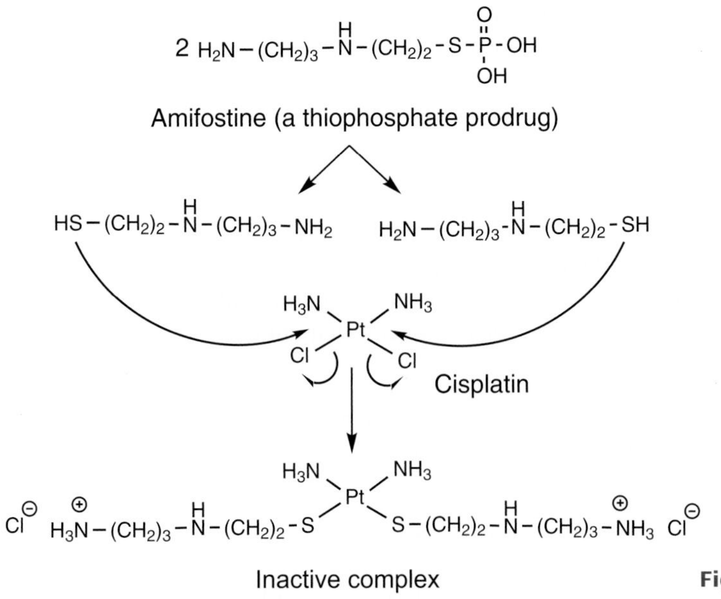

2 $H_2N-(CH_2)_3-\overset{H}{N}-(CH_2)_2-S-\overset{O}{\underset{OH}{\overset{\parallel}{P}}}-OH$

Amifostine (a thiophosphate prodrug)

$HS-(CH_2)_2-\overset{H}{N}-(CH_2)_3-NH_2$ $H_2N-(CH_2)_3-\overset{H}{N}-(CH_2)_2-SH$

$\underset{Cl}{\overset{H_3N}{}}Pt\underset{Cl}{\overset{NH_3}{}}$ **Cisplatin**

$Cl^{\ominus}$ $\overset{\oplus}{H_3N}-(CH_2)_3-\overset{H}{N}-(CH_2)_2-S$ Pt $S-(CH_2)_2-\overset{H}{N}-(CH_2)_3-\overset{\oplus}{NH_3}$ $Cl^{\ominus}$

Inactive complex

Figure 26.13 Cisplatin inactivation by amifostine.

Topoisomerase Poisons (Fig. 26.14)

MOA

- Topoisomerase is essential to DNA replication.
  - During replication, double-stranded DNA is cleaved and a Tyr is tethered to DNA 5′-phosphate.
  - Once replication is complete, the phosphate bond is cleaved and DNA is resealed.

Camptothecins

Epipodophyllotoxins

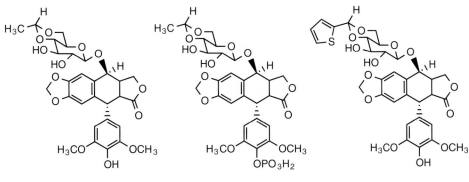

Irinotecan hydrochloride
(Camptosar)

Topotecan hydrochloride
(Hycamtin)

Etoposide
(VePesid)

Etoposide phosphate
(Etopophos)

Teniposide
(Vumon)

Figure 26.14 Camptothecins and epipodophyllotoxins.

- Topoisomerase poisons stimulate DNA cleavage but inhibit resealing. DNA is irreversibly lysed and unable to replicate.

Camptothecins

MOA
- Camptothecins target topoisomerase I (topI). They bridge the enzyme and nicked DNA to form a stable ternary complex.
- The highly conjugated camptothecin ring system is planar, allowing the drug to intercalate DNA at the site of cleavage like a complementary DNA base.

Resistance Mechanisms
- Downregulation of topI and/or activating enzymes.
- P-Glycoprotein (P-gp) mediated cellular efflux.

SAR
- The intact pentacyclic camptothecin ring system is required for optimum activity.
  - Hydrolysis of the E-ring lactone yields a less active topI poison.
- The free C10-phenol is essential. Esters or amides are prodrugs.
- The two semisynthetic camptothecans have basic side chains at C9 (topotecan) or C10 (irinotecan) of the quinoline ring.
  - Allows IV administration of water-soluble salts.
  - Topotecan is also available as capsules.

Metabolism
- The 10-carbamate of irinotecan must be hydrolyzed in the liver to liberate the essential O10-phenol.
  - Hydrolysis is sluggish but the active metabolite's half-life is prolonged when compared to the parent drug (11.5. vs. 5 to 9.6 hours).
- Topotecan is active intact and its half-life is short (2 to 3.5 hours).
- Both camptothecans undergo cyclic hydrolysis–lactonization at pH 7.4. The hydroxy acid hydrolysis product is less active than the parent lactone.
  - The preferential binding of the irinotecan lactone to serum albumin shifts the equilibrium to favor formation of the lactone, explaining its higher potency.
- Inactivating reactions include:
  - Irinotecan: piperidine hydroxylation (CYP3A4) followed by amide hydrolysis, O10-glucuronidation.
  - Topotecan: N-dealkylation (CYP3A4), O10-glucuronidation.
- Dosage adjustments may be required when CYP3A4 inducers or inhibitors are coadministered.

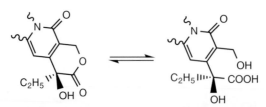

Camptothecan
(active lactone)

Camptothecan hydroxy acid
(less active)

Camptothecin hydrolysis–lactonization cycle

Clinical Applications
- Irinotecan, in combination with the antimetabolite fluorouracil, is used in metastatic colorectal cancer.
- Topotecan is used in cervical, ovarian, and small cell lung cancer.

⚠ ADVERSE EFFECTS

Camptothecans

Topotecan
- Neutropenia (Black Box Warning)
- Thrombocytopenia, anemia
- Interstitial lung disease

Irinotecan
- Acute and delayed diarrhea (Black Box Warning)
  - UGT1A1*6 (poor metabolizer) patients have increased toxicity risk.
  - Anticholinergic pretreatment and loperamide therapy ease symptoms
- Neutropenia, leukopenia
- Interstitial lung disease
- Nausea, vomiting requiring antiemetic pretreatment

Epipodophyllotoxins

MOA

- Epipodophyllotoxins inhibit topIIα by binding only to the enzyme (not to DNA). The binary complex stimulates enzyme-induced DNA cleavage and inhibits resealing.

Resistance Mechanisms

- Downregulation of topIIα.
- P-gp cellular efflux.
- Development of novel mechanisms of repair of drug-induced DNA damage.

SAR

- All functional groups/rings of the epipodophyllotoxin system are essential.
- Epipodophyllotoxins differ only at C2 of the β–D-glucopyranosyl moiety.

Physicochemical and Pharmacokinetic Properties (Table 26.3)

- The higher lipophilicity of teniposide (log P 1.24), compared to etoposide (log P 0.60) is responsible for many activity differences.
  - Teniposide's higher tumor cell concentration secondary to passive diffusion results in a 10-fold potency increase.
  - Teniposide's extensive first-pass metabolism demands IV administration. Etoposide's lower hepatic extraction allows it to be given orally as well as IV.
  - Epipodophyllotoxins must be solubilized for IV administration with oil-based solubilizers.
    - Teniposide requires solubilization with Kolliphor EL while etoposide can be solubilized with Tween. Compared to Tween, Kolliphor EL increases the risk of severe hypersensitivity reactions.
  - The risk of spontaneous drug precipitation from IV lines is significantly greater with teniposide.
- A water-soluble prodrug analog of etoposide is marketed as Etopophos.
  - The phosphate ester is cleaved by phosphatases in the bloodstream to release active etoposide.
  - No hypersensitivity-inducing solubilizer is needed so higher doses can be given.

Metabolism (Fig. 26.15)

- Lactone hydrolysis to the hydroxy acid and CYP3A4-mediated O-dealkylation to the catechol are inactivating.
  - Doses should be adjusted when 3A4 inducers, inhibitors, or competing substrates are coadministered.
- Oxidation of the catechol to an electrophilic *ortho*-quinone puts patients (particularly children) at risk for drug-induced acute myeloid leukemia within 2 years of treatment.

Clinical Applications

- Etoposide is used in small cell lung cancer and, in combination with other agents, in refractory testicular cancer.
- Teniposide's sole indication is refractory childhood acute lymphoblastic leukemia.

> **⚠ ADVERSE EFFECTS**
> **Epipodophyllotoxins**
> - Drug-induced leukemia
> - Solubilizer-induced hypersensitivity (bronchoconstriction, hypotension)
> - Mucositis
> - Myelosuppression (teniposide: anemia, neutropenia, leukopenia; etoposide: leukocytopenia; both drugs: thrombocytopenia).
> - Nausea and vomiting.
> - Alopecia (primarily with etoposide)

Table 26.3 Pharmacokinetic Properties of Epipodophyllotoxins

	Protein Binding (%)	Terminal Half-Life (h)	Renal Elimination (%)	Oral Activity
Etoposide	97	4–11	45	Approximately 50% of IV
Teniposide	>99	5–40	21	None

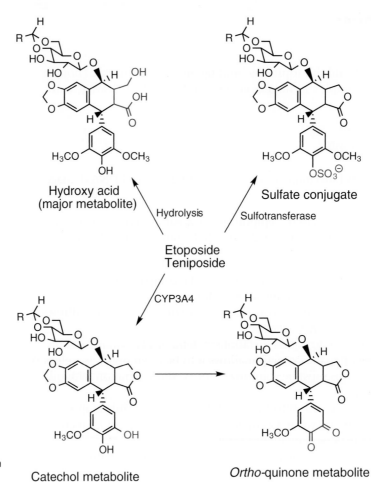

Figure 26.15 Epipodophyllotoxin metabolism.

Antibiotics (Fig. 26.16)

Anthracyclines and Anthracenediones

MOA

- Most anthracyclines and anthracenediones are topIIα inhibitors. They bind to both the enzyme and DNA to form a stable ternary complex that cleaves, but does not reseal, DNA.
- DNA intercalation precedes ternary complex formation. The flat anthracylinone ring system is the intercalating moiety.
- Valrubicin, a unique analog of doxorubicin, acts as a DNA and RNA polymerase inhibitor.

Resistance Mechanisms

- Downregulation of topIIα.
- Cellular efflux by P-gp and multidrug resistance proteins (MRPs).

Anthracycline SAR

- Anthracyclines are glycosides that contain a substituted anthracyclinone ring system and a daunosamine sugar. Both structural segments are required.
- Anthracycline SAR is summarized in Fig. 26.17. Relative potencies predicted by SAR are reflected in Table 26.4.

Anthracycline Metabolism and Related Myocardial Toxicity

NADPH/CYP450 Reductase (CYP450 Reductase)

- CYP450 reductase reduces the quinone (C) ring to a hydroquinone, releasing superoxide radical anions.

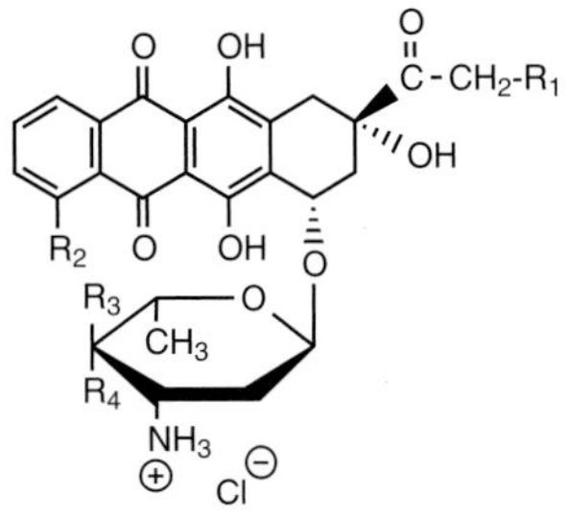

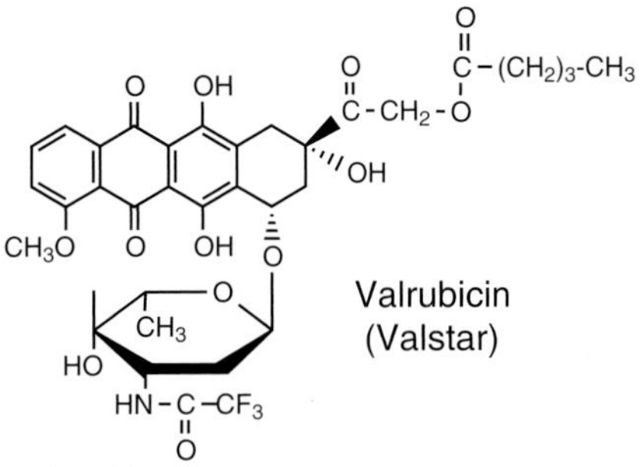

Valrubicin
(Valstar)

Drug	Trade Name	R_1	R_2	R_3	R_4
Doxorubicin hydrochloride	Adriamycin	OH	H_3CO	H	OH
Epirubicin hydrochloride	Ellence	OH	H_3CO	OH	H
Daunorubicin hydrochloride	Cerubidine	H	H_3CO	H	OH
Idarubicin hydrochloride	Idamycin PFS	H	H	H	OH

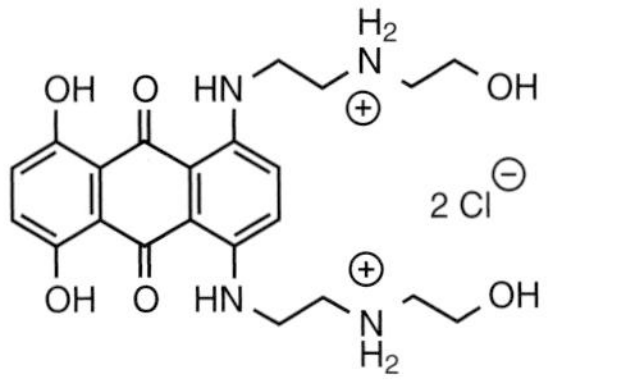

Mitoxantrone hydrochloride
(Novantrone)

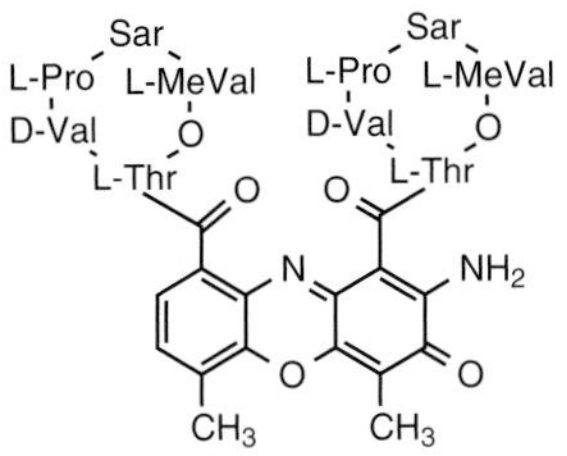

Dactinomycin
(Cosmegen)

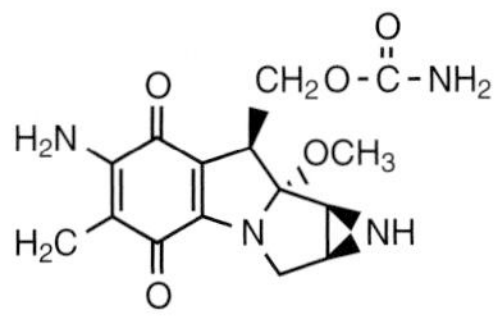

Mitomycin
(Mutamycin)

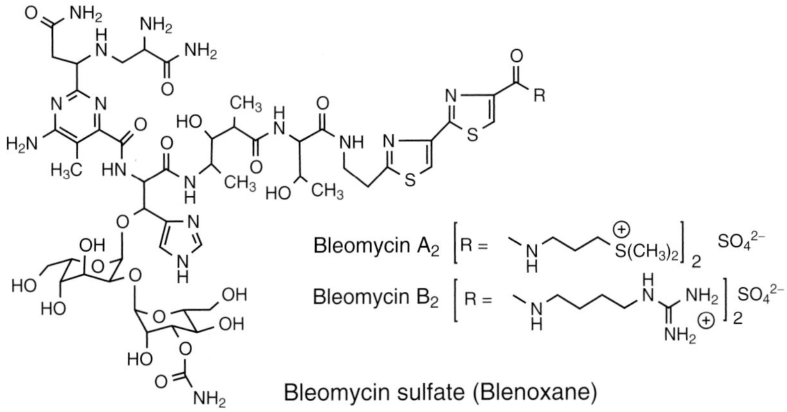

Bleomycin A$_2$ $\left[R = \right]$ 2 SO_4^{2-}

Bleomycin B$_2$ $\left[R = \right]$ 2 SO_4^{2-}

Bleomycin sulfate (Blenoxane)

Figure 26.16 Antibiotic antineoplastics.

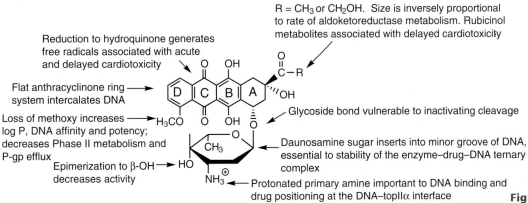

R = CH$_3$ or CH$_2$OH. Size is inversely proportional to rate of aldoketoreductase metabolism. Rubicinol metabolites associated with delayed cardiotoxicity

Reduction to hydroquinone generates free radicals associated with acute and delayed cardiotoxicity

Flat anthracyclinone ring system intercalates DNA

Loss of methoxy increases log P, DNA affinity and potency; decreases Phase II metabolism and P-gp efflux

Epimerization to β-OH decreases activity

Glycoside bond vulnerable to inactivating cleavage

Daunosamine sugar inserts into minor groove of DNA, essential to stability of the enzyme–drug–DNA ternary complex

Protonated primary amine important to DNA binding and drug positioning at the DNA–topIIα interface

Anthracycline

Figure 26.17 Anthracycline SAR.

Table 26.4 Selected Therapeutic-Related Properties of Anthracyclines

	Indications	Half-Life (h)	Log P (Calc)[a]	Standard Dose (mg/m²)	Active Rubicinol Metabolite?
Doxorubicin	Leukemias, lymphomas, solid tumors, Kaposi's sarcoma (liposomal)	48	0.92	60–75	No
Epirubicin	Breast cancer (adjuvant therapy)	33	0.92	100–120	No
Daunorubicin	Leukemias, Kaposi's sarcoma (liposomal)	19	1.73	30–45	No
Idarubicin	Acute myeloid leukemia	22	1.9	12	Yes

[a]ChemAxon, reported by DrugBank (http://www.drugbank.ca/drugs)

- Superoxide radical anions generate hydrogen peroxide (H_2O_2) as shown in Fig. 26.18.
- Catalase converts H_2O_2 to water and oxygen. In cells without catalase, H_2O_2 reacts with intracellular Fe^{2+} to form the highly destructive hydroxyl radical (Fenton reaction).
- The myocardium is deficient in catalase. Anthracyclines readily chelate intracellular Fe^{2+}, promoting the formation of destructive hydroxyl radicals in the heart, particularly with doxorubicin.
- Drug-induced free radical damage to the myocardium can be acute or delayed. Delayed toxicity manifests as drug-resistant heart failure involving systolic and diastolic functions.
- Pretreatment with the iron chelating agent dexrazoxane 30 minutes before doxorubicin decreases the risk of free radical-induced myocardial toxicity.
  - Dexrazoxane hydrolyzes intracellularly to generate the active dianion known as ADR-925
  - The affinity of ADR-925 for intracellular Fe^{2+} is higher than doxorubicin's
  - Dexrazoxane can also prevent serious tissue injury after doxorubicin extravasation.

Dexrazoxane   ADR-925

Aldoketoreductase (Fig. 26.19)

- Aldoketoreductase converts the C13 ketone of anthracyclines to a secondary alcohol (rubicinol metabolite).
  - In all cases except idarubicin, this rubicinol metabolite is inactive or significantly less active than the parent.
  - Idarubicinol is equipotent with idarubicin.
- Anthracyclines where C14 is CH_3 are metabolized faster than when C14 is CH_2OH.
- Rubicinol metabolites are chronically cardiotoxic. They are sequestered in the heart and form a "long-lived reservoir" of toxicity which can induce drug-resistant heart failure that is fatal to approximately 60% of those diagnosed.
- Prior to excretion, unsequestered rubicinol metabolites are cleaved to the aglycone. O-dealkylation at C4 is followed by Phase II conjugation.
  - Idarubicin lacks the 4-OCH_3 and is excreted primarily as unconjugated idarubicinol.

Anthracenediones

- Mitoxantrone's aromatic ring system intercalates DNA. The protonated secondary amines mimic the daunosamine cation.

Anthracycline chelation of Fe^{2+}

ADVERSE EFFECTS

Anthracyclines
- Acute and delayed cardiotoxicity (cumulative dose-dependent)
- Severe myelosuppression (especially leukopenia)
- Mucositis
- Nausea and vomiting
- Alopecia
- Skin necrosis (upon extravasation)

CHEMICAL NOTE

Colorful Conjugation: The highly conjugated structure of anthracyclines imparts a reddish-orange color to these compounds that is implied in the name "rubicin." This red color is maintained in excreted metabolites. Patients should be counseled that the reddish urine and/or feces they will see is not hemorrhagic but, rather, the result of the highly conjugated chemistry of this class of anticancer drugs.

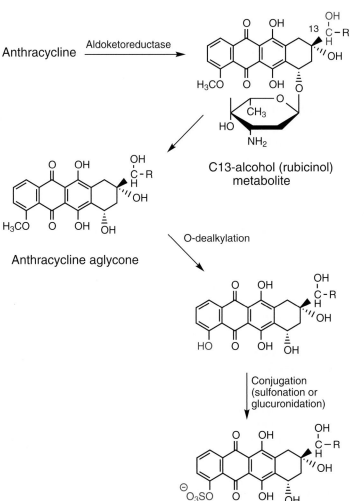

Figure 26.18 Anthracycline-mediated free radical formation.

Figure 26.19 Anthracycline metabolism.

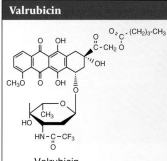

Valrubicin

Valrubicin

- Valrubicin is an orphan drug used to treat BCG-refractory bladder cancer.

- Unlike other anthracyclines, valrubicin has no ionizable functional group and must be solubilized with Kolliphor EL for intravesicular administration.

- Patients retain the drug in the bladder for 2 hours, then void in the normal fashion. Unlike other anthracyclines, valrubicin is not necrotic to skin.

- High lipophilicity (log P 4.49) ensures drug trapping in bladder cells. The lack of systemic exposure makes serious toxicities (including myocardial toxicity) rare.

- The risk of serious toxicity increases if administered to patients with disease-damaged bladders. These patients should not receive valrubicin.

- Mitoxantrone's quinone oxygen atoms are stabilized by intramolecular hydrogen bonding involving two adjacent hydrogen donating groups. This may confer resistance to CYP450 reductase.
- Mitoxantrone is inactivated by N-dealkylation, deamination, and cytosolic oxidation to the carboxylic acid. Glucuronic acid conjugation at one COOH precedes biliary excretion.

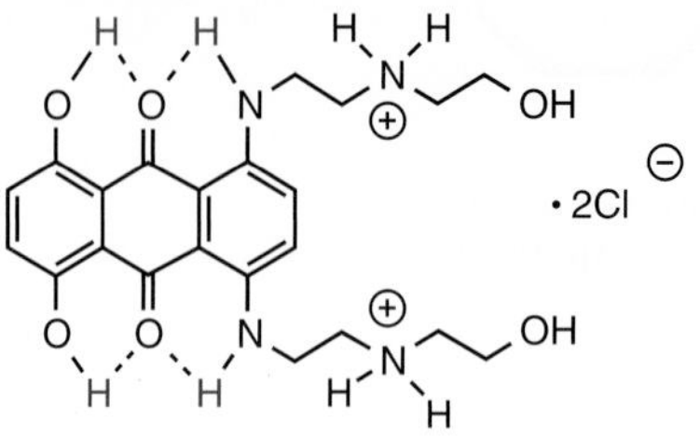

Mitoxantrone intramolecular
hydrogen bonding

Sites of potential
glucuronide conjugation

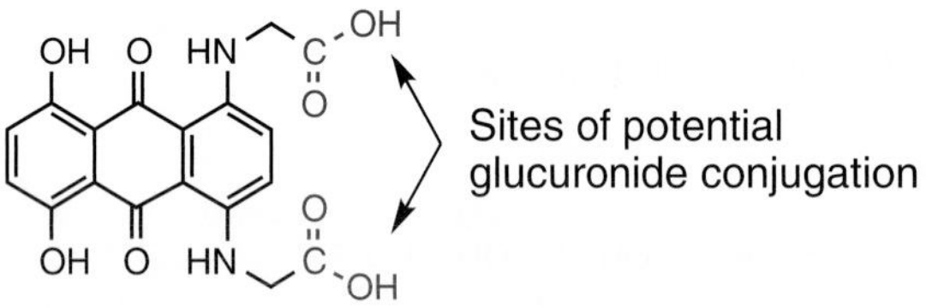

Mitoxantrone metabolite

ADVERSE EFFECTS

Mitoxantrone
- Myelosuppression
- Heart failure (Black Box Warning)
  - Free radicals not believed to be involved
- Alopecia and GI distress less than anthracyclines
- Bluish-green coloration of body fluids, skin, and whites of eyes due to highly conjugated structure

Clinical Applications
- Acute nonlymphocytic leukemia and prostate cancer.
- Progressing/relapsing multiple sclerosis.

Miscellaneous Antibiotic Antineoplastics
- Properties of three structurally unique antibiotic antineoplastics are provided in Table 26.5. All are administered IV.
- Additional details on the chemistry and mechanism of these anticancer agents can be found in Chapter 37 in *Foye's Principles of Medicinal Chemistry, Seventh Edition*.

Antimetabolites (Fig. 26.20)

Pyrimidine Antagonists: dTMP Biosynthesis Inhibitors

MOA
- dTMP is produced through the methylation of dUMP (Fig. 26.21).
- The rate-limiting enzyme is the sulfhydryl-containing thymidylate synthase. The methyl donor is 5,10-methylenetetrahydrofolate.

Table 26.5 Properties of Miscellaneous Antibiotic Antineoplastics

	Mechanism	Indications	Black Box Warnings
Dactinomycin	RNA polymerase inhibition	Solid tumors, muscle-related cancer	Highly necrotic to skin
Mitomycin	DNA alkylation	Gastric, pancreatic, and superficial bladder cancers	Hemolytic uremic syndrome (hemolytic anemia, thrombocytopenia, renal failure)
Bleomycin	Free radical DNA destruction	Squamous cell head and neck cancer, genital carcinomas, lymphomas	Pulmonary fibrosis

Pyrimidine antagonists: **Purine antagonists:**

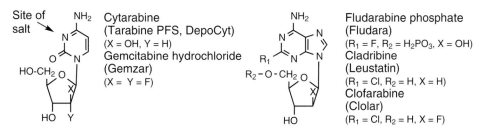

Folate antagonists:

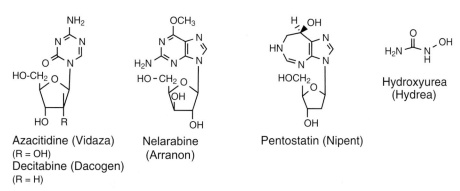

Methotrexate (Trexall)   Pemetrexed disodium (Alimta)
(X = N-CH₃)

Pralatrexate (Folotyn)
(X = CH-CH₂-C≡CH

DNA polymerase and chain elongation inhibitors:

Site of salt → Cytarabine
(Tarabine PFS, DepoCyt)
(X = OH, Y = H)
Gemcitabine hydrochloride
(Gemzar)
(X = Y = F)

Fludarabine phosphate
(Fludara)
(R_1 = F, R_2 = H_2PO_3, X = OH)
Cladribine
(Leustatin)
(R_1 = Cl, R_2 = H, X = H)
Clofarabine
(Clolar)
(R_1 = Cl, R_2 = H, X = F)

DNA Methyltransferase inhibitors: **Miscellaneous antimetabolites:**

Azacitidine (Vidaza)   Nelarabine   Pentostatin (Nipent)   Hydroxyurea
(R = OH)   (Arranon)   (Hydrea)
Decitabine (Dacogen)
(R = H)

Figure 26.20 Antimetabolites and related antineoplastics.

- Thymidylate synthase attacks dUMP at electrophilic C6 to form an unstable ternary complex that breaks down to release dTMP and dihydrofolate (DHF).
  - The key step is abstraction of the C5-H of the substrate by N10 of the folate cofactor.
- DHF is reduced to tetrahydrofolate (THF) by DHF reductase (DHFR) and NADPH. The cofactor is regenerated by enzymatic insertion of a CH_2 donated by Ser.
- dTMP biosynthesis inhibitors (1) generate a stable ternary complex that cannot release product (fluorouracil analogs) or (2) block cofactor regeneration via inhibition of DHF reduction (antifolates).

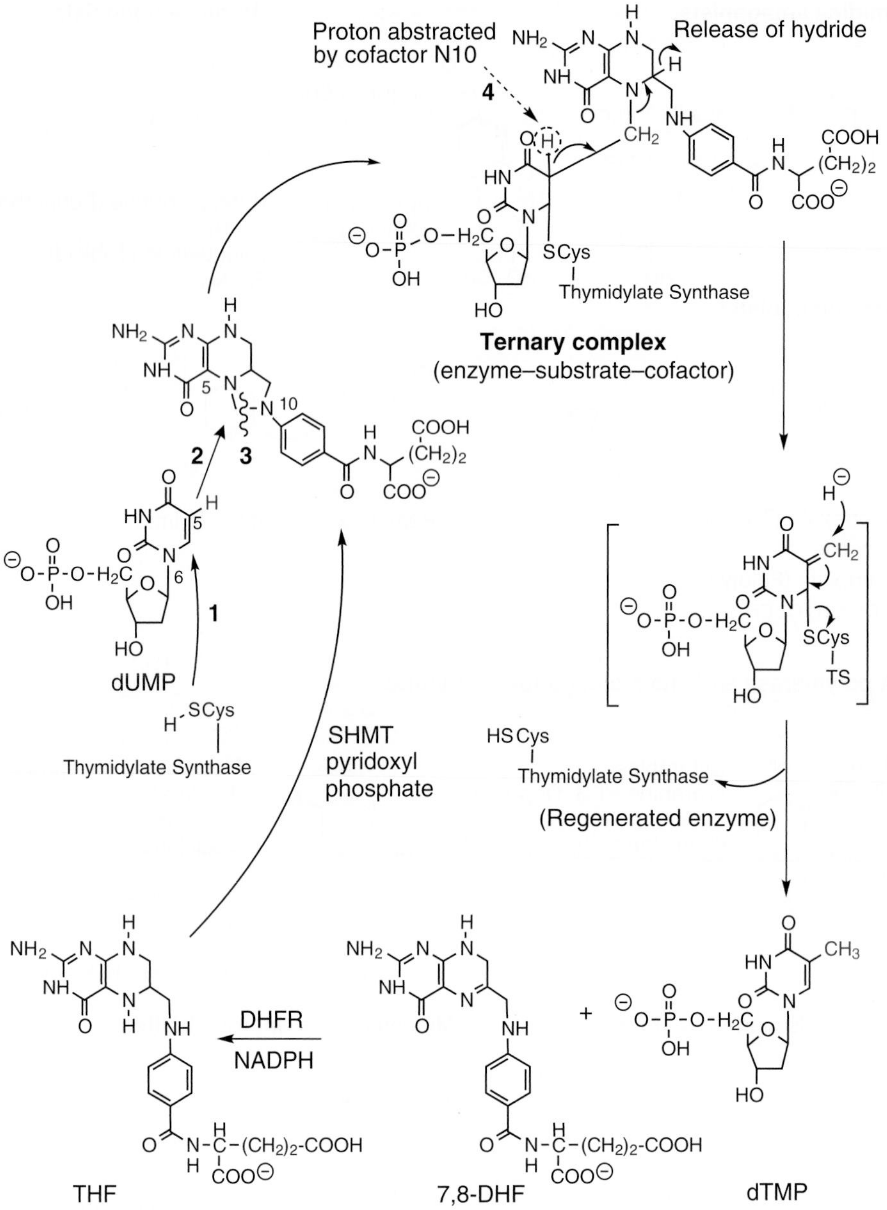

Figure 26.21 Deoxythymidine monophosphate (dTMP) biosynthesis.

Fluorouracil Analogs (Table 26.6)

MOA

- 5-F-dUMP is the active form.
- The 5-F increases the δ^+ character of C6, leading to an accelerated attack by the enzyme's SH group. The fluorinated false substrate is attacked preferentially over the endogenous substrate.

Table 26.6 Selected Therapeutic-Related Properties of Fluorouracil Analogs

	Indications	Dosage Form	Decrease Dose in Poor DHD Metabolizers?
Fluorouracil	Colorectal, breast, stomach, pancreatic cancer (palliative)	IV solution	Yes
Floxuridine	GI adenocarcinoma with hepatic metastasis (palliative)	Solution for intraarterial injection	No
Capecitabine	Colorectal cancer, metastatic breast cancer	Tablets	Yes

- The false ternary complex is stable, as the 5-F cannot be abstracted by cofactor N10. The rate limiting enzyme is irreversibly inhibited. The cell dies a "thymineless death."

Bioactivation and Metabolism (Fig. 26.22)

- Three marketed prodrugs differ only in the mechanism of activation to 5-F-dUMP
  - Floxuridine, a deoxyribonucleoside, is phosphorylated at 5' via dUMP kinase.
  - 5-fluorouracil (5-FU), a base, requires conversion to the deoxyribonucleotide via orotate phosphoribosyltransferase and ribonucleotide reductase. Kinases and phosphatases are involved in the bioconversion.
  - Capecitabine, a false cytidine nucleoside, requires conversion of cytidine to uridine and cleavage of the 5'-deoxyribose to generate 5-FU (Fig 26.23).
    - Loss of 5'-deoxyribose is catalyzed by thymidine phosphorylase, an enzyme found in higher concentration in tumor cells.
    - Capecitabine is more selectively toxic to tumor cells than 5-FU or floxuridine.

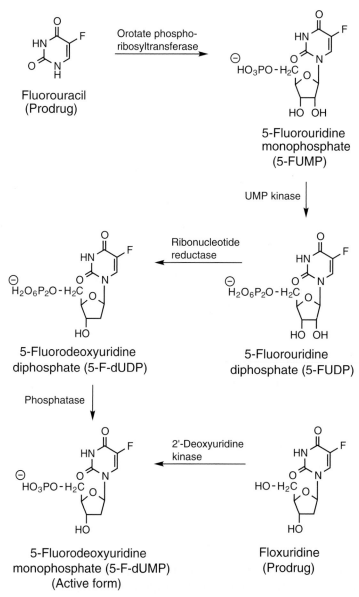

Figure 26.22 5-Fluorouracil and floxuridine activation.

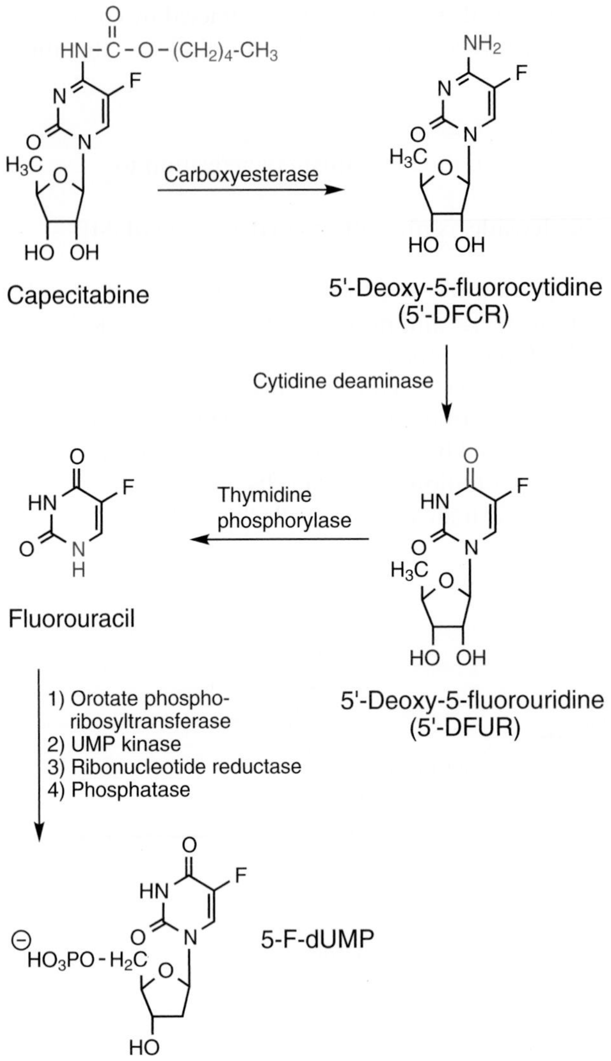

Figure 26.23 Capecitabine activation.

⚠ **ADVERSE EFFECTS**

Fluorouracil Analogs
Fluorouracil and Floxuridine
- Myelosuppression (leukopenia)
- GI ulceration, stomatitis
- Nausea, vomiting, diarrhea

Capecitabine
- Diarrhea
- Hand and foot syndrome
- Myelosuppression (neutropenia, thrombocytopenia)
- Nausea

DRUG–DRUG INTERACTIONS

Capecitabine–Warfarin Interaction: Capecitabine augments warfarin-induced anticoagulation and can result in potentially fatal hemorrhage. Capecitabine's ability to inhibit CYP2C9 may be, in part, responsible. The interaction is sufficiently severe to warrant a Black Box Warning. Patient INR or PT should be monitored frequently and warfarin doses titrated accordingly.

- 5-FU and capecitabine are inactivated by dihydropyrimidine dehydrogenase (DPD).
  - Patients genetically deficient in DPD are at risk for use-limiting toxicity unless doses are decreased.
- Floxuridine does not generate 5-FU in significant amounts and is not impacted by DPD phenotype.

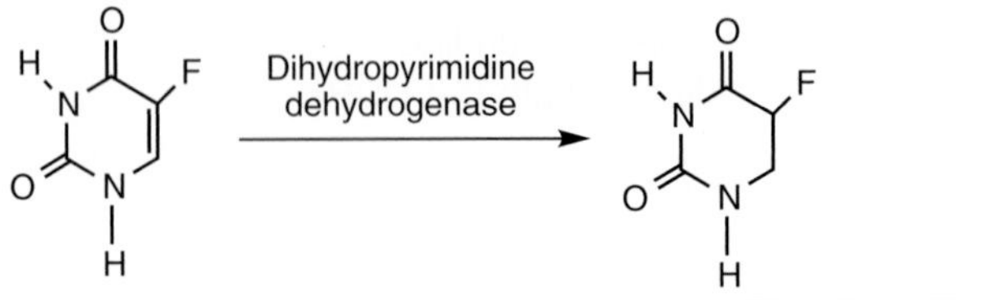

5-Fluorouracil   5-Fluoro-5,6-dihydrouracil

Fluorouracil inactivation

Folate Antagonists

MOA

- Antifolates enter tumor cells via reduced folate carrier-1 (RFC1). They are trapped in cells through polyglutamation by folylpolyglutamate synthase (FPGS).

- Trapping is the result of multiple anionic charges.
- Polyglutamation is more extensive in tumor cells, leading to selective toxicity.
- Antifolates bind with higher affinity to DHFR and, in some cases, thymidylate synthase, than endogenous folate substrates.
- Antifolates that inhibit DHFR bind pseudoirreversibly to the enzyme. A strong ion–ion bond between protonated N1 of the drug and DHFR Asp^{27} is crucial.
  - The endogenous substrate (DHF) anchors to Asp^{27} through protonated N5.
  - Drug misorientation at DHFR inhibits reduction by NADPH.
  - DHF builds up in the cell and exerts feedback inhibition on thymidylate synthase.
- Antifolates also inhibit GAR transformylase, a key enzyme in purine biosynthesis.

Resistance Mechanisms

- Downregulation of RFC1 and FPGS.
- Upregulation of γ-glutamyl hydrolase (which cleaves glutamate residues).
- Upregulation of target enzymes.
- Cellular efflux by P-gp and other MDR proteins.

SAR

- An antifolate SAR overview is provided in Figure 26.24.
- Methotrexate has the highest affinity for DHFR.
- Pemetrexed, in polyglutamated form, has the highest affinity for thymidylate synthase.
- Pralatrexate has the highest affinity for FPGS and RFC1.

Metabolism

- Antifolates are primarily excreted unchanged in urine.
- A small percentage of a methotrexate dose is hydroxylated at C7 to produce a less water-soluble metabolite. Accumulation of this metabolite, along with the parent drug, can result in crystalluria.

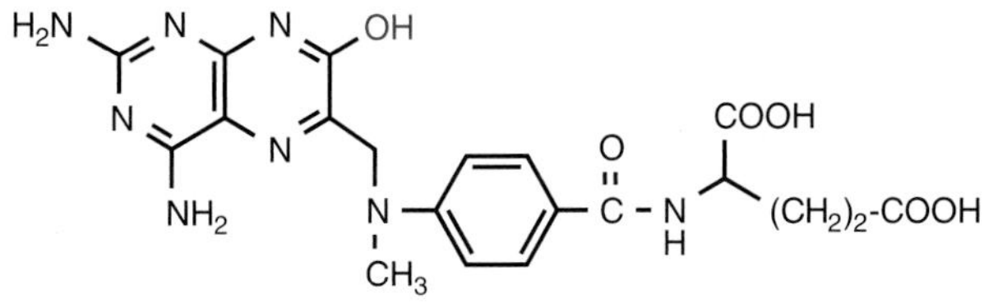

7-Hydroxymethotrexate

Clinical Applications

- Hydration while on methotrexate is critical due to the risk of life-threatening crystalluria. Urine should be basified to at least pH 7 to minimize the risk of drug precipitation.

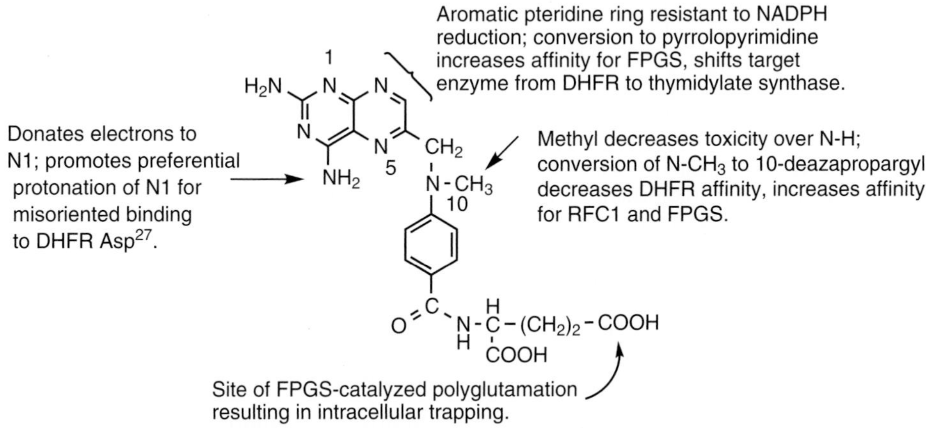

Aromatic pteridine ring resistant to NADPH reduction; conversion to pyrrolopyrimidine increases affinity for FPGS, shifts target enzyme from DHFR to thymidylate synthase.

Donates electrons to N1; promotes preferential protonation of N1 for misoriented binding to DHFR Asp^{27}.

Methyl decreases toxicity over N-H; conversion of N-CH$_3$ to 10-deazapropargyl decreases DHFR affinity, increases affinity for RFC1 and FPGS.

Site of FPGS-catalyzed polyglutamation resulting in intracellular trapping.

Figure 26.24 Antifolate SAR.

- Coadministration of Vitamin B_{12} and folic acid can decrease risk of adverse effects.
- Folate replacement therapy with 5-formyl THF (leucovorin) can rescue healthy cells in the case of severe antifolate toxicity that is anticipated or experienced.
- Indications for use include:
  - Methotrexate: breast, head and neck, squamous and small cell lung cancers, non-Hodgkin's lymphoma, choriocarcinoma, and chorioadenoma.
  - Pemetrexed: nonsquamous nonsmall cell lung cancer, malignant pleural mesothelioma.
  - Pralatrexate: peripheral T-cell lymphoma.

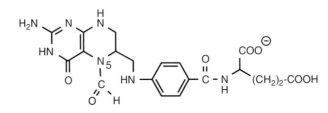

5-Formyltetrahydrofolate
(Leucovorin)

> **⚠ ADVERSE EFFECTS**
>
> **Antifolates**
> - Mucositis, ulcerative stomatitis, hemorrhagic enteritis
> - Myelosuppression (neutropenia, thrombocytopenia, anemia, leukopenia)
> - Dermatitis
> - Methotrexate-induced lung disease
> - Crystalluria (methotrexate)

Purine Antagonists: AMP and GMP Biosynthesis Inhibitors

MOA

- Purine antagonists are 6-thio analogs of guanine and purine (inosine). They are orally active prodrugs that require conversion to active ribonucleotides.
- The active form of 6-mercaptopurine (6-thioinosinate) and its 6-CH$_3$ metabolite inhibit the rate-limiting enzyme of purine nucleotide biosynthesis, hypoxanthine-guanine phosphoribosyltransferase (HGPRT) (Figure 37.38, Chapter 37 in *Foye's Principles of Medicinal Chemistry, Seventh Edition*).
- The active form of 6-thioguanine (6-thioguanylate) and its 6-CH$_3$ metabolite act as DNA and RNA polymerase inhibitors. This apoptotic process is a secondary mechanism for 6-thioinosinate.

Metabolism (Fig. 26.25)

- 6-Thioinosinate is methylated to a more active metabolite by thiopurine S-methyltransferase (TPMT). S-adenosylmethionine (SAM) is the methyl-donating cofactor.
- 6-Thioguanylate is not extensively methylated by TPMT.

6-Thioinosinate ($R_1 = R_2 = H$)
S-Methyl-6-thioinosinate ($R_1 = H, R_2 = CH_3$) ⎫ Active
6-Thioguanylate ($R_1 = NH_2, R_2 = H$) ⎭

6-Thiouric acid
(inactive)

S-Methyl-6-thiopurine
(inactive)

S-Methyl-6-thioguanine
(inactive)

Figure 26.25 Thiopurine metabolites.

- The two prodrugs can be directly methylated by TPMT but they are poor substrates for the activating enzyme HGPRT.
- TPMT is polymorphic.
  - Poor metabolizers inactivate fewer molecules of prodrug. Patients may be at increased risk for adverse effects unless the dose is decreased.
  - Extensive TPMT metabolizers on mercaptopurine, but not thioguanine, may show enhanced drug sensitivity since more highly active S-methyl-6-thioinosinate forms.
- Xanthine oxidase (XO) converts mercaptopurine to inactive 6-thiouric acid. The dose of mercaptopurine, but not thioguanine, can be halved if the XO inhibitor allopurinol is coadministered.

Clinical Applications

- Mercaptopurine is used in combination with other agents in acute lymphoblastic leukemia.
- Thioguanine is used to treat acute nonlymphocytic leukemias.

> ⚠️ **ADVERSE EFFECTS**
> **Thiopurines**
> - Myelosuppression (anemia, leukopenia, thrombocytopenia)
> - Secondary malignancy
> - Hepatotoxicity

DNA Polymerase/Chain Elongation Inhibitors

MOA

- These nucleosides are prodrugs that are activated by phosphorylation.
- In prodrug form, they are actively transported into tumor cells.
- Once activated to the triphosphate nucleotide by kinases, they are erroneously incorporated into DNA and arrest further polymerization.

Resistance Mechanisms

- Alterations in expression of activating kinase enzymes.
- Downregulation of active carrier proteins.
- Cellular efflux.

SAR

- All but one DNA polymerase inhibitor (fludarabine) are nucleosides, which are actively transported into tumor cells.
- Fludarabine's 5′-phosphate enhances water solubility, but it is cleaved in vivo for binding to nucleoside-specific transporting proteins.

Bioactivation and Metabolism

- Whether purine or pyrimidine based, all DNA polymerase inhibitors are initially phosphorylated by deoxycytidine kinase. Mono- and diphosphate kinases convert the monophosphate nucleotide to the active triphosphate form.
- Deaminase metabolizes purine-based inhibitors cytarabine and gemcitabine to inactive uracil analogs.
- The 2-halogen on all pyrimidine-based inhibitors (fludarabine, cladribine, and clofarabine) renders them resistant to inactivating deaminase.

Clinical Applications

- DNA polymerase inhibitors cytarabine, fludarabine, cladribine, and clofarabine are used IV to treat a variety of leukemias. Gemcitabine's indications include breast, pancreatic, and nonsmall cell lung cancer.
- Gemcitabine renal elimination is more efficient in men than women, putting women at higher risk for toxicity.

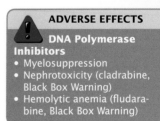

> ⚠️ **ADVERSE EFFECTS**
> **DNA Polymerase Inhibitors**
> - Myelosuppression
> - Nephrotoxicity (cladribine, Black Box Warning)
> - Hemolytic anemia (fludarabine, Black Box Warning)

DNA Methyltransferase Inhibitors

- Three nucleoside DNA methyltransferase inhibitors are related in mechanism to DNA polymerase inhibitors.
- These agents are converted to triphosphorylated nucleotides and, after incorporation into DNA, irreversibly inhibit the DNA methyltransferase enzyme.

- The therapeutic goal of DNA hypomethylation is restoration of normal gene function and apoptosis of cells that have become insensitive to homeostatic proliferation control processes.
- All DNA methyltransferase inhibitors undergo inactivating deamination.

Clinical Applications

- Azacitidine and decitabine are used in myelodysplastic syndrome.
- Nelarabine's indication is T-cell acute lymphoblastic leukemia or lymphoma.

Mitosis Inhibitors (Fig. 26.26)

MOA

- Microtubules assemble during cellular mitosis to form the mitotic spindle.

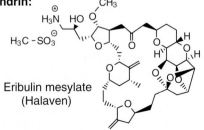

Figure 26.26 Mitosis inhibitors.

- Microtubules are comprised of α and β tubulin proteins that polymerize (elongate) and depolymerize (erode) in a process called dynamic instability. This is essential to the formation of the mitotic spindle.
- Mitosis inhibitors disrupt dynamic instability by promoting elongation at the expense of erosion (taxanes, epothilones) or erosion at the expense of elongation (vinca alkaloids, eribulin).
- Mitosis inhibitors bind to specific sites on β tubulin.

Taxanes

MOA

- The β tubulin binding site for taxanes has been identified (see Table 37.8, Chapter 37 in *Foye's Principles of Medicinal Chemistry, Seventh Edition*).
- The "southern face" functional groups interact with β tubulin residues while the "northern face" groups primarily maintain the conformation required for high affinity binding.

Resistance Mechanisms

- Cellular efflux via P-gp and MRP.
- Mutated β tubulin with decreased taxane affinity.
- Overexpression of tubulin isotypes that circumvent taxane toxicity (e.g., Class III β tubulin).
- Altered expression of microtubule associated proteins.

SAR

- A taxane SAR overview is provided in Fig. 26.27.
- Taxanes are unionized neutral molecules. No water-soluble salt forms are possible.
- The most polar taxane (docetaxel) has free OH groups at C7 and C10. It can be solubilized with Tween, as can the 7,10-dimethoxy analog cabazitaxel.
- Paclitaxel has a free OH at C7 and an acetate ester at C10. It requires Kolliphor EL for solubilization, which increases hypersensitivity risk.
- The 7,10-dimethoxy structure of cabazitaxel decreases affinity for P-gp.
  - Drug concentration in tumor cells increases.
  - Drug concentration in enterocytes and nerve cells increases, leading to adverse effects (diarrhea, sensory and motor neuropathy).
  - Resistance to therapy decreases.

Metabolism (Fig. 26.28)

- Taxanes with a *t*-butoxycarboxamido group (docetaxel, cabazitaxel) are metabolized preferentially by CYP3A4.
  - Docetaxel undergoes hydroxylation at a *t*-butyl CH_3. The primary alcohol metabolite is known as hydroxydocetaxel.
  - Cabazitaxel is O-dealkylated at C7 and/or C10. All three metabolites are active. The didesmethyl metabolite is docetaxel.

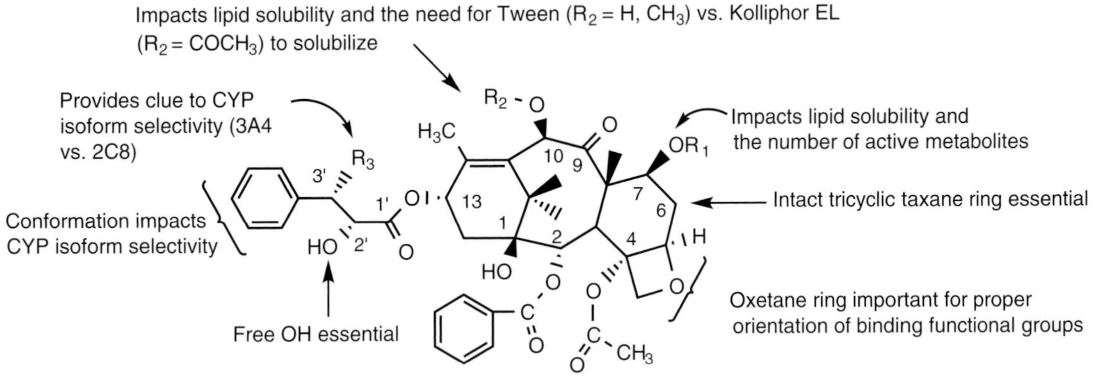

Figure 26.27 Taxane SAR.

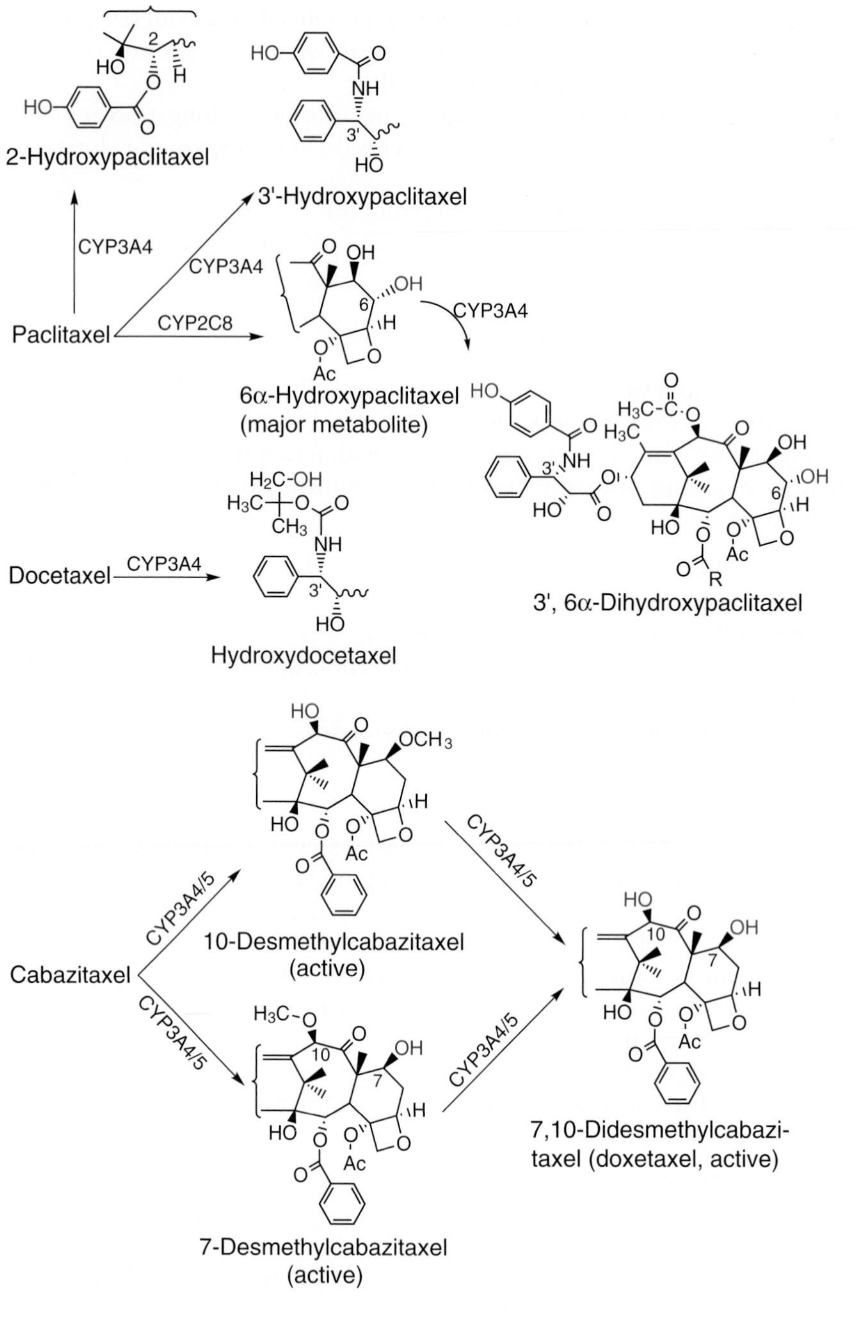

Figure 26.28 Taxane metabolism.

- Paclitaxel's 3′-benzamido moiety predicts preferential CYP2C8 metabolism. Hydroxylation provides 6α-hydroxypaclitaxel, which is 30-fold less active.
- Dosage adjustments may be needed with coadministered therapies that compete for metabolizing CYP isoforms.

Clinical Applications

- Taxane indications include:
  - Paclitaxel: advanced ovarian cancer, metastatic breast cancer, nonsmall cell lung cancer.
  - Docetaxel: breast, nonsmall cell lung, head and neck, and metastatic prostate cancer.
  - Cabazitaxel: docetaxel-resistant metastatic prostate cancer.
- Pretreatment with antihistamines and a corticosteroid minimizes hypersensitivity from oil-based solubilizers.
- A water-soluble, albumin-bound paclitaxel formulation without Kolliphor EL is available as Abraxane.

- Taxanes upregulate thymidine phosphorylase, an activating enzyme of the antimetabolite capecitabine. Paclitaxel-capecitabine combination therapy is common in anthracycline-resistant breast cancer.

Epothilone (Ixabepilone)

- The β tubulin binding site of epothilones overlaps that of taxanes and they act by the same mechanism (stimulating microtubule elongation at the expense of erosion).
- Like paclitaxel, ixabepilone requires solubilization by Kolliphor EL. Hypersensitivity prophylaxis is required.
- Like cabazitaxel, ixabepilone has a low affinity for P-gp. This increases potency and neuropathy risk. The incidence of severe diarrhea is about half that of cabazitaxel.
- The amide linkage of ixabepilone protects against inactivating hydrolysis. CYP3A4 metabolism results in many inactive metabolites. Doses should be adjusted when CYP3A4 substrates, inhibitors, or inducers are coadministered.

> **⚠ ADVERSE EFFECTS**
>
> **Taxanes and Epothilones**
>
> Neutropenia
>
> Fluid retention (docetaxel)
>
> Sensory and/or motor neuropathy (cabazitaxel, ixabepilone, albumin-bound paclitaxel)
>
> Diarrhea (cabazitaxel, ixabepilone)

Vinca Alkaloids

MOA

- Like taxanes, vinca alkaloids bind to the luminal surface of β tubulin at a specific binding site. Unlike taxanes, they promote microtubule erosion at the expense of polymerization.
- Vinca alkaloids are highly effective antimitotic agents. Occupation of 1% to 2% of the binding sites can result in up to 50% inhibition of microtubule assembly.

SAR

- Vincristine and vinblastine differ in structure only at N1 of the vindoline segment. Vinblastine's N-CH$_3$ group increases lipophilicity over vincristine's N1-formyl moiety.
- Vinorelbine and vinblastine differ in structure only in the number of carbons connecting positions 6' and 9' on the catharanthine segment and in the nature of the piperidine ring within that segment.
  - Vinblastine and vincristine have a 4'-tertiary alcohol while vinorelbine has been dehydrated.
  - Esterification of the 4'-OH of vincristine and vinblastine destroys activity.
  - Reduction of the 3',4'-olefin of vinorelbine is inactivating.
- The C18'-methoxycarbonyl moiety and S configuration is essential to activity.

Metabolism

- Hydrolysis of the 4-acetate ester provides active metabolites.
- CYP3A4 oxidizes vinca alkaloids to inactive metabolites that are excreted in bile.

Physicochemical Properties

- Vinca alkaloids contain ionizable amines and are administered as water-soluble salts. There is no need for hypersensitizing solubilizers.
- A liposomal formulation of vincristine is available.
- The less lipophilic vinblastine penetrates cells more sluggishly than vinblastine and vinorelbine, but binds with higher affinity to the β tubulin binding site.
- Vincristine is cleared more slowly than other vinca alkaloids, resulting in a more prolonged action within tumor cells.

Clinical Applications

- The indications of the vinca alkaloids are:
  - Vincristine: leukemias, lymphomas.
  - Vinblastine: leukemias, lymphomas, advanced testicular carcinoma, Kaposi's sarcoma.
  - Vinorelbine: nonsmall cell lung cancer.

- Administration of vinca alkaloids is IV only. Intrathecal administration can be fatal (Black Box Warning).

Eribulin

- Eribulin, a natural product, is classified chemically as a halichondrin. Referred to as an "end poison," it binds to β tubulin within the vinca binding domain on the end of the microtubule.
- Eribulin inhibits microtubular elongation without impacting erosion. The impact on dynamic instability is the same as elicited by vinca alkaloids.
- Hydrogen bonding is crucial in holding eribulin to its binding site.
- Like vinca alkaloids, eribulin contains a basic amine and forms water-soluble salts. Oil-based solubilizers are not required.
- Like ixabepilone, eribulin retains activities in cancer cells that overexpress Class III β tubulin. Unlike ixabepilone, it is a P-gp substrate and resistant to metabolism.

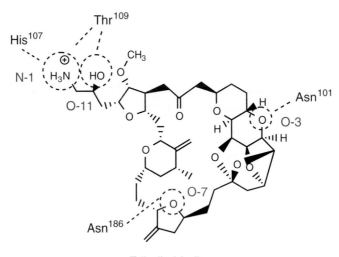

Eribulin binding

Estramustine

- Although estramustine contains a bis-β-chloroalkylamine (nitrogen mustard) moiety normally associated with DNA alkylation, it functions as a mitosis inhibitor.
- Estramustine binds to microtubule associated proteins and inhibits the formation of the mitotic spindle.
- Estramustine is used for the palliative treatment of metastatic prostate cancer. The estrogen component of the structure was incorporated to facilitate uptake into hormone-dependent prostate cells.
- The estrogen moiety is also believed responsible for the drug's ability to induce cardiovascular adverse effects (e.g., thrombosis, myocardial infarction).
- The sodium phosphate ester promotes water solubility of this orally administered drug.

Tyrosine Kinase Inhibitors (Fig. 26.29)

MOA

- Deregulated tyrosine kinase (TK) activity is associated with many neoplastic diseases. Inhibiting the enzyme halts TK-induced cell division and antiapoptotic actions.
- TKs are classified as receptor-associated and cellular (nonreceptor). TK targets for antineoplastic tyrosine kinase inhibitors (TKIs) include EGFR, VEGFR, HER2, PDGRF, and Bcr-Abl.
  - Bcr-Abl is a cellular TK.

ADVERSE EFFECTS

Vinca Alkaloids and Eribulin
Vinca Alkaloids
- Severe tissue damage on extravasation (vinca alkaloids, apply heat to affected areas 1 to 4 times daily for 3 to 5 days)
- Alopecia
- Peripheral neuropathy (vincristine)
- Constipation (vincristine)
- Leukopenia (vinblastine)
- Granulocytopenia (vinorelbine)
- Interstitial pulmonary disease (vinorelbine)
Eribulin
Peripheral neuropathy
- Neutropenia, anemia
- QT prolongation
- Peripheral neuropathy

Bcr-Abl kinase inhibitors:

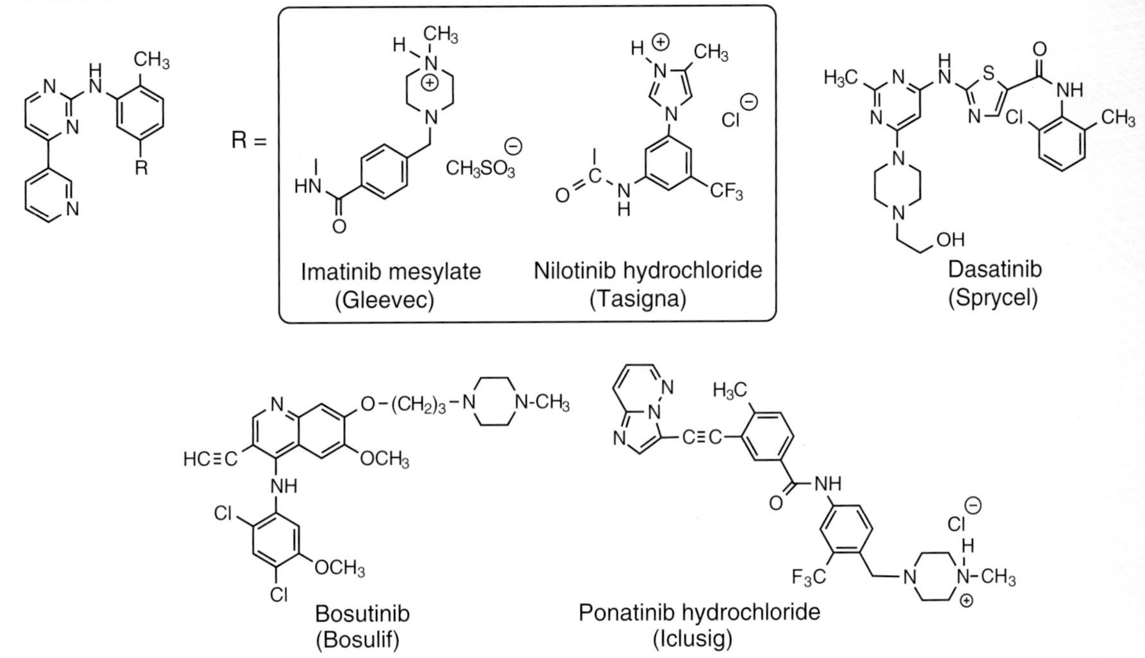
R =
Imatinib mesylate
(Gleevec)
Nilotinib hydrochloride
(Tasigna)
Dasatinib
(Sprycel)
Bosutinib
(Bosulif)
Ponatinib hydrochloride
(Iclusig)

EGFR kinase inhibitors:

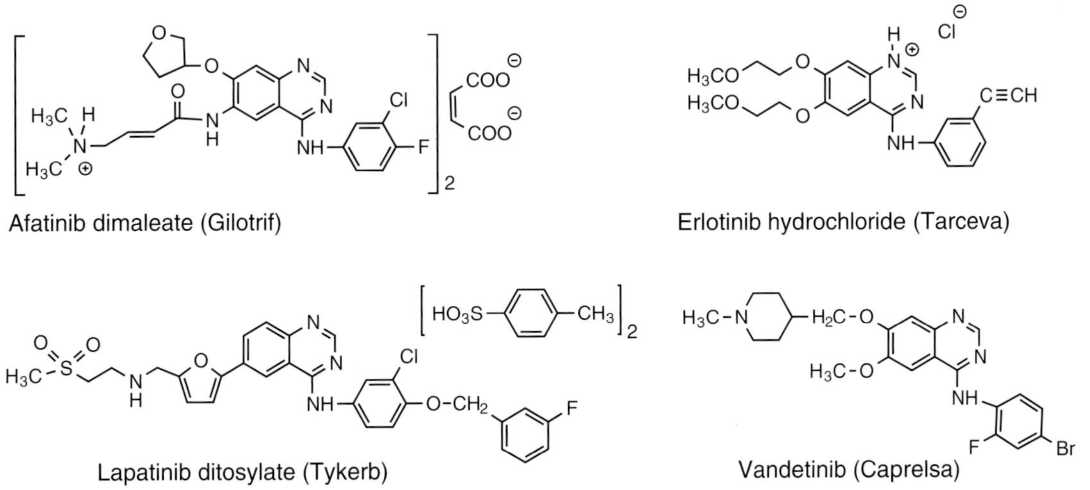
Afatinib dimaleate (Gilotrif)
Erlotinib hydrochloride (Tarceva)
Lapatinib ditosylate (Tykerb)
Vandetinib (Caprelsa)

Multikinase Inhibitors:

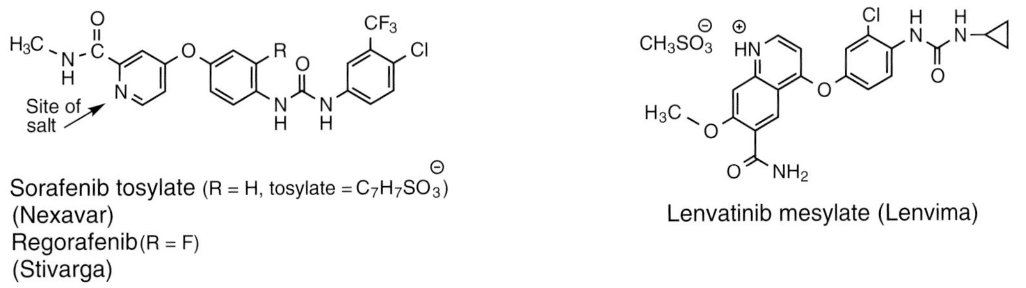
Site of salt
Sorafenib tosylate (R = H, tosylate = $C_7H_7SO_3^-$)
(Nexavar)
Regorafenib (R = F)
(Stivarga)
Lenvatinib mesylate (Lenvima)

Figure 26.29 Tyrosine kinase inhibitors. (*Continued*)

VEGFR kinase inhibitors:

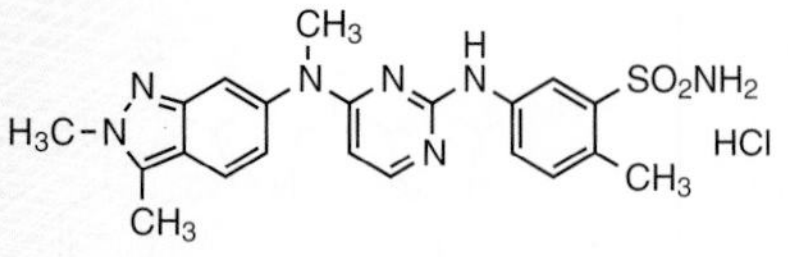

Pazopanib hydrochloride
(Votrient)

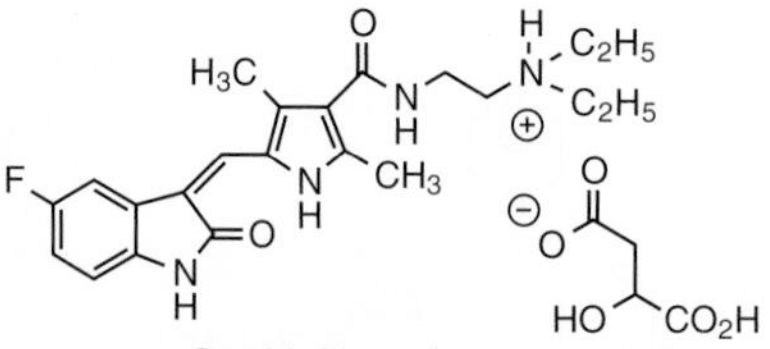

Sunitinib malate
(Sutent)

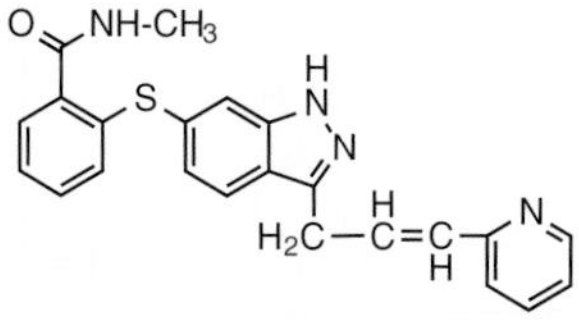

Axitinib (Inlyta)

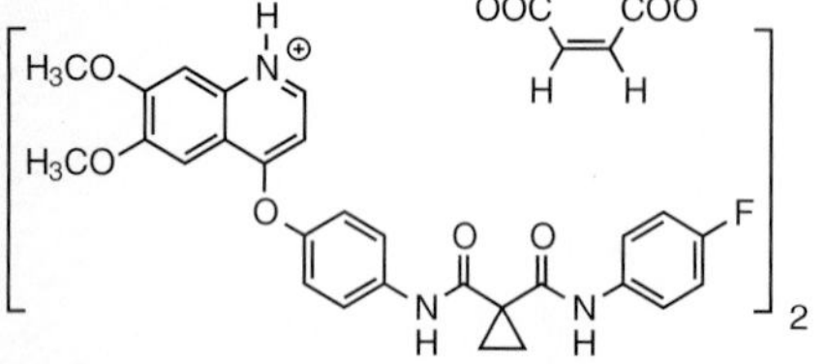

ALK Inhibitors:

Cabozantinib maleate (Exelixis)

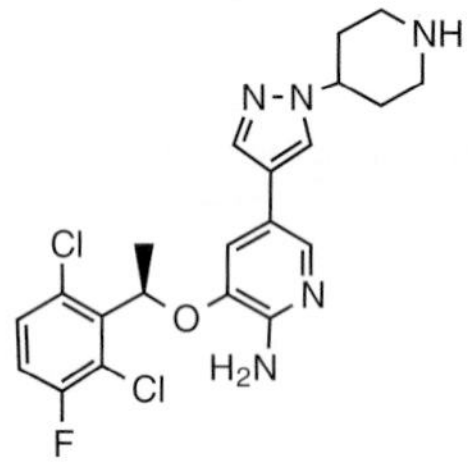

Crizotinib (Xalkori)

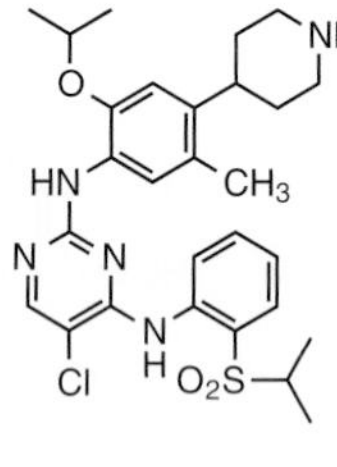

Ceritinib (Zykadia)

Other TKIs:

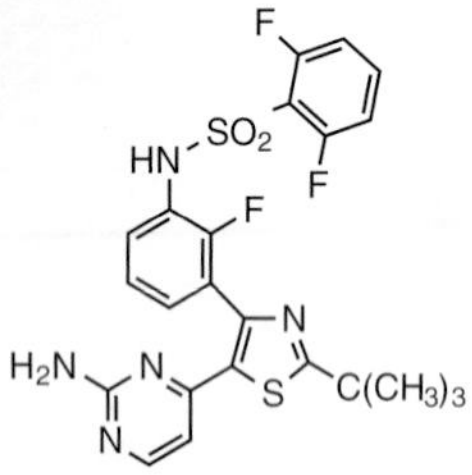

Dabrafenib (Tafinlar)
B-RAF TKI

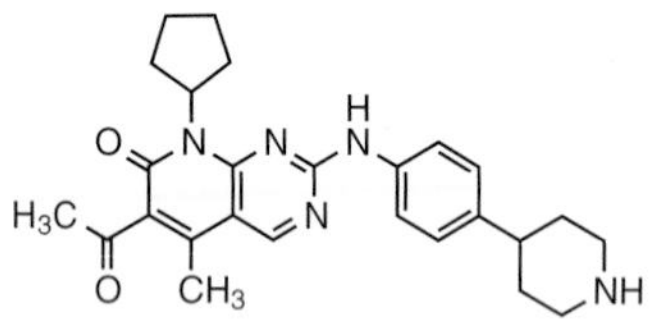

Palbociclib (Ibrance)
Cyclin-Dependent Inhibitor

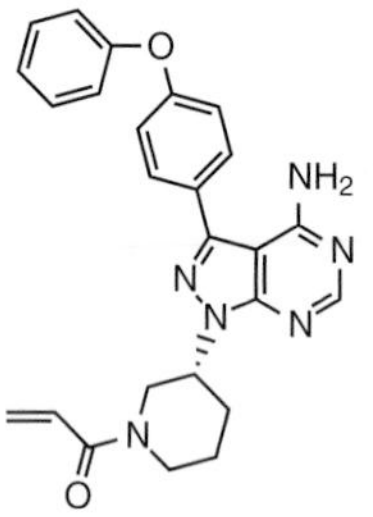

Ibrutinib (Imbruvica)
Bruton TKI

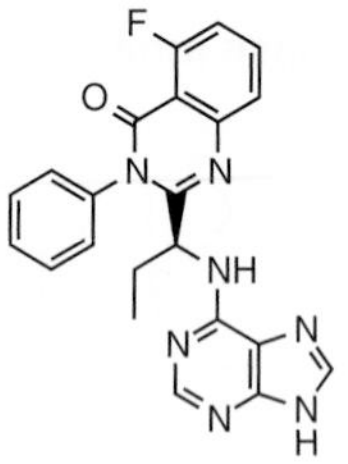

Idelalisib (Zydelig)
PI3K Inhibitor

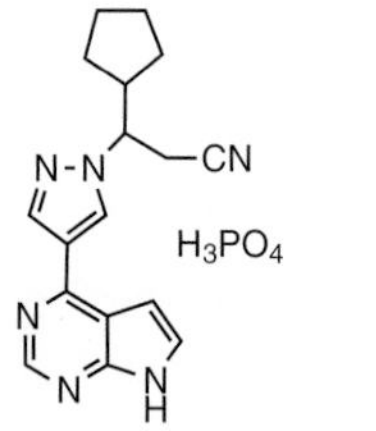

Ruxolitinib phosphate (Jakafi)
JAK1 & JAK2 Inhibitor

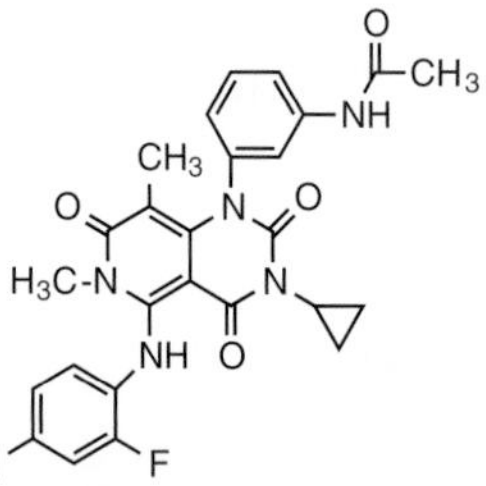

Trametinib (Mekinist)
MEK Inhibitor

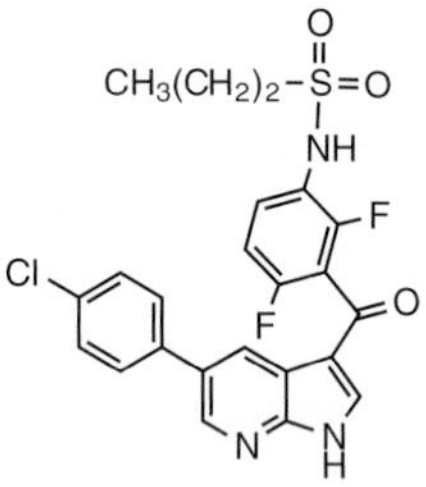

Vemurafenib (Zelboraf)
B-RAF TKI

Figure 26.29 (*Continued*)

- TKIs bind to the ATP-binding domain of their specific TK enzyme.
  - Type 1 inhibitors bind to the active conformation.
  - Type 2 inhibitors have highest affinity for the inactive conformation.
  - Type 2 inhibitors show the greater TK selectivity.
- The TKI binding site is primarily hydrophobic and spans a "hinge region" that connects the N and C termini of the protein.
- The binding site residues of several TKIs have been identified. (For specifics, refer to Figure 37.49, Chapter 37 in *Foye's Principles of Medicinal Chemistry, Seventh Edition*.)

Resistance Mechanisms

- Point mutations in the TK target (e.g., T315I in Bcr-Abl).
- P-gp overexpression.
- Downregulation of organic cation active transporting proteins.

Table 26.7 CYP Mediated Metabolism of Selected Tyrosine Kinase Inhibitors

	Reaction (Site)	CYP Isoform	Active Metabolite?
Axitinib	Sulfoxidation	3A4	No
Bosutinib	N-dealkylation (piperazine), oxidative dechlorination (chlorinated phenyl)	3A4	No
Cabozantinib	N-oxidation	3A4	No
Ceritinib	O-dealkylation (isopropoxy)	3A4	Yes, questionable significance
Crizotinib	Oxidation to lactam (piperidine), O-dealkylation	3A4	Yes (lactam)
Dasatinib	Aromatic hydroxylation (phenyl), benzylic hydroxylation (phenyl), N-dealkylation (piperazine), N-oxidation (piperazine N4)	3A4	Yes (questionable clinical relevance)
Erlotinib	O-dealkylation (terminal methoxy), aromatic hydroxylation (acetylenated phenyl)	3A4, 1A1, 1A2	Yes (desmethyl-erlotinib)
Imatinib	N-dealkylation (piperazine)	3A4	Yes
Lapatinib	Dealkylation (fluorobenzyl)	3A4	No
Nilotinib	Methyl hydroxylation (imidazole)	3A4	No
Pazopanib	Benzylic hydroxylation	3A4	Yes
Ponatinib	N-dealkylation (piperazine)	3A4	No
Regorafenib	N-oxidation (pyridine), N-demethylation	3A4	Yes
Sorafenib	N-oxidation (pyridine)	3A4	Yes
Sunitinib	N-deethylation to secondary amine	3A4	Yes
Vandetanib	N-dealkylation	3A4	Yes

Metabolism (Table 26.7)

- CYP3A4 is a common metabolizing isoform in all TKIs.
  - Decrease dose if coadministered with 3A4 substrates or inhibitors. Patients on TKIs should not drink grapefruit juice.
  - Increase dose by as much as 50% if coadministered with 3A4 inducers.
- Several TKIs are metabolized to *p*-aminophenols. If the phenyl ring has one or more electron withdrawing substituents, these metabolites are vulnerable to oxidation to electrophilic hepatotoxic quinoneimines.
- TKIs are known inhibitors of CYP isoforms (e.g., 3A4, 2D6, 2C9, 2C8) and P-gp. Potentially serious drug–drug interactions are possible

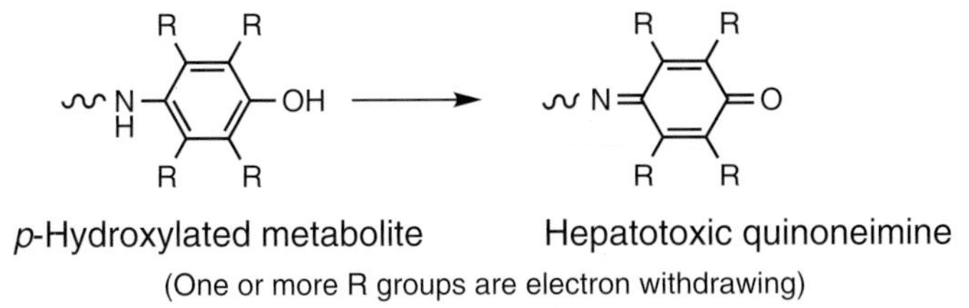

p-Hydroxylated metabolite   Hepatotoxic quinoneimine

(One or more R groups are electron withdrawing)

TKI hepatotoxic quinoneimine metabolite

Clinical Application

- Indications and therapeutic properties of selected TKIs are provided in Table 26.8.

Bcr-Abl Inhibitors (Bosutinib, Dasatinib, Imatinib, Nilotinib, Ponatinib)

SAR

- The 2-phenylaminopyrimidine moiety is common to imatinib and nilotinib.
  - The 4-pyridyl substituent increases TK affinity through hydrogen bonding.

Table 26.8 Indications and Therapeutic Properties of Selected Tyrosine Kinase Inhibitors

	Indications	Dosage Form	Therapeutic Issues
Bcr-Acl Inhibitors			
Bosutinib	Imatinib-resistant chronic myelogenous leukemia (CML)	Tablets	Administer with food. GI and hepatotoxicity possible
Dasatinib	Ph+ CML, imatinib-resistant acute lymphoblastic leukemia (ALL)	Tablets	May take with food and water if GI distress occurs. Fluid retention, hemorrhage, QT interval prolongation possible.
Imatinib	Ph+ CML, ALL, gastrointestinal stromal tumors (GIST)	Tablets	Take with food and water to minimize GI distress. Edema and hepatotoxicity possible
Nilotinib	Imatinib-resistant CML	Capsules	Black Box Warning: QT interval prolongation leading to sudden death
Ponatinib	Imatinib-resistant CML and ALL	Tablets	Black Box Warning: vascular occlusion, heart failure, hepatotoxicity
EGFR Inhibitors			
Afatinib	Nonsmall cell lung cancer	Tablets	GI, hepatic, and dermatologic toxicity possible
Erlotinib	Nonsmall cell lung cancer, pancreatic cancer	Tablets	24% increase in clearance in smokers demands dose increase. Food increases bioavailability from 60–100%
Lapatinib	HER2 positive breast cancer	Tablets	Black Box Warning: Potentially fatal hepatotoxicity
Vandetanib	Medullary thyroid cancer	Tablets	Black Box Warning: QT interval prolongation leading to sudden death
VEGFR Inhibitors			
Axitinib	Advanced renal cell carcinoma	Tablets	Hypertension common, hemorrhage possible
Cabozantinib	Medullary thyroid cancer	Capsules	Black Box Warning: perforations, fistulas, hemorrhage
Pazopanib	Advanced renal cell carcinoma	Tablets	Black Box Warning: potentially fatal hepatotoxicity
Sunitinib	Advanced renal cell carcinoma, advanced neuroendocrine tumors, imatinib-resistant GIST	Capsules	Black Box Warning: potentially fatal hepatotoxicity
Multikinase Inhibitors			
Regorafenib	GIST, metastatic colorectal cancer	Tablets	Black Box Warning: potentially fatal hepatotoxicity
Sorafenib	Advanced renal cell carcinoma, thyroid and hepatocellular carcinoma	Tablets	Cardiac ischemia, hemorrhage, hypertension and dermatologic toxicities possible
Lenvatinib	Iodine-refractory differentiated thyroid cancer	Capsules	Hypertension and cardiac decompensation possible
AKL Inhibitors			
Ceritinib	Nonsmall cell lung cancer	Capsules	Bioavailability may decrease at high gastric pH. GI, hepatotoxicity and pulmonary toxicity possible.
Crizotinib	Nonsmall cell lung cancer	Capsules	Bioavailability may decrease at high gastric pH. GI, hepatotoxicity and pulmonary toxicity possible.

- The O-methyl confers selectivity for cytosolic Bcr-Abl.
- The benzamide *para* to the CH_3 allows some inhibition of PDGFR, resulting in fluid retention.
- Imatinib and nilotinib have ionizable nitrogens in the same structural area.
  - Imatinib's higher pKa (piperazine) permits binding to organic cation transporting proteins (OCTP).
  - Nilotinib's lower pKa (imidazole) decreases affinity for OCTP and bypasses one resistance mechanism.
- Dasatinib and bosutinib's ionizable piperazine ring ensures water solubility.
- Nilotinib's CF_3 moiety augments potency through hydrophobic bonding compared to imatinib. Ponatinib also has an aromatic CF_3.

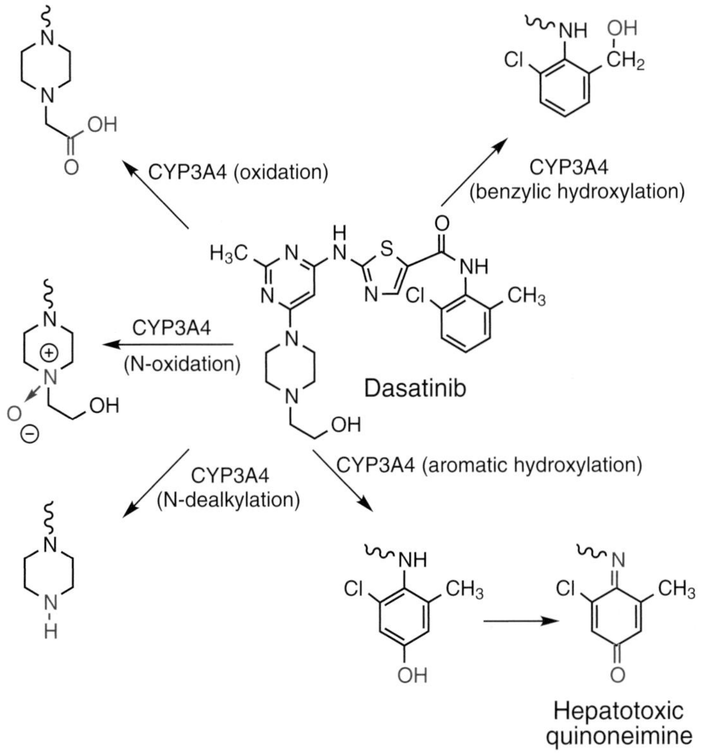

Hepatotoxic quinoneimine **Figure 26.30** Dasatinib metabolism.

- Dasatinib's thiazole ring shifts binding mode at the active site. Hydrogen and hydrophobic bonding is still of critical importance.
  - The unique binding confers mixed action (type 1 and type 2), which increases potency and therapeutic utility in imatinib- and nilotinib-resistant patients.
- Ponatinib's acetylene moiety provides rigidity that permits kinase binding in T315I mutants resistant to imatinib.

Metabolism

- The metabolic pathway of dasatinib is provided by way of example of Bcr-Abl TKIs (Fig. 26.30).
- Dasatinib has the potential to induce hepatotoxicity. The electron withdrawing Cl on the *p*-aminophenol metabolite enhances the electrophilicity and reactivity of the hepatotoxic quinoneimine.
- The primary CYP3A4-generated metabolites of imatinib, nilotinib, bosutinib, and ponatinib are shown in Figure 26.31. The oxydeschloro bosutinib metabolite is capable of generating a quinoneimine.
- Only desmethylimatinib retains clinically relevant activity.

EGFR and EGFR/HER2 Inhibitors (Afatinib, Erlotinib, Lapatinib, Vandetanib)

SAR

- The pharmacophore for both EGFR and HER2 inhibitors is a 4-anilinoquinazoline substituted at C6 or C7 with an oxygen-containing moiety.
- EGFR kinase selectivity is enhanced with an electron-withdrawing *m*-substituent on the 4-anilino ring system (erlotinib). The *p*-position tolerates only very small substituents (e.g., F).
- EGFR TKIs bind to the active conformation of the kinase (type 1 TKI).
- Increasing the size of the *p*-substituent (lapatinib) broadens activity to include inhibition of the inactive conformation of HER2 (type 2). EGFR inhibition is also shifted to type 2.
  - Type 2 inhibition increases drug-EGFR kinase dissociation half-life 10-fold.

Omacetaxine Mepesuccinate for TKI-Resistant CML and B-ALL

Omacetaxine mepesuccinate (Synribo) has proven effective in treating CML and B-cell acute lymphoblastic leukemia resistant to Bcr-Abl TKIs. This drug targets enzyme-positive leukemia stem cells, inhibiting a Bcr-Abl stabilizing protein known as HSP90 and resulting in kinase degradation. The rates of malignant cell apoptosis and patient survival increase. Omacetaxine is given by subcutaneous injection. Myelosuppression, diarrhea, and nausea are common adverse effects. The drug is also known as homoharringtonine, a reference to its natural origin (*Cephalotaxus harringtonia* an evergreen tree native to China).

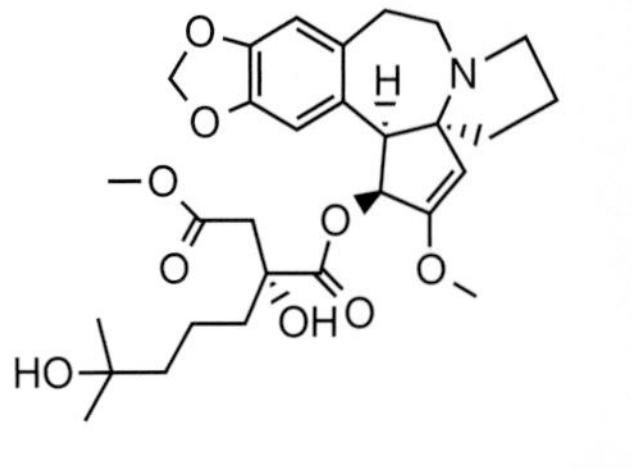

Omacetaxine mepesuccinate (homoharringtonine)

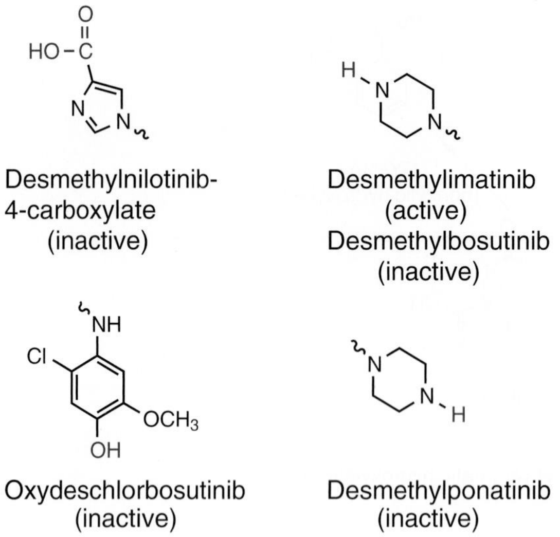

Figure 26.31 Major metabolites of selected Bcr-Abl TKIs.

Desmethylnilotinib-4-carboxylate (inactive)

Desmethylimatinib (active)
Desmethylbosutinib (inactive)

Oxydeschlorbosutinib (inactive)

Desmethylponatinib (inactive)

Physicochemical Properties

- All EGFR/HER2 TKIs can be made into water-soluble salts at N4.
- Polar groups incorporated in C6 substituents of lapatinib (methylsulfone), afatinib (dimethylamine), and vandetanib (methylpiperidine) enhance water solubility.
- Erlotinib lacks polar functional groups in the corresponding molecular area, so its water solubility is the lowest of the four.

Mechanisms of Resistance

- T790M mutation within EGFR results in an unfavorable drug–ATP binding affinity ratio.

Metabolism

- Erlotinib is O-demethylated by CYP3A4 and 1A1 (Fig. 26.32).
  - Desmethyl erlotinib's primary alcohol can be oxidized to the carboxylic acid.

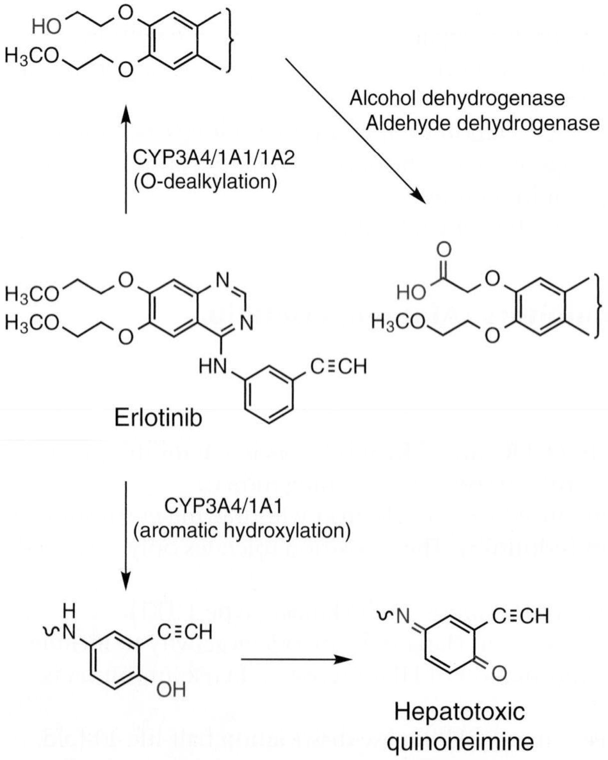

Figure 26.32 Erlotinib metabolism.

- Erlotinib generates an electrophilic quinoneimine through aromatic hydroxylation.
- Lapatinib loses the *m*-fluorophenyl substituent in a CYP3A4-catalyzed process. The phenol can oxidize to an electrophilic quinoneimine.
  - The close proximity of the electron withdrawing Cl increases the reactivity of this hepatotoxic metabolite.
- Vandetanib is N-dealkylated to an active nor metabolite.
- Afatinib is minimally metabolized by CYP.

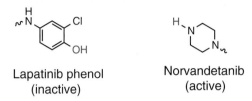

Lapatinib phenol
(inactive)

Norvandetanib
(active)

Lapatinib and vandetanib major metabolites

VEGFR Inhibitors (Axitinib, Cabozantinib, Pazopanib, Sorafenib, Sunitinib)

SAR

- These TKIs bind to the active, inactive, or both forms of VEGFR kinase.
- All structures contain a potentially ionizable amine. In some structures, the unionized conjugate binds within the hinge region of the kinase.
- Geometric isomerism can be important to activity. The *Z* isomer of sunitinib is 100 times as active as the *E* isomer.
- Sorafenib, along with regorafenib and lenvatinib, has affinity for other closely related enzymes, including PDGFR and c-kit. These three TKIs are classified as multikinase inhibitors.

Metabolism

- The major metabolites of VEGFR and multikinase inhibitors are provided in Figure 26.33.

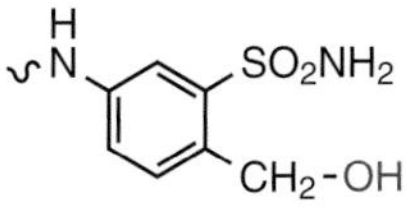

Hydroxypazopanib
(active)

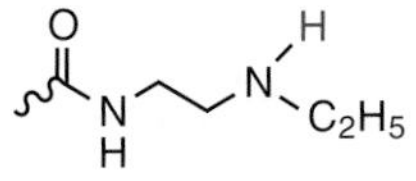

Desethylsunitinib
(active)

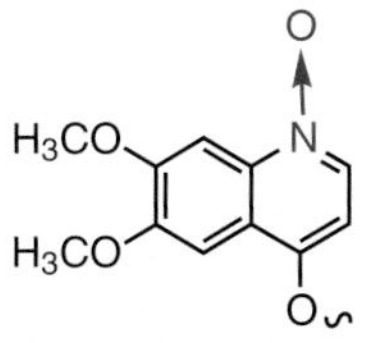

Cabozantinib N-oxide
(inactive)

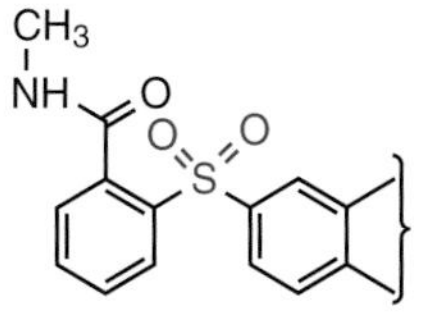

Axatinib sulfoxide
(inactive)

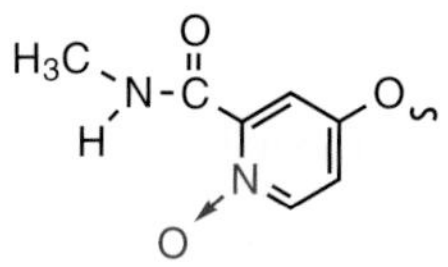

Sorafenib N-oxide
(active)

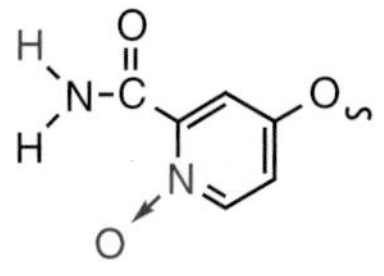

Desmethylregorafenib
N-oxide
(active)

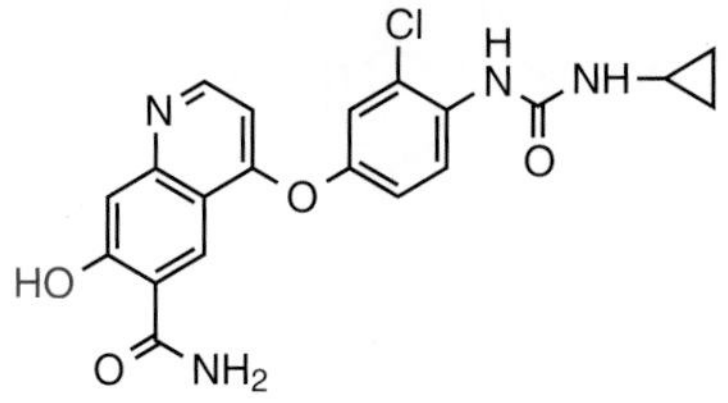

Desmethyllenvatinib **Figure 26.33** Major metabolites of selected VEGFR and multikinase TKIs.

ALK Inhibitors (Ceritinib, Crizotinib)

MOA

- Anaplastic lymphoma kinase (ALK) inhibitors suppress abnormal *ALK* in patients with nonsmall cell lung cancer.

Metabolism

- Crizotinib and ceritinib undergo CYP3A4 metabolism but the majority of the dose is excreted unchanged in feces.

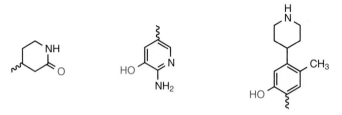

Crizotinib lactone   Dealkylated crizotinib   Dealkylated ceritinib

Crizotinib and ceritinib major metabolites

⚠ ADVERSE EFFECTS

TKIs (also see Table 26.8)
- Edema
- Diarrhea, abdominal pain
- Nausea, anorexia, weight loss
- Rash (sunlight sensitive)
- Hand-foot syndrome
- Fatigue
- Pulmonary, myocardial, endocrine dysfunctions

mTOR Inhibitors (Fig. 26.34)

MOA

- mTOR stands for mammalian target of rapamycin. It is a serine/threonine kinase regulated through the action of phosphatidylinositol. Activation accelerates the synthesis of kinases that promote neoplastic growth.

mTOR Inhibitors:

Temsirolimus (Torisel)   Everolimus (Afinitor)

Proteasome Inhibitors:

Bortezomib (Velcade)   Carfilzomib (Kyprolis)

Figure 26.34 mTOR and proteasome inhibitors.

- mTOR inhibitors bind deep within a 100-residue binding domain within the kinase. When interacting with two proteins (FKBP12 and FRB), kinase function is inhibited.

SAR

- Van der Waals and H-bonding interactions occur between mTOR inhibitors and aromatic residues on FKBP12 protein.
- Hydrophobic bonding is important in holding inhibitor to FRB.
- Polarity at O13 (ester, alcohol) promotes water solubility.

Metabolism

- Both mTOR inhibitors are metabolized by CYP3A4.
  - Temsirolimus is converted to rapamycin, an active metabolite.
  - Everolimus undergoes inactivating hydroxylation, O-dealkylation, and lactone ring cleavage.
  - Doses must be adjusted if CYP3A4 inhibitors, inducers or substrates are coadministered.

Clinical Application

- mTOR inhibitors are used in the treatment of advanced TKI-resistant renal cell carcinoma.

> **⚠ ADVERSE EFFECTS**
>
> **mTOR Inhibitors**
> - Immunosuppression
> - Blood glucose and serum triglyceride elevation
> - Angioedema
> - Stomatitis/mucositis
>
> ***Temsirolimus***
> - Interstitial lung disease
> - Bowel perforation
> - Renal failure
> - Cerebral hemorrhage
>
> ***Everolimus*** *(Black Box Warning)*
> - Immunosuppression nephro-toxicity
> - Graft thrombosis
> - Mortality in heart transplant patients

Proteasome Inhibitors (Fig. 26.34)

MOA

- Inhibition of proteasome induces cell death by halting the action of several anti-apoptotic proteins, including factor-κB and IκBα.
- Proteasome inhibitors bind to the 20S core of the 26S protein (specifically, the "chymotrypsin-like site" β5 of the proteasome), and halt normal cleavage of regulatory proteins. The build-up of proteins inside the cell initiates the apoptotic process.
- Binding selectivity and affinity are very high. Carfilzomib acts irreversibly by alkylating the N-terminal Thr of "chymotrypsin-like site" β5 of the proteasome.
- Malignant hematologic cells are particularly responsive to the lethal effects of these drugs.
- The higher β5 selectivity of carfilzomib may explain the decreased incidence of adverse effects, specifically serious sensory neuropathy that is use-limiting with bortezomib.

Metabolism

- CYP3A4 is the primary isoform that biotransforms bortezomib. Deboronation and hydroxylation are the principle reactions.
- Carfilzomib undergoes peptidase-mediated cleavage. CYP metabolism is minor.

Clinical Application

- Both proteasomes are given IV in the treatment of multiple myeloma.

Histone Deacetylase Inhibitors (Fig. 26.35)

MOA

- Deacetylation of Lys residues associated with histone proteins regulates gene transcription. Inhibition of this hydrolytic reaction allows acetylated histones to accumulate, stimulating cell cycle arrest and apoptosis.
- Histone deacetylase inhibitors (HDACi) also increase production of antioncogene *p*21.

> **⚠ ADVERSE EFFECTS**
>
> **Proteasome Inhibitors**
> - Myelosuppression
> - Cardiac and pulmonary complications
> - Sensory neuropathy (bortezomib)
> - Nausea, diarrhea, fever (carfilzomib)

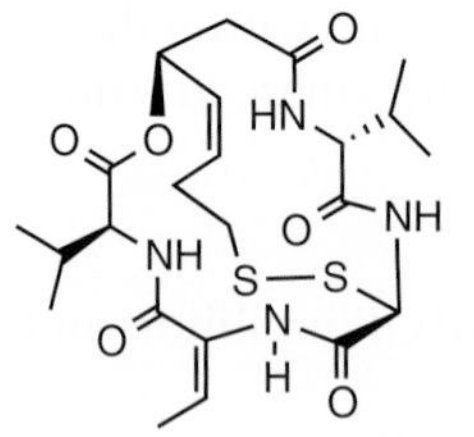

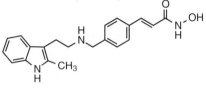

Romidepsin (Istodax)

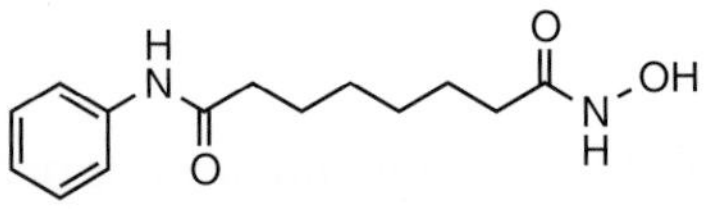

Vorinostat (Zolinza)

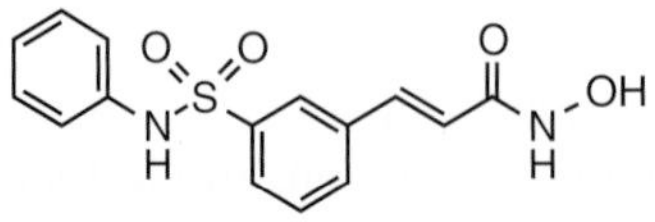

Belinostat (Beleodaq)

Panobinostat (Farydak)

Figure 26.35 Histone deacetylase inhibitors.

SAR

- The four marketed HDACi drugs have varied structures.
- Romidepsin is the original macrolide HDACi. Once reduced to the active thiol, it sequesters Zn^{2+} associated with deacetylase enzymes. This drug inhibits primarily Class I deacetylase enzymes which regulate cellular proliferation. There is a minor inhibitory effect on Class II (malignant cell apoptosis suppressor) enzymes.
- Belinostat, vorinostat, and panobinostat are hydroxamic acid analogs that terminate in an aromatic moiety (phenyl or indole). The long connecting chain between these groups is predominantly lipophilic, and may contain an aromatic ring (belinostat, panobinostat) or a straight chain alkyl moiety (varinostat).
- The hydroxamic acid-based HDACi drugs inhibit both Class I and Class II deacetylase enzymes. As such, they are sometimes classified as "pan inhibitors."
- Panobinostat, a cinnamoyl hydroxamic acid, also inhibits Class III and Class IV deacetylase enzymes. Class III HDACi regulate normal cell growth and proliferation. Little is known about Class IV enzymes beyond the possible ability to regulate immune responses. Panobinostat has been estimated to be 10-fold more effective as an HDACi than vorinostat.

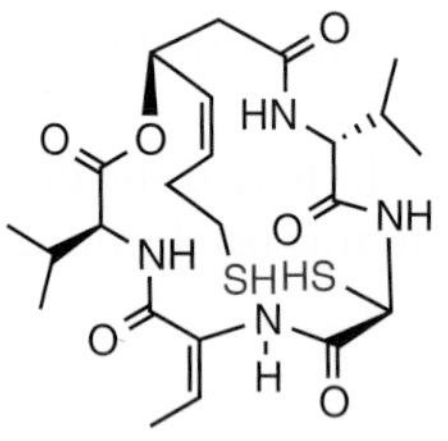

Romidepsin thiol metabolite
(active)

Metabolism

- Romidepsin is reduced intracellularly by glutathione to the active thiol.
- All HDACi drugs except vorinostat are vulnerable to CYP-mediated metabolism, with CYP3A4 playing a major biotransformation role.
- Vorinostat is metabolized via amide hydrolysis with subsequent β-oxidation to 4-anilino-4-oxobutanoic acid. Direct glucuronidation can also occur.

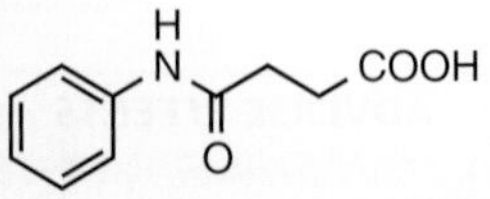

Clinical Applications

- Belinostat and vorinostat are used in peripheral and cutaneous T-cell lymphoma, respectively. Romidepsin is utilized in both disease states.
- Panobinostat is currently indicated for the treatment of multiple myeloma. It is the only HDACi with a Black Box Warning related to severe diarrhea and potentially fatal cardiac arrhythmia, ischemic episodes, and EEG abnormalities.

4-Anilino-4-oxobutanoic acid

Selected Miscellaneous Antineoplastic Agents (Fig. 26.36)

- Olaparib is a poly (ADP-ribose) polymerase (PARP) inhibitor. It induces double strand breaks in DNA in BRCA1/2 deficient cancer cells, leading to cell destruction.
- Tretinoin is *trans*-retinoic acid. It is used for the induction of remission of selected types of acute promyelocytic leukemia (APL).
- Arsenic trioxide is also used to treat APL patients. It acts by destroying the APL-inducing fusion protein PML-RARα which promotes chromosomal translocation through the joining of the promyelocytic gene and retinoic acid receptor α.
- Thalidomide, lenalidomide, and pomalidomide are immunomodulators that increase natural killer cells, interleukin-2, and/or interferon-γ in patients with multiple myeloma.
  - Lenalidomide is used in doses 32-fold lower than thalidomide, and it is considered less toxic.
  - Pomalidomide is the most potent of the three and is used at a dose 1/200 that of thalidomide.
  - All three drugs induce serious dose-independent teratogenicity and are contraindicated in pregnancy.
- Bexarotene activates retinoid X receptors, resulting in controlled cellular proliferation. This drug is used in the treatment of cutaneous T-cell lymphoma.
- Vismodegib inhibits the hedgehog signaling pathway crucial to the regulation of early-stage tissue growth.
  - The specific target for vismodegib is a transmembrane protein called smoothened which, when active, promotes neoplastic cell growth.
  - Vismodegib has found value in the treatment of basal cell carcinoma.
- While the exact mechanism of mitotane in the treatment of adrenal cortical carcinoma is unknown, it is believed to involve adrenal cortex suppression and alterations in the metabolism of hydrocortisone and steroids.

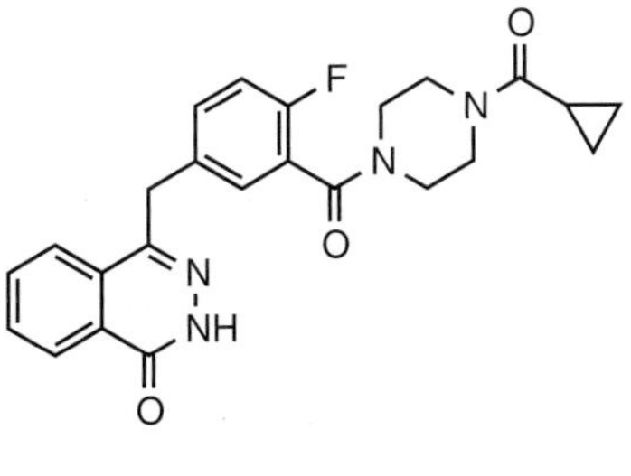

Olaparib (Lynparza)

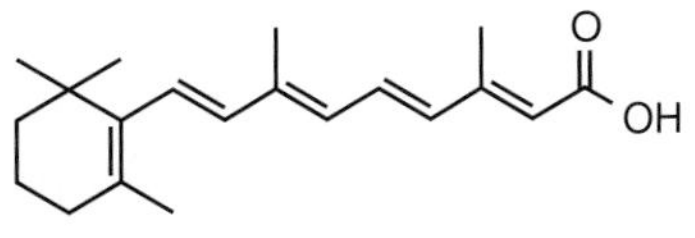

Tretinoin

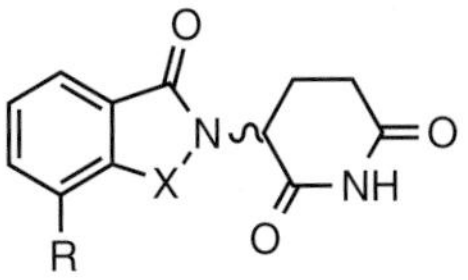

Thalidomide (Thalomid)
(R = H, X = C = O)
Lenalidomide (Revlimid)
(R = NH_2, X = CH_2)
Pomalidomide (Pomalyst)
(R = NH_2, X = C = O)

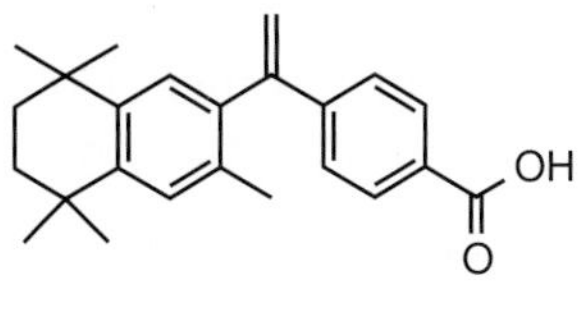

Bexarotene (Targretin)

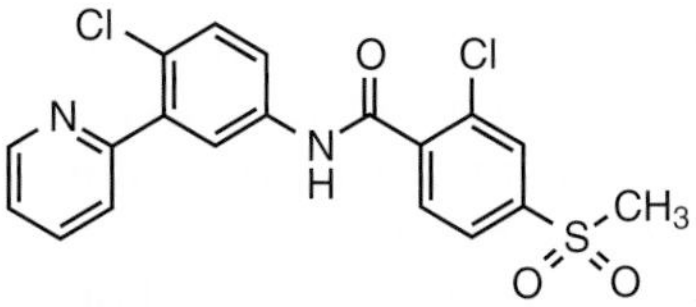

Vismodegib (Erivedge)

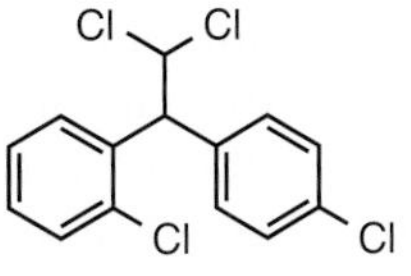

Mitotane (Lysodren)

As_2O_3

Arsenic trioxide
(Trisenox)

Figure 26.36 Miscellaneous antineoplastic agents.

Chemical Significance

Unlike some disorders where the pharmacist has the luxury of guiding primary therapeutic decision making, the drugs used to treat specific neoplastic orders are used because they are known to work. Most commonly, specific regimens that harness the power of several antineoplastic agents with differing mechanistic and side effect profiles are employed to maximize dose while minimizing adverse effects.

Pharmacists' understanding of antineoplastic drug mechanism, physicochemical properties, and toxicity risk make them invaluable members of the oncology health care team. Pharmacists who are not board certified in oncology also have a critical role to play, for they are sure to serve patients with cancer who come to them for prescription and other medication therapy management services. They may be the first to hear about an adverse effect that requires attention. Some examples of where a pharmacist's expertise can be life-saving or risk-attenuating follow.

- Does an antineoplastic drug being given IV require Kolliphor EL or Tween for solubilization? Recognizing the potential for hypersensitivity, be sure to pretreat with antihistamines and corticosteroids, and have epinephrine ready to administer if blood pressure drops or bronchi constrict.
- Is a colorectal cancer patient about to begin therapy on capecitabine? Recommend that his DPD phenotype be assessed, and cut the doses back if he is a poor DPD metabolizer.
- Is an ovarian cancer patient initiating a regimen that contains cisplatin? Minimize her risk of permanent hearing loss by administering amifostine to react with any organoplatinum complex that distributes to cochlea.
- Is a breast cancer patient about to receive her first dose of doxorubicin? Make sure that the iron chelating agent dexrazoxane is on board so that her risk of life-threatening myocardial toxicity is lowered. Let her know the red color of the drug will be apparent in her urine.
- Is a Hodgkin's lymphoma patient on cyclophosphamide experiencing painful and hemorrhagic urination? Alert the oncologist that the sulhydryl rescue agent mesna should be administered ASAP, and encourage the patient to force fluids to rid the urinary track of drug and its urotoxic acrolein byproduct.
- Is a BCG-refractory bladder cancer patient on valrubicin experiencing adverse reactions more commonly found with IV anthracyclines (e.g., doxorubicin)? His disease may have progressed to the point where the bladder wall has lost integrity. The drug should be discontinued.
- Is a lung cancer patient about to have methotrexate added to his regimen? Be sure his urinary pH is at least 7.0 to avoid crystalluria from precipitated free acid, and counsel him to hydrate well during therapy.
- Check the patient profile of any cancer patient being administered a TKI for potential DDIs. These drugs are metabolized by 3A4 and inhibit that isoform and many others. Counsel the patient not to drink grapefruit juice at any time during their therapy.

Cancer takes a huge toll on a patient's physical and emotional well-being, and that of their loved ones. The importance of having faith in a health care team that fully understands that the drugs being administered are intentional poisons, and takes every precaution to keep the patient safe and functional under these challenging circumstances, is paramount. Pharmacists know more about how these drugs work on a molecular level than anyone, and they have a mandate to keep current on new drugs coming on the market at an unprecedented pace. They must bring every bit of their chemical expertise to bear when treating patients facing this routinely debilitating and too often fatal disease. They must also bring all of the compassion and respectful sensitivity for which pharmacists are known to the patient (and family) care environment.

Review Questions

1 Which DNA cross-linking agent would put patients at highest risk for urotoxicity?

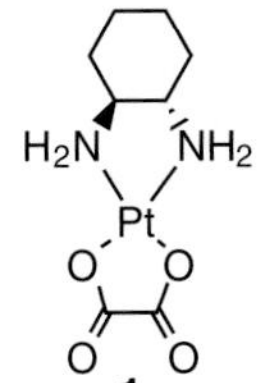

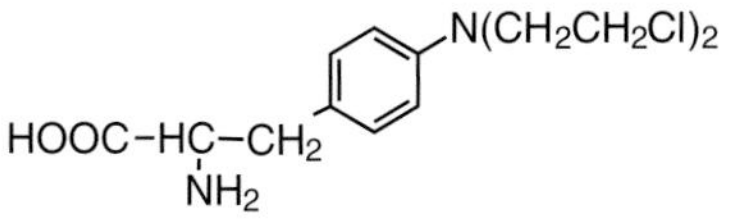

A. Cross-linker **1**
B. Cross-linker **2**
C. Cross-linker **3**
D. Cross-linker **4**

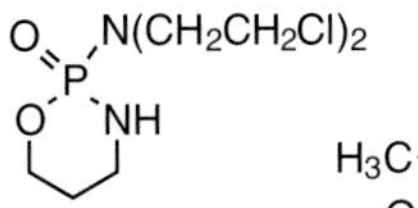

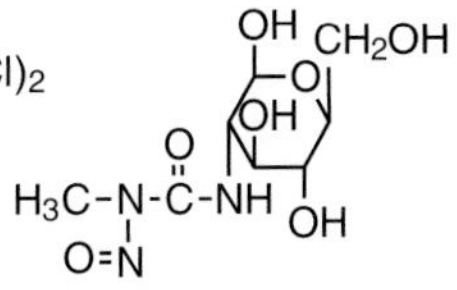

2 Which antineoplastic would put patients at highest risk for serious hypersensitivity?

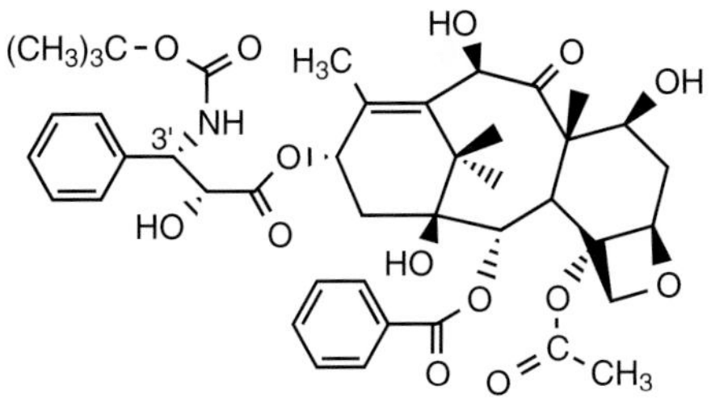

A. Antineoplastic **1**
B. Antineoplastic **2**
C. Antineoplastic **3**
D. Antineoplastic **4**

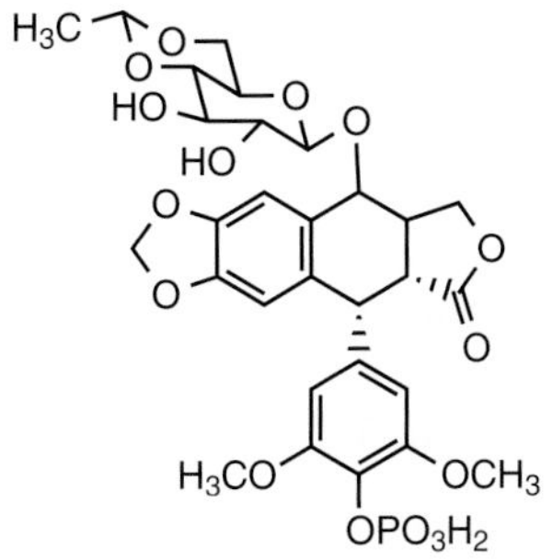

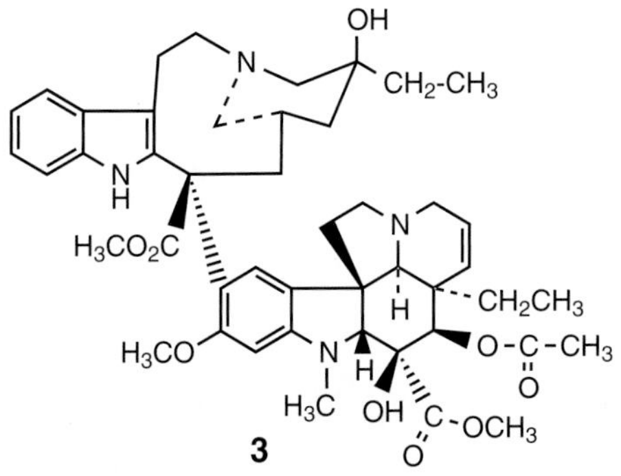

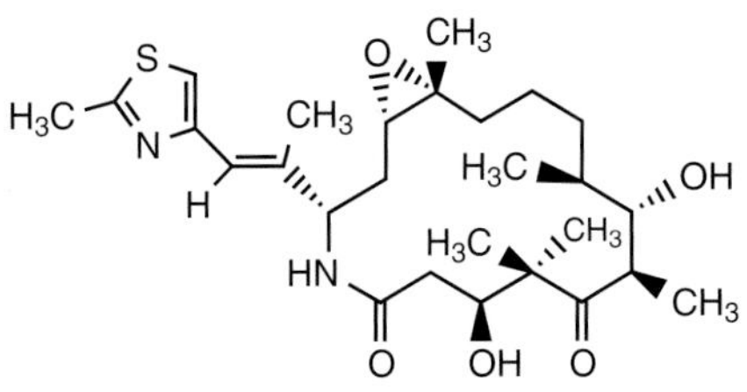

3 In which class of anticancer agents is grapefruit juice a strict contraindication?
A. Tyrosine kinase inhibitors
B. Fluorouracil antimetabolites
C. Organoplatinum DNA cross-linking agents
D. Anthracycline topoisomerase poisons
E. Antifolate antimetabolites

4 Which anthracycline/anthracenedione metabolite is associated with the generation of cytotoxic free radicals that can acutely damage the heart?

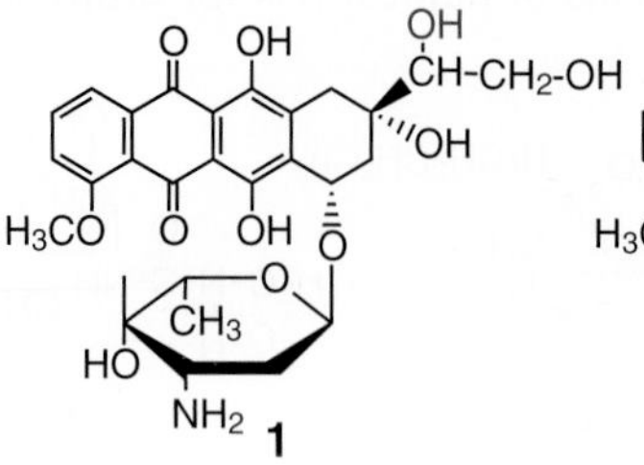
1

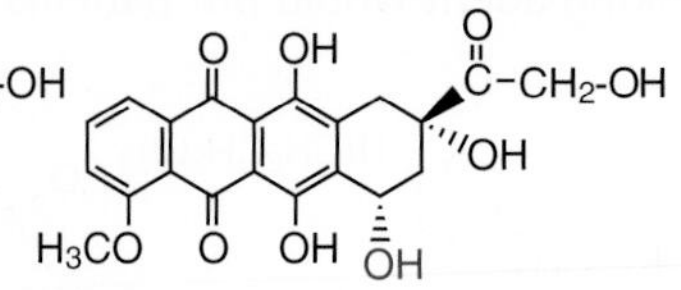
2

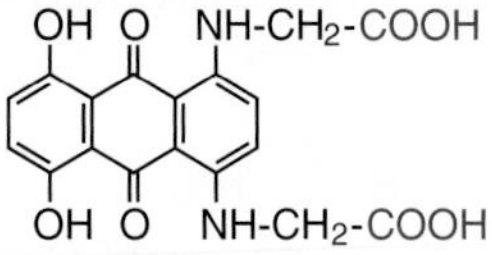
3

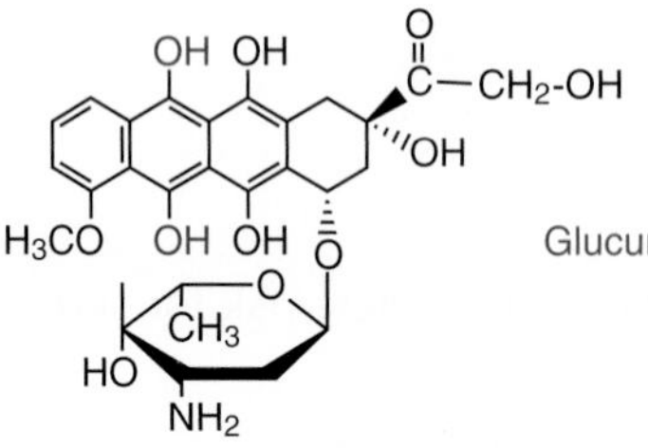
4

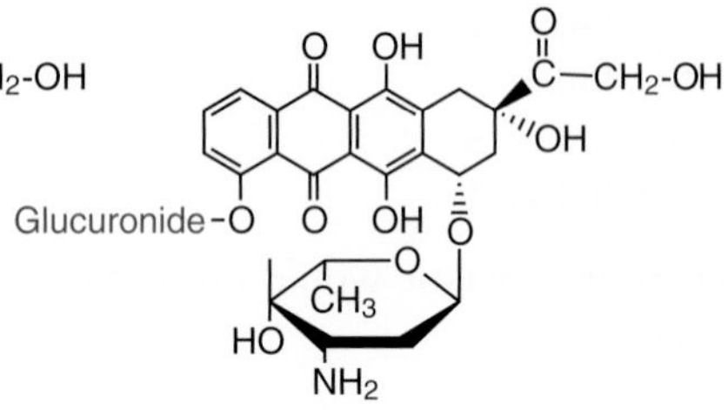
5

A. Metabolite **1**
B. Metabolite **2**
C. Metabolite **3**
D. Metabolite **4**
E. Metabolite **5**

5 Which tyrosine kinase inhibitor would be least likely to generate a hepatotoxic quinoneimine metabolite?

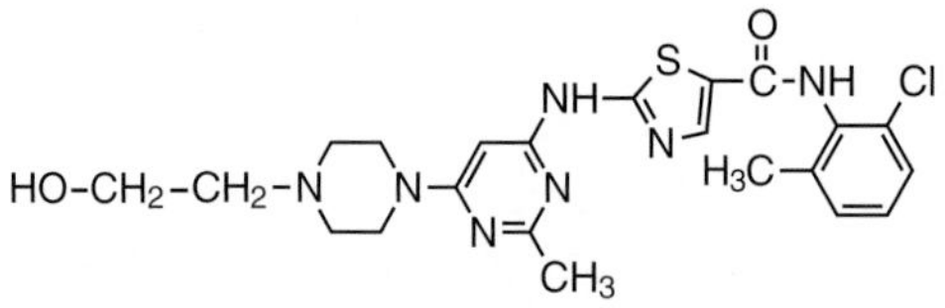
1

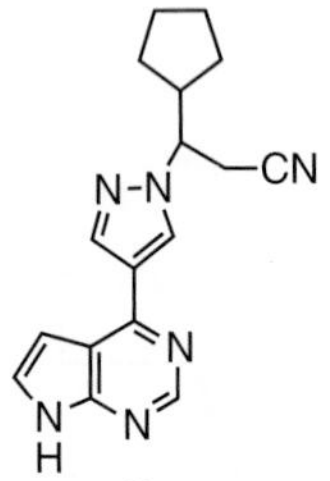
2

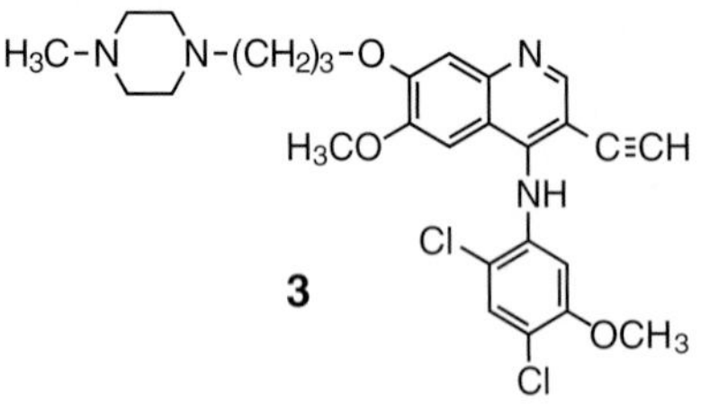
3

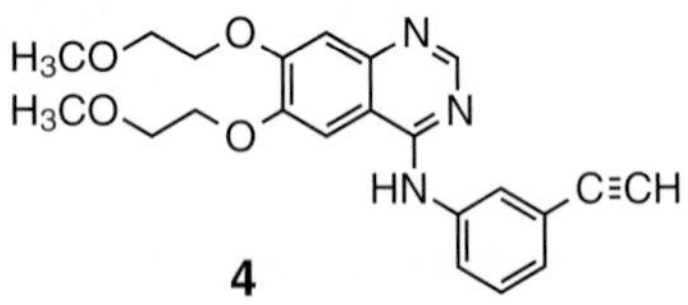
4

A. TKI **1**
B. TKI **2**
C. TKI **3**
D. TKI **4**

Chemotherapy: Antiviral and Antiretroviral Agents

27

Learning Objectives

Upon completion of this chapter the student should be able to:

1 Identify the site and describe the mechanisms of action of the antiviral drugs for the condition of:

a. HIV
  - adsorption/fusion inhibitors
  - nucleoside and nonnucleoside reverse transcriptase inhibitors
  - protease inhibitors
  - integrase inhibitors

b. Viral hepatitis B and C
  - vaccines
  - nucleosides
  - protease inhibitors
  - polymerase inhibitors

c. Herpes virus

d. Influenza

2 List structure–activity relationships (SAR) of key drug classes where presented.

3 Identify clinically significant physicochemical and pharmacokinetic properties.

4 Discuss, with chemical structures, the clinically significant metabolism of discussed drugs and how the metabolism affects their activity.

5 Discuss the clinical application of the various drug classes for the treatment of the specific viral infection.

Introduction

Viruses are associated with diseases, which are as common as the "common cold" but are also as deadly as HIV/AIDs and cancer. Viruses are infectious particles made up of genetic material, DNA or RNA, and associated proteins and come in various shapes and sizes. As obligatory organisms virus are intracellular organisms, which require the machinery of the host cell to carry out their infectious processes. The basic unit of a virus is either DNA or RNA, which may be either single or double strand and appear as either linear or circular forms. The virus has associated nucleoproteins and may be surrounded by a symmetrical protein known as a capsid. The capsid is made up of repeating structural units known as protomers, which themselves are made up of nonidentical protein subunits. In some cases the nucleocapsid is surrounded by a lipid-containing membrane or envelope. When present the envelope contains virus-encoded specific proteins. The virus particle, with or without an envelope, is referred to as the virion. Viral architecture consists of one of three basic types: cubic (icosahedral) symmetry, helical symmetry, or a complex structure. In addition, viruses are classified by morphology, properties of the genome (i.e., DNA vs. RNA, single vs. double strand, linear vs. circular, sense vs. antisense), physicochemical properties, structure of associated proteins, and replication process. Viruses are classified into major families, with names ending with the suffix -viridae, and then into genera that end in -virus. A listing of these families can be found in Chapter 38, Table 38.1 in *Foye's Principles of Medicinal Chemistry, Seventh Edition.*

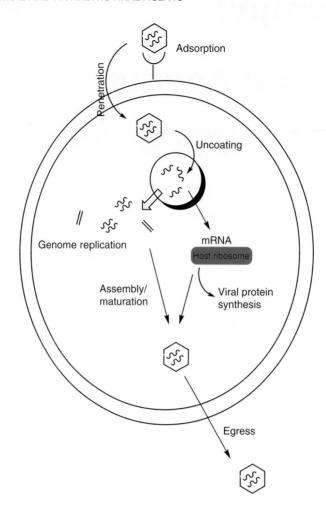

Figure 27.1 A schematic drawing of viral life cycle involving adsorption and penetration of the host cell, uncoating of genetic material, genome replication and viral protein synthesis, assembly and maturation, and finally egress from the host cell to continue infection. (Modified 36.1 from Principles of Pharmacology, 2nd ed.)

Viral Replication

In order for a virus to replicate the organism is required to produce usable mRNA which in turn is necessary for production of all of the macromolecules required by the virus. This is done in a highly organized sequence. The process begins with adsorption of the virus to a specific receptor site on the host cell. The sequence of events is depicted in Figure 27.1 as described in Table 27.1. The array of drugs used to treat the various viral infections take advantage of one or more of the key steps involved in the life cycle of the virus. As indicated above, viral infections require the human cell and make use of various biochemical processes of the host cell to

Table 27.1 **RNA and DNA Life Cycle**	
Stages in Replication	**Characteristics**
Adsorption/attachment of virion	Complementarity between virion surface and cell surface receptors.
Penetration/entry	Virion crosses cell membrane to gains access to host cell.
Uncoating	Nucleic acid or nucleocapsid exposed.
Transcription/translation	Viral nucleic acid enters host cell nucleus: viral DNA is transcribed into mRNA or serves as a template for additional DNA synthesis. Viral RNA can be transcribed directly to mRNA, viral RNA can be replicated to new infectious RNA, or can be transcribed into DNA with the aid of viral reverse transcriptase. mRNA leaves nucleus and with host ribosome is translated into viral protein.
Genome replication	Synthesis of the appropriate viral DNA or RNA
Assembly/maturation	Viral proteins are synthesized and infectiousness of virus occurs (for HIV infectiousness occurs outside cell).
Egress	Release of virion from host cell.

propagate and spread to additional cells. As a result, the chemotherapy of viruses based upon the unique biochemistry of the virus is challenging since many of steps and stages in viral infections involve normal host cell machinery. Drug categories will be based upon the key steps shown in Figure 27.1 namely: agents inhibiting attachment and penetration of the host cell, uncoating of the nucleic material, transcriptional/translational processes, genomic replication, viral protein synthesis, assembly/maturation processes, and egress from the infected cell.

HIV/AIDS

Because of the epidemic nature of the human immunodeficiency virus (HIV-1) and the worldwide efforts directed toward an understanding and treatment of the resulting disease, acquired immunodeficiency syndrome (AIDS), chemotherapeutic agents used to treat this condition will be covered separately. It is estimated that as of 2012 there are 35.3 million people worldwide living with HIV/AIDS. In 2012, more than 1.6 million people died of AIDS with 2.3 million new cases including more than 640,000 children were diagnosed with the disease. While the disease cannot be cured, the selection of drugs continues to grow despite rapid drug resistance and virus mutation.

The life cycle of HIV RNA virus appears similar to that of the DNA viruses, but with several significant changes (Fig. 27.2). Similar to the stages reported for DNA viruses shown in Table 27.1 the following process occurs with HIV:

- Binding and adsorption occurs between the viral extracellular protein gp120 and CD4 receptors on T-lymphocytes of the host cell.
- Fusion and uncoating gives rise to single-stranded RNA (ssRNA) which serves as a substrate for reverse transcriptase (RT) forming proviral double-stranded DNA.
- The proviral double-stranded DNA enters the nucleus where it is incorporated into host DNA with the aid of viral integrase.

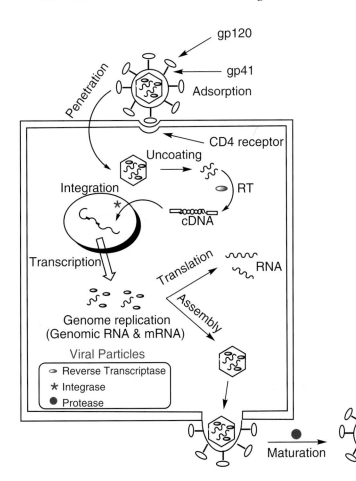

Figure 27.2 Schematic drawing of the HIV life cycle. Significant to drug therapy are the adsorption/penetration, reverse transcriptase (RT), integration of viral cDNA into the host DNA (integrase catalyzed), and maturation (protease catalyzed) processes.

- Transcription by host RNA polymerase II results in viral RNA, which is released from the nucleus where genome replication occurs.
- Viral assembly occurs in which viral RNA, integrase, transcriptase, and proteases are incorporated into new virions ready to egress and maturation.

Antiretroviral Drug Therapy

Adsorption/Fusion Inhibitors

The attachment of HIV to the host cell membrane is associated with embedded proteins in the cell membrane of HIV, gp120 and gp41, and CD4 receptor in the host membrane (Fig. 27.2). Through a complex interaction of these proteins and host receptor, fusion of the two membranes occurs allowing penetration of the HIV virion into the host cell. Theoretically drugs can be designed that inhibit attachment or fusion of the membranes thus preventing entry of HIV into human host cells.

Enfuvirtide

N-acetyl-Tyr-Thr-Ser-Leu-Ile-His-Ser-Leu-Ile-Glu-Glu-Ser-Gln-Asp-Gln-Gln-Glu-Lys-Asp-Glu-Gln-Glu-Leu-Leu-Glu-Leu-Asp-Lys-Try-Ala-Ser-Leu-Try-Asp-Try-Phe-NH$_2$

Enfuvirtide (Fuzeon)

Enfuvirtide is a synthetic 36 amino acid peptide the design of which is based upon the structure of a portion of the HIV surface protein gp41.

MOA

- Normally, gp41 viral protein undergoes a conformational change, which allows gp120 protein to bind to the CD4 host receptor protein. Fusion of the HIV membrane to the host membrane results.
- Enfuvirtide binds to the tryptophan-rich gp41 protein of HIV-1 to prevent the conformational change thus preventing fusion.
- Unfortunately, the virus can mutate any of 10 amino acids in a specific domain of the gp41, thus leading to resistance to enfuvirtide.

Physicochemical and Pharmacokinetic Properties

- As a peptide enfuvirtide must be administered via subcutaneous injection twice daily.
- The drug is highly bound to plasma protein (~92%).
- Enfuvirtide is prone to proteolytic metabolism.
- Most patients experience local injection site reactions (i.e., erythema, cysts, pruritus).
- Cost of therapy is nearly prohibitive (~$25,000/year).

Maraviroc

The binding of HIV-1 cell surface gp41 and gp120 to CD4 cells in host T-cells is assisted in some individuals by CCR5 protein. The CCR5 protein is a chemokine found on the host cell, which functions as a coreceptor for HIV fusion and penetration of the HIV virion. A second coreceptor associated with HIV-1 penetration into host cells is the CXCR4 chemokine protein, which may be present on some T-cells.

Maraviroc (Selzentry)   Maraviroc metabolite

MOA

- Maraviroc selectively binds to CCR5 inhibiting the binding of HIV-1 to the CCR5~gp120/gp41 complex.
- Inhibition of HIV-1 binding to the host cell prevents fusion and penetration into the host cells.
- Individuals with a genetic mutation in the *CCR5* gene (*CCR5-δ32*) may be unresponsive to maraviroc.
- Maraviroc is not effective on cells exhibiting CXCR4 coreceptors and therefore a viral tropism assay is required and the drug is not recommended for patients with dual/mixed or CXCR4-topic HIV-1 infections.

Physicochemical and Pharmacokinetic Properties

- Orally administered and effectively absorbed.
- Maraviroc's poor bioavailability (~23%) is possibly due to P-glycoprotein transport for which it is a substrate.
- Highly bound to plasma protein (~76%).

Metabolism

- Maraviroc is metabolized by CYP3A4 giving rise to the inactive N-dealkylated product shown above. The potential for drug–drug interactions with other drugs metabolized by CYP3A4 should be considered.
- Maraviroc does not appear to clinically inhibit any of the CYP450 enzymes.

Clinical Applications

- Maraviroc should be used in combination with other antiretroviral therapy regiments when treating HIV-1 patients.
- Combination therapy should include at least three different drugs from at least two different classes: nucleoside reverse transcriptase inhibitors, nonnucleoside reverse transcriptase inhibitors, protease inhibitors, and the entry inhibitor enfuvirtide.
- Care should be taken with atazanavir, saquinavir, and ritonavir, which are CYP3A4 inhibitors.
- Allergic reactions and potential liver diseases are associated with the use of maraviroc.

Nucleoside Reverse Transcriptase Inhibitors (NRTIs)

As indicated in Figure 27.2 the uncoated HIV virion is encoded by reverse transcriptase, an RNA-dependent DNA polymerase to proviral DNA, which is then incorporated into a host DNA cellular chromosome. The structures of the NRTIs are shown in Figure 27.3.

MOA

- All of the NRTIs are prodrugs, which must be phosphorylated at the 5′ position to the triphosphate, which compete with natural substrates for incorporation into the proviral DNA (Fig. 27.4A).
- The NRTIs are either purine (didanosine, abacavir) or pyrimidine (zidovudine, zalcitabine, stavudine, lamivudine, emtricitabine) derivatives of natural DNA monomers.
- Only tenofovir disoproxil is unique in that this prodrug, is hydrolyzed to a phosphonic acid which is then converted into a "diphosphate" of the phosphonic acid (actually a triphosphate) which is a competitive inhibitor of the natural substrate, deoxyadenosine triphosphate (Fig. 27.4B).
- All of the NRTIs lack a 3′-hydroxyl group and will thus inhibit propagation of the DNA leading to termination of elongation.

Physicochemical and Pharmacokinetic Properties (Table 27.2)

- As indicated in Table 27.2 all NRTIs are administered via the oral route although zidovudine is also available in an injectable form.

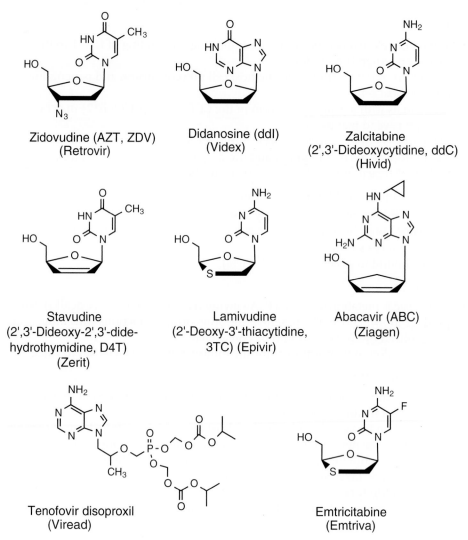

Figure 27.3 Nucleoside reverse transcriptase inhibitors (NRTIs).

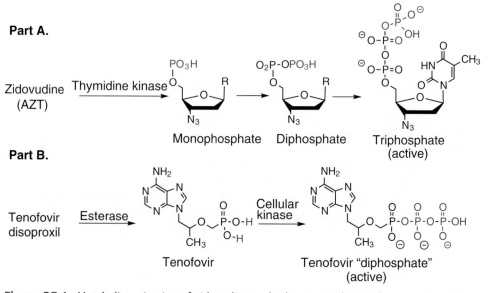

Figure 27.4 Metabolic activation of zidovudine and other pyrimidine and purine NRTIs (**A**) and of tenofovir disoproxil to active NRTIs (**B**).

Table 27.2 Pharmacokinetic Properties of HIV Reverse Transcriptase Inhibitors

Generic Name	Oral Bio-Availability	Plasma Protein Binding	Dosage Forms	Additional Properties
Nucleoside Reverse Transcriptase Inhibitors (NRTIs)				
Zidovudine	65%	30%	Tab, Cap, Syrup, Inj	Store at 15-25 degrees/protect from light–azide is unstable; enters CNS.
Didanosine	25%	<5%	Chewable/buffered Tab,	
Zalcitabine	87%	~11%	Tab	Food reduces bioavailability; enters CNS
Stavudine	85%	negligible	Cap, Powder for Oral Sol	
Lamivudine	86%	<35%	Tab, Sol	
Abacavir	83%	50%	Tab, Sol	Enters CNS
Tenofovir disoproxil	25%	<0.7%	Tab	
Emtricitabine	93%	<4%	Cap	(−) Isomer> activity (+) isomer
Non-Nucleoside Reverse Transcriptase Inhibitors (NNRTIs)				
Nevirapine	95%	60%	Tab	Crosses placenta, enters breast milk
Delavirdine	85%	98%	Tab	Inhibitor of CYP3A4
Efavirenz	~45%	~99%	Cap	Avoid high fat meals
Etravirine	Unknown	~96%	Tab	Both drugs show increase absorption with food
Rilpivirine	Unknown	<99.7%	Tab	

Cap, capsule; *Tab*, tablet; *Sol*, solution; *Inj*, injectable.

- NRTIs in general show good bioavailability and low binding to plasma protein.
- Tenofovir is an exception which is poorly absorbed while the disoproxil derivative improves absorption and bioavailability.

Metabolism

- Zidovudine undergoes significant glucuronidation presumable at the 5′ position and eliminated via the urine.
- The major inactive metabolite of didanosine is hypoxanthine (removal of the dideoxy sugar) followed by normal purine metabolism (i.e., to xanthine and uric acid).
- The major metabolic product of zalcitabine is dideoxyuridine, an inactive metabolite.
- The metabolism of stavudine is unknown.
- The majority of lamivudine is excreted unchanged with a small amount of the sulfoxide also being reported while tenofovir disoproxil is also primarily excreted unchanged.
- Abacavir is extensively metabolized to inactive metabolites through glucuronidation and oxidation (Fig. 27.5).

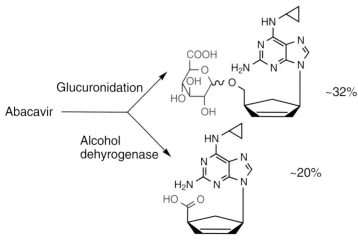

Figure 27.5 Metabolic inactivation of abacavir.

> **⚠ ADVERSE EFFECTS**
>
> **Zidovudine:** Common adverse effects include fatigue, malaise, myalgia, nausea, anorexia, headache, and insomnia. Less common effects include bone marrow suppression and skeletal muscle myopathy.
>
> **Stavudine:** Peripheral neuropathy is common especially with high doses of the drug. Peripheral neuropathy is much more likely if the drug is combined with other agents known to produce the same effect (i.e., several antituberculin drugs, and ddI). The combination of stavudine and ddI is also suspected of causing fatal pancreatitis.
>
> **Abacavir:** A fatal hypersensitivity has been reported with abacavir and should the patient be removed from the drug it should NOT be restarted only removed.
>
> **Didanosine:** Peripheral neuropathy possibly associated with mitochondrial toxicity has been reported. Acute pancreatitis which is potentially fatal is rare, but has been reported as has hepatic toxicity.

Table 27.3 Fix Combination Antiretroviral Drugs Used in the Treatment of HIV Infections

Fixed Combinations	Abacavir	Efavirenz	Elvitegravir	Emtricitabine	Lamivudine	Lopinavir	Rilpivirine	Ritonavir	Tenofovir Disoproxil Fumarate	Zidovudine
Atripla		✓		✓					✓	
Combirvir					✓					✓
Complera				✓			✓		✓	
Epzicom	✓				✓					
Kaletra						✓		✓		
Stribild			✓	✓					✓	
Trizivir	✓				✓					✓
Truvada				✓					✓	

NRTIs shown in bold.

- Emtricitabine is metabolized leading to the sulfoxide and a glucuronidation product.

Clinical Applications

- In nearly all cases the NRTIs are used in combination with other HIV drugs for the treatment of AIDS and Aids Related Complex (ARC).
- Combinations are meant to delay resistance to an individual drug.
- Approved FDA fixed combinations are shown in Table 27.3.

Nonnucleoside Reverse Transcriptase Inhibitors (NNRTIs)

The NNRTIs are represented by four chemically distinct classes of molecules as shown in Figure 27.6. Etravirine and rilpivirine are second-generation NNRTI which joins the first-generation NNRTIs, nevirapine, delavirdine, and efavirenz. Etravirine and rilpivirine have recently been approved (etravirine in 2008 and rilpivirine in 2011) for use in combination with other antiretroviral drugs in the treatment of human immunodeficiency virus type 1 (HIV-1) especially those patients who have shown resistance to other reverse transcriptase inhibitors.

Nevirapine, NVP
(Viramune)

Delavirdine, DLV
(Rescriptor)

Efavirenz, EFV
(Sustiva)

Etravirine
(Intelence)

Rilpivirine
(Edurant)

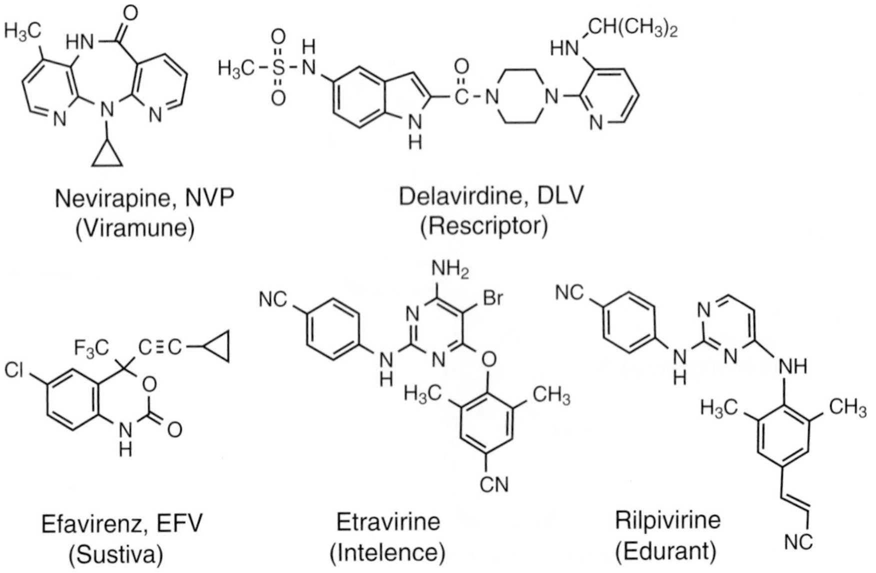

Figure 27.6 Nonnucleoside RT Inhibitors (NNRTI).

MOA

- The first-generation NNRTIs bind directly to RT and disrupt the catalytic site.
- The result of binding to RT blocks RNA- and DNA-dependent polymerase.
- Both etravirine and rilpivirine are considered to be diarylpyrimidine (DAPY) derivatives and as such demonstrate increased conformational flexibility over the first-generation agents.
- Because of the conformational flexibility, the DAPYs are able to bind to the non-nucleoside inhibitory binding-pocket (NNIBP) of native as well as drug-resistant mutant HIV-1 virus.
- The cyanovinyl functional group of rilpivirine increases binding to the RT.

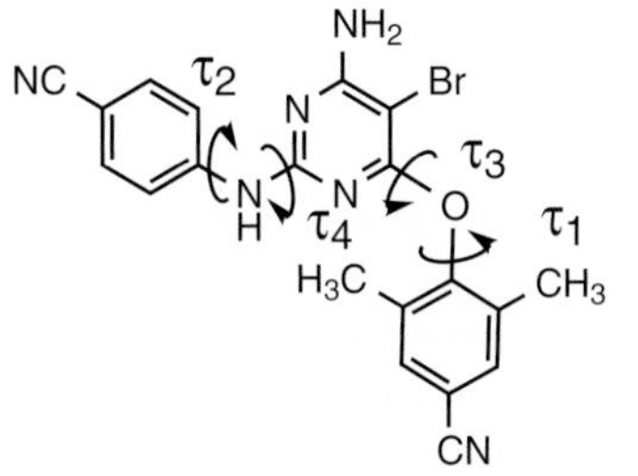

Conformational flexibility as indicated by rotation (τ) around indicated bonds.

Physicochemical and Pharmacokinetic Properties

- As indicated in Table 27.2, all NNRTIs are administered via the oral route.
- Unlike NRTIs, the NNRTIs are highly bound to plasma protein.

Metabolism

- Nevirapine is extensively oxidized by CYP3A4 and CYP2D6 leading to C2 or C12 hydroxylation or C3 or C8 hydroxylation, respectively (Fig. 27.7).
- A "quinone intermediate" is an activated intermediate which, account for the rash, hepatotoxicity, and inactivation most likely by CYP3A4.
- The major route of metabolism and inactivation of delavirdine is oxidative removal of the N-isopropyl by CYP3A4.

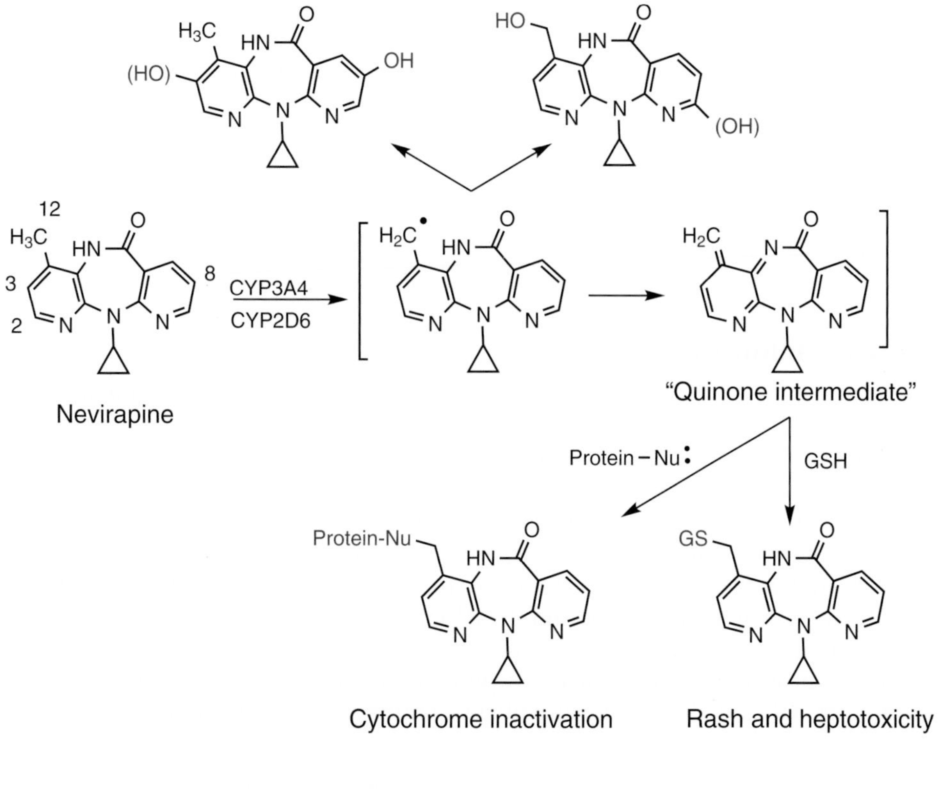

Figure 27.7 Metabolic oxidation of nevirapine by CYP3A4 leading to C2 or C12 hydroxylation or by CYP2D6 leading to C3 or C8 hydroxylation and an explanation for side effects associated with a quinone intermediate.

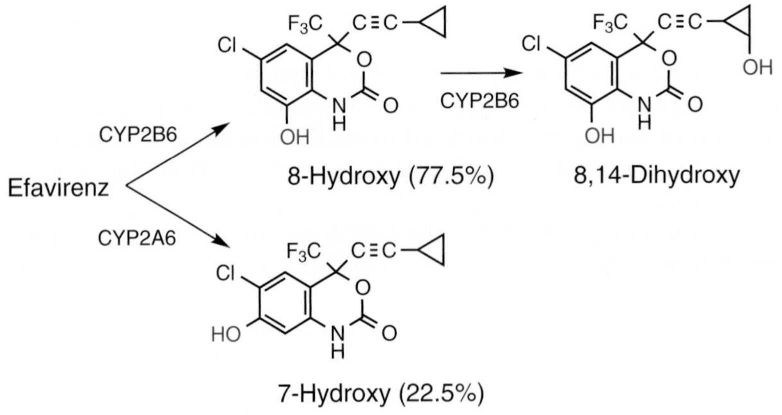

Figure 27.8 Metabolism of efavirenz.

⚠ ADVERSE EFFECTS

NNRTIs: A common adverse effect of first-generation drugs is rash, which in some cases results in patients discontinuing the drugs. Hepatic transaminase elevation has also been reported. Efavirenz is also a teratogenic in primates and women must avoid pregnancy. Second-generation NNRTIs have also been reported to cause rash. Rilpivirine's adverse effects parallel those of efavirenz and include depressive disorders, insomnia, headache, and rash. The rash with second-generation NNRTIs appears to be milder than that seen in the first-generation drugs.

- Efavirenz is metabolized to 7-hydroxyefavirenz and 8-hydroxyefavirenz by CYP2A6 and CYP2B6, respectively (Fig. 27.8).
- A secondary metabolite is the 8,14-dihydroxyefavirenz.
- Etravirine is metabolized by various CYP450 isoforms leading to monomethyl hydroxylation, dimethyl hydroxylation, and aromatic hydroxylation (Fig. 27.9).
- Etravirine induces formation of CYP3A4 suggesting the possibly for drug–drug interactions.

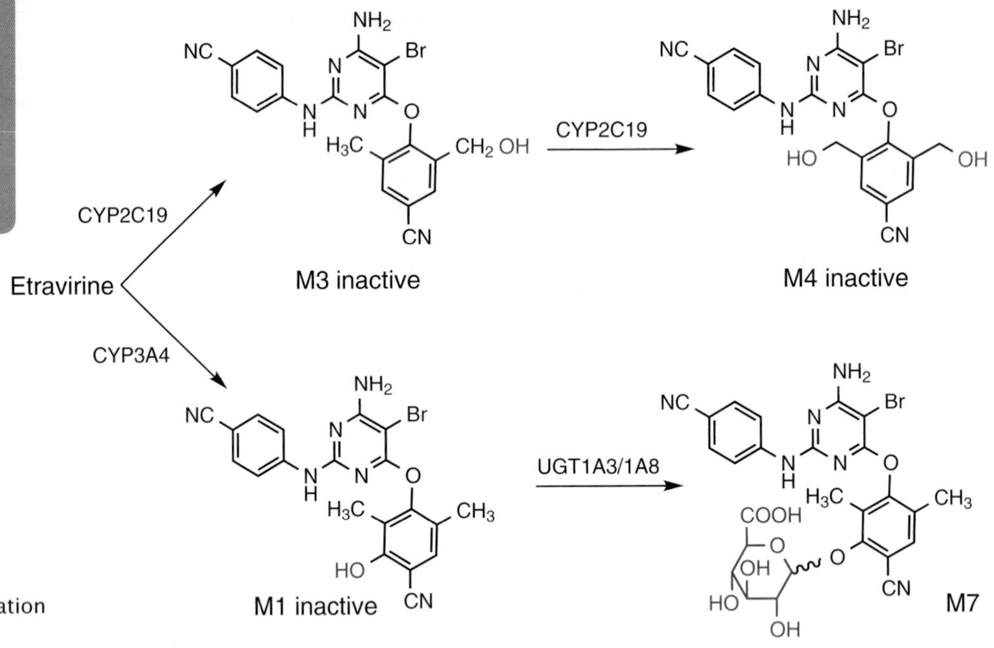

Figure 27.9 Metabolic inactivation of etravirine.

Protease Inhibitors (PIs)

The final step in the HIV-1 life cycle involves assembly of the key components of the virus including the viral RNA and structural and regulator proteins within the infected cell membrane. Budding occurs with the release of an immature virion. At this point HIV-1 protease begins the maturation process (Fig. 27.2). The HIV-1 protease is composed of a homodimer of 99 amino acids with a C-terminal sequence of Asp[25], Thr[26], and Gly[27] (aspartyl protease). The two Asp[25] residues of the homodimer protease serve as catalytic sites for the cleavage of key virus proteins (several structural proteins and the RT enzyme). It is the cleavage of these proteins that leads to the infectious nature of the HIV and thus the maturation. The HIV-1 protease prefers to cleave the virus structural and regulator proteins at the N-terminal side of a proline and phenylalanine sequence. It should be noted

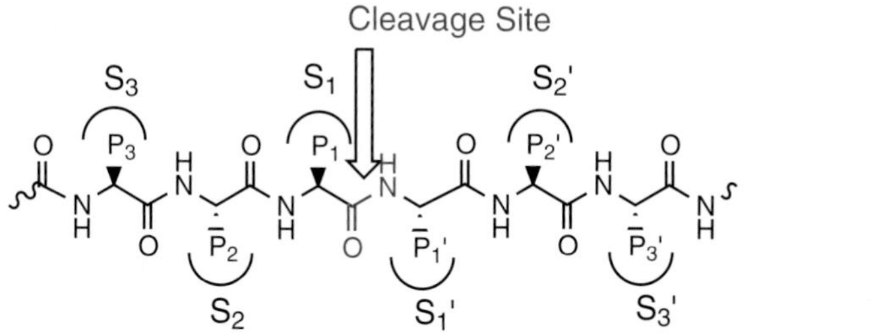

A. Nomenclature of amino acid side chains (P) and binding pockets (S) surrounding a protease cleavage site.

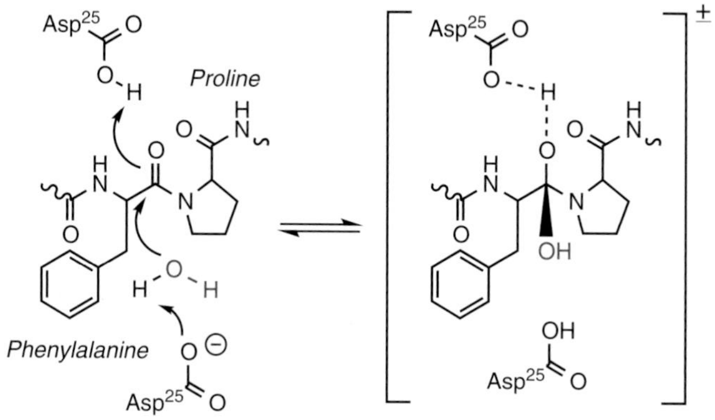

B. Role of two Asp^{25} residues in formation of the hydrolytic transition state.

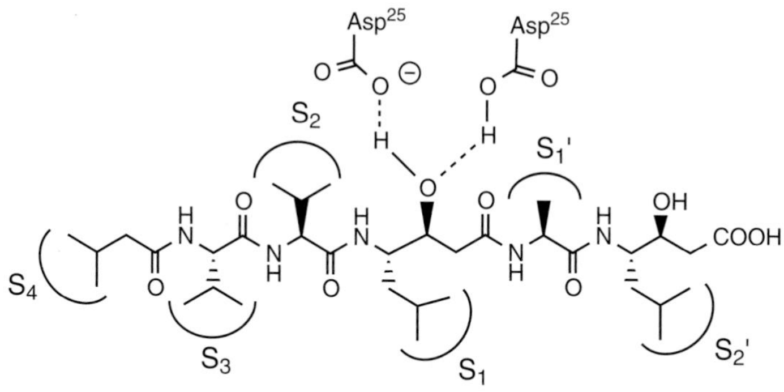

C. Coordination of the Asp^{25} residues by the transition state analog pepstatin.

Figure 27.10 Assembly of structural and regulator proteins on HIV protease (**A**), cleavage of a protein catalyzed by the protease dimer (**B**), and the occupancy of the protease by a protease inhibitor (**C**).

that human asparyl proteases have only a single aspartyl catalytic site thus allowing for drug selectivity between human and viral proteases. The assembly and cleavage of a protein is pictured in Figure 27.10A and B. While one aspartyl residue coordinates with a carbonyl of the protein being cleaved the second aspartyl reside coordinates with water which becomes the nucleophile in this hydrolysis.

MOA

- The PIs were designed to inhibit HIV protease by mimicking the transition state of normal peptide cleavage (Fig. 27.10B).
- Pepstatin, a natural occurring aspartyl protease, was a model protease inhibitor as shown in Figure 27.10C, which represents the types of bonding that may occur between the protease inhibitor and HIV protease. Note that both aspartyl residues of HIV protease bind to the alcohol functional group of the transition state mimics and that lipophilic substituents bind to pockets within the protease.

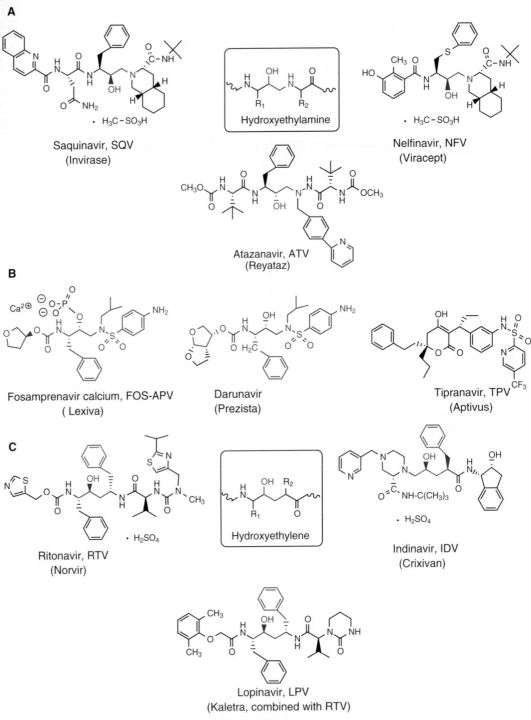

Figure 27.11 Structures of HIV protease inhibitors that are used clinically (**A–C**). In addition, **A** represents structures build upon the hydroxyethylamine motif. **B** represents the relationship of fosamprenavir to darunavir. **C** represents structures build upon the hydroxyethylene motif. Tipranavir is a structurally unique HIV protease inhibitor.

- Figure 27.11 shows the clinically available HIV protease inhibitors. The majority of the protease inhibitors fit either the hydroxyethylamine or hydroxyethylene motif (Fig. 27.11).
- With the exception of the nonpeptide nelfinavir, fosamprenavir, darunavir, and tipranavir the other HIV protease inhibitors are peptidomimetic agents.

Physicochemical and Pharmacokinetic Properties

- Saquinavir was the initial approved PI and suffered from poor bioavailability (Table 27.4).

Table 27.4 Pharmacokinetic Properties of HIV Protease Inhibitors

Drug Name	Oral Bioavailability (%)	Protein Binding (%)	Half-Life (h)	Absorption: with Meal[a]
Atazanavir	NA	86	6.5/8.6[b]	Increase
Darunavir	30–80	95	15[b]	Increase
Fosamprenavir	NA	90	7.5/15–23[b]	None
Indinavir	~65	60	1.8	Decrease
Lopinavir	NA	>98	5–6[b]	Minimal
Nelfinavir	20–80	>98	3.5–5	Increase
Ritonavir	~60	~99	3–5	Minimal
Saquinavir	13	98	1.8	None
Tipranavir	NA	~100	~6	None

[a]Variable dependent on food intake;
[b]Half-life when combined with ritonavir
NA, Not available.

- Food in the stomach generally improves absorption and area under the curve (AUC)(Table 27.4).
- All PIs are substrates for CYP3A4 (see Metabolism below).
- All PIs inhibit and several enhance CYP450 metabolism.
- All PIs are available in dry dosage forms (i.e., capsules, tablets, or powders) while ritonavir, lopinavir, and tipranavir are also available in solution. The lopinavir/ritonavir solution contains 42% alcohol with potential drug–drug interactions with disulfiram or metronidazole.

Metabolism

- As indicated above, all PIs are substrates for CYP450 enzymes. The prominent metabolizing isoform is CYP3A4 except for nelfinavir where CYP2C19 is the prominent isoform.
- Examples of protease inhibitor metabolism are shown in Figures 27.11 to 27.13.
- Darunavir, lopinavir, ritonavir, and tipranavir are strong inhibitors of CYP3A4 and dosage modification is often necessary when administered with drugs known to be 3A4 substrates or 3A4 enhancers (see Chapter 4, Tables 4.10–4.14, in *Foye's Principles of Medicinal Chemistry, Seventh Edition*).

> **DRUG–DRUG INTERACTIONS**
>
> Rifampins, clarithromycin, sulfamethoxazole/trimethoprim, and azole antifungal drugs (substrates, inducers, inhibitors of CYP) are commonly administered to AIDS patients to treat secondary infections. Dose adjustment may be necessary when used in combination with CYP3A4 PI inhibitors.

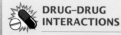

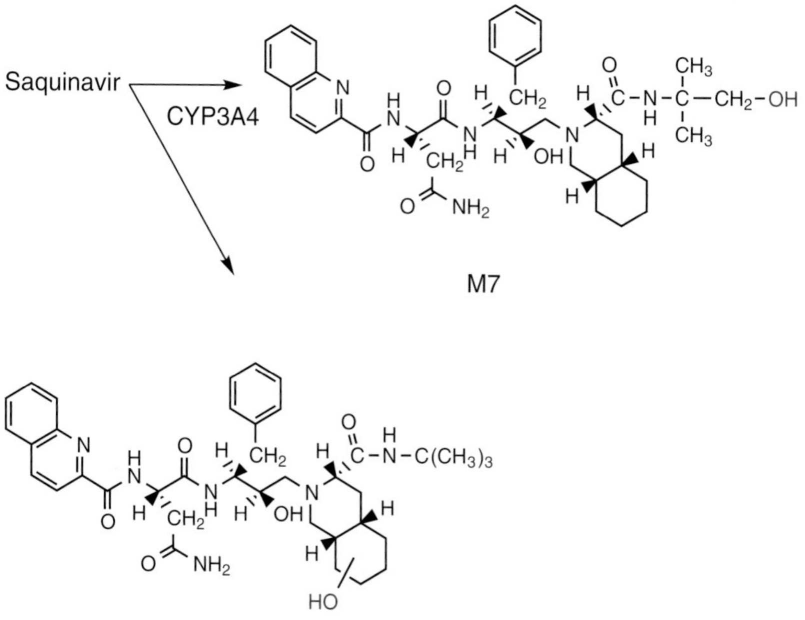

Figure 27.12 Major metabolic products from saquinavir.

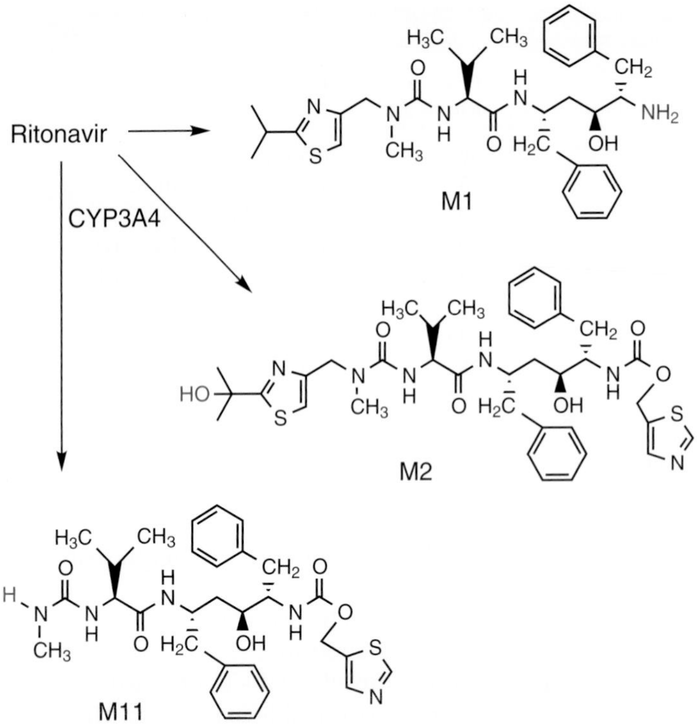

Ritonavir → M1

CYP3A4

M2

M11

Figure 27.13 Major metabolic products from CYP3A4 metabolism of ritonavir.

- Ritonavir is often administered in combination with a second protease inhibitor. Ritonavir inhibits CYP3A4 (irreversibly) allowing the second protease inhibitor to be used at a reduced dose or frequency thus reducing the potential of adverse effects and "pill burden" (i.e., saquinavir, darunavir, and tipranavir). Lopinavir is only available in a fixed combination with ritonavir (4:1 ratio) resulting in a significant concentration of lopinavir appearing in the serum (~90%).
- Nelfinavir is metabolically oxidized by CYP2C19 along with CYP3A4. It is the only PI in which of the metabolite is still active.
- Fosamprenavir is a prodrug which is activated via phosphate metabolic hydrolysis (Fig. 27.14)

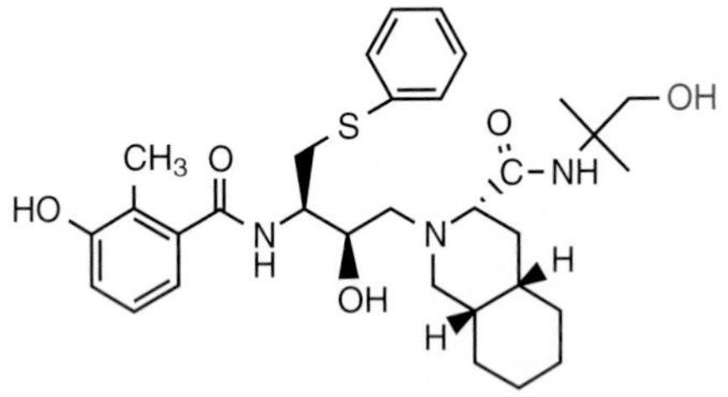

Nelfinavir's active metabolite

⚠ ADVERSE EFFECTS

Protease Inhibitors: Most of the PIs cause GI disturbances (nausea, vomiting, and diarrhea). Combination therapy may lead to a reduced dose of these drugs and reduce the degree of GI disturbance. Fosamprenavir, tipranavir, and darunavir are reported to cause skin eruption and rash. In some cases this leads to discontinuation of drug. This effect is associated with the sulfonamide moiety present in the drugs.

UNIQUE DRUG–DRUG INTERACTIONS

Atazanavir requires and acidic pH for successful absorption and should not be administered with proton pump inhibitors while patients on indinavir may experience crystalluria and nephrolithiasis because of poor water solubility of the drug. Buffering agents should not be used with indinavir. Lower pH and an increased water consumption is recommended.

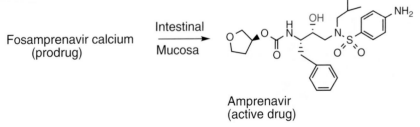

Fosamprenavir calcium (prodrug) → Intestinal Mucosa → Amprenavir (active drug)

Figure 27.14 Fosamprenavir calcium (prodrug) is activated via phosphate metabolism in the intestinal lining.

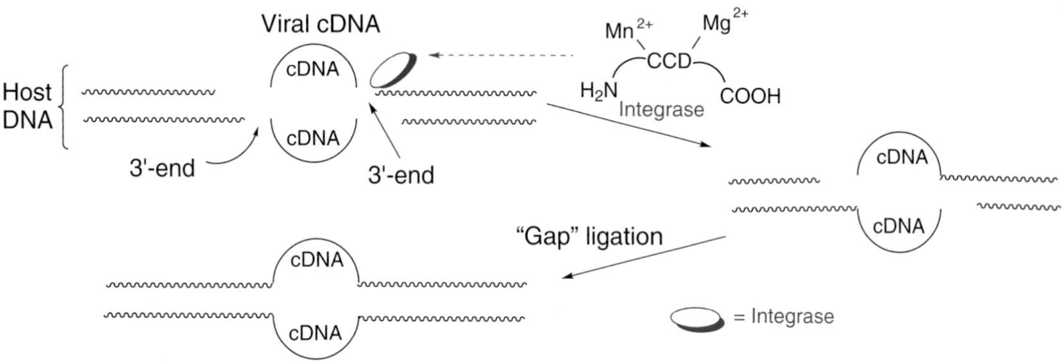

Figure 27.15 Schematic representation of the integration of viral cDNA into host DNA through the action of integrase.

HIV Integrase Inhibitors

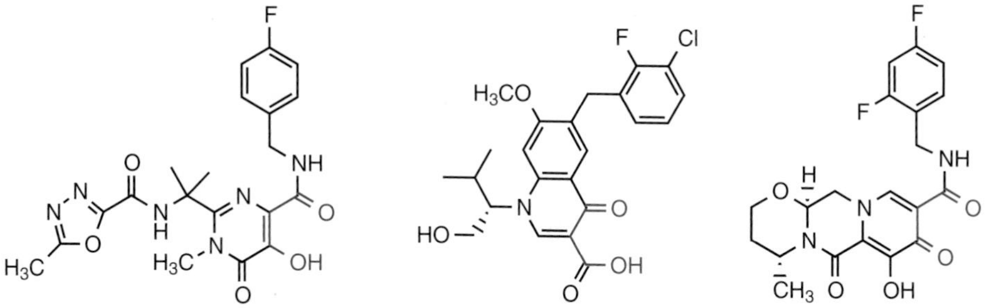

Raltegravir
(Isentress)

Elvitegravir
(Stribild)

Dolutegravir
(Tivicay)

(Pharmacophore highlighted in red)

Among the HIV particles are a complex of 40 to 100 integrase molecules the role of which is to process and integrate viral complementary DNA (cDNA) into the host cell genome. Following the action of RT and production of cDNA in the cytoplasm the cDNA must be processed by trimming of the 3′-ends of the cDNA. This cDNA along with viral and cellular proteins (preintegration complex) migrates into the nucleus where the integration into host double-stranded DNA occurs. The HIV integrase (32-kDa protein) is composed of three regions: amino-terminal domain, the catalytic core domain (CCD), and the carboxy-terminal domain. The amino-terminal domain binds to zinc while the CCD, the active portion of the integrase, binds to manganese (Mn^{2+}) and magnesium (Mg^{2+}) (Figs. 27.2 and 27.15).

MOA

- It is proposed that the integrase inhibitors act by chelation of a divalent cation at the CCD site of HIV integrase (Fig. 27.16).
- The result of binding is inhibition of cDNA integration into the host DNA.
- Similar pharmacophores, capable of metal chelation through the electron-rich oxygen atoms, are present in the integrase inhibitors.

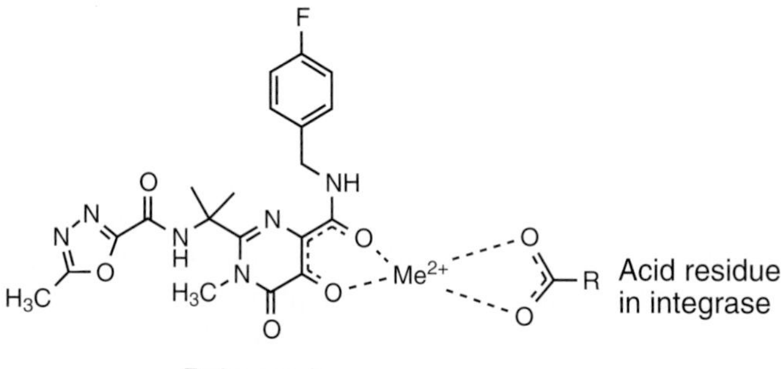

Raltegravir

Figure 27.16 Chelation complex between raltegravir and integrase.

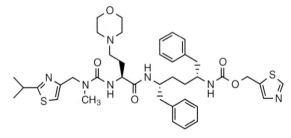

Figure 27.17 Metabolism of integrase inhibitor by UGT1A1 and CYP3A.

Raltegravir   Elvitegravir   Dolutegravir

⟹ Site of glucuronidation   ⤷ Site of hydroxylation

- Coplanarity is important between the ketone and the β-carbonyl (either a second ketone or a carboxyl oxygen) to improve chelation of the metal (probably Mg^{2+}).

Physicochemical and Pharmacokinetic Properties

- Because of the chelation properties of the integrase inhibitors these drugs should not be taken with di- and trivalent cations.

Metabolism

- Raltegravir is metabolized primarily by O-glucuronidation catalyzed by UGT1A1 (Fig. 27.17).
- A combination of raltegravir and atazanavir will greatly increase the AUC of raltegravir since atazanavir is a moderate inhibitor of UGT1A1.
- Elvitegravir is metabolized by both CYP3A4 oxidation and O-glucuronidation (Fig. 27.17).
- Elvitegravir is available in a fixed combination product (QUAD, Table 27.3). The combination includes two NRTIs (emtricitabine and tenofovir disoproxil fumarate) and cobicistat (a nonantiviral CYP3A4 inhibitor).
- Cobicistat has favorable pharmacokinetics in that it is highly selective for CYP3A4, is rapidly cleared yet produces prolonged inhibition of the CYP3A4 and inhibits CYP3A4 in an irreversible manor.

Cobicistat

- Cobicistat is a competitive inhibitor of P-glycoprotein efflux increasing overall absorption of tenofovir disoproxil fumarate.
- Dolutegravir (DTG) is reported to be metabolized by CYP3A (minor) and UGT1A1 (major) although 53% of the drug is excreted unchanged in the feces (Fig. 27.17).

Clinical Application

Antiretroviral therapy (ART) for the treatment of HIV-infected individuals varies considerably and is updated constantly based upon available drug therapy. In addition, coinfections, drug cost, men verses women, drug–drug interactions, treatment-naïve, and treatment-experienced adults and children influence the types of combination therapy called for. An overriding principle is the need to reduce the risk of disease progression with recommendations being based upon CD4 cell count. It is recommended that the reader refer to the internet for current

HIV combination guidelines (http://www.aidsifo.nih.gov/Guidelines). In general, combination therapy will be called for with drugs from various mechanism-based classes being used. As indicated in Table 27.3 various fixed combinations are available, but single drugs combined in individualized combinations are commonly used. A new fixed combination of the integrase inhibitor dolutegravir with two transcriptase inhibitors, abacavir and lamivudine is now available (Triumeq).

Adjunct Therapy

Crofelemer

Crofelemer (Fulyzaq)
(n = 1–28; R = H, OH)

Crofelemer, a nature product isolated from the red sap of the South American plant *Croton lechleri*. The herbal extract is commonly referred to as Dragon's blood and is a proanthocyanidin and has been found to alleviate the diarrhea symptoms as a delayed-release tablet for use in HIV/AIDS patients.

MOA

- Crofelemer has been shown to inhibit two mechanisms involving chloride ion (Cl^-) loss:
  - Chloride ion loss through the cystic fibrosis transmembrane regulator conductance (CFTR), a cAMP-stimulated Cl^- channel.
  - Calcium-activated Cl^- channels (CaCCs).
- The inhibition occurs at these channels in the luminal membrane of enterocytes.
- Crofelemer is selective for the Cl^- channels.
- Crofelemer acts as a concentration-dependent partial antagonist of CFTR Cl^- conductance.

Physicochemical and Pharmacokinetic Properties

- Crofelemer is administered as a mixture of oligomers with an average molecular weight of 2.3 kDa.
- Because of the size of crofelemer very little of the drug is absorbed when administered orally.

Clinical Significance

- The inhibition of CaCCs is significant in that some antiretrovirals and chemo-therapeutics cause secretory diarrhea, which crofelemer is beneficial in treating.
- Crofelemer has been shown to be effective in treating acute diarrhea caused by *E. coli* and *V. cholera* infections. These organisms have been reported to be seen in many AIDS patients.

ADVERSE EFFECTS

Integrase Inhibitors: Minimal adverse effects have been reported with raltegravir while the "quad combination" (Stribild; Table 27.3) has been reported to cause life-threatening side effects. Acid buildup in the blood can lead to lactic acidosis and serious liver toxicity has been reported. New or worsening renal impairment, decreased bone mineral density and immune reconstitution syndrome are additional warnings. Dolutegravir has been reported to produce life-threatening adverse effects, which include rash, fever, extreme tiredness, redness or swelling of eyes, and trouble breathing.

Viral Hepatitis

Viral hepatitis is presented as an inflamed liver caused by any one of the hepato-tropic viruses which include Hepatitis A (HAV), Hepatitis B (HBV), Hepatitis C (HCV), and less commonly Hepatitis D and E (HDV and HEV, respectively). The diseases can be acute with a rapid onset or chronic.

- HAV is a member of the picornavirus family and has a single-stranded RNA genome. The disease is spread commonly through contaminated raw sea food or drinking contaminated water. The patient's immune system produces antibodies which confer immunity and usually the patient recovers within 2 months. HAV causes an acute infection, which does not progress to a chronic infection.
- HBV is a member of the hepadnavirus family and causes both an acute and chronic infection. The virus has a double-stranded circular DNA genome. The disease is spread through various means including blood transfusion, tattoos, and bodily fluids including breast milk. Chronic HBV infections are reported to be responsible for 500,000 to 1.2 million deaths a year worldwide and globally the condition is the 10th leading cause of death.
- HCV is a member of the flaviviridae family and leads to a chronic hepatitis. This virus is a positive-stranded RNA virus, which is spread through contact with blood and can cross the placenta. It has been estimated that as many as 150 to 200 million people worldwide are infected with HCV. An HCV patient is also susceptible to HAV and HBV infections.

Viral Hepatitis Drug Therapy

Vaccines

- Hepatitis A vaccine (Havrix, Vaqta)
  - Sterile suspension of inactivated virus used IM.
  - Virus is grown in MRC-5 human diploid cells, lysed, and inactivated with formalin.
  - The duration of immunity is unknown
- Hepatitis B vaccine (Engerix-B, Recombivax)
  - Surface antigen gene of hepatitis B virus is cultivated genetically in *Saccharomyces cerevisiae* cultures.
  - It is a noninfectious recombinant DNA vaccine.
  - Product is a suspension of the antigen used IM.
  - Administered as three vaccinations with a duration of protection for a lifetime.
- HepA–HepB (Twinrix)
  - Combination of both Havrix and Engerix B.
  - Administered as three vaccinations similar to hepatitis B vaccine.

Nucleoside Therapy for HBV

Adefovir Dipivoxil (PMEA) (Fig. 27.18)

MOA

- PMEA is a prodrug, which is activated by esterase hydrolysis in the intestine and blood to adefovir (a phosphonic acid) and then phosphorylated to the active "diphosphate" (Fig. 27.19).
- PMEA inhibits HBV DNA polymerase and is incorporated into the viral DNA leading to DNA chain termination.
- PMEA is reported to have high affinity for HBV DNA polymerase.

Physicochemical and Pharmacokinetic Properties

- Adefovir has low oral bioavailability (~59%), but PMEA is readily absorbed and is not affected by food in the stomach.
- The adefovir is excreted without further change.

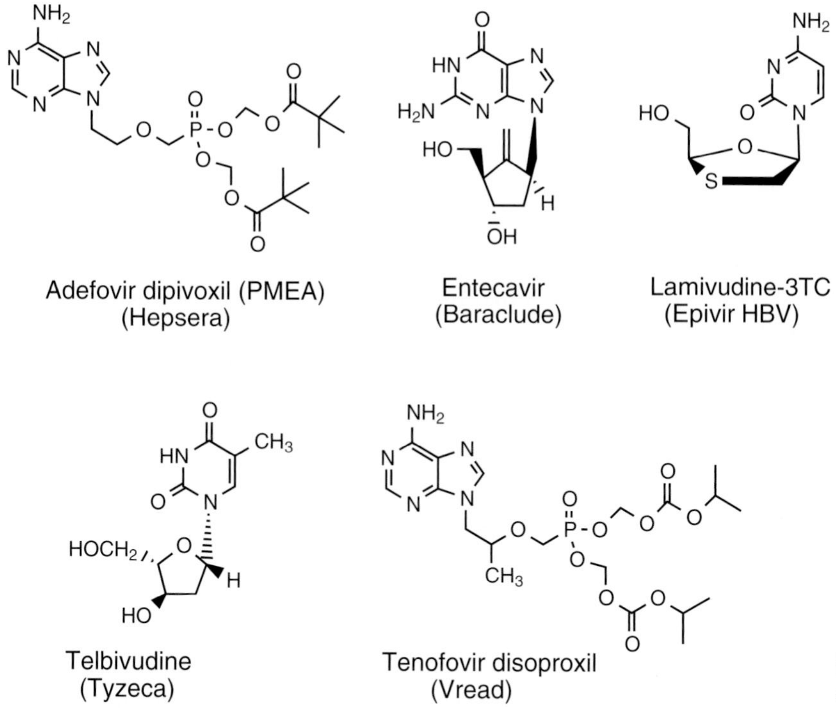

Adefovir dipivoxil (PMEA)
(Hepsera)

Entecavir
(Baraclude)

Lamivudine-3TC
(Epivir HBV)

Telbivudine
(Tyzeca)

Tenofovir disoproxil
(Vread)

Figure 27.18 Low-molecular-weight drugs used for the treatment of HBV infections.

Entecavir (Fig. 27.18)

MOA

- Entecavir is a prodrug, which is rapidly phosphorylated to the 5'-triphosphate.
- The triphosphate competes for incorporation with deoxyguanosine triphosphate into the DNA of the HBV.
- Entecavir containing DNA inhibits the protein linked priming of the HBV polymerase, RNA directed first-strand DNA synthesis or reverse transcription, and second-strand DNA-directed DNA synthesis.
- The 3'-hydroxy group of entecavir may not lead to immediate termination of DNA polymerase activity in that additional nucleotides may be incorporated into the DNA chain before the termination.
- The drug is know as a nonobligate or pseudo-terminator.

Physicochemical and Pharmacokinetic Properties

- Orally administered entecavir is rapidly absorbed with a bioavailability of near 100%.
- The drug is weakly bound to plasma protein (13%).

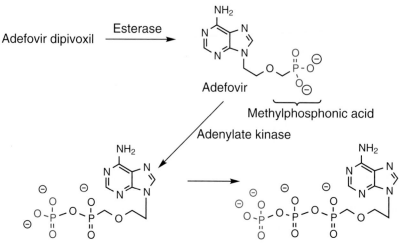

Adefovir dipivoxil → (Esterase) → Adefovir + Methylphosphonic acid → (Adenylate kinase) → Adefovir phosphate → Adefovir "diphosphate"

Figure 27.19 Activation of the prodrug adefovir dipivoxil by esterase and adenylate kinase.

- The absorption is hindered by meals and it is recommended that entecavir be administered on an empty stomach.
- Entecavir is not a substrate for CYP450 or acetylation. Only minor amounts of phase 2 glucuronated or sulfated metabolites have been found.

Telbivudine (Fig. 27.18)

MOA

- Telbivudine is the L enantiomer of thymidine and is a prodrug. The active drug is the 5'-triphosphate
- The triphosphate is a DNA polymerase inhibitor or is incorporated into the developing viral DNA and leads to termination of the polymerization.
- While the drug is reported to inhibit the first strand of DNA (RNA dependent) it appears to have a preference for the second strand of DNA synthesis (DNA dependent).
- Telbivudine is exclusively active against hepadnavirus with no activity against other viruses.

Physicochemical and Pharmacokinetic Properties

- Orally administered telbivudine is readily absorbed and rapidly achieves peak plasma levels (3 hours) with low plasma protein binding (~3.3%).
- Telbivudine is excreted via urine and feces with nearly all of the drug appearing unchanged.
- The drug is not affected by food in the stomach.

ADVERSE EFFECTS

HBV Nucleoside Therapy: Generally few adverse effects are noticed beyond headache, nausea, and diarrhea. Exacerbation of the hepatitis has occasionally occurred upon discontinuation of the therapy.

Lamivudine and Tenofovir Disoproxil (Fig. 27.18)

- As discussed earlier (see Nucleoside Reverse Transcriptase Inhibitors) these drugs are prodrugs that require metabolic activation where they act as NRTIs (Fig. 27.4).
- They also inhibit HBV DNA polymerase causing chain termination.
- The pharmacokinetic properties are reported in Table 27.2.

Interferons

The interferons are a group of glycoproteins with molecular weights of 20 to 160 kDa. These proteins are produced by various human cells with the property of inhibiting viral growth, cell multiplication, and immunomodulatory activities. Human interferons are classified as leukocyte interferon (IFN-α) produced by lymphocytes and macrophages; fibroblast interferon (IFN-β) produced by fibroblasts, epithelial cells, and macrophages; and immune interferon (IFN-γ), synthesized by CD4+, CD8+, and natural killer lymphocytes. IFN-α and IFN-β are known as type I interferons while IFN-γ is a type II interferon.

To improve the potency and the pharmacokinetic properties of the interferons they are often attached covalently to a polyethylene glycol (PEG) molecule. This complex is referred to as pegylated (PEG-IFN). Two PEGs are used to prepare the PEG-IFN: monomethyl PEG with an average molecular weight of 12 kDa or the larger branched PEG with a molecular weight of 40 kDa are combined with the interferon (i.e., PEG-IFN alpha-2b [12 kDa], PEG-IFN alfa-2a [40 kDa], respectively) improves the half-life of the interferon significantly and improves blood levels of the drug.

PEG-IFN Alfa-2a [40 kDa] for HBV (Pegasys)

- The interferon appears to stimulate effector proteins leading to a decline in DNA and polymerase activity.
- Sustained effects are less valuable in HBV than with HCV infections.
- The IFN-α is of recombinant human source and is administered via SC injection.
- Serious adverse effects are common and include: depression, myelosuppression with granulocytopenia and thrombocytopenia. The neuropsychiatric effects may be life-threatening.
- The PEG-IFN appears to be superior to IFN-α.

Clinical Applications of Antihepatitis B Drugs

- HBV is a chronic disease resulting from viral DNA being incorporated into the host chromosomal DNA.
- The goal for treatment of chronic HBV is: prevent cirrhosis of the liver, hepatic failure, and hepatocellular carcinoma. Success is observed when alanine aminotransferase (ALT) serum levels return to normal, HBV DNA levels are decreased, and hepatitis B e-antigen (HBeAg) is not detected.
- Interferons (IFN-α and IFN-β, and PEG-IFN) are administered for a defined period of time (several weeks up to 6 months).
- The nucleoside analogs are used singly or in combination to meet the treatment goals.
- PEG-IFN alfa-2a, entecavir, and tenofovir disoproxil fumarate are considered first-line agents.
- The most recent WHO Guidelines for treatment of HBV recommends treatment with entecavir, and tenofovir disoproxil fumarate for adults and children aged 12 years and older and recommends against the use of lamivudine, adefovir, or telbivudine which can lead to drug resistance (for more details see WHO Guidelines for the prevention, care, and treatment of persons with chronic hepatitis B infections March 2015).

Drugs for Hepatitis C

PEG-IFN Alfa-2a [40 kDa](Pegasys) and PEG-IFN Alfa-2b [12 kDa] (Peg-Intron, Redipen)

- As previously indicated, the IFN bind to cellular receptors leading to inhibition of transcription and translation of mRNA into viral nucleic acid and protein.
- Inhibitor proteins are produced by the host cell, which terminate viral enzymes, nucleic acid, and structural proteins.
- IFN-α–2a is a product of human recombinant DNA while IFN-α–2b is a product of *Escherichia coli* recombinant DNA technology and both are used in their pegylated form.
- These products are used SC in combination with ribavirin for treatment of chronic HCV infections.

Ribavirin

MOA

- Ribavirin is phosphorylated to a triphosphate and as such imitates either adenosine or guanosine and is incorporated into RNA.
- Base pairing with uracil or cytosine leads to mutations in RNA-dependent replication in RNA virus such as HCV (Note: its action in DNA virus may be different).

Physicochemical and Pharmacokinetic Properties

- Ribavirin is used orally in combination with injectable IFNs (see above).
- Available as capsule, tablet, and injectable solution.
- Bioavailability ~45%.
- Gastrointestinal adverse effects are minor consisting of nausea, vomiting, and diarrhea.

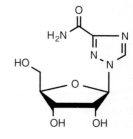

Ribavirin
(Rebetol, Copegus,
Ribasphere,
Pegintron/Rebetol)

HCV Protease Inhibitors – Nonnucleoside Inhibitors (Fig. 27.20)

HCV as a single-stranded positive sense RNA genome binds to the host ribosome where it is translated into a large polyprotein (~3,000 amino acids). This polyprotein in turn is converted, via a proteolytic process into structural proteins (required for viral capside and envelope) and nonstructural proteins (NS). The NSs proteins

NS3/NS4A protease inhibitors:

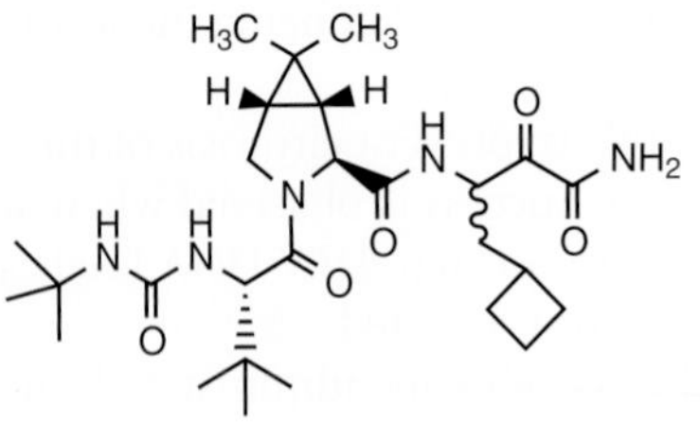

Boceprevir
(Victrelis)

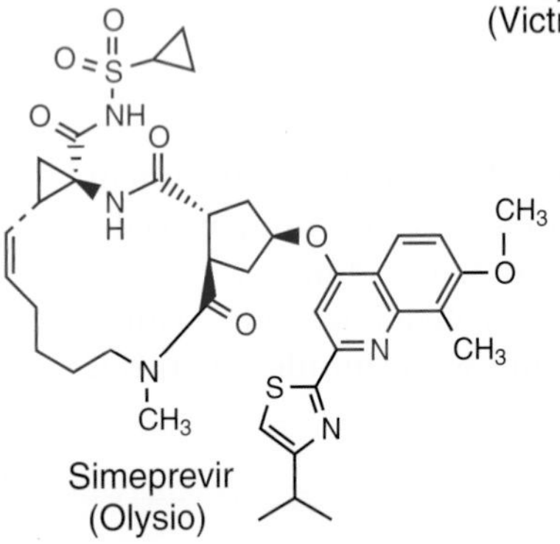

Simeprevir
(Olysio)

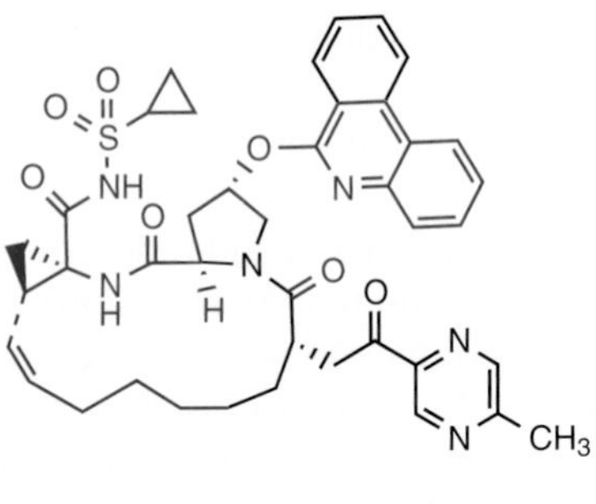

Paritaprevir
(component of Viekira Pak)

NS5A/NS5B protease inhibitors:

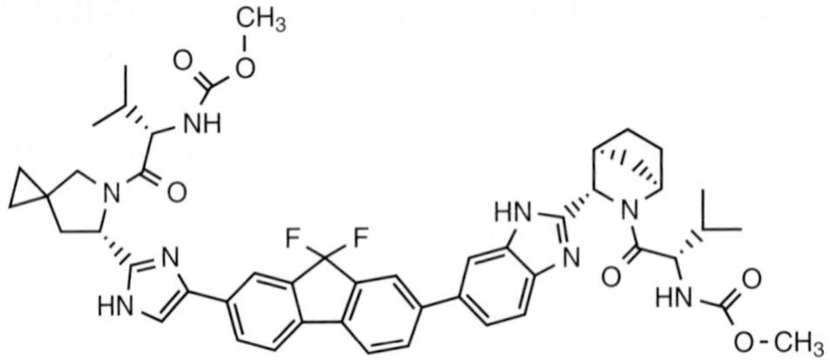

Ledipasvir
(a component of Harvoni)

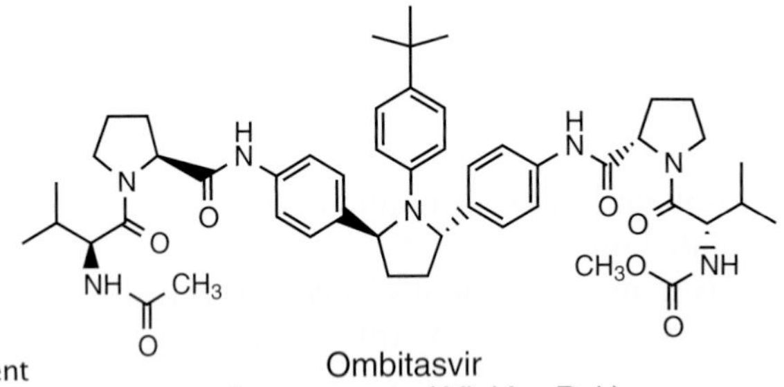

Ombitasvir
(component of Viekira Pak)

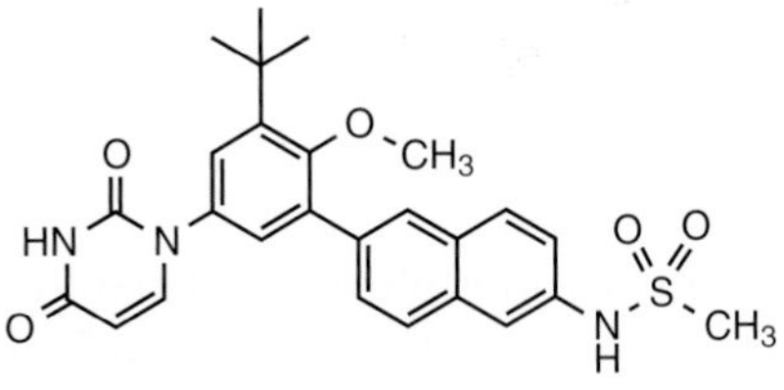

Dasabuvir
(component of Viekira Pak)

Figure 27.20 Low-molecular-weight drugs used for the treatment of HCV infections.

NS2, NS3, NS4A, NS4B, NS5A, and NS5B and are found in the endoplasmic reticulum of the infected cell. The roles for the NSs are several and involve replicating HCV RNA genome (RNA polymerase), mediating HCV resistance to interferon, inhibiting apoptosis, and possibly promoting tumorigenesis. Of these nonstructural proteins NS3, a serine proteinase; NS5A, a polymerase modulator; and NS5B, the RNA-dependent RNA polymerase, represent enzyme targets for newer HCV drugs.

MOA

- Boceprevir (α-ketoamide) and simeprevir or paritaprevir (macrocyclic molecules) represent two classes of NS3/4A inhibitors (Fig. 27.20).

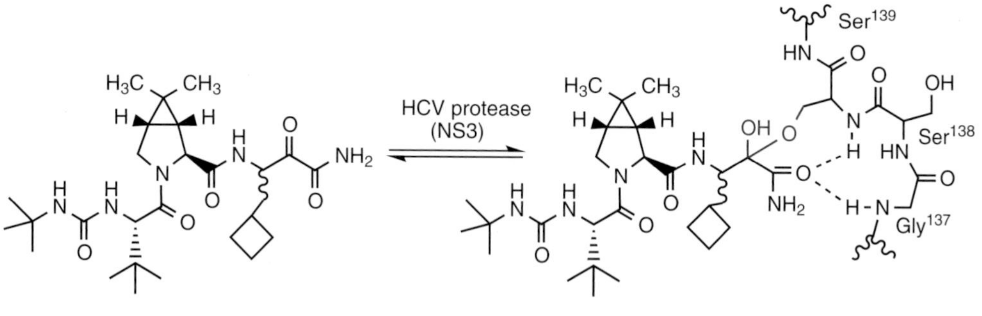

Boceprevir

Figure 27.21 Representation of boceprevir reacting with the serine OH of HCV protease. This model is based upon a X-ray co-crystal structure of telaprevir (discontinued from US market) bound to HCV NS3–4A protease complex.

- NS3 has protease and helicase activity while NS4A is a cofactor protein, which is essential for the functioning of NS3.
- Boceprevir reversibly binds to NS3/4A through a hemiketal bond formed at the α-keto of the α-keto amide group and the serine OH of NS3/4A (Fig. 27.21).
- This inhibition prevents replication of HCV RNA genome.
- Evidence suggests that simeprevir, a 14-membered macrocycle, binds to the same site as the α-ketoamides, but does so via electrostatic bonding rather than covalent binding.
- Paritaprevir, a 15-membered macrocycle, presumably binds to the same site as simeprevir due to its structural similarity (Fig. 27.20), although literature supporting this assumption has not been published at this time.
- Paritaprevir is coadministered with ritonavir (a CYP3A4 inhibitor which increases the concentration of paritaprevir—see below).
- Ledipasvir and ombitasvir are reported to act as NS5A inhibitors while dasabuvir is a NS5B inhibitor (Fig. 27.20).
- NS5A, a 447 amino acid protein, exists in a phosphorylated (p56) and a hyperphosphorylated (p58) form which are thought to play a role in RNA-dependent RNA polymerase (RdRp).
- The NS5A drugs may regulate or inhibit the p56/p58 phosphorylations.
- NS5A inhibitors result in redistribution of NS5A from the endoplasmic reticulum (ER) to lipid droplets, which correlates with inhibition of HCV replication.
- Dasabuvir (Fig. 27.20) is a selective inhibitor of the viral protein of HCV genotype 1 RNA polymerase (NS5B).
- Binding by the drug likely occurs at the palm I allosteric inhibitory site blocking the RNA chain initiation steps.

Physicochemical and Pharmacokinetic Properties

- Boceprevir exhibits moderate plasma protein binding (~68%).
- Simeprevir has good bioavailability when taken with food (especially a high fat meal).
- Simeprevir is highly bound to plasma protein (~100%).
- Paritaprevir, ombitasvir, and dasabuvir combination (Viekira Pak) are reported to exhibit high plasma protein binding (~97% to 99%).
- The combination shows increased absorption if taken with a moderate fat meal and should always be taken with a meal.
- All of the nonnucleoside protease inhibitors are eliminated in the feces.

Metabolism

- Boceprevir undergoes extensive metabolism consisting of primarily keto reduction and CYP oxidation as a minor metabolic route (Fig. 27.22).
- Simeprevir is oxidized by CYP3A4 with aromatic hydroxylation and O-demethylation of the methoxyl group.
- Paritaprevir undergoes CYP3A4/5 catalyzed metabolism leading to M2 as the major metabolite as well as hydrolysis to M29 (Fig. 27.23).
- Ledipasvir and ombitasvir are eliminated primarily unchanged.

Telaprevir Market Status

Telaprevir (Incivek) was approved by the USFDA for the treatment of HVC in May, 2011. This drug is an analog of Boceprevir with the same mechanism of action. On August 12, 2014 Vertex Pharmaceutical announced that it will discontinue its sales and distribution in the United States as a result of diminished demand for the product caused by competition from newer HCV treatments. In addition, telaprevir's adverse effects of severe skin reactions including Stevens–Johnson Syndrome and toxic epidermal necrolysis are significant problems.

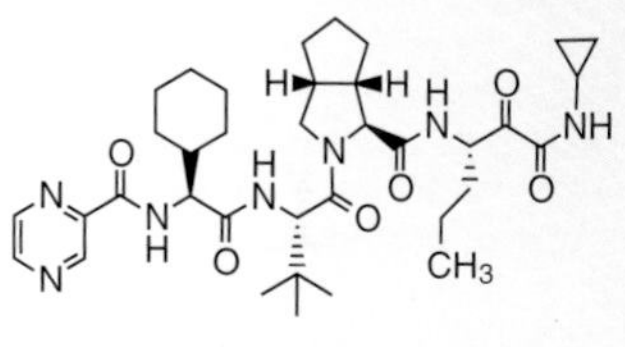

Telaprevir
(Incivek)

Boceprevir $\xrightarrow[\text{Major}]{\text{Aldo-ketoreductase}}$

Mixture of 4 isomer at
positions 2 & 3 (inactive)

Figure 27.22 Metabolic inactivation
of boceprevir.

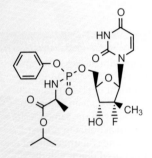

Sofosbuvir
(Sovaldi)

⚠ **ADVERSE EFFECTS**

**Interferon/Ribavirin
Therapy:** Adverse effects are
said to be almost universal with
this combination. The effects
consist of fatigue, headache,
myalgias, fever, and arthralgias.
GI effects of nausea, anorexia,
and diarrhea and psychiatric
effects of depression, irritability,
and insomnia are quite common.
Rash, pruritus, and alopecia
occur along with dyspnea and
cough are commonly reported.
Dose modification is often
needed, but since combination
therapy is quite important the
patient should be encouraged to
continue their therapy.

- Dasabuvir's metabolism involves CYP2C8 oxidation to *t*-butylhydroxylate followed by sulfate and glucuronide conjugation.

HCV Protease Inhibitors – Sofosbuvir

MOA

- Sofosbuvir is a nucleoside prodrug, which is activated by hydrolysis and phosphorylation to the triphosphate, which then inhibits viral RNA polymerase (Fig. 27.24).
- Other nucleosides have proven ineffective because of the slow rate of monophosphorylation, which is not important with sofosbuvir.
- The triphosphate is a potent inhibitor of HCV NS5B, which acts as the viral RNA-dependent RNA polymerase.
- Sofosbuvir acts as a nonobligate chain terminator.

Clinical Applications of Antihepatitis C Drugs

- HCV caused hepatitis, often undiagnosed, is the most common blood-borne pathogen and is the leading cause of death in the United States (surpassing HIV).
- Drug choices depend upon the HCV genotype with type 1 being the most common worldwide (of the 6 genotypes), but with a response rate to therapy of 42% to 52%.
- Successful therapy is measured by alanine aminotransferase (ALT) and HCV RNA levels.
- The standard of care for chronic HCV has been injectable PEG-IFN alfa-2a plus oral ribavirin (~50% effective in type 1 and ~80% effective in types 2 and 3 genotypes). This therapy commonly is used for 48 weeks.
- Improved effectiveness has been reported when boceprevir or simeprevir is added to the IFN-ribavirin combination (~80% effectiveness). Normally the triple combination is used (IFN-ribavirin for several weeks followed by addition of PI).
- Sofosbuvir has been reported to be effective in treatment of HCV as a single agent, but a fixed combination product (Harvoni – containing ledipasvir,

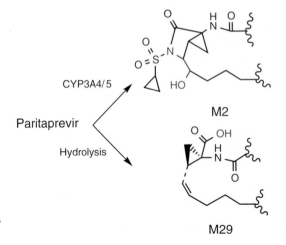

Figure 27.23 Plasma metabolites
of paritaprevir.

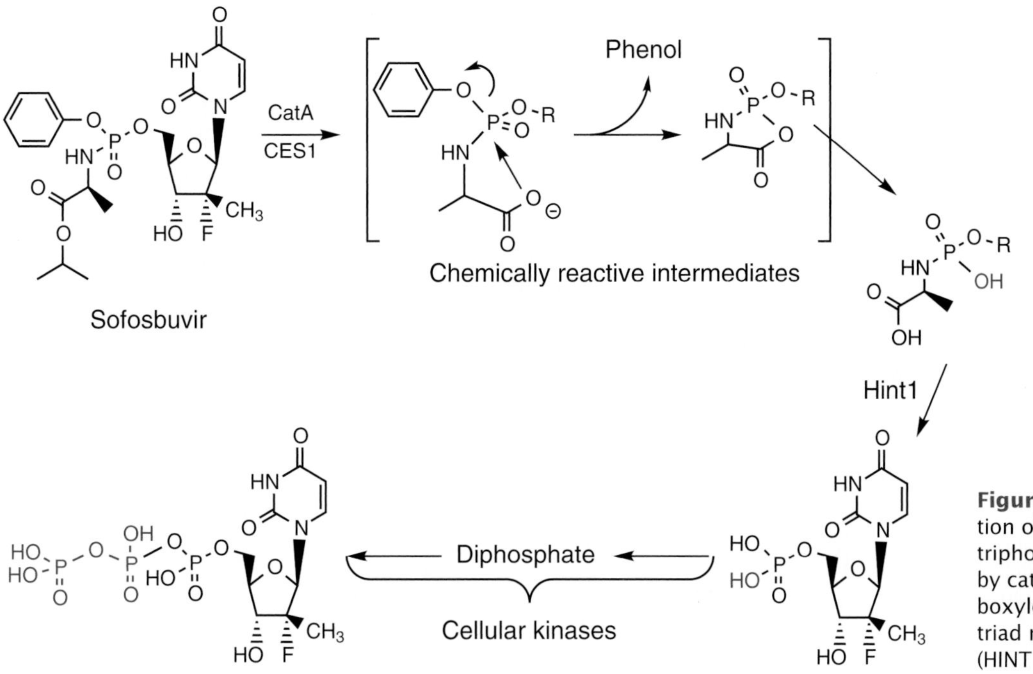

Figure 27.24 Metabolic activation of sofosbuvir to the active triphosphate; a reaction catalyzed by cathepsin A (CatA) and carboxylesterase 1 (CES1), histidine triad nucleotide-binding protein 1 (HINT1).

Fig. 27.20) has been approved by the FDA. Combination of sofosbuvir with ribavirin and/or PEG-IFN for treatment HCV is not necessary.

- Viekira Pak is a fixed dose combination of paritaprevir with ritonavir (PVT/r) and ombitasvir (OBV) copackaged with dasabuvir (DSV). This formulation is referred to as the 3-dimensional (3D) regimen (three direct acting antiviral agents).
- This combination is administered for 12 weeks and early clinical studies suggest high cure rates (93% to 100% depending on genotype HCV infection) without treatment with interferon.
- Ritonavir greatly increased peak and trough concentrations of paritaprevir thus allowing once daily dosing of this drug.

Herpesvirus

The herpesviruses are members of the Herpesviridae family which are composed of a wide spectrum of viruses causing a significant number of human diseases. Included within this family are: herpes simplex virus (HSV) 1 and 2 (causing cold sores and genital herpes, respectively), varicella-zoster virus (causing chicken-pox and shingles), cytomegalovirus (CMV) (causing mononucleosis-like condition), Epstein–Barr virus (EBV) (causing infectious mononucleosis), human herpesvirus 6 (HHV-6) and herpesvirus 7 (HHV-7), and Kaposi sarcoma-associated herpesvirus (HHV-8). The herpesviruses are linear double-stranded DNA viruses coated by a protein with the genome encoding for more than 100 different proteins. A characteristic of the herpesvirus infections is that they are quite common (>90% of the population has experienced an infection) and the pathogen establishes a lifelong persistence. These organisms target a variety of tissue including mucoepithelial, B cells, monocytes, and lymphocytes.

Drugs Used in the Treatment of Herpesvirus Infections

Nucleoside Inhibitors (Fig. 27.25)

MOA

- All of the nucleoside antiherpes virus drugs are prodrugs that must be activated by phosphorylation (Table 27.5).

ADVERSE EFFECTS

Nonnucleoside and Nucleoside HCV Therapy: The newness of these therapeutic treatments of HCV plus the use of these drugs in combination with IFN/ribavirin make it difficult to isolate adverse effects associated exclusively with any of the seven new drugs. Boceprevir appears to increase the incidence of anemia while other effects may be caused by IFN/ribavirin. Associated with simeprevir is the reported effects of rash, itching, sun sensitivity, muscle pain, and shortness of breathing. Sofosbuvir is well-tolerated with common adverse effects usually resulting from additional drugs used in combination with sofosbuvir. Viekira Pak associated adverse effects include fatigue, headache, and nausea which were reported in >70% of the patients.

Cost of Therapy for HCV

Most insurance plans provide coverage for the HCV drugs, but the amount of coverage may vary and the patient should be aware of their particular coverage before beginning treatment. The previous gold standard of treatment for HCV has been PEG-IFN/ribavirin combination used for up to 48 weeks at a cost of approximately $30,000. Boceprevir is recommended for use for 24 to 48 weeks at a dose of four tablets three times a day. Normally boceprevir is proceeded by 4 weeks of treatment with PEG-IFN/ribavirin. With the 24-week regimen of boceprevir costing approximately $66,360 plus 4 weeks of PEG-IFN/ribavirin at a cost of $2,700 the total cost could easily approach $69,000. The use of sofosbuvir, commonly used in a fixed combination with ledipasvir (Harvoni) for 12 weeks is estimated to cost $94,500, while the newest product Viekira Pak costs approximately $84,000 for 12 weeks of therapy.

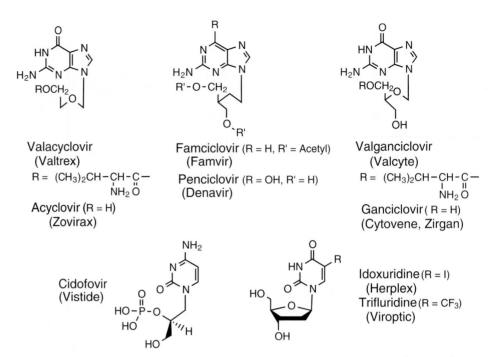

Figure 27.25 Drugs used to treat herpes viral infections which act by interfering with nucleic acid replication.

- An example of the metabolic activation is shown for valacyclovir and acyclovir in Figure 27.26.
- With one exception the triphosphates are inhibitors of viral DNA polymerase. Trifluridine triphosphate inhibits thymidine triphosphate incorporation into DNA.
- Incorporation of the false monophosphate via the triphosphate into viral DNA gives rise to a nucleotide that blocks DNA elongation and death of the pathogen.
- The diphosphate of Cidofovir is an active drug blocking DNA synthesis by acting as a substrate for DNA polymerase and serving as a competitive inhibitor of dCTP.
- Trifluridine monophosphate is an irreversible inhibitor of thymidylate synthase while the triphosphate inhibits thymidine triphosphate incorporation into DNA and is incorporated into the viral DNA.

Physicochemical and Pharmacokinetic Properties
- The pharmacokinetic properties are reported in Table 27.6.
- It should be noted that the bioavailability of oral acyclovir is significantly improved by use of valacyclovir, a prodrug to acyclovir.

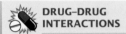

With the 3D Regimen: Drug–drug interactions between Viekira Pak drugs (OBV, PTV/r, and DSV) were minimal with the exception of a Combination with carbamazepine which resulted in a substantial decreased PVT, ritonavir, and DSV. Likewise, Gemfibrozil increased DSV exposures while coadministration with ethinyl estradiol contraceptives resulted in an elevated alanine aminotransferase level, while no change was seen with progestin-only contraceptives. Therefore the 3D regimen is contraindicated with gemfibrozil, carbamazepine, or ethinyl estradiol contraceptives.

Table 27.5 Metabolic Activation of Herpesvirus Nucleosides

Generic Name	Common Name	Intermediate	Active Drug
Valacyclovir		Acyclovir	Acyclovir triphosphate
Acyclovir			Acyclovir triphosphate
Famciclovir	FCV	Penciclovir	Penciclovir triphosphate
Penciclovir	PCV		Penciclovir triphosphate
Valganciclovir		Ganciclovir	Ganciclovir triphosphate
Ganciclovir	DHPG		Ganciclovir triphosphate
Cidofovir	CDV		Cidofovir diphosphate
Idoxuridine	IDUR		Idoxuridine triphosphate
Trifluridine	TFT	Trifluridine monophosphate[a]	Trifluridine triphosphate

[a]Active intermediate.

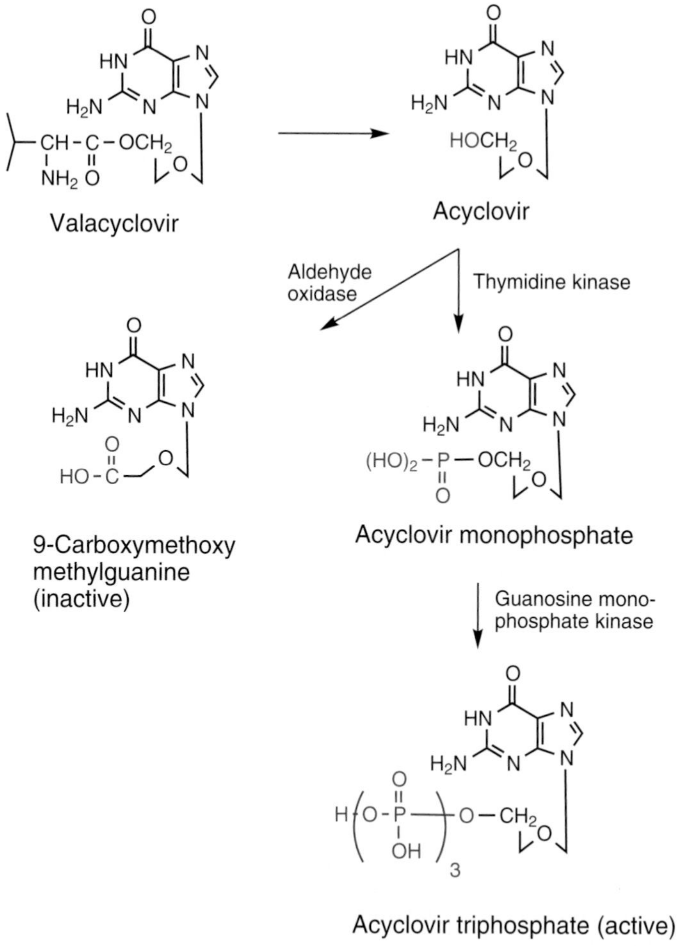

Figure 27.26 Metabolic activation of valacyclovir and acyclovir.

- Idoxuridine and trifluridine are used topically (ophthalmic use) and therefore their bioavailability and elimination routes are less important.
- Valacyclovir/acyclovir is the only drug with reported significant oxidative metabolism (aldehyde oxidase – Fig. 27.26).

Foscarnet Sodium

MOA

- Foscarnet is a selective inhibitor of viral DNA polymerase.
- The drug reversible binds to the site normally occupied by the pyrophosphate of deoxynucleotide triphosphates needed by DNA polymerase.

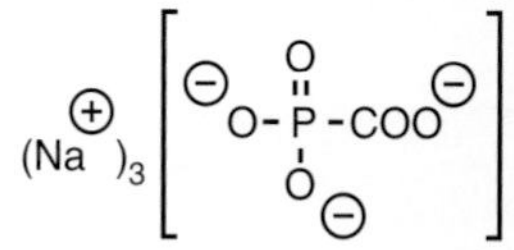

Physicochemical and Pharmacokinetic Properties

- Foscarnet is poorly absorbed orally, but is available as an IV injectable.
- The drug is distributed into vitreous fluid and the CSF as well as being sequestered in bone.

Foscarnet sodium
(Foscavir)

Table 27.6 **Pharmacokinetic Properties of Orally Administered Nucleosides**		
Generic Name	**Bioavailability (%)**	**Elimination Route/Unchanged (%)**
Valacyclovir	54	
Acyclovir	20	Urine/~70
Famciclovir	75	
Penciclovir	<5	Urine/~90
Valganciclovir	61 (>food in stomach)	
Ganciclovir	6–9	Urine/~90
Cidofovir	Poor (2–26)	Urine/>90

ADVERSE EFFECTS

Systemic Antiherpes Nucleosides: The adverse effects reported for the nucleosides vary considerably, but appear to be most significant for the systemic agents and especially for those injected. *Acyclovir:* renal insufficiency and nephrotoxicity being the most serious although reversible and possibly associated with other complicating conditions. CNS effects have also been reported including somnolence. These effects are less likely with oral *valacyclovir.* Similarly, *ganciclovir* associated myelosuppression and neutropenia are reported with IV administration. With oral *valganciclovir* diarrhea, nausea, vomiting, and abdominal pain are likely as are headache, behavioral changes, and coma. Patients on valganciclovir may experience some of the same effects as reported for the injectable product. With *cidofovir* nephrotoxicity as indicated by proteinuria and elevated creatinine as well as neutropenia have been reported. In some instances drug withdrawal is called for.

ADVERSE EFFECTS

Foscarnet: Hydration is especially important when foscarnet sodium is used in AIDS patients with CMV retinitis as indicated by a large percentage of patients experiencing renal impairment. When hydrated sufficiently those experiencing this problem is greatly reduced. Foscarnet also causes changes in serum electrolytes leading to hypocalcemia, hypo- and hyperphosphatemia, hypomagnesemia, and hypokalemia. CNS effects include seizures, tremors, and headache. Additional common adverse effects include nausea, vomiting, diarrhea, dizziness, loss of appetite, and increased sweating.

Table 27.7   Clinical Use of Antiherpesvirus Drugs

Generic Name	Spectrum of Activity	Dosage Forms
Valacyclovir	HSV-1; VZV	Tab
Acyclovir	HSV-1; HSV-2; VZV	Tab; Cap; Top; Inj; Susp
Famciclovir	HSV; VZV	Tab
Penciclovir	HSV	Top
Valganciclovir	CMV	Tab; Sol
Ganciclovir	CMV retinitis	Tab; Top; Inj
Cidofovir	CMV (in patients with AIDS)	Inj (intraocular)
Idoxuridine	HSV keratitis	Top
Trifluridine	HSV-1; HSV-2	Sol
Foscarnet Sodium	CMV retinitis; HSV; VZV	Inj

Tablets (Tab); Topical (Top - creams and ointments); Injectable (Inj); Suspension (Susp); Sol (solution).

Clinical Application of Drugs Used for Treatment of Herpesvirus Infections
- The clinical application of the herpesvirus drugs is shown in Table 27.7.
- It should be noted that there is no approved drug therapy for EBV.

Influenza

Influenza is caused by a single-stranded, negative-sense RNA genome, and is a member of the orthomyxoviridae family. Three immunologic types of influenza viruses are known and are identified as influenza A, B, and C. The virion contains nine different structural proteins. Among these proteins is the ribonucleoprotein making up the viral nucleocapsid, three proteins bound to viral ribonuclear protein involved in RNA transcription and replication. Underneath the viral lipid envelope is a shell of matrix protein. The envelope contains two viral surface glycoproteins called hemagglutinin (HA) and neuraminidase (NA). Mutations cause antigenic changes in these structural proteins and changes in HA is responsible for continual evolution of new strains. Upon replication influenza virus bind to the interior cell surface of the host though HA and sialic acid groups (N-acetylneuraminic acid). In order for the virus to be released from the infected cell NA removes sialic acid from surface glycoproteins.

Anti-Influenza Drugs
Amantadine and Rimantadine (Fig. 27.27)

MOA
- These drugs are thought to be involved in preventing uncoating of virus thus inhibiting viral replication.
- Additionally by blocking viral transmembrane ion channel protein (M2) viral replication may be effected.
- The amantadine type drugs inhibit penetration of the virus into the host cell.

Physicochemical and Pharmacokinetic Properties
- Readily absorbed following oral administration.
- Rimantadine is found in high concentrations in nasal fluid.
- Amantadine does not undergo significant metabolism while rimantadine is extensively metabolized with ring hydroxylation being the major oxidized metabolite.

Clinical Applications
- Of the two drugs amantadine's use in influenza has been significantly reduced while rimantadine continues limited use.
- Rimantadine has some value as prophylaxis treatment of influenza including: influenza A, H1N1, H2N2, and H3N2.
- Beneficial effects are associated with a reduction of the duration of symptoms.

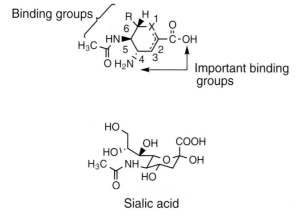

Amantadine
(generic)

Rimantadine
(Flumadine)

Zanamivir
(Relenza)

Oseltamivir phosphate
(Tamiflu)

Peramivir
(Rapivab)

Sialic acid

Figure 27.27 Drugs used in the treatment of influenza.

Neuraminidase Inhibitors (Zanamivir, Oseltamivir, and Peramivir)(Fig. 27.27)

Without NA and its ability to hydrolyze sialic acid from the HA the influenza viron cannot be released from the infected cell. The action of NA is shown in Fig. 27.28.

MOA

- Zanamivir, oseltamivir, and peramivir are transition state mimics of sialic acid and by binding to the active site of NA they inhibit this enzyme.
- Blocking NA leads to an accumulation of influenza virus at the cell surface of the infected host cell and prevents the spread of the virus.

SAR

Binding groups

Important binding groups

Sialic acid

> **ADVERSE EFFECTS**
>
> **Amantadine and Rimantadine:** It is reported that the adverse effects are much less with rimantadine than with amantadine and normal consist of minor GI effects and dose-related CNS effects. Nervousness, insomnia, dizziness, headaches, and fatigue are the most common reported.

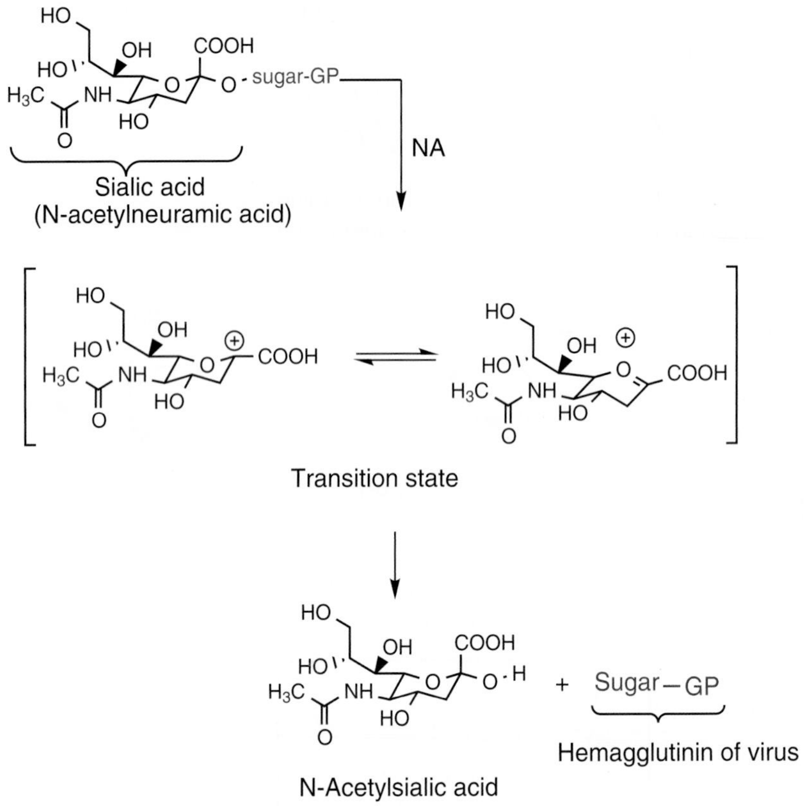

Figure 27.28 Neuraminidase-catalyzed removal of a sialic acid residue from a glycoprotein chain (NA, neuraminidase; GP, glycoprotein).

- Groups important for binding to the NA include: a carboxyl at C2, a basic group at C4 (amino or guanidine), N-acetyl at C5 and R at C6.
- Carboxyl ester at C2 is orally active, but must be hydrolyzed for activity.
- Maximum binding to NA occurs when R is a 3-pentyloxy side chain.

Physicochemical and Pharmacokinetic Properties

- Oseltamivir is a prodrug, which is orally active but must be hydrolyzed by esterase to the active drug (Fig. 27.29).
- Zanamivir and peramivir are not orally active, but zanamivir is used via oral inhalation and peramivir is given by IV.
- Zanamivir has poor bioavailability (~17%), peramivir is weakly bound to plasma protein and both drugs are excretion in urine unchanged.
- Oseltamivir is well absorbed via the oral route (absolute bioavailability of ~80%) and excreted in the urine as the carboxyoseltamivir. While a minor inactive side chain carboxylic acid has been reported in rats no addition metabolism has been reported in humans.

Clinical Applications

- The neuraminidase inhibitors zanamivir and oseltamivir are useful for the treatment and prevention of influenza A and B infections, while peramivir is indicated for the treatment of acute uncomplicated influenza in adults.

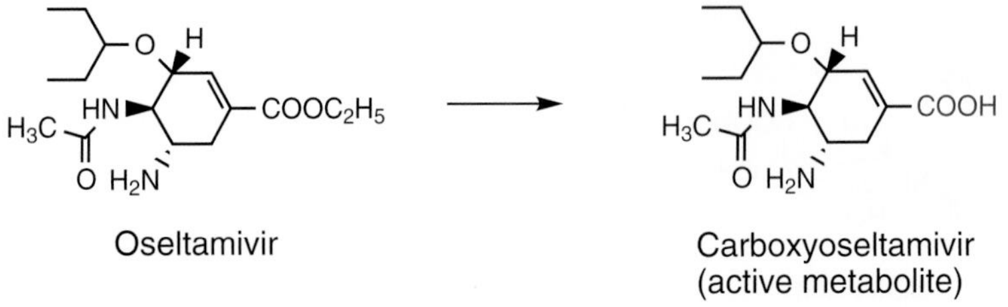

> **ADVERSE EFFECTS**
>
> **Zanamivir, Oseltamivir, and Peramivir:** In general the adverse effects of zanamivir and oseltamivir are quite mild and typically resolve themselves. On the other hand, zanamivir is not recommended for use in patients with underlying airway disease, which includes asthma or chronic obstructive pulmonary disease. Potentially hazardous bronchospasm has been reported which may require hospitalization. Peramivir may cause rare but serious skin hypersensitivity reactions.

Figure 27.29 Metabolism of oseltamivir by hydrolysis to the active drug.

Oseltamivir

Carboxyoseltamivir
(active metabolite)

- When used for prophylaxis these drugs have a high success rate of reducing the infection.
- When used in infected individuals these drugs shorten the time for recovery.

Chemical Significance

Viral infections are responsible for a high incidence of morbidity and mortality worldwide as well as in the United States. These infections are as common as the common cold or as deadly as HIV/AIDS. HIV effects up to 35 million individuals worldwide with an estimated death rate of 1.6 million, HBV is estimated to cause 1.2 million deaths/year while HCV is reported to infect as many as 160 million worldwide with 2.9 million of these individuals being US citizens. Chronic HCV can lead to cirrhosis and hepatocellular carcinoma. In the period of 1999 to 2007 chronic HCV was responsible for more deaths than HIV. Herpesvirus the cause of herpes simplex, chicken-pox, shingles, and mononucleosis is said to infect >90% of the world population with the HCV often remaining in the individual for a lifetime. And finally influenza effects millions of individuals each year and during pandemic years millions of people will die from the disease and its complications. With numbers like these, the pharmacist is likely to become the health care professional to whom the patient and their caretakers go for information.

Whether it be working with other health care providers or educating the patient an understanding of the selection of therapy to treat viral infections can utilize various aspects of medicinal chemistry. Examples include:

- The use of combination therapy for the treatment of HIV and HCV infections is based upon the selection of drugs with different mechanisms of action. The mechanisms of action arise from specific chemical structures within distinct drug classes.
- Drug metabolism is a key factor with many of the antiviral agents. Within the class of herpesvirus nucleosides there are three pairs of drugs, which actually represent only three distinct drugs (valacyclovir–acyclovir; famciclovir–penciclovir; valganciclovir–ganciclovir). The prodrugs nature and interconversion of one member of the pair to the other drug is easily recognized from the chemical structures, but not from their trade names.
- The chemical structures help identify the HIV nucleoside reverse transcriptase inhibitors and herpesvirus nucleosides as prodrugs which undergo activation to competitive inhibitors of DNA polymerase or mimic natural substrates resulting in incorporation into viral DNA leading to death of the virus.
- Ritonavir is used at a subtherapeutic dose in combination with another protease inhibitor. When used as such the ritonavir is not used for its therapeutic effects, but rather to inhibit CYP metabolism of the co-administered drug, which results in a reduced dose and side effects of the second PI.

These and many other effects can be directly traced to the chemical structures of drug molecules. When fully utilized, medicinal chemistry can serve as an additional resource available to the pharmacist for the understanding and application of clinical and pharmacokinetic properties of a drug to best serve the patient.

Review Questions

1 At which step in the life-cycle of HIV is the drug Tenofovir disoproxil active?
  A. Adsorption/fusion
  B. Reverse transcriptase
  C. Integrase
  D. Protease
  E. Transcriptase

2 Ritonavir binds to the viral protease enzyme at which of the labeled locations?

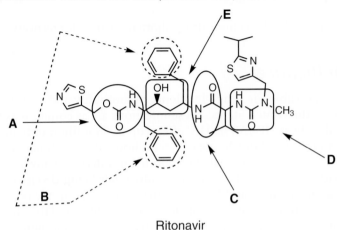

Ritonavir

3 Lopinavir is normally administered in combination with ritonavir. What is the rationale for this combination?
A. The two drugs bind to adjacent sites on the viral protease and act synergistically.
B. Ritonavir inhibits the efflux of lopinavir increasing the AUC for lopinavir.
C. Ritonavir inhibits the CYP3A4 metabolism of lopinavir which is the active drug.
D. Ritonavir has inhibiting effects on two critical steps in the virus reproduction and therefore the combination blocks sequential steps.
E. By combining the two drugs the dose of each can be reduced thus reducing adverse effects.

4 What portion of the zanamivir/oseltamivir is responsible for the drugs' ability to bind to the neuraminidase enzyme?

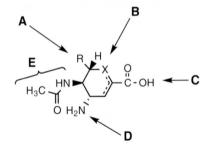

Zanamirvir/oseltamivir nucleus

A. C and D functional groups through ionic bonds.
B. B, C functional groups through chelation.
C. A, C, D, and E through a combination of bonding interactions.
D. Ion–ion bonding through a protonated ammonium ion to a carboxyl group on the enzyme.
E. All of the indicated portions of the drugs.

5 The metabolism of valacyclovir is shown below. Which of the structures actually represents the ACTIVE drug?

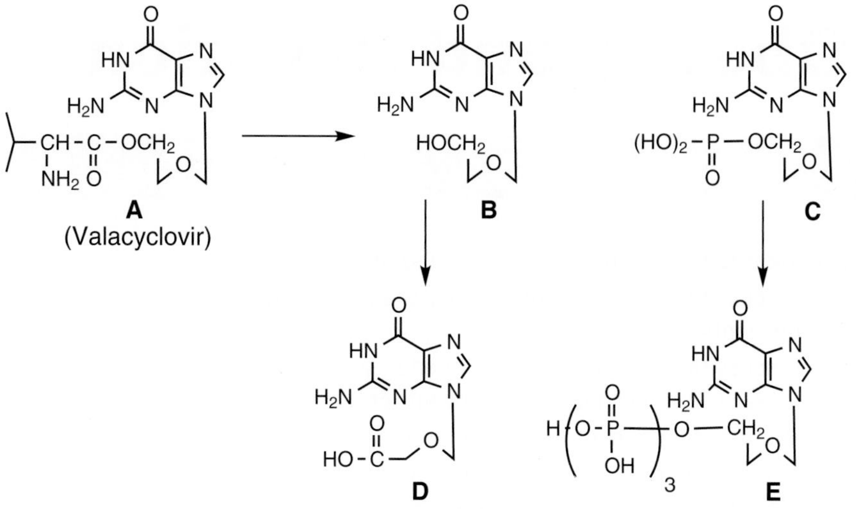

A
(Valacyclovir)

B

C

D

E

28

Obesity and Nutrition

Learning Objectives

Upon completion of this chapter the student should be able to:

1. Identify the macronutrients utilized by man.
2. Discuss the processes utilized by the body to produce energy from nutrients.
3. Discuss the units of measure of energy and state the amount of energy derived from each of the macronutrients.
4. Discuss the concept of obesity and calculate body mass index (BMI).
5. Discuss the role of each macronutrient, their terminology, structural identification, absorption, transportation, storage, and metabolism.
6. Identify and discuss the characteristics of macronutrient replacements for control of body weight (i.e., nonnutritive sweeteners, fats and fatty acid replacements).
7. Identify and discuss anorexiants used in weight control.
8. Identify and discuss the micronutrients essential for human health.

Introduction

Without a doubt, nutrition is essential since all living cells require nutrients to remain alive. The chemical ingredients, which are essential to the body are composed of both organic and inorganic chemicals and are normally referred to as the nutrients. Poor nutrition can result from both insufficient quantities of nutrients as well as over consumption of nutrients. Examples of diseases associated with low levels of nutrients consist of blindness and vitamin A, osteoporosis and vitamin D, calcium and phosphate, scurvy and vitamin C, and kwashiorkor and protein deficiencies. Far more important to U.S. consumer is the excessive intake of nutrients leading to obesity which in turn has been associated with heart diseases and hypertension, sleep apnea and respiratory problems, type 2 diabetes, and some types of cancer. It is estimated that 69% of the adult U.S. population are overweight or obese and today obesity is recognized as a chronic disease by the American Medical Association. The intent of this chapter is to identify the nutrients, both macronutrients (those required in large quantities) and micronutrients (those needed in small quantities), their role in meeting the body's needs, and in the case of macronutrients how excess intake can lead to overweight and obesity. The available drug therapy for these diseases will be presented followed by a discussion of the role of micronutrients as nutritional agents.

Macronutrients

- The macronutrients consist of carbohydrates, fats, and proteins.
- Quantities of macronutrients change with age.
- From conception until the mid teens the body undergoes considerable growth and the quantities of nutrients for good health will be high.
- From teenage throughout much of the remainder of life nutrients are expected to maintain the structure of the body.

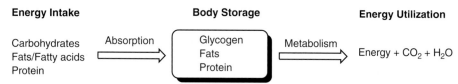

Figure 28.1 Diagrammatic representation of macronutrient absorption, storage, and utilization needed to run the body.

- During much of our adult life (20 and older) nutrition serves the roll of maintaining the body composition by replacement of essential chemicals lost due to normal turnover and to supply energy for running the body.
- The energy supplied to the body is primarily met through the macronutrients (i.e., carbohydrates and fats).
- Excessive body weight or the medical condition of obesity occurs when the intake of macronutrients exceeds the maintenance and energy needs of the body and the nutrients are put into storage.
- Carbohydrates are stored in the form of glycogen; amino acids, the absorbable form of protein are stored in the form of protein; and fats are stored as fats (triglycerides) or lipids (Fig. 28.1).

Body Energy Needs

- Carbohydrates, fats, and proteins serve as potential sources of energy.
- Oxidative breaking of carbon-carbon and carbon-hydrogen bonds convert the potential energy into the body's energy carrier adenosine triphosphate (ATP) (Fig. 28.2).
- ATP is utilized to carry out biochemical processes in the body (i.e., muscle contraction, vascular and heart contraction for blood flow, active transport of molecules and ions, biosynthesis of macromolecules, and a host of other cellular functions).
- Nearly half of the potential energy from macronutrient foods is lost in the form of heat.
- The unit of measure of potential energy is the calorie (Cal or kcal).

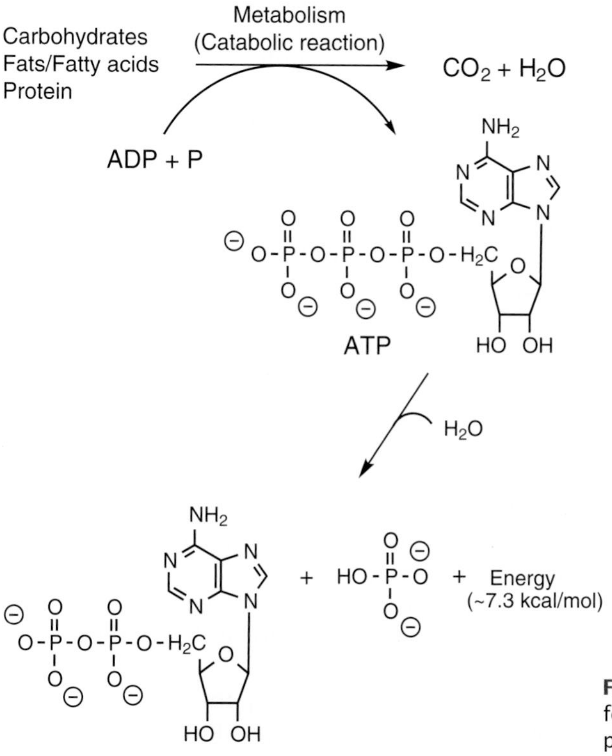

Carbohydrates
Fats/Fatty acids
Protein

Metabolism
(Catabolic reaction) $\longrightarrow$ $CO_2 + H_2O$

ADP + P

ATP

HO OH

H_2O

+ HO$-$P$-$O + Energy
(~7.3 kcal/mol)

HO OH

Figure 28.2 Conversion of foodstuff to ATP (body energy currency) followed by energy release involving ATP hydrolysis to ADP plus phosphate.

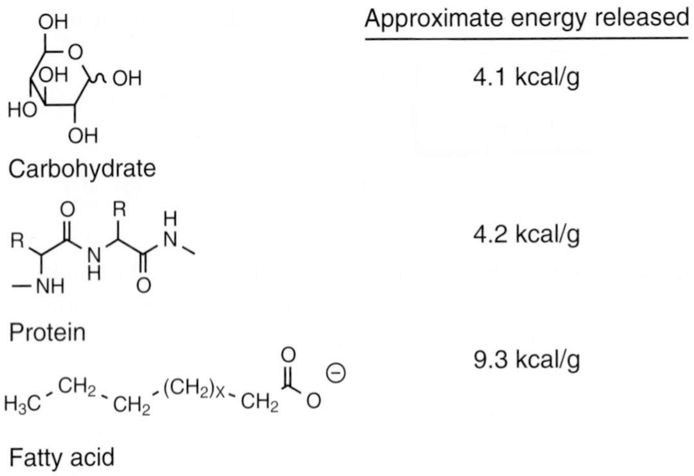

Figure 28.3 Potential energy derived from various foodstuffs.

Approximate energy released

Carbohydrate — 4.1 kcal/g

Protein — 4.2 kcal/g

Fatty acid — 9.3 kcal/g

- Calorie is defined as the amount of energy required to heat 1 kg of water 1°C from 15° to 16°C (Note: Cal [with an upper case "C"] replaces the older unit terminology cal [with a lower case "c"] which is 1/1,000 of a kcal).
- Nutritional labels report caloric content of foods in the form of calories (Cal). One may also see energy reported in joules or kilojoules (kj) where 1 kcal = 4.184 kj a standard set by International System (IS) of Units.
- The hydrolysis of ATP, a phosphoric acid anhydride, releases approximately 7.3 kcal/mol and results from the exothermic hydrolysis of an anhydride bond (Fig. 28.2).
- The potential energy present in any particular foodstuff is determined in a calorimeter. The average energy potentials are shown in Figure 28.3.
- Complete combustion of the carbon atoms in these molecules gives rise to carbon dioxide and water (NOTE: Nitrogen within the structure of proteins is not oxidized nor it is a source of energy).
- The carbons in carbohydrates are already in a partially oxidized form (C-OH) as are several of the carbons in a protein (C-NH, C = O).
- A fatty acid has a large number of carbon-carbon and carbon-hydrogen bonds that are available for metabolism and as a result fatty acids would be expected to have a significantly larger potential for generating energy.
- Generally carbohydrates are the first source of energy utilized by the body followed by fats/fatty acids, while protein is only used when other sources of energy are depleted.
- The amount of energy being generated from the macronutrients should equal the amount of energy needed to run and maintain the body.
- There are two sources of macronutrient calories:
  - Those arising from ingestion of the external foods (i.e., exogenous calories).
  - Those calories that are stored in the body (i.e., endogenous calories [Fig. 28.4]).
- An approximate measure of calories needed to maintain the body at a resting state (i.e., energy needed for cellular metabolism plus energy for blood circulation and respiration) is the basal metabolic rate (BMR).
- Combining the BMR with the energy expenditure of activity (EEA) also known as the total energy expenditure (TEE) and the diet-induced thermogenesis (DIT) can give an estimate of daily energy needs.
  - EEA is associated with lifestyle activities and can range from 400 Cal/d to several thousand based upon labor activities.
  - DIT is associated with nutrient absorption, the steps involved in metabolism, and storage of the absorbed nutrients. It is estimated to be 10% of the total amount of energy ingested.
- Combined with an estimate of calories consumed will help the individual determine their balance between calorie intake and calorie expenditure.

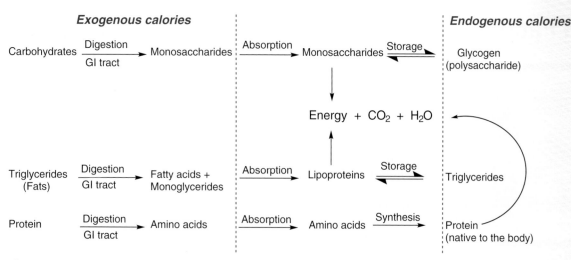

Figure 28.4 Digestion, absorption, and storage of macromolecules and conversion to energy.

- BMR calculations can be done with the formula shown in Equation 28.1.

$$\text{BMR (in Cal/d for women)} = 655 + (9.6 \times W) + (1.8 \times H) - (4.7 \times A)$$
$$\text{BMR (in Cal/d for men)} = 66 + (13.7 \times W) + (5 \times H) - (6.8 \times A) \qquad \text{[Eq. 28.1]}$$

Where W = weight in kg of the individual (1 kg = 2.2 lb), H = height in cm of the individual (1 in = 2.54 cm), A = age of the individual.

Overweight and Obesity

By definition, obesity is defined as excessive body fat while overweight is defined as excessive body weight composed of bone, muscle, fat, and water. It is generally agreed that men with more than 25% body fat and women with more than 30% body fat are obese. There are various methods for measuring body fat both directly and indirectly.

The most commonly used method of measuring body weight is the body mass index (BMI) calculation and waist circumference measurement which give a good, but indirect estimate of total body fat. The BMI is a simple calculation useful for adults and is shown in Equation 28.2.

$$\text{BMI} = \text{Weight (kg)/Height (m)}^2$$
$$\text{BMI} = \text{Weight (lb)/Height (in)}^2 \times 703 \qquad \text{[Eq. 28.2]}$$

- Individuals with a BMI of <25 are considered normal; a BMI of 25 to 29.9 are classified as overweight; and a BMI ≥30 are said to be obese. With a BMI ≥40 the person is considered morbidly obese.
- In general the BMI gives a good estimate of total body fat. The relationship of the BMI to potential health risks is shown in Table 28.1.

Table 28.1   Relationship of BMI to Health Risk

BMI Category	Health Risk	Risk Adjusted to Comorbidity
<25	Minimal	Low
25–<27	Low	Moderate
27–<30	Moderate	High
30–<35	High	Very high
35–<40	Very high	Extremely high

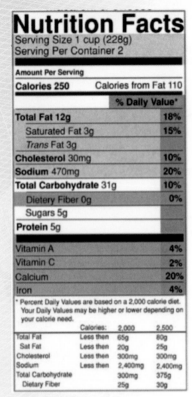

Nutrition Facts

Serving Size 1 cup (228g)
Serving Per Container 2

Amount Per Serving

Calories 250	Calories from Fat 110

	% Daily Value*
Total Fat 12g	18%
Saturated Fat 3g	15%
Trans Fat 3g	
Cholesterol 30mg	10%
Sodium 470mg	20%
Total Carbohydrate 31g	10%
Dietary Fiber 0g	0%
Sugars 5g	
Protein 5g	

Vitamin A	4%
Vitamin C	2%
Calcium	20%
Iron	4%

* Percent Daily Values are based on a 2,000 calorie diet. Your Daily Values may be higher or lower depending on your calorie need.

	Calories:	2,000	2,500
Total Fat	Less then	65g	80g
Sat Fat	Less then	20g	25g
Cholesterol	Less then	300mg	300mg
Sodium	Less then	2,400mg	2,400mg
Total Carbohydrate		300mg	375g
Dietary Fiber		25g	30g

Figure 28.5 Typical food label.

- Excessive abdominal fat can also be utilized as an indicator of health risk and this is done through the measuring of waist circumference. The waist circumference gives a good measure of abdominal fat mass.
- The circumference is measured as a horizontal plane at the iliac crest. A high-risk waist circumference is >40 in in a male and >35 in in a female.

Food Labels

- Food labels are meant to help the individual manage their caloric and nutrient intake. Figure 28.5 illustrates an example of a typical food label.
- The label indicates a "normal" serving size and the number of calories obtained from this size serving.
- The actual content of the nutrients includes macronutrient: fats, cholesterol, carbohydrates, and protein; micronutrients: minerals (sodium, calcium, and iron) and vitamins (vitamin A and C).
- The recommended calorie intake for individuals over the age of 4 years is presently set at 2,000 calories, which serves as the basis for listing the % of daily value (DV) for nutrients which is also shown on the typical label.
- The DVs are actually based upon the highest Recommended Dietary Allowance (RDA) for the age or gender groups listed for RDAs and therefore would be a high end recommendation. The RDAs are given in gram or milligram quantities representing the average daily nutrient intake level required to meet the daily needs of an individual for good health while the DV values are a percent of the total daily requirements.
- In 2014 the United States Food and Drug Administration (USFDA) announced that package nutritional labeling will change significantly in the future including: more prominent caloric counts, realistic serving size, and replacement of vitamin A and C with vitamin D and potassium.
- For information about dietary recommendations see Table 28.2.

Specific Macronutrients

Carbohydrates

- Carbohydrates are compounds composed of carbon, hydrogen, and oxygen (Fig. 28.6) and consist of:
  - Simple monosaccharides (i.e., glucose, galactose, and fructose). Glucose is the major source of energy for the CNS.

Table 28.2 Dietary Recommendations for Various Nutrients

Macronutrients	RDI[a]	Micronutrients			
		Vitamins	**RDI**	**Minerals**	**RDI**
Carbohydrates	300 g	A	5,000 IU	Sodium	2.3 g
Fiber	25 g	D	400 IU	Potassium	4.7 g
Total fat	65 g	E	30 IU	Calcium	1.0 g
Saturated fat	20 g	K	80 µg	Iron	18 mg
Protein	50 g	C	60 mg	Phosphorus	1.0 g
		Thiamine	1.2 mg	Iodine	150 µg
		Riboflavin	1.7 mg	Magnesium	400 mg
		Niacin	20 mg	Zinc	15 mg
		Pyridoxine	2 mg	Selenium	70 µg
		Folate	20 mg	Copper	2 mg
		Cobalamine	400 µg	Manganese	2 mg
		Biotin	300 µg	Chromium	120 µg
		Pantothenic acid	10 mg	Chloride	3.4 g

[a]RDAs, also referred to as the Recommended Daily Intakes (RDIs), have become part of a broader dietary guidelines known as the Dietary Reference Intake (DRIs) which is revised and updated continually and serves as the basis for determination of the % DV. Based upon the most up-to-date scientific findings the RDI may be greater than, lesser than, or equal to the older RDA numbers.

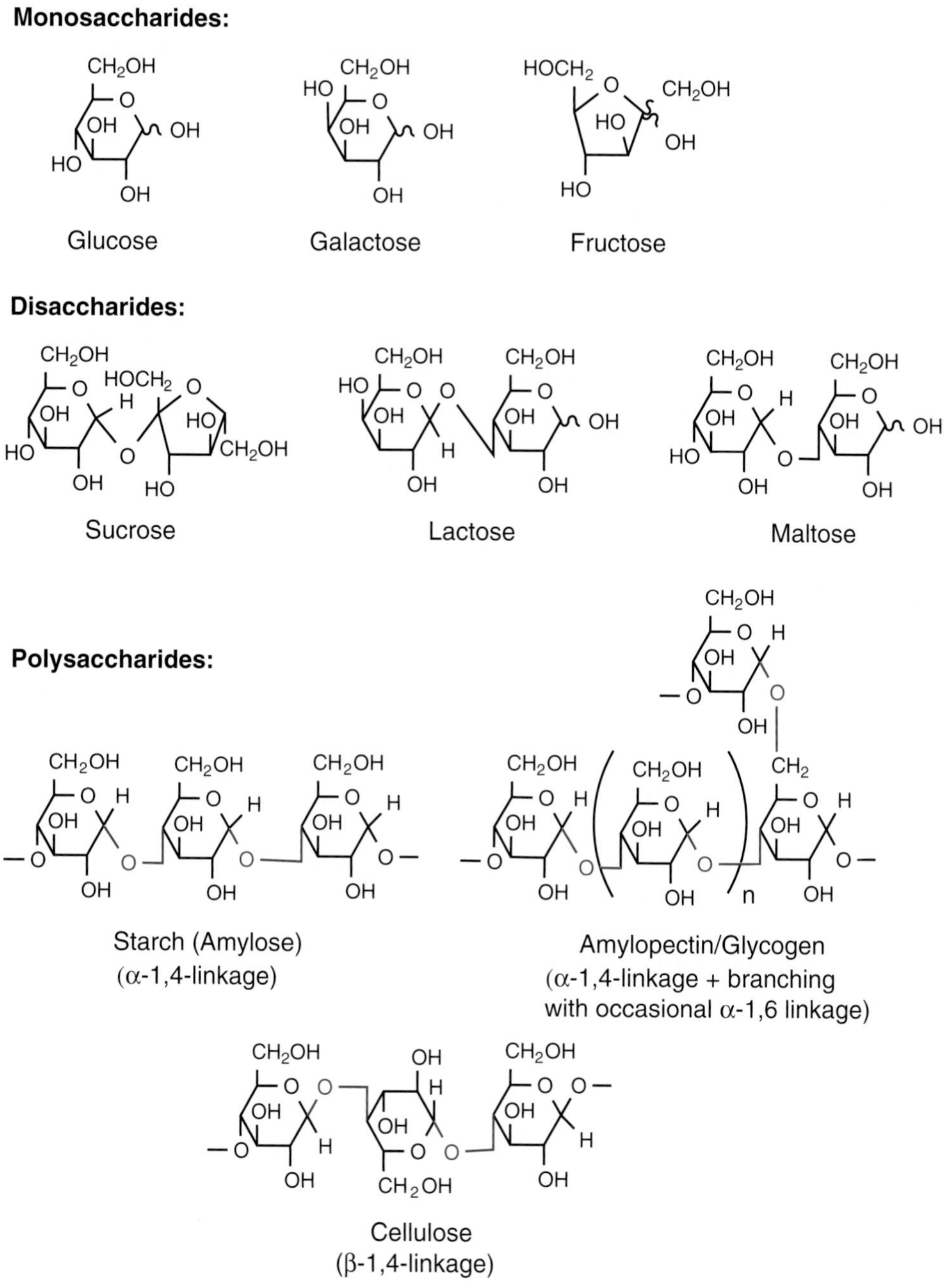

Monosaccharides:

Glucose   Galactose   Fructose

Disaccharides:

Sucrose   Lactose   Maltose

Polysaccharides:

Starch (Amylose)   Amylopectin/Glycogen
(α-1,4-linkage)   (α-1,4-linkage + branching
          with occasional α-1,6 linkage)

Cellulose
(β-1,4-linkage)

Figure 28.6 Structures of common carbohydrates.

- Disaccharides (i.e., sucrose [table sugar], maltose and lactose [milk sugar]).
- Polysaccharides (i.e., starch, glycogen, and cellulose).
- The absorbable form of a carbohydrate is the monosaccharide.
- Disaccharides and polysaccharides must first be hydrolyzed/metabolized to monosaccharides in the GI tract before absorption.

Monosaccharide Absorption

- Absorption of the very hydrophilic monosaccharides requires facilitated absorption/transport from the intestine and body fluids into the cytoplasm of cells.
- Two families of membrane proteins involved in monosaccharide transport are the glucose transporter proteins (GLUT 1–GLUT 5) and the Na^+-glucose cotransporters (SGLT 1–SGLT 3).
- GLUTs are 12 membrane-spanning proteins of approximately 500 amino acids found in red blood cells, the blood–brain barrier, liver cells, fat cells and brush borders of the intestine.

- GLUT transporters facilitate the absorption of glucose and fructose.
- SGLTs are 12 membrane-spanning proteins (unrelated to the GLUTs) exhibit high affinity for glucose following binding to sodium ion.
- SGLT 1 is primarily expressed in the small intestine and increases the absorption of glucose.
- SGLT 2 is expressed in the kidney and increases reabsorption of glucose from the kidney tubules. SGLTs also show affinity for galactose transport.

Disaccharide Metabolism

- Disaccharides require hydrolysis to monosaccharides prior to absorption.
- Enzymes present in the brush border of the intestine are involved in the hydrolytic process (i.e., sucrase-isomaltase, lactase-phlorizin hydrolase, maltase-glucoamylase, and trehalase).

Polysaccharides – Metabolized and Unmetabolized

Starch

- A polysaccharide of glucose exists as a straight chain polymer (amylose) with three or more glucose monomers linked via α-1,4-linkages and amylopectin, a highly branched polymer consisting of α-1,4-linkages together with α-1,6-linkages (Fig. 28.6).
- Amylose exists in a helix configuration.
- Starch is a major constituent of plants with various ratios of amylose to amylopectin (i.e., potato is 20% amylose and 80% amylopectin; rice is 18.5% amylose and 81.5% amylopectin).
- The hydrolytic metabolism of starch begins in the oral cavity with α-amylase, which attacks the α-1,4-linkages, and is secreted by the salivary glands and later by α-amylase secreted by the pancreas into the duodenum.
- Glucoamylase in the small intestine is also involved in starch hydrolysis. The result of the amylase metabolism is that maltose and maltotriose (a three α-1,4 glucose unit molecule) are formed.

Glycogen

- Human storage form for glucose with storage occurring in the liver and muscle tissue.
- Absorbed glucose is phosphorylated to glucose-6-phosphate and is then incorporated into glycogen through the action of glycogen synthase and glycogen branching enzyme.
- The structure of glycogen is similar to starch with both α-1,4-links and α-1,6-links (Fig. 28.6).
- Glycogen hydrolysis occurs when the body requires energy and involves a process of phosphorylation, transferase, and a debranching enzyme leading to the release of glucose into the blood stream.

Cellulose

- A major polysaccharide found in plants (Fig. 28.6).
- A linear polymer of glucose with β-1,4-linkages which increases the stability of the polymer through internal hydrogen bonding.
- This configuration prevents hydrolytic metabolism by human enzymes. The β-1,4 linkage can be hydrolyzed by cellulase found in some bacteria, protozoa, and fungi and in bovine animals.

Fiber

- Dietary fiber is made up of nondigestible oligosaccharides usually of plant origin such as the polysaccharides cellulose and pectin (polysaccharide containing 1,4-α-D-galactosyluronic acid monomeric units).
- Fiber may contain complex material found in wood and cell walls of plants such as lignin (Table 28.3). There is no USFDA definition for dietary fiber. Fibers may

Table 28.3	Classes of Dietary Fibers
	Source

Oligosaccharides

Cellulose	Bran, gegumes, peas, vegetable, cabbage, seeds, and apple covers
Hemicellulose	Bran and whole grains
Polyfructose	Oligofructans and Inulin
Gums	Oatmeal, barley, and legumes
Pectin	Apples, strawberries, and citrus fruits
Resistant starches	Ripe bananas and potatoes

Nonsaccharide

Lignin	Root vegetables, wheat, fruits, strawberries

be further classified as soluble fibers (water soluble) such as pectin or insoluble fibers such as cellulose, hemicellulose, and lignin.

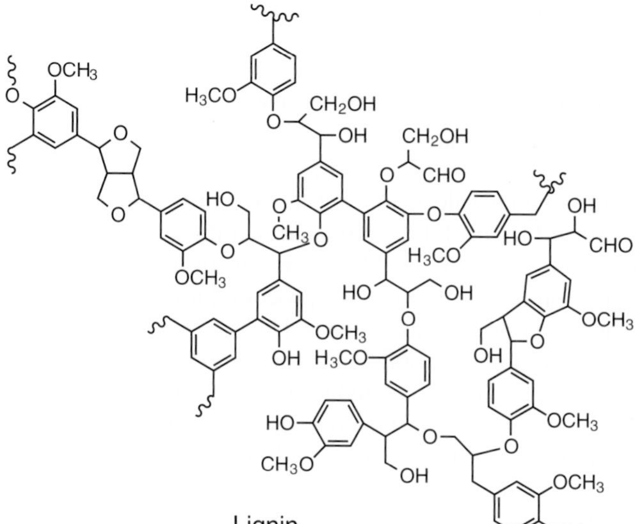

Lignin

- Soluble fibers absorb and dissolve in water forming gels, which are prone to fermentation by bacteria in the colon generating short-chain fatty acids (i.e., butyric acid).
- Insoluble fibers create bulk, which is suggested to be beneficial via increased stomach transit time and a reduction of intestinal transit time leading to a decreased urge to eat and reduction in the ability for chemical absorption to occur from the intestine.
- It is generally agreed that fiber in the diet can:
  - lower the incidence of obesity (creates a feeling of satiety), reduce insulin resistance and the risk of type 2 diabetes.
  - reduce the risk of heart disease (decrease cholesterol absorption from the intestine).
  - reduce the risk of colon cancer (decrease absorption of potential carcinogens from the gut).
- Insoluble fibers also act as bulk laxatives. The average American diet contains 12 to 15 g of fiber compared to 40 to 150 g of fiber in the diets of Africans and Indians.
- Recommendations for fiber content in the diet: 38 g/d for men under the age of 50, 30 g/d for men over the age of 50; 25 g/d for women under the age of 50 and 20 g/d over the age of 50.
- While nondigestible fibers have little or no caloric value, soluble fiber has a caloric value of 4 kcal/g.

Nonnutritive Sweeteners

Sugar is defined as a sweet tasting crystalline carbohydrate, which includes sucrose (table sugar), glucose, galactose, lactose, and fructose. Normally carbohydrates

Table 28.4 Comparison of the Sweetness of Nutritive and Nonnutritive Sweeteners

	% of Sweetness	Cal/100 g
Nutritive Sweetener		
Sucrose	100	All mono- and disaccharides ~370
Glucose	60–70	
Fructose	110–180	
Maltose	33–50	
Galactose	15–40	
Xylitol	100	~192
Nonnutritive Sweetener		
Saccharin	300 × sucrose	0
Sodium cyclamate[a]	30 × sucrose	0
Aspartame	200 × sucrose	~0.05
Neotame	7000 × sucrose	0
Acesulfame-K	200 × sucrose	0
Sucralose	600 × sucrose	0
Rebaudioside A	200 × sucrose	0
Mogrosides	300 × sucrose	<5 Cal/serving

[a]Sodium cyclamate is no longer available in the United States.

make up 40% to 70% of the human daily calorie intake (i.e., primarily starch, but also sizable quantities of sucrose and lactose). It is estimated that the average American adult consumes 64 lb of sucrose/year while children may consume as much as 140 lb/y. The majority of sugar is found in commercially processed food, mostly as high fructose corn syrup. The perception of sweetness is detected by G-protein coupled receptors (GPCRs) located in the taste buds on the tongue. Excessive intake of sucrose or high fructose corn syrup is directly related to the development of tooth caries and indirectly related to the development of overweight and obesity as well as the onset of type 2 diabetes. With the mounting concern with overweight and obesity and the need to reduce calorie intake and yet the desire to meet the craving for sweet substances, nonnutritive sweeteners have gained universal acceptance.

Saccharin (Sweet'N Low)

- Saccharin is 300-fold sweeter than sucrose with zero calories (Table 28.4).
- Saccharin has a bitter aftertaste.
- Unstable to prolonged heating.

Aspartame (Equal, NutraSweet) and Neotame

Saccharin

Aspartame   Neotame

- Aspartame a dipeptide composed of the methyl ester of two natural amino acids, L-phenylalanine and L-aspartic acid.
- 200-fold sweeter than sucrose with no bitter aftertaste (Table 28.4).
- Unstable to prolonged heating.
- Prone to hydrolytic metabolism either in the intestine or mucus membrane of the intestinal lining producing phenylalanine, aspartic acid, and methanol (Fig. 28.7).

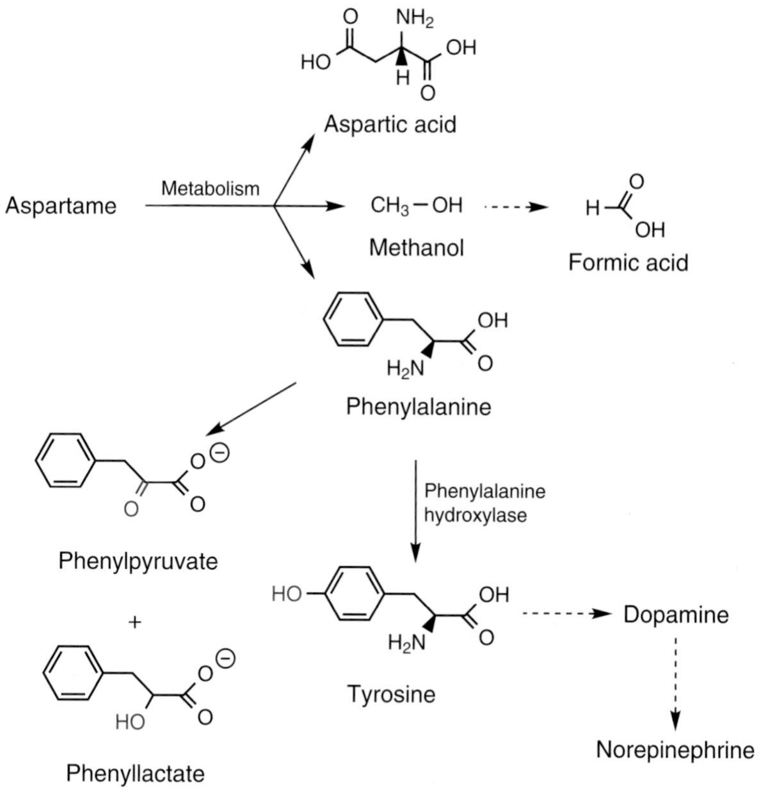

Figure 28.7 Metabolism of aspartame and its metabolites.

- Methanol, or wood alcohol, is a potential toxin when absorbed leading to formaldehyde and then formic acid. The quantity of formic acid detected is lower than that found after consuming some fruit juices.
- Aspartame containing foods contain a warning label indicating its presence and its potential for harm in phenylketonuria (PKU) patients (both pregnant women and infants) due to the metabolic products of phenylalanine, phenylpyruvate, and phenyllactate.
- In 2002, an analog of aspartame, neotame was approved for marketing in the United States.
- Neotame is 7,000- to 13,000-fold sweeter than sucrose and is used in a very low dose.
- Metabolism of neotame leads to formation of methanol, phenylalanine, and the N-dimethylbutylaspartic acid (decreased hydrolysis due to the nitrogen substitution).
- Only 20% to 30% of the neotame is absorbed presumably as the individual amino acids.

Acesulfame-K (Sunette)

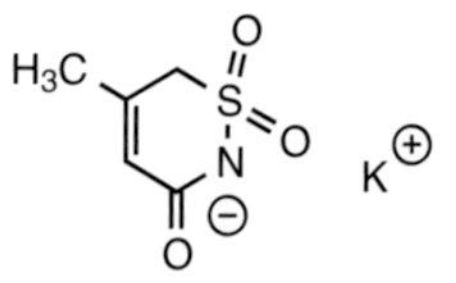

- Nonnutritive sweetener, which is ~200 times sweeter than sucrose.
- Heat stable and therefore can be used in cooking and baking (Table 28.4) and is not metabolized.
- Used in combination with other sweeteners because of its bitter aftertaste.
- Used in dry foods and as a general purpose sweetener in carbonated and noncarbonated beverages.

Acesulfame-K

Sucralose (Splenda)

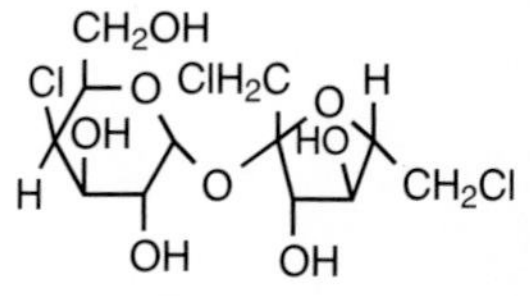

- A chlorinated synthetic derivative of sucrose resulting in a compound that is not metabolized.
- Approximately 600-fold sweeter than sucrose.
- The chloride groups prevent the metabolism of sucralose and increase stability of the chemical to heat. Sucralose can be used in baking and cooking.

Sucralose

Steviol Glycosides (Stevioside/Rebaudioside A mixture) (Truvia, Rebiana or Reb-A)

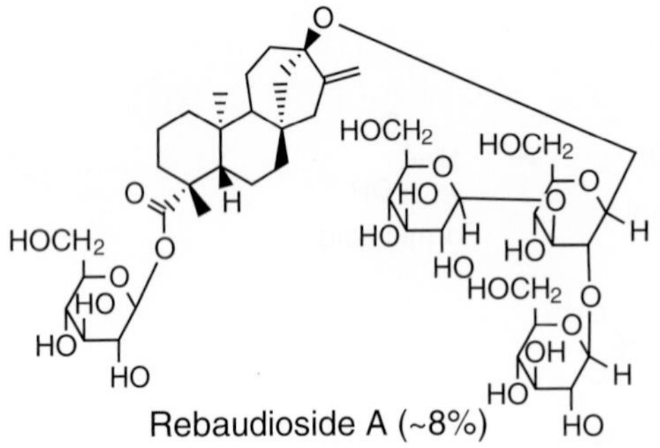

Rebaudioside A (~8%) 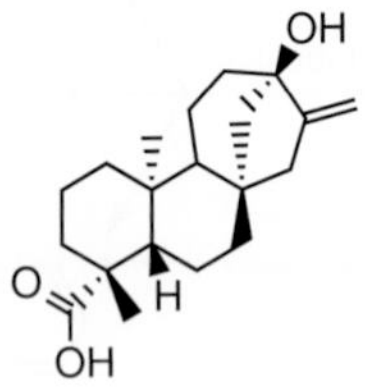   Steviol

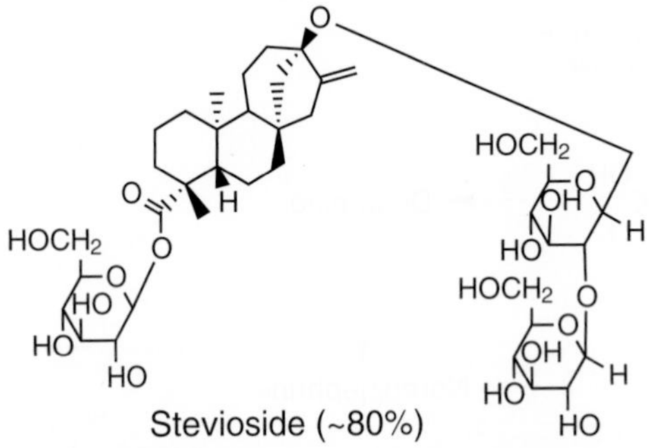

Stevioside (~80%)

- Isolated from the leaves of the stevia plant (*Stevia rebaudiana* Bertoni).
- Two major glycosides possess much of the sweet taste: stevioside and rebaudioside A.
- The diterpene glycosides possess sweetness of ~200-fold greater than sucrose (Table 28.4) with zero calories.
- The glycoside is prone to acid-catalyzed hydrolysis. In a low pH environment stevioside can undergo hydrolysis by intestinal bacterial flora leading to the formation of steviol, the aglycone component of the steviol glycosides.
- Trace amounts of stevioside can be detected in blood but no steviol is detected.
- The steviol glycosides are USFDA approved for inclusion in the Generally Recognized as Safe (GRAS) list as well as for use in foods and beverages as a nonnutritive sweetener.
- Erythritol is added as a sweetener to mask the licorice aftertaste, which appears to be associated with rebaudioside A. Erythritol itself has a caloric content of 0.2 kcal/g.

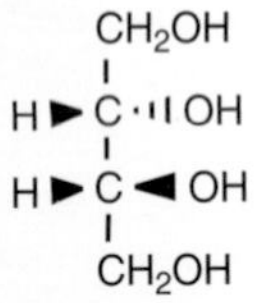

Erythritol

Mogrosides (Nectresse)

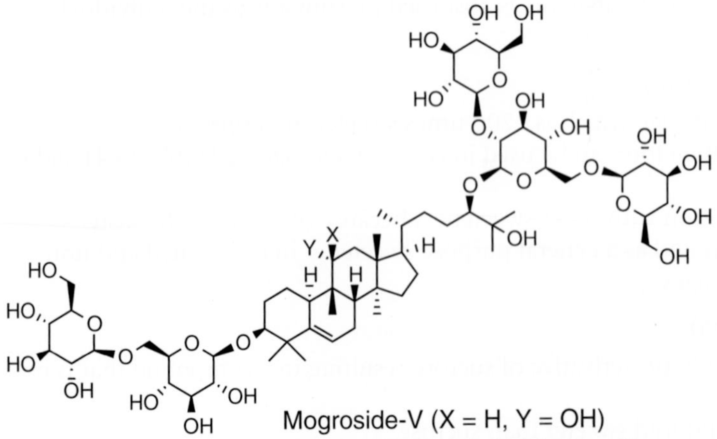

Mogroside-V (X = H, Y = OH)
11-Oxo-mogroside V (X = Y = O)
(major components of monk fruit extract)

- Isolated from the plant *Siraitia grosvenorii* which produces a fruit known as luo han guo or the monk fruit. Contains the triterpene glycosides mogroside V and 11-oxo-mogroside V.
- Approximately 300 times sweeter than sucrose.
- The fruit extract has been approved by the USFDA for inclusion on the GRAS list.
- Both ingredients have been reported to exhibit antioxidant properties and act as scavengers of reactive oxygen species (ROS).
- Nectresse contains monk fruit extract (mogrosides), erythritol, molasses, and sugar with <5 calories/serving.

Xylitol

- Xylitol is a polyol with structural similarity to glucose and erythritol.
- Commonly found in chewing gums because it does not promote the development of dental caries or plaque formation, and is not metabolized to polyhydric acids.
- Can have a laxative effect due to its poor absorption and water osmotic effect.

$$CH_2OH$$
$$HC \blacktriangleleft OH$$
$$HO \blacktriangleright CH$$
$$HC \blacktriangleleft OH$$
$$CH_2OH$$

Xylitol

Fats and Fatty Acids

Chemistry of Fats and Fatty Acids

A fat is a neutral molecule composed of three fatty acids attached to the alcohol glycerol, referred to as triglycerides, and represents a storage and transport form for fatty acids.

Glycerol

Triglyceride
(Fat)

Role of Fatty Acids

- Component of the cell wall in the form of phospholipids, glycolipids, and fatty acid acetylated proteins.
- High energy source of calories.
- Hormone and intracellular messenger.

Classification of Fatty Acids

- Saturated fatty acids of 12 to 24 carbon atoms with an even number of carbon atoms due to their biosynthesis from two carbon acetate units (Fig. 28.8).
- Monounsaturated fatty acids contain one double bond with a *cis* or *Z* configuration. Examples include palmitoleic acid (9Z-hexadecenoic acid or 9-*cis*-hexadecenoic acid) and oleic acid (9Z-octadecenoic acid or 9-*cis*-octadecenoic acid).
- Polyunsaturated fatty acids (PUFA) contain multiple double bonds with a *cis* or *Z* configuration. Examples include linoleic and linolenic acid (9Z,12Z-octadecadienoic acid and 9Z,12Z,15Z-octadecatrienoic acid, respectively).
- Additional fatty acids of nutritional significance are arachidonic acid (C20:4 *Z* or *cis* double bonds), eicosapentaenoic acid (EPA) (C20:5 *Z* or *cis* double bonds), and docosahexaenoic acid (DHA) (C22 with 6 *Z* or *cis* double bonds) (Fig. 28.8).
- Linolenic, EPA, and DHA constitute the ω-3 fatty acids (double bond three carbon atoms from the end of the fatty acid chain).

Physical–Chemical Properties

- Saturated fatty acids tend to be low melting solids while the unsaturated and PUFAs are liquids (Table 28.5).

Saturated Fatty Acids:

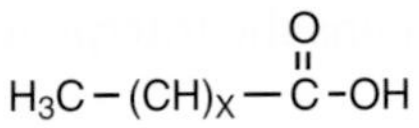

$$H_3C-(CH)_x-\overset{\overset{O}{\|}}{C}-OH$$

x	Common name	Official name
10	Lauric acid	n-Dodecanoic acid
12	Myristic acid	n-Tetradecanoic acid
14	Palmitic acid	n-Hexadecanoic acid
16	Stearic acid	n-Octadecanoic acid
18	Arachidic acid	n-Eicosanoic acid
20	Behenic acid	n-Docosanoic acid
22	Lignoceric acid	n-Tetracosanoic acid

Unsaturated Fatty Acids:

Palmitoleic acid

Oleic acid

Linoleic acid

Arachidonic acid

Linolenic acid

Eicosapentaenoic acid (EPA)

Docosahexaenoic acid (DHA)

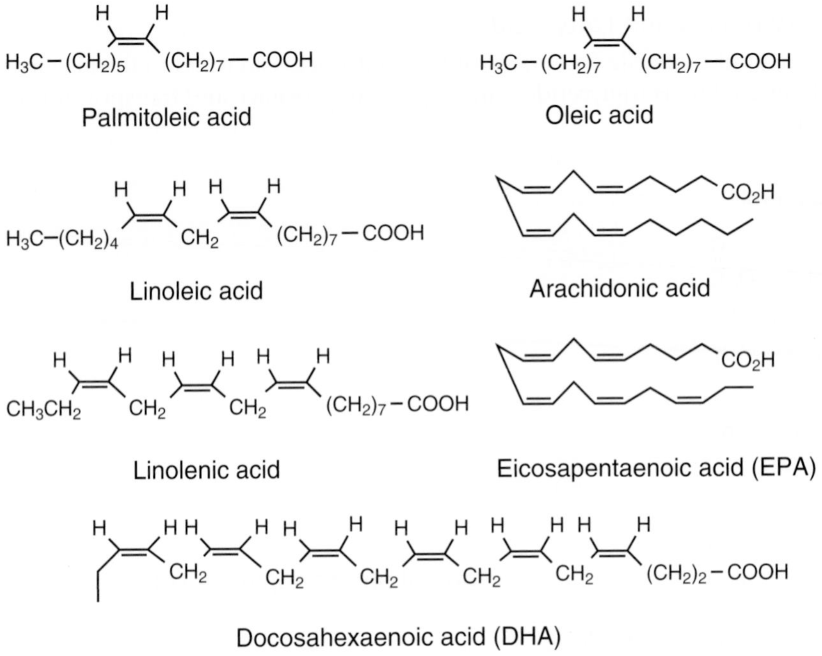

Figure 28.8 Structures of saturated, monounsaturated, and polyunsaturated fatty acids.

Table 28.5 Melting Points of Saturated and Unsaturated Fatty Acids

$$H_3C-(CH)_x-\overset{\overset{O}{\|}}{C}-OH \qquad H_3C-(CH_2)_y\left(CH_2\right)_x(CH_2)_w-COOH$$

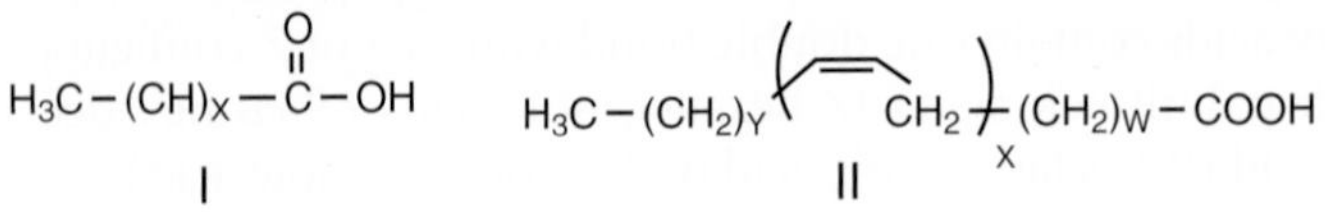

Common Name	X			mp (°C)
Saturated (I)				
Lauric acid	10			44
Myristic acid	12			58
Palmitic acid	14			63
Stearic acid	16			70
Unsaturated (II)	w	x	y	
Palmitoleic acid	6	1	5	−1
Oleic acid	6	1	7	4
Linoleic acid	6	2	4	−5
Linolenic acid	6	3	1	−12
Arachidonic acid	2	5	1	−49.5

Table 28.6 Approximate Percent of Fatty Acids in Various Food[a]

Food	Saturated Fatty Acids (%)	Monounsaturated Fatty Acids (%)	Polyunsaturated Fatty Acids (%)
Butter fat	60	36	4
Pork fat	59	39	2
Beef	53	44	2
Chicken fat	39	44	21
Margarines[a]	20	40	40
Salmon oil	24	39	36
Corn oil	15	31	53
Soybean oil	14	24	53
Olive oil	14	75	11
Sunflower	10	22	66
Canola	7	61	22

[a]Fatty acid content shows considerable variations between percentage of saturated fatty acids (palmitic acid and stearic acid), monounsaturated fatty acid (oleic acid), and polyunsaturated fatty acid (linoleic acid) as well as *cis/trans* ratio in various brand named margarines.

- Saturated fatty acids tend to be found in higher concentrations in animal fats while the unsaturated and PUFAs are found in high concentrations in vegetable oils (Table 28.6).
- Saturated fatty acids generally have a stable shelf life and are stable at elevated temperatures to air oxidation.
- Unsaturation within a fat is susceptible to air oxidation especially at elevated temperature (baking and frying temperatures).
- Oxidative reactions lead to rancidity, darkening in color, increased viscosity, and polymer formation.
- Reducing the concentration of the PUFA is through partial hydrogenation (i.e., hydrogenated oils) improves shelf life by increasing stability to oxidation.
- Catalytic hydrogenation involves a reversible addition of hydrogen and catalyst to the double bond followed by the irreversible addition of a second hydrogen to the initial double bond (Fig. 28.9). The reversibility of the initial step in the hydrogenation leads to formation of a new alkene with the more stable *trans* or *E* stereochemistry. Thus formation of a *trans* fatty acid (TFA) impurity.

> **⚠ ADVERSE EFFECTS**
>
> ***Trans* Fatty Acids (TFAs):** TFA consumption is reported to increases the levels of low-density lipoprotein (LDL) cholesterol, reduces the level of high-density lipoprotein (HDL) cholesterol, and increases the ratio of total cholesterol to HDL cholesterol, measures that are associated with increased risk of coronary heart disease (CHD). In addition, TFAs lead to changes in triglyceride levels, promotion of inflammatory reactions, and endothelial dysfunction. *Trans* fats are those in which one or more TFA is attached to the glycerin backbone (these normally are not present in natural foods). Commercially deep-fried foods, bakery products, packaged snack foods, margarines, and crackers prepared with *trans* fat containing oils may increase the account of *trans* fats in our diet leading to 2% to 3% of the total caloric intake being TFAs. Legislation outlawing the use of *trans* fat containing oils is being developed. The USFDA allows the reporting of zero grams of *trans* fats/serving if the food contains <500 mg/serving.

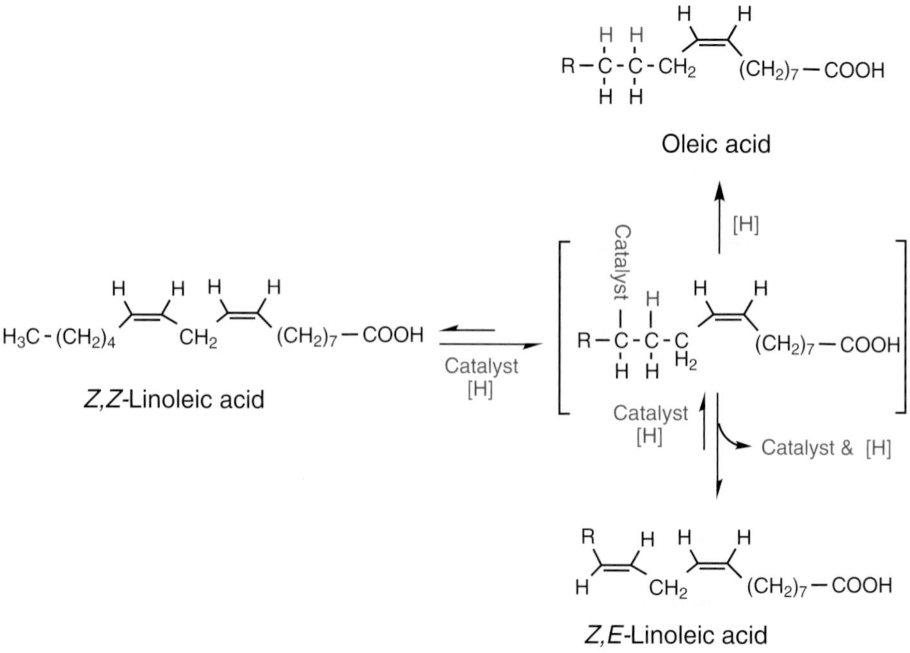

Figure 28.9 Catalytic hydrogenation of a polyunsaturated fatty acid to a monounsaturated fatty acid.

Nutritional Properties of Fats

Approximately 35% to 40% of the caloric intake in the United States comes in the form of triglycerides (>90%), free fatty acids, cholesterol, and phospholipids. This amounts to 100 to 150 g/d of triglycerides with a recommended dietary intake (RDI) for fat (based upon a 2,000 Cal diet). Fats are a major source of body energy (80% to 90%), fatty acids are an endogenous source of cholesterol via acetyl CoA units, and cholesterol is the precursor to hormones of the body (i.e., bile salts, vitamin D, sex hormones, and adrenocortical hormones). Phospholipids and unsaturated fatty acids are component of the cell membrane and subcutaneous fats give the body thermal insulation.

⚠ ADVERSE EFFECTS

Fats: Evidence points to dietary fats as a major contributor to:
1. coronary heart disease (CHD) (i.e., ischemic heart disease, myocardial infarction, and atherosclerosis, angina pectoris).
2. colorectal cancer.
3. insulin resistance leading to type II diabetes.
4. although debatable, decreased insulin sensitivity.
5. obesity when dietary fats make up a high level of caloric intake.

Fat and Fatty Acid Absorption

- Fats and fatty acids are high density sources of energy following absorption, transportation via the blood stream, and deposition in cells for immediate energy or to storage sites in adipose cells for future energy needs.
- Because of their low hydrophilicity a sequence of hydrolysis and ester formation occurs during the stages in the absorption–transport–storage–utilization processes.
- Fats travel to the small intestine where they are emulsified with the aid of bile acids and are then hydrolyzed through the action of pancreatic lipase.
- The resulting fatty acids and glycerol are absorbed into the intestinal mucosa.
- The fatty acid is converted into chylomicrons (lipoprotein particles consisting of 85% to 92% triglycerides) and transported to storage sites in adipocytes or muscle cells for utilization as an energy source.
- Transport into adipocytes or muscle cells required fat hydrolysis by lipases, absorbed across the cell membrane, and reesterified to the triglycerol in adipocytes or metabolized in the mitochondria of the muscle cell (myocytes), respectively.
- Mobilization of fats from adipocytes is a G-protein innervated process involving β-adrenergic receptors and various hormones (i.e., glucagon, epinephrine, or β-corticotropin).
- Transport to myocytes involves blood albumin and ultimately passive absorption into the myocytes.
- Transport into the mitochondria from the cytoplasm requires esterification to a fatty acyl-CoA, transesterification to fatty acylcarnitine, and transesterification back to fatty acyl-CoA followed by metabolism in the Krebs cycle to carbon dioxide, water, and ATP.

Fats and Fatty Acid Substitution

Skimmed Milk

- Skimmed milk is a dairy product in which nearly all of the butterfat is removed (whole milk contains approximately 3.5 g of a mixture of fats in 100 g of fluid).
- The fatty acid content of whole milk consists of saturated fatty acids: palmitic acid (31%), myristic acid (12%), and stearic acid (11%) and unsaturated fatty acids: oleic acid (24%), palmitoleic acid (4%), and linoleic acid (3%).
- Skimmed milk's caloric content is approximately 34 calories/100 g (whole milk caloric content is 64 calories).
- Frozen yogurt (undefined by the FDA) if made from skimmed milk has very little fat (ice cream prepared from cow's milk contains from 10% to 19% fat and ~97 mg of cholesterol/100 g of ice cream).

Olestra (Olean)

Olestra

- Olestra is composed of a sucrose nucleus with the hydroxyl groups esterified with stearic acids.
- The properties of olestra include a high boiling point, stability to baking and cooking, and a taste and feel similar to fat.

- Olestra is not readily metabolized because of poor absorption and produces zero grams of fat (products containing it are considered "fat free").
- Olestra-prepared potato chips have 70 calories while an equivalent quantity of normally prepared chips have 10 g of fat and 160 calories.

Protein

- Proteins are essential to the body as a source of amino acids.
- Proteins are polymers of amino acids that are not absorbed as such but must be metabolized by peptidases in the GI tract to the monomeric amino acids as well as di- and tripeptides, which are absorbed via active transport.
- Protein makes up 12% to 14% of our normal diet.
- Protein, in the form of muscle mass, accounts for approximately 6 kg of body weight and is a source of potential energy via metabolism to amino acids, which are converted into glucose through the biochemical process called gluconeogenesis.
- Proteins are used for body structure (muscle), hormones and enzymes, and as a source of nitrogen for the synthesis of key chemical components such as purines for RNA and DNA synthesis and only as a last resort will the body utilize protein as a source of energy.
- Twenty amino acids are required for the human body. Some can be synthesized in the human body (nonessential), while others must be obtained from the proteins in the diet (essential).
- Amino acids are utilized for protein synthesis under the direction of DNA and RNA.
- While not precisely known generally accepted RDA for males is 45 to 63 g of protein/day, while females have an RDA of 44 to 50 g/d (the RDA is affected by pregnancy and lactation increasing by as much as 50% and stress can greatly affect the RDA).

Pharmacotherapy of Overweight and Obesity

Overweight and obesity are complex chronic conditions for which successful treatment is not easily attained. Obesity and overweight results from a chronic intake of calories in excess to what the body needs resulting in the storage of the excess calories commonly in the form of fat. Dieting and physical activity can normally produce a loss of weight, but the long-term maintenance of the desired body weight is extremely difficult. Health hazards are associated with both excessive body weight as well as weight cycling.

The anorexiants, also known as anorectics or anorexigenics, antiobesity and appetite suppressants, constitute the starting place of pharmacologic therapy for obesity. These drugs are used as adjuncts in weight reduction programs for morbidly obese patients (BMI >30 kg/m). Anorexiants/antiobesity drugs are not a practical long-term solution for patients who are overweight and are best used under a doctor's supervision with an approved diet and exercise program. Anorexiants are generally divided into sympathomimetic amine and serotonergic agents.

Sympathomimetic Amines

The sympathomimetic anorexiants shown in Figure 28.10 are structurally and pharmacologically related to amphetamine and its closely related analogs. The sympathomimetic anorexiants are related to each other through the phenylpropylamine pharmacophore. Anorexiants are primarily intended to suppress the appetite, but most of the drugs in this class appear not to have a primary effect on appetite, but their appetite suppressant actions appear to be secondary to CNS stimulation, and as CNS stimulants have a potential for abuse or addiction. As a result of the abuse potential amphetamine itself and many of its early analogs are no longer on the market for use as anorexiants. Lisdexamfetamine, a recently approved drug, is a prodrug, which releases dextroamphetamine, but its use is limited to treatment of binge eating disorders (BEDs).

⚠ ADVERSE EFFECTS

Olestra: As a fatty material olestra has the potential for dissolving the fat-soluble vitamins A, D, E, and K, increasing the potential for hypovitaminosis of these vitamins. Additionally olestra causes cramping and loose stools. Olestra containing foods commonly are fortified with the small amounts of the fat-soluble vitamins to reduce the potential for hypovitaminosis.

Nonessential and Essential Amino Acids

Nonessential – alanine, aspartic acid, asparagine, glycine, glutamic acid, glutamine, proline, serine, and tyrosine; Essential – arginine, histidine, isoleucine, leucine, lysine, methionine, phenylalanine, threonine, tryptophan, and valine.

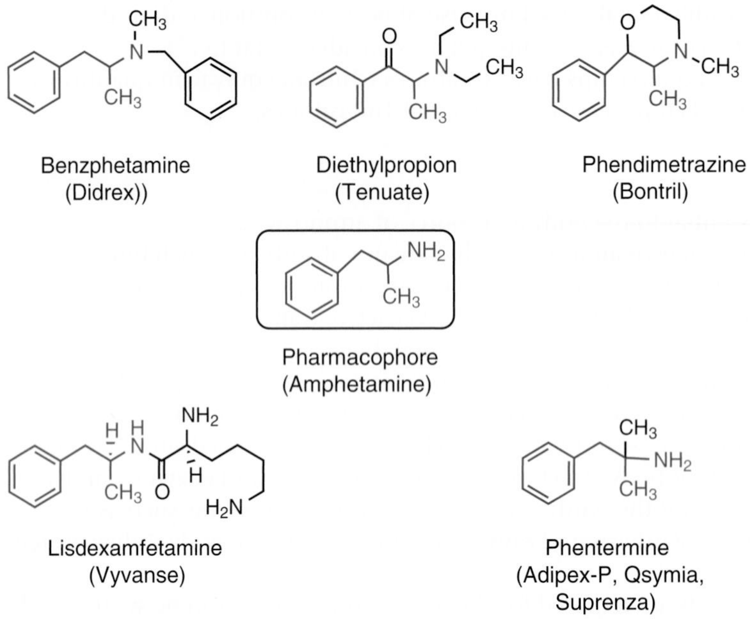

Benzphetamine
(Didrex))

Diethylpropion
(Tenuate)

Phendimetrazine
(Bontril)

Pharmacophore
(Amphetamine)

Lisdexamfetamine
(Vyvanse)

Phentermine
(Adipex-P, Qsymia,
Suprenza)

Figure 28.10 Sympathomimetic anorexiants.

ADVERSE EFFECTS

Phenmetrazine: Phenmetrazine, the demethylated metabolite of phendimetrazine, was previously marketed as an anorexiant, but was withdrawn from the market because of its high potential for abuse and addiction. Phendimetrazine has been judged to have a low abuse potential and is only slowly metabolized to phenmetrazine and thus remains available as a prescription drug.

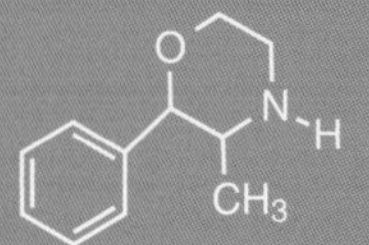

Phenmetrazine
(Preludin)

ADVERSE EFFECTS

Sympathomimetic Amines: Common adverse effects include cardiovascular effects (i.e., palpitation, tachycardia, elevated blood pressure), CNS effects (i.e., restlessness, dizziness, insomnia, headaches), and gastrointestinal effects (i.e., dry mouth, nausea, diarrhea). These drugs may cause withdrawal reactions, especially if they have been used regularly for a long time or in high doses. In such cases, withdrawal symptoms (such as depression, severe tiredness) may occur when these medications are suddenly stopped. The abuse and addiction liability is high with many of the phenylpropylamines.

MOA

- The action of sympathomimetic amines in treatment of obesity is not fully known, but the following effects are seen with these drugs.
- Release norepinephrine from storage vesicles in the adrenergic neuron.
- Block norepinephrine reuptake from the synapse.
- Inhibit reuptake of dopamine (which is linked to abuse and addiction potential).
- None of these effects occur on the appetite brain center.

Physicochemical and Pharmacokinetic Properties

- All of the sympathomimetic anorexiants are available as water-soluble salts (i.e., hydrochlorides, tartrates, dimesylate).
- These compounds are readily absorbed and often extensively metabolized.
- Lisdexamfetamine is a prodrug composed of L-lysine (lis)-dextroamphetamine.

Metabolism

- Common metabolism consists of N-dealkylation.
  - N-Demethylation of benzphetamine and phendimetrazine to benzylamphetamine and phenmetrazine, respectively.
  - N-Deethylation of diethylpropion to ethocathrinone.
- The N-dealkylated products retain biologic activity.
- Lisdexamfetamine is metabolized to L-lysine and dextroamhetamine the active drug (Fig. 28.11).

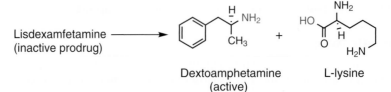

Lisdexamfetamine
(inactive prodrug)

Dextoamphetamine
(active)

L-lysine

Figure 28.11 Metabolism of sympathomimetic anorexiants.

Clinical Applications

- Sympathomimetic appetite suppressants are used in the short-term treatment of obesity (i.e., no more than 12 weeks).
- Their appetite-reducing effect tends to decrease after a few weeks of treatment.
- These drugs should not be taken with other appetite suppressants due to the possibility of serious adverse effects.

- Lisdexamfetamine is specifically approved for treatment of BEDs and for treatment of attention deficit hyperactivity disorder (ADHD) in children age 6 or older.

Serotonergic Agents

Lorcaserin

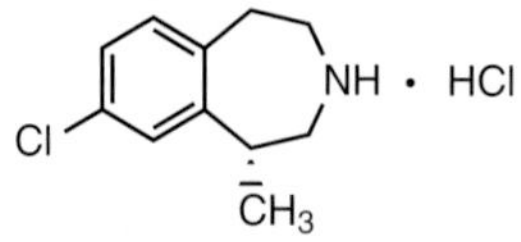

Lorcaserin (Belviq)

MOA

- Lorcaserin acts as a specific agonist at $5HT_{2C}$ receptors thus promoting satiety and a decrease in food consumption (100-fold selectivity over $5HT_{2B}$ receptors).
- The site of action is thought to be the hypothalamus.
- Lorcaserin does not produce significant agonist activity at $5HT_{2A}$ or $5HT_{2B}$ receptors. $5HT_{2B}$ receptor stimulation has been associated with valvular heart disease, an adverse effect previously seen with some antiobesity drugs.
- Stimulation of $5HT_{2A}$ receptors has been associated with psychological disturbance leading to hallucinations.

Physicochemical and Pharmacokinetic Properties

- Lorcaserin is readily absorbed following oral administration and is moderately bound to plasma protein (~70%).
- The drug reaches maximum steady state plasma concentrations within 2 hours and maximum CSF concentration within 6 hours.

Metabolism

- Lorcaserin is metabolized via oxidation (various CYP isoforms) and conjugation (UDP-glucuronosyltransferase [UGT – various isoforms] and sulfotransferase [SULT]) (Fig. 28.12).

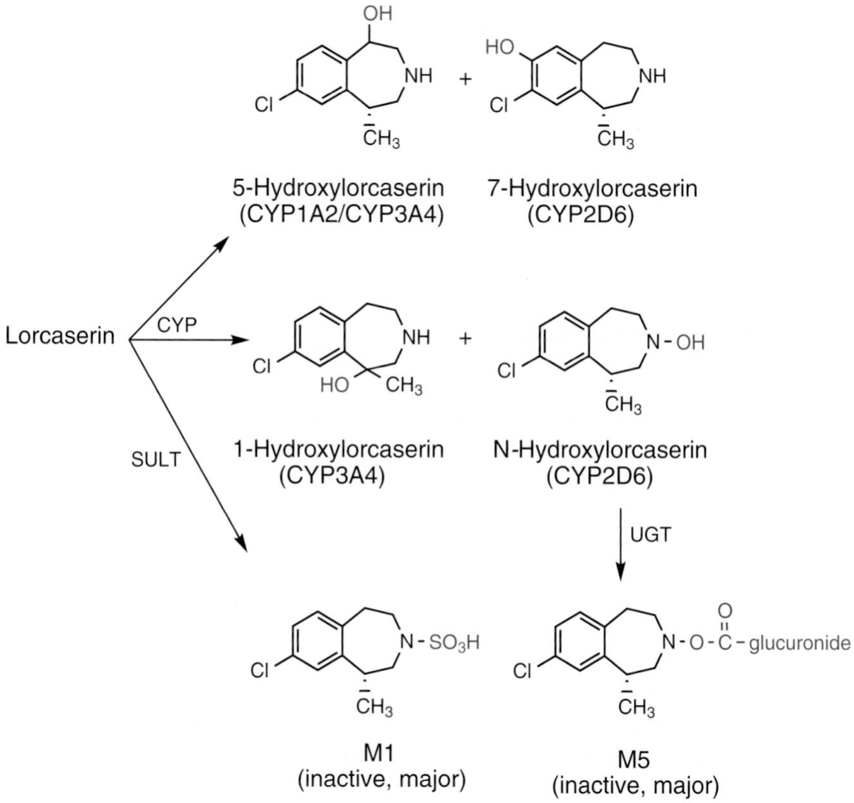

Figure 28.12 Metabolism of lorcaserin.

Clinical Applications

- Clinical studies have shown weight loss of >5% in patients after 1 year.
- Improvement in fasting glucose, fasting insulin, total cholesterol, and a reduction in waist circumference were also reported.
- No major increases in mood-related adverse events were reported.

Orlistat (Alli, Xenical)

Miscellaneous Class of Anorexiants

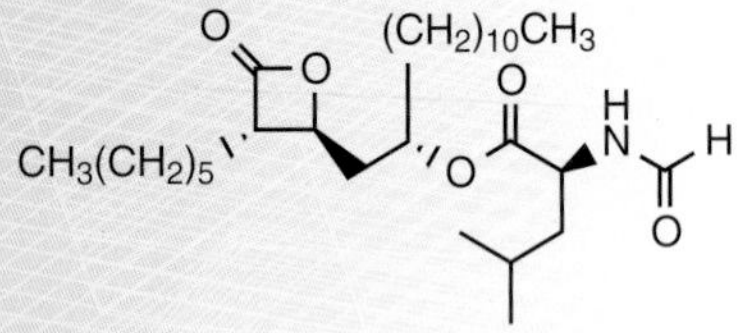

Orlistat

Orlistat is a semisynthetic derivative of the natural product lipstatin produced by *Streptomyces toxytricini.*

MOA

- The compound is an irreversible inhibitor of pancreatic lipase as well as several other lipases.
- Acylation of serine[152] is responsible for this inhibition of lipase activity (Fig. 28.13).
- It prevents the metabolic breakdown of fats leading to a reduced absorption of fatty acids from the GI tract.
- Fecal fat loss (undigested fats) increases by as much as 30% after treatment.

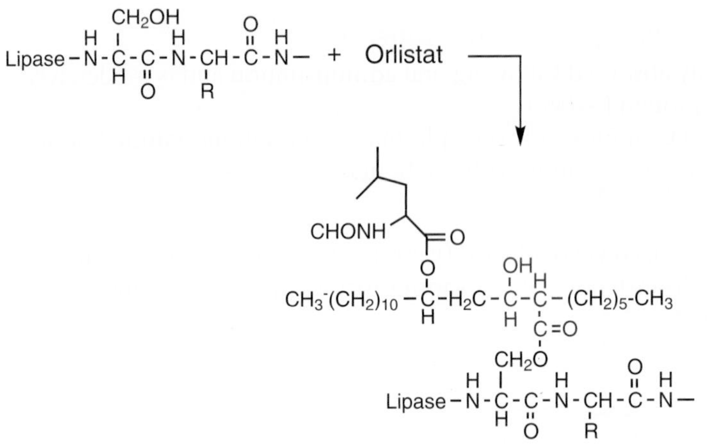

Figure 28.13 Irreversible acylation of a serine in pancreatic lipase by orlistat.

Acylated pancreatic lipase

⚠ **ADVERSE EFFECTS**

Orlistat: The most common adverse effects are associated with the lack of absorption of fats and consist of oily stools, spotting bowel movements, stomach pain, and flatulence. These events occurred in approximately 5% of the patients receiving orlistat at 30 to 360 mg/day. Additionally it was noted that while on orlistat obese patients showed a decrease in LDL and total cholesterol levels and that HbA_{1c} levels decreased in patients with type 2 diabetes.

Physicochemical and Pharmacokinetic Properties

- Orlistat itself is not absorbed to any appreciable extent and it is lost from the body via the GI tract.
- Orlistat does not interfere with the absorption of a wide spectrum of drugs nor with the fat-soluble vitamins with the exception of vitamin E.

Clinical Applications

- Orlistat is available in a nonprescription product (Alli, 60 mg taken with each fat containing meal up to three doses/day).
- The prescription product (Xenical, 120 mg taken with each fat containing meal up to three doses/day).
- Orlistat is used as an adjunct to diet and exercise and can be used for long-term therapy.

Internet Sold Anorexiants

- Sold as dietary supplements.
- Often are herbal or so-called natural remedies.
- None of these products are USFDA-approved products.
- Effectiveness and safety has not been proven.

Micronutrients

Vitamins

Vitamins are organic chemicals found in microorganisms, plants, and animals, but not man, which are essential to man and higher animals for the maintenance of growth, physiological function, metabolism, and reproduction. The role of vitamins in the body is to act as catalysts for the biochemical reactions of the body. Vitamins have a normal turn over and elimination, and therefore must be continuously resupplied to the body.

The vitamins are divided into fat-soluble and water-soluble vitamins based upon these physiochemical properties. Diseases associated with the vitamins occur due to a deficiency (primary or secondary hypovitaminosis) (Table 28.7), or an excessive quantity of a vitamin (hypervitaminosis).

- Primary hypovitaminosis – inadequate diet because of a lack of foods containing the vitamins.
- Secondary hypovitaminosis – result of poor health or chronic disease states (i.e., hyperthyroidism, chronic diarrhea, liver disease, or diseases of the GI track).

Table 28.7 Vitamins, Outward Disease(s), and Common Sources and Roles

Vitamin	Specific Disease(s)	Source	Role
Fat Soluble			
Vitamin A	Xerophthalmia Nyctalopia	Yellow and green leafy vegetables, carrots, yellow fruits, fish oils, milk, fats and eggyolk, apricots, peaches, and sweet potatoes	Necessary for normal eye function, integrity of the epithelium, synthesis of adrenocortical steroids
Vitamin D	Rickets Osteomalacia	Egg yolks, dairy products, fish, and UV light	Calcium absorption and bone formation
Vitamin E	Spinocerebellar ataxia Skeletal myopathy	Cereals, nuts, unsaturated oils, leafy green and yellow vegetables, milk, muscle meats, and butter	Health of sensory neurons leading to peripheral neuropathy if absent
Vitamin K	Pigmented retinopathy Hypoprothrombinemia	Synthesized in the intestinal tract by bacteria and found in leafy green vegetables	Involved in the blood clotting process
Water Soluble			
Vitamin C (Ascorbic acid)	Scurvy	Citrus fruits, cabbage, peppers, berries, melons, and salad greens	Prevent scurvy, necessary for proper bone and teeth formation
Thiamine (Vitamin B_1)	Wernicke encephalopathy Beriberi	Yeast, beans, brown rice, lean pork, whole wheat bread, liver, and outer layers of seeds of plants or unrefined cereal grains	Prevents beriberi (leg edema, paralysis, muscle atrophy)
Riboflavin (Vitamin B_2)	Anemias	Liver, kidney, milk, yeast, heart, anaerobic fermenting bacteria, and many vegetables	Deficiency leads to lip lesions, seborrheic dermatitis
Niacin (Vitamin B_3)	Pellagra	Liver, kidney, lean meat, soybeans, wheat germ, poultry products, yeast	Prevent pellagra (dermatitis, diarrhea, dementia)
Pantothenic acid (Vitamin B_5)	Rare, no specific disease	Muscle meats, outer layer of whole grains, broccoli and avocados	Deficiencies are rare, but may appear as numbness, paresthesia, and muscle cramps
Pyridoxine (Vitamin B_6)	Homocystinuria	Wheat germ, milk, yeast, liver	Prevents dermatitis sicca in adults and convulsions in infants, essential to biochemical reactions
Cobalamine (Vitamin B_{12})	Pernicious anemia	Liver, kidney, muscle meat	Prevents pernicious anemia
Folic acid	Megaloblastic anemia Neural tube defects (NTDs)	Leafy vegetables, beans, peas, liver, kidney, baker's yeast, grains, various nuts – sunflower seeds	Important for prevention of neural tube defects (NTDs), necessary to DNA and RNA synthesis and other biochemical reactions
Biotin	Rare–no specific disease	Tomatoes, kidney, egg yolk, peanuts, chocolates, yeast, and liver	Deficiency causes myodermatitis, lassitude, gastrointestinal symptoms, and hyperesthesias including muscle pain

Fat-Soluble Vitamins

Vitamin A

trans-Retinol   11-*cis*-Retinal

β-Carotene

α-Carotene

Vitamin A as a generic term is generally considered to be a family of chemicals containing retinol, retinal as well as the poorly absorbed carotenes (see Table 28.7 for source and role). α- and β-carotene are provitamin A chemicals which are oxidatively converted into all *trans* retinal, which in turn can be converted into the all-*trans* retinol, the active form of vitamin A. Retinol is commonly found as a lipid-soluble ester with either palmitic acid or acetic acid. The carotenoids, a combination of all sources of vitamin A, have their biological activity measured in the retinol activity equivalent (RAE), which is defined as 1 μg of all-*trans* retinol or 2 μg of *trans*-β-carotene, 12 μg of food-based *trans*-β-carotene, or 24 μg of other food-based all-*trans* provitamin A carotenoids. See Table 28.8 for the RDI.

PHYSIOLOGICAL ACTION OF VITAMIN A

- An outline of the role of vitamin A in vision is shown in Figure 28.14. The functional form of vitamin A is rhodopsin.
- The absorption of a photon of light by the rhodopsin results in Na^+ conductance at a GPCR, which leads to signaling in the visual center in the cortex via the optic nerve.
- Opsin and *trans* retinal are released and recycled.
- A deficiency of vitamin A (i.e., hypovitaminosis A) leads initially to night blindness (nyctalopia) or total blindness (xerophthalmia)

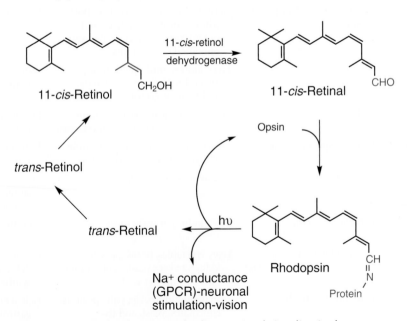

Figure 28.14 Role of vitamin A leading to visual signaling in the cortex.

- Retinoic acid may also be involved in the development of epithelial tissue, mucus-producing tissue, reproduction, and bone growth.
- Weak evidence suggests that the carotenoids may also function as antioxidants with a preventive role in heart disease and cancer.

ABSORPTION

- The absorbable form of vitamin A is retinol (obtained by hydrolysis of retinol esters in the intestine).
- Vitamin A from β-carotene requires oxidative cleavage to retinal, which in turn is reduced to all-*trans* retinol.
- Approximately 70% to 90% absorption of vitamin A occurs from the small intestine with storage occurring in the liver.

Vitamin D (Sunshine Vitamin, Calcitriol, Calciferols)

Vitamin D is classified as a vitamin as well as a hormone since it is synthesized in the skin via a series of reactions, which include ultraviolet stimulated ring cleavage (UV B radiation) and hydroxylation reactions that occur in the liver and kidney (Fig. 28.15). In the absence of sufficient sunlight vitamin D associated diseases can occur.

Since the amount of sunlight that an individual receives can vary considerably, the daily needs for an individual will vary (see Tables 28.7 and 28.8 for sources and doses).

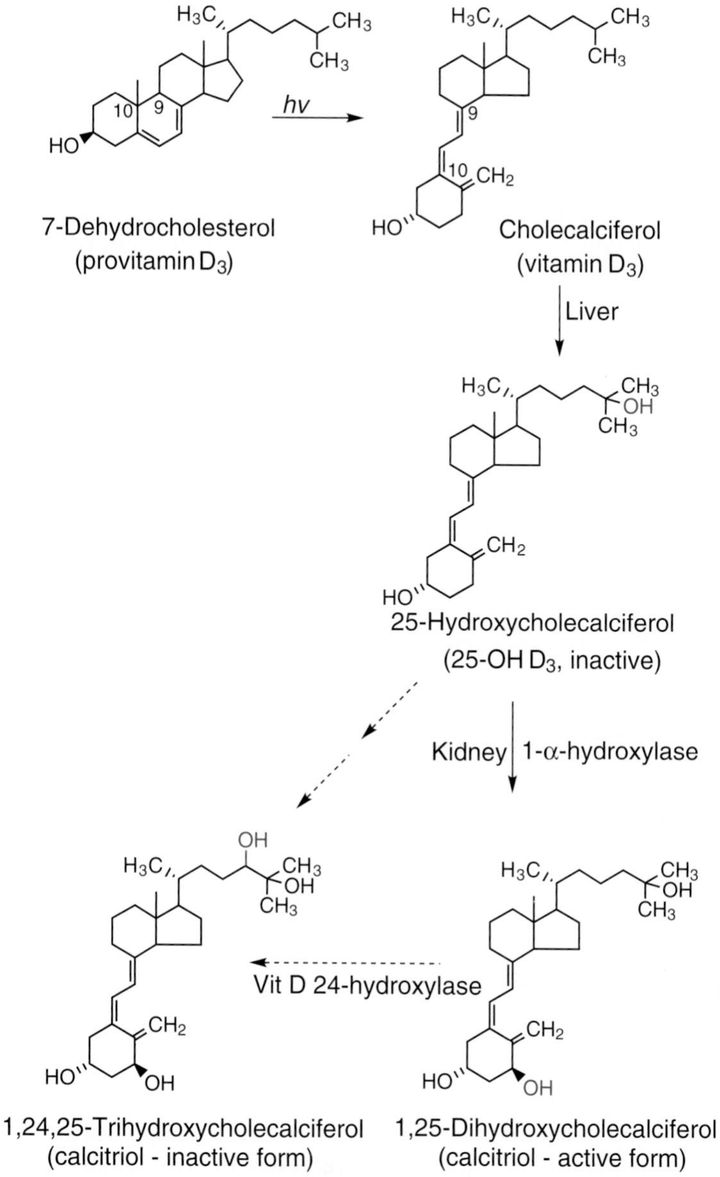

ADVERSE EFFECTS

Vitamin A Overdose: Hypervitaminosis A signs and symptoms include headache, fatigue, insomnia, cracked bleeding lips, cirrhosis-like damage of the liver, and yellow skin coloration. The upper safe limit for vitamin A consumption is considered to be 3,000 µg. Birth defects are associated with the intake of 13-*cis*-retinoic acid (Accutane) a known teratogen if consumed during the first trimester of pregnancy.

Figure 28.15 Bioactivation and inactivation of vitamin D.

Table 28.8 Recommended Daily Intake (RDI) of Micronutrients

Nutrient	Male	Female	Female During Pregnancy/Lactation
Vitamin A	900 µg	700 µg	770–1,300 µg
Vitamin D	200 IU/5 µg up to age 50, 400 IU 51–70; >71 400 IU		
Vitamin E	30 IU for all adults		
Vitamin K	80 µg for all adults (1 µg/kg body weight/d)		
Vitamin C	60 mg for all adults		75 mg
Vitamin B_1	1.5 mg for all adults		
Vitamin B_2	1.7 mg for all adults		
Vitamin B_3	20 mg for all adults		
Vitamin B_5	10 mg for all adults		
Vitamin B_6	2 mg for all adults		
Vitamin B_{12}	6 µg for all adults		
Folic Acid	400 µg for all adults		600 µg/500 µg
Biotin	300 µg for all adults		

ADVERSE EFFECTS

Vitamin D Overdose: Hypervitaminosis D resulting from consumption of high doses of vitamin D supplements (10,000 IU/day) can cause constipation, bone pains and stiffness, and confusion. High serum calcium levels increase the likelihood for precipitation of calcium salts in soft tissue such as kidney and heart. This could lead to a decrease in renal function and kidney stones.

PHYSIOLOGICAL ACTION OF VITAMIN D

- Vitamin D regulates serum levels of ionic calcium and phosphate, which in turn affects bone ossification and neuromuscular function.
- Calcitriol governs the transcription of mRNA leading to synthesis of proteins, which increase the absorption of calcium from the gut or resorption of calcium from the bones (see Chapter 20).
- The termination of the action of calcitriol occurs through hydroxylation (Fig. 28.15).
- Hypovitaminosis D leads to bones resorption resulting in rickets or osteomalacia.
- Women who are breast feeding should supplement their diet with vitamin D.

Vitamin E (Tocopherols)

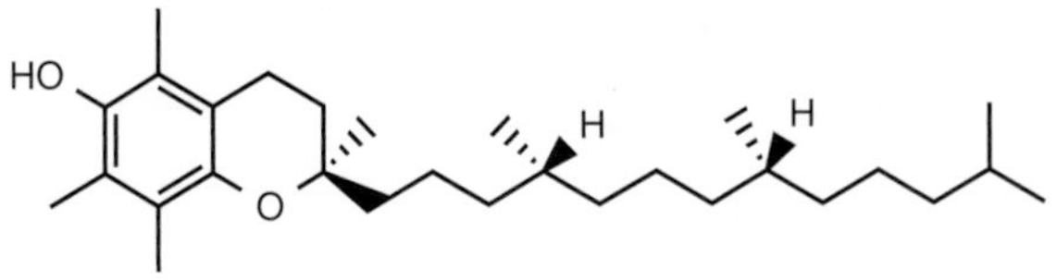

α-Tocopherol

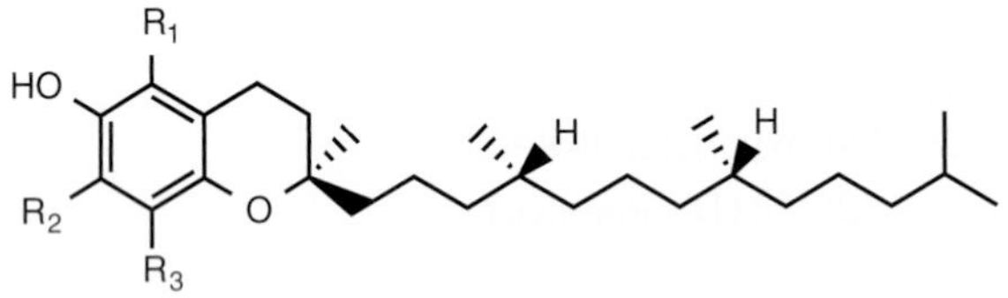

β-Tocopherol (R_2 = H, R_1 = R_3 = CH_3)
γ-Tocopherol (R_1 = H, R_2 = R_3 = CH_3)
δ-Tocopherol (R_1 = R_2 = H, R_3 = CH_3)

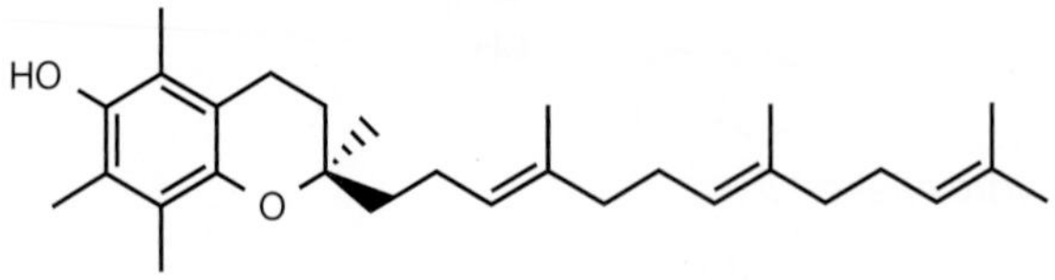

α-Tocotrienol

Vitamin E consists of a group of chemicals composed of α, β, γ, δ-tocopherol and α, β, γ, δ-tocotrienol. α-Tocopherol and γ-tocopherol represent the two major tocopherols. Various stereoisomers are possible for each of the tocopherols all of which are normally called vitamin E. Commercially available vitamin E may be a single isomer of one of the tocopherols or a mixture of tocopherols although it is generally considered that α-tocopherol or R,R,R-α-tocopherol is the more active form of vitamin E. Often the tocopherols come in the form of an ester of the phenol. The quantization of vitamin E is based upon a direct analytical measure or an indirect biologic rat assay method.

PHYSIOLOGICAL ACTION OF VITAMIN E

- Vitamin E is ubiquitous in nature (Table 28.7) and therefore human deficiency of this vitamin is uncommon.
- It is difficult to identifying a role for vitamin E.
- The lay press contains many claims as to the value for vitamin E which includes: improving wound healing without scaring; slowing down the ageing process; protection against various cancers; prevention of heart disease; improving the performance of athletes.
- Vitamin E has the chemical property of being a fat-soluble antioxidant acting by virtue of its high reactivity toward ROS thus protecting cellular components from oxidation (antioxidant).
- The tocopherols may protect PUFA from the damaging effects of ROS-catalyzed reactions.

MECHANISM OF ANTIOXIDANT ACTIVITY

- Vitamin E (especially α-tocopherol) scavenger lipid peroxyl radicals.
- The mechanism and products of this reaction are shown in Figure 28.16.
- ROS can readily oxidize PUFA leading to a dienyl radical and lipid peroxyl radicals which can damage cellular components (i.e., addition PUFA, DNA, RNA, etc.).
- Vitamin E can form a relatively stable tocopherol radical thus effectively chain-breaking the autoxidation of the PUFA.

Resonance stabilization of α-tocopherol radical

ABSORPTION

- The absorption of vitamin E occurs in the intestine.
- Ester of vitamin E must be hydrolyzed first by intestinal esterase.
- Absorption is bile acid dependent. Bile acids favor micelle formation, which is essential for absorption.

METABOLISM

- Vitamin E, the tocopherols and the tocotrienols, is oxidized to more water-soluble chemicals that can be readily excreted from the body (i.e., ω-oxidation followed by progressive shortening of the side chain) (Fig. 28.17).
- As a result of the metabolism the danger of hypervitaminosis E is low.

Vitamin K (Menadiones)

Vitamin K derives its name from the German word Koagulation for which there is no single chemical entity possessing the biological activity attributed

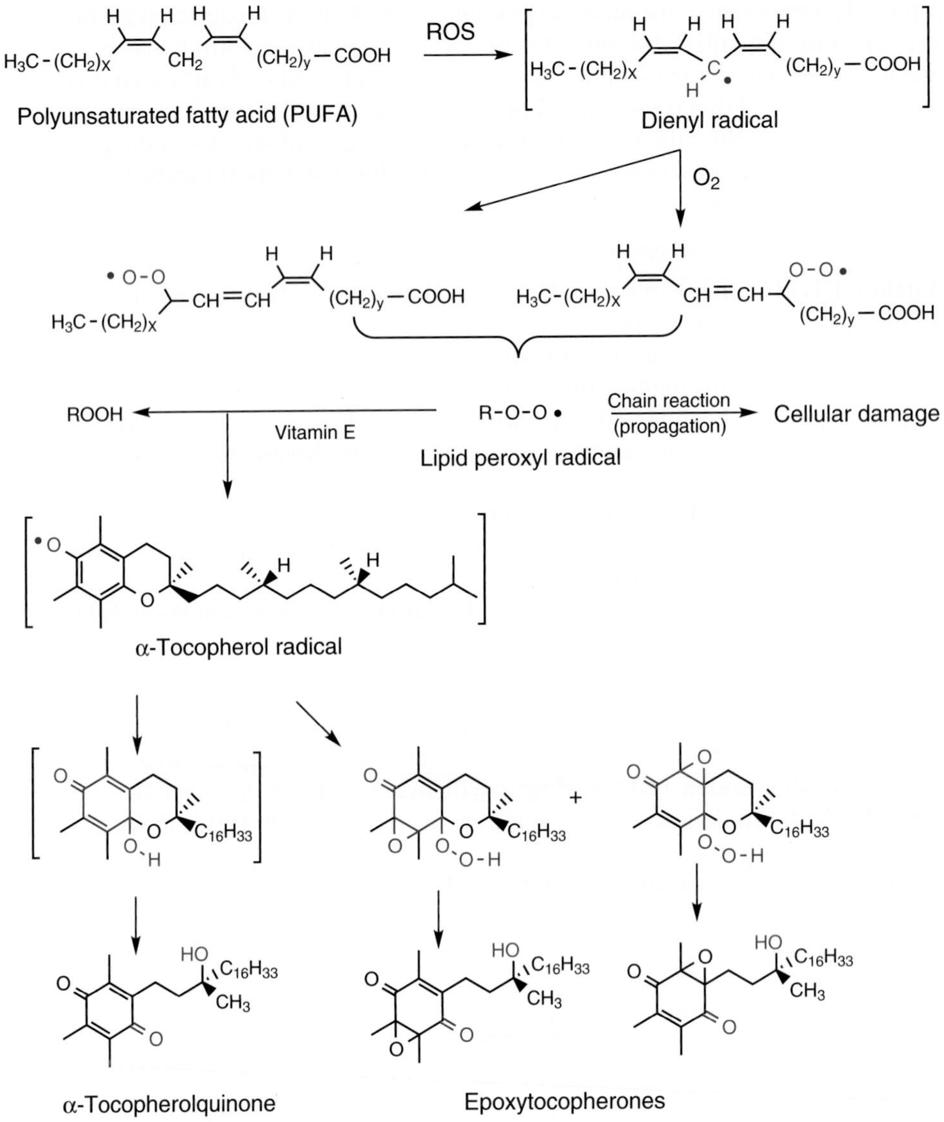

Figure 28.16 The action of vitamin E in decreasing cellular damage via reaction with PUFA-free radicals.

to vitamin K. The basic nucleus of all vitamin K compounds is the 2-methyl-1,4-naphthoquinone to which various chain lengths of the 5-carbon isoprenoid unit are attached at the 3-position.

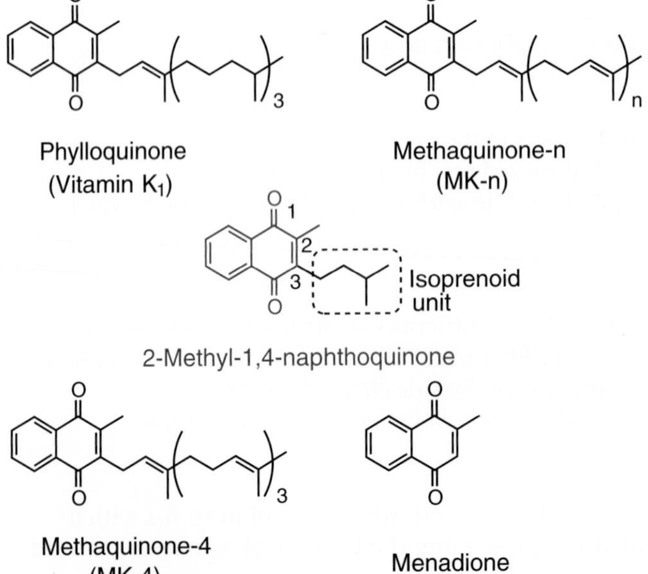

Phylloquinone
(Vitamin K$_1$)

Methaquinone-n
(MK-n)

2-Methyl-1,4-naphthoquinone

Methaquinone-4
(MK-4)

Menadione

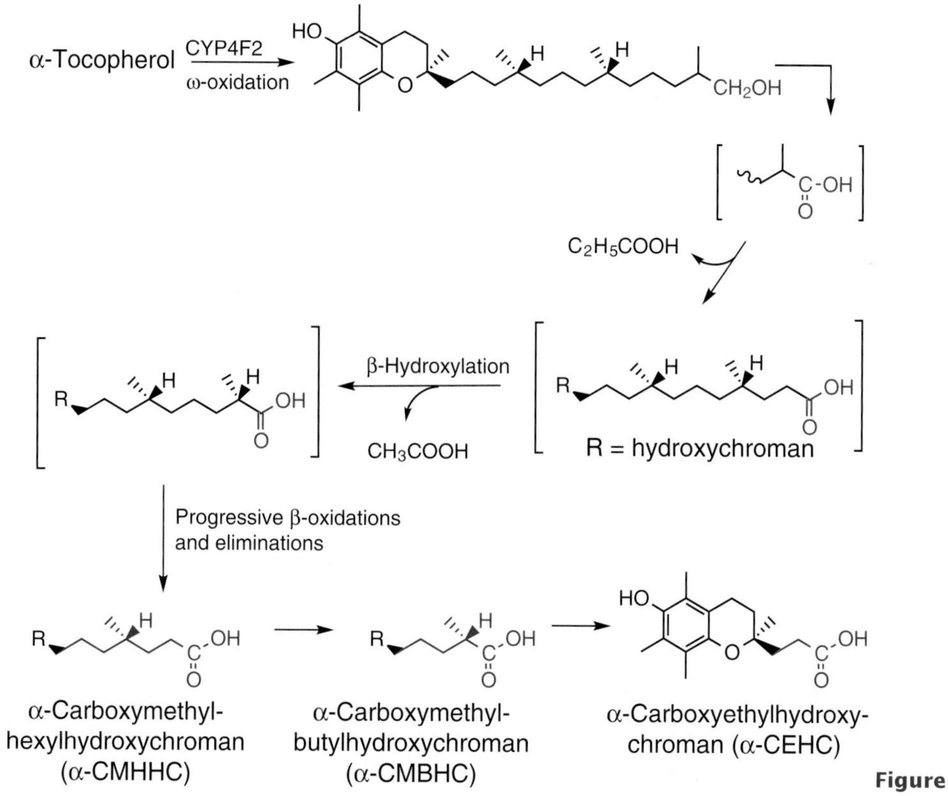

Figure 28.17 Metabolism of α-tocopherol.

PHYSIOLOGICAL ACTION OF VITAMIN K

- Vitamin K catalyzes posttranslation carboxylation of several proteins that are involved in blood coagulation (see Table 28.7 and 28.8 and Chapter 16 for details of biological function).
- The absorption of dietary vitamin K occurs via the lymphatic system.

Water-Soluble Vitamins

The water-soluble vitamins consist of the B complex of vitamins (i.e., thiamine [B_1], riboflavin [B_2], niacin [B_3], pantothenic acid [B_5], pyridoxine [B_6], cobalamine [B_{12}], folic acid, and biotin) and vitamin C. Diseases, sources, and roles for the water-soluble vitamins are found in Table 28.7. Vitamin C and folic acid are discussed below while the reader is referred to a biochemistry text for additional information about the remaining water-soluble vitamins.

Vitamin C (Ascorbic Acid)

Vitamin C or ascorbic acid has a long history dating back to 3000 BC. It was not until 20th century that Albert Szent-Gyorgyi identified and named the chemical in fruit juices as ascorbic acid. Vitamin C is synthesized in plants from glucose and fructose and most mammals can also synthesize this substance with the exception of humans, nonhuman primates, including guinea pigs, and several other animal species.

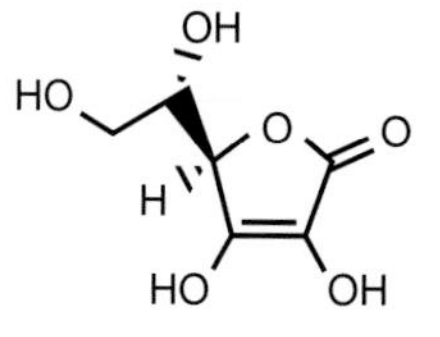

L-Ascorbic Acid

PHYSIOLOGICAL ACTION OF VITAMIN C

- Vitamin C is essential for the integrity of blood vessels due to its role in the synthesis of collagen.
- Vitamin C (L-ascorbic acid) is a cofactor in the hydroxylation of proline and lysine, which are essential components of collagen.
- The hydroxylated amino acids have a role in stabilizing the helix structure of collagen.
- Deficiency is indicated by the development of scurvy a condition characterized by subcutaneous hemorrhage due to blood vessel rupture, commonly seen in the gingiva, skin, and GI mucosa.

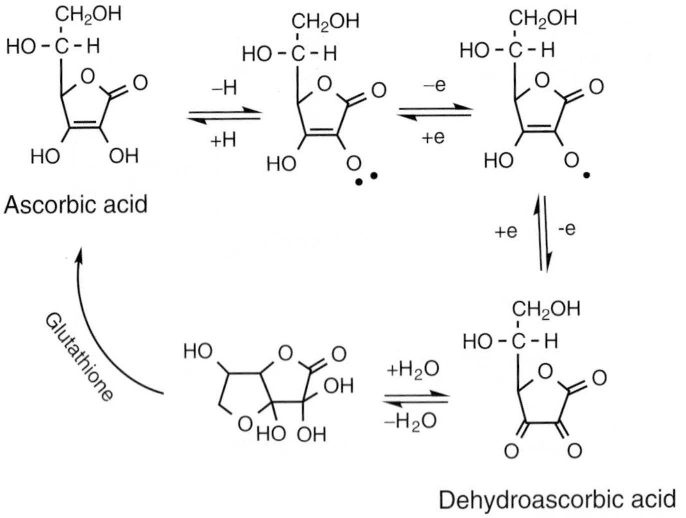

Figure 28.18 Antioxidant property of ascorbic acid.

⚠️ **ADVERSE EFFECTS**

Mega Doses of Vitamin C: High doses of vitamin C have been associated with increased oxalate excretion with the potential for oxalate-based kidney stones, and will lower urine pH affecting drug excretion. High doses, in the range of 3 g/d may cause diarrhea and bloating. Vitamin C may prevent intestinal oxidation of ferrous ion to ferric ion, the latter is poorly absorbed from the GI tract.

- Additional features of hypovitaminosis C include loss of teeth, bone damage, internal bleeding, and infections leading ultimately to death.
- Vitamin C (both isomers) is a water-soluble antioxidant by virtue of its ability to give up two electrons, an oxidative process (Fig. 28.18).
- The antioxidant property suggests a role in the prevention of cardiovascular diseases by inhibiting LDL oxidation and blood vessels plaque formation.
- Vitamin C has also been touted as valuable for prevention and/or hastening the recovery from the common cold via increasing the body's production of interferons.

Folic Acid

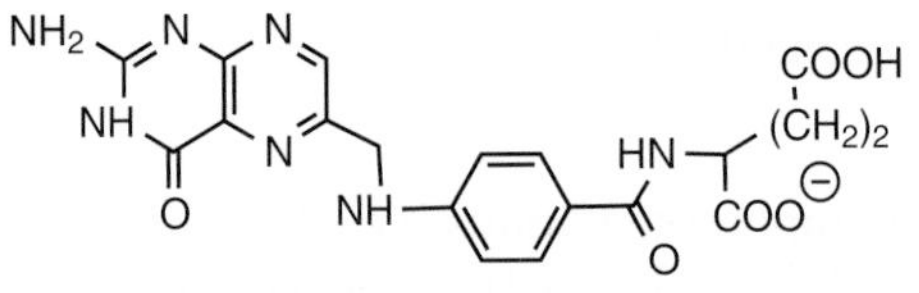

Folic acid

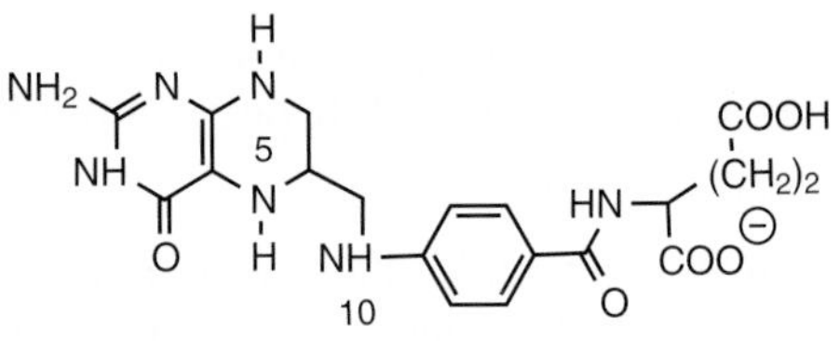

5,6,7,8-Tetrahydrofolic acid

PHYSIOLOGICAL ACTION OF FOLIC ACID

- Folic acid or more specifically 5,6,7,8-tetrahydrofolic acid functions as a coenzyme involved in one carbon transfers to key biochemical compounds.
- The single carbon is held by the N^5, N^{10}, or both the N^5-N^{10} nitrogens in various oxidative states prior to transfer to substrates such as:
  - Homocysteine to be converted into methionine.
  - Deoxyuridylate to be converted into thymidylate for DNA base synthesis.
  - Formiminoglutamine to glutamic acid where tetrahydrofolic acid acts as a carbon acceptor.
  - Glycinamide ribonucleotide and 5-aminoimidazole-4-carboxamide ribonucleotide involved in purine synthesis and thus DNA and RNA bases (see Chapter 26 for the role of folic acid in purine and pyrimidine synthesis).

ABSORPTION AND METABOLISM

- Folic acid in food is readily absorbed in the small intestine.
- In the mucosa folic acid is reduced by folate reductase and methylated to active forms of the vitamin.
- Tetrahydrofolic acid is transported to tissue via plasma proteins and excreted tetrahydrofolate is reabsorbed via enterohepatic recycling involving bile.

CLINICAL APPLICATIONS

- Deficiencies may occur in acute and chronic alcoholism.
- Folate deficiencies have been implicated in the development of neural tube defects (i.e., spina bifida, encephaloceles, and anencephaly) in infants born to women not supplementing their diet with folic acid (see Table 28.8).

Essential Electrolytes

The essential minerals include calcium, chromium, copper, iodine, iron, magnesium, manganese, molybdenum, phosphorus, selenium, and zinc along with the essential electrolytes of sodium, potassium, and chloride. This discussion will be limited to sodium and chloride ions.

Sodium Chloride (Table Salt, NaCl)

DIETARY ROLE OF SODIUM ION AND CHLORIDE ION

- Sodium ion is essential for nerve conduction and fluid movement.
- Sodium ion is most commonly consumed in the form of NaCl.
- It is estimated that Americans consume approximately 10 to 12 g/d of NaCl.
- It is not known what the minimum amount of NaCl for good health is although it is thought that an intake of <1,000 mg/d of NaCl would meet the needs for a normal adult (1.5 to 2.3 g/d of sodium ion and 3.8 to 9.1 g/d of chloride ion).
- The 2005 dietary guidelines for NaCl consumption was established at 5.8 g/d (~1 teaspoon) or 2.3 g/d of sodium ion.

PHYSIOLOGICAL/PATHOLOGICAL ROLE OF SODIUM ION AND CHLORIDE ION

- The human appetite for NaCl has both a physiological and an educated basis.
- Taste receptors in the oral cavity can detect a salty taste presumably as a means of encouraging and assuring the intake of NaCl.
- Commercially processed food commonly has added sodium, usually in the form of NaCl, as a preservative but also to improve the taste (estimates are that as much as 65% to 85% of the daily consumed NaCl is present as "hidden salt" in processed food).
- Restaurant foods are also reported to be high sources of NaCl.
- Reports suggest Americans need to reduce their intake of sodium ion in the form of NaCl and that high sodium ion intake is associated with elevated blood pressure, heart disease, stroke, congestive heart failure, and renal disease.
- In 2009 the Institute of Medicine Committee on Strategies to Reduce Sodium Intake convened and issued their report suggesting that the USFDA work to progressively decrease sodium content of processed food and restaurant food.

Chemical Significance

The major health concern among many Americans today is weight control, obesity, and the associated diseases of diabetes, cardiovascular diseases, and cancer. The lay press and television alert the public to this issue and the public has become quite proficient at using the vocabulary associated with weight control and obesity, but do they really know what these words mean? As examples the public now reads food labels but may not understand serving size. They can talk calories, but do they know the difference between carbohydrate, fat, and protein calories? Do they understand the differences between mono-, di-, and polysaccharides? Do they understand how saturated, monounsaturated, and polyunsaturated fats differ? And what is a TFA or fats versus fatty acids? These and many other terms used in discussions of nutrition are chemical terms and define specific chemical structures.

The applications of medicinal chemistry to clinical situations involving nutrition are many and can be utilized in daily conversations between the pharmacist and the patient as well as many health professionals. The following are but a few examples of where medicinal chemistry can be utilized directly or indirectly while discussing nutritional issues:

- Explaining that TFAs are not common to nature, but arise through the partial hydrogenation of unsaturated and polyunsaturated fatty acids to saturated fatty acids.
- Discussing the essential nature of fats (triglycerides) and fatty acids to the human body and what a PUFA is requires an understanding of basic organic chemistry. Explaining to an interested patient what a ω-fatty acid is requires an understanding of chemistry.
- The recognition that fructose does not differ significantly from the chemistry of glucose, but that the concentration of fructose in a commercially processed food does make a difference in caloric content and can best be justified through its chemistry.
- Fat-soluble versus water-soluble antioxidants and the role of antioxidants in protecting biopolymer from destructive oxidative reactions can be explained by reference to the medicinal chemistry of these chemicals.

These and many more examples represent practical applications of how medicinal chemistry can help bring understanding to the sometimes confusing topics related to nutrition. The public is quite aware of the importance of nutrition to their personal health yet complicated chemical terminology can lead to misunderstanding and frustration. The pharmacist stands as an important interpreter of nutritional information and by utilizing their chemical knowledge can serve their patients well.

Review Questions

1 Deficiency of vitamin A in children can cause:
   A. lack of sex drive
   B. anorexia and hyperirritability
   C. soft bones
   D. scurvy
   E. blindness

2 Label each of the following sweeteners as to whether they are derived from natural occurring chemicals or natural occurring as such (N) or synthetic sweeteners (S).
   A. Aspartame
   B. Mogrosides
   C. Acesulfame-K
   D. Sucralose
   E. Rebaudioside A

3 Which statement concerning the macronutrients of the body is incorrect?
   A. Carbohydrates are converted to monosaccharides prior to absorption and then stored as glycogen.
   B. Fats are stored as triglycerides
   C. Fats are absorbed through the action of GLUT and SGLT enzymes.
   D. Protein consumed must be converted to amino acids prior to absorption and are then converted to protein in the body.
   E. Energy is released when organic material is converted to carbon dioxide and water.

4 If your body needed energy the highest source of energy per gram of weight is:
   A. Carbohydrates
   B. Fat
   C. Protein
   D. Vitamins
   E. Body electrolytes such as sodium

5 What is the relationship between aspartame and phenylketonuria (PKU)?
  A. Aspartame is a source of phenylalanine which cannot be properly metabolized by a PKU patient.
  B. A PKU patient will develop cancer if they consume large quantities of aspartame.
  C. Aspartame causes allergic reactions in the PKU patient.
  D. The PKU patient will metabolize aspartame leading to nyctalopia.
  E. There is no relationship between aspartame and phenylketonuria.

Abbreviations

AA-NAT	arylalkylamine N-acetyltransferase
ABC	ATP-binding cassette
ACE	angiotensin-converting enzyme
ACEI	angiotensin-converting enzyme inhibitor
ACh	acetylcholine
AChE	acetylcholinesterase
AChEI	acetylcholinesterase inhibitor
AChRs	acetylcholine receptors
ACT	artemisinin-based combination therapy
ACTH	adrenocorticotropic hormone
AD	Alzheimer's disease
ADH	alcohol dehydrogenase
ADHD	attention deficit hyperactivity disorder
ADP	adenosine diphosphate
AF-2	activation factor-2
AG	arabinogalactan
AIDS	acquired immunodeficiency syndrome
AKRs	aldo-keto reductases
ALDH	aldehyde dehydrogenase
ALK	anaplastic lymphoma kinase
AMP	adenosine monophosphate
AMPA	L-α-amino-3-hydroxy-5-methyl-4-isoxazole propionic acid propionate
AO	aldehyde oxidase
6-APA	6-aminopenicillanic acid
APC	antigen-presenting cell
APL	acute promyelocytic leukemia
aPTT	activated partial thromboplastin time
AR	androgen receptor
ARB	angiotensin receptor blocker
ARC	AIDS-related complex
ARE	androgen-response element
AT	antithrombin III
ATP	adenosine triphosphate
AUC	area under plasma concentration–time curve
AV	atrioventricular
BAS	bile acid sequestrant
BCRP	breast cancer resistance protein
BMD	bone mineral density
BMI	body mass index
BMR	basal metabolic rate
BPH	benign prostatic hyperplasia
BZ	benzodiazepine
BZR	benzodiazepine receptor
CAD	coronary artery disease
cAMP	cyclic adenosine monophosphate
CaSRs	calcium-sensing receptors
CD	collecting duct
cDNA	complementary DNA
cGMP	cyclic guanosine monophosphate

CHD	coronary heart disease
CHF	congestive heart failure
CKD	chronic kidney disease
CML	chronic myelogenous leukemia
CMV	cytomegalovirus
CNS	central nervous system
CNTs	concentrative nucleoside transporters
CO₂	carbon dioxide
CoA	coenzyme A
CoASH	coenzyme A
COMT	catechol-O-methyltransferase
COPD	chronic obstructive pulmonary disease
COX	cyclooxygenase
CRF	corticotropin-releasing factor
CRP	C-reactive protein
CRPC	castration-resistant prostate cancer
CSF	cerebrospinal fluid
CT	calcitonin
CTZ	chemoreceptor trigger zone
CYP	cytochrome P450
D₃	cholecalciferol
D5W	5% dextrose in water
DA	dopamine
DAG	1,2-diacylglycerol
DAGAT	diacylglycerol acyltransferase
DAO	diamine oxidase
DAT	dopamine reuptake transporter
DBD	DNA-binding domain
DCT	distal convoluted tubule
DDI	drug–drug interaction
DdRp	DNA-dependent RNA polymerase
DHEA	dehydroepiandrosterone
DHF	dihydrofolate
DHFR	dihydrofolate reductase
1,4-DHPs	1,4-dihydropyridines
DHT	dihydrotestosterone
DMARDs	disease-modifying antirheumatic drugs
DMF	dimethyl fumarate
DOPA	dihydroxyphenylalanine
DOT	direct observation therapy
DPD	dihydropyrimidine dehydrogenase
DPI	dry powder inhaler
DPP-IV	dipeptidyl peptidase-IV
DSM-5	Diagnostic and Statistical Manual of Mental Disorders, Fifth Edition
DTI	direct thrombin inhibitor
dTMP	deoxythymidine monophosphate
dUMP	deoxyuridine monophosphate
DVT	deep vein thrombosis
E	entgegen
ED	erectile dysfunction
EE	ethinyl estradiol
EEG	electroencephalogram
EGFR	epidermal growth factor receptor
EMB	ethambutol
ENL	erythema nodosum leprosum
ENT	equilibrative nucleoside transporter

Epi	epinephrine (Chapters 3,4,6)
EPR-3	Guidelines for the Diagnosis and Management of Asthma
ER	estrogen receptor
ERE	estrogen-response element
ERT	estrogen replacement therapy
ET	endothelin
FXa	direct factor Xa
FAD	flavin adenine dinucleotide
FDA	U.S. Food and Drug Administration
FFAs	free fatty acids
FMOs	flavin monooxygenases
FPGS	folylpolyglutamate synthase
FSH	follicle-stimulating hormone
GABA	γ-aminobutyric acid
GERD	gastroesophageal reflux disease
GFJ	grapefruit juice
GFR	glomerular filtration rate
GI	gastrointestinal
GIP	gastric inhibitory polypeptide
GLP-1	glucagon-like peptide
Glu	glutamate
GLUT	glucose transporter proteins
GLUT4	glucose transporter 4
GnRH	gonadotropin-releasing hormone
GP	glycoprotein
GPCR	glycoprotein-coupled receptor
GPCR	G-protein–coupled receptor
GR	glucocorticoid receptor
G_s	G-stimulatory proteins
GSH	glutathione
GST	glutathione S-transferase
GTP	guanosine triphosphate
HV	hepatitis virus (e.g., A, B, C, E)
H₂O₂	hydrogen peroxide
HA	hemagglutinin
HAT	histone acetyltransferase
HbA₁c	glycosylated hemoglobin
H-bond	hydrogen bond
Hct	hypocretin
HDAC	histone deacetylase
HDACis	histone deacetylase inhibitors
HDL	high-density lipoprotein
HER2	human epidermal growth factor 2
HGPRT	hypoxanthine-guanine phosphoribosyl transferase
HIOMT	5-hydroxyindole–O-methyltransferase
HIV	human immunodeficiency virus
HMG	3-hydroxy-3-methyl-glutaryl
HMGR	HMG-CoA reductase
HoFH	homozygous familial hypercholesterolemia
HPA	hypothalamus–pituitary–adrenal axis
HPETE	5-hydroperoxyeicosatetraenoic acid
HPPH	5-hydroxyphenyl-5-phenylhydantoin
HPT	hypothalamic–pituitary–testicular
11β-HSD1	11β–hydroxysteroid dehydrogenase
HRE	hormone-response element

Hst	histamine
HSV	herpes simplex virus
5-HT	serotonin
I_f	"funny" current
IFN	interferon
IL	interleukin
IM	intramuscular
INH	isoniazid
INR	international normalized ratio
IP$_3$	inositol-1,4,5-triphosphate
IRSP	insulin receptor substrate peptides
ISA	intrinsic sympathomimetic activity
ITP	idiopathic thrombocytopenia purpura
IUD	intrauterine device
IV	intravenous
KA	kainic acid
L-AAD	L-amino acid decarboxylase
LABAs	long-acting β$_2$-adrenergic agonists
LBD	ligand-binding domain
LDL	low-density lipoprotein
L-DOPA	levo-dihydroxyphenylalanine
LH	luteinizing hormone
LHRH	luteinizing hormone–releasing hormone
LMWH	low–molecular-weight heparin
Log P	log$_{10}$ partition coefficient
LTs	leukotrienes
MAC	minimum alveolar concentration (Chapter 9)
MAC	*Mycobacterium avium*-intracellulare complex (Chapter 23)
mAChR	muscarinic acetylcholinergic receptor
MAGAT	monoacylglycerol acyl transferase
MAO	monoamine oxidase
MAOIs	monoamine oxidase inhibitors
MAPKs	mitogen-activated protein kinases
MATEs	multidrug and toxin extrusion transporters
m-CPP	m-chlorophenylpiperazine
MCTs	monocarboxylate transporters
MDI	metered dose inhaler
MDR	multidrug resistance
MDR-TB	multiple drug resistant TB
MEN 2	multiple neoplasia syndrome type 2
MI	myocardial infarction
MMF	monomethyl fumarate
MMRs	mismatch repair proteins
MOA	mechanism of action
MPO	myeloperoxidase
MR	mineralocorticoid receptor
MRPs	multidrug resistance proteins
MRSA	methicillin-resistant *Staphylococcus aureus*
MTC	medullary thyroid cancer
mTOR	mammalian target of rapamycin
MTTP	microsomal triglyceride transfer protein
MTR	melatonin receptors
MTT	methylthiotetrazole
N$_2$O	nitrous oxide
NA	neuraminidase

nAChR	nicotinic acetylcholine receptor
NAD	nicotinamide adenine dinucleotide
NAPQI	N-acetyl-*p*-benzoquinoneimine
NaSSAs	noradrenergic and serotonergic specific antidepressants
NAT	N-acyltransferase
NCEs	new chemical entities
NDRIs	norepinephrine and dopamine reuptake inhibitors
NE	norepinephrine
NER	nucleotide excision repair
NET	norepinephrine reuptake transporter
NIH	National Institutes of Health
NMDA	N-methyl-D-aspartate
NO	nitric oxide
NREM	nonrapid eye movement sleep
Nrf2	nuclear factor erythroid 2-related factor 2
NRTIs	nucleoside reverse transcriptase inhibitors
NS	nonstructural
NSAIDs	nonsteroidal anti-inflammatory drugs
NSRIs	norepinephrine and serotonin reuptake inhibitors
NTZ	nitazoxanide
O$_2$	molecular oxygen
OAB	overactive bladder
OATP	organic anion transporting protein
OATs	organic anion transporters
OC	oral contraceptive
OCTP	organic cation transporting protein
ODC	ornithine decarboxylase
OTC	over the counter
OXR	orexin receptors
PABA	*p*-aminobenzoic acid
PAH	polyaromatic hydrocarbons (Chapter 2)
PAH	pulmonary arterial hypertension (Chapter 13)
PAI-1	plasminogen activator inhibitor-1
PAP	phosphatidic acid phosphatase (Chapter 15)
PAP	prostatic acid phosphatase (Chapter 18)
PAPS	3′-phosphoadenosine-5′-phosphosulfate
PAR-1	protease-activated receptor-1
PARP	poly (ADP-ribose) polymerase
PAS	*p*-aminosalicylic acid
PBP	penicillin-binding protein
PCP	pneumocystis pneumonia
PCT	proximal convoluted tubule
PD	Parkinson's disease
PDE	phosphodiesterase
PE	pulmonary embolism
PEPTs	peptide transporters
PG	prostaglandin
PGI$_2$	prostacyclin
P-gp	P-glycoprotein
PIP$_2$	phosphatidylinositol 4,5-diphosphate
PKA	protein kinase A
PLA$_2$	phospholipase A$_2$

PPAR	peroxisome proliferator-activated receptor	**TALH**	thick ascending limb of the loop of Henle
PPIs	proton pump inhibitors	**TB**	tuberculosis
PSA	prostate-specific antigen	**TBG**	thyroxine-binding globulin
PT	prothrombin time	**TCAs**	tricyclic amines
PTH	parathyroid hormone	**TF**	tissue factor
PTSD	posttraumatic stress disorder	**TFA**	*trans* fatty acid
PTU	*n*-propyl thiouracil	**Tg**	thyroglobulin
PUFA	polyunsaturated fatty acids	**THF**	tetrahydrofolate
RANKL	receptor activator of nuclear factor-kB	**TIZ**	tizoxanide
RAS	reticular activating system	**TK**	tyrosine kinase
RDA	recommended dietary allowance	**TKI**	tyrosine kinase inhibitor
REM	rapid eye movement sleep	**TM**	transmembrane
RFC1	reduced folate carrier-1	**TMH**	transmembrane helix
RIF	rifampin	**TMN**	tuberomammillary nucleus
RIMAs	reversible inhibitors of MAO-A	**TNFα**	tumor necrosis factor alpha
ROS	reactive oxygen species	**tPA**	tissue plasminogen activator
rRNA	ribosomal ribonucleic acid	**TPMT**	thiopurine S-methyltransferase
RT	reverse transcriptase	**TPO**	thyroperoxidase
S1PR	sphingosine 1-phosphate receptor	**TRH**	thyrotropin-releasing hormone
SA	sinoatrial	**TRT**	testosterone replacement therapy
SABAs	short-acting β_2-adrenergic agonists	**TSH**	thyroid-stimulating hormone
SAM	S-adenosylmethionine	**TXA$_2$**	thromboxane A$_2$
Sar	sarcosine	**UFH**	unfractionated heparin
SAR	structure–activity relationship	**UGT**	UDP-glucuronosyltransferase (Chapters 2,17)
SARIs	serotonin-2 antagonists/serotonin reuptake inhibitors	**SUR1**	sulfonylurea receptor type 1
SC	subcutaneous	**URAT1**	urate anion transporting protein
SCN	suprachiasmatic nucleus	**UTIs**	urinary tract infections
SERM	selective estrogen receptor modulator	**VEGFR**	vascular endothelial growth factor receptor
SERT	serotonin reuptake transporter	**VLDL**	very low-density lipoprotein
SGLT	sodium glucose cotransporter	**VRSA**	vancomycin-resistant *Staphylococcus aureus*
SLC	solute carrier	**VSM**	vascular smooth muscle
SNRIs	selective norepinephrine reuptake inhibitors	**VTE**	venous thromboembolism
SRM	serotonin receptor modulators	**vWF**	von Willebrand factor
SSRIs	selective serotonin reuptake inhibitors	**WHO**	World Health Organization
SULT	sulfotransferase	**XO**	xanthine oxidase
T$_3$	triiodothyronine	**XTR-TB**	extensively drug-resistant tuberculosis
T$_4$	thyroxine	**Z**	zusammen

Answers to Review Questions

Chapter 2
1. B
2. E
3. A
4. B
5. D

Chapter 3
1. C
2. B
3. E
4. B
5. A

Chapter 4
1. E
2. B
3. D
4. A
5. C

Chapter 5
1. C
2. A
3. B
4. B
5. C

Chapter 6
1. C
2. B
3. E
4. C
5. A

Chapter 7
1. A
2. D
3. C
4. E
5. B

Chapter 8
1. C
2. A
3. A
4. B
5. C

Chapter 9
1. E
2. B
3. B
4. B
5. B

Chapter 10
1. D
2. B, C
3. A
4. C
5. B, E

Chapter 11
1. C
2. E
3. C
4. D
5. A

Chapter 12
1. B
2. A
3. E
4. D
5. C

Chapter 13
1. B
2. D
3. A
4. C
5. E

Chapter 14
1. B
2. D
3. C
4. E
5. D

Chapter 15
1. B
2. D
3. A
4. B
5. C

Chapter 16
1. A
2. B
3. D
4. C
5. A

Chapter 17
1. B
2. C
3. C
4. B
5. B

Chapter 18
1. D
2. B
3. E
4. A
5. C

Chapter 19
1. C
2. D
3. D
4. B
5. A

Chapter 20
1. C
2. B
3. A
4. B
5. B

Chapter 21
1. B
2. C
3. E
4. A
5. C

Chapter 22
1. B
2. D
3. C
4. D
5. A

Chapter 23
1. B
2. C
3. E
4. C
5. D
6. C
7. A
8. D

9. B
10. C

Chapter 24
1. C
2. A
3. B
4. B
5. A

Chapter 25
1. Enterobiasis   C
   Malaria   A
   Giardiasis   A
   Lice   B
2. A
3. B
4. C
5. B

Chapter 26
1. C
2. D
3. A
4. D
5. B

Chapter 27
1. B
2. E
3. C
4. C
5. E

Chapter 28
1. E
2. Aspartame   S
   Mogrosides   N
   Acesulfame-K   S
   Sucralose   S
   Rebaudioside A   N
3. C
4. B
5. A

Subject Index

Note: Page numbers followed by "f" indicate illustrations and chemical structures; those followed by "t" indicate to tables. Drugs are listed under the generic name.

Drug Index

Page numbers followed by "f" indicate illustrations and chemical structures those followed by "t" indicate to tables. Drugs are listed under the generic name.

A

Abacavir, 563, 564f, 565, 565f, 565t, 566t
Abatacept, 400
Abciximab, 285, 285t
Abiraterone acetate, 340
Acarbose, 303, 304
Acebutolol, 57f, 209f, 210f, 210t
Acesulfame-K, 601
Acetaminophen, 13, 20f, 395–396, 396f
Acetazolamide, 203f
Acetohexamide, 297f, 298t, 299
Acetylsalicylic acid, 381f
Aclidinium, 39f, 40, 80f
Acrivastine, 414f, 420, 420f
Acyclovir, 584f, 584t, 585f, 585t, 586t
Adalimumab, 400t
Adefovir dipivoxil, 576–578, 577f
Afatinib, 545f, 548t, 549–551
Albendazole, 504, 505t
Albiglutide, 305
Albuterol, 76t
Alcaftadine, 421, 421f, 421t
Alclometasone, 320t, 324f
Alendronate, 370f, 370t
Alfentanil, 192, 192t, 193
Alfuzosin, 55f, 334f, 335t, 337
Alirocumab, 2t
Aliskiren, 246
Allopurinol, 402–403, 402f
Alloxanthine, 403
Alogliptin, 306, 307f, 308t
Alprazolam, 109f
Alprostadil, 378
Alteplase, 287f, 288, 289t
Altretamine, 514f
Amantadine, 69, 586–587, 587f
Ambrisentan, 232f, 233
Amcinonide, 320t, 324f
Amfenac, 389
Amifostine, 525f
Amikacin, 470
Amiloride, 206f
ε-Aminocaproic acid, 289–290
p-Aminosalicylic acid, 16f, 20f, 471
Amiodarone, 218t, 223, 223f, 223t, 224f
Amitriptyline, 160t, 162f, 166t, 168t
Amlodipine, 215f, 215t
Amobarbital, 117f, 117t
Amoxapine, 160t, 162f, 163f
Amoxicillin, 444f, 445t
Amphetamine, 50f, 66f
Amphotericin B, 479f, 480t
Ampicillin, 443f, 444f, 445t
Anakinra, 400

Anastrozole, 354f
Angiotensin, 244f
Anidulafungin, 485f
Anisindione, 268f
Apixaban, 275f, 276–277, 277f
Apomorphine, 68, 68t
Apraclonidine, 52
Apremilast, 399–400
Arformoterol, 74f
Argatroban, 272f, 273t, 274f
Aripiprazole, 106, 106f, 107t, 160t, 172f, 172t
Arsenic oxide, 497
Arteflene, 502f
Artemether, 502f, 503t
Artemisinin, 502f
Artesunate, 502f, 503t
Articaine, 140t, 141t
Ascorbic acid (Vitamin C), 618f
Asenapine, 103f, 103t
Aspartame, 295t, 600–601, 601f
Aspirin, 278–279, 279f, 380t, 381–383, 383f
Atazanavir, 570f, 571t, 572, 574
Atenolol, 209f, 210t
Atomoxetine, 160t, 162f, 163f, 164, 164t, 165f
Atorvastatin, 26t, 256f, 258f, 259t
Atovaquone, 495
Atracurium, 41f, 42f, 42t
Atripla, 566t
Atropine, 38f, 69f, 80f
Auranofin, 396–397, 396f
Avanafil, 332f, 333f
Avibactam, 446–447
Axitinib, 547t, 548t, 551
Azacitidine, 533f, 540
Azelastine, 421f, 421t
Azilsartan, 244f, 245t
Azithromycin, 455
Azitinib, 546f
Aztreonam, 452, 453
Azulfidine, 16f

B

Bacitracin A, 462–463
Baclofen, 70
Bazedoxifene, 352f, 352t, 367–368, 368f, 369t
Beclomethasone, 89t, 320t, 324f, 325f, 326f, 326t
Bedaquiline, 472–473, 473f
Belinostat, 554f
Benazepril, 241f, 242f, 243
Bendamustine, 514f

Benoxinate, 139t, 141t
Benzisoxazole, 104–105, 104f, 105t
Benznidazole, 497f
Benzocaine, 139t, 141t
Benzothiazole, 104f, 105t
Benzoyl alcohol, 510
Benzphetamine, 608f, 609f
Benztropine, 39f, 69
Benzyl alcohol, 510
Benzylpenicillin, 444f, 445t
Bepotastine, 411f, 412
Besifloxacin, 440f
Betamethasone, 323t
Betaxolol, 57f, 58
Bethanechol, 34f, 35t, 44
Bicalutamide, 339f, 340t, 341f
Bismuth subsalicylate, 431
Bisoprolol, 209f, 210t
Bivalirudin, 272f, 273f, 273t, 274f
Bleomycin, 529f, 532t
Blinatumomab, 2t
Boceprevir, 580, 580f, 581, 581f, 582f
Bortezomib, 552f
Bosentan, 232f, 233, 233f
Bosutinib, 545f, 547–549, 547t, 548, 548t
Brimonidine, 52, 52f
Brinzolamide, 203f
Bromocriptine, 68, 68t
Brompheniramine, 418–419
Budesonide, 89t, 320t, 325f, 326f, 326t
Buformin, 301
Bumetanide, 208f, 208t
Bupivacaine, 140t, 141t
Buprenophine, 184, 188, 189, 189f
Bupropion, 160t, 170–171, 171f
Busulfan, 514f
Butabarbital, 117f, 117t
Butamben, 139t
Butenafine, 484
Butoconazole, 481t
Butorphanol tartrate, 189, 189t

C

Cabazitaxel, 341–342, 540f, 542, 542f
Cabozantinib, 363–364, 363f, 364, 546f, 547t, 548t, 551
Cafditoren pivoxil, 451f, 452–453
Caffeine, 13, 83f
Calcitonin (CT), 366, 371, 371f
Canagliflozin, 309f, 310–311, 311f
Candesartan, 244f, 245, 245t
Capecitabine, 533f, 534t, 535, 536f
Capreomycin, 472
Captopril, 28f, 240f, 241f, 242f, 242t

CCS0416